Little, Brown and Company
34 Beacon Street
Boston, Massachusetts 02106

Basic
Neurochemistry

Basic Neurochemistry

Second Edition

Edited by

George J. Siegel, M.D.
University of Michigan Medical School,
Ann Arbor

R. Wayne Albers, Ph.D.
National Institute of Neurological
and Communicative Disorders and Stroke,
Bethesda

Robert Katzman, M.D.
Albert Einstein College of Medicine
of Yeshiva University, Bronx

Bernard W. Agranoff, M.D.
University of Michigan Medical School,
Ann Arbor

Little, Brown and Company   Boston

Preface

In the past half century, neurochemistry has emerged as a distinct, if hybrid, discipline. Its unifying objective is the elucidation of biochemical phenomena that subserve activity of the nervous system or are associated with neurological diseases. For neurochemistry as for all of experimental biology, this unifying objective generates certain subsidiary goals: (1) isolation and identification of components, (2) determination of the functional interactions of the components, and (3) development of integrating hypotheses that account satisfactorily for the activities of the intact organ in terms of molecular events.

A comprehensive description of nervous system function in terms of molecular events presumably would supply intellectually satisfying and socially useful explanations of the neural responses that ultimately mediate mentation and behavior and their pathologies. Advances in neurochemistry already have led to a remarkable increase in our understanding of many of the inherited neurological disorders and to effective diagnostic techniques that may be employed both prenatally and postnatally. In another area, the identification of a deficiency in a specific neurotransmitter associated with Parkinson's disease has led to a useful therapy in this common disorder.

The validity and vitality of neurochemistry are proportionate to the successes of its practitioners in providing biochemical explanations for neural functions. These explanations cannot be derived from chemical analyses alone, and those who wish to contribute to this endeavor must be prepared to pursue various levels of structural and functional organization. The scope of neurochemistry is determined by the junctures that develop between the field of biochemistry and the fields of neurobiology, neurology, and the behavioral sciences. Neurochemists are crucially dependent upon data from these diverse subjects if they are to formulate functionally meaningful molecular hypotheses.

As a result, the student of neurochemistry must become familiar with concepts and information that are widely dispersed in the scientific literature. A number of neurochemistry courses have been organized in medical and graduate schools within recent years, and the organizers have become acutely aware of the difficulties in selecting the most significant material to place within the perspective of neurochemistry. Few existing books are at once comprehensive and sufficiently concise to be practical as texts in these courses.

A conference on neurochemistry curriculum considered these problems in June 1969 and made recommendations for the scope of the subject matter that subsequently was developed into the first edition of *Basic Neurochemistry*. It was anticipated that the experience gained through the construction and use of this text would initiate a continuing reappraisal of the field that could contribute to the evolution of later editions.

v

To this end, the editors organized a conference, attended by most of the contributing authors, preceding the 1974 meeting of the American Society for Neurochemistry. This conference resulted in this second edition. More extensive material on cell structure and ultrastructure has been included (Chap. 1). Areas of recent research emphasis are reflected in the separate chapters on photoreception (Chap. 7), cyclic nucleotides (Chap. 12), prostaglandins (Chap. 13), axoplasmic flow (Chap. 21), and endocrine effects on the brain (Chap. 35). Certain deficiencies in the coverage of clinical problems have been corrected by new chapters on abnormal brain states (Chap. 31) and the mucopolysaccharidoses (Chap. 27). Discussion of the neuromuscular diseases has also been expanded (Chap. 23).

Needless to say, the second edition is larger than the first, although the editors have deleted some topics and consolidated others. Consideration of special senses other than vision has been omitted; amino acid and carbohydrate metabolism have been combined (Chap. 14); and discussions of analytical techniques have been curtailed. Consideration of biochemical theories of the origin of psychoses has been condensed and combined with psychopharmacology (Chap. 34).

The editors wish to thank the contributing authors for their diligence, promptness, and cooperation in editorial problems. The excellent editorship of Mrs. Helene Jordan Waddell has been a key factor in the production of this edition. We thank Little, Brown and Company and their staff, especially Mrs. Lin Richter and Mrs. Anne Merian, for their cooperation in the rapid and careful publication of this volume.

G. J. S.
R. W. A.
R. K.
B. W. A.

Contents

Contributing Authors

Leo G. Abood, Ph.D.
Professor, Center for Brain Research and
Biochemistry, University of Rochester School of
Medicine and Dentistry, Rochester, New York
Chapter 5

Bernard W. Agranoff, M.D.
Professor of Biological Chemistry, University of
Michigan Medical School, Ann Arbor
Chapter 36

R. Wayne Albers, Ph.D.
Chief, Section on Enzymes, Laboratory of
Neurochemistry, National Institute of Neurological
and Communicative Disorders and Stroke,
National Institutes of Health, Bethesda, Maryland
Chapter 3

Stanley H. Appel, M.D.
Professor and Chief, Division of Neurology, Duke
University School of Medicine, Durham, North
Carolina
Chapter 2

Samuel H. Barondes, M.D.
Professor of Psychiatry, University of California,
San Diego, School of Medicine at La Jolla
Chapter 16

Joyce A. Benjamins, Ph.D.
Assistant Professor of Neurology, Wayne State
University School of Medicine, Detroit, Michigan
Chapter 18

Murray B. Bornstein, M.D.
Professor, Department of Neurology, Albert
Einstein College of Medicine of Yeshiva University,
Bronx, New York
Chapter 28

xi

Roscoe O. Brady, M.D.
Chief, Developmental and Metabolic Neurology
Branch, National Institute of Neurological and
Communicative Disorders and Stroke, National
Institutes of Health, Bethesda, Maryland
Chapter 26

Donald D. Clarke, Ph.D.
Professor of Chemistry, Fordham University,
New York
Chapter 14

Maynard M. Cohen, M.D., Ph.D.
Professor and Chairman, Department of
Neurological Sciences, and Professor of
Biochemistry, Rush Medical College, Chicago
Chapter 33

Eugene D. Day, Ph.D.
Professor of Immunology, Duke University
Medical Center, Durham, North Carolina
Chapter 2

Pierre M. Dreyfus, M.D.
Professor of Neurology, University of California,
Davis, School of Medicine
Chapter 29

Thomas E. Duffy, Ph.D.
Associate Professor of Biochemistry in Neurology
and Assistant Professor of Biochemistry, Cornell
University Medical College, New York
Chapter 31

Joseph S. Eisenman, Ph.D.
Associate Professor of Physiology, Mount Sinai
School of Medicine of The City University of
New York
Chapter 22

Stanley E. Geel, Ph.D.
Adjunct Assistant Professor of Neurochemistry,
Department of Neurology, University of California,
Davis, School of Medicine
Chapter 29

J. Gergely, M.D., Ph.D.
Associate Professor, Department of Biological
Chemistry, Harvard Medical School, Boston
Chapter 23

Paul Greengard, Ph.D.
Professor of Pharmacology, Yale University School
of Medicine, New Haven, Connecticut
Chapter 12

Richard Hammerschlag, Ph.D.
Associate Research Scientist, Division of
Neurosciences, City of Hope National Medical
Center, Duarte, California
Chapters 8 and 11

Francis C. G. Hoskin, Ph.D.
Professor of Biology, Illinois Institute of
Technology, and Department of Neurological
Sciences, Rush Medical College, Chicago
Chapter 33

Wayne Hoss, Ph.D.
Assistant Professor, Center for Brain Research and
Biochemistry, University of Rochester School of
Medicine and Dentistry, Rochester, New York
Chapter 5

Y. Edward Hsia, B.M., M.R.C.P., D.C.H.
Associate Professor, Departments of Human
Genetics and Pediatrics, Yale University School
of Medicine, New Haven, Connecticut
Chapter 24

Robert Katzman, M.D.
Professor and Chairman of Neurology, and
Professor of Neuroscience, Albert Einstein College
of Medicine of Yeshiva University, Bronx,
New York
Chapter 20

Abel Lajtha, Ph.D.
Director, New York State Research Institute for
Neurochemistry and Drug Addiction, Ward's Island,
New York
Chapter 14

F. C. MacIntosh, Ph.D.
Professor of Physiology, McGill University Faculty
of Medicine, Montreal, Quebec, Canada
Chapter 9

Henry R. Mahler, Ph.D.
Research Professor of Chemistry, Center for
Neural Sciences, Indiana University, Bloomington
Chapter 17

Howard S. Maker, M.D.
Associate Professor of Neurology, Mount Sinai
School of Medicine of The City University of
New York, New York
Chapter 14

David B. McDougal, Jr., M.D.
Professor, Department of Pharmacology,
Washington University School of Medicine,
St. Louis, Missouri
Chapter 25

Bruce S. McEwen, Ph.D.
Associate Professor of Neurobiology, Rockefeller
University, New York
Chapter 35

Guy M. McKhann, M.D.
Kennedy Professor of Neurology, and Chairman,
Department of Neurology, The Johns Hopkins
University School of Medicine, Baltimore
Chapter 18

Pierre Morell, Ph.D.
Associate Professor, Department of Biochemistry
and Nutrition, University of North Carolina School
of Medicine, Chapel Hill
Chapter 28

James A. Nathanson, M.D., Ph.D.
Pharmacology Research Associate, Laboratory of
Neuropharmacology, National Institute of Mental
Health, St. Elizabeth's Hospital, Washington, D.C.
Chapter 12

Elizabeth F. Neufeld, Ph.D.
Chief, Section on Human Genetics, National
Institute of Arthritis, Metabolism, and Digestive
Diseases, National Institutes of Health, Bethesda,
Maryland
Chapter 27

William T. Norton, Ph.D.
Professor of Neurology (Neurochemistry) and
Neuroscience, Albert Einstein School of Medicine
of Yeshiva University, Bronx, New York
Chapter 4

Sidney Ochs, Ph.D.
Professor of Physiology, University of Indiana
School of Medicine, Indianapolis
Chapter 21

Fred Plum, M.D.
Professor and Chairman, Department of Neurology,
Cornell University Medical College, New York
Chapter 31

Cedric S. Raine, Ph.D., D.Sc.
Associate Professor of Pathology (Neuropathology),
Albert Einstein College of Medicine of Yeshiva
University, Bronx, New York
Chapter 1

Eugene Roberts, Ph.D.
Director, Division of Neurosciences, City of Hope
National Medical Center, Duarte, California
Chapters 8 and 11

Frederick J. Samaha, M.D.
Associate Professor of Pediatrics and Neurology,
The University of Pittsburgh School of Medicine,
Pittsburgh, Pennsylvania
Chapter 23

Larry J. Shapiro, M.D.
Assistant Professor of Pediatrics, University of
California, Los Angeles, School of Medicine
Chapter 27

Hitoshi Shichi, Ph.D.
Research Chemist, National Eye Institute,
National Institutes of Health, Bethesda, Maryland
Chapter 7

George J. Siegel, M.D.
Professor of Neurology, University of Michigan
Medical School, Ann Arbor
Chapter 22

Solomon H. Snyder, M.D.
Professor of Pharmacology and Psychiatry,
The Johns Hopkins University School of Medicine,
Baltimore
Chapter 10

Louis Sokoloff, M.D.
Chief, Laboratory of Cerebral Metabolism,
National Institute of Mental Health,
Bethesda, Maryland
Chapter 19

Theodore L. Sourkes, Ph.D.
Professor of Psychiatry and Biochemistry, McGill
University Faculty of Medicine, Montreal, Quebec,
Canada
Chapters 32 and 34

William L. Stahl, Ph.D.
Associate Professor of Physiology and Biophysics
and of Medicine (Neurology), University of
Washington School of Medicine, Seattle
Chapter 6

Kunihiko Suzuki, M.D.
Professor of Neurology and Neuroscience, Albert
Einstein College of Medicine of Yeshiva University,
Bronx, New York
Chapter 15

Phillip D. Swanson, M.D., Ph.D.
Professor of Medicine, and Head, Division of
Neurology, University of Washington School of
Medicine, Seattle
Chapter 6

Donald B. Tower, M.D., Ph.D.
Director, National Institute of Neurological and
Communicative Disorders and Stroke, National
Institutes of Health, Bethesda, Maryland
Chapter 30

Leonhard S. Wolfe, M.D., Ph.D.
Professor, Department of Neurology and
Neurosurgery and Department of Biochemistry,
McGill University Faculty of Medicine, Montreal,
Quebec, Canada
Chapter 13

Part One

General Neurochemistry

Section I

Morphological Basis of Neurochemistry

Chapter 1

Neurocellular Anatomy

Cedric S. Raine

Although our understanding of the functional relationships of the central nervous system (CNS) components is still in its infancy, particularly in the area of neurotransmitters and synaptic modulation, the fine structure of most elements is relatively well worked out [1–7]. The purpose of this chapter is to collate available information and present it in a manner relevant to neurochemistry. The excellent neuroanatomical atlases of Peters et al. [4] and Palay and Chan-Palay [1] should be consulted for a more detailed ultrastructural analysis of specific cell types, particularly of neurons with their diverse forms and connections. For the sake of simplicity, the present section is subdivided into a section on general organization and then according to major cell types.

GENERAL CELLULAR ORGANIZATION

Central nervous system parenchyma is made up of nerve cells and their afferent and efferent extensions (dendrites and axons), all closely enveloped by glial cells. A highly diagrammatic representation of the major CNS elements is shown in Figure 1-1. The entire central nervous system is bathed both internally and externally by cerebrospinal fluid (CSF), which circulates throughout the ventricular and leptomeningeal spaces. This fluid, a type of plasma ultrafiltrate, plays a significant role in protecting the CNS from mechanical trauma, in the balance of electrolytes and protein, and in the maintenance of ventricular pressure. This subject is reviewed in Chapter 20. The outer surface of the CNS is invested by a triple membrane system, the meninges, composed of flattened fibrous or elastic connective tissue, collagen, and blood vessels. The outermost of the three, the dura mater, is applied tightly to the inner surfaces of the calvania and has the arachnoid membrane closely applied to its inner surface. The innermost pia mater loosely covers the CNS surface. The pia and arachnoid together are called *leptomeninges.* Cerebrospinal fluid occupies the subarachnoid space (between the arachnoid and the pia) and the ventricles. The CNS parenchyma is overlaid by a layer of subpial astrocytes, which, in turn, is covered on its leptomeningeal aspect by a continuous basement membrane material (see Fig. 1-1). On the inner (ventricular) surface, the CNS parenchyma is separated from the CSF by a layer of ciliated ependymal cells, which are thought to facilitate the movement of CSF. In some species, areas of the CNS are lined by nonciliated ependymal cells, e.g., the central canal of the human spinal cord. The production and circulation of CSF is maintained by specialized mesodermal elements, the choroid plexus, which are grapelike collections of vascular tissue and cells that protrude into the ventricles. Resorption of CSF is effected by vascular structures known as *arachnoid villi,* located in the leptomeninges over the surface of the brain.

5

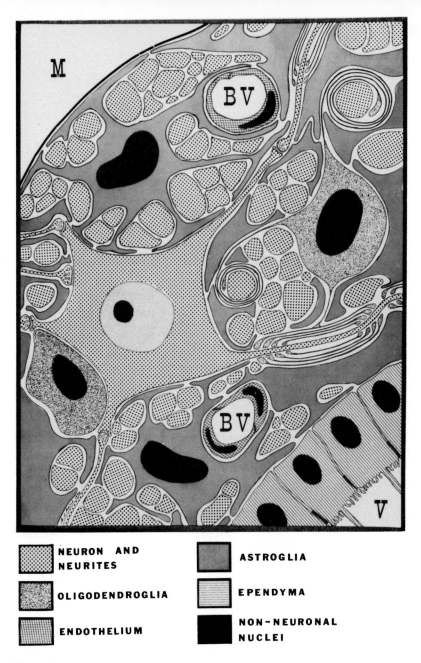

NEURON AND NEURITES   ASTROGLIA

OLIGODENDROGLIA   EPENDYMA

ENDOTHELIUM   NON-NEURONAL NUCLEI

Figure 1-1

Diagrammatic representation of the major components of the CNS and their interrelationships. Microglia are not depicted. In this simplified situation, the CNS extends from its meningeal surface (M), through the basement lamina over the CNS parenchyma (solid black line), subpial astrocytes, parenchyma proper, subependymal astrocytes, to the ciliated ependymal cells lining the ventricular space (V). Note how the astrocyte provides the covering between external factors and CNS parenchyma and also invests blood vessels (BV), neurons, cell processes, etc. One neuron is seen (center) with synaptic contacts on its soma and dendrites. Its axon emerges to the right and is myelinated by an oligodendrocyte (above). Other axons are shown in transverse section, some of which are myelinated. The oligodendrocyte to the lower left of the neuron is of the nonmyelinating, satellitic type.

The ependymal cells abut layers of astrocytes, which, in turn, envelop neurons, neurites, and vascular components. In addition to neurons and glial cells (astrocytes and oligodendrocytes), the CNS parenchyma contains blood vessels, macrophages (pericytes), and microglial cells.

The peripheral and autonomic nervous systems consist of bundles of myelinated and nonmyelinated axons enveloped by Schwann cells — the peripheral counterparts of the oligodendrocytes. The nerve bundles are enclosed by the perineurium and the epineurium — tough, fibrous, elastic sheaths. Between individual nerve fibers are isolated connective tissue (endoneurial) cells and blood vessels. The ganglia (e.g., dorsal-root and sympathetic ganglia), located peripherally to the CNS, are made up of large neurons, usually uni- or bipolar, surrounded by satellite cells that are specialized Schwann cells. A dendrite and an axon, both of which can be of great length (up to several feet), arise from each neuron.

THE NEURON

Functional Features

From a historical standpoint, no other cell type has attracted as much attention or caused as much controversy as the nerve cell. It is impossible, within the confines of a single chapter, to delineate comprehensively the tremendous structural, topographical, and functional variation achieved by this cell type. Consequently, despite an enormous literature, the neuron still defies precise understanding, particularly from the functional aspect. It is known that the neuronal population is usually established shortly after birth, that mature neurons do not divide, and that in man there is a daily dropout of neurons amounting to about 20,000 cells. These facts alone make the neuron unique. Development and maturation of neurons are discussed in Chapter 18.

Neurons can be either excitatory, inhibitory, or modulatory in their effect, motor, sensory, or secretory in function. They can be influenced by a large battery of neurotransmitters and hormones (see Chap. 35). Naturally, this enormous repertoire of functions is associated with different developmental influences upon different neurons, largely reflected in the variations of dendritic and axonal outgrowth. Specialization also occurs at axonal terminals, where a variety of junctional complexes (synapses) exists. The subtle synaptic modifications are best visualized ultrastructurally, although immunofluorescent staining for light microscopy also permits distinctions among synapses on the basis of the type of transmitter released.

General Structural Features of Neurons, Dendrites, and Axons

Although one tends to think of the neuron as a stellate cell with broad dendrites and a fine axon emerging from one pole, an impression gained from the older work of Purkinje, who first described the nerve cell in 1839, and of Deiters, Cajal, and Golgi, this appearance does not hold true for many neurons. The neuron is the most polymorphic cell in the body and defies formal classification on the basis of shape, location, function, fine structure, or transmitter substance. Early workers described the neuron as a globular mass suspended between nerve fibers, but the teased preparations by Deiters and his contemporaries soon proved this was not the case. Later work, utilizing impregnation staining and culture techniques, elaborated on Deiters' findings. For a long time, before the work of Deiters and Cajal, neurons were believed to form a syncytium with no intervening membranes, a postulation that also was proposed for neuroglia. Sherrington suggested that the manner in which neurons interact with each other is mediated via specialized junctions, which he named synapses. Today, of course, we are familiar with the enormous variety of neuron shapes and sizes; they range from the small, globular, cerebellar granule cells with a perikaryal diameter of about 6μ to 8μ to the pear-shaped Purkinje cells and star-shaped anterior horn cells, both of which may reach diameters of 60μ to 80μ in man. The perikaryal size, however, is generally a poor index of cell volume, because it is a general rule in neuroanatomy that the neurites occupy a greater percentage of the cell surface area than does the soma. For example, the pyramidal cell of the somatosensory cortex has a cell body that accounts for only 4 percent of the total cell surface area, whereas from the dendritic tree, the dendritic spines alone claim 43 percent (Mungai, quoted by Peters et al. [4]). Hydén [2] quotes the work of Scholl (1956), who calculated that the perikaryon of a "cortical cell" represents 10 percent of the neuronal surface area. In the feline reticular formation, certain giant cells possess ratios between soma and dendrites of about 1:5. A single axon is the usual rule, but some cells, e.g., the Golgi cell of the cerebellum, are endowed with several axons, some of which may show branching.

The extent of the branching displayed by the dendrites is a useful index to their functional importance. Dendritic trees represent the expression of the receptive fields, and large fields can receive inputs from multiple origins. A cell with less-developed dendritic ramification, e.g., the cerebellar granule cells, synapses with a more homogeneous population of afferent sources.

The axon emerges from a neuron as a slender thread and frequently does not branch until it nears its target. In contrast to the dendrite and the soma (with very few exceptions), the axon is frequently myelinated, thus increasing its efficiency as a conducting unit. Myelin, a spirally wrapped membrane (see Chap. 4), is laid down in segments (internodes) by oligodendrocytes in the CNS and in the peripheral nervous system (PNS) by Schwann cells. The naked region of axon between adjacent internodes is known as a *node of Ranvier,* and it is across this region of the axon that saltatory conduction (Chap. 5) is effected.

Cellular Organization of Neurons

No unique cytoplasmic features of the neuronal soma serve to characterize this cell as different from any other. Neurons have all the morphological counterparts of other cell types, the structures are similarly distributed, and some of the commonest, e.g., Golgi apparatus and mitochondria, were first described in neurons (Fig. 1-2).

Outer Limiting Membrane

As seen by electron microscopy, the neuron often stands out poorly from the background neuropil, most of which is composed of nonmyelinated neurites (axons and

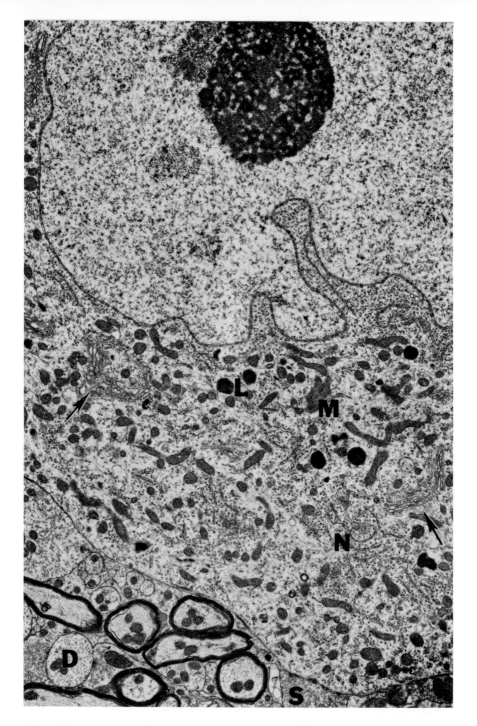

Figure 1-2

A motor neuron from the spinal cord of an adult rat shows a nucleus containing a nucleolus, clearly divisible into a pars fibrosa and a pars granulosa, and a perikaryon filled with organelles. Among these, Golgi apparatus (arrows), Nissl substance (N), mitochondria (M), and lysosomes (L) can be seen. An axo-somatic synapse (S) occurs below and two axo-dendritic synapses abut a dendrite (D). × 8,000.

dendrites), synaptic complexes, and glial processes – profiles with similar electron density at low power. Closer inspection shows that, like all cells, the neuron is delineated by a typical triple-layered unit membrane about 75 Å wide, consisting of two dense leaflets about 25 Å thick enclosing a clear space. Some fixation procedures, such as glutaraldehyde followed by postosmication, render the inner cytoplasmic leaflet slightly more osmiophilic, giving the unit membrane an asymmetrical appearance.

Nucleus

A large, usually spherical, nucleus containing a prominent nucleolus is typical of many neurons. The nucleochromatin is invariably pale, with little dense heterochromatin. In some neurons, the karyoplasm can be more differentiated and contain dense heterochromatin, e.g., cerebellar granule cells. The nucleolus is vesiculated, clearly delineated from the rest of the karyoplasm, and usually contains two textures, the pars fibrosa (fine bundles of filaments) and the pars granulosa, in which dense granules predominate. An additional juxtaposed structure, found in neurons of females of some species, is the nucleolar satellite, or sex chromatin, which consists of dense, but loosely-packed, coiled filaments. The nucleus is enclosed by the nuclear envelope, made up on the cytoplasmic side by the inner membrane of the perikaryon (sometimes seen in continuity with the endoplasmic reticulum) and a more regular membrane on the inner, nuclear aspect. Between the two is a clear channel of between 200 and 400 Å. Periodically, the inner and outer membranes of the envelope come together to form a single diaphragm – a nuclear pore (Fig. 1-3). In tangential section, nuclear pores are seen as empty vesicular structures, about 700 Å in diameter. In some neurons, that segment of the nuclear envelope which faces the dendritic pole is deeply invaginated, as in Purkinje cells.

The Perikaryon

The body of the neuron, called the perikaryon, is rich in organelles (Fig. 1-2). Among the most prominent features of the cytoplasm is a system of membranous cisternae, divisible into rough endoplasmic reticulum (ER), which forms part of the Nissl substance; smooth (agranular) ER; subsurface cisternae (the hypolemmal system); and the Golgi apparatus. Although these various components are structurally interconnected, and some are even connected to the nuclear envelope, each possesses distinct enzymologic properties. Also present within the cytoplasm are abundant lysosomes, lipofuscin granules (aging pigment), mitochondria, multivesicular bodies, neurotubules, neurofilaments, and ribosomes.

Nissl Substance

The intracytoplasmic, basophilic masses that ramify loosely throughout the cytoplasm and are typical of most neurons are known collectively as Nissl substance (Figs. 1-2 and 1-3). The distribution of Nissl substance in certain neurons is characteristic and is used as a criterion for identification. By electron microscopy (EM), this substance is seen to comprise regular arrays or scattered portions of flattened cisterns of rough ER, surrounded by clouds of free polyribosomes (Chap. 17). The membranes of the rough

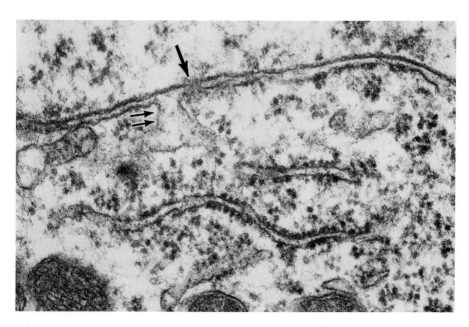

Figure 1-3
Detail of the nuclear envelope showing a nuclear pore (arrow) and the outer leaflet connected
to smooth ER (double arrows). Two cisterns of rough ER with associated ribosomes are also
present. × 80,000.

ER are studded with rows of ribosomes. A spacing of 200 to 400 Å is maintained
within cisternae. Sometimes, cisternal walls meet at fenestrations. Unlike the rough
ER of glandular cells or other protein-secreting cells (e.g., plasma cells), the rough
ER of neurons probably produces proteins for its own use, a feature imposed by the
extraordinary functional demands placed upon this cell. Nissl substance does not
penetrate axons but extends along dendrites.

Smooth ER
Most neurons contain at least a few cisternae or tubules of smooth (agranular) ER,
sometimes difficult to differentiate from rough ER, due to disorderly arrangement
of ribosomes. Ribosomes are not attached to these membranes, and the cisterns
usually assume a meandering, branching course throughout the cytoplasm. In some
neurons, smooth ER is quite a prominent component, e.g., in Purkinje cells. Individual cisterns of smooth ER extend along axons and dendrites. There is some discussion of whether the pockets of smooth ER within axons may play a role in the
transport or packaging of neurotransmitters or other proteins.

Subsurface Cisternae
Although not a constant feature of all neurons, a system of smooth, membrane-bound,
flattened cisterns can be found in many neurons. These structures, referred to as

hypolemmal cisternae by Palay and Chan-Palay [1], abut the plasmalemma of the neuron and, in such areas, constitute a secondary membranous boundary within the cell. The distance between these cisternae and the plasmalemma is usually in the range of 100 to 120 Å and, on occasion, a mitochondrion may be found in close association with the innermost leaflet (e.g., in Purkinje cells). Similar cisternae have been described beneath synaptic complexes, but the functional significance of this is not clearly understood. Some authors have suggested that such a system may play a role in the uptake of metabolites.

Golgi Apparatus

Undoubtedly, the most impressive demonstration of the Golgi system, a highly specialized form of agranular reticulum, is achieved by using the metal impregnation techniques of Golgi. Ultrastructurally, it consists of aggregates of smooth-walled cisternae and a variety of vesicles, surrounded by an heterogeneous assemblage of organelles, including mitochondria, lysosomes, and multivesicular bodies. In most neurons, the Golgi apparatus encompasses the nucleus, extends into dendrites, but is absent from axons. A three-dimensional analysis of the system reveals that the stacks of cisternae are pierced periodically by fenestrae. Tangential sections of these fenestrations show them to be circular. A multitude of vesicles is associated with each segment of Golgi apparatus, in particular "coated" vesicles which are proliferated from the lateral margins of flattened cisternae (Fig. 1-4). Such structures have been variously named, but the term *alveolate vesicle* seems to be generally accepted. Histochemical staining reveals that these bodies are rich in acid hydrolases, and they are believed to represent primary lysosomes. Acid phosphatase is also found elsewhere in the cisterns but in lesser amount than in alveolate vesicles.

Lysosomes

Lysosomes are organelles that are common constituents of all nerve cells and can often be seen at various stages of development (Fig. 1-2). They range in size from about 0.1μ to several microns in diameter. The matrix of a mature lysosome is made up of an electron-dense, finely granular material, which gives a positive reaction for acid phosphatase when stained histochemically. A single limiting membrane can be shown in favorable sections, and paracrystalline arrays of lamellae are not uncommon amid the granular matrix. A brown pigment (lipofuscin), characteristic of aging neurons, is accumulated in old (tertiary) lysosomes (Fig. 1-5).

Multivesicular Bodies

Multivesicular bodies are usually found in association with the Golgi apparatus. They are visualized by EM as small, single, membrane-bound sacs about 0.5 micron in diameter, containing several minute spherical profiles, sometimes regularly arranged about the periphery. Currently, they are believed to represent secondary lysosomes because they contain acid hydrolases and form from the alveolate vesicles of the Golgi apparatus.

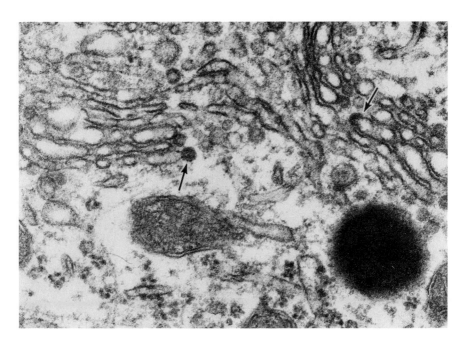

Figure 1-4
A portion of a Golgi apparatus. The smooth-membraned cisternae appear beaded. The many
circular profiles represent tangentially-sectioned fenestrae and alveolate vesicles. Two of the
latter can be seen budding from Golgi saccules (arrows). Mitochondria and a dense body
(lysosome) are also present. × 60,000.

Neurotubules

The neurotubule has been the subject of intensive research in recent years. Neurotubules usually are arranged haphazardly throughout the perikaryon of neurons.
They are aligned longitudinally in axons and dendrites and, in the latter, form the
major filamentous component. Each neurotubule consists of a dense walled structure enclosing a clear lumen in the middle of which may be found an electron-dense
dot. Sometimes axonal neurotubules display 50 Å, filamentous, interconnecting side
arms. The diameter of neurotubules varies between 220 and 240 Å. High-resolution
studies seem to indicate that each neurotubule wall consists of thirteen filamentous
subunits arranged helically around the lumen. These structures and the neurofilaments (below) can be seen in the photographs discussed later in connection with
astrocytes.

Neurofilaments

Neurofilaments usually are found in association with neurotubules. The function of
the two organelles has been debated for some time, and, although it seems reasonable
to assume that they play a role in the maintenance of form, their putative role in
axoplasmic transport remains to be clarified (see Chap. 21). Neurofilaments have a
diameter of about 100 Å, are of indeterminate length, and frequently occur in

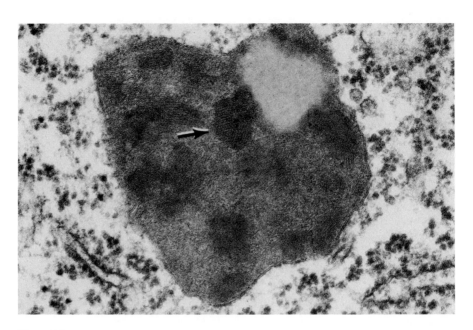

Figure 1-5
A lipofuscin granule from a cortical neuron shows membrane-bound lipid (dense) and a soluble component (gray). The denser component is lamellated. The lamellae appear as paracrystalline arrays of tubular profiles when sectioned transversely (arrow). The granule is surrounded by a single unit membrane. Free ribosomes can also be seen. × 96,000.

bundles. They are constant components of axons, but are rarer in dendrites. In the axon, individual filaments can be seen to possess a minute lumen and to be interconnected by floccular, proteinaceous side arms, thereby forming a meshwork. Because of these cross-bridges, they do not form tightly packed bundles in a normal axon, in distinction to filaments within glial processes (i.e., astroglia, see Fig. 1-12). which lack cross-bridges. Neurofilaments within neuronal somata do not always display cross-bridges and can be found in tighter bundles. The biochemistry of neurotubules and neurofilaments has been covered by Shelanski and Weisenberg [8]. A form of filamentous structure finer than neurofilaments is seen at the tips of growing neurites, particularly in the growth cones of developing axons. These 50 Å structures, known as *microfilaments,* facilitate movement and growth, for axonal extension can be arrested pharmacologically by treatment with compounds which depolymerize these structures.

Mitochondria
Mitochondria are organelles that are the centers for oxidative phosphorylation. They occur ubiquitously throughout the neuron and its processes (Figs. 1-2 and 1-4). Their overall shape may change from one type of neuron to another, but their basic morphology is identical to that of other cell types. They consist morphologically

of double-membraned sacs surrounded by protuberances or cristae extending from the inner membrane into the matrix space. The recent review by Novikoff and Holtzman [9] discusses in more detail the ultrastructure and enzymatic properties of mitochondria and the above cellular components.

The Axon

As the axon egresses, it is physiologically and structurally divisible into distinct regions – the axon hillock, the initial segment, and the axon proper. These segments are discussed in detail by Peters et al. [4]. Basically, the segments differ ultrastructurally on the basis of membrane morphology and the content of rough and smooth ER. The axon hillock may contain fragments of Nissl substance, including abundant ribosomes, which diminish as the hillock continues into the initial segment. Here the various axoplasmic components begin to align longitudinally. A few ribosomes and smooth ER still persist and some axo-axonic synapses occur. More interestingly, however, the axolemma of the initial segment, the region for the generation of the action potential, is underlaid by a dense granular layer similar to that seen at the node of Ranvier. Also present in this region are neurotubules, neurofilaments, and mitochondria. The neurotubules in the initial segment, unlike their scattered pattern in the distal axon, occur in fascicles and are interconnected by side arms [4]. Beyond its initial segment, the axon maintains a relatively uniform morphology. It contains an axolemma without any structural modification, microtubules sometimes cross-linked, neurofilaments interconnected by granular proteinaceous strands, mitochondria, and tubulo-vesicular profiles probably derived from smooth ER. Myelinated axons show granular densifications beneath the axolemma at nodes of Ranvier, and synaptic complexes may also occur at the same regions. The terminal portions of axons arborize and enlarge at their synaptic regions, where they contain synaptic vesicles lying beneath the specialized presynaptic junction.

The Dendrite

Dendrites, the afferent components of neurons, are frequently arranged around the neuronal soma in a stellate fashion. In some neurons they may arise from a single trunk from which they arborize into a dendritic tree. Unlike axons, they are generally lacking in neurofilaments, although they may contain fragments of Nissl substance. Larger branches of dendrites, however, in close proximity to neurons, may contain small bundles of neurofilaments. Some difficulty may be encountered in distinguishing small unmyelinated axons, terminal segments of axons, and small dendrites. In the absence of synaptic evidence, however, they often can be assessed by the content of neurofilaments. The synaptic regions of dendrites occur either along the main stems (Fig. 1-6) or at small protuberances known as *dendritic spines* or thorns. Axon terminals abut these structures.

The Synapse

The axons and dendrites that emerge from different neurons intercommunicate by means of specialized junctional complexes known as *synapses,* a fact and name first proposed by Sherrington in 1897. Their existence was immediately demonstrable

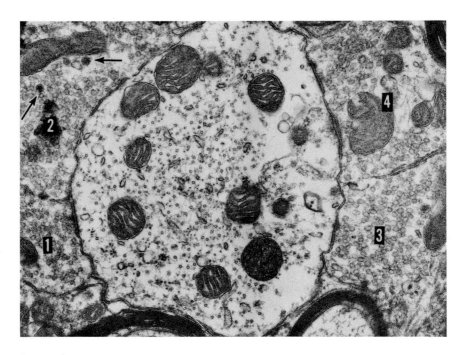

Figure 1-6
A dendrite emerging from a motor neuron in the anterior horn of the spinal cord of a rat is contacted by 4 axonal terminals. Terminal 1 contains clear spherical synaptic vesicles, terminals 2 and 3 contain both clear spherical and dense core vesicles (arrows), and terminal 4 contains many clear flattened (inhibitory) synaptic vesicles. Note also the synaptic thickenings and (within the dendrite) the mitochondria, neurofilaments, and neurotubules. × 33,000.

by EM and today can be recognized in a dynamic fashion by Nomarski optics for light microscopy and by scanning EM. With the development of neurochemical approaches toward neurobiology (see Chap. 2 for synaptosomes and Chap. 8 for synaptic transmission), an understanding of synaptic form and function becomes of fundamental importance. As was noted in the first ultrastructural study on synapses (Palade and Palay in 1954, quoted in [5]), synapses display interface specialization and are frequently polarized or asymmetrical. The asymmetry is due to the unequal distribution of electron-dense (osmiophilic) material or thickening applied to the apposing membranes of the junctional complex and the heavier accumulation of organelles within the presynaptic (usually axonal) component. The closely applied membranes constituting the synaptic site are overlaid on the pre- and postsynaptic aspects by an osmiophilic material similar to that seen in desmosomes (see later under Ependymal Cell), and are separated by a gap (cleft) of a constant spacing between 150 and 200 Å. The presynaptic component usually contains a collection of clear, 400- to 500-Å synaptic vesicles and various numbers of small mitochondria about 0.2 to 0.5 micron in diameter (Figs. 1-6 to 1-8). Occasional 240-Å microtubules, coated vesicles, and cisternae of smooth ER are not uncommon in this region. On the postsynaptic side

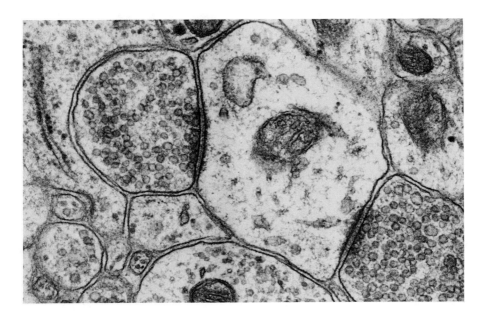

Figure 1-7
A dendrite (center) is flanked by two axonal terminals packed with clear spherical synaptic vesicles. Details of the synaptic region are clearly shown. × 75,000.

is a greater amount of thickening, referred to as the subsynaptic web, but apart from an infrequent, closely applied packet of smooth ER (subsurface cisterna) belonging to the hypolemmal system, there are no aggregations or organelles in the dendrite.

At the neuromuscular junction, the morphological organization is somewhat different. Here the axon terminal is greatly enlarged and is ensheathed by Schwann cells, and the postsynaptic (sarcolemmal) membrane displays less densification and is deeply infolded.

Before elaborating further on synaptic diversity, it might be helpful to outline briefly other ways in which synapses have been classified in the past. Using the light microscope, Cajal (quoted by Bodian [10]) was able to identify 11 distinct groups of synapses. Nowadays, most neuroanatomists apply a more fundamental classification schema to synapses, depending upon the profiles between which the synapse is formed, i.e., axo-dendritic, axo-somatic, axo-axonic, dendro-dendritic, somato-somatic, and somato-dendritic. Unfortunately, such a list totally disregards the type of transmission (chemical or electrical) or, in the case of chemical synapses, the neurotransmitter involved.

In terms of physiological typing, three groups of synapses are recognized: excitatory, inhibitory, and modulatory. Some neuroanatomical studies (Walberg, 1965, and Uchinozo, 1965, quoted by Bodian [10]) on excitatory and inhibitory synapses have claimed that the former possess spherical synaptic vesicles, whereas inhibitory

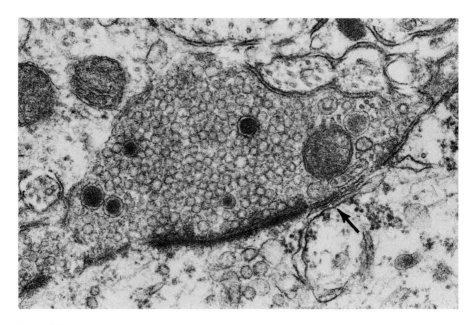

Figure 1-8
An axonal terminal at the surface of a neuron from the dorsal horn of a rabbit spinal cord con-
tains both dense-core and clear-spherical synaptic vesicles lying above the membrane thickenings.
A subsurface cisterna (arrow) is also seen. × 68,000.

synapses contain a predominance of flattened vesicles (Fig. 1-6). Other studies, e.g.,
Gray [11], have correlated this synaptic vesicular diversity with physiological data.
In his study on cerebellum, Gray showed that neurons with a known predominance
of excitatory input on dendrites and an inhibitory input on the cell body possess two
corresponding types of synapses. However, although the above interpretation fits
well in some loci of the CNS, it does not hold true for all regions. Furthermore, some
workers feel that the differences between flat and spherical vesicles may reflect an
artifact of aldehyde fixation or a difference in physiological state at the time of sampling.
In the light of these criticisms, it is clear that further confirmation as to the correlation
between flattened vesicles and inhibitory synapses is required.

Another criterion for the classification of synapses by EM was introduced in 1959
by Gray [11]. Briefly, certain synapses in the cerebral cortex can be grouped into
two types, depending upon the length of the contact area between synaptic membranes and the amount of postsynaptic thickening. Relationships have been found
between type 1, the membranes of which are closely apposed for long distances and
have a large amount of associated postsynaptic thickening, and excitatory axo-dendritic
synapses. Type 2 synapses, on the other hand, are mainly axo-somatic, show less close
apposition and thickening at the junction, and are believed to be inhibitory. This
broad grouping has been confirmed in the cerebral cortex by a number of workers, but
it does not hold true for all centers of the CNS.

Most of the data gained from studies on synapses in situ or on synaptosomes (see Chap. 2) have been on cholinergic transmission. Understanding of the vast family of chemical synapses belonging to the autonomic nervous system which utilize biogenic amines (Chap. 10) as neurotransmitter substances is still in its infancy. Morphologically, catecholaminergic synapses are similar but possess, in addition to clear vesicles, dense-core or granular vesicles of variable and slightly larger dimensions (Figs. 1-6 and 1-8). These vesicles were first identified as synaptic vesicles by Grillo and Palay [12], who segregated classes of granular vesicles based on vesicle and core size, but no relationship was made between granular vesicles and transmitter substances. About the same time, EM autoradiographic techniques were being employed and, by using tritiated norepinephrine, Wolfe et al. [13] were able to localize the label to granular vesicles within axonal terminals.

Since this work, numerous other labeling techniques for aminergic synapses have been developed. Several of the methods and requirements for detecting such transmitters have been reviewed by Bloom [14]. Catecholaminergic vesicles are generally classified on a size basis, and not all have dense cores.

Another, as yet unclassified category of synapses may be the so-called silent synapses observed in both in vitro and in vivo CNS tissue. These synapses are morphologically identical to functional synapses but are physiologically dormant.

Finally, with regard to synaptic types, there is the well-characterized electrical synapse [15], in which current can pass from cell to cell across regions of membrane apposition that essentially lack the associated collections of organelles present at the chemical synapse. In the electrical synapse (Fig. 1-9), the unit membranes are closely apposed; the outer leaflets sometimes fuse to form a pentalaminar structure. However, in most places, a gap of about 20 Å exists producing a so-called gap junction. Not infrequently, such gap junctions are separated by desmosomelike regions [4]. Sometimes electrical synapses exist at terminals which also display typical chemical synapses — in such a case the structure is referred to as a *mixed synapse*. The comparative morphology of electrical and chemical synapses has been recently reviewed by Pappas and Waxman [16].

THE NEUROGLIA

Classification
Virchow (1846) first recognized the existence in the CNS of a fragile, nonnervous, interstitial component made up of stellate or spindle-shaped cells, morphologically distinct from neurons, which he named neuroglia (nerve glue). It was not until the early part of this century that this interstitial element was classified into distinct cell types [4]. Today, we identify two broad groups of glial cells: (1) the macroglia, embracing astrocytes and oligodendrocytes, all of ectodermal origin; and (2) the smaller microglia, of mesodermal origin. Macroglial cells develop from a common stem cell, the spongioblast. Microglia invade the CNS at the time of vascularization via the pia mater, the walls of blood vessels, and the tela choroidea.

All glial cells differ from neurons in that they possess no synaptic contacts, and all retain the ability to divide throughout life, particularly in response to injury. The

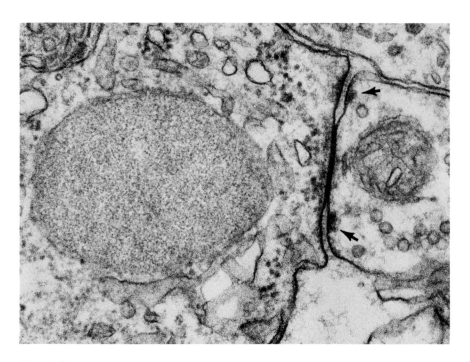

Figure 1-9
An electrotonic synapse is seen at the surface of a motor neuron from the spinal cord of a toad-fish. Between the neuronal soma (left) and the axonal termination (right) a tight junction flanked by desmosomes is visible. Photograph, courtesy of Drs. G. D. Pappas and J. S. Keeter. × 80,000.

rough schema represented by Figure 1-1 demonstrates the interrelationships between the macroglia and other CNS components.

The Astrocyte

The complex packing achieved by the processes and cell bodies of astrocytes heralds the potential involvement of this cell type in brain metabolism because virtually nothing can enter the CNS parenchyma without being confronted by an astrocytic interphase.

Although astrocytes traditionally have been subdivided into protoplasmic and fibrous astrocytes [5], these two forms probably represent the opposite ends of a large spectrum of variation within the same cell type. The morphological components of fibrous and protoplasmic astrocytes are identical; the differences are quantitative. In the early days of electron microscopy, structural differences between the two vari-ants were more apparent, due to artifactual changes, but, with the development of better fixation procedures, it became apparent that the differences were not so great.

Protoplasmic astrocytes range in size from 10 to 40 microns, are frequently located in gray matter in relation to capillaries, and have a clearer cytoplasm than do fibrous astrocytes (Fig. 1-10). Within the perikaryon are scattered 90-Å filaments and 240-Å microtubules (Fig. 1-10), glycogen granules, lysosomes, and lipofuscinlike bodies,

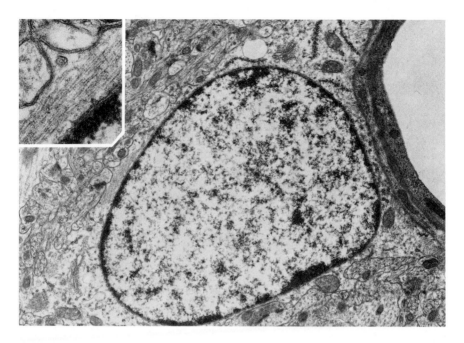

Figure 1-10
A protoplasmic astrocyte abuts a blood vessel in the cerebral cortex of a rat. The nucleus shows a rim of denser chromatin, and the cytoplasm contains many organelles including Golgi and rough ER. × 10,000. Inset: Detail of perinuclear cytoplasm showing filaments. × 44,000.

isolated cisternae of rough ER, a small Golgi apparatus opposite one pole of the nucleus, and small elongated mitochondria, often extending, together with loose bundles of filaments, along cell processes. A centriole is not uncommon. Characteristically, the nucleus is ovoid and the nucleochromatin homogeneous, except for a narrow rim of dense chromatin and one or two poorly defined nucleoli. The fibrous astrocyte occurs in white matter (Fig. 1-11). Its processes are twiglike, being composed of large numbers of glial filaments oriented in tight bundles. The filaments within these cell processes can be distinguished from neurofilaments by their close packing and the absence of cross-bridges (Figs. 1-11 and 1-12). Desmosomes are often between adjacent astrocytic processes.

In addition to protoplasmic and fibrous forms, regional specialization occurs among astrocytes. The outer membranes of those located in subpial zones and adjacent to blood vessels possess a specialized thickening [17]. Also, desmosomes between astrocytic processes are common in these regions. In the cerebellar cortex, protoplasmic astrocytes can be segregated into three classes — the Golgi epithelial cell, the lamellar or velate astrocyte, and the smooth astrocyte [1], each ultrastructurally distinct.

Astrocytic Function
The functions of astrocytes have long been debated. One of their major roles is related to their connective tissue or skeletal function since they invest, possibly

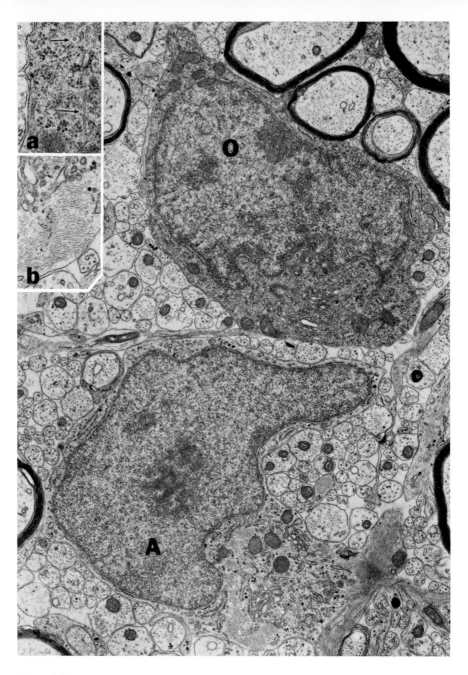

Figure 1-11

A section from the myelinating white matter of a kitten contains a fibrous astrocyte (A) and an oligodendrocyte (O). The nucleus of the astrocyte has homogeneous chromatin with a denser rim and a central nucleolus. That of the oligodendrocyte is denser and more heterogeneous. Note the denser oligodendrocytic cytoplasm and the prominent filaments within the astrocyte. × 15,000. Inset: (a) Detail of the oligodendrocyte showing microtubules (arrows) and absence of filaments. × 30,000. (b) Detail of astrocytic cytoplasm showing filaments, glycogen, rough ER, and Golgi. × 30,000.

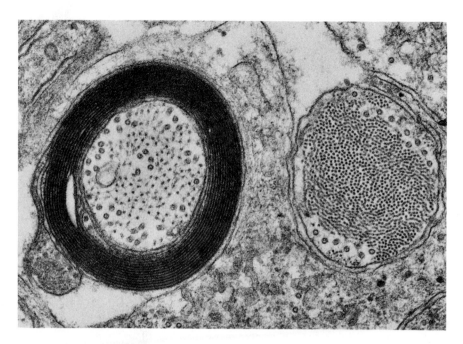

Figure 1-12
A myelinated axon (left) and the process of a fibrous astrocyte (right) from the spinal cord of an adult dog are sectioned transversely. The axon contains scattered neurotubules and loosely packed neurofilaments interconnected by proteinaceous strands. The astrocytic process contains a bundle of closely packed filaments with no cross-bridges, flanked by several microtubules. Sometimes a lumen can be seen within a filament. × 60,000.

sustain, and provide a packing for other components. In the case of the astrocytic ensheathment seen around synaptic complexes and the bodies of some neurons, e.g., Purkinje cells, it may be speculated that the astrocyte serves to isolate these structures.

One well-known function of the astrocyte is concerned with reparation. Subsequent to trauma, astrocytes invariably proliferate, swell, accumulate glycogen, and undergo fibrosis by the accumulation of filaments. This state of gliosis may be total where all other elements are lost, leaving a glial scar, or it may be a generalized response occurring in a background of regenerated or normal CNS parenchyma. Such a phenomenon can occur in both the gray and white matter thereby indicating common links between protoplasmic and fibrous astrocytes. With age, both fibrous and protoplasmic astrocytes accumulate filaments. In some diseases, astrocytes have been shown to become macrophagous. It is interesting to note that the astrocyte is probably the most disease-resistant component in the CNS as very few diseases (one of them is alcoholism) can cause depletion or degeneration of astrocytes.

Another putative role for the astrocyte is its involvement in transport mechanisms and in the blood-brain barrier system. It was believed for some time that the transport of water and electrolytes is effected by the astrocyte, a fact never definitively demonstrated and largely implied from pathological or experimental evidence. It is known, for example, that damage to the brain vasculature, the production of local heat or cold injuries, or inflammatory changes produce a focal swelling of astrocytes, presumably due to a disturbance in fluid transport. Large molecules can be transferred, however, across the CNS without the involvement of astrocytes. The astrocytic investment of blood vessels might suggest a role in the blood-brain barrier system, but the studies of Reese and Karnovsky [18] and Brightman [19] indicate that the astrocytic end feet provide little resistance to the movement of molecules and that blockage of passage of material into the brain occurs at the endothelial cell lining blood vessels (see Chap. 20).

The Oligodendrocyte

The ultrastructural studies of Schultz et al. and Farquhar and Hartman in 1957 [5] were among the first to contrast the EM features of oligodendrocytes with astrocytes (Fig. 1-10). The study by Mugnaini and Walberg [5] more explicitly laid down morphological criteria for identification of these cells and, apart from technical improvements apropos quality of preservation, our EM understanding of these cells has changed little since that time. Like the astrocyte, the oligodendrocyte is of ectodermal origin but has fewer cell processes and is spindle shaped.

From a number of EM studies [20], it is apparent that, as with the astrocyte, oligodendrocytes are highly variable, differing in location, morphology, and function but definable by some morphological criteria. The cell soma ranges from 10 to 20 microns, is roughly globular, and is more dense than that of an astrocyte. The margin of the cell is irregular and compressed against the adjacent neuropil. Few cell processes are seen. Within the cytoplasm, many organelles are found. Parallel cisterns of rough ER and a widely dispersed Golgi apparatus are common. Free ribosomes occur, scattered amidst occasional multivesicular bodies, mitochondria, and coated vesicles. Serving to distinguish the oligodendrocyte from the astrocyte is the apparent lack of glial filaments but the constant presence of 240-Å microtubules (Fig. 1-11), most common at the margins of the cell, in the occasional cell process, and in the cytoplasmic loops around myelin sheaths. This absence of filaments, first noticed by Mugnaini and Walberg (1964), has been confirmed by several workers. Oligodendrocytic filaments have been observed only in experimental situations. Dense bodies also occur in oligodendrocytes. Most of these are related to lipid droplets, lysosomes, and pigment granules. The nucleus is usually ovoid but slight lobation is not uncommon. The nucleochromatin stains heavily and contains clumps of denser heterochromatin; the whole structure is sometimes difficult to discern from the background cytoplasm. Desmosomes are known to occur between interfascicular oligodendrocytes.

Ultrastructural studies on the developing nervous system [21, 22] and labeling studies [20] have demonstrated the variability in oligodencrocyte morphology and activity. Mori and Leblond [20] segregated oligodendrocytes into three groups based upon location, stainability, and DNA turnover. The three classes corresponded to satellite, intermediate, and interfascicular (myelinating) oligodendrocytes. Satellite oligodendrocytes are small (about 10 microns), are restricted to the gray matter, and are closely applied to the surface of neurons. They are assumed to play a role in the maintenance of the neuron. Interfascicular oligodendrocytes are large (about 20 microns) during myelination but, in the adult, range from 10 to 15 microns with the nucleus occupying a large percentage of the volume of the cell soma. Intermediate oligodendrocytes are regarded as potential satellite or myelinating forms. The nucleus in these cells is small, the cytoplasm occupying the greater area of the soma.

The Oligodendrocyte and Myelin

Myelinating oligodendrocytes have been studied extensively [21, 23, 24]. Examination of the CNS during myelinogenesis (Fig. 1-13) reveals connections between the cell body and the myelin sheath [21]. However, connections between the cell body and the myelin sheath have never been demonstrated in a normal adult animal, unlike its PNS counterpart, the Schwann cell. In contrast to the Schwann cell (see below), the oligodendrocyte is capable of producing many internodes of myelin simultaneously. It is estimated that oligodendrocytes in the optic nerve might produce between 30 and 50 internodes of myelin [23]. In addition to this heavy structural commitment, the oligodendrocyte is known to possess a slow mitotic rate and a poor regenerative capacity. Damage to only a few oligodendrocytes, therefore, can be expected to produce an appreciable area of primary demyelination. Indeed, in most CNS diseases in which myelin is a target, the oligodendrocytes are known to be most vulnerable and among the first CNS elements to degenerate.

Somewhat analogous to the neuron, the relatively small oligodendrocytic cell body produces and supports many more times its own volume of membrane and cytoplasm. For example, take an average $12\,\mu$ oligodendrocyte producing 30 internodes of myelin (the lowest number quoted by Peters and Proskauer; see [23]), each axon with a diameter of $3\,\mu$ (small for CNS fibers) and covered by 6 lamellae of myelin (a conservative estimate), each lamella representing two fused layers of unit membrane. By statistical analysis, taking into account the length of myelin internode (possibly 500 microns) and the length of the membranes of the cell processes connecting the sheaths to the cell body (c. $12\,\mu$), the ratio between the surface area of the cell soma and the myelin it sustains is approximately 1:620. The formation of myelin is more completely discussed in Chapter 4.

In some rare instances, oligodendrocytes have been shown to elaborate myelin around structures other than axons. Rosenbluth (1966) was among the first to report the presence of aberrant myelin around neuronal somata and nonaxonal profiles. Since that time, numerous reports have appeared in which myelinated neuronal or oligodendrocytic somata are described [25].

THE EPENDYMAL CELL

Ependymal cells are arranged in single palisade arrays and line the ventricles and central canal. These cells are usually ciliated, their cilia extending from their free (apical) surfaces into the ventricular cavity. Their fine structure has been elucidated by Brightman and Palay [26]. They possess several features which clearly differentiate them from any other CNS cell. The cilia emerge from the apical pole of the cell where they are attached to a blepharoplast, the basal body (Fig. 1-14), which is anchored in the cytoplasm by means of rootlets and a basal foot. The basal foot is the contractile component which apparently determines the direction of the ciliary

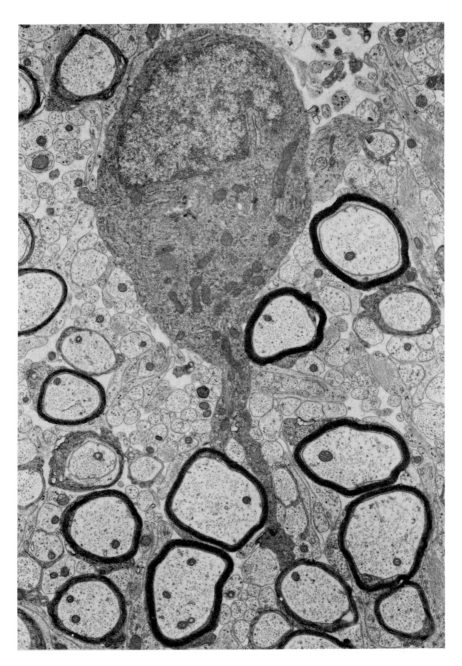

Figure 1-13
A myelinating oligodendrocyte from the spinal cord of a two-day-old kitten shows cytoplasmic connections to at least two myelin sheaths. Other myelinated and unmyelinated fibers, as well as glial processes, are seen in the surrounding neuropil. × 12,750.

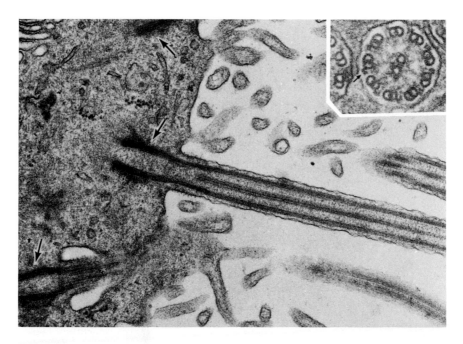

Figure 1-14
The surface of an ependymal cell contains basal bodies (arrows) connected to the microtubules
of cilia, seen here in longitudinal section. Several microvilli are also present. × 37,000. Inset:
Ependymal cilia in transverse section possess a central doublet of microtubules surrounded by
nine pairs, one of each pair having a characteristic hooklike appendage (arrow). × 100,000.

beat. Like all flagellar structures, the cilium contains the 9 + 2 microtubule arrangement (Fig. 1-14). In the vicinity of the basal body, the arrangement is of nine triplets,
and at the tip of each cilium the pattern is of haphazardly organized single tubules.
Also extending from the free surface of the cell are numerous microvilli containing
microfilaments (Fig. 1-14). The cytoplasm stains intensely, having an electron density about equal to that of the oligodendrocyte while the nucleus is similar to that
of the astrocyte. Microtubules, large whorls of filaments, coated vesicles, rough ER,
Golgi apparatus, lysosomes, and abundant small, dense mitochondria are also present.
The base of the cell is reflected into tortuous processes which interdigitate with the
underlying neuropil. The lateral margins of each cell characteristically display long
compound junctional complexes (Fig. 1-15) made up of desmosomes (zonulae
adhaerentes) and tight junctions (zonulae occludentes). The biochemical properties
of these structures are well known. Desmosomes display protease sensitivity, divalent
cation dependency, and osmotic insensitivity, and the membranes are mainly of the
smooth type. In direct contrast to desmosomes, the tight junction (and also gap junctions and synapses) displays no protease sensitivity and no divalent cation dependency,
is osmotically sensitive, and morphologically is made up of complex membranes.
These facts have been utilized in the development of techniques to isolate purified
preparations of junctional complexes.

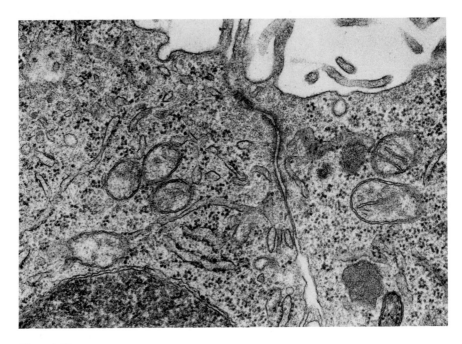

Figure 1-15
A typical desmosome and gap junction complex between two ependymal cells. Microvilli and pinocytic vesicles are also seen. × 35,000.

THE MICROGLIAL CELL

Of the few remaining types of CNS cells, the most important, and probably the most enigmatic, is the microglial cell, a cell of mesodermal origin, located in the normal brain in a resting state and purported to become a very mobile, active macrophage in times of need. While these cells can be selectively stained and demonstrated by light microscopy using Hortega's silver carbonate method, no comparable definitive technique exists for their ultrastructural demonstration. The cells have spindle-shaped bodies and a thin rim of densely staining cytoplasm difficult to distinguish from the nucleus. The nucleochromatin is homogeneously dense, and the cytoplasm does not contain an abundance of organelles although representatives of the usual components can be found. During normal wear and tear, some CNS elements degenerate, and microglia can phagocytose the debris (Fig. 1-16). Their identification and numbers (as determined by light microscopy) differ from species to species. The rabbit CNS is known to be richly endowed. In a number of disease instances, e.g., trauma, microglia are known to be stimulated and migrate to the area of injury where they phagocytose debris. The relatively brief mention of this cell type in EM text books [4], and the conflicting EM descriptions [27, 28], are indicative of the uncertainty attached to their identification. Pericytes, the pericapillary macrophages, are believed to be a resting form of microglial cell.

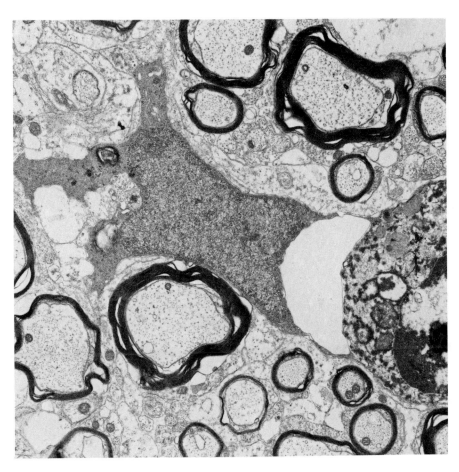

Figure 1-16
A microglial cell (center) has elaborated two cytoplasmic arms to encompass a degenerating oligodendrocyte (right) in the spinal cord of a three-day-old kitten. The microglial cell nucleus is difficult to distinguish from the narrow rim of densely staining cytoplasm. × 10,000.

THE SCHWANN CELL

When axons leave the CNS, they lose their neuroglial interrelationships, and traverse a short transitional zone, after which they become invested by Schwann cells. These are the axon-ensheathing cells of the peripheral nervous system, equivalent function-ally to the oligodendrocytes of the CNS. Not all PNS fibers are myelinated but, in contrast to nonmyelinated fibers in the CNS, unmyelinated fibers in the PNS are suspended in groups within Schwann cell cytoplasm, each axon connected to the extracellular space by a short channel, the mesaxon, formed by the invaginated Schwann cell plasmalemma. Myelinated fibers of the PNS, on the other hand, occur singly, are clearly separated, and each internode of myelin is elaborated by one Schwann cell. This ratio of one internode of myelin to one Schwann cell (1:1) is a fundamental distinction between this cell type and its CNS analog, the oligodendro-cyte, which is able to proliferate internodes in the ratio of 1:30 or greater. Furthermore, the Schwann cell body always remains in intimate contact with its myelin internode (Fig. 1-17). Periodically, myelin lamellae open up into Schwann cell cytoplasm producing bands of cytoplasm around the fiber — Schmidt-Lanterman incisures, reputed to be the stretch points along PNS fibers. These incisures are usually lacking in the CNS. The PNS myelin period is 119 Å in preserved specimens (some 30 percent less than in the fresh state), in contrast to the 106 Å of central myelin. In addition to these structural differences, PNS myelin is known to differ biochemically and antigenically from that of the CNS (Chap. 4).

Ultrastructurally, the Schwann cell is unique and distinct from the oligodendrocyte. Each Schwann cell is surrounded by a basement lamina about 200 to 300 Å thick that does not extend into the mesaxon and is presumably made up of mucopolysaccharide (Fig. 1-17). The basement laminae of adjacent myelinating Schwann cells at nodes of Ranvier are continuous, and the Schwann cell processes interdigitate so that the PNS myelinated axon is never in direct contact with the extracellular space. The Schwann cells of nonmyelinated PNS fibers overlap, and nodes of Ranvier are lacking. The cytoplasm of the Schwann cell is rich in organelles. A Golgi apparatus is located near to the nucleus, and cisterns of rough ER occur throughout the cell. Lysosomes, multivesicular bodies, glycogen granules, and lipid granules (pi granules) can also be seen. The cell is rich in filaments (in contrast to the oligodendrocyte), and many microtubules are present. The plasmalemma frequently shows pinocytic vesicles. Small round mitochondria are scattered throughout the soma. The nucleus is flattened, oriented along the nerve fiber, and stains intensely. Aggregates of denser heterochromatin are arranged peripherally. The various features of the Schwann cell are outlined in greater detail by Peters et al. [4].

The Schwann Cell During Disease

In sharp contrast to the oligodendrocyte, the Schwann cell has a vigorous response to most forms of injury. An active phase of mitosis occurs following traumatic insult, and the cells are capable of local migration. Studies on their behavior after primary demyelination have shown that in some cases they are able to phagocytose actively damaged myelin. They possess remarkable reparatory properties and can begin to lay down new myelin about one week after a fiber loses its myelin sheath. Studies on PNS and CNS remyelination by Raine et al. [29] have shown that, by three months after primary demyelination, PNS fibers are well remyelinated while similarly affected areas in the CNS show relatively little proliferation of new myelin. Under circumstances of severe injury, e.g., transection, axons degenerate, and the Schwann cells form tubes (Bungner bands) containing cell bodies and processes surrounded by a single basement lamina. These structures provide channels along which regenerating axons grow. The presence and integrity of the Schwann cell basement lamina is an essential factor for reinnervation.

OTHER PNS ELEMENTS

The extracellular space between PNS nerve fibers is occupied by bundles of collagen fibrils, blood vessels, and endoneurial cells. The endoneurial cells are elongated spindle-shaped cells with tenuous processes relatively poor in organelles except for

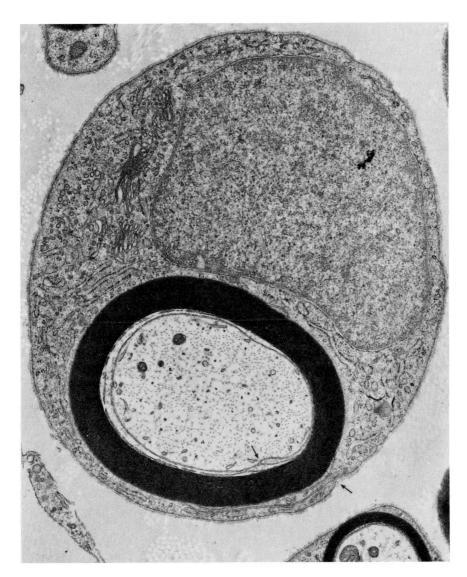

Figure 1-17
A myelinated PNS axon is surrounded by its Schwann cell. Note the fuzzy basement lamina
around the cell, the rich cytoplasm, the inner and outer mesaxon (arrows), the close proximity
of the cell to its myelin, and the one-to-one (cell-to-myelin internode) relationship. A process
of an endoneurial cell is seen (lower left) and unstained collagen lies in the extracellular space
(white dots). X 20,000.

large cisterns of rough ER. There is some evidence that these cells proliferate collagen fibrils. Sometimes mast cells, the histamine producers of connective tissue, can be seen. Bundles of nerve fibers are arranged in fascicles that are emarginated by flattened connective-tissue cells forming the perineurium, an essential component in the blood-nerve barrier system. Fascicles of nerve fibers are aggregated into nerves and invested by a tough elastic sheath of cells known as the *epineurium.*

IN CONCLUSION

This chapter has been compiled to provide a concise description of the major cyto-architectural features of the nervous system and to provide an entrance into the relevant literature. Although the fine structures of organelles of the CNS and PNS are not peculiar to these tissues, the interrelations of all types, e.g., the synaptic contracts and myelin, are unique. These specializations and those that sequester the CNS from the general circulation, viz., blood-brain barrier and absence of lymphatics, are major factors in normal and disease processes in the nervous system.

ACKNOWLEDGMENTS

The author thanks Dr. Robert D. Terry for his helpful discussion. The excellent technical assistance of Everett Swanson, Howard Finch, and Miriam Pakingan is appreciated. I thank Mrs. Mary Palumbo for her secretarial assistance.

The work represented by this chapter was supported in part by grants NS 08952 and NS 03356 from the National Institutes of Health and a grant from the Alfred P. Sloan Foundation.

The author is the recipient of a Research Career Development Award from the NIH — grant NS 70265. Figures 1-11 and 1-13 are from a study by the author and Dr. P. S. Spencer on the developing nervous system, currently in preparation.

REFERENCES

*1. Palay, S. L., and Chan-Palay, V. *Cerebellar Cortex: Cytology and Organization.* New York: Springer, 1974.
*2. Hydén, H. The Neuron. In Brachet, J., and Mirsky, A. E. (Eds.), *The Cell,* Vol. 5. New York: Academic, 1960.
*3. Windle, W. F. *Biology of Neuroglia.* Springfield, Ill.: Thomas, 1958.
*4. Peters, A., Palay, S. L., and Webster, H. deF. *The Fine Structure of the Nervous System: The Cells and Their Processes.* New York: Harper & Row, 1970.
*5. Mugnaini, E., and Walberg, F. Ultrastructure of neuroglia. *Ergebn. Anat. Entwickl. Gesch.* 37:194, 1964.
 6. Sabatini, D. D., Bensch, K., and Barrnett, R. Cytochemistry and electron microscopic preservation of cellular ultrastructure and enzymatic activity by aldehyde fixation. *J. Cell Biol.* 17:19, 1963.
 7. Palay, S. L., McGee-Russell, S. M., Gordon, S., Jr., and Grillo, M. A. Fixation of neural tissues for electron microscopy by perfusion with solutions of osmium tetroxide. *J. Cell Biol.* 12:385, 1961.

*Asterisks denote key references.

*8. Shelanski, M. L., and Weisenberg, R. C. Cytochemical Methods for the Study of Microtubules and Microfilaments. In Glick, D., and Rosenbaum, R. M. (Eds.), *Techniques of Biochemical and Biophysical Cytology*, Vol. 1. New York: Wylie, 1972, p. 25.

*9. Novikoff, A. B., and Holtzman, E. *Cells and Organelles.* New York: Holt, Rinehart and Winston, 1970.

*10. Bodian, D. Synaptic Diversity and Characterization by Electron Microscopy. In Pappas, G. D., and Purpura, D. P. (Eds.), *Structure and Function of Synapses.* New York: Raven, 1972, p. 45.

11. Gray, E. G. Electron microscopy of excitatory and inhibitory synapses: A brief review. *Prog. Brain Res.* 31:141, 1969.

12. Grillo, M. A., and Palay, S. L. Granule-Containing Vesicles in the Autonomic Nervous System. In Breese, S. S. (Ed.), *Proc. Fifth Int. Cong. Electron Microscopy.* New York: Academic, 1962, p. U-1.

13. Wolfe, D. E., Potter, L. T., Richardson, K. C., and Axelrod, J. Localizing tritiated norepinephrine in sympathetic axons by electron microscopic auto-radiography. *Science* 138:440, 1962.

*14. Bloom, F. E. Localization of Neurotransmitters by Electron Microscopy. In *Neurotransmitters, Proc. ARNMD,* Vol. 50. Baltimore: Williams & Wilkins, 1972, p. 25.

*15. Bennett, M. V. L. Electrical Versus Chemical Neurotransmission. In *Neurotransmitters, Proc. ARNMD,* Vol. 50. Baltimore: Williams & Wilkins, 1972, p. 58.

*16. Pappas, G. D., and Waxman, S. Synaptic Fine Structure: Morphological Correlates of Chemical and Electrotonic Transmission. In Pappas, G. D., and Purpura, D. P. (Eds.), *Structure and Function of Synapses.* New York: Raven, 1972, p. 1.

17. Cook, R. D., Raine, C. S., and Wisniewski, H. M. On perivascular astrocytic membrane specializations in monkey optic nerve. *Brain Res.* 57:491, 1973.

18. Reese, T. S., and Karnovsky, M. J. Fine structural localization of a blood-brain barrier to exogenous peroxidase. *J. Cell Biol.* 34:207, 1967.

19. Brightman, M. The distribution within the brain of ferritin injected into cerebrospinal fluid compartments. II. Parenchymal distribution. *Am. J. Anat.* 117:193, 1965.

20. Mori, S., and Leblond, C. P. Electron microscopic identification of three classes of oligodendrocytes and a preliminary study of their proliferative activity in the corpus callosum of young rats. *J. Comp. Neurol.* 139:1, 1970.

*21. Bunge, R. P. Glial cells and the central myelin sheath. *Physiol. Rev.* 48:197, 1968.

22. Caley, D. W., and Maxwell, D. S. An electron microscope study of the neuroglia during postnatal development of the rat cerebrum. *J. Comp. Neurol.* 133:45, 1968.

*23. Davison, A. N., and Peters, A. *Myelination.* Springfield, Ill.: Thomas, 1970.

24. Hirano, A., and Dembitzer, H. A structural analysis of the myelin sheath in the central nervous system. *J. Cell Biol.* 34:555, 1967.

25. Raine, C. S., and Bornstein, M. B. Unusual profiles in organotypic cultures of central nervous tissue. *J. Neurocytol.* 3:313, 1974.

26. Brightman, M., and Palay, S. L. The fine structure of ependyma in the brain of the rat. *J. Cell Biol.* 19:415, 1963.

27. Mori, S., and Leblond, C. P. Identification of microglia in light and electron microscopy. *J. Comp. Neurol.* 135:57, 1969.

28. Blakemore, W. F. Microglial reactions following thermal necrosis of the rat cortex: An electron microscopic study. *Acta Neuropath.* (Berlin) 21:11, 1972.

29. Raine, C. S., Wisniewski, H., and Prineas, J. An ultrastructural study of experimental demyelination and remyelination. II. Chronic experimental allergic encephalomyelitis in the peripheral nervous system. *Lab. Invest.* 21:316, 1969.

Chapter 2

Cellular and Subcellular Fractionation

Stanley H. Appel
Eugene D. Day

One of the significant aims of neurochemistry is to describe neuronal and glial behavior at the molecular level. Past investigative efforts have attempted to isolate relatively pure fractions of subcellular components that retain metabolic activity and structural integrity. However, such fractions are necessarily heterogeneous because their organelles are derived from a widely diverse population of neurons and glia. Decreasing the numbers of neurons and glia by isolating specific regions of the brain may help, but heterogeneity would still exist in the smaller population of cells as, no matter what degree of purity might be achieved, such organelles reflect the diverse population of cells from which they were derived. It is gratifying that, since the last edition of this text, much effort has been directed toward limiting such heterogeneity through the isolation of specific cells and the characterization of their components. In this chapter, we shall emphasize these recent techniques and include the methods for both cellular separation and subcellular fractionation.

GENERAL SCHEME OF SUBCELLULAR FRACTIONATION

Cells usually are disrupted by such various means as shearing, shocking, swelling, and sonication. Depending upon the method chosen, one sacrifices the yield of the more sensitive particles in order to dissociate the units that are sounder structurally. No single disruption technique will yield all particles intact. It is remarkable that the simple method of tissue homogenization in buffered isoosmotic sucrose or saline works as well as it does. The tolerance between the Teflon pestle and glass tube may be important for one particle (e.g., a loose fit for synaptosomes); the number of up-and-down strokes may be critical for some tissues (several for lung, few for brain); the viscosity of the medium can be significant for others (high for tumor cells); and the adjustment of bivalent cation concentrations and pH has an optimal, and sometimes critical, effect upon certain particles (e.g., whether they disperse or aggregate, swell or shrink, absorb or discharge components). The usual subcellular fractionation methods attempt to treat tissues mildly and separate their parts into a few major categories (Fig. 2-1, right ordinate). Before separating tissues in the centrifuge, it may be necessary to treat them by osmotic shock in order to release a particular component, e.g., mitochondria, from their parent synaptosomes prior to gradient centrifugation. In general, the size and density of the subcellular particles are the most critical features in the design of separation. Usually there are enough differences among cell constituents with respect to these two parameters to make it theoretically possible to define each type uniquely. One may order the placement of subcellular constituents of neurons and glia within a typical suspension of disrupted cells according to sedimentation coefficient (one aspect of size) and intrinsic density, as shown in Figure 2-1. From a practical point of view, however, it is often difficult to separate

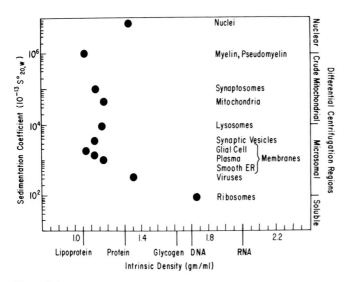

Figure 2-1
Size and density of subcellular particles in a typical brain homogenate is ordered by sedimentation coefficient and intrinsic density.

organelles with closely related sizes and densities. As a result, recent efforts have concentrated on both antigenic and other chemical features of cells and subcellular fractions that permit separation by properties independent of size and density. These approaches utilize immunological and partitioning techniques and are proliferating at such a rapid rate that they should have significant effects on all cell separation methods within the next few years.

The process of homogenization and release of subcellular particles into a suspending medium will change particle properties through solvation, osmotic action, and ion effects.

1. Solid particles (e.g., membranes) that carry a solvation mantle have a density that is intermediate between the medium and the particles themselves until the isopycnic point is reached, i.e., the point at which the medium and the particles have the same density. At that point the "anhydrous," or intrinsic, density of the particles can be measured.

2. Vesicular particles that exhibit osmotic behavior may eventually become equilibrated with the medium in which they move as they approach an isopycnic density. They, too, will then exhibit an intrinsic density value independent of fluid content or mantle. Pure osmotic behavior, however, is rarely observed among subcellular components, perhaps because the investigator cannot often wait long enough to achieve equilibrium and because other effects are also involved; nevertheless, mitochondria and synaptosomes both approach the ideal as osmometers.

3. Vesicular particles, such as cell nuclei, that respond less readily to changes in osmotic pressure may exhibit volume changes that are contrary to expectation (e.g.,

swelling when taken from hypotonic 0.01 M $CaCl_2$ into hypertonic 0.88 M ion-free sucrose) [1]. Such a response may be related to the ionic environment and the ion-exchange capacity of the particles. In such cases, there may be considerable difficulty in obtaining intrinsic densities or even a stabilized extrinsic one that is independent of the operational procedure.

In none of these situations will the sedimentation coefficient be descriptive of a particle devoid of environmental factors, and the coefficient must be interpreted as being more operational than definitive. However, the purpose of subcellular fraction-ation procedures is not so much to duplicate the intracellular milieu as it is to reach sufficient artifactual stability to permit experiments to be conducted and meaningful questions asked.

The classic method of differential centrifugation involves only sedimentation velocity. Particles are dissociated and separated in a medium of relatively low and fixed density, usually isotonic salt or sucrose solutions. Four fractions are obtained by sequential centrifugation steps of increasing rotational velocity and time: a low-speed, short-term nuclear pellet; a medium-speed, medium-time, crude mitochondrial pellet; a high-speed, prolonged-time, microsomal pellet; and a supernatant fluid of soluble components. This technique has been widely used although it has a serious defect in that there is an inevitable contamination of low-speed pellets with high-speed components. As a result, pellets must be washed extensively by repeated centrifugation. Sufficient dilution of suspensions of both original and homogenates and subsequent washings may limit the number of contaminants already at the bottom of the tube before centrifugation starts.

Isopycnic or buoyant density centrifugation orders the separation of constituents solely on the basis of initial extrinsic particle density and final intrinsic particle den-sity. Usually this is accomplished by equilibrium density-gradient centrifugation in which each constituent will either sediment down to or float up to the density of the medium at which it is isopycnic and at which it is weightless. Sucrose or iso-osmotic Ficoll-sucrose is commonly used (Fig. 2-2).

The relative amount of four main constituents — RNA, DNA, protein, and lipid — determines the intrinsic density of any one particle. The intrinsic density is a fixed property and is not subject to much change. The actual extrinsic density is subject to fluctuation because it depends upon the volume and concentration of fluids as well as upon the nature of surface-adsorbed components. To obtain a true density value experimentally (e.g., in CsCl) one usually must sacrifice metabolic activity. From a practical standpoint, most separation techniques depend more upon extrinsic or oper-ationally defined densities than upon fixed parameters.

Rate-zonal centrifugation makes use of the velocity-density differential among components. It is based on the fact that the sedimentation coefficient decreases as the density difference between particle and medium narrows when the particle moves into denser regions of a gradient. Thus, although both dense and light particles are slowed down, the ratio between their respective sedimentation velocities continuously widens and makes the clear separation of one from the other much easier to obtain than by differential procedures.

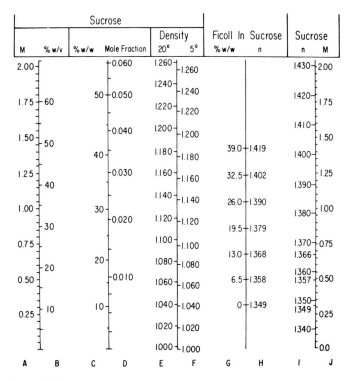

Figure 2-2
Sucrose solutions. Comparison of methods of expression: (A) and (J) molarity, and (B) weight-volume per 100 at 25° C; (C) weight-weight per 100; (D) mole fraction; (E) density at 20° C; (F) density at 5° C. Comparison with isoosmotic Ficoll-sucrose (of 0.19 g per liter osmosity or 0.32 M sucrose): (G) weight-weight per 100; (H) refractive index measured at 20° C; (I) refractive index of sucrose alone. Line up straight edge for the same values in (A) and (J).

Some of the practical advantages of density-gradient centrifugation procedures are:

1. When the sample is layered on top of the medium, the desired purified component is sedimented away from the initial sample (by contrast, see differential centrifugation, above).
2. Multiple washings are not required.
3. Several components are separated in a single centrifugation step.
4. Improvement of separation can be made according to sedimentation velocity, density, and time.

CRITERIA OF PURITY

It is still not possible to set up absolute criteria of purity for isolated particles because the heterogeneity of cell types permits statements only of quantitative enrichment rather than qualitative separation and mutual exclusion of contaminating organelles. It no longer is sufficient to distinguish biochemical characteristics of brain nuclei, mitochondria, plasma membranes, endoplasmic reticulum, etc., from each other. At best, such characterizations represent average statements, for considerable differences exist in the composition of membranes or organelles in neurons as compared to glia, and among various types of neurons and glia. For example, the specific activity of cytochrome oxidase varies widely from brain region to brain region as well as from cell type to cell type, e.g., it may be high in cerebral cortex, medium in the thalamus, and low in the corpus callosum. Furthermore, gangliosides that once were thought to be specific markers for neuronal membranes are now recognized as markedly enriched in astrocytes. Myelin constituents may also be present in oligodendroglia.

The morphological and biochemical markers can be presented as a guide to particle and cell enrichment, but not as specific criteria of absolute purity. The biochemical markers provided in Table 2-1 should be used in this context. As yet, there are no comparable immunochemical markers of cellular and subcellular membrane components, although Schachner [2] has shown that "the catalogue of antigens demonstrable on the surface of neuroblastoma C1300 cells, which are of a sympathetic nervous system origin, is similar to that of normal brain tissue." Specific components within glial and neuronal cells can be analyzed immunochemically or by immunohistochemical techniques. Several different antigenic constituents have been employed to separate neuronal and glial cell types, and these should be of increasing importance in the future.

Glial Fibrillary Acidic Protein

The glial fibrillary acidic protein (GFAP) reported by Eng et al. [3] is a 43,000-dalton major constituent of neuroglial fibers and is astrocyte specific according to Bignami and Dahl [4]. It shows evidence of a high degree of evolutionary stability [5] and is easily measured as an extract by immunodiffusion techniques and in situ in tissues by immunofluorescence techniques. The antigens that were used for raising antisera were obtained from gliotic regions of multiple sclerosis plaques from autopsied human brains. Rabbits received 6 mg purified GFAP in complete Freund's adjuvant.

Table 2-1. Selected Properties of Isolated Cell Types[a]

	Rat		Bovine
	Neurons	Astrocytes	Oligodendroglia
Yield 10^6 cells/g fresh tissue	17	3.5	11.4
Dry weight pg/cell	178	590	25
RNA pg/cell	24.2	29.1	2.0
DNA pg/cell	8.2	11.2	5.2
Total lipid % of dry weight	24.1	38.9	29.5
Gangliosides % of dry weight	0.23	0.60	0.25
Lipids % of total lipid			
Cholesterol	10.6	14.0	14.1
Total glycolipid	2.1	1.8	9.9
Total phospholipid	72.3	70.9	62.2

[a]Data from Poduslo and Norton [19].

S-100 Protein
The acidic protein reported by Moore [6], although quantitatively localized within glia, may also be detected by immunofluorescent staining in neurons [7]. In the pyramidal neurons of the hippocampus in rat brain, the staining is strongly nuclear without involving the nucleoli. In cerebellar granular cells, the S-100 is localized to the cytoplasmic or perinuclear regions [8]. The 25,000-dalton, S-100 protein appears also to be stable evolutionarily and its production can be noted by cultured glioma cells.

14-3-2 Protein
This protein of Moore and Perez [9] is predominantly neuronal in location but has been noted to a limited extent in glial cells [10] by employing the immunofluorescent technique.

Basic Protein
Myelin basic protein is a predominantly cytoplasmic constituent in myelin. It is located asymmetrically along the inner cytoplasmic surface of the membrane [11]. Antibodies directed against basic protein do not react with the outermost lamellar surface of the native intact myelin sheath. Basic protein is detectable in radioimmunoassay, therefore, only after extraction of myelin or in situ after sectioning by immunohistochemical techniques. There are no known immunological studies concerned with the intrinsic proteins of myelin such as the proteolipid Folch-Lee protein, the Triton X-100 insoluble Wolfram protein, and the DM-20 Agrawal protein.

SEPARATION OF NEURONS AND GLIA
The availability of homotypic cell populations of isolated neurons and glia has become one of the pressing needs of modern neurobiology. Such populations are required for characterizing specific surface antigens, neurotransmitter receptors, membrane characteristics, and other metabolic features. They also are critical for isolating more homogeneous subcellular fractions. Only with procedures that can efficiently and harmlessly resolve such tissue into its component cells will we be able to identify the cellular properties responsible for biochemical uniqueness. The relative proportions of neurons and glia change significantly with age of the brain. In immature rats, glia represent a small percentage of cortical cells, whereas in adult animals, glia outnumber neurons. Attempts have been made to isolate various fractions by hand dissection, micromanipulation, and macroscale bulk preparations. Neuronal perikarya were first isolated by Deiters in 1865. Subsequently, the studies by Lowry in 1953, Chu in 1954, and Hydén in 1959 demonstrated the usefulness of microdissection techniques for sampling neuronal and glial populations.

Several methods have been described for isolating purified cell fractions in bulk. In general these depend upon maintaining the integrity of the cells by selecting the appropriate medium, disrupting the tissue into single cells, either through mechanical or enzymatic means, then sieving and collecting the desired neuronal and glial cells by centrifugation in discontinuous gradients of sucrose or Ficoll-sucrose. The first of the recent procedures for cell fractionation was developed by Rose [12]. Subsequently, procedures have been presented by Satake et al. [13], Flangas and Bowman [14], Blomstrand and Hamburger [15], Sellinger et al. [16], and Norton and Poduslo [17].

In the technique developed by Poduslo and Norton [18], rat brain is removed from the animal, weighed, and the cerebellum discarded. The tissue is minced and incubated in a 0.1 percent trypsin solution at 37° C for 90 minutes. The tissue is then cooled in ice and a 0.1 percent trypsin inhibitor is added. The tissue is centrifuged at 500 g for 5 minutes and washed two times in a cell medium consisting of 5 percent glucose, 5 percent fructose, 1 percent albumin in a 10-mM KH_2PO_4:NaOH buffer, pH 6.0. All steps, except for the 37° C incubation, are performed at 4° C. The washed tissue is resuspended in cell medium and stroked through a nylon screen of 100 to 150 mesh. The nylon is stretched across a funnel with a deep insert that has been placed in a vacuum filtration setup. The tissue is stroked with a blunt glass rod while a minimum of vacuum pulls the dissociated cells through the screen. The cell suspension is then passed several times through a stainless-steel screen of 200 mesh or 74-μ openings. The cell-suspension volume is adjusted with a cell medium, and aliquots are layered onto discontinuous sucrose gradients. The gradients consist from the bottom of 2.0, 1.55, and 0.9 M sucrose and are centrifuged at 4,100 g for 10 minutes.

A. *Neurons* layer on 2 M sucrose; a mixed-cell layer with red blood cells and endothelial cells is on 1.55 M sucrose, a crude astrocyte layer is on 1.35 M, and myelinated axons, broken-cell processes, clumped tissue, and subcellular debris are found on 0.9 M. The neurons can be diluted gently at least fivefold with medium and concentrated by centrifugation at 630 g for 10 minutes.

B. *Astrocytes* can be purified further by relayering onto another discontinuous sucrose gradient. The 1.35 M astrocyte layer is diluted gently with cell medium to twice the volume, and aliquots are layered onto gradients consisting of 2.0, 1.35, and 1.0 M sucrose. These are centrifuged at 5,000 g for 20 minutes. The astrocytes collect at the 1.35 M interface. They are concentrated by diluting gently with medium at least fivefold and centrifuging the suspension at 630 g for 10 minutes.

C. *Oligodendroglia* are isolated by Poduslo and Norton [19] from calf white matter. From the corpus callosum and centrum semiovale of one calf brain that has been extracted, placed in a plastic bag, packed in ice, and transported as quickly as possible to the laboratory, 30 to 40 g of white matter is dissected. The tissue is minced as well as possible with a sharp blade and incubated in 0.1 percent trypsin solution at 37° C for 90 minutes. After incubation, the tissue is cooled, and a 0.1 percent solution of soybean trypsin inhibitor is added. The tissue is centrifuged at 630 g for 5 minutes and washed twice with cell medium. The white matter is suspended in 0.9 M sucrose and cell medium and is stroked through the 100-to-150-mesh nylon screen as noted above for neuronal and astrocytic isolation from rat brain. Often higher vacuum and more stroking are required to completely dissociate the tissue. The suspension is then passed three times through a stainless-steel screen of 200-mesh. After the volume is adjusted, aliquots are layered onto discontinuous sucrose gradients consisting of 1.55, 1.4, and 0.9 M sucrose and are centrifuged at 4,100 g for 20 minutes. The thick myelinated-axon layer at the top of the gradient, as well as other cellular constituents in the reddish 1.4 M sucrose layer, is discarded. The remainder of the gradient, including the pellet, consists of oligodendroglia. If gray matter persists during the dissection, neurons are found in the pellet fraction. Oligodendroglial cells can be concentrated by diluting them fivefold with cell medium and centrifuging them at 630 g for 10 minutes.

Cell fractions are routinely monitored by phase microscopy, using a 200- or a 400-power objective. Cell counts can be made on an aliquot by adding a 10 percent formalin solution and counting with a phase hemocytometer. To remove trypsin activity, the cells may be washed three times in cell medium made 50 percent with calf serum and centrifuged at 500 g for 5 minutes at 4° C. Freshly isolated cells are able to maintain their capacity for protein and nucleic-acid synthesis, and presently are being investigated intensely for their metabolic characteristics. Some of their biochemical characteristics are summarized in Table 2-2. Neurons have been found to possess the neurotransmitter enzymes — dopamine-β-hydroxylase, tyrosine hydroxylase, choline acetyltransferase, glutamine decarboxylase, and phenylethanolamine-N-methyltransferase — whereas these same enzymes are absent from astrocytes and oligodendroglia. The enzyme 2′,3′-cyclic nucleotide-3′-phosphohydrolase is present predominantly in oligodendroglia although approximately 10% of the activity can be demonstrated in neurons and in astrocytes. Similarly, the activity of galactocerebroside phosphotransferase can be localized in oligodendroglia with much smaller amounts of activity being detected in neurons and astrocytes. The presence of this enzyme parallels the relative distribution of galactocerebroside and sulfatide in these three types of cells.

The ultrastructure of these isolated neurons, astroglia, and oligodendroglia is presented in Figure 2-3 [20]. All three cell types in these samples retained most of their characteristic features. A triple-layered plasma membrane was constantly present and showed only minor interruptions. Neurons ranged from 15 to 40 μm in diameter and contained abundant Nissl substance. Astroglia varied between 10 and 30 μm and proved difficult to obtain in a well-preserved state. There was a paucity of filaments possibly because of precipitation or dissolution during processing. Oligodendroglial preparations were greater than 90 percent pure and contained rounded cells, 8 to 20 μm in diameter, each with a narrow rim of dense cytoplasm usually containing microtubules, surrounding an equally dense rounded nucleus. The ultrastructural features of all these cell types are in accord with their counterparts in vivo.

The technique Sellinger and his coworkers [21] use for the bulk separation of neuronal and glial cells differs from the scheme detailed above primarily by eliminating added digestive enzymes. In addition, temperatures are maintained close to 0° C throughout the preparation.

Cerebral cortex from immature rats (3 to 30 days old) is dissected free of underlying white matter, chopped into a mince for 1 minute, and transferred at 25° C into a beaker that contains 106 ml per cortex of ice-cold solution 1 — 4.7 pH, 7.5% (w/v) polyvinylpyrrolidone (PVP), 1 percent (w/v) fraction-V bovine serum albumin (BSA), and 10 mM $CaCl_2$. The mince is decanted into a 20-ml plastic syringe, the tip of which is sawed off and has a 333-μm nylon cloth stretched over its end. The tissue mince is pushed through the nylon cloth by steady hand pressure. Two additional sievings with rinses of the filter between passages are carried out. The resulting suspension is then successfully sieved six more times, three times through a double layer of 333-μm and the 110-μm filters and three more times through the 333-μm and the 73-μm combination. In the final filtrate, this suspension is adjusted to 60 ml with solution 1, and 20 ml is layered on a two-step gradient of 1.0 M sucrose plus 1 percent BSA over 1.75 M sucrose plus 1 percent BSA. The tubes are centrifuged in the swinging bucket (SW) 25.1 rotor at 41,000 g for 30 minutes. Neuronal cell bodies are recovered in pure form in the pellet. Material banding at the interface between the sample solution and the 1.0 M sucrose solution is discarded as it contains chiefly impure myelin. Material banding at the 1.0 M to 1.75 M sucrose interface contains a mixture of glial cells, nerve-cell bodies, capillaries, nerve endings, and fibers. The material at this interface is suspended

Table 2-2. Biochemical Markers of Brain Cells and Subcellular Fractions

Fraction Subfraction	Marker	Author and Reference
Cytosol	lactate dehydrogenase (EC 1.1.1.27)	Kornberg [70]
Microsomes	NADPH$_2$:cytochrome c oxidoreductase (EC 1.6.2.3)	Miller and Dawson [76]
	NADPH:cytochrome c reductase (EC 1.6.99.1)	Sottocasa et al. [85]
Rough ER	RNA	Fleck and Begg [65]
Mitochondria	cytochrome c oxidase	Miller and Dawson [76] Cooperstein and Lazarow [58]
	succinic dehydrogenase	Earl and Korner [60]
Inner membrane	rotenone-sensitive NADH: cytochrome c reductase (EC 1.6.99.3)	Schnaitman et al. [84]
	D-3-hydroxybutyrate: NAD$^+$ oxidoreductase (EC 1.1.1.30)	Fitzgerald et al. [64]
Outer membrane	rotenone-insensitive NADH: cytochrome c reductase (EC 1.6.99.3 & 1.6.2.2)	Sottocasa et al. [85]
	Monoamine oxidase (EC 1.4.3.4)	Weissbach et al. [88] Krajl [71] Robinson et al. [82]
Synaptosomes	acetylcholinesterase (EC 3.1.17)	Ellman et al. [61]
Surface membranes	Na$^+$+K$^+$-activated ouabain-sensitive ATPase (EC 3.6.1.4)	Verity [87]
Catecholamine storage vesicles	dopamine-β-hydroxylase (EC 1.14.2.1)	Molinoff et al. [77]
Lysosomes	acid phosphatase (EC 3.1.3.2)	Jongkind [69] Cotman and Matthews [59]
	β-glucuronidase	Fishman and Bernfeld [63] Robins et al. [80, 81]
	β-glucosidase and β-galactosidase	Robins et al. [80, 81]
	N-arylamidase	Boer [56]
Nuclei	DNA and DNP	Steele and Busch [86]
Nuclear membrane	RNA polymerase	Roeder et al. [83]
Myelin	2',3'-cyclic nucleotide- 3'-phosphohydrolase (EC 3.1.4d)	Olafson et al. [78]

Table 2-2 (*Continued*)

Fraction Subfraction	Marker	Author and Reference
Cell membranes in general	total ATPase (EC 3.6.1.0) minus Mg^{2+}-ATPase (EC 3.6.1.4) equals Na^+,K^+-ATPase	Lowry and Lopez [74]
Brain particles in general	phosphatidic acid phosphatase	McCaman et al. [75]
Oligodendroglial cells	cerebroside sulphotrans- ferase	Farrell and McKhann [62] Benjamins et al. [55]
	glyceride galactosyl- transferase	Brenkert and Radin [57] Radin et al. [79]
Neuronal cells	guanyl cyclase	Goridis et al. [67, 68]
	glutamate decarboxylase	Fonnum [66]
	tyrosine hydroxylase	Kuczenski and Mandell [72]

in a medium of 7.5 percent PVP, 5 percent Ficoll, and 1 percent BSA and sieved three times through the 73-μm filter. The crude glial suspension is relayered onto a gradient of 30 percent Ficoll over 1.2 M sucrose over 1.65 M sucrose all containing 1 percent BSA. After centrifugation for 30 minutes at 41,000 g, neurons and capillaries are found in the pellet, and the glia are collected on 1.65 M sucrose. The glia are diluted with 0.32 M sucrose and pelleted at 480 g for 20 minutes. The glia are resuspended in 0.32 M sucrose, sieved through one thickness of 73-μm nylon filter, and placed on a third gradient of 1.3 M sucrose over 1.65 M sucrose. Following centrifugation at 2500 g for 20 minutes in the SW 25.1 rotor, the glial cells are recovered at the 1.65 M sucrose interface.

With cerebral cortex, cell yield was found to be maximal 8 days postnatally, whether expressed as protein or as RNA. The yield from neuronal cell bodies was a maximum of 6.5 mg per g at 5 days, falling to 0.9 mg per g at 18 days. The neuronal and glial cells can then be used to study amino-acid incorporation into protein or nucleotide incorporation into RNA [22]. In addition, they may form the basis for isolating subcellular organelles or membrane fractions.

It is important to note that none of these cell-isolation procedures may be reproduced easily the first time it is attempted. It is important to choose media containing high-molecular-weight substances in order to minimize cell breakage, to select appropriate sieving or enzymatic digestion procedures, and to choose isolation media with appropriate pH and osmotic characteristics. However, apparently many steps still cannot be defined, and these may be responsible for the variation in purity and yield found from laboratory to laboratory. Furthermore, many of the cell processes are lost in cellular isolation.

As our understanding of biochemical markers for various types of neurons and glia becomes clarified, we will be able to employ these characteristics for purification. It is also important to note that surface antigens of brain cells may possess fine differences

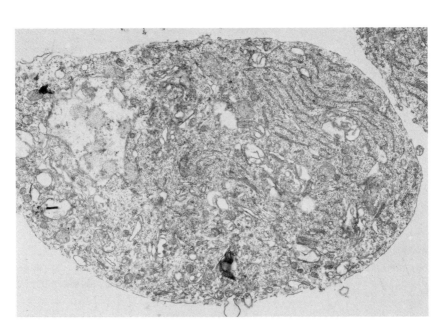

A

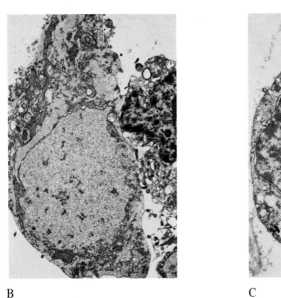

B   C

Figure 2-3
(A) Large neuron isolated from rat brain. Note abundant Nissl substance, Golgi apparatus, and good membrane preservation. × 7,000. (B) Isolated astroglial cell from rat brain. The cell has retained most of its features. Note pale cytoplasm and homogenous nucleochromatin. × 4,000. (C) Oligodendroglial cell isolated from bovine brain. The cell has retained most of its in vivo features, including a narrow rim of dense cytoplasm around the dark nucleus containing hetero-chromatin. Mitochondria are clear and swollen. × 6,000. (From Raine et al., *Brain Res.* 27:11, 1971.)

both between neurons and glia and among different cell species within each major class. With a greater understanding of these characteristics, the differential adhesion properties of viable cells to specified surfaces may aid in the bulk preparation of desired fractions. Such an approach has been reported by Schachner and Hammerling [23]. Red blood cells are caused to form rosettes on selected cells (via immunochemical or other specific ligands); the rosette-loaded cells are readily sedimented out and subsequently freed of the attached erythrocytes. An alternative source of bulk amounts of neuronal or glial materials is provided by neoplastic tissue. Tumors of neural as well as nonneural origin do retain some of differentiated characteristics of their normal counterparts and may be successfully adapted to grow in vitro and exhibit such characteristics. These cells may also be useful in the future for defining neural and glial properties to be employed for bulk isolation of nonneoplastic tissue.

ISOLATION OF CELL NUCLEI

Attempts to isolate nuclei from the nervous system [24] have been complicated by the heterogeneous populations of cells present in the brain and the varied array of particles produced by homogenization. Generally, preparations of neuronal and glial nuclei are contaminated with endothelial cells because of their similar densities. The major difficulty in isolating nuclei from whole brain is that both neuronal and glial nuclei will be included in the populations. Depending upon the length of time after death, the percentage of neuronal and glial nuclei isolated from whole brain tissue will vary. The hallmark of the neuronal nucleus is the single, dense, spherical nucleolus and the diffuse chromatin network. Astrocytic and oligodendroglial nuclei differ from each other when examined in vivo, but morphological differentiation is difficult in isolated nuclear fractions. As a result, to obtain such nuclei, purified populations of astrocytes or oligodendroglia must be used as the starting material.

The isolation of nuclei is facilitated by their high DNA and RNA content with the resulting high intrinsic density. Other structures are less dense, and thus should be separated easily. The major problems are to prevent aggregation or disruption by their ionic constituents and the pH of the homogenization and isolation media and to minimize those changes that would alter the extrinsic density of the nuclei in unpredictable ways. It has been claimed that homogenization and isotonic sucrose preserve the integrity of nuclei if calcium or magnesium ions at approximately 1 mM and pH between 6.2 and 6.7 are used [25].

Siakotos et al. [26] employ a technique of homogenization in 0.4 M sucrose with a loose-fitting (0.01-inch clearance) Potter-Elvehjem homogenizer, 5 to 6 passes at 1,200 rpm. The homogenate is passed through a 10-mesh sieve and centrifuged at 1,000 g for 30 minutes. Thereafter the crude nuclear fraction is purified by suspension and pelleting through higher-density sucrose and by passage through a Sephadex G-25 column. It is then resuspended and centrifuged through 0.83 percent NH_4Cl solutions and, finally, through 1.2 M sucrose. The purity is monitored by phase-contrast microscopy, and this technique yields 1 to 2 g of nuclei from 600 g of crude gray matter [27].

Cortical neuronal nuclei have also been purified starting with a neuronal fraction [21]. A crude nuclear fraction is isolated by homogenization and sedimentation through 0.5 M sucrose, followed by sedimentation through buffered (pH 6.5) 2.0 M and 2.40 M sucrose. The preparation has an RNA-to-DNA ratio of 0.54 and a protein-to-DNA ratio of 5.0, which is similar to results obtained with nuclei obtained from adult rabbit or cat brain.

MYELIN

Myelin may be defined morphologically as the lamellar structure that results from the concentric layers of "glial-cell membranes" that are deposited around axons. In the central nervous system, the cell membranes are derived from the oligodendroglial cell; in the peripheral nervous system, from the Schwann cell. These membranes condense into a compact laminated structure that may include small amounts of cytoplasm in both the innermost and outermost layers. Because of the strong interaction between the glial-cell membranes and the axonal membranes, many of the isolation procedures yield myelin that includes axonal membranes.

Myelin comprises about 50 percent of the white matter of the brain [28] and has been extensively studied because of its stability in isolation procedures and the high yield that may be obtained. It is a lipid-rich membrane; its protein, lipid, carbohydrate, and fatty-acid compositions are known and are described in greater detail in Chapter 4.

The methods of preparing myelin from the central nervous system usually begin with brain white matter or spinal-cord homogenate. Myelin is separated from other subcellular constituents either by differential centrifugation or by density-gradient centrifugation. The separation is enhanced by the high lipid content and low density of the membranes and by the large size of the vesicles produced during the initial homogenization.

By differential centrifugation, myelin can be found predominantly in the crude mitochondrial fraction. This material is then centrifuged on sucrose gradients, which yield crude myelin, synaptosomes, and mitochondrial fractions [29, 30], and the myelin usually will band above 0.85 M sucrose. The advantage of this technique is that other membrane fractions, such as smooth endoplasmic reticulum, are removed at an early stage in the isolation procedure. Its disadvantage, however, is that the total yield of the myelin might not be realized, and material might be lost in the crude nuclear pellet or the microsomal pellet.

An alternative approach has been utilized by Norton's laboratory [31]. A crude myelin fraction is isolated by direct sedimentation on a discontinuous sucrose gradient, after which it is purified to remove microsomal material. The major contamination of this procedure appears to be axoplasm as well as axonal membranes. This can be decreased by omitting ions from the initial homogenization procedure and shocking the myelin fractions osmotically in water and removing the myelin debris by centrifugation.

The criteria for purity are discussed in Chapter 4. With purified fractions, morphological techniques are inadequate to define the degree of contamination of the myelin by other membrane vesicles. The polypeptide pattern, as noted by SDS polyacrylamide gel profiles, or the specific enzymatic activities of such membranes are better indices of purity. However, the recently developed methods for purifying oligodendroglia make it apparent that constituents such as myelin-basic protein, proteolipid protein, and 2′,3′-nucleotide phosphohydrolase may be constituents of the oligodendroglial plasma membrane as well as of the myelin sheath. As a result, further work is needed to characterize constituents that may be present in mature myelin and absent in the oligodendroglial cell membranes.

MYELIN-FREE AXONS

Past studies of axons have depended on the use of cranial and peripheral nerves or the isolation of limited numbers of unmyelinated axons from cat cerebellum and basal ganglia. With the availability of isolation procedures for CNS myelinated axons from bovine white matter, it has become possible to remove the myelin sheath to obtain axons in sufficient quantities for morphological and biochemical analysis (Fig. 2-4) [32].

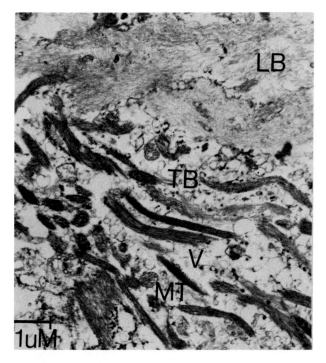

Figure 2-4
Electron micrograph of final rat central-nervous-system myelin-free axons. The preparation is essentially filamentous with loose bundles (LB) and tight bundles (TB) of filaments as well as membrane vesicles (V) and occasional mitochondria (MT) (De Vries, G. H., Hadfield, M. G., and Cornbrooks, C. J., used by permission).

In this procedure, the brain-stem areas from adult rats are minced and homogenized in 0.85 M sucrose, 0.10 M NaCl, 0.05 M KH_2PO_4 buffer at pH 6.0 in a loose and tight Dounce homogenizer. Following centrifugation at 9,500 g for 30 minutes, the floating layer is rehomogenized and centrifuged at 81,500 g for 15 minutes. Floating layers are again homogenized and are recentrifuged until no capillaries or nuclei are noted in the pellet. The final floating layers are suspended in 0.05 M KH_2PO_4 buffer at 6.0 and stirred for 12 hours. Under these conditions, the myelin sheath swells, forms vesicles, and is stripped from the axon. Following centrifugation at 81,500 g for 15 minutes, the axons remain in the pellet fraction and the myelin floats to the surface. The procedure is repeated until no myelin is seen floating above the axonal pellet.

This technique results in axons that contain 15.2 percent lipid, including 19.4 percent cholesterol, 56.9 percent phospholipid, and 23.7 percent galactolipid, with a molar ratio of cerebroside to sulfatide of 4:1. The phospholipids consist primarily of choline phosphatides and, to a lesser extent, of ethanolamine phosphatides. Electron microscopy reveals no myelin. Polyacrylamide gel electrophoresis shows the absence of a characteristic ratio of myelin-specific proteins. The myelin marker enzyme, $2',3'$-cyclic nucleotidase, is present only in trace amounts. No axolemma is evident in the outer edges of the bundle of filaments. There is no evidence of glial fibrillary acidic protein, as assayed by an immunoradiometric assay. Neurotubules are absent, but this is not surprising because they are known to be cold labile, and the entire procedure is carried out at 4° C.

The availability of preparations of myelinated axons and of myelin-free axons should enable a further investigation of the metabolic relationships between the axons and the myelin sheath. The high specific activity of the acetylcholinesterase in myelinated axon fractions, and its much lower specific activity in the myelin-free axons, would suggest that the majority of axolemma is removed from the axon with the stripping of the myelin. Perhaps the most surprising finding is the high amount of galactolipid found in axons. The presence of this lipid cannot be explained by myelin contamination, but it may be a constituent of axonal filaments themselves.

NERVE-ENDING PARTICLES (SYNAPTOSOMES)

Homogenization of brain tissue results in nerve terminals shearing from their axonal connections and from surrounding glial elements. The synaptic terminals are relatively resistant to mechanical stresses and can reseal to form *synaptosomes*. Often the intersynaptic cleft and a postsynaptic process can be isolated as a portion of this same particle. The presynaptic terminals may be used to study transmitter metabolism, macromolecular synthesis, presynaptic terminal mitochondria, vesicles, and the structure and function of plasma membrane. A synaptic-membrane fraction may then be utilized to isolate synaptic junctional complexes and postsynaptic densities.

The synaptosomal fraction has presented a unique opportunity to study neuronal properties in isolation with minimal glial contamination. Before purified preparations of neuronal and glial cell fractions were available, synaptosomes were the only means of approaching biochemical characteristics of neuronal membranes. However, drawbacks of this fraction are its lack of complete purity and the heterogeneous population of cell synapses included. It has been an extremely powerful tool, nonetheless, in the study of neurotransmitter metabolism and synaptic biochemical specializations.

Initially, synaptosomes were purified on a sucrose gradient by Whittaker et al. [29] and deRobertis et al. [33]. The particles were found to be made up primarily of a plasma membrane with inclusions of vesicles, mitochondria, and assorted soluble components. The basic procedure for isolation uses differential centrifugation, followed by further resolution on a sucrose or Ficoll-sucrose gradient. Homogenization of brain tissue has been performed with a Teflon-and-glass homogenizer that has the clearance of 0.025 cm, and uses a 10 percent (w/v) suspension in 0.32 M sucrose. A variety of techniques have been described for isolating the synaptosomal fraction from crude brain homogenates. These techniques take advantage of the fact that synaptosomes are predominantly in the crude mitochondrial fraction.

Attempts to sacrifice speed for increased yield of fractions may increase the disruption of synaptosomes. A contamination of the mitochondrial fraction by myelin may be minimized by using younger animals (10 to 20 days old) or by attempting to separate gray from white matter before homogenization [34]. After a spin at 1,000 g for 10 minutes, the supernatant is sedimented at 20,000 g for 20 minutes. At these low speeds, the yield of synaptosomes will be decreased by the loss of smaller-sized particle into the "microsomal fraction." On the other hand, the microsomal contamination of this fraction would thereby be minimized. Other sedimentations of the crude mitochondrial fraction (as high as 17,000 g for 55 minutes) have been used to maximize the yield, but these also increased the microsomal contamination in the crude mitochondrial fractions. Fractionation of the crude mitochondrial pellet has been performed in two ways.

One is the standard technique devised by Whittaker [29] and deRobertis [33]: Sedimentation is carried out in a discontinuous sucrose gradient made up in equal volumes of 0.8 and 1.2 M sucrose. The crude mitochondrial pellet is suspended in 0.32 M sucrose and centrifuged in a gradient at 53,000 g for 2 hours. With this gradient, small myelin fragments have been found above the 0.8 M fraction. Synaptosomes are isolated at the 0.8 and 1.2 M sucrose-fraction interface, and mitochondria are recovered in the pellet below 1.2 M sucrose. With this procedure, synaptosomal constituents are found in the mitochondrial pellet. Furthermore, mitochondria are present in the 0.8 to 1.2 M interface.

More recent procedures are based upon differential centrifugation and density-gradient centrifugation in Ficoll-sucrose [34–36]. After homogenization and dilution to a 10 percent (w/v) in 0.32 M sucrose, the material is centrifuged at 1,000 g for 5 minutes. The supernatant is pelleted at 11,000 g for 20 minutes and is washed twice in 0.32 M sucrose and applied to a Ficoll-sucrose gradient of 7.5 percent, 13 percent, 17 percent (w/v) Ficoll in 0.32 M sucrose. The synaptosomal fractions are located at the 7.5 to 13 percent and 13 to 17 percent interfaces after 45 minutes of centrifugation at 68,580 g.

The Ficoll-sucrose density gradients require less centrifugation time and maintain isosmotic conditions throughout the isolation procedure. Purity of the synaptosome fractions is determined both by morphology (Fig. 2-5) and by the specific activity of various enzymes, such as NaK ATPase and acetylcholinesterase. The difficulty in using

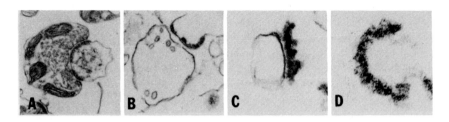

Figure 2-5
Synaptic structures that can be purified from brain. (A) Synaptosome with presynaptic element containing vesicles and mitochondria. (B) Synaptic plasma membranes. (C) Synaptic junctions. (D) Postsynaptic densities. (Courtesy of Dr. Carl Cotman [40].)

Ficoll is its variability from batch to batch, depending upon the ion concentration and the particle size. To study ion flux it is necessary to remove the high salt content by dissolving the Ficoll in a small amount of water and reprecipitating it from 95 percent ethanol. The ethanol is removed by lyophilization for prolonged periods of time.

SUBFRACTIONATION OF NERVE-ENDING PARTICLES
Nerve-ending particles can be disrupted easily by changing the osmotic pressure of the medium. The most popular technique for separating subsynaptosomal constituents is that devised by Whittaker et al. [37]. In this procedure, the washed crude mitochondrial pellet is subjected to osmotic shock in water, and the preparation is centrifuged at 53,000 g for 2 hours in a discontinuous sucrose density gradient that consists of equal volumes of 0.4, 0.6, 0.8, 1.0, and 1.2 M sucrose fractions. The synaptic plasma membrane layers at the 0.8 to 1.0 M interface; the mitochondria are found in the pellet, and synaptic vesicles are found at the 0.4 and 0.6 M fractions.

In recent modifications, introduction of a 0.7 and 1.1 M sucrose layer has helped to minimize mitochondrial contamination from the synaptic plasma membranes isolated at the 0.7 to 1.0 M and the 1.0 to 1.1 M interfaces. Contamination by mitochondria is also minimized by increasing the buoyant density of mitochondria by reducing iodonitroneotetrazolium violet by NADH [36, 38]. With this method, mitochondria are found entirely in the pellet, and synaptic plasma membranes can be isolated at the 0.7 to 1.0 M and 1.0 to 1.1 M interfaces. The other major change from the original Whittaker technique involves the hypoosmotic treatment of the synaptosomal fraction itself, rather than the crude mitochondrial fraction. Once again the NaK ATPase and acetylcholinesterase provide measures of the purity of the membrane fraction. In addition, the protein polypeptide profile provides a sensitive assessment of both relative purity and reproducibility.

ISOLATION OF SYNAPTIC COMPLEXES
The role of the synaptic complex in synaptic function is now unknown, but recent techniques for its isolation present the opportunity for detailed investigation of both structure and function. The isolation of the complex was made possible by the

demonstration that Triton X-100 selectively releases the synaptic complex from adjoining pre- and postsynaptic membranes. This work was confirmed and extended by Davis and Bloom [38] and by Cotman et al. [39], who demonstrated that the synaptic complexes could be identified by staining with ethanolic-phosphotungstic acid both in synaptic plasma-membrane fractions and after treatment with Triton X-100. Inclusion of calcium ions was found to aid in optimal preservations of these complexes.

The synaptic plasma-membrane fraction was isolated by the technique given in the previous section. Thereafter it was treated by the slow addition of Triton X-100 (4 mg per ml Triton X-100, 2 mM EDTA, 2 mM Bicine, pH 7.5) [40]. After 10 minutes, the suspension was applied to a sucrose density gradient of 1.0 M, 1.2 M, 1.4 M, and 1.5 M sucrose, 50 μM in $CaCl_2$, pH 7.4. It was centrifuged at 63,580 g for 1.25 hours. Synaptic complexes could be demonstrated in all fractions banding below 1.0 M sucrose. The optimal Triton-protein ratio was 1:1 or 2:1, and the yield of synaptic complexes was approximately 0.10 to 0.15 mg of protein per gram brain wet weight.

After isolation, synaptic complexes contain a section of pre- and postsynaptic plasma membranes separated by the synaptic cleft. A density is situated on the inner surface of the postsynaptic membrane (see Fig. 2-5). Proteolytic enzymes at high concentrations removed this postsynaptic density (PSD) and opened the synaptic cleft. At low concentrations, the PSD was destroyed selectively. The appearance of the PSD was not changed by treatment with NaCl, EGTA, and low concentrations of urea. Cotman et al. [40] concluded that polypeptides are major constituents of the PSD and may determine its structural integrity.

ISOLATION OF POSTSYNAPTIC DENSITIES
Isolation of the postsynaptic densities has been based on their buoyant density and their insolubility in n-lauroyl sarcosinate (NLS) [40]. Treatment of synaptic membranes with this detergent solubilizes most plasma membranes and detaches PSDs from the plasma membrane for future purification on density gradients. Sodium NLS was used either as a 0.5 percent (w/v) or 3.9 percent (w/v) solution in 10 mM Bicine, pH 7.5. With the 3.9 percent NLS, 20 ml is added to 5 to 8 ml of the membrane suspension that contains 40 to 80 mg of synaptic plasma membrane protein, so that the final detergent concentration is approximately 3 percent. After it becomes soluble, the suspension is applied to a discontinuous sucrose gradient consisting of 1.0, 1.4, and 2.2 M sucrose with 0.05 mM CaCl and at pH 7.0. After being centrifuged at 63,600 g for 75 minutes, the material that collects at the 1.4 to 2.2 M sucrose interface is found to be predominantly the postsynaptic densities, as determined by electron microscopy and by staining with ethanolic phosphotungstic acid (Fig. 2-5). Such fractions also stain with bismuth iodidine-uranyl lead, and contain cyclic 3',5'-phosphodiesterase activity. Quantitative analysis of the PSD fraction by electron microscopy assigns an estimated purity of better than 85 percent. These PSDs are primarily associated with dendritic excitatory synapses and therefore

represent the isolation of a distinctive structure that should provide an opportunity for us to understand its biochemical characteristics and its possible contribution in intercellular communication.

VESICLES

As mentioned above, the gradients devised by Whittaker [37] enabled the separation of synaptic vesicles in his "fraction D" at a particle-density equivalent to 0.4 M sucrose. Because of their relative resistance to changes in osmotic pressure, vesicles can be purified further by relayering them on a continuous gradient of 0.4 to 0.6 M sucrose above a layer of 1.6 M sucrose.

Small vesicles of approximately 500 Å diameter, isolated at a sucrose density of 0.4 to 0.5 M, contain approximately 50 percent of the acetylcholine found in the synaptosome fraction. The remainder of the acetylcholine is found in the supernatant fraction as the so-called labile pool. That the acetylcholine is within the synaptic vesicles is evident because it can be liberated by osmotic shock. During the initial osmotic shock to the synaptosomes, the synaptic vesicles apparently are protected by the synaptic membrane. However, if the vesicles are subjected directly to osmotic shock in the presence of distilled water, they will release their constituents into the medium.

There is similar evidence that nerve endings from tissues rich in catecholamines possess vesicles with high catecholamine content. According to Whittaker's data, yield of synaptic vesicles is approximately 3.8×10^{12} vesicles per gram of original tissue (guinea-pig cortex). The fraction is claimed to represent a 15 to 16 percent yield of the vesicles present in the initial homogenate. The membranes of the synaptic vesicles have different protein constituents from other membranes, as indicated by the studies of Cotman, Mahler, and Hugli [41], as well as by those conducted in our own laboratory [42]. In addition, they are thought to have a Mg^{2+}-activated ATPase. Markers of the various fractions include gangliosides, which are relatively rich in the synaptic plasma membrane and poor in the other fractions, and glycoproteins, which also are relatively rich in the synaptic plasma membrane.

LYSOSOMES

Brain lysosomes, like those of other tissues, are membrane-limited cytoplasmic organelles with acid hydrolase activity [43]. They comprise a heterogeneous population of granules that vary widely in size, structure, physical properties, and chemical composition. The reason for presenting a brief discussion of lysosomes in this chapter is to point out that such organelles contaminate almost all fractions and may impair the integrity of isolated particles.

Lysosomes contain enzymes, usually glycoproteins, with optimal pHs in the acid range. They sediment in most of the particulate fractions and usually must be disrupted by various techniques in order to release optimal enzyme activity. Koenig used acid hydrolases as markers and demonstrated that an enriched lysosomal fraction could be isolated from a crude mitochondrial fraction as a 1.4 M sucrose pellet after being centrifuged through a discontinuous gradient of 0.8, 1.0, 1.2, and 1.4 M sucrose. Some less dense lysosomal particles were isolated at the 1.2 to 1.4 M sucrose interface, especially when the microsomal fraction was used as starting material. Histochemical and electron-microscopic studies have confirmed this isolation procedure [44].

BRAIN MITOCHONDRIA

Mitochondria isolated from brain tissue possess both qualitative and quantitative differences from mitochondrial preparations of heart and liver. The methods used for isolating brain mitochondria cannot distinguish between glial and neuronal mitochondria, and neither type is homogeneous in all of its functional and biochemical characteristics [45, 46]. Brain tissue is a rich source of mitochondria, because the brain utilizes 25 percent of the total body oxygen and glucose, and the function of mitochondria is oxidative metabolism and the production of ATP. In fact, 15 percent of the total content of protein in brain is mitochondrial [47]. There is a greater proportion of mitochondria in neurons than in glia. Therefore the gray matter of the brain contains more mitochondria than does the white matter. Mitochondria may be found throughout the cytoplasm, axons, and dendrites of neurons, but are particularly abundant near the synaptic regions. Most of the mitochondria in cells derived from glia are found in the oligodendrocytes, with relatively few in the astrocytes and microglia. Schwann cells are also rich in mitochondria, with mitochondrial clusters around the nuclei.

Procedures for the isolation of neural mitochondria invariably start with a brain homogenate. The isolation and purification procedures are facilitated because mitochondria possess tough outer membranes that are not easily destroyed by osmotic and other physical forces, whereas the cellular plasma membranes are fragile and easily disrupted. Most of the isolation procedures for neural mitochondria have utilized rat, mouse, rabbit, or guinea-pig brains, but bovine, frog, and chicken brains have also been used.

The isolation procedures have included both differential and density-gradient centrifugation. The earliest procedures utilized simple differential centrifugation of the brain homogenate in isotonic (0.25 M) sucrose solutions [47, 48]. The next refinement was introduced by Gray and Whittaker [29] and deRobertis et al. [33]. Their technique was to use sucrose step gradients ranging from 5 to 30 percent sucrose. The mitochondrial preparations resulting from these procedures were still contaminated with myelin and synaptic material. Also, the increased osmotic properties of the higher concentrations of sucrose disrupted many of the mitochondrial membranes, thus increasing contamination and decreasing the yield.

Tanaka and Abood [49] introduced the use of Ficoll in conjunction with a continuous sucrose gradient as an improved method for isolating mitochondria. Ficoll can impart increasing densities to the sucrose while maintaining isoosmotic or near isoosmotic conditions. Abdel-Latif [35] used a Ficoll-sucrose step gradient, instead of a continuous gradient, in a simple method for isolating nerve-ending particles. His technique was adapted by Autilio et al. [34] to isolate other subcellular particles, including mitochondria.

Individual rat brains are hand homogenized in a 0.32 M sucrose solution (containing 5×10^{-5} M $CaCl_2$ and 2.5×10^{-3} M Tris, adjusted to pH 7.4) in the proportion of one volume of brain to nine volumes of sucrose solution. Isoosmotic Ficoll-sucrose [50] is used to prepare step gradients with densities of 1.049, 1.067, 1.080, and 1.120. Samples of the homogenate are centrifuged on this gradient for 20 minutes at 68,108 g and allowed to decelerate without braking. The fraction of the homogenate that collects as a band on top of 1.120 Ficoll-sucrose is removed, diluted with 0.32 M sucrose to a density of 1.062, and layered on another Ficoll-sucrose step gradient with densities of 1.080, 1.120, and 1.156. After centrifugation for 20 minutes at 68,108 g, the material that bands at density 1.156 is removed; this is the mitochondrial fraction.

Because the intrinsic isopycnic point of mitochondria is 1.176, the banding of mitochondria at 1.156 may represent another population of mitochondria. The purity of these brain mitochondria has been tested by biochemical, immunological, and structural methods. Apparently they are not contaminated with synaptic membranes, because cycloheximide, an inhibitor of ribosomal and microsomal protein synthesis, did not inhibit uptake of labeled amino acids into protein. On the other hand, puromycin, a potent inhibitor of mitochondrial protein synthesis, greatly inhibited the uptake of labeled amino acid in this mitochondrial preparation. As an immunological test of purity, antibodies were produced in rabbits against the preparation and, by adsorption experiments, the antibody was shown to react specifically with brain mitochondria. As a structural check for purity, samples were taken for electron microscopy, and their sections were shown to be composed primarily of brain mitochondria.

PURIFICATION OF MICROTUBULES AND FILAMENTS
FROM BRAIN

In the nervous system, microtubules are fibrillar constituents with high concentrations in neurons, Schwann cells, and immature or reactive astrocytes [51]. The microtubule is composed of a 110,000-dalton protein subunit called *tubulin*. The tubulin is present not only in microtubules, but also appears to represent a significant percentage (11 to 40 percent) of soluble protein in brain. Colchicine binds directly and reversibly to the tubulin dimer and results in depolymerization of the microtubules. Vinblastine binds to the tubulin molecule at a site other than the colchicine-binding site in a less reversible fashion than does colchicine. In vitro vinblastine causes aggregation of tubulin, whereas in vivo it induces large paracrystalline arrays. Calcium inhibits the assembly of tubulin into microtubules. Thus, EGTA can be used to promote tubule assembly, and this effect can be reversed by the addition of calcium.

The purification of microtubules and tubulin can take advantage of these biochemical properties indicated above [51]. Kirkpatrick's [52] technique for purifying microtubules depends upon maintaining the labile microtubular structure intact for sufficient periods of time to purify the samples by differential centrifugation. Conditions include the use of hexylene glycol, together with low temperature.

A 16.7 percent homogenate of brain tissue in 1.0 M hexylene glycol buffered with 20 mM KH_2PO_4, pH 6.4, at 1° C is centrifuged at 48,000 g for 30 minutes.

The supernatant is then sedimented in a discontinuous sucrose gradient that consists of sucrose dissolved in buffered hexylene glycol to obtain solutions with density of 1.16, layered over a solution of density 1.19. The most highly purified microtubules were obtained after centrifugation at 150,000 g for 1 hour from the interface of the 1.16 and 1.19 fractions.

According to Kirkpatrick's data, the final product was 100 times more pure than the starting material. Interestingly, this fraction showed similar bands on sodium dodecyl sulfate and acrylamide-gel electrophoresis, as does tubulin prepared according to the techniques of Weisenberg et al. [53].

Filaments have been isolated from brain by floating myelinated axons and fractionating the axons on a series of density gradients [54]. Myelinated axons were prepared as indicated in a previous section. Myelin was then stripped off by suspension in hypotonic 30 mM phosphate buffer pH 6.5, homogenized, incubated at 4° C for one hour, and pelleted through 1.0 M sucrose. Two types of filaments, loose discrete bundles and short densely-staining bundles, were isolated. Both filaments had a diameter of 90 Å after density-gradient centrifugation in a discontinuous gradient of 1.0, 1.5, and 2.0 M sucrose. The filaments could be isolated at the 1.5 to 2.0 M sucrose interface and to a lesser extent in the pellet. Further purification could be obtained by isopycnic banding in CsCl at 1.3 g per cm^3.

The purified filaments do not bind colchicine or nucleotides and thus differ from tubulin. However, the effect of CsCl on filament structure is not known. In polyacrylamide-gel electrophoresis, the most prominent component migrates at a molecular weight of 60,000, whereas tubulin is a dimer with two distinct subunits with molecular weight between 55,000 and 60,000. Thus, in almost all biochemical and histological respects, the constituents of neurofilaments appear to differ from those of neurotubules.

ACKNOWLEDGMENTS

This work was supported in part by Grants NS07872, NS05417, NS12213, NS06233, and NS10237 from the National Institutes of Health, and Grants 558-D-5 and 883-C-3 from the National Multiple Sclerosis Society. We would like to thank Drs. C. Raine, C. Cotman, and G. DeVries for the electron micrographs of isolated fractions, Drs. S. Poduslo, W. Norton, C. Cotman, and O. Sellinger for providing descriptions of their isolation procedures, and Mrs. E. Chapman for secretarial assistance.

REFERENCES

1. Anderson, N. G., and Wilbur, K. M. Studies on isolated cell components. IV. The effect of various solutions in the isolated rat liver nucleus. *J. Gen. Physiol.* 35:781, 1952.
2. Schachner, M. Serologically demonstrable cell surface specificities on mouse neuroblastoma C1300. *Nature [New Biol.]* 243:117, 1973.
3. Eng, L. F., Vanderhaegen, J. J., Bignami, A., and Gerstl, B. An acidic protein isolated from fibrous astrocytes. *Brain Res.* 28:351, 1971.

*Asterisks denote key references.

4. Bignami, A., and Dahl, D. Astrocyte-specific protein and neuroglial differentiation. An immunofluorescence study with antibodies to the glial fibrillary acidic protein. *J. Comp. Neurol.* 153:27, 1974.

5. Dahl, D., and Bignami, A. Immunochemical and immunofluorescence studies of the glial fibrillary acidic protein in vertebrates. *Brain Res.* 61:279, 1973.

6. Moore, B. W. A soluble protein characteristic of the nervous system. *Biochem. Biophys. Res. Commun.* 19:739, 1965.

7. Hydén, H., and McEwen, B. A glial protein specific for the nervous system. *Proc. Natl. Acad. Sci. U.S.A.* 55:354, 1966.

8. Mahadik, S. P., Laev, H., Graf, L., and Rapport, M. M. Immunofluorescent localization of S-100 in rat brain neurons. *Trans. Am. Soc. Neurochem.* 5:161, 1974.

*9. Moore, B. W., and Perez, V. J. Specific Acidic Proteins of the Nervous System. In Carlsson, F. D. (Ed.), *Physiological and Biochemical Aspects of Nervous Integration.* Englewood Cliffs, N.J.: Prentice-Hall, 1968, p. 343.

10. Perez, V. J., Olney, J. W., Cicero, T. J., Moore, B. W., and Bahn, B. A. Wallerian degeneration in rabbit optic nerve: Cellular localization in the central nervous system of the S-100 and 14.3.2 proteins. *J. Neurochem.* 17:511, 1970.

11. Braun, P. E., and Golds, E. E. Asymmetric location of the basic protein in the intact myelin sheath. *Trans. Am. Soc. Neurochem.* 5:142, 1974.

12. Rose, S. P. R., and Sinha, A. K. Some properties of isolated neuronal cell fractions. *J. Neurochem.* 16:1319, 1969.

13. Satake, M., Hasegana, S., Abe, S., and Tanaka, R. Preparation and characterization of nerve cell perikaryon from pig brain stem. *Brain Res.* 11:246, 1968.

14. Flangas, A. L., and Bowman, R. F. Neuronal perikarya of rat brain isolated by zonal centrifugation. *Science* 161:1025, 1968.

15. Blomstrand, C., and Hamburger, A. Protein turnover in cell-enriched fractions from rabbit brain. *J. Neurochem.* 16:1401, 1969.

16. Sellinger, O. Z., Azcurra, J. B., Johnson, D. E., Ohlsson, W. G., and Lodin, Z. Independence of protein synthesis and drug uptake in nerve cell bodies and glial cells isolated by a new technique. *Nature [New Biol.]* 230:253, 1971.

17. Norton, W. T., and Poduslo, S. E. Neuronal soma and whole neuroglia of rat brain: A new isolation technique. *Science* 167:1144, 1970.

*18. Poduslo, S. E., and Norton, W. T. Isolation of Specific Brain Cells. In Fleischer, S., and Packer, L. (Eds.), *Methods in Enzymology.* Vol. 31, Part B. New York: Academic, 1975.

19. Poduslo, S. E., and Norton, W. T. Isolation and some chemical properties of oligodendroglia from calf brain. *J. Neurochem.* 19:727, 1972.

20. Raine, C. S., Poduslo, S. E., and Norton, W. T. The ultrastructure of purified preparations of neurons and glial cells. *Brain Res.* 27:11, 1971.

*21. Sellinger, O. Z., and Azcurra, J. M. Bulk Separation of Neuronal Cell Bodies and Glial Cells in the Absence of Added Digestive Enzymes. In Marks, N., and Rodnight, R. (Eds.), *Research Methods in Neurochemistry.* Vol. II. New York: Plenum, 1974, p. 3.

22. Azcurra, J. M., Sellinger, O. Z., and Carrasco, A. E. *In vivo* labeling of cytoplasmic RNA in neurons of the immature brain cortex. *Brain Res.* 86:144, 1975.

23. Schachner, M., and Hammerling, V. The postnatal development of antigens on mouse brain cell surfaces. *Brain Res.* 73:362, 1974.

*24. Rappaport, D. A., Maxey, P., Jr., and Daginawala, H. F. Nuclei. In Lajtha, A. (Ed.), *Handbook of Neurochemistry.* Vol. 2. New York: Plenum, 1969, p. 241.

25. Lovtrup-Rein, H., and McEwen, B. S. Isolation and fractionation of rat brain nuclei. *J. Cell Biol.* 30:405, 1966.

26. Siakotos, A. N., Rousen, G., and Fleischer, S. Isolation of highly purified human and bovine brain endothelial cells and nuclei and their phospholipid composition. *Lipids* 4:234, 1969.

*27. Siakotos, A. N. The Isolation of Nuclei from Normal Human and Bovine Brain. In Fleischer, S., and Packer, L. (Eds.), *Methods in Enzymology*. Vol. 31, Part A. New York: Academic, 1974, p. 452.

28. Norton, W. T., and Autilio, L. A. The lipid composition of bovine brain myelin. *J. Neurochem.* 13:213, 1966.

29. Gray, E. G., and Whittaker, V. P. The isolation of nerve endings from brain: An electron microscopic study of the cell fragments of homogenation and centrifugation. *J. Anat.* 96:79, 1962.

30. deRobertis, E., deIraldi, A. P., Rodriguez de Lores Arnaiz, G., and Salganicoff, L. Cholinergic and non-cholinergic nerve endings in rat brain. I. Isolation and sub-cellular distribution of acetylcholine and acetyl cholinesterase. *J. Neurochem.* 9:23, 1962.

*31. Norton, W. T. Isolation of Myelin from Nerve Tissue. In Fleischer, S., and Packer, L. (Eds.), *Methods in Enzymology*. Vol. 31, Part A. New York: Academic, 1974, p. 435.

32. DeVries, G. H., and Norton, W. T. The lipid composition of axons from bovine brain. *J. Neurochem.* 22:259, 1974.

33. deRobertis, E., deIraldi, A. P., Rodriguez de Lores Arnaiz, G., and Gomez, C. On the isolation of nerve endings and synaptic vesicles. *J. Biophys. Biochem. Cytol.* 9:229, 1961.

34. Autilio, L. A., Appel, S. H., Pettis, P., and Gambetti, P. Biochemical studies of synapses in vitro. I. Protein synthesis. *Biochemistry* 7:2615, 1968.

35. Abdel-Latif, A. A. A simple method for isolation of nerve ending particles from rat brain. *Biochim. Biophys. Acta* 121:403, 1966.

*36. Cotman, C. W. Isolation of Synaptosomal and Synaptic Plasma Membrane Fractions. In Fleischer, S., and Packer, L. (Eds.), *Methods in Enzymology*. Vol. 31, Part A. New York: Academic, 1974, p. 445.

37. Whittaker, V. P., Michaelson, I. A., and Kirkland, R. J. A. The separation of synaptic vesicles from nerve ending particles. *Biochem. J.* 90:293, 1964.

38. Davis, G., and Bloom, F. E. Isolation of synaptic junctional complexes from rat brain. *Brain Res.* 62:135, 1973.

39. Cotman, C. W., Levy, W., Banker, G., and Taylor, D. An ultrastructural and chemical analysis of the effect of Triton X-100 on synaptic plasma membranes. *Biochim. Biophys. Acta* 249:406, 1971.

40. Cotman, C. W., Banker, G., Churchill, L., and Taylor, D. Isolation of post-synaptic densities from rat brain. *J. Cell Biol.* 63:441, 1974.

41. Cotman, C. W., Mahler, H. R., and Hugli, T. E. Isolation and characterization of insoluble proteins of the synaptic plasma membrane. *Arch. Biochem. Biophys.* 126:821, 1968.

42. Day, E. D., and Appel, S. H. The biologic half-life of brain-localized antisynapse radioantibodies. *J. Immunol.* 3:710, 1970.

*43. deDuve, C. The separation and characterization of subcellular particles. *Harvey Lect.* 59:49, 1965.

*44. Koenig, H. The Isolation of Lysosomes from Brain. In Fleischer, S., and Packer, L. (Eds.), *Methods in Enzymology*. Vol. 31, Part A. New York: Academic, 1974, p. 457.

45. Blokhuis, G. G., and Veldstra, H. Heterogeneity of mitochondria in rat brain. *FEBS Let.* 11:197, 1970.

46. Hamberger, A., Blomstrand, C., and Lehninger, A. L. Comparative studies on mitochondria isolated from neuron-enriched and glia-enriched fractions of rabbit and beef brain. *J. Cell Biol.* 45:221, 1970.

*47. Abood, L. G. Brain Mitochondria. In Lajtha, A. (Ed.), *Handbook of Neurochemistry*. Vol. 2. New York: Plenum, 1969, p. 303.

48. Brody, T. M., and Bain, J. A. A mitochondrial preparation from mammalian brain. *J. Biol. Chem.* 195:685, 1952.

49. Tanaka, R., and Abood, L. G. Studies on adenosine triphosphatase of relatively pure mitochondria and other cytoplasmic constituents of rat brain. *Arch. Biochem. Biophys.* 105:554, 1964.

50. Day, E. D., McMillan, P. N., Mickey, D. D., and Appel, S. H. Zonal centrifuge profiles of rat brain homogenates, instability in sucrose, stability in iso-osmotic Ficoll-sucrose. *Anal. Biochem.* 39:29, 1971.

*51. Shelanski, M. Methods for the Neurochemical Study of Microtubules. In Marks, N., and Rodnight, R. (Eds.), *Research Methods in Neurochemistry*. Vol. 2. New York: Plenum, 1974, p. 281.

52. Kirkpatrick, J. B., Haynes, L., Thomas, V. L., and Howley, P. M. Purification of intact microtubules from brain. *J. Cell Biol.* 47:384, 1970.

53. Weisenberg, R. C., Borisy, C. G., and Taylor, E. W. The colchicine binding protein of mammalian brain and its relation to microtubules. *Biochemistry* 7:4466, 1968.

54. Shelanski, M., Albert, S., DeVries, G. H., and Norton, W. T. Isolation of filaments from brain. *Science* 174:1242, 1971.

55. Benjamins, J. A., Guarnieri, M., Miller, K., Sonneborn, M., and McKhann, G. M. Sulfatide synthesis in isolated oligodendroglial and neuronal cells. *J. Neurochem.* 23:751, 1974.

56. Boer, G. J. A microassay for arylamidase activity. *Anal. Biochem.* 59:410, 1974.

57. Brenkert, A., and Radin, N. S. Synthesis of galactosyl ceramide and glucosyl ceramide by rat brain: Assay procedures and changes with age. *Brain Res.* 36:183, 1972.

58. Cooperstein, S. J., and Lazarow, A. A microspectrophotometric method for the determination of cytochrome oxidase. *J. Biol. Chem.* 189:665, 1951.

59. Cotman, C. W., and Matthews, D. A. Synaptic plasma membranes from rat brain synaptosomes: Isolation and partial characterizations. *Biochim. Biophys. Acta* 249:380, 1971.

60. Earl, D. C. N., and Korner, A. The isolation and properties of cardiac ribosomes and polysomes. *Biochem. J.* 94:721, 1965.

61. Ellman, G. L., Courtney, K. D., Andres, V., Jr., and Featherstone, R. M. A new and rapid colorimetric determination of acetylcholinesterase activity. *Biochem. Pharmacol.* 7:88, 1961.

62. Farrell, D. F., and McKhann, G. M. Characterization of cerebroside sulfotransferase from rat brain. *J. Biol. Chem.* 246:4694, 1971.

63. Fishman, W. H., and Bernfeld, P. Glucuronidases. In Colowick, S. P., and Kaplan, N. O. (Eds.), *Methods in Enzymology*. Vol. I. New York: Academic, 1955, p. 262.

64. Fitzgerald, G. G., Kaufman, E. E., Sokoloff, L., and Shein, H. M. D(-)-B-hydroxybutyrate dehydrogenase activity in closed cell lines of glial and neuronal origin. *J. Neurochem.* 22:1163, 1974.

65. Fleck, A., and Begg, D. The estimation of ribonucleic acid using ultraviolet absorption measurements. *Biochim. Biophys. Acta* 108:333, 1965.

66. Fonnum, F. The distribution of glutamate decarboxylase and aspartate transaminase in subcellular fractions of rat and guinea-pig brain. *Biochem. J.* 106:401, 1968.

67. Goridis, C., Massarelli, R., Sensenbrenner, M., and Mandel, P. Guanyl cyclase in chick embryo brain cell cultures: Evidence of neuronal localization. *J. Neurochem.* 23:135, 1974.
68. Goridis, C., Virmaux, N., Urban, P. F., and Mandel, P. Guanyl cyclase in a mammalian photoreceptor. *FEBS Let.* 30:163, 1973.
69. Jongkind, J. F. Quantitative histochemistry of hypothalamus. II. Thiamine pyrophosphatase, nucleoside diphosphatase and acid phosphatase in the activated supraoptic nucleus of the rat. *J. Histochem. Cytochem.* 17:23, 1969.
70. Kornberg, A. Lactic Dehydrogenase of Muscle. In Colowick, S. P., and Kaplan, N. O. (Eds.), *Methods of Enzymology*. Vol. I. New York: Academic, 1955, p. 441.
71. Krajl, M. A rapid microfluorimetric determination of monoamine oxidase. *Biochem. Pharmacol.* 14:1684, 1965.
72. Kuczenski, R. T., and Mandell, A. J. Regulatory properties of soluble and particulate rat brain tyrosine hydroxylase. *J. Biol. Chem.* 247:3114, 1972.
73. Kurihara, T., and Tsukada, Y. The regional and subcellular distribution of 2', 3'-cyclic nucleotide 3'-phosphohydrolase in the central nervous system. *J. Neurochem.* 14:1167, 1967.
74. Lowry, O. H., and Lopez, J. A. The determination of inorganic phosphate in the presence of labile phosphate esters. *J. Biol. Chem.* 162:421, 1946.
75. McCaman, R. E., Smith, M., and Cook, K. Intermediary metabolism of phospholipids in brain tissue. II. Phosphatidic acid phosphatase. *J. Biol. Chem.* 240:3513, 1965.
76. Miller, E. K., and Dawson, R. M. C. Can mitochondria and synaptosomes of guinea-pig brain synthesize phospholipids? *Biochem. J.* 126:805, 1972.
77. Molinoff, P. B., Weinshilboum, R., and Axelrod, J. A sensitive enzymatic assay for dopamine-β-hydroxylase. *J. Pharmacol. Exp. Ther.* 178:425, 1971.
78. Olafson, R. W., Drummond, G. I., and Lee, J. F. Studies on 2', 3'-cyclic nucleotide-3'-phosphohydrolase from brain. *Can. J. Biochem.* 47:961, 1969.
79. Radin, N. S., Brenkert, A., Arora, R. C., Sellinger, O. Z., and Flangas, A. L. Glial and neuronal localization of cerebroside-metabolizing enzymes. *Brain Res.* 39:163, 1972.
80. Robins, E., Fisher, H. K., and Lowe, I. P. Quantitative histochemical studies of the morphogenesis of the cerebellum. II. Two β-glycosidases. *J. Neurochem.* 8:96, 1961.
81. Robins, E., Hirsch, H. E., and Emmons, S. S. Glycosidases in the nervous system. I. Assay, some properties, and distribution of β-galactosidase, β-glucuronidase, and β-glucosidase. *J. Biol. Chem.* 243:4246, 1968.
82. Robinson, D. S., Lovenberg, W., Keiser, H., and Sjoerdsma, A. Effects of drugs on human blood platelet and plasma amine oxidase activity in vitro and in vivo. *Biochem. Pharmacol.* 17:109, 1968.
83. Roeder, R. G., and Rutter, W. J. Multiple forms of DNA-dependent RNA polymerase in eukaryotic organisms. *Nature* 224:234, 1969.
84. Schnaitman, C., Erwin, V. G., and Greenswalt, J. W. The submitochondrial localization of monoamine oxidase. An enzymatic marker for the outer membrane of rat liver mitochondria. *J. Cell Biol.* 32:719, 1967.
85. Sottocasa, G. L., Kuylenstierna, B., Ernster, L., and Bergstrand, A. An electron-transparent system associated with the outer membrane of liver mitochondria. *J. Cell Biol.* 32:415, 1967.
86. Steele, W. J., and Busch, H. Studies on acidic nuclear proteins of the Walker tumor and liver. *Cancer Res.* 23:1153, 1963.
87. Verity, M. A. Cation modulation of synaptosomal respiration. *J. Neurochem.* 19:1305, 1972.
88. Weissbach, H., Smith, T. E., Daly, J. W., Witkop, B., and Udenfriend, S. A rapid spectrophotometric assay of monoamine oxidase based on the rate of disappearance of kynuramine. *J. Biol. Chem.* 235:1160, 1960.

Chapter 3

Biochemistry of Cell Membranes
R. Wayne Albers

The integrating function of the nervous system depends upon the ability of combinations of neurons to generate differential responses to various stimuli and to communicate these responses to other neurons. At the molecular level, the outer cell membrane (plasmalemma) is primarily involved in these interactions. A single neuron is characterized typically by one or more receptor specializations that are usually confined to the dendritic or somal regions of the plasmalemma (see Chap. 1). Activation of these receptors initiates a propagated membrane depolarization, which, in turn, initiates a presynaptic response that usually culminates in the release of a chemical neurotransmitter (see Chaps. 8 and 9). This chain of events involves, in addition to specific receptors, the ion transport mechanism which generates the electrical membrane potential (Chap. 6), the specific ion channels which propagate the electrical impulse (Chap. 5), and presynaptic mechanisms for extrusion of neurotransmitters (Chap. 8). In addition, the outer cell membrane contains other transport mechanisms and receptors which participate in metabolic and regulatory processes.

This chapter considers some biological membrane features which appear to be either general principles or to be particularly relevant to neural functions. A number of recent reviews of membrane biochemistry are available [1–5].

Membranes are composed predominantly of proteins and complex lipids (Table 3-1). The lipids are well characterized, and studies of membrane protein have advanced to the point at which some generalizations are possible [3, 4].

PHYSICAL PROPERTIES OF AMPHIPHILIC LIPIDS
The structural organization of membranes is closely related to the physicochemical properties of membrane lipids because the forces of molecular interactions in membranes primarily involve the formation of noncovalent bonds. These may be classified as ionic, hydrogen, and lyophobic bonds (or, in aqueous solutions, hydrophobic bonds). Such interactions are weak relative to covalent bonds, but a particular complementary orientation between molecules, which maximizes the number of these interactions, may produce a stable or favored association.

Molecules are soluble if their interactions with the solvent are stronger than their interactions with other solute molecules. Ionic and polar groups of a molecule will become hydrated, i.e., they will interact with hydrogen or oxygen dipoles of water. If such polar constituents predominate, the molecule will dissolve. Regions of large molecules that lack such polar groups will associate with nonpolar regions of other molecules to form micelles. Within such micelles, molecules assume an orientation that minimizes the exposure of nonpolar groups to water or, conversely, maximizes the interactions of polar groups with water.

60

TABLE 3-1. Composition of Cell Membranes

	Liver			Neural		
	Microsomes [1, 4, 41]	Golgi [1, 41]	Plasma [3, 41]	Axolemma [5]	Myelin [18, 33]	Microsomes [18]
% Solids						
Protein	70	60	68	37	25	10
Cholesterol	1.5	5	8	9	18	10
Phospholipids	28	21	15	29	32	44
Glycolipids	0	....	....	....	20	7
% Total Phospholipids						
Phosphatidylcholine	65	45	35	....	22	47
Phosphatidylethanolamine	20	17	22	....	36	26
Phosphatidylserine	4	4	4	....	22	12
Phosphatidylinositol	12	9	8	....	2	3
Sphingomyelin	4	12	20	....	11	11
% Unsaturation						
Phosphatidylcholine	49	41	30	....	....	....
Phosphatidylethanolamine	34	34	39	....	....	....
Phosphatidylserine	75	55	14	....	....	....
Sphingomyelin	30	10	11	....	....	....

Both complex lipids and proteins may be classified as amphiphilic molecules, i.e., molecules that contain both polar and nonpolar moieties. Amphiphilic molecules in aqueous solution will form aggregates or, in the case of macromolecules, will fold in such a way as to sequester the hydrophobic regions in minimal volumes. Hydrophobic bonding is now recognized as a major factor in determing the stability of associations of both proteins and lipids [6].

Formation of Lipid Bilayers

The complex lipid constituents of membranes are characterized by distinct segregation of polar and nonpolar moieties within a single molecule. Thus, phosphatides can be represented as having a polar "head" composed of a glycerophosphoryl ester moiety and a "tail" composed of two fatty-acid chains. Aqueous emulsions of phospholipids consist of a few to several hundred molecules that form more or less spherical aggregates (micelles), wherein the polar heads contact water and sequester the hydrocarbon tails in the interior. With increasing concentration, these micelles will coalesce further to form large, multilayered complexes that sometimes are called liposomes. Similar structures may be formed by adding small amounts of water to solid phospholipids. These have been called myelin figures. Their basic structural unit is a bilayer sheet of phospholipid molecules oriented with polar heads outward on either side in contact with water and with the hydrocarbon tails directed inward, the hydrocarbon chains being more or less perpendicular to the plane of the bilayer.

Single bilayer membranes may be formed from various mixtures of amphiphiles by a process of spontaneous thinning, which takes place after a solution in organic solvent is painted across a small aperture immersed in aqueous solution. A single bilayer membrane is too thin to diffract light and is often called a black lipid membrane [2].

The experimental use of liposomes and black lipid membranes has been providing extensive correlations between their chemical and physical properties. Their permeability to small, uncharged molecules is increased by increasing the degree of hydrocarbon-chain unsaturation, by decreasing the chain length, or by decreasing the cholesterol content.

Physical State of Lipids in Biological Membranes

The existence of some form of lipid bilayer in natural membrane as a barrier to the entrance of small nonelectrolytes was recognized by Overton in 1895 [7]. In 1925, Fricke deduced that the hydrocarbon barrier should have a thickness of about 33 Å from electrical impedance measurements [8]. In the same year, Gorter and Grendel estimated the surface area of red cells and, from the lipid content, deduced that there was precisely enough lipid to cover the surface with a bilayer [9]. More recently, the existence of bilayers in myelin [10] and the plasma membranes of *Mycoplasma laidlawii* [11] has been established by x-ray diffraction.

MEMBRANE PROTEINS

Tertiary and Quaternary Protein Structure

Current reviews of this subject are readily available [12, 13]. The chief point to be emphasized here is that hydrophobic forces are major determinants of higher-order structure in proteins, as well as in lipids. Proteins often associate to form polymeric structures. These range from simple dimerization of similar proteins to the formation of highly ordered complexes of several unlike subunits. The one case in which the interfacial contacts between subunits are known is hemoglobin. Here the contacts between unlike subunits ($\alpha\beta$) are chiefly nonpolar, whereas $\alpha\alpha$ and $\beta\beta$ contacts are polar [14, 15].

Physical State of Proteins in Biological Membranes

Early workers felt that the low surface tension and elasticity of plasmalemma could not be properties of a lipid film. Danielli [16] found that proteins were effective in lowering surface tension at an oil-water interface, and proposed that a layer of "unrolled" protein might be adsorbed to the polar surfaces of the lipid bilayer. This would lead to the expectations that cell membranes should have fairly constant ratios of protein to lipid and that much of the "extended" protein might exhibit β secondary structure [13]. Neither expectation has been upheld. Protein-lipid ratios seem to vary over a range of nearly tenfold from cell to cell [17]. Proteins containing peptide chains in β conformation have a characteristic infrared absorption band at 1630 cm^{-1}. This has not been detected in plasma membrane proteins. On the other hand, optical rotatory dispersion and circular dichroism measurements are compatible with as much as 60 percent α-helical content, which is higher than that of many globular proteins.

Chemical Composition of Membrane Proteins

As procedures are developed for solubilizing and resolving membrane proteins, it becomes clear that many different proteins are involved. Amino acid analyses are available for membranes from a variety of cells and they have been discussed with respect to the content of nonpolar amino acids. These are, in fact, not much higher than in many soluble proteins; nevertheless, these proteins have a marked tendency to form large aggregates. From studies of red-cell membrane proteins, Maddy and Kelly [18] have concluded that at least three classes of proteins can be separated on the basis of their dominant aggregative forces: (1) bivalent cation bridges; (2) direct electrostatic bridges; and (3) hydrophobic bonding. Categories of "integral" and "peripheral" proteins have been defined on the basis of their ease of solubilization [3].

Several proteins which appear to be integral membrane constituents have a distinctly amphipathic structure. Among the best-characterized of these is cytochrome b$_5$, which is associated with endoplasmic reticulum [19]. This protein attaches to membranes through a "handle" constructed predominantly of hydrophobic amino acids. This segment may be cleaved proteolytically from the remainder that becomes a soluble form of cytochrome b$_5$, which is now incapable of binding to membranes. Other integral membrane proteins may span the lipid bilayer and contain a hydrophobic binding segment in the midregion. Evidence has been obtained for a protein of this structure associated with erythrocyte membranes [20].

LIPID-PROTEIN INTERACTIONS IN MEMBRANES

There are few covalent bonds between membrane proteins and lipids. An exception appears to be the class of "proteolipids" that contains, in addition to a high proportion of hydrophobic amino acids, covalently bound fatty acids [21]. Hypothetical models for membrane structure differ with respect to the relative importance of polar and hydrophobic interactions. Models that suppose that membranes contain a continuous core of lipid bilayer must, explicitly or by exclusion, rely chiefly on polar lipid-protein bonding. However, membrane proteins are not commonly dissociated from lipids at high ionic strength, which seems to eliminate electrostatic binding as a dominant force. The most effective solvents for membrane proteins are certain organic solvents, and solubilization is promoted by low ionic strength, low pH, and divalent metal chelating agents. Anionic and neutral detergents are often effective in bringing membrane proteins into aqueous solution. A number of "chaotropic" anions, such as SCN^- and ClO_4^-, have strong solubilizing effects, which appear to arise from their ability to disrupt the hydrogen-bonded structure of water and thereby to reduce the relative stability of hydrophobic associations [6, 22].

This type of information indicates that hydrophobic forces are generally dominant in binding membrane components together. Recent work has suggested that membrane proteins may exist as globular units, or arrays, which "float" in the lipid bilayer [3] (Fig. 3-1). The composition of the hydrocarbon part of the bilayer of biological membranes is such that individual lipid molecules may have a high degree of mobility within the plane of the bilayer [11]. The fluidity of the hydrocarbon phase will depend upon its chemical composition: chain length, unsaturation, branching. All these influence the phase-transition temperature, or "melting-point," of bilayers and could be factors in metabolic regulation of membrane processes (Chapman in [1]).

Functional requirements dictate that certain proteins, e.g., those of transport systems and the external-receptor—internal-enzyme systems such as adenyl cyclase, must extend through the lipid bilayer. There has been limited confirmation of this by use of the freeze-fracture method of preparing membranes for electron microscopy. However, this technique shows that most of the globular protein units appear to be associated with the inner half of the lipid bilayer [23, 24].

The examples of freeze-etched membrane topography that have been published to date reveal a rather low density of such globular subunits, far too few to fulfill a structural role. Other techniques of electron microscopy have supported quite a different concept. The usual trilaminar image of thin-sectioned membranes stained with heavy metals is still evident in mitochondria after about 95 percent of the lipid has been extracted [25], indicating that protein is a major determinant of this image. The negative-staining technique presents an image of closely spread globular subunits attached to the membrane surface in the mitochondrial inner membrane [26] and sarcoplasmic reticulum [27]. Subunit structure has been seen within membranes in special cases, such as retinal photoreceptor discs (see Chap. 4) and Mauthner-cell synapses [28].

Evidence about the modes of association between proteins and lipids indicates considerable diversity [17], as it seems likely that the molecular organization of

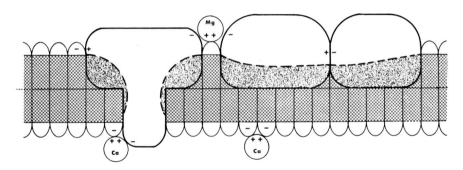

Figure 3-1
Binding forces in cell membranes. Some current concepts of structure are illustrated. The lipid bilayer is interrupted by proteins inserted primarily (but not exclusively) into the cytoplasmic half of the bilayer, and, in some cases, extending through the bilayer. Hydrophobic bonding is a dominant force (indicated by adjacent shaded areas). This is supplemented by electrostatic bonds directly between lipids and proteins as well as by divalent cation bridges.

membranes is at least as variable as the range of associations of soluble proteins. Satisfactory models can be derived only from the intensive study of specific cases.

MEMBRANE BIOSYNTHESIS

The composite nature of membranes introduces a number of interdependent questions relating to the synthesis of individual components and their assembly into a functional structure. How are newly synthesized lipids and proteins assembled to form new membranes? Are membranes synthesized only de novo, by addition of supramolecular lipoprotein subunits, or by the insertion of new molecules into old membranes? What controls the type of membrane formed? Is there genetic control at the supramolecular level, or is this a "spontaneous" assembly determined by intrinsic characteristics of individual protein and lipid components?

Ribosomes in combination with messenger RNA, i.e., polyribosomes, are found in close association with endoplasmic reticulum (ER). In some cases they appear to be closely adherent to these membranes. Correlations have been made between the appearance and structure of the rough endoplasmic reticulum (RER) and the rate of synthesis of certain membrane-bound enzymes. The evidence suggests that, at least in some instances, nascent proteins may move directly from ribosomes into the membranes of the endoplasmic reticulum. Such studies must be extended to more membrane proteins before the generality of this mechanism can be established. If most membranes originate as part of the endoplasmic reticulum, other questions arise. How are they transformed into other types of membranes? How is the structural and functional differentiation of membranes regulated?

Almost all of the complex lipids of cells are associated with membranes. The enzymes for phospholipid synthesis are largely bound to endoplasmic reticulum [29]. Little is known, however, about the relation between the lipid biosynthetic process and the assembly of lipids into membranes.

Different hypotheses of membrane biosynthesis are current. One might be termed the sequential transformation hypothesis [30]. It is possible to view the nuclear, endoplasmic reticulum, Golgi, and plasma membranes (Chap. 1) as forming a continuum, and indeed instances of physical continuity have been demonstrated between nuclear and rough endoplasmic membranes, between endoplasmic reticulum (rough and smooth) and Golgi, and between smooth endoplasmic reticulum (SER) and plasma membrane. One may envision the polyribosomes assembling at the nuclear membrane to form rough endoplasmic reticulum, which will generate new membrane proteins, including those which catalyze phospholipid synthesis. The Golgi apparatus may be a region into which these newly synthesized membranes flow and which, to a large extent, mediates the transition from rough to smooth endoplasmic reticulum. This Golgi region is also a site of protein modification reactions, particularly the addition of carbohydrates to form glycoproteins. Furthermore, in secretory cells, elements of SER appear to convey secretory precursor material into the interiors of the Golgi membranes preparatory to the formation of membrane-bound secretory granules [31].

A functional continuity is postulated between the Golgi membranes and the plasma membrane [32], perhaps via a type of smooth endoplasmic reticulum [33]. In certain cases, Golgi membranes can be shown to have enzyme profiles that are intermediate between those of isolated endoplasmic reticulum and plasma membranes. This has been interpreted as evidence for the sequential transformation of ER to Golgi to plasma membrane [34].

A somewhat different, but not incompatible, view arises from the work of Palade, Siekevitz, and co-workers, who have studied the development of membranes in rat liver cells during a period of rapid development [34, 35]. Nuclear, ER, and plasma-membrane proteins appear to be synthesized at about the same average rate. On the other hand, there are differences in the rates of synthesis of particular enzymes. Even within the ER, lipids turn over more rapidly than do proteins and, within the lipids, the glycerol moiety turns over more rapidly than do the fatty acids. They proposed that ER membranes may be "mosaics of functionally different patches, each bearing a characteristic enzyme or enzyme set . . . assembled in a single-step operation from structural protein, phosphatides, and corresponding enzymes, but various types [of patches] produced in different proportions at different times" [36, 37].

Both of the preceding hypotheses envisage the RER as the primary site of membrane protein synthesis. However, Glick and Warren have reported that plasma membranes isolated from mouse fibroblasts contain 60 to 80 μg RNA per milligram of protein, most of which appeared to be ribosomal. These membranes incorporate amino acids into protein at a rate comparable to that of microsomes from the same cells, and these authors propose that "association of the protein-synthesizing machinery with the surface membranes could facilitate production of proteins for membrane synthesis or for export" [38]. Some mechanism of protein incorporation into membranes other than into RER must take place, because the enzyme content of plasma membrane differs markedly from that of RER [32, 37].

Neuronal membrane synthesis poses several unique questions. Because ribosomal protein synthesis is largely excluded from the axon (Chap. 16), the problems related to assembly and transport into the axon are especially obvious. Not only must membrane proteins be transported the length of the axon, but specialized synaptic structures [39] must be assembled. Moreover, the axolemma itself must contain an exceedingly complex array of specific proteins.

Mahler and co-workers have studied the rate of in vivo protein turnover in sub-cellular brain fractions from rats [40]. The average rates of turnover of proteins were all very similar in the following fractions: soluble, synaptic vesicle, mitochon-drial, and plasma-membrane components of synaptosomes. The initial rate of incor-poration of radioactive amino acid was much more rapid into the microsome fraction than into synaptic plasma membranes. This finding is consistent with a temporal separation of the processes of membrane protein synthesis and synaptic-membrane assembly; possibly this delay corresponds to the time required for axoplasmic trans-port (Chap. 21). A large fraction of the radioactive protein that moves actively down axons is insoluble. It is not known how much of this is in the form of membranes or precursors of membranes.

Membrane biosynthesis is also an important aspect of the process of quantal release of neurotransmitters (Chap. 9). The origin and disposition of the synaptic-vesicle membranes is speculative. Mahler and co-workers find that the qualitative electrophoretic pattern of vesicle membranes is distinct from that of synaptic plasma membranes, although their average turnover rates are similar. This information is not consistent with theories that vesicles either arise from or merge with undifferentiated synaptic plasma membrane.

Several workers have investigated the possibility that there may be protein syn-thesis within the axon, despite the general absence of ribosomes. These studies are extremely difficult, because all axons are closely invested with glial or Schwann cells, which do contain ribosomes. There are suggestions that the axolemma may contain bound RNA and that some axonal and synaptic-membrane protein might be synthesized locally. This would require some transport of genetic information to those sites rather than the transport of precursor protein. The interchange of macromolecules between glia and neurons is also a possibility that has been discussed but not demonstrated.

Regardless of whether there may be some potential for local synthesis, there is no doubt that a large increase in protein synthesis takes place within the neuronal peri-karyon during the process of axonal regeneration [33]. During the outgrowth of axons, their tips become enlarged and project long, membranous extensions. Although the axon proper contains neurofilaments and neurotubules, only a type of SER extends into this "growth cone" region where most of the new membrane synthesis appears to occur.

COOPERATIVE EFFECTS OF MEMBRANE PROTEINS

The nerve impulse is a process localized in the plasma membrane and seems to involve several temporally and spatially coordinated changes in membrane structure. These changes are characterized by abrupt transitions, e.g., with respect to cation permea-bilities as a function of transmembrane potential. As discussed elsewhere (Chap. 5), the molecular basis of these effects is still speculative.

From a biochemical viewpoint, it is attractive to consider whether these changes may be consistent with the known ability of proteins to mediate interactions between diverse molecules and ions. In a fundamental sense, this role of proteins is simply a molecular example of Newton's third law. If molecules A and B each react with a protein, P, then an interaction will occur between A and B which depends largely upon the properties of P.

If A and B bind to distinct sites on P, some physical effects are necessarily propagated between the sites. These may be sufficient to cause changes in the protein folding (conformation). Detailed studies of these changes are available for hemoglobin, oxygen, and 2,3-diphosphoglycerate [41]. Oxygen-free hemoglobin contains six cationic groups between the two β-subunits that form a site complementary to the diphosphoglycerate anion. The oxygen-free conformation of hemoglobin can be established by combination with this "effector" molecule. Upon combination with oxygen, electrostatic bonds between hemoglobin subunits are broken, H^+ is absorbed, and the binding site for diphosphoglycerate is distorted. Thus, diphosphoglycerate acts to hinder the binding of oxygen and, conversely, it binds only weakly to oxyhemoglobin.

Figure 3-2 is a qualitative illustration of effects that may accrue from the interaction of multiple binding sites for the same effector. Such effects are common occurrences, usually resulting from the interaction of several identical or similar protein subunits, each with one effector binding site. If the protein can exist in only two states, P and Q, and the fraction, q, in state Q is a function of the concentration of free effector molecule, m, then

$$q = [(K/m)^n + 1]^{-1}$$

This is an approximation proposed by Hill [42] to describe this function. When $n = 1$, this becomes the Michaelis equation. As n increases, the response becomes increasingly sigmoidal, and as n approaches ∞, there is an increasingly abrupt transition between states P and Q. In this form of the equation, K is the concentration of m at which the protein molecules are equally distributed between states P and Q. The physical interpretation of n is ambiguous, as it reflects the number of binding sites per protein molecule, the number of interacting protein molecules, and the strength of the interactions between binding sites. However, n can never be greater than the total number of strongly interacting binding sites. Usually there will be only one binding site per protein subunit, so an abrupt transition between two states requires a large array of strongly interacting proteins. In the Hodgkin-Huxley description (Chap. 5), the factors controlling Na^+ and K^+ permeabilities are related to these permeabilities by exponents of 3 and 4, respectively, which might reflect high values of n in the Hill formulation.

Although a number of membrane-bound enzymes exhibit marked cooperative effects, the extent to which these concepts can be extended to such phenomena as impulse conduction, transmitter release, or neuroreceptor actions is uncertain. All of these processes involve highly specific interactions and nonlinear, or threshold, responses. In impulse conduction, tetrodotoxin and DDT appear to interact

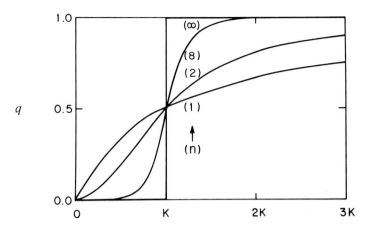

Figure 3-2
The influence of an effector molecule on the conformational or functional state of a protein as depicted by the Hill approximation (see text).

specifically with membrane Na^+ channels at two distinct sites [43]; DDT and veratrine prolong the action potential by interfering with the closure of Na^+ channels. The "cooperative" nature of DDT action is indicated by the fact that the Na^+ channels must be in the open state before DDT binding can occur.

Membranes could accommodate an "infinite" two-dimensional array of interacting subunits [44]. Morphological evidence of large arrays of globular subunits is seen in mitochondrial inner membranes, chloroplasts, certain elements of sarcoplasmic reticulum, photoreceptor discs, and some junctional areas of plasma membranes. In general, however, freeze-etching and negative-staining techniques demonstrate rather widely spaced units or small clusters of globular units in plasma membranes [23, 24].

Calculations for the node of Ranvier in frog indicate a density of Na^+ channels of about 160 per square micron. If each channel is associated with a protein of 100,000 molecular weight, only about 10^{-4} of the membrane area would be occupied by that protein [43]. Thus, even such a specialized area does not require a high density of specialized protein molecules.

Coupling in membranes is undoubtedly mediated to a large extent by the membrane potential. This is a large force (about 100,000 volts per centimeter), and the effect of membrane potentials on membrane proteins has not been studied intensively. The effect of adding various proteins to black lipid membranes has been reported in terms of changes in membrane conductances [2], and more studies are now appearing concerning enzymes artificially bound to membranes, but the field still is relatively unexplored.

Another possible factor in the propagation of effects through membranes is the lipid bilayer. Bilayers may undergo state transitions from one packing geometry to another [45]. This could lead to protein transformations and the change of properties of the bilayer itself, as well.

PREPARATION OF PURIFIED PLASMA MEMBRANE FRACTIONS

The general aspects of subcellular fractionation (Chap. 2) and the preparation of synaptosomes and myelin (Chap. 4) are discussed elsewhere in this volume. An extensive compendium of techniques is available [46]. Biochemical studies of membranes are almost totally contingent upon the adequacy of these methods.

Several of the more useful methods depend on a strategic choice of starting material. Mammalian erythrocyte membranes are probably the most readily prepared and the most intensively studied. The red cells are easily separated from serum and white cells by low-speed centrifugation in isotonic media. Cytoplasmic proteins, predominantly hemoglobin, are then removed in a second stage of washing in hypotonic media. Because these cells are essentially devoid of internal organelles, such preparations can be pure. Even in this simple case, however, there are some uncertainties when individual components are studied: some artifactual adsorption of soluble protein may occur or some "loosely" associated components may be removed.

Myelin is another relatively rich source of plasma membranes, and purification strategies have been based upon the possibility of preparing multilaminar vesicles of low density and large size (Chaps. 2 and 4).

Heterogeneity of more complex tissues is a major problem at both the histological and subcellular levels. In some cases, pure cell lines can be grown in culture, but the amount of material is usually limited and the properties of the plasma-membrane structure of cultured cells may differ importantly from that of normal cells. Accordingly, considerable effort has been directed to preparing plasma-membrane fractions from complex tissues.

The strategies may involve preliminary dissociation of the tissue and fractionation of cell types; mild disruptive techniques designed to obtain large, readily sedimenting fragments of membranes; the use of low ionic-strength media and chelating agents to minimize aggregation artifacts; and the use of density gradients to separate plasma membranes from the somewhat more dense membranes of nuclei and mitochondria. These methods require certain criteria for purity in terms of morphology, chemical composition, enzymatic activity, and so on. Adenyl cyclase, (Na^+, K^+)-ATPase, and $5'$-nucleotidase are frequently used as "marker enzymes" for plasma membranes.

Brain and electric-organ synaptosomes are advantageous starting points, and several subfractionation procedures have been published (see Chap. 2) with the goal of investigating synaptic receptors.

Preliminary work has appeared on the isolation of axolemma [47]. The strategy here is the converse of that involved in preparing myelin. The large, unmyelinated, retinal nerve bundles of squid were chosen as starting tissue because up to 10 g of nerve fibers per squid is available. These membranes have a high protein-lipid ratio relative to myelin. However, no satisfactory way of distinguishing axolemma from Schwann cell membranes has yet been applied.

REFERENCES

*1. Oseroff, A., Robbins, P., and Burger, M. The cell surface membrane: Biochemical aspects and biophysical probes. *Annu. Rev. Biochem.* 42:647, 1973. (Discusses the freeze cleaving E. M. technique, membrane isolation, biosynthesis and composition, biophysical probes.)

*2. Bangham, A. D. Lipid bilayers and biomembranes. *Annu. Rev. Biochem.* 41:753, 1972.

*3. Singer, S. J. The molecular organization of membranes. *Annu. Rev. Biochem.* 43:805, 1974. (Reviews recent studies on the interactions between membrane components; membrane "fluidity.")

*4. Guidotti, G. Membrane proteins. *Annu. Rev. Biochem.* 41:731, 1972.

*5. Finean, R., Coleman, R., and Michell, R. A. *Membranes and Their Cellular Functions.* Oxford: Blackwell, 1974. (Overview of roles of membranes in cellular activities. Excellent illustrations of concepts at introductory level.)

*6. Tanford, C. *The Hydrophobic Effect.* New York: Wiley, 1973. (A unified presentation of the subject, its physical chemical basis, and its role in macromolecular and supramolecular structures.)

 7. Overton, E. Ueber die osmotischen Eigenschaften der lebenden Pflanzen und Tiergelle. *Vjschr. Naturforsch. Ges. Zurich* 40:159, 1895.

 8. Fricke, H. A mathematical treatment of the electric conductivity and capacity of disperse systems. II. The capacity of a suspension of conducting spheroids surrounded by a nonconducting membrane for a current of low frequency. *Phys. Rev.* 26:678, 1925.

 9. Gorter, E., and Grendel, F. On bimolecular layers of lipoids on the chromocytes of the blood. *J. Exp. Med.* 41:439, 1928.

10. Blaurock, A. E. Structure of the nerve myelin membrane: Proof of the low-resolution profile. *J. Mol. Biol.* 56:35, 1971.

11. Engelman, D. M. Structure in the membrane of *Mycoplasma laidlawii. J. Mol. Biol.* 58:153, 1971.

*12. Klotz, I. M., Langerman, N. R., and Darnell, D. W. Quaternary structure of proteins. *Annu. Rev. Biochem.* 39:25, 1970.

*13. Rich, A. Molecular Configuration of Synthetic and Biological Polymers. In Oncley, J. L. (Ed.), *Biophysical Science: A Study Program.* New York: Wiley, 1959.

14. Perutz, M. F., and Lehmann, H. Molecular pathology of human hemoglobin. *Nature* 219:902, 1968.

*15. Dickerson, R. E. X-ray studies of protein mechanisms. *Annu. Rev. Biochem.* 41:815, 1972. (Contains a lucid description of the conformational changes which occur in hemoglobin.)

16. Danielli, J. F. Protein films at the oil-water interface. *Quant. Biol.* 6:190, 1938.

*17. Korn, E. D. Structure of biological membranes. *Science* 153:1491, 1966.

*18. Maddy, A. H., and Kelly, P. G. Factors affecting the interactions of proteins isolated from erythrocyte membranes. *Biochem. J.* 122:62P, 1971.

19. Strittmatter, P., Rogers, M. J., and Spatz, L. The binding of cytochrome b_5 to liver microsomes. *J. Biol. Chem.* 247:7188, 1972.

20. Marchesi, V., Tillack, T., Jackson, R., Segrest, J., and Scott, R. *Proc. Natl. Acad. Sci. U.S.A.* 69:1445, 1972.

21. Stoffyn, P., and Folch-Pi, J. *Biochem. Biophys. Res. Commun.* 44:157, 1971.

22. Hatefi, Y., and Hanstein, W. G. Solubilization of particulate proteins and nonelectrolyte by chaotropic agents. *Proc. Natl. Acad. Sci. U.S.A.* 62:1129, 1969.

23. Pinto da Silva, P., and Branton, D. Membrane splitting in freeze-etching covalently bound ferritin as a membrane marker. *J. Cell Biol.* 45:598, 1970.

*Asterisks denote key references.

24. Tillack, T. W., and Marchesi, V. T. Demonstration of the outer surface of freeze-etched red blood cell membranes. *J. Cell Biol.* 45:649, 1970.

25. Fleischer, S., Fleischer, F., and Stoeckenius, W. Fine structure of whole and fragmented mitochondria after lipid depletion. *Fed. Proc.* 24:296, 1965.

26. Parsons, D. F. Mitochondrial structure: Two types of subunits on negatively stained mitochondrial membranes. *Science* 140:985, 1963.

27. Greaser, M. L., Cassens, R. J., Hoekstra, W. G., and Briskey, E. J. Purification and ultrastructural properties of the calcium accumulating membranes in isolated sarcoplasmic reticulum preparations from skeletal muscles. *J. Cell Physiol.* 74:37, 1969.

28. Robertson, J. D. The occurrence of a subunit pattern on the unit membranes of club endings in Mauthner cell synapses in goldfish brains. *J. Cell Biol.* 19:201, 1963.

29. Willgram, G. F., and Kennedy, E. P. Intracellular distribution of some enzyme catalyzing reactions in the biosynthesis of complex lipids. *J. Biol. Chem.* 238:2615, 1963.

30. Grove, S. N., Bracker, C. E., and Morré, D. J. Cytomembrane differentiation in the endoplasmic reticulum—Golgi apparatus—vesicle complex. *Science* 161:171, 1968.

31. Claude, A. Growth and differentiation of cytoplasmic membranes on the course of lipoprotein granule synthesis in the hepatic cell. I. Elaboration of elements of the Golgi complex. *J. Cell Biol.* 47:745, 1970.

32. Cheetham, R. D., Morré, D. J., and Yunghans, W. N. Isolation of a Golgi apparatus — Rich fraction from rat liver. II. Enzymatic characterization and comparison with other cell fractions. *J. Cell Biol.* 44:492, 1970.

33. Bunge, M. B., and Bray, D. Fine structure of growth cones from cultured sympathetic neurons. *J. Cell Biol.* 47:241a, 1970.

34. Omura, T., Siekevitz, P., and Palade, G. E. Turnover of constituents of the endoplasmic reticulum membranes of rat hepatocyte. *J. Biol. Chem.* 242:2389, 1967.

35. Widnell, C. C., and Siekevitz, P. The turnover of the constituents of various rat liver membranes. *J. Cell Biol.* 35:142A, 1967.

36. Dallner, G., Siekevitz, P., and Palade, G. E. Biogenesis of ER membranes. II. Synthesis of cinstitute microsomal enzymes in developing rat hepatocyte. *J. Cell Biol.* 30:97, 1966.

*37. Meldolesi, J. Membranes and membrane surfaces. *Philos. Trans. R. Soc. Lond. [Bio* 268:39, 1974. (Reviews evidence in support of "membrane patch" hypothesis.)

38. Glick, M. C., and Warren, L. Membranes of animal cells. III. Amino acid incorporation by isolated surface membranes. *Proc. Natl. Acad. Sci. U.S.A.* 63:563, 1969.

*39. Whittaker, V. P. The Synaptosome. In Lajtha, A. (Ed.), *Handbook of Neurochemistry.* Vol. 2. New York: Plenum, 1969, p. 327.

*40. Lajtha, A. (Ed.). *Protein Metabolism of the Nervous System.* New York: Plenum, 1970. (Chapters of particular relevance are those by Austin et al. [Protein biosynthesis in axons], Koenig [Membrane protein synthesis in axons], and Mahler and Cotman [Proteins in synaptic plasma membranes].)

41. Perutz, M. F., and Lehmann, H. Molecular pathology of human haemoglobin. *Nature* 219:902, 1968.

42. Hill, A. V. The combinations of haemoglobin with oxygen and with carbon monoxide. *Biochem. J.* 7:471, 1913.

43. Hille, B. Pharmacological modifications of the sodium channels of frog nerve. *J. Gen. Physiol.* 51:199, 1968.

44. Changeaux, J., Thiery, J., Tung, Y., and Kittel, C. On the cooperativity of biological membranes. *Proc. Natl. Acad. Sci. U.S.A.* 57:335, 1967.

45. McFarland, B. G., and McConnell, H. M. Bent fatty acid chains in lecithin bilayers. *Proc. Natl. Acad. Sci. U.S.A.* 68:1274, 1971.

*46. Fleischer, S. F., and Packer, L. (Eds.). Biomembranes, Parts A and B. In *Methods in Enzymology*, Vol. 31 (Part A) and Vol. 32 (Part B). New York: Academic, 1974. (Part A contains methods for preparing membranes and subcellular organelles. Part B contains methods for isolating membrane components and application of biophysical techniques to membranes.)

47. Fischer, S., Cellino, M., Zambrano, F., Zampighi, G., Tellez-Nagel, M., Marcus, D., and Canessa-Fischer, M. The molecular organization of nerve membranes. I. Isolation and characterization of plasma membranes from retinal axons of the squid: An axolemma-rich preparation. *Arch. Biochem. Biophys.* 138:1, 1970.

Chapter 4

Formation, Structure, and Biochemistry of Myelin

William T. Norton

This chapter presents a survey of the structure, cellular formation, and biochemistry of the myelin sheath. For a fuller treatment of some subjects, the reader is referred to a number of recent books and reviews [1–13]. An introduction to the histology of oligodendrocytes and myelin is presented in Chapter 1.

The morphological distinction between white matter and gray matter is one that is also useful for the neurochemist. White matter is composed of myelinated axons, glial cells, and capillaries. Gray matter contains, in addition, the nerve cell bodies with their extensive dendritic arborizations, and quite different ratios of the other elements. The predominant element of white matter is the myelin sheath, which comprises about 50 percent of the total dry weight. Myelin is mainly responsible for the gross chemical differences between white and gray matter. It accounts for the glistening white appearance, high lipid content, and relatively low water content of white matter.

The myelin sheath is a greatly extended and presumably modified plasma membrane which is wrapped around the nerve axon in a spiral fashion. The myelin membranes originate from, and are part of, the Schwann cell in the peripheral nervous system (PNS), and the oligodendroglial cell in the central nervous system (CNS). In the mature myelin sheath, these membranes have condensed into a compact, paracrystalline structure in which each unit membrane is closely apposed to the adjacent one; the protein layers of each unit membrane are fused with, or bound to, the proteins of the unit in apposition. The sheath is not continuous for the whole length of an axon because each myelin-generating cell furnishes myelin for only a segment of the axon. Between these segments, short portions of the axon are left uncovered. These periodic interruptions, called nodes of Ranvier, are critical for the function of myelin.

PATTERNS OF MYELINATION

Myelination does not proceed in all parts of the nervous system at the same time, but follows the order of phylogenetic development [14, 15]. Portions of the PNS myelinate first, then the spinal cord, and the brain last. In all parts of the nervous system there are, of course, many small fibers that never myelinate. Even within the brain, different areas myelinate at different rates, the intracortical association areas being the last to do so. It is generally true that pathways in the nervous system become myelinated before they become completely functional. This relationship may be reciprocal, however, and function also appears to stimulate myelination. The relationship of myelination to function is somewhat clouded, because the period of maximum myelination also coincides with many other, less-known, changes in the nervous system. It is not seriously doubted, although somewhat difficult to prove, that much

74

of the loss of function in demyelinating diseases is a result of loss of myelin. (The reader will find diseases of myelin discussed in Chap. 28.)

The importance of myelin to proper performance of the nervous system can be inferred by comparing the different capabilities of newborns of various species. The period at which rapid myelination takes place varies considerably among different species. For instance, the CNS of such nest-building animals as rats myelinates largely postnatally and the animals are quite helpless at birth. Grazing animals, such as horses, cows, and sheep, have considerable myelin in the CNS at birth and a correspondingly much higher level of complex activity immediately postnatally.

The maximal rate of myelination in humans takes place during the perinatal period. The motor roots begin to myelinate in the fifth fetal month. The brain is almost completely myelinated by the end of the second year of life, although apparently myelination continues in the human neocortex through the end of the second decade, or even longer (see Chap. 18).

Although it is easy to ascertain when myelination begins, it is difficult to determine when the process of accumulation stops. In the rat, myelin is still being deposited in the brain up to 425 days of age and possibly longer [16]. The rat, however, continues to grow in body size and brain weight for many months, and such a prolonged period of myelin accumulation may not occur in all species.

FUNCTION

The conventional view that myelin acts as an electrical insulator that surrounds axons much as insulation surrounds a wire is probably true, at least in part. However, the main function of this insulating sheath appears to be to facilitate conduction in axons rather than to prevent "short circuiting" between adjacent fibers. The nature of this facilitation has no exact analogy in electrical circuitry. In unmyelinated nerves, impulse conduction is propagated by local circuits that flow between the resting and active nerve in and out of the axon through the axonal membrane (see Chap. 5). The electrical resistance of the myelin sheath is 10 to 20 times higher than extracellular salt solutions, and this prevents local current flow. Conduction in myelinated fibers depends upon the sheath being interrupted periodically at the nodes of Ranvier. The bioelectrical current generated at the low-resistance region of the node acts through the medium external to the sheath to activate the axonal membrane at the next node, rather than causing continuous sequential depolarization of the membrane through the high-resistance sheath. This results in "saltatory" conduction; that is, the impulse jumps from node to node. Current flow by saltatory conduction is approximately six times faster than current flow in a comparably sized unmyelinated fiber. Saltatory conduction requires only 1/300th of the sodium-ion flux required for conduction in an unmyelinated nerve of the same diameter, and an equivalent or greater reduction in energy.

Myelin also lowers the capacitance per unit length of axon, which results in increased speed of local circuit-spreading. The calculated ratio of axon diameter to myelinated fiber diameter for optimal current flow between nodes is 0.6 to 0.7. This ratio is close to that observed in myelinated peripheral nerves. Conduction velocity can also be increased by increasing the diameter of the unmyelinated axon. In unmyelinated fibers, however, conduction velocity is proportional to the square root of the diameter, whereas in myelinated fibers the velocity is proportional to the diameter of the fiber (including the myelin sheath). Thus, the main function of myelin appears to be to facilitate conduction and, at the same time, to conserve space and energy.

There is an exception to the rule that myelinated fibers conduct impulses more rapidly than do unmyelinated ones. If myelinated fibers are smaller than one μ in diameter (according to theoretical calculations), their rate of impulse transmission falls below that of unmyelinated fibers of the same diameter. This is because the axons of these myelinated fibers are smaller than those of unmyelinated fibers of equal diameter. This critical one-μ dimension is close to the observed minimum limit of myelinated fibers in peripheral nerves. In the CNS, the minumum diameter of myelinated fibers is about 0.3 μ − considerably less than this critical size. (See [17] for a review of the function of the myelin sheath.)

ULTRASTRUCTURE

The existence of a sheath surrounding nerve fibers has been known since the early days of light microscopy (Fig. 4-1). Myelin, as well as many of its structural features, such as nodes of Ranvier and Schmidt-Lantermann clefts, can be seen readily in the

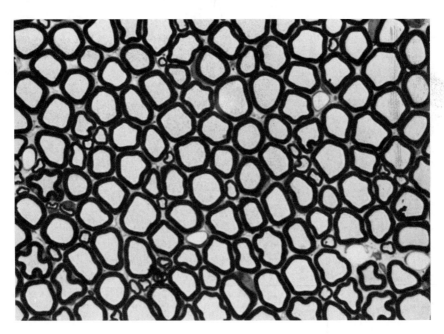

Figure 4-1
Light micrograph of a one-micron epon section of rabbit peripheral nerve (anterior root), stained with toluidine blue. The myelin sheath appears as a thick black ring around the pale axon. X 600, before 30% reduction. (Courtesy of Dr. Cedric Raine.)

light microscope. Before the 1930s, sufficient chemical and histological work had been done to indicate that myelin was primarily lipoidal in nature but had a protein component, as well. Our current view of myelin as a system of condensed plasma membranes with alternating protein-lipid-protein lamellae is derived mainly from studies by three physical techniques: polarized light, x-ray diffraction, and electron microscopy. The earliest physical studies, in the latter half of the nineteenth century, showed that myelin was birefringent when examined by polarized light. This property indicates a considerable degree of long-range order. As early as 1913, Göthlin showed that there was both a lipid-dependent and a protein-dependent birefringence, and that the lipid-dependent type predominated (see [18] for review). Further work with polarized light by Schmidt and Schmitt and co-workers in the 1930s established that myelin was built up of layers. They also found that the lipid components of these layers were oriented radially to the axis of the nerve fiber, whereas the protein component was oriented tangentially to the nerve (see [18] for review). Danielli and Davson had already formulated their concept of the cell membrane as a bimolecular lipid leaflet, coated on both sides with protein. So the results of the latter studies were interpreted with the awareness of the possible membrane nature of myelin.

During the same period, in pioneering studies with x-ray diffraction, Schmitt and co-workers [19] found that peripheral-nerve myelin had a radial repeating unit of 170 to 180 Å, a distance sufficient to accommodate two bimolecular leaflets of lipid together with the associated protein. In 1939, Schmitt and Bear [19] concluded that the configuration of the lipid and protein in the myelin sheath was as follows: "The proteins occur as thin sheets wrapped concentrically about the axon, with two bimolecular layers of lipoids interspersed between adjacent protein layers," a description nearly consistent with our current view.

The x-ray diffraction studies were extended and elaborated by a series of investigations by Finean and co-workers (reviewed in [20]). Low-angle diffraction studies of peripheral nerve myelin provided an electron density plot of the repeating unit that showed three peaks and two troughs, with a repeat distance of 180 Å. The peaks were attributed to protein plus lipid polar groups and the troughs to lipid hydrocarbon chains. The dimensions and appearance of this repeating unit were consistent with a protein-lipid-protein-lipid-protein structure, in which the lipid portion is a bimolecular leaflet and adjacent protein layers are different in some way.

Similar electron density plots of mammalian optic nerve showed a repeat distance of 80 Å (Fig. 4-2); i.e., adjacent protein layers reacted identically to the x-ray beam. Because 80 Å can accommodate one bimolecular layer of lipid (about 50 Å) and two protein layers (about 15 Å each), this represents the width of one unit membrane, and the main repeating unit of two fused unit membranes is twice this figure, or 160 Å.

The electron-density plot reproduced in Figure 4-2, and similar ones obtained for PNS myelin, were obtained by using only the first five diffraction orders. More recent high-resolution studies, using more diffraction orders, reveal a fine structure in the repeat period that could not be obtained in low-resolution work. These plots have not yet been definitively interpreted in terms of molecular structure [21].

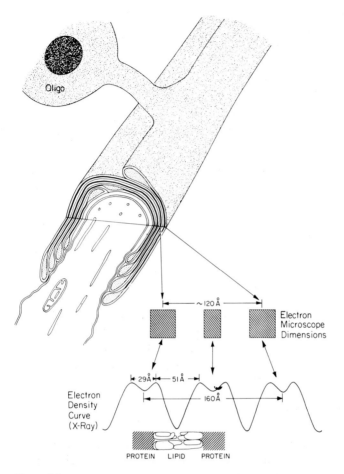

Figure 4-2
A composite diagram summarizing some of the ultrastructural data on CNS myelin. At the top
an oligodendroglial cell is shown connected to the sheath by a process. The cutaway view of the
myelin and axon illustrates the relationship of these two structures at the nodal and paranodal
regions. (Only a few myelin layers have been drawn for the sake of clarity.) At the internodal
region, the cross section reveals the inner and outer mesaxons and their relationship to the inner
cytoplasmic wedges and the outer loop of cytoplasm. Note that in contrast to PNS myelin,
there is no full ring of cytoplasm surrounding the outside of the sheath. The lower part of the
figure shows roughly the dimensions and appearance of one myelin repeating unit as seen with
fixed and embedded preparations in the electron microscope. This is contrasted with the dimensions of the electron density curve of CNS myelin obtained by x-ray diffraction studies in fresh
nerve. The components responsible for the peaks and troughs of the curve are sketched below.
(Reprinted courtesy of Lea & Febiger, publishers.)

The conclusions regarding myelin ultrastructure derived from these two techniques are fully supported by electron microscope studies. Myelin is now routinely seen in electron micrographs as a series of alternating dark and less dark lines separated by unstained zones (Figs. 4-3 to 4-6). The stained or osmiophilic lines are thought to represent the protein layers and the unstained zones the lipid hydrocarbon chains (see Fig. 4-2). The asymmetry in the staining of the protein layers results from the way the myelin sheath is generated from the cell plasma membrane (see following section and Fig. 4-7). The less dark, or intraperiod line represents the closely apposed outer protein coats of the original cell membrane; the dark, or major period line is the fused, inner protein coats of the cell membrane.

The x-ray diffraction data and the electron microscope data correlate very well. Myelin in the PNS, when swollen in hypotonic solutions, is found to split only at the intraperiod line, and the electron density plots show that the broadening occurs at the wider of the three peaks in the repeating unit. This combined approach shows the continuity of the membrane junction of the minor period with the extracellular space and proves that the wide electron density peak in peripheral nerve plots corresponds to the intraperiod lines seen in electron micrographs [22]. It also indicates

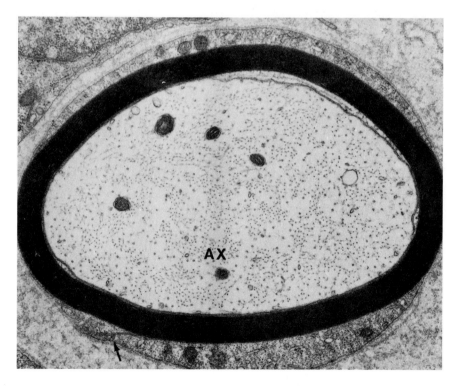

Figure 4-3
Electron micrograph of a single peripheral nerve fiber from rabbit. AX = axon. Note that the myelin sheath has a lamellated structure and is surrounded by Schwann cell cytoplasm. The outer mesaxon (arrow) can be seen in lower left. × 18,000. (Courtesy of Dr. Cedric Raine.)

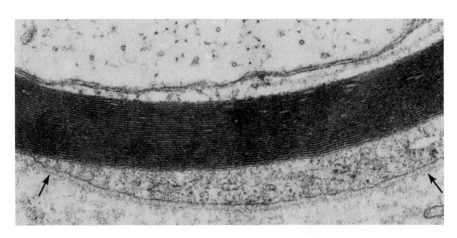

Figure 4-4
Higher magnification of Figure 4-3 to show the Schwann cell cytoplasm covered by basal lamina (arrows). × 50,000.

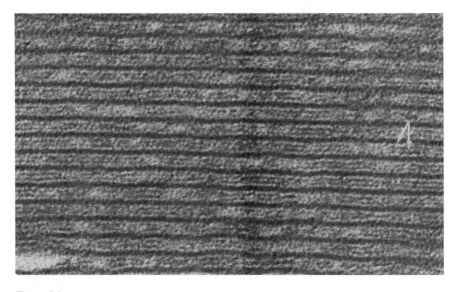

Figure 4-5
Magnification of the myelin sheath of Figure 4-3. Note that the intraperiod line (arrows) at this high resolution is a double structure. × 350,000. (Courtesy of Dr. Cedric Raine.)

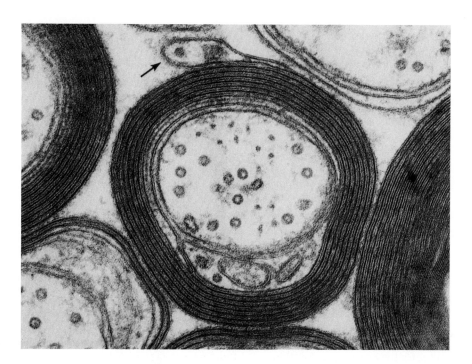

Figure 4-6
A typical CNS myelinated fiber from the spinal cord of an adult dog. Contrast this figure with the PNS fiber in Figure 4-3. The course of the flattened oligodendrocytic process, beginning at the outer tongue (arrow), can be traced. Note that the fiber lacks investing cell cytoplasm and a basal lamina — as is the case in the PNS. The major dense line and the paler, double intraperiod line of the myelin sheath can be discerned. The axon contains neurotubules and neurofilaments. X 135,000.

that the troughs correspond to the light zones between the two dark lines in electron micrographs (see Fig. 4-2). From both x-ray and electron microscope data, it can be seen that the smallest radial subunit that can be called myelin is a five-layered structure of protein-lipid-protein-lipid-protein. This unit, comprising the center-to-center distance between two major period lines, has a spacing in fixed and imbedded preparations of about 120 Å. This repeat distance is lower than the 160- to 180-Å period given by x-ray data because of the considerable shrinkage that takes place after fixation and dehydration.

Although a significant difference was found between central and peripheral myelin in the low-angle x-ray diffraction data, the electron micrographs were much the same for both; each showed a major repeat period of about 120 Å with the intermediate, minor period line of low electron density (compare Figs. 4-5 and 4-6). Recently it has been shown that peripheral myelin has an average repeat distance of 119 Å and the central myelin of 107 Å, confirming the x-ray data for the differences in the size of the period. However, the detailed appearance of electron micrographs is highly dependent on the processing procedures, and some fixation procedures favor retention of the relatively large period seen in x-ray diffraction. Also, with improved methods of perfusion fixation and tissue staining, the intraperiod line is now routinely seen as a double line rather than as a single one (see Figs. 4-5 and 4-6). This ultrastructural appearance indicates that the extracellular sides of the unit membranes probably are not fused.

So far, we have discussed only the ultrastructure of the major portion of the sheath in the internodal region. As we have said, two adjacent segments of myelin on one axon are separated by a node of Ranvier, and in this region the axon is uncovered. These nodes are present in both central and peripheral myelin and apparently have a similar structure. At the paranodal region, the membrane sheaths open up at the major period line and loop back upon themselves, enclosing Schwann or glial-cell cytoplasm within the loop. The myelin loops ending at the node form membrane complexes with the axolemma, whereas myelin in the internodal region is separated from the axon by a gap of extracellular space. These membrane complexes are believed to be helical structures that seal the myelin to the axolemma but provide, by the intervening spaces, a path from the extracellular space to the periaxonal space. The morphology of this region is complex and can be best understood by referring to Figure 4-2. Peripheral nerves also have Schmidt-Lantermann clefts, which are cone-shaped separations of myelin layers that occur at major period lines. The best way to visualize the structure of both the nodes and the Schmidt-Lantermann clefts is to refer to the diagrams by Hirano and Dembitzer [23], which show that, if myelin were unrolled from the axon, it would be a flat sheet surrounded by a tube of cytoplasm. These diagrams show clearly how the nodal structures and the clefts can arise from the cytoplasmic tubes.

CELLULAR FORMATION OF MYELIN

Myelination in the PNS is preceded by Schwann cells invading the nerve bundle, rapid multiplication of these cells, and segregation of the individual axons by Schwann cell processes. Smaller axons (less than 1 μ), which will remain unmyelinated, are segregated and several may be enclosed in one cell, each within its own pocket, similar to the structure shown in Figure 4-7A. Large axons (greater than 1 μ) destined for myelination are enclosed singly; one cell per axon per internode. These cells line up along the axons with intervals between them; these intervals become the nodes of Ranvier.

Geren [24] showed that before myelination the axon lies in an invagination of the Schwann cell (see Fig. 4-7A). The plasmalemma of the cell then surrounds the axon and joins to form a double membrane structure that communicates with the cell surface. This structure, previously noted in unmyelinated fibers and called the "mesaxon," then elongates around the axon in a spiral fashion (see Fig. 4-7). Geren postulated that mature myelin is formed in this jelly-roll fashion; the mesaxon winds about the axon, and the cytoplasmic surfaces condense into a compact myelin sheath.

In early electron micrographs, cell membranes appeared as single dense lines, the mesaxon as two lines, and myelin as a series of repeating dense lines 120 Å apart. Robertson [22] was later able to show that the Schwann cell membrane is composed

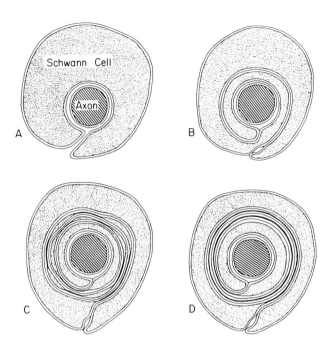

Figure 4-7
Myelin formation in the peripheral nervous system. (A) The Schwann cell has surrounded the axon but the external surfaces of the plasma membrane have not yet fused in the mesaxon. (B) The mesaxon has fused into a five-layered structure and spiralled once around the axon. (C) A few layers of myelin have formed but are not completely compacted. Note the cytoplasm trapped in zones where the cytoplasmic membrane surfaces have not yet fused. (D) Compact myelin showing only a few layers for the sake of clarity. Note that Schwann cell cytoplasm forms a ring both inside and outside of the sheath. (Reprinted courtesy of Lea & Febiger, publishers.)

of two dense lines, as illustrated in Figure 4-7. When the two portions of this membrane come together to form the mesaxon, the two external surfaces appear to join to form a single line that eventually becomes the myelin intraperiod line. As a mesaxon spirals into the compact myelin layers, the cytoplasmic surfaces of the mesaxon fuse to form the major dense line. It was thus shown beyond doubt that peripheral myelin is morphologically an extension of the Schwann cell membrane. The mesaxon is thus the smallest myelin subunit and has the five-layered structure previously described.

It was reasonable to assume that myelin in the CNS was formed in a similar fashion by the oligodendroglial cell. However, these nerve fibers are not separated by connective tissue nor are they surrounded by cell cytoplasm, and specific glial nuclei are not obviously associated with particular myelinated fibers. In 1960 it was shown independently by Maturana and Peters that central myelin is a spiral structure similar to peripheral myelin; it has an inner mesaxon and an outer mesaxon that ends in a loop or tongue of glial cytoplasm (Fig. 4-2; see also [4] for a review of this work).

Unlike peripheral nerve, where the sheath is surrounded by Schwann cell cytoplasm, the cytoplasmic tongue in the CNS is restricted to a small portion of the sheath. It was assumed that this glial tongue was eventually connected in some way to the glial cell, but confirmation was difficult. Finally, Bunge [1] and colleagues showed that the central myelin sheath is continuous with the plasma membrane of the oligodendroglial cell through slender processes. They also showed that one glial cell apparently can myelinate more than one axon. Peters has recently calculated that, in the rat optic nerve, one oligodendroglial cell myelinates, on the average, 42 separate axons [4].

The actual mechanism of myelin formation is still obscure. In the PNS, a single axon may have up to 100 myelin layers, and it is therefore improbable that myelin is laid down by a simple rotation of the Schwann cell nucleus around the axon. In the CNS, such a postulate is precluded by the fact that one glial cell can myelinate several axons. During myelination, there are increases in the length of the internode, the diameter of the axon, and the number of myelin layers. Myelin is therefore expanding in all planes at once, and any mechanism to account for this growth must assume the membrane system is flexible. It must be able to expand and contract, and layers probably slip over each other.

ISOLATION

Much biochemical information has been inferred from studies of organized tissue. For more definitive studies, myelin can readily be isolated in high yield and high purity by conventional methods of subcellular fractionation [25]. It is the only homogeneous, well-defined, pure membrane fraction of any tissue that can be obtained easily in large amounts. Methods of isolation and fractionation also are described in Chapter 2.

Myelin is usually isolated by centrifugation of tissue homogenates in sucrose solutions. During homogenization, myelin peels off the axons and re-forms in spherical vesicles of the size range of nuclei and mitochondria. (It is important to homogenize the tissue in media of low ionic strength, otherwise much of the myelin remains bound to the axon and the intact myelinated axon fragments present in the homogenate will contaminate the crude myelin preparations.) Because of their high lipid content, these myelin vesicles have the lowest intrinsic density of any membrane fraction of the nervous system. Most isolation methods utilize both of these properties — large vesicle size and low density.

One general class of methods involves the isolation, by differential centrifugation, of a crude mitochondrial fraction that contains mitochondria, synaptosomes, and myelin. This fraction is re-suspended in isotonic (0.3 M) sucrose and layered over 0.8 M sucrose. Myelin collects at the interface during centrifugation, whereas mitochondria and synaptosomes sediment through the dense layer. Unless the nuclear fraction is processed similarly, the myelin fragments in that fraction are lost. The other general class of methods bypasses the initial differential centrifugation step. A homogenate of nervous tissue in isotonic sucrose is layered directly onto 0.85 M sucrose and centrifuged at high speed. A crude myelin layer collects at the interface.

The crude myelin layer obtained by either of these methods is a varying purity, depending on the tissue from which the myelin is isolated. White matter from adult brain yields reasonably pure myelin; myelin from the whole brain of a very young animal might be quite impure. The major impurities are microsomes and axoplasm trapped in the vesicles during the homogenization procedure. Further purification is generally achieved by subjecting the myelin to osmotic shock in distilled water. This opens up the myelin vesicles, releasing trapped material. The larger myelin particles can then be separated from the smaller membranous material by low-speed centrifugation or by repeating the density gradient centrifugation on continuous or discontinuous gradients of sucrose or CsCl. On sucrose gradients, myelin forms a band centering at approximately 0.65 M sucrose, equivalent to a density of 1.08 g per milliliter. In CsCl gradients, however, the myelin layers out at a higher density and in sucrose-Ficoll gradients (see Chap. 2) at a lower density. For comparison, mitochondria have a density of 1.2 (equivalent to about 1.55 M sucrose).

Criteria of purity have been difficult to set for myelin preparations because there is no a priori way of knowing what the intrinsic myelin constituents are. The types of criteria for myelin should be the same as for any other subcellular fraction: typical ultrastructure, the absence or minimization of markers characteristic of other particles, and the maximization of markers characteristic of myelin. Most investigators have used the obvious criterion of electron-microscopic appearance. Isolated myelin retains the typical five-layered structure and repeat period of about 120 Å seen in situ. The difficulty of identifying small membrane vesicles of microsomes in a field of myelin membranes and the well-known sampling problems inherent in electron microscopy make this characterization unreliable after a certain purity level has been reached. Useful markers for contamination are succinic dehydrogenase (for mitochondria), Na^+, K^+-activated ATPase, glucose 6-phosphatase, 5'-nucleotidase, and nucleic acid (for cell membranes, microsomes, ribosomes, and nuclei), all of which are very low in purified myelin.

Markers characteristic of myelin are fewer, and some used previously should now be regarded with suspicion. Purification by a final density-gradient centrifugation step establishes, within limits, the density and, therefore, the lipid-protein ratio. Many people have used the solubility, or near solubility, of the final product in 2:1 chloroform-methanol as an indicator of purity. It now appears that one class of myelin proteins is insoluble in this solvent; even so, most myelin preparations of the CNS are about 95 percent soluble. Therefore, complete or nearly complete solubility is a good sign, but may be unreliable if new species or very young animals are being examined or if there is any reason to suspect that the protein composition may be altered. In recent years the enzyme 2',3'-cyclic nucleotide 3'-phosphohydrolase has been shown to be myelin-specific. This enzyme should prove very useful in devising new myelin isolation methods and in assaying myelin contamination in other fractions. With increasing knowledge of myelin proteins and with the development of new methods for their separation and measurement, determination of the myelin protein pattern by, for example, acrylamide-gel electrophoresis should be an excellent indicator of myelin purity.

The foregoing discussion is concerned mainly with the isolation of myelin from the CNS. Peripheral nerve myelin can be isolated by similar techniques, but special homogenization conditions are required because of the large amount of connective tissue present in the nerve. Because PNS myelin has different proteins from CNS myelin, the solubility in $CHCl_3:CH_3OH$ is not a reliable indicator of purity. Also, the enzyme 2',3'-cyclic nucleotide 3'-phosphohydrolase does not seem to be a component of PNS myelin.

COMPOSITION

In contrast with other subcellular fractions, myelin is characterized by a low amount of water, low protein, and a high lipid content. The solids of myelin are 70 to 85 percent lipid and 15 to 30 percent protein; the lipids of mammalian CNS myelin are composed of 25 to 28 percent cholesterol, 27 to 30 percent galactosphingolipid, and 40 to 45 percent phospholipid. Early inferences about myelin composition were made from three types of indirect measurements: comparative analyses of gray and white matter, the measurement of brain constituents during the period of rapid myelination, and studies of brain and nerve composition during experimental demyelination. From such studies it became generally accepted that proteolipid protein, cerebrosides, and sulfatides were exclusively myelin constituents; that sphingomyelin and the plasmalogens were predominantly myelin constituents; that cholesterol and phosphatidylserine were major, but not exclusively, myelin lipids; and that lecithin was probably not a myelin lipid. These suppositions have now been shown to be only partially correct. Even so, in 1949 Brante [26] calculated that myelin sheath lipids were 25 percent cholesterol, 29 percent galactolipids, and 46 percent phospholipids, figures very close to those obtained by direct analysis of isolated myelin.

No direct determination of water can be made on myelin, although obviously myelin is a relatively dehydrated structure. The low water content of white matter (72 percent) as opposed to gray matter (82 percent) is largely due to the high myelin content of white matter. From x-ray diffraction studies on nerve tissue during drying, Finean determined the water content to be about 40 percent. This is probably a fairly accurate calculation, and all the data on yields of myelin and the composition of myelin and white matter are consistent with myelin having about 40 percent water and the nonmyelin portions of white matter having about 80 percent water, similar to the rest of the nervous system.

CNS Myelin Lipids

Table 4-1 lists the composition of bovine, rat, and human myelin compared to bovine and human white matter, human gray matter, and rat whole brain. (The classification and metabolism of brain lipids are discussed in Chap. 15.) It can be seen that all the lipids found in whole brain are also present in myelin; that is, there are no lipids localized exclusively in some nonmyelin compartment except, possibly, cardiolipin. We also know that the reverse is true; that is, there are no myelin lipids which are not also found in other subcellular fractions of the brain. Even though there are no "myelin-specific" lipids, cerebroside is the most typical of myelin. During development,

TABLE 4-1. Composition of CNS Myelin and Brain[a]

Substance[b]	Myelin			White Matter		Gray Matter (Human)	Whole Brain (Rat)
	Human	Bovine	Rat	Human	Bovine		
Protein	30.0	24.7	29.5	39.0	39.5	55.3	56.9
Lipid	70.0	75.3	70.5	54.9	55.0	32.7	37.0
Cholesterol	27.7	28.1	27.3	27.5	23.6	22.0	23.0
Cerebroside	22.7	24.0	23.7	19.8	22.5	5.4	14.6
Sulfatide	3.8	3.6	7.1	5.4	5.0	1.7	4.8
Total galactolipid	27.5	29.3	31.5	26.4	28.6	7.3	21.3
Ethanolamine phosphatides	15.6	17.4	16.7	14.9	13.6	22.7	19.8
Lecithin	11.2	10.9	11.3	12.8	12.9	26.7	22.0
Sphingomyelin	7.9	7.1	3.2	7.7	6.7	6.9	3.8
Phosphatidylserine	4.8	6.5	7.0	7.9	11.4	8.7	7.2
Phosphatidylinositol	0.6	0.8	1.2	0.9	0.9	2.7	2.4
Plasmalogens[c]	12.3	14.1	14.1	11.2	12.2	8.8	11.6
Total phospholipid	43.1	43.0	44.0	45.9	46.3	69.5	57.6

[a]All average figures obtained on adults in the author's laboratory.
[b]Protein and lipid figures in % dry weight; all others in % total lipid weight.
[c]Plasmalogens are primarily ethanolamine phosphatides.

the concentration of cerebroside in brain is directly proportional to the amount of myelin present [16]. There are only minor differences between the lipid composition of myelin and the corresponding white matter, although myelin lipids tend to have somewhat more cholesterol, cerebrosides, and ethanolamine phosphatides than white matter lipids, and somewhat less sulfatides and lecithin.

Figures expressed in this way give no information about lipid concentrations in either the dry or wet tissue. Thus, although human myelin and white matter lipids have similar galactolipid contents, the total galactolipid is 19.3 percent of the myelin dry weight but 14.5 percent of white matter dry weight. These differences are much greater if a wet-weight reference is used. Then galactolipid is 11.6 percent of myelin but only 4.1 percent of fresh white matter. As another example, total phospholipids are a larger percentage of gray matter lipid than of either myelin or white matter lipids. However, on a total dry-weight basis, phospholipids are 30.2 percent of myelin, 25.2 percent of white matter, and 22.7 percent of gray matter.

Figures in Table 4-1 show that many of the suppositions of the earlier deductive work are true. The major lipids of myelin are cholesterol, cerebrosides, and ethanolamine phosphatides in the plasmalogen form. However, lecithin is seen to be a major myelin constituent and sphingomyelin a relatively minor constituent. If the data for lipid composition are expressed in mole percent, most of the preparations analyzed so far contain cholesterol, phospholipid, and galactolipid in a molar ratio of 4:3:2. Thus, cholesterol constitutes the largest proportion of lipid molecules in myelin, although the galactolipids are usually a greater proportion of the lipid weight.

The composition of brain myelin from all mammalian species studied is very much the same. However, there are some obvious species differences. For example, rat myelin has less sphingomyelin than does ox or human (Table 4-1). As one goes down the phylogenetic scale, the differences in myelin become much more apparent. Myelin from amphibians and fish apparently has less sphingolipid and more glycerylphospholipid than does myelin from mammals.

Besides the lipids listed in the table, there are several others of importance. If myelin is not extracted with acid organic solvents, the polyphosphoinositides remain tightly bound to the myelin protein, and therefore are not included in the lipid analysis. There is good evidence that the brain triphosphoinositide, which is stable to postmortem degradation, is localized mainly in myelin and may, therefore, have some status as a myelin marker. These lipids are of considerable interest because triphosphoinositide has the highest turnover rate of any phospholipid of the brain, yet one of the main characteristics of myelin is its low metabolic activity. Triphosphoinositide accounts for between 4 and 6 percent of the total myelin phosphorus, and diphosphoinositide for 1 to 1.5 percent of the myelin phosphorus. Although low ganglioside levels have been used as an indicator of myelin purity, it is now apparent that there is an irreducible amount of ganglioside associated with myelin; this is in the order of 40 to 50 μg of sialic acid per 100 mg of myelin (about 0.15 percent ganglioside). In the mature rat and cow, the myelin gangliosides have a pattern completely unlike the pattern of gangliosides extracted from whole brain; the major monosialoganglioside, G_{M1}, accounts for 80 to 90 mole percent of the total myelin ganglioside [28]. Sialosylgalactosylceramide, an unusual ganglioside, has been shown to be a major component of human myelin gangliosides, although it is less abundant in other species. This ganglioside is derived from galactocerebroside and has a fatty-acid composition typical of that galactolipid rather than of the other brain gangliosides [29].

The long-chain fatty residues of myelin are characterized by a very high proportion of fatty aldehydes. These fatty aldehydes, which are derived primarily from phosphatidalethanolamine and, to a lesser extent, from phosphatidalserine, constitute one-sixth of the total glycerylphosphatide fatty residues and 12 mole percent of the total hydrolyzable fatty chains of the myelin lipids. The phospholipid fatty acids differ considerably from one phospholipid to another, but are generally characterized by a high oleic acid content and a low level of polyunsaturated fatty acids. The glycosphingolipids, cerebrosides, and sulfatides have two classes of fatty acids — unsubstituted and α-hydroxy — both of which can be saturated or monounsaturated, whereas sphingomyelin has only unsubstituted fatty acids. The sphingolipid acids are primarily long-chain (22 to 26 carbon atoms) with varying amounts of 18:0. For example, human myelin glycosphingolipids have very little α-hydroxystearic acid but significant amounts of stearic acid, whereas bovine glycosphingolipids have both. Cerebrosides with unsubstituted acids and hydroxy fatty acids correspond to cerasine and phrenosine, respectively, of the older literature.

The previous discussion on composition refers primarily to myelin isolated from the brain. There is some evidence that myelin isolated from the spinal cord has a considerably higher lipid-protein ratio than that isolated from the brain of the same species. The differences between myelins isolated from different parts of the CNS are not well documented and deserve much further study. We do know, however, that myelin from the PNS has a different composition from myelin of the CNS. Peripheral nerve myelin has not received the same extensive documentation primarily because of the technical difficulty of homogenizing peripheral nerve. The few analyses that have been made show that PNS myelin has less cerebroside and sulfatide and considerably more sphingomyelin than CNS myelin. (For general references on CNS myelin lipids, see [2–4, 6–13, 27].)

Nonlipid Constituents
Myelin does not contain any substantial amount of mucopolysaccharides, although it does contain glycoproteins. Little enzymatic activity has been detected in myelin, with a few exceptions. Some workers have reported reasonably high levels of various peptidases in isolated myelin. Other workers, however, claim they can dissociate the peptidase activity from myelin. The enzyme 2',3'-cyclic nucleotide 3'-phosphohydrolase appears to be definitely established as a CNS myelin-specific enzyme, but apparently it is not in PNS myelin. Sixty percent of the enzyme activity of whole rat brain can be recovered in the purified myelin fraction. The increase of this enzyme in brain and spinal cord during development parallels the course of myelination, and very low levels have been found in the two mouse mutants, quaking and jimpy, that are characterized by deficient myelination [30]. This enzyme could have considerable utility as a myelin marker in future studies. Recently, a second enzyme has been discovered that seems to be myelin specific. This is the cholesterol ester hydrolase, with a pH optimum of 7.2; it is one of three separate cholesterol ester hydrolases of brain [31]. Nothing is yet known of the true function of either enzyme in myelin.

It was once generally believed that myelin was an inert tissue that contained little or no enzymatic activity. Now that enzymes have been proved to be present, it is likely that other enzymatic activities will be discovered. There is recent evidence for the presence of a protein kinase that can phosphorylate myelin basic protein, and for UDP-galactose:ceramide galactosyl transferase, the enzyme that catalyzes the last step in cerebroside synthesis.

With the availability of purified myelin preparations, the once rather confused picture of myelin proteins began to be somewhat clarified. The studies of myelin proteins were slow in getting started because, with the exception of the "basic protein," the experimental allergic encephalomyelitis (EAE) antigen, myelin proteins are insoluble in aqueous media and thus are not amenable to study by conventional protein techniques. In recent years, rapid advances have been made in this area. These have been due largely to the application of the technique of polyacrylamide gel electrophoresis of proteins dissolved in sodium dodecylsulfate. This technique permits the separation of proteins according to molecular weight and has made possible the routine analysis of myelin proteins, studies of their synthesis and turnover, and the discovery of new myelin proteins. Myelin proteins can also be partially separated by solvent fractionation and by differential solubility in salt solutions or salt solutions containing detergents. As far as the gross composition goes, these various techniques give reasonably consistent results, and it is generally agreed that there are three major proteins in myelin: 30 to 50 percent proteolipid protein; 30 to 35 percent basic protein, and a lower percentage of a higher molecular-weight protein doublet that is soluble in acidified $CHCl_3:CH_3OH$ (Wolfgram protein).

The electrophoresis studies indicate that the protein composition is more complex than this. In addition to these three well-known proteins, some rodents have an additional smaller basic protein, and all mammals have: a proteolipid-type protein doublet of lower molecular weight than proteolipid protein; at least one glycoprotein; and a family of high molecular-weight proteins. These high molecular-weight proteins appear to vary in amount, depending on the species, and comprise a higher percentage of the total in mouse and rat myelin than in bovine and human myelin. For this reason, there is some doubt whether all of these high molecular-weight proteins are intrinsic myelin components.

The major protein, the classic Folch-Lees proteolipid protein, can be extracted from whole brain with chloroform-methanol and can be solubilized in either chloroform or chloroform-methanol, even though its lipid content is reduced to less than 10 percent. It contains about 40 percent polar amino acids and 60 percent nonpolar amino acids. This protein has been the subject of much study over the years because of its unusual solubility properties (for a historical review see [32]). However, it remains somewhat of an enigma. The proteolipid protein fraction, as isolated from whole brain by solvent extraction, is heterogeneous, although the major component appears to be a single protein of molecular weight about 24,000. Part of the heterogeneity probably arises from aggregation of subunits. There is a myelin protein, actually usually seen as a doublet, which runs between basic protein and proteolipid protein on SDS-polyacrylamide gels and which has been called DM-20 by Agrawal et al. [33]. This fraction accompanies the proteolipid protein during extraction procedures but has not yet been well characterized.

Central nervous system myelin contains a small amount of glycoprotein which can be labeled with fucose, glucosamine, or N-acetylmannosamine. This protein runs in the high-molecular-weight region of SDS-polyacrylamide gels and undergoes an apparent decrease in molecular weight during development [34].

The basic protein of myelin has been the most extensively studied [35, 36], because, when it is injected into an animal, it elicits a cellular antibody response that produces an autoimmune disease of the brain called experimental allergic encephalomyelitis. This disease involves focal areas of inflammation and demyelination that resemble multiple sclerosis in some respects. The basic protein cannot be extracted from whole brain with organic solvents, but it dissolves when isolated myelin is treated with chloroform-methanol (2:1). It can be extracted readily from myelin or from whole brain by dilute acid or salt solutions and, when so extracted, is soluble in water.

The bovine protein is a highly basic protein (isoelectric point greater than pH 12), and it is highly unfolded, with essentially no tertiary structure. It has a molecular weight of around 18,000 and contains approximately 54 percent polar amino acids and 46 percent nonpolar amino acids. It has no cysteine and has one mole of tryptophan per mole of protein. This is in contrast to proteolipid protein, which is high in cysteine and methionine, as well as being rich in tryptophan.

The complete amino acid sequences of both human and bovine basic protein have been elucidated by Eylar and co-workers [35], and the sequence of the human basic protein has been determined independently by Carnegie [37]. These two proteins differ from each other by only 11 residues.

Considerable effort has been devoted to determining which portions of the sequence are necessary for encephalitogenic activity. A small peptide sequence containing the tryptophan is necessary for EAE production in guinea pigs. Another region (residues 45 to 90) that does not contain tryptophan is encephalitogenic in rabbits but inactive in guinea pigs. Other regions are apparently active in monkeys but not in either guinea pigs or rabbits. It has also been shown that the antibody-combining site does not necessarily correspond to any encephalitogenic site. This is an active and still-controversial area of investigation.

As noted above, mice and rats have a second basic protein, smaller than the encephalitogen. This second protein has been found in some other rodents, but no clear-cut evolutionary pattern is evident. The small basic protein has the same N- and C-terminal sequences as the larger, but differs by a deletion of 40 residues [38].

The third class of myelin proteins is composed of acid-soluble proteolipid proteins that have been called "Wolfgram proteins" after their discoverer [39]. This protein fraction is water insoluble and is also apparently insoluble in neutral chloroform-methanol. It contains about 53 percent polar amino acids and 47 percent nonpolar amino acids. On disc gel electrophoresis, this fraction appears to consist of two or three major bands and possibly several minor ones, all of considerably higher molecular weight than the other two myelin protein fractions.

PNS myelin proteins are different from those of CNS myelin. It has been known for many years that peripheral nerve has little proteolipid protein. Recent electrophoretic studies show that PNS myelin has one major nonpolar protein, two lower molecular-weight basic proteins, and 3 or 4 other prominent bands. The principal protein, P_0, has a molecular weight of about 30,000 and has been shown to be a glycoprotein. The larger of the two basic proteins, P_1, has a molecular weight of 18,000 and is identical to the CNS-myelin basic protein. The smaller basic protein, P_2, has a molecular weight of 12,000 and does not have a counterpart in the CNS (see, e.g., [40]).

COMPOSITIONAL CHANGES

Developing Brain

The developing nervous system is marked by several overlapping periods, each defined by one major event in brain growth and structural maturation (see Chaps. 17 and 18). These periods can be determined by following the concentration of a specific marker. For example, the period of cellular proliferation can be followed by measuring the amount of DNA per whole brain, and the period of myelination by following a myelin marker such as cerebroside. In the rat, whose CNS undergoes considerable development postnatally, the maximal rate of cellular proliferation occurs at 10 days. The period of rapid myelination overlaps with this period of cellular proliferation and is one of the most dramatic in nervous system development. The rat brain begins to form myelin at about 10 to 12 days postnatally. At 15 days of age, about 4 mg of myelin can be isolated from one brain. This amount increases sixfold during the next 15 days, and at 6 months of age, 60 mg of myelin can be isolated from one brain. This represents an increase of about 1,500 percent over 15-day-old animals. During the same 5½ month period, the brain weight increases by 50 to 60 percent [16].

It has been proved convincingly by several groups that the myelin that is first deposited has a very different composition from that of the adult. (For reviews, see [4, 10, 16].) As the rat matures, the myelin galactolipids increase by about 50 percent and lecithin decreases by a similar amount. The very small amount of desmosterol declines, but the other lipids remain relatively constant. In addition, the polysialogangliosides decrease and the monosialoganglioside, G_{M1}, increases to become the predominant ganglioside. These changes are not complete until the rat is about 2 months old. There is, however, a change in the composition of the protein portion as well as in the lipid portion. Earlier work suggested that the ratio of basic protein to proteolipid protein in human myelin increases during development. More recent studies, using disc gel electrophoresis, suggest that these protein changes, at least in rodents, may be complex, but that both protein and proteolipid protein increase in the myelin sheath during development whereas the amount of higher molecular-weight protein decreases.

These studies of myelin during development are complicated by the presence of the fraction called the "myelinlike fraction," which is found in crude myelin from young animals. The myelinlike fraction can be separated from true myelin by a density-gradient centrifugation technique, and, when separated, it has physical properties similar to those of microsomes. It differs from myelin in lipid, protein, and enzyme composition, and its possible role in myelination is still obscure. The main reasons for the interest in this fraction are that it is closely associated with myelin from young animals but it is not present in crude myelin fractions from mature animals, and its enzyme profile is quite different from that of microsomes. It has been suggested that the myelinlike fraction may represent the transition form from the plasma membrane of an oligodendroglial cell to myelin, or perhaps even represents the first few layers that the glial cell forms during myelination. Brain development is described further in Chapter 18.

Disease

Besides the normal changes seen during development, myelin composition is also altered in certain neurological diseases (see Chap. 28). The myelin abnormalities in human diseases have been found to include both nonspecific and specific changes.

Abnormal myelin was first reported in a case of subacute sclerosing panencephalitis, a disease caused by an atypical measles virus infection of the brain. Analysis of the brain white matter showed a picture typical of many demyelinating diseases: a loss of major myelin constituents and a very high concentration of cholesterol esters, which are myelin degradation products. Grossly, the isolated myelin had a normal ultrastructural appearance. However, it was quite abnormal chemically, with very high cholesterol, low cerebroside, and low ethanolamine phosphatide (plasmalogen) contents.

Subsequently, similar abnormalities have been found in myelin from human brains with spongy degeneration, G_{M1} gangliosidosis, Tay-Sachs disease, and infantile Neimann-Pick disease (see Chap. 26). These diseases, although they all involve some form of demyelination (usually secondary), are quite different both in etiology and in pathology. Therefore, it is probable that the abnormal myelin is produced by a nonspecific degradative process.

Specific myelin changes have been found in metachromatic leukodystrophy and Refsum's disease. The former is caused by a lack of a sulfatase that degrades sulfatide to cerebroside, leading to abnormal accumulations of sulfatide in the brain and other organs. It is accompanied by a striking deficiency of myelin. Myelin isolated from several cases of this disease reflects the chemical defect seen in the brain as a whole; it is very high in sulfatide and low in cerebroside content. In Refsum's disease, the victims lack the ability to degrade phytanic acid. This tetraisoprenoid acid, a normal dietary product, has been found to accumulate in myelin phospholipids as it does in other tissues of the body. (For reviews see [9, 17, and 41].)

BIOCHEMISTRY

The principal biochemical features of myelin are its high rate of synthesis during the early stages of myelination and its relative metabolic stability in the adult. Although myelin is one of the most stable structures of the body when it is once formed, it is not by any means a completely inert tissue, and there is evidence that its constituents have widely varying metabolic activities. Before myelination begins, the immature brain has a relatively high concentration of cholesterol and phospholipids, but the amount of cerebroside, the lipid typical of myelin, is extremely low, as is the activity of the enzyme which synthesizes cerebrosides from UDP-galactose and ceramide. In the mouse, the activity of the cerebroside-synthesizing system reaches a peak at 10 to 20 days, coinciding very well with the maximal rate of myelination. Most other lipid synthesizing systems also become most active at this time.

A remarkable amount of synthetic work is done by the oligodendroglial cell during the time of maximum myelination. It has been shown that a 20-day-old rat brain synthesizes about 3.5 mg of myelin per day. A rough calculation shows that there are probably fewer than 20×10^6 oligodendroglia in such a brain, with each cell body having a dry weight of $\sim 50 \times 10^{-12}$ g. Thus each cell makes about 175×10^{-12} g of myelin per day, which is greater than threefold its own weight.

The concept of the metabolic stability of myelin in the adult has been largely developed by the studies of Davison and co-workers [3, 4]. This concept originated in classic studies of Waelsch, Sperry, and Stoyanoff in the 1940s (for a review, see [3]). They found that heavy water is incorporated slowly into adult brain cholesterol and fatty acids, but it is incorporated much more rapidly in young myelinating rats. Davison and colleagues confirmed these observations in greater detail, using various isotopic labels and several precursors of myelin constituents. Because these constituents were isolated from whole brain, the biochemical behavior of myelin itself had to be determined by inference. Davison and his group then pioneered in the biochemical study of isolated myelin, using similar tracer techniques, and the earlier results were found to be valid, for the most part. The conclusions from this work were that myelin shows considerably more long-term metabolic stability than do other structures in the nervous system, and that all of the myelin lipids turn over at about the same rate; therefore, the myelin membrane may be metabolized as a unit.

Experiments by Smith and colleagues [13] confirmed the relative metabolic stability of myelin but cast some doubt on the idea that myelin is metabolized as a unit. They showed that lipid precursors are incorporated into myelin and mitochondria at similar rates in young animals but that they are lost much more rapidly from the mitochondrial lipids. This finding is in accord with Davison's earlier work. However, long-term experiments show that the radioactivity is lost from individual myelin lipids at different rates. Three myelin lipids — phosphatidylinositol, lecithin, and phosphatidylserine — have half-lives of 5 weeks, 2 months, and 4 months, respectively, whereas the ethanolamine phospholipids — cholesterol, spongomyelin, cerebrosides, and sulfatides — have half-lives of 7 months to more than 1 year. By contrast, the half-lives of the lipids in mitochondria ranged from 11 days (phosphatidylinositol) to 59 days (cerebroside).

The rates of synthesis of these lipids, as determined in short-term experiments, are in agreement with the rates of turnover, phosphatidylinositol and lecithin being labeled more than the others. These observations have been extended to long-term studies of adult rats, as well as to in vitro studies of the myelin of rat spinal cord slices, using ^{14}C-glucose as a precursor in both instances [42, 43]. These experiments also confirm the earlier data: phosphatidylinositol and lecithin have the most active metabolism, and sulfatide, cerebroside, sphingomyelin, and cholesterol the least. The incorporation of precursors into myelin in vitro is, as might be expected, age-dependent: higher activities are present in actively myelinating animals, but considerable uptake can still be measured in slices from 120-day-old animals.

In the past few years, some possible explanations for the discrepancies in turnover times obtained by different investigators have been explored. In particular, the nature of the precursor and route of administration seem to be important variables. One of the more interesting recent findings is that apparently there is considerable reutilization of myelin constituents and exchange of constituents between myelin and other brain membranes. For example, it has been shown that in Wallerian degeneration in the PNS, cholesterol from the degenerating sheath is retained in the myelin debris and reused for the formation of new myelin [44]. Spohn and Davison [45], in a study of long-term cholesterol metabolism of brain, showed that eventually there is a relatively uniform distribution of labeled sterol in all subcellular fractions. They explained these results by suggesting that there is a single pool of cholesterol with which all membrane structures, including myelin, readily exchange. A similar conclusion was reached by Jungalwala and Dawson [46] with regard to phospholipid molecules, although a small pool of slowly exchangeable material may also exist. The data on the long-term persistence of sulfatide [47] in brain membranes indicates that membranes other than myelin have a very slow turnover after long periods, which also suggests exchange, but not complete equilibrium. These concepts are rather different from the usual way of looking at brain metabolism and may explain the disparate results obtained for myelin-protein turnover. If proteins are degraded, some precursors may be retained and reutilized, whereas others go into pools that turn over faster and on to other metabolic pathways. The implications are, of course, that a slow incorporation or turnover of a particular precursor does not necessarily reflect the true rate of renewal of molecules in the sheath.

Methods for separating myelin proteins have now made possible metabolic studies of these individual species. Although the turnover studies make it clear that, in general, the half-lives of myelin proteins are greater than those of other brain proteins, the half-lives reported vary considerably. For example, the half-life of basic protein in the rat has been reported as 14 to 21 days, 21 days, and 42 to 44 days, and, in the mouse, as 95 to >100 days, depending on the age at injection (for references, see [48]). Proteolipid protein has generally been found to have about the same half-life as basic protein, whereas Wolfgram protein has a shorter half-life. The rate of synthesis of these proteins is also roughly in the same order: basic protein and proteolipid protein are labeled more slowly than the high-molecular-weight proteins of myelin.

Myelin assembly and site of synthesis have been the subject of considerable speculation. It has generally been assumed that synthesis would proceed in an orderly manner, with the parts of the sheath formed early in development segregated from parts formed late. Because it was believed myelin had little metabolic turnover, the inner layers might represent early myelin and the outer layers late myelin. Recent autoradiographic studies by Rawlins show that this static picture cannot be true, at least for cholesterol [49]. Young mice were injected with ^{3}H-cholesterol and the sciatic nerves examined after short labeling times. After 20 minutes, cholesterol was mainly at the outer and inner edges of the sheath, with little at the midzone. After three hours, the cholesterol was homogeneously distributed throughout the sheath. These results are consistent with the idea discussed above — that lipids of the sheath continuously exchange with other membranes and with blood, and certainly do not support the concept of a sequential deposition of myelin into unchanging layers. Rawlins [49] suggested that the cholesterol entering the sheath from the inner edge might actually be supplied by the axon. There is now evidence that some myelin proteins in the optic pathway are labeled if precursors are injected into the eye [50]. The current feeling is that any label in the nerve and tract, when corrected for systemic labeling in the noninjected side, represents synthesis in the ganglion cells only. Thus it is speculated that label in the myelin would come from proteins synthesized in the nerve cell and transported through the axon. The exchange of lipid molecules from axon to myelin seems feasible, but the idea that the neuron synthesizes myelin protein strikes most as a heretical idea. This problem is being pursued actively by several groups, and so far the most likely conclusion is that the proteins labeled by axoplasmic flow are probably contaminants, rather than intrinsic myelin proteins.

It is possible that the relative inertness of myelin is an evolutionary adaptation that serves a useful purpose in the adult nervous system. It may contribute to stability of function and help buffer the effects of minor or short-lived traumas. The actual reasons for the greater metabolic stability of myelin compared with other brain constituents are unknown. However, we may need to look no farther than the peculiar geometry of the Schwann and oligodendroglial cells. Consider a typical large PNS axon of 5 μ diameter, with a myelin sheath 1.0 μ thick and an internode length of 1000 μ. Such an axon will have about 50 layers of myelin. A simple calculation shows that the total volume of myelin generated by one Schwann cell is 18,840 cu μ. The total area of a myelin double-unit membrane (assuming a thickness of 180 Å or 0.018 μ), if it were unrolled from the axon, is about 1×10^6 sq μ or 1 sq mm. This enormous amount of membrane is maintained by the remainder of the cell, which may be two orders of magnitude smaller. The situation with respect to the oligodendroglial cell is similar. Although CNS sheaths are thinner and internodes shorter, one cell having a perikaryon volume of about 500 cu μ myelinates an average of about 40 such internodes. It can easily be seen that the combination of the physical isolation of most of the sheath from enzymatic systems and the enormous ratio of membrane to cytoplasm may be sufficient to explain the average low metabolic activity.

MOLECULAR MODELS

Although myelin is an extension of a cell plasma membrane, it is probably quite different from it. The low protein content and low enzymatic activity indicate that, in the process of differentiation to become myelin, the original membrane must have lost much of its specialized components, producing a sort of skeletal or minimal membrane. Nevertheless, its ultrastructural appearance suggests it is very much like most membranes, and evidence obtained from myelin studies has been used to construct detailed molecular models that may resemble other membranes as well.

Given the dimensions of myelin from x-ray data and the complete chemical composition, there are two main problems in constructing a model: the fatty acid chains are too long to fit in the bilayer space and there is not enough protein to cover the bilayers. The most detailed of such models [51] solves the first problem by allowing the long sphingolipid fatty acid chains to interdigitate across the bilayer. It solves the second by proposing an unorthodox extended β-keratin structure for the proteins. The large amount of cholesterol in the structure is complexed with both sphingolipids and phosphatides.

It has been suggested that lipids with long-chain fatty acids, which produce physical stability, are also metabolically more stable, and it is true that the sphingolipids turn over more slowly than do the phosphatides, which have shorter fatty acids.

In the last few years much dissatisfaction has arisen about the universality of the unit-membrane model, and a plethora of new membrane theories and models have been proposed. In the midst of this, myelin has stood out as being the one unshakable example, so far, of the protein-lipid-protein sandwich type proposed by Danielli and Davson so many years ago.

ACKNOWLEDGMENT

I wish to thank Dr. Cedric Raine for supplying the elegant microphotographs that illustrate this chapter.

REFERENCES

*1. Bunge, R. P. Glial cells and the central myelin sheath. *Physiol. Rev.* 48:197, 1968.

2. Davison, A. N. Myelination. In Linneweh, F. (Ed.), *Fortschritte der Paedologie,* Band II. Berlin: Springer, 1968.

*3. Davison, A. N., and Dobbing, J. *Applied Neurochemistry.* Philadelphia: Davis, 1968.

*4. Davison, A. N., and Peters, A. *Myelination.* Springfield, Ill.: Thomas, 1970.

*5. Davison, A. N. Biosynthesis of the Myelin Sheath. In *Lipids, Malnutrition and the Developing Brain* (A Ciba Foundation Symposium). New York: Elsevier, 1972.

6. Mokrasch, L. Myelin. In Lajtha, A. (Ed.), *Handbook of Neurochemistry.* Vol. I. New York: Plenum, 1969.

*7. Mokrasch, L. C., Bear, R. S., and Schmitt, F. O. Myelin. *Neurosci. Res. Program Bull.* 9:439, 1971.

*8. Morell, P. (Ed.). *Myelin.* New York: Plenum, in press.

*9. Norton, W. T. The Myelin Sheath. In Shy, G. M., Goldensohn, E. S., and Appel, S. H. (Eds.), *The Cellular and Molecular Basis of Neurologic Disease.* Philadelphia: Lea & Febiger, 1976.

*10. Norton, W. T. Recent Developments in the Investigation of Purified Myelin. In Paoletti, R., and Davison, A. N. (Eds.), *Chemistry and Brain Development.* New York: Plenum, 1971.

*11. O'Brien, J. S. Lipids and Myelination. In Himwich, H. E. (Ed.), *Developing Brain.* Springfield, Ill.: Thomas, 1970.

*12. O'Brien, J. S. Chemical Composition of Myelinated Nervous Tissue. In Vinken, B., and Bruyn, G. (Eds.), *Handbook of Neurology.* Vol. 7. Amsterdam: North-Holland Publishing Co., 1970.

13. Smith, M. E. The metabolism of myelin lipids. *Adv. Lipid Res.* 5:241, 1967.

*14. Rorke, L. B., and Riggs, H. E. *Myelination of the Brain in the Newborn.* Philadelphia: Lippincott, 1969.

*15. Yakovlev, P., and Lecours, A. R. The Myelogenetic Cycles of Regional Maturation of the Brain. In Minkoski, A. (Ed.), *Regional Development of the Brain in Early Life.* Oxford: Blackwell, 1967.

16. Norton, W. T., and Poduslo, S. E. Myelination in rat brain: Changes in myelin composition during brain maturation. *J. Neurochem.* 21:759, 1973.

*17. Hodgkin, A. L. *The Conduction of the Nervous Impulse.* Springfield, Ill.: Thomas, 1964.

*18. Schmitt, F. O. Ultrastructure of Nerve Myelin and Its Bearing on Fundamental Concepts of the Structure and Function of Nerve Fibers. In Korey, S. R. (Ed.), *The Biology of Myelin.* New York: Hoeber-Harper, 1959, p. 1.

*19. Schmitt, F. O., and Bear, R. S. The ultrastructure of the nerve axon sheath. *Biol. Rev.* 14:27, 1939.

*20. Finean, J. B. Biophysical contributions to membrane structure. *Q. Rev. Biophys.* 2:1, 1969.

*21. Worthington, C. R. X-Ray Diffraction Studies on Biological Membranes. In Sanadi, D. R., and Packer, L. (Eds.), *Current Topics in Bioenergetics.* Vol. 5. New York: Academic, 1973.

22. Robertson, J. D. Design Principles of the Unit Membrane. In Wolstenholme, G. E. W., and O'Connor, M. (Eds.), *Principles of Biomolecular Organization* (A Ciba Foundation Symposium). London: Churchill, 1966.

23. Hirano, A., and Dembitzer, H. M. A structural analysis of the myelin sheath in the central nervous system. *J. Cell Biol.* 34:555, 1967.

*Asterisks denote key references.

24. Geren, B. H. The formation from the Schwann cell surface of myelin in the peripheral nerves of chick embryos. *Exp. Cell Res.* 7:558, 1954.

25. Norton, W. T. Isolation of Myelin from Nerve Tissue. In Fleischer, S., and Packer, L. (Eds.), *Methods in Enzymology.* Vol. 31. New York: Academic, 1974, p. 435.

26. Brante, G. Studies on lipids in the nervous system with special reference to quantitative chemical determinations and topical distribution. *Acta Physiol. Scand.* 18:(Suppl. 63), 1949.

*27. Eichberg, J., Hauser, G., and Karnovsky, M. L. Lipids of Nervous Tissue. In Bourne, G. H. (Ed.), *The Structure and Function of Nervous Tissue.* Vol. 3. New York: Academic, 1969.

28. Suzuki, K., Poduslo, J. F., and Poduslo, S. E. Further evidence for a specific ganglioside fraction closely associated with myelin. *Biochim. Biophys. Acta* 152:576, 1968.

29. Ledeen, R. W., Yu, R. K., and Eng, L. F. Gangliosides of human myelin: Sialosylgalactosylceramide (G_7) as a major component. *J. Neurochem.* 21:829, 1973.

30. Kurihara, T., Nussbaum, J. L., and Mandel, P. 2′,3′-cyclic nucleotide 3′-phospho-hydrolase in brains of mutant mice with deficient myelination. *J. Neurochem.* 17:933, 1970.

31. Eto, Y., and Suzuki, K. Cholesterol ester metabolism in rat brain. A cholesterol ester hydrolase specifically localized in the myelin sheath. *J. Biol. Chem.* 248:1986, 1973.

32. Folch, J., and Stoffyn, P. Proteolipids from membrane systems. *Ann. N.Y. Acad. Sci.* 195:86, 1972.

33. Agrawal, H. C., Burton, R. M., Fishman, M. A., Mitchell, R. F., and Prensky, A. L. Partial characterization of a new myelin protein component. *J. Neurochem.* 19:2083, 1972.

34. Matthieu, J.-M., Brady, R. O., and Quarles, R. H. Change in a myelin-associated glycoprotein in rat brain during development: Metabolic aspects. *Brain Res.* 86:55, 1975.

35. Eylar, E. H. The structure and immunologic properties of basic proteins of myelin. *Ann. N.Y. Acad. Sci.* 195:481, 1972.

36. Kies, M. W., Martenson, R. E., and Deibler, G. E. Myelin Basic Protein. In Davison, A. N., Mandel, P., and Morgan, I. G. (Eds.), *Functional and Structural Proteins of the Nervous System.* New York: Plenum, 1972, p. 201.

37. Carnegie, P. R. Amino acid sequence of the encephalitogenic basic protein from human myelin. *Biochem. J.* 123:57, 1971.

38. Martenson, R. E., Deibler, G. E., Kies, M. W., McKneally, S. S., Shapira, R., and Kibler, R. F. Differences between the two myelin basic proteins of the rat central nervous system: A deletion in the smaller protein. *Biochim. Biophys. Acta* 263:193, 1972.

39. Wolfgram, F. A new proteolipid fraction of the nervous system. I. Isolation and amino acid analyses. *J. Neurochem.* 13:461, 1966.

40. Brostoff, S. W., Karkhanis, Y. D., Carlo, D. J., Reuter, W., and Eylar, E. H. Isolation and partial characterization of the major proteins of rabbit sciatic nerve myelin. *Brain Res.* 86:449, 1975.

41. Suzuki, K. Lipid composition of purified myelin in various white matter diseases. A hypothesis of chemical abnormality of myelin in nonspecific demyelination. In Tourtellotte, W. W. (Ed.), *Proceedings of Symposium on Selected Topics in Human Chemical Neuropathology. Riv. Patol. Nerv. Ment.* 1971, p. 87.

42. Smith, M. E. The turnover of myelin in the adult rat. *Biochim. Biophys. Acta* 164:285, 1968.
43. Smith, M. E. An *in vitro* system for the study of myelin proteins. *J. Neurochem.* 16:83, 1969.
44. Rawlins, F. A., Hedley-Whyte, E. T., Villegas, G., and Uzman, B. G. Reutilization of cholesterol-1,2-^{3}H in the regeneration of peripheral nerve. An autoradiographic study. *Lab Invest.* 22:237, 1970.
45. Spohn, M., and Davison, A. N. Cholesterol metabolism in myelin and other subcellular fractions of rat brain. *J. Lipid Res.* 13:563, 1972.
46. Jungalwala, F. B., and Dawson, R. M. The turnover of myelin phospholipids in the adult and developing rat brain. *Biochem. J.* 123:683, 1971.
47. Jungalwala, F. B. Synthesis and turnover of cerebroside sulfate of myelin in adult and developing rat brain. *J. Lipid Res.* 15:114, 1974.
48. Fischer, C. A., and Morell, P. Turnover of proteins in myelin and myelin-like material of mouse brain. *Brain Res.* 74:51, 1974.
49. Rawlins, F. A. A time-sequence autoradiographic study of the *in vivo* incorporation of [1,2-^{3}H] cholesterol into peripheral nerve myelin. *J. Cell Biol.* 58:42, 1973.
50. Giorgi, P. P., Karlsson, J. O., Sjostrand, J., and Field, E. J. Axonal flow and myelin protein in the optic pathway. *Nature [New Biol.]* 244:121, 1973.
51. Vandenheuvel, F. A. The structure of myelin in relation to other membrane systems. *J. Am. Oil Chem. Soc.* 42:481, 1965.

Section II
Function of Neural Membranes

Chapter 5
Excitation and Conduction in the Neuron

Leo G. Abood
Wayne Hoss

The problem of nerve conduction and its fundamental significance for neural function first emerged in 1850 when Helmholtz conducted his measurements on the velocity of the nerve impulse in frogs. The present concept of the membrane potential originated with Ostwald in 1890 with the demonstration that colloidal membranes, by virtue of their ion selectivity, can promote the development of electrochemical forces of considerable magnitude. During the fifty years that preceded the definitive work of Bernstein [1], the fundamental theories of electricity and electrochemistry were developed by Arrhenius, Nernst, Planck, and many others; these theories were to be of primary importance for the subsequent development of hypotheses for neural excitation.

EXCITABILITY

In its broadest sense the term *excitability* can be defined as the ability of a cell or an organelle to respond to a stimulus. Unless this definition is narrowed, however, it would include virtually all processes of the cell — including even metabolic ones — that interact with the cell's immediate environment. Historically the term has been restricted to a description of the transient, rapid responsiveness of cells to a physicochemical stimulus in a way that involves a reversible change in the cellular membrane potential. Nonexcitable tissues, such as skin, erythrocytes, glandular cells, and even plant cells, exhibit a transient voltage change when they are hyperpolarized. As will be discussed later, a number of artificial membrane preparations, both organic and inorganic, exhibit excitatory properties. Although the physicochemical events of such systems are not unlike those that occur in so-called excitable tissues, the term *excitation* is primarily used to describe bioelectrical processes in muscle (including smooth, skeletal, and heart muscle) and nerve cells. Nevertheless, the possibility exists that bioelectrical phenomena may somehow participate in a variety of other cellular functions (e.g., secretion, mitosis, and permeability).

There are generally two aspects of excitability as applied to neurons: one is the initiation of a nerve impulse; the other, its propagation or transmission. The initiation of the nerve impulse takes place within the neuron and may involve the summation of the activities of innumerable synapses. Propagation refers to the cablelike transmission of an action potential along an axon or through any conductive medium of excitable tissues.

Originally the phenomenon of excitation was assumed to be the result of an explosive release of chemical energy. With an increasing understanding of bioenergetics and membrane biophysics, the excitatory membrane has come to be regarded as a dynamic metastable unit. One of the crucial questions concerning excitation is whether the rhythmic instability, characteristic of excitatory tissues, is caused by a biochemical reaction or by a purely physical one. Is the sudden change in membrane permeability triggered by a chemical event, such as the splitting of ATP, or is it the result of the interplay of such physical forces as electrochemical potential, fixed charges, and osmotic gradients? Much of the discussion in this chapter will focus on these and related questions.

NATURE OF THE MEMBRANE POTENTIAL

The membrane potential of a living cell is an electrical potential between two different electrolyte systems which are separated by a semipermeable cellular membrane. The resting membrane potential is a steady-state potential, which means that it is a time-invariant property of a system not at equilibrium. Because all systems seek to maintain electroneutrality, the membrane potential balances the net forces resulting from all ionic fluxes. The tendency of the two solutions to mix provides the free energy of the system, which, in the case of an ideally permselective membrane, is totally conserved or utilizable. (A *permselective* membrane is an ion-exchange membrane that is selective only for counter ions, whereas a *semipermeable* membrane is one that is permeable to one type of ion or molecule in preference to another.) Among the factors that determine the magnitude of an ionic flux across the semipermeable cellular membrane are the electrolyte concentration difference, the electrical potential difference, the permeability characteristics of the membrane barrier, and the energy-linked ion transport systems.

Interfacial (Donnan) Potentials

When colloid science developed near the turn of the century, it was recognized that polyvalent colloids, including proteins, in contact with an electrolyte will establish an equilibrium potential even in the absence of a semipermeable membrane. This "Donnan" potential arises because the electrical charge of the colloidal matrix is fixed (immobile) and the diffusible ions become distributed unequally to attain electrical neutrality. Biological membranes have a negative surface charge at neutral and alkaline pH because of the presence of ionized carboxyl and phosphate groups in their constituent proteins and lipids. These anionic groups tend to be oriented at the aqueous interfaces (membrane surfaces), because they are more easily solvated by water molecules (lower free energy) than the nonpolar interior of the membrane.

The distribution of counterions between a fixed-charge compartment and an external compartment has been shown to follow the Boltzmann relation:

$$C_i^+ = C_0^+ \, e^{-ZE/RT}$$

to within experimental error, where C_i^+ and C_0^+ are the total counterion activities in the fixed charge and external compartments respectively, Z refers to the counterion charge, and E is the resulting potential (Donnan potential). In the present case of a negatively charged membrane surface, C_i would refer to the total interfacial activity of cations. Techniques are not yet available to make direct measurements of the interfacial potentials at membrane surfaces, although measurement of the fluorescence properties of certain cyanine dyes is a promising technique [2]. The potential of the outer surface of frog and squid axons in normal bathing solutions is probably 60 to 70 millivolts (mv) [3] and, at the inner surface of the squid axon, about 20 mv [4]. Details of the charge distribution at nerve membrane interfaces have not been worked out; however, recent studies of surface-potential data from monolayers of hydrophobic polyacid spread at the air-water interface suggest that most interfacial carboxylate ions are paired in the presence of Na^+ or Ca^{2+} [5], or both. The apparent dissociation constant of the carboxyl group as well as the Ca^{2+}/Na^+ exchange-equilibrium constant was found to vary with the surface potential, and it was inferred that the effective pH at the interface was different from that of the bulk. The relationships between interfacial properties and bulk pH may have functional significance. Indeed, a positive increment of 0.2 pH unit from the normal internal value in the perfused axon preparation results in spontaneous repetitive nerve activity [6].

Diffusion Potentials

The concentration differences of various ions across living membranes can be expressed as electrochemical potential gradients, $\Delta\bar{\mu}_j$. These gradients are the forces that produce diffusional flows within the membrane, and they are related to the resulting potential difference across the membrane, E_{diff}, by the expression

$$\Delta\bar{\mu}_j = \Delta\mu_j + Z_j F \bar{E}_{diff}$$

where $\Delta\mu_j$ is the difference in chemical potential of ion j, Z refers to valence, and F is the Faraday constant. The membrane potential, E_m, is the sum of two interfacial potentials and the diffusion potential (Fig. 5-1).

$$E_m = \bar{E}_{diff} + E_d'' - E_d'$$

If the steady-state condition of zero current flow is imposed, then ideally

$$E_m = E_d'' - E_d' + \frac{RT}{F} \sum_j \frac{t_j}{Z_j} \ln \frac{C_j^0}{C_j^i}$$

where R and T have their usual significance, t is the transference number (the fraction of current carried by ion j), and C_j represents the concentration of ion j outside and inside the cellular membrane. Although the meaning of this expression is conceptually straightforward, the determination of E_d and t_j values are not. Therefore, its application now is limited.

The Action Potential

In the resting state, the membrane potential, E_r, across the squid giant axon membrane is about 60 to 90 mv (positive outside). When E_r is reduced (depolarized) below a certain critical value, referred to as *threshold*, a transient change in E_m occurs, including a reversal of potential with subsequent return to its resting value. This phenomenon, which is several milliseconds in duration, is called the *action potential*

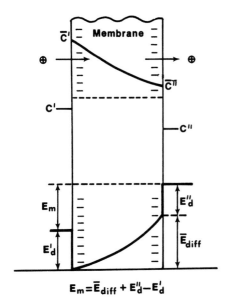

$$E_m = \overline{E}_{diff} + E''_d - E'_d$$

Figure 5-1
Profile of a cation exchange membrane demonstrating origin of the membrane potential, E_m. Essentially E_m is a concentration potential resulting from the unequal distribution of a cation (e.g., K^+). Because the membrane is more permeable to cations than to anions, the cations diffuse across more readily; the momentary excess cation transfer creates an electrical potential. A steady state is attained in which the diffusion potential enforces equivalence of cation and anion fluxes. Here, $\overline{E}_{diff}$ = diffusion potential, E_d = Donnan equilibrium potential, C = concentration of cation, $(-)$ indicates membrane, and $(')$ and $('')$ internal and external components. The membrane potential E_m is the sum total of the diffusion and the two Donnan potentials.

(Fig. 5-2). It is propagated along the axon at constant velocities in the range of 20 to 120 m per sec.

In recent years considerable attention has focused on the mechanisms responsible for maintenance of the electrochemical balance of excitable tissues because this balance is crucial for excitation. It should be stressed, however, that the ionic movements associated with excitation take place independently of those involved in active ion transport (see Chap. 6). Among other differences, the ionic movements that take place during depolarization are from zones of higher to those of lower concentration gradients. They require no energy and involve considerably greater fluxes than do active transport processes.

ELECTROPHYSIOLOGICAL BASIS OF EXCITATION
It once was thought that the resting potential was a simple K^+ equilibrium potential describable by the Nernst equation:

$$E_K = \frac{RT}{F} \ln \frac{[K^+]_i}{[K^+]_o}$$

where R = is the gas constant (8.31 joules/mole/degree)

T = degrees K

F = Faraday's constant (96,500 coulombs per mole)

Although this equation does yield a reasonably correct value for E, it does not take into account the differential Na^+ (and Cl^-) distribution. If the ratio $[Na^+]_i/[Na^+]_o$ is incorporated in the equation, the value for the potential would be 0. The addition of the Cl^- ratio does not rectify the problem. In order to circumvent this difficulty, Hodgkin and Huxley [7, 8] proposed a steady-state model in which the permeability (P) differences of the various ions were considered. In the squid axon, it was determined experimentally that the relative permeabilities of K^+, Na^+, and Cl^- in the resting state were 1.0 to 0.04 to 0.45. The revised equation for E_R was then

$$E_R = (RT/F) \ln \frac{P_K [K^+]_i + P_{Na} [Na^+]_i + P_{Cl} [Cl^-]_o}{P_K [K^+]_o + P_{Na} [Na^+]_o + P_{Cl} [Cl^-]_i}$$

$$E_R = 58 \log_{10} \frac{[K^+]_i + (P_{Na}/P_K) [Na^+]_i + (P_{Cl}/P_K) [Cl^-]_o}{[K^+]_o + (P_{Na}/P_K) [Na^+]_o + (P_{Cl}/P_K) [Cl^-]_i}$$

at 20° C. Substitution of the P ratios and $[Na^+]_o = 120$ mM, $[Na^+]_i = 4$ mM, $[K^+]_o = 2.5$ mM, $[K^+]_i = 140$ mM, and $[Cl^-]_o = 120$ mM yields the correct value for E, which is -70 mv.

The term *permeability* refers to the sum of interactions at both membrane interfaces and within the membrane; in contrast to the thermodynamic parameters discussed in the previous section, permeabilities are experimentally measurable.

If a driving force (electrical or chemical) that increases P_{Na} is applied, the ionic permeabilities are altered so that, at the peak of the action potential, the ratio P_K to P_{Na} to P_{Cl} is 1.0 to 0.20 to 0.45. The inward Na^+ current (I_{Na}) could then be expressed as a function of the electrochemical potential difference $E - E_{Na}$ by a variant of the Goldman equation [9]:

$$I_{Na} = P_{Na} \frac{F^2 E}{RT} [Na^+]_o \frac{e^{(E - E_{Na}) F/RT} - 1}{e^{EF/RT} - 1}$$

Based on the supposition that the initial unit event in nerve conduction involves a sudden change in P_{Na}, a series of empirical differential equations were formulated. These described in quantitative terms the permeability changes that presumably were responsible for the action potential.

Hodgkin-Huxley Hypothesis

Briefly, the ionic hypothesis can be described as follows: At rest, the excitatory membrane exhibits some permeability to K^+ and Cl^-, but it is virtually impermeable to Na^+. When activated, either chemically or electrically, the membrane potential decreases to a critical threshold, at which point the Na^+ permeability increases abruptly. A transient, inward-moving Na^+ current results from the electrochemical gradient. The magnitude and rate of depolarization are functions of the influx of Na^+. With diminution of the transmembrane potential, the driving (electrochemical) force of Na^+ influx slows and ceases. At the same time, there is a compensatory outward current, which is the result of K^+ outflux (and Cl^- influx), so that, at the height of the action potential, the outward and inward currents are balanced. The increased Na^+ permeability is only transient. So the transmembrane potential decreases until the resting equilibrium potential is attained. The net result of the single action potential is the exchange of equal quantities of Na^+ for K^+, and both the shape and magnitude of the action potential are described mainly in terms of the time course of the Na^+ and K^+ permeability changes (see Fig. 5-2).

By employing the ingenious technique of the voltage clamp, which permits the maintenance of the transmembrane potential at a fixed value, Hodgkin and Huxley demonstrated that the time courses of the mechanisms that determine ionic fluxes

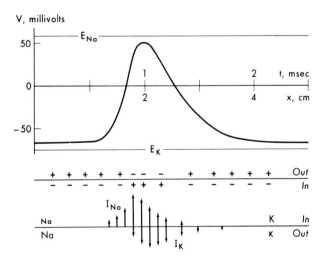

Figure 5-2

Descriptions of an impulse in a squid axon propagating to the left as a function of time t at one point or distance x at one time. At rest (left), the potential (top) is near the potassium potential E_K; there is a deficit of positive charges inside the axon (center), and the ion currents (bottom) are negligible. The increase of potential as the impulse approaches allows an inward sodium current I_{Na} which further increases V and I_{Na}. As the potential approaches E_{Na}, I_{Na} decreases; I_K increases until, after the peak of the spike, the outward I_K dominates to return V to the rest potential. (Reproduced from [52], published in 1968 by the University of California Press; reprinted by permission of the University of California Press.)

across the membrane are expressible by rate constants which depend only upon voltage and temperature. They were thus able to solve the following formulations:

$$I = C(dV/dt) + P_K (V - V_K) + P_{Na} (V - V_{Na}) + P_\lambda (V - V_\lambda) \qquad (5\text{-}1)$$

$$P_{Na} = \bar{g}_{Na} m^3 h \qquad (5\text{-}2)$$

$$P_K = \bar{g}_K n^4 \qquad (5\text{-}3)$$

$$dm/dt = \alpha_m (1 - m) - \beta_m m \qquad (5\text{-}4)$$

$$dh/dt = \alpha_h (1 - h) - \beta_h h \qquad (5\text{-}5)$$

$$dn/dt = \alpha_n (1 - n) - \beta_n n \qquad (5\text{-}6)$$

where I = total ionic current density, C = capacitance per cm^2, and V = transmembrane potential. The terms P_{Na} and P_K are the permeability coefficients for Na$^+$ and K$^+$; V_{Na} and V_K are the resting potentials for Na$^+$ and K$^+$. The subscript λ refers to similar parameters for other ions involved. The ionic permeabilities are expressed in terms of $\bar{g}_{Na}$ and $\bar{g}_K$, the conductance due to Na$^+$ and K$^+$, respectively. The letters m, h, and n represent the mole fractions of unspecified factors that regulate ionic permeabilities and α and β, the empirical rate constants. By fixing the voltage (i.e., V = constant), equation (5-1) becomes

$$I = P_{Na} (V - V_{Na}) + P_K (V - V_K) + P_\lambda (V - V_\lambda) \qquad (5\text{-}7)$$

or

$$I = I_{Na} + I_K + I_\lambda \qquad (5\text{-}8)$$

where I_{Na} is the inward Na$^+$ current and $I_K + I_\lambda$ the outward currents due to K$^+$ and other ions. Under these conditions, the total current is only a function of the combined ionic permeabilities. The rate constants are monotonic functions of the transmembrane potential, and their values are obtained over the range V = resting potential to V = Na equilibrium potential

$$V_{Na} = (RT/F) \ln [Na^+]_0/[Na^+]_i = -50 \text{ mv}$$

After determining the values for the constants α and β, the four differential equations (first order, nonlinear) are solved simultaneously for the boundary condition V = constant. A comparison is then made between the calculated current values and the data from the voltage-clamp experiments to determine the fit of the α and β functions and the necessity for making adjustments. A plot of the data obtained from the solution of the equations reveals excellent agreement between the computed and experimental curves for the action potential with respect to size and shape, as well as to threshold and subthreshold responses to the stimuli.

The overall equation describing the action potential is obtained by substituting equations (5-2) and (5-3) into (5-1):

$$I = C(dV/dt) + \bar{g}_{Na}m^3h(V - V_{Na}) + \bar{g}_K n^4(V - V_K) + \bar{g}_\lambda(V - V_\lambda) \tag{5-9}$$

where the first term refers to the current contribution of the membrane capacitance and the subsequent terms to the current contributions made by Na^+, K^+, and non-specified conductances.

The time course of the Na^+ permeability change measured by a radiotracer technique during a single action potential was in excellent agreement with that predicted by this formulation [10]. From the standpoint of neurochemistry, one of the most challenging and important problems is to define the chemical nature of the dimensionless parameters m, h, and n, which regulate ionic permeability during activity. Conceivably, the parameters m and n involve certain metastable structural components of the membrane, such as phospholipids and protein, which undergo reversible rearrangements that affect both the spatial and charge distribution of the components. Because the rate constants of these processes change dramatically with slight changes in the membrane potential, a cooperative effect probably prevails so that a rearrangement of one molecule is transferred to a number of neighboring molecules. It is apparent, therefore, that a comprehension of the molecular aspects of nerve conduction must await an understanding of the physical and chemical nature of the excitable membrane. Apart from describing their rate constants, the ionic hypothesis provides no hint as to their nature or if one or more mechanisms are involved. A number of plausible mechanisms will be discussed in later sections of this chapter.

Na^+ Conductances and Ion Channels

Since the early observation of Lorente de Nó that quaternary ammonium compounds can be substituted for Na^+ in the excitatory process, the question has arisen as to the dispensability of Na^+. Although there is a wide variety of Na^+ substitutes, including guanidine, hydrazine, and hydroxylamine, only hydrazine yields action potentials similar to those obtained with Na^+ [11]. More recently, it has been shown that in some tissues, such as the molluscan giant nerve and the crustacean muscle, the inward current is produced by Ca^{2+} and not by Na^+. Such observations have led to the hypothesis that within the excitatory membrane the channels that are opened (or activated) during the initial phase of the action potential have limited or varying specificities. Small currents presumably associated with the opening and closing of the sodium channels (gating currents) have now been measured for the squid giant axon by a signal-averaging technique [12]. They evidently arise from charge movement within the membrane and support the concept of potential-dependent conformational changes within excitable membranes. Presumably, the various ionic channels are distinct; for example, in general, the transient currents of inward and outward Na^+, but not of Ca^{2+} and K^+, are blocked by low concentrations of tetrodotoxin [13]. In some species of fish, however, Na^+ conductance is not blocked by the toxin, although Ca^{2+} conductance is blocked. A recent observation that tetrodotoxin interacts with cholesterol and not with other membranous lipids has led to the suggestion that cholesterol may be a constituent of the Na^+ channel [14].

CHEMICAL ASPECTS OF EXCITATION

The Role of Lipids

Since the observation by Langmuir about fifty years ago that a monolayer of lipids assumes a particular molecular orientation, the interactions of all surface-active substances at water-lipid interfaces have received considerable attention by both physical chemists and biologists. Such interactions, which can be described for the most part in terms of Gibbsian adsorption (i.e., as a lowering of surface free energy), are found in biological systems as well as in pure lipid-water systems.

Among the unique characteristics of lipid-water systems is their ability to undergo reversible phase transformations, ranging from a solid form to a liquid crystalline (anisotropic) form to various lipid isotropic forms. Of interest to the problems of membrane phenomena is that such phase transformations can occur readily, even under physiological conditions, and may account for such characteristics as ionic selectivity, hydrophilic-lipophilic shifts (i.e., dielectric changes), permeability changes, variations in equivalent pore size, and, possibly, even excitability. A number of dynamic membrane models have been proposed based on the phase transitions of lipids between micellar and lamellar forms [15, 16]. Many drugs and detergents have been shown to increase conductivity and selectivity to cations and anions, respectively, presumably by promoting lamellar-to-micellar transitions in the lipid bilayer membrane.

The lipid bilayer preparation of Rudin and Mueller has been employed for the study of membrane phenomena associated with excitability [17]. In addition to having the molecular configuration and dimensions of biological membranes, the lipid bilayer has comparable electrical and permeability characteristics and exhibits excitability characteristics. A variety of lipophilic substances, including an anti-biotic polypeptide (alamethicin) and certain enzymes and antigens, can produce transient decreases in the membrane resistance of the bilayer by inducing ionic con-duction pores. If certain heterocyclic antibiotics are introduced into the membrane, the membranes become K^+ selective.

One of the shortcomings of the lipid bilayer is that, in addition to phospholipids, it requires the presence of relatively high concentrations of lipid solvents (e.g., n-decane) and cholesterollike materials to maintain the stability and fluidity needed for carrying out experiments. It has been estimated that the ratio of n-decane to phospholipid in the membrane preparation may be as high as ten to one [18]. Con-sequently, many of the electrical and permeability characteristics of the membrane may be largely attributable to the hydrocarbon. A further difficulty in utilizing it as a membrane model is that it does not account for the role of proteins in determining the structural and functional characteristics of membranes, which is felt by many investigators to be the major one.

Recent evidence has indicated that most of the hydrophobic interior of lipid bilayer regions is not directly involved in excitation. Electron spin-resonance spectra of lipophilic spin labels incorporated into axonal membranes has demonstrated that the motion of the labels did not change during excitation, as would be expected if a lipid phase transition were involved [19].

Small changes in lipid head-group vibrational frequencies between resting and active axons have been observed by infrared difference spectroscopy [20]. It was inferred that these changes were caused by the action of the changing electric field on lipid dipoles of the nerve membrane.

Apart from its questionable resemblance to natural excitatory phenomena, the modification of the electrical properties of the lipid bilayer by alamethicin and the peptide isolated from *Enterobacter cloacae* is of interest from the standpoint of the role of proteins in regulating ion transport. Presumably, the peptides develop cation-selective channels within the membrane, whereas the addition of the protamine, a basic peptide, renders some of these channels anion selective. Under certain critical conditions, the transmembrane potential can be made to open and close either type of channel, causing an oscillatory change in ion conductance. Research in this area is currently directed toward the isolation and characterization of peptide ionophores from membranous ATP-hydrolyzing enzyme systems [21]. Such cation-selective agents — including many cycloethers and cyclopeptides which facilitate ion transport and agents such as tetrodotoxin which specifically block ion transport — are of considerable importance in elucidating the structural and functional role of the excitatory membrane.

Another simple artificial lipid model for excitable membranes has been developed [22]. This consists of a thin, flexible membrane (1 μ thick) formed by spreading a drop of an unsaturated glyceride (e.g., linseed oil) on a solution of 0.1 percent $KMnO_4$. Because a film of this type exhibits such characteristics as cation selectivity, biionic potentials, electrical parameters comparable to biological membranes, and "excitatory" responses, a mechanism was proposed based on the notion that cation transfer requires the breaking of the hydrogen bonds between water and the hydroxyl groups of lipids [22]. Conceivably, a depolarizing electrical field could initiate the cleavage of hydrogen bonds and thereby promote cation transfer.

The Role of Proteins

It has now become recognized that the membranous proteins play an important role in both structure and function of membranes. In addition to their catalytic, conformational, contractile, and ion-exchange properties, they possess a strong hydrophobic character that permits them to interact with one another, as well as with lipids. One of the unique characteristics of proteins is their ability to undergo cooperativity, a process whereby an interaction at one point of a macromolecule (e.g., with a substrate, ATP, or a metal) can induce a change (e.g., proton transfer or other energy change) elsewhere [23]. It is highly probably that such macromolecular characteristics, which involve intramolecular energetic changes, are of prime significance to the problem of excitation.

Metabolism and Excitation

The question of the manner in which respiratory metabolism participates in excitatory activity has preoccupied neurochemists for almost a half century. If respiratory metabolism is accelerated during neuronal excitation (this point remains problematical), it is not known if the metabolic increment is a cause or a consequence of the activity or if it is associated entirely with the recovery phase. It is conceivable that the generalized increase in membrane permeability during activity accelerates the exchange of substrates, cofactors, and metabolites and thus leads to increased metabolism. Until neurochemists are able to understand fully the biophysical events of excitation in terms of energetics, it is futile to attempt to relate energetic metabolism to neuronal activity. At present it is more appropriate to consider the energetic requirements of such functions as Na^+ extrusion, neurotransmitter metabolism and storage, and the maintenance of membrane configuration.

One approach to the problem of relating metabolic activity to excitability is to examine the chemical changes occurring in excitable tissues under the influence of stimulating and depressant drugs. Although the respiratory metabolism of the brain in vivo is diminished during surgical anesthesia, the content of ATP, phosphocreatine, and other energy-rich metabolites is essentially unchanged. It is not yet known whether the reduced metabolism is entirely the result of a specific suppression of brain centers that control cerebral blood flow and respiratory rate or whether the anesthetics affect metabolic activity that is more directly associated with excitation. In the isolated ganglion preparation of the cat, a direct relationship was found between the inhibition of oxidative metabolism by various anesthetics and the degree of synaptic transmission, particularly postsynaptic [24].

The major difficulty with this approach is that anesthetics, particularly at the concentrations used, could affect the physicochemical characteristics of the excitatory membrane in a variety of ways, and the diminished respiratory metabolism may not necessarily be responsible for diminished excitatory activity. Anesthetics and other lipophilic drugs could alter the hydrophilic-lipophilic balance, electrical charge distribution, and molecular configuration of the membrane, all of which are critical for excitation. The reader is referred elsewhere for discussions of this problem [25].

An impressive body of evidence indicates that nerve conduction is not necessarily accompanied by an increase in respiratory metabolism, insofar as it proceeds in the absence of oxygen or in the presence of oxidative and glycolytic inhibitors [25, 27]. In such tissues as frog spinal ganglia, spinal cord, and sciatic nerve, there is no increase in the rate of incorporation of $^{32}P_i$ into ATP during excitation. There is a stage, however, after metabolic inhibition (e.g., by inhibitors of oxidative phosphorylation), when excitation begins to fail; at this point the cellular concentration of ATP falls below 70 percent of its normal value. At this stage the membrane potential has fallen below a critical value (about −60 mv). Upon hyperpolarization, however, the nerves and ganglia are once again excitable. Such studies suggest that the concentration of ATP is a critical factor for the maintenance of the membrane potential above a critical level necessary for excitability [26, 28].

Role of Ca^{2+} in Excitability

Since the work of Ringer in 1890, the importance of Ca^{2+} in the maintenance of excitability has been recognized (see [11, 29] for reviews). It is believed that Ca^{2+} is responsible for the organization of the membranous components directly involved in the ion permeability changes accompanying the action potential. In addition to regulating the Na^+ and K^+ conductance during excitation, Ca^{2+} serves as a charge carrier in some tissues [11]. Still another role of Ca^+ is its participation in the storage and release of neurotransmitters at synaptic junctions [30, 31].

MOLECULAR MODELS FOR EXCITATION

Possible Mechanisms for Modifying Membrane Structure

If it is assumed that the initial event in excitation is an alteration of the configuration and physicochemical characteristics of the excitatory membrane, it is appropriate to consider the ways in which membrane structure can be reversibly altered electrically, mechanically, or chemically. Among the more apparent mechanisms are the following: (1) those causing the alteration of the conformation (i.e., tertiary or quaternary structure) of the integral proteins by affecting electrical charge distribution or through combination with a divalent cation or other mobile substance, (2) those modifying the state (phase transition) or charge of membrane lipids, (3) those affecting the equivalent pore radius of the membrane, (4) those affecting the configuration or charge characteristics of a sialopolysaccharide, (5) those initiating an enzymic or chemical reaction within the membrane, or (6) those triggering a purely electrochemical process controlling ion conductance. The various theories of excitation are based on one or more of these mechanisms, and several representative ones will be discussed. The phenomena they attempt to describe include the action potential, threshold, the variation in conductance (permeability) with membrane potential, and chemical synaptic transmission. A few common elements have emerged from the diverse experimental and theoretical approaches (discussed below).

1. Excitability is a property of the macromolecular complex that makes up functional neuronal membranes. Indeed, the protoplasm in giant axons can be removed by the technique of internal perfusion, and bioelectric properties remain [32]. But internal digestion with proteolytic enzymes or phospholipases quickly suppresses the ability of the axon to produce action potentials. This concept has been carried one step further by the demonstration [33] that microsacs (vesicular membrane fragments) prepared from the innervated face of the eel electroplax retain the ability to respond to acetylcholine agonists by an increase in membrane permeability.

2. Excitation is a transition between discrete membrane states: resting → active. This can be viewed as an extension to the membrane macromolecular complex of the well-established concept of transitions between discrete electronic, vibrational, and conformational states within atoms and molecules. In perfused axon preparations, the active state can be maintained during extended periods (>1 sec). In addition to electrical stimulation, transitions between membrane states can be induced by changes in electrolyte composition and temperature, both of which are known to affect conformational transitions in proteins and lipid bilayers. The observation of electromagnetic emission (infrared) from the external surface of unmyelinated crab nerves during stimulation [34] is at least consistent with a picture of excitation as a transition between membrane states of different energies.

3. Calcium plays a key role in excitable membranes. It is generally agreed that Ca^{2+} is bound to fixed negative sites in the resting state and that it is released during the active state.

Two-Channel Theory

The Hodgkin and Huxley analysis of the membrane events that give rise to the action potential is entirely in terms of the time and voltage dependencies of ionic permeabilities. It does not say anything explicitly about the nature of the molecular events that determine these permeabilities. It does say that, in the case of squid axons, the major ionic currents are normally Na^+ and K^+.

The implicit experimental question, then, is whether the two cations traverse the same or distinct pathways through the membrane during the action potential [35]. The Na^+ permeability changes are effectively eliminated in the presence of tetrodotoxin, leaving the K^+ conductance changes intact. Conversely, tetraethylammonium ion selectively eliminates the K^+ permeability changes. Internal perfusion of squid axons with certain proteolytic enzymes can destroy the inactivation phase of Na^+ conductance without modifying the K^+ conductance. Thus, axons so treated will conduct large currents of Na^+ and K^+ simultaneously during the latter phase of the action potential. Taken together, these experiments strongly suggest that Na^+ and K^+ traverse physically distinct pathways during the action potential.

Two principal alternatives exist for selective membrane transport: channel mechanisms and mobile carrier mechanisms (see Chap. 6). Several lines of evidence favor the channel mechanism. These include the estimated conductance per "channel" and the low temperature coefficient of the process (about 1.2). A variety of pharmacological and electrical techniques have been employed to characterize these postulated channels as to shape and size and associated charges [35, 36]. The dissociation constant of tetrodotoxin is sufficiently small to permit an estimation of the Na^+-channel density from the amount of binding of the radioactive compound to various excitable membranes [37]. Current noise-frequency analysis under voltage-clamping conditions appears to permit significant estimates of the density of both Na^+ and K^+ channels and their conductances [38]. Information that appears to bear upon the properties of the Na^+ activation "gate" [and the parameter m in equation (5-2), above] is appearing from studies of the shape and size of the initial capacitative current after depolarization [39]. Comparison of the estimate of gating charge per square centimeter of membrane with the estimates of Na^+-channel density is compatible with the original suggestion of Hodgkin and Huxley that several charged "particles" act cooperatively in the gating of each channel.

In contrast to several of the theories discussed below, proponents of the two-channel theory do not consider that Ca^{2+} participates directly in the gating processes, but rather that its effects on Na^+ conductance arise secondarily to its general effect on the membrane surface charge [35].

Two-Stable-States Theory

A theory of excitation proposed by Tasaki is that the action potential results from a reversible conformational change in the macromolecular complex that comprises the excitable membrane. This reversible change is triggered by the exchange of ions [32]. In the resting (stable) state, the negatively charged membrane is occupied largely by Ca^{2+}. An increase in the external ratio of univalent to divalent cations (produced, for example, by an outward stimulating current or K^+ outflux) causes a corresponding ion exchange at the external membrane surface. At a critical concentration ratio of univalent to divalent cations in the membrane, a discontinuous conformational transition takes place from the resting to the active state. The cooperative conformational change is accompanied by an increase in the intramembrane diffusion of cations, especially Ca^{2+}, in addition to an increase in the selectivity of membrane external surface for univalent cations. The transient changes in membrane impedance and potential that accompany excitation are the combined effect of the flux of ions and water and fluctuations in the fixed charges of the membrane.

Until the introduction of the perfused axon preparation, the theory had little experimental support other than the observation that Na^+ was not indispensable. After the axoplasm has been removed from an excised squid axon by proteolytic digestion (pronase), it can be perfused with various media while electrical and other measurements are being made. When the axon is immersed in artificial sea water (300 to 450 mM NaCl, 50 to 200 mM $CaCl_2$) and perfused with potassium fluoride or other appropriate potassium salts, excitability can be maintained for hours. The ability of the various potassium salts to maintain excitability conforms to the relative lyotropic number of the anionic species. Fluoride is the most favorable and thiocyanate (SCN^-) is the least. Because the affinity of anions for the $-NH_3^+$ groups of proteins also conforms to the lyotropic series but in the opposite direction (i.e., SCN^- is greater in affinity than F^-), proteins are believed to be the major constituents involved in excitation. Presumably, anions with the greatest affinity for ammonium would more readily disrupt salt linkages in proteins, thereby disrupting the appropriate protein conformations. Just as a variety of monovalent cations could substitute for Na_0 in an intact axon, it is possible to make substitutions in the internal perfusate. The order for maintaining excitability is $Cs > Rb > K > NH_4 > Na$ and is in agreement with that found for the affinity to phosphoryl groups of various colloids.

Abrupt transitions from the resting to the active state can be elicited in axons perfused internally with dilute monovalent cation salts and externally with $CaCl_2$ by gradually increasing the external concentration of KCl. Influx of Ca^{2+} in squid axon during excitation has been demonstrated by internal perfusion of aequorin, a protein that emits light in the presence of Ca^{2+} [40]. The difference between the fluorescence spectrum of axons stained internally with 2-p-toluidinylnaphthalene-6-sulfonate in the resting and active states can be explained by hypothesizing an invasion of water molecules during excitation [32].

The significance of water in nerve excitation has just begun to be investigated. Some penetration of water into lipid bilayer regions of biological membranes has been demonstrated by a study of solvent effects on electron spin-resonance spectra of spin-labeled lipids incorporated into myelin and other membranes [41]. By varying the label along the hydrocarbon chain, it was concluded that water penetrates to at least the C-2 position (adjacent to the ester carbonyl) and perhaps infrequently as far as C-9. The proton magnetic-resonance absorptions arising from nerve water can be resolved into a sharp signal originating from intracellular water superimposed on a broad extracellular contribution [42]. Upon depolarization of frog sciatic nerve, the data indicated that the internal water content is reduced from

65 to 35 percent of the total and that the membrane permeability to water is decreased. In addition, more intracellular water is apparently bound in the depolarized state. Other studies on natural and artificial membranes have shown that the immobilized water in the vicinity of phospholipid molecules is released in the presence of Ca^{2+}, and the mobility of the phospholipid molecules is increased. Despite the importance of resonance spectroscopy in detecting molecular changes in pure materials, the problem of interpretation poses considerable difficulties in more complex systems.

The Tasaki theory attempts to explain nerve conduction as well as excitation. As a region of the membrane is converted to the active state, a local current is established between the active and adjacent resting areas. This current, which flows inward in the active area and outward in the resting area, drives the internal cations into the resting membrane and thus establishes a self-propagating reaction along the axon. Meanwhile, immediately behind the excited region, the reversible Ca^{2+} exchanges for the monovalent ion, restoring the membrane to the resting or stable state.

A Model Based on a Ca^{2+}-ATP Membranous Complex

A molecular model of an excitatory membrane has been proposed, in which the functional unit is a macromolecular complex of Ca-ATP-phospholipid-protein [26, 29]. The Ca^{2+} is linked to ATP and the phosphate group of the phospholipid (e.g., phosphatidylserine), which in turn interacts hydrophobically with a structural protein moiety. In addition to having a high affinity for ATP, the protein may have certain catalytic functions, such as ATP-ADP exchange and ATP hydrolysis.

The initial chemical event in excitation is believed to be the displacement (electrophoretically) of Ca^{2+} followed by the release of ATP and its eventual hydrolysis. Associated with the chemical events are the conformational changes in membrane structure responsible for the increased Na^+ permeability. Restoration of the membranous structure requires the resynthesis and recombination of ATP, which in turn promotes the reuptake of Ca^{2+}. (For a detailed discussion of the kinetic and chemical aspects of the model, see [26].)

In addition to interfacial and other studies demonstrating the interaction of Ca^{2+}, ATP, and phospholipids, the model derives some support from studies on synaptic and other neural membranes that demonstrate the presence of high concentrations and combinations of the various constituents. Such membranes contain a high concentration of ATP which is bound directly to the protein, whereas the Ca^{2+} seems to be associated largely with phospholipids. According to the model, the state of membranous Ca^{2+} is dependent upon ATP, which, by virtue of its chelating and sequestering roles as well as its catalytic role, determines the conformational state of the membrane. The model provides an explanation for the change in membrane permeability, and it also conforms to the notion that excitation results from fluctuations in the fixed charges of the membrane.

Because Ca^{2+} evidently plays a critical role in regulating ion permeability during excitation, some consideration should be given to possible mechanisms involved in its regulation. If the surface layer of the excitatory membrane behaves like a surface (cationic) film, the equilibrium concentration of Ca^{2+} within the surface can be expressed as

$$a' = Ka e^{(-2F\psi/RT)}$$

where a' is the surface concentration of Ca^{2+}, a the concentration in the external solution, ψ the phase boundary (surface) potential, K a constant, and the other symbols have their customary designation.

Electrical depolarization may produce a change in the dipole orientation of phospholipid groups (e.g., CH_2-O-P, $P=O$, and $P-O-CH_2CH_2N$) attached to a Ca^{2+}. As a result, a change takes place in ψ followed by a change in the surface concentration of Ca^{2+}. Depolarization may also result in the hydrolysis of a complex of Ca^{2+} with ATP by promoting hydrolysis of the Ca-phosphate bond. Although such mechanisms are intended to be no more than examples, there is some experimental evidence in support of them.

A Thermodynamic Approach

As often emphasized by Katchalsky, the living cell functions in an asymmetric, nonequilibrium environment. As a consequence, it cannot be understood in terms of models based on reversible processes at equilibrium [43].

Of the thermodynamic functions, it is the entropy, S, which distinguishes between reversible-equilibrium processes and irreversible-nonequilibrium ones. There are two contributions to the entropy change associated with a given process; thus,

$$dS = d_iS + d_eS$$

where d_eS refers to the entropy exchanged with the surroundings and d_iS to internal entropy produced in the system. The second law of thermodynamics can be expressed as

$$\frac{d_iS}{dt} \geqslant 0$$

where the equality refers to the equilibrium condition.

Irreversible processes can be described in terms of flows, J_i, and their conjugate forces, X_i. In addition to diffusional flows and their conjugate forces (the electrochemical potential gradients, the role of which in membrane potential has already been mentioned), heat flows caused by temperature gradients and chemical-reaction flows produced by the difference between the chemical potentials of the reactants and products are familiar examples. Forces and flows are related to the internal entropy production by the dissipation function, Φ:

$$\Phi = T\frac{d_iS}{dt} = \Sigma_i J_i X_i \geqslant 0$$

where T refers to Kelvin temperature. Only the sum of the products of forces and flows need be positive; a particular flow can proceed in a direction opposite to its conjugate force. A well-known example is active transport in which a diffusional flow proceeds against its own force, $\Delta\tilde{\mu}_i$, because it is coupled to metabolic reactions that provide the necessary dissipation.

Systems in flow tend toward steady states where Φ is a minimum and certain parameters become time independent. The resting state of a nerve membrane can be described as a steady state in which diffusional and chemical reaction flows are present (and therefore the system is not at equilibrium), but Φ is minimized and the membrane potential is constant. In this context, excitation is viewed as a transition from one steady-state flow pattern to another.

This approach has been extended by Blumenthal and his associates [44], who have derived an explicit expression for a simple case of the flow of a permeant through a membrane in terms of the conformational state of the membrane in addition to other constants and variables. For certain values of the parameters, including the requirement for a high degree of membrane cooperativity (see below), three possible steady states exist for a single value of the overall gradient. Examination of the three steady states has revealed that the two extreme states were stable (restorable after a small perturbation) and the third unstable. If the resting state of a nerve membrane can be represented by one of the stable steady states, then after a small perturbation the resting state will be restored. But if the perturbation is large enough so that the unstable state is reached or exceeded (threshold), the change in the system parameters will continue until the second stable steady state (active state) is reached. In addition to excitation and threshold, the theory has also predicted the experimental conductance-voltage curves and dose-response relationship observed in axons.

Cooperativity of membrane transitions, which is an essential feature of this model, is introduced in the following manner: The membrane is conceived as a lattice system which contains equivalent lipoprotein units (protomers) specialized in the translocation of some membrane permeant. Two conformations, S and R (more permeable), are available to each protomer. The free energy of the transition from S to R, ΔF, is related to the fraction of protomers that have already made the transition, $<r>$, by the expression

$$\Delta F = \epsilon - \eta <r>$$

where ϵ is the energy required for the transition when all protomers are in the S conformation and η is a positive constant. There is no experimental evidence for long-range structural coupling in excitable membranes; however, cooperativity among a small number of units could prevail.

Irreversible thermodynamics is rapidly providing a unifying framework for the understanding of the oscillatory phenomena typical of nerve cells [43]. If the simple system described above includes the membrane-surface layers proposed by Tasaki, the development of concentration oscillations can be shown theoretically.

Some of the earliest theories of excitation held that the action potential resulted from oscillatory phenomena associated with electrochemical reactions at the membrane interface. Early studies included the work of Lillie, whose "iron-nerve" model generated a propagating, oscillatory wave of polarization and depolarization along an iron wire coated with a film of iron oxide as it underwent oxidation and reduction.

More recently, a nonmetallic model has been developed by Teorell [45]. This consists of a porous membrane with fixed charges (e.g., a fritted glass disc) that separates two stirred solutions of different electrical conductance. When a constant electrical current is passed through the membrane, electrophoresis or electroendosmosis occurs, which tends to increase the fluid volume in the compartment with higher conductance. The electroosmotic flow establishes a hydrostatic pressure difference, p, across the membrane and regulates the velocity of flow, v, so that

$$v = q \; (dp/dt)$$

where q is a geometry coefficient. With a variation in v (which is a function of the applied voltage and p), a corresponding variation takes place in the concentration profile of the electrolyte across the membrane. This, in turn, affects the electrical resistance, R, of the membrane. As v is varied, a time lag is required for R to reach a steady state, and the rate of change of the membrane resistance is found to be directly proportional to the deviation of the resistance from the steady state, R^0. Therefore

$$dR/dt = k(R - R^0)$$

where k is a rate constant dependent upon the nature of the membrane and electrolyte. A set of differential equations is then derived expressing the oscillatory, time-dependent variation for the fluid velocity, v, and the drive electrical potential, E (from Ohm's law). The solution of the equations yields integral curves, which may be damped, undamped, or "growing" sinusoidal oscillations, depending on the "damping-factor" in the equations. For a given fixed-charge porous membrane and electrolyte, the model can describe accurately the cyclical changes in R (or E) induced by electroosmosis. This experimental model incorporates many electrochemical and physical concepts relevant to bioelectrical phenomena and so serves as an excellent teaching device. Like many such physical models, however, the chemical structural nature of the model membrane (fritted glass) bears little resemblance to a biological system.

Quantum Dipole Theory

The negative charge known to exist at the surface of nerve membranes may be expressed in terms of dipoles, each of which can assume one of two quantum states with populations N_1 and N_2 [46]. The resting state is characterized by a certain distribution ratio between the two states, r, separated in energy by ΔE, so that

$$r = N_2/N_1 = e^{-\Delta E/kT}$$

Most of the dipoles are in the lower energy state (state 1 with the negative end directed outward). Thus they present a barrier against the penetration of the membrane by cations. The resulting outward force, f, on a cation will be given by

$$f = qQ(N_1 - N_2)/\epsilon$$

where q is the charge on each pole, Q is the charge on the cation, and ϵ refers to the permittivity. Such cations are also subject to diffusional and other molecular forces. Excitation results when the net force on an external cation (normally Na^+) is in the inward direction. The nerve can be excited by reducing N_1, thus reducing the outward dipole force, f. One way to reduce N_1 is to apply a negative (depolarizing) potential near the negative ends of the dipoles, which will flip to the upper state if the stimulus exceeds threshold.

The theory, which has been developed by L. Y. Wei [46], derives its major support from its ability to quantitatively predict the birefringence [47] and heat changes [48] that accompany the action potential.

The heat change associated with the action potential of various nerves is diphasic, and the time course of the resulting temperature change is similar to that of the action potential. During the rising phase of the compound action potential, rabbit vagus nerve liberates 24 μcal per g impulse, which is mostly reabsorbed during the falling phase [48].

Both proteins and lipids have the ability to undergo changes in dipole orientation when they are spread as insoluble monolayers and subjected to a moderate electrostatic field. This distortion polarizability, which is especially high for zwitterions, may serve as a triggering phenomenon for the membrane conformational change required for the increased permeability during the action potential [49]. Polarization distortion is associated with entropy changes and could also account for some of the heat changes associated with the action potential.

Neumann and Katchalsky have demonstrated electric-field-induced conformational changes in solutions of biopolymers [50]. The electric fields, which were comparable in magnitude to those which occur during nerve impulses (20 kv per cm), apparently acted by displacing the counterion atmosphere from the charged polymer. This counterion shift resulted in a large dipole moment leading to the conformational change. Although the polymers investigated were of the RNA type, the phenomenon may be applicable to charged membranes.

ROLE OF GLIA IN NERVE EXCITATION

Among the most poorly understood, yet most important, problems for the neuroscientist is the role of glia in neuronal function. Apart from their apparent role in transporting substances between capillaries and neurons, they are believed to play a role in the regulation of the ionic environment of the neuron [51]. The Schwann cell of the squid giant axon is known to have a membrane potential of -40 mv (a K^+ equilibrium potential), which remains unchanged during axonal conduction. Insofar as the Schwann cell contains six times $[Na^+]_0$, it has been suggested that the cell may be responsible for maintaining the ionic composition of the axolemma-Schwann cell space. A ouabain-sensitive K^+ transport system may be involved in the maintenance of the $[K^+]_i$ at the level of the membrane potential. In the absence of a K^+ transport system, the K^+ equilibrium potential would be about -80 mv.

CONCLUSION

In spite of all the available biophysical and biochemical data concerning bioelectrical phenomena, the mechanisms underlying excitation are still largely obscure. Although the ionic hypothesis appears to be the most adequate theory to explain the electrical events, it has not provided any insight into the biochemical nature or physicochemical characteristics of the excitatory membrane. One of the most important areas for neurochemical investigation is that concerned with membrane biochemistry, particularly the dynamic aspects associated with the action potential. Apart from the need to determine if specific lipids or proteins play a primary role, it is necessary to determine how a depolarizing stimulus or chemical agent can initiate the change in ionic permeability or whatever other physicochemical events may be responsible for excitatory potentials. As Cole [52] so aptly stated in the conclusion of a classic treatise on the ionic hypothesis: "It is most unfortunate, yet most challenging, that a possible and reasonable explanation of the most powerful facts of ion permeabilities may be completely wrong."

REFERENCES

*1. Bernstein, J. *Electrobiologie*. Braunschweig: Wieweg und Sohn, 1912.
  2. Sims, P. J., Waggoner, A. S., Wang, C.-H., and Hoffman, J. F. Studies on the mechanism by which cyanine dyes measure membrane potential in red blood cells and phosphatidyl choline vesicles. *Biochemistry* 13:3315, 1974.
  3. Ehrenstein, G., and Gilbert, D. L. Evidence for membrane surface charge from measurement of potassium kinetics as a function of external divalent cation concentration. *Biophys. J.* 13:495, 1973.
  4. Chandler, W. K., Hodgkin, A. L., and Meves, H. The effect of changing the internal solution on sodium inactivation and related phenomena in giant axons. *J. Physiol.* (Lond.) 180:821, 1965.
  5. Ter-Minassian-Saraga, L., and Thomas, C. Spread monolayer of a hydrophobic polyacid II. Surface potential and calcium binding by ion exchange. *J. Coll. Interface Sci.* 48:42, 1974.
*6. Bass, L., and Moore, W. J. A Model of Nervous Excitation Based on the Wien Dissociation Effect. In Rich, A., and Davidson, N. (Eds.), *Structural Chemistry and Molecular Biology*. San Francisco: W. H. Freeman, 1968.
  7. Hodgkin, A. L., and Huxley, A. F. Currents carried by sodium and potassium through the membrane of the giant axon of *Lodigo. J. Physiol.* (Lond.) 116:449, 1952.
*8. Hodgkin, A. L., and Huxley, A. F. A quantitative description of membrane current and its application to conduction. *J. Physiol.* (Lond.) 117:500, 1952.
  9. Goldman, D. E. Potential, impedance, and rectification in membranes. *J. Gen. Physiol.* 27:37, 1943.

*Asterisks denote key references.

10. Atwater, I., Benzanilla, F., and Rojas, E. Time course of the sodium permeability change during a single membrane action potential. *J. Physiol.* (Lond.) 211:753, 1970.

*11. Koketsu, K. Calcium and the Excitable Cell Membrane. In Ehrenpreis, S., and Solnitzky, O. C. (Eds.), *Neurosciences Research.* New York: Academic, 1969.

12. Benzanilla, F., and Armstrong, C. M. Gating currents of the sodium channels: Three ways to block them. *Science* 183:753, 1974.

13. Watanabe, A., Tasaki, I., Singer, I., and Lerman, L. Effects of tetrodotoxin on excitability of squid giant axons in sodium-free media. *Science* 155:95, 1967.

14. Villegas, R., Barnola, F. V., and Camejo, G. Ionic channels and nerve membrane lipids. *J. Gen. Physiol.* 55:548, 1970.

*15. Parsons, D. F. Ultrastructural and Molecular Aspects of Cell Membranes. *Canadian Cancer Conference.* Vol. 7. New York: Pergamon, 1966, p. 193.

16. Torch, W. C., and Abood, L. G. Effect of calcium, sodium, and aluminum ions on electrical and physical properties of bimolecular lipid membranes. *Int. J. Neurosci.* 5:143, 1973.

*17. Mueller, P., and Rudin, D. O. Bimolecular Lipid Membranes: Techniques of Formation, Study of Electrical Properties and Induction of Ionic Gating Phenomena. In Passow, H., and Stämpfli, R. (Eds.), *Laboratory Techniques in Membrane Biophysics.* Berlin: Springer-Verlag, 1969, p. 141.

*18. Henn, F. A., and Thompson, T. E. Synthetic lipid bilayer membranes. *Annu. Rev. Biochem.* 38:693, 1969.

19. Calvin, M., Wang, H. H., Etine, C., Gill, D., Ferruti, P., Herrolod, M. A., and Klein, M. P. Biradical spin labeling for nerve membranes. *Proc. Natl. Acad. Sci. U.S.A.* 63:1, 1969.

20. Sherebrin, M. H., MacClement, B. A. E., and Franko, A. J. Electric-field-induced shifts in the infrared spectrum of conducting nerve axons. *Biophys. J.* 12:977, 1972.

21. Shamoo, A. E. Isolation of sodium-dependent ionophore from (Na^+, K^+)-ATPase preparations. *Ann. N. Y. Acad. Sci.* 242:389, 1974.

*22. Monnier, A. M. Experimental and theoretical data on excitable artificial lipidic membranes. *J. Gen. Physiol.* 51:265, 1968.

23. Wyman, J. Regulation in macromolecules as illustrated by hemoglobin. *Q. Rev. Biophys.* 1:35, 1968.

24. Larrabee, M. G., and Pasternak, J. M. Selective actions of anesthetics on synapses in mammalian sympathetic ganglia. *J. Neurophysiol.* 15:91, 1952.

*25. Seeman, P. M. Membrane stabilization by drugs, tranquilizers, steroids, and anesthetics. *Int. Rev. Neurobiol.* 9:145, 1966.

*26. Abood, L. G. Interrelationships between phosphates and calcium in bioelectric phenomena. *Int. Rev. Neurobiol.* 9:223, 1966.

*27. Ling, G. N. *A Physical Theory of the Living State.* New York: Blaisdell, 1962.

*28. Caldwell, P. C. Factors governing movement and distribution of inorganic ions in nerve and muscle. *Physiol. Rev.* 48:1, 1968.

*29. Abood, L. G. Calcium-ATP-Lipid Interactions and Their Significance in the Excitatory Membrane. In Ehrenpreis, S., and Solnitzky, O. C. (Eds.), *Neuroscience Research.* New York: Academic, 1969.

*30. Smith, A. D., and Winkler, H. Fundamental Mechanisms in the Release of Catecholamines. In Blaschko, H., and Muscholl, E. (Eds.), *Handbook of Experimental Pharmacology.* Vol. 33. Berlin: Springer-Verlag, 1972, p. 538.

31. Berl, S., Puszkin, S., and Nicklas, W. J. Actomyosin-like protein in brain. *Science* 179:441, 1973.

*32. Tasaki, I. Energy transduction in the nerve membrane and studies of excitation processes with extrinsic fluorescent probes. *Ann. N. Y. Acad. Sci.* 227:247, 1974.

33. Kasai, M., and Changeux, J.-P. Excitability in vitro. *J. Membr. Biol.* 6:1, 1971.

34. Fraser, A., and Frey, A. H. Electromagnetic emission at micron wavelengths from active nerves. *Biophys. J.* 8:731, 1967.

*35. Armstrong, C. M. Ionic pores, gates and gating currents. *Q. Rev. Biophys.* 7:179, 1975.

36. Hille, B. An essential ionized acid group in sodium channels. *Fed. Proc.* 34:1318, 1975.

37. Almers, W., and Levinson, S. R. Tetrodotoxin binding to normal and depolarized frog muscle and the conductance of a single sodium channel. *J. Physiol.* (Lond.) 247:483, 1975.

38. Conti, F., de Felice, L. J., and Wanke, E. Potassium and sodium ion current noise in the membrane of the squid giant axon. *J. Physiol.* (Lond.) 248:45, 1975.

39. Keynes, R. D., and Rojas, E. Kinetics and steady-state properties of the charged system controlling sodium conductance in the squid giant axon. *J. Physiol.* (Lond.) 239:393, 1974.

40. Hallett, M., and Carbone, E. Studies of calcium influx into squid giant axons with aequorin. *J. Cell Physiol.* 80:219, 1972.

41. Griffith, O. H., Dehlinger, P. J., and Van, S. P. Shape of the hydrophobic barrier of phospholipid bilayers: Evidence for water penetration in biological membranes. *J. Membr. Biol.* 15:159, 1974.

42. Dea, P., Chan, S. I., and Dea, F. J. High-resolution proton magnetic resonance spectra of rabbit sciatic nerve. *Science* 175:206, 1972.

*43. Katchalsky, A. K., Rowland, V., and Blumenthal, R. Dynamic patterns of brain cell assemblies. *Neurosci. Res. Program Bull.* 12:1, 1974.

44. Blumenthal, R., Changeux, J.-P., and Lefever, R. Membrane excitability and dissipative instabilities. *J. Membr. Biol.* 2:351, 1970.

45. Teorell, T. Oscillatory phenomena in a porous, fixed-charge membrane. In Passow, H., and Stämpfli, R. (Eds.), *Laboratory Techniques in Membrane Biophysics.* Berlin: Springer-Verlag, 1969, p. 130.

46. Wei, L. Y. Dipole mechanisms of electrical, optical and thermal energy transductions in nerve membrane. *Ann. N. Y. Acad. Sci.* 227:285, 1974.

*47. Cohen, L. B. Changes in neuron structure during action potential propagation and synaptic transmission. *Physiol. Rev.* 53:373, 1973.

48. Abbott, B., and Howarth, J. V. Heat studies in excitable tissues. *Physiol. Rev.* 53:120, 1973.

49. Pethica, B. A. Structure and physical chemistry of membranes. *Protoplasma* 63:147, 1967.

50. Neumann, E., and Katchalsky, A. Long-lived conformational changes induced by electric impulses in biopolymers. *Proc. Natl. Acad. Sci. U.S.A.* 69:993, 1972.

*51. Kuffler, S. W., and Nicholls, J. G. Physiology of neuroglia cells. *Ergeb. Physiol.* 57:1, 1966.

*52. Cole, K. S. *Membranes, Ions, and Impulses.* Berkeley: University of California Press, 1968.

Chapter 6

Ion Transport

Phillip D. Swanson
William L. Stahl

GENERAL CONSIDERATIONS

Different cellular and subcellular compartments of the body are separated from each other and from the external environment by a series of membranes. In order to maintain the compartmental concentrations of essential nutrients and ions at levels necessary for normal cellular activity, membrane transport processes are essential. Each compartment has a reasonably characteristic composition. Steep chemical and electrical gradients often exist across the membranes. These are often important links in cellular function. For example, sodium and potassium ions are involved in the generation and propagation of the action potential in nerve and muscle tissues.

Movement of molecules through membranes can, in some instances, be described in terms of simple forces that are sufficient to bring about movement without the addition of metabolic energy. In these passive transport processes, a net flux of material arises due to dissipation of the total free energy of the system. The energy of the system decreases until thermodynamic equilibrium is reached and all net fluxes are zero. In other instances, movements of molecules across membranes require expenditure of energy (active transport).

The rate of passage of a molecule through a membrane depends on the magnitude of forces responsible for the movement (concentration or electrical potential gradients) and the relative ease with which the molecule passes through the membrane, the latter being expressed in terms of permeability of the membrane. Mechanisms of movement remain largely unproved, but may involve direct movement through the membrane, passage through membrane pores, and pinocytosis.

Stein [1] has carefully examined data on the permeation of a large number of substances through membranes. Passive diffusion involves movement of a solute across a membrane as a result of random molecular motion. Membrane charge may alter the rate of diffusion. The model of a simple bimolecular leaflet with substances directly diffusing through the hydrophobic lipid phase accounts for much of the assembled permeability data for nonelectrolytes as based on considerations of the permeant's size, the number of potential hydrogen-bonding groups, and, in some cases, the number of bare methylene groups in relation to the oil-water partition coefficient. However, this model does not account for the anomalous permeability of water, urea, and ions in most natural membrane systems. Alternatively, a model that postulates movement of the permeant within channels penetrating through the hydrophobic environment of the membrane predicts that diffusion might be determined exclusively by the size of the penetrating species. This view is not wholly adequate because no single value for the pore radius can account completely for the available data. Penetration of many molecules may occur through nonrigid lattice spaces between the lipoprotein chains of the membrane matrix. There is probably a range of such pore sizes in any given membrane.

It is well known that a number of molecules — glucose and glycerol, for example — pass through cell membranes faster than is predicted on the basis of the above considerations. This mediated transport of specific molecules imparts great functional flexibility to the membrane. Systems controlling the rate at which specific molecules enter the cell down a preexisting concentration gradient have been termed *facilitated diffusion* systems. These systems do not require metabolic energy. The kinetics of such systems have been interpreted successfully in terms of membrane carriers. Reversible binding of a permeating substance P with a membrane component C, designated as the carrier, is postulated (Fig. 6-1). The complex then traverses the membrane and dissociates, thereby releasing the permeant into the cell. The carrier then returns to the opposite face of the membrane to complete the cycle. Assuming the presence of carriers enables one to explain many of the known kinetic characteristics of membrane-transport processes, including saturation, specificity, and competition. Coupling of inward movement of one substance with outward movement of another can also be encompassed in such models. Although the mechanism shown in Figure 6-1 indicates movement of the carrier through the membrane, the carrier molecule could be viewed as rotating or changing in conformation in order to permit movement of the permeant through the membrane (see Fig. 6-4).

Cells are able to maintain steady-state concentrations of many substances far from equilibrium by supplying energy derived from metabolic processes. Therefore, active transport of a substance is transport against its electrochemical or osmotic gradient, requiring a source of energy to bring about vectorial movement of particles through a membrane. Energy might be supplied at one or more of the steps shown in Figure 6-1 in such a way that transfer of permeant is more effective in one direction than in the other. A useful example is the movement of sodium ions. During the initial phase of a nerve action potential, sodium moves rapidly into the cell by passive diffusion, due to forces that result from its concentration and electrical potential gradients. During the recovery phase, however, sodium moves out of the cell against these gradients. Thus, the transport during the recovery phase is an "active" process requiring energy for the transport.

Active transport of sodium out of the cell is important for maintenance and restoration of ion gradients that are necessary for initiation of an action potential,

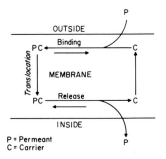

Figure 6-1
Carrier-mediated transport.

as well as for regulation of intracellular cation concentrations and for maintenance of a constant cell volume. The concentrations of Na^+ and K^+ within the cell undoubtedly influence the rates of a variety of enzyme reactions. Complex systems, such as mitochondrial oxidative phosphorylation, are inhibited by high Na^+ or low K^+ concentrations.

The necessity of active transport for cell volume stabilization has been discussed in theoretical terms by Post and Jolly [2] and by Stein [1]. Animal cells are surrounded by water-permeable, nonrigid membranes that vary in volume if placed in media of varying tonicity. The steady-state volume will change if a cell is prevented from metabolizing appropriate substrates. The membrane is permeable to some solutes (e.g., inorganic ions) and not to others (e.g., certain proteins). If there were no active transport, the permeable constituents would become distributed in equilibrium with equal internal and external concentrations or, in the case of ions, distributed according to a Donnan equilibrium ratio (see Chap. 5). Because of the presence of impermeable substances within the cell, water would move into the cell, thus increasing the volume. As developed by Post and Jolly [2], the three factors that control the volume of a cell without a rigid cell wall are: (1) the amount of fixed intracellular material; (2) the concentration of external permeable material; and (3) the ratio of the rate constants of the "leak" to the "pump." Increasing leakiness or decreasing pump rates will bring about an increased cell volume.

ISOLATION AND IDENTIFICATION OF TRANSPORT SYSTEMS

Transport Proteins

Although many of the properties of transport systems have been described in great detail, little is known about them at the molecular level. A major goal has been to isolate and to characterize transport components and, it is hoped, to reconstitute activity in model systems by reassociation of membrane components, as in the synthetic membranes described below. Study of transport proteins has been hampered because the methods used to isolate the transport species necessarily lead to dissolution of the cell membrane, with concomitant loss of the transport function. The properties of the transporting system and the isolated transport protein must be compared to establish a relationship between the two. For example, in both the bacterial sulfate-transporting system and a sulfate-binding protein discussed below, they: (1) are identically inhibited by a series of anions and a series of protein reagents; (2) are lost from cells to a similar degree by osmotic shock; (3) are simultaneously lost by mutation, are regained by reversion or transduction, and are lacking in certain transport-negative mutants; and (4) show a similarity in affinity constants for binding and transport.

Membrane-transport proteins have been successfully isolated in pure form, mainly from microorganisms. Bacterial systems offer several advantages. Many transport proteins are liberated easily by osmotic shock, which simplifies further purification, and specific transport function can often be induced by growing the organisms in the presence of substrate.

Beta-Galactoside Transport Protein

In the now-classic experiments of Fox and Kennedy [cf. 3], evidence was presented for the involvement of a protein (M protein) in the facilitated entrance of β-galactosides into *Escherichia coli*. This component is distinct from previously characterized proteins of the lactose system, is localized in the membrane fraction, and has a high affinity for certain β-galactosides. Organisms that were induced or noninduced for the transport system were used. *N*-Ethylmaleimide (NEM), which combines irreversibly with the sulfhydryl groups (–SH) of proteins, blocks transport of β-galactosides. The bacteria were first treated with nonradioactive NEM in the presence of thiodigalactoside. This improved the specificity of isotopic labeling in a subsequent step by first permitting all –SH groups not protected by thiodigalactoside to react with unlabeled NEM. The excess NEM and the thiodigalactoside were then removed. The induced organisms were treated with ^{14}C-NEM and the uninduced organisms with ^{3}H-NEM. The labeled bacteria were then mixed and the proteins were extracted and fractionated. It was hoped that the transport species (induced) would be selectively labeled with ^{14}C-NEM. Indeed, a membrane fraction that had a higher ratio of ^{14}C to ^{3}H was isolated and the component was purified to yield a protein with a molecular weight of 30,000.

Sulfate-Binding Protein

Transport of sulfate into *Salmonella typhimurium* occurs against a concentration gradient and requires energy. Utilizing substrate recognition (binding of $^{35}SO_4{}^{2-}$) as a means of testing for the transport protein, Pardee and his associates [3] have isolated and crystallized a protein with a molecular weight of 32,000. One sulfate ion appears to be bound per protein molecule, with a dissociation constant of about 0.1 μM. Thus far, attempts to reconstitute the transport system by adding pure sulfate-binding protein to shocked organisms (which are unable to transport sulfate) have been unsuccessful. It is likely that both the sulfate-binding protein and M protein are involved in the initial binding step of the transport process (Fig. 6-1).

Calcium-Binding Proteins

Calcium plays an important role in many physiological processes including release of neurotransmitters at synapses. A number of calcium-binding proteins have now been isolated [4–10], and a partial list is given in Table 6-1. In general, they are low molecular-weight proteins with high glutamic and aspartic acid compositions, and they possess no known enzymatic activity. With the exception of calsequestrin [7], all of the proteins are easily extracted from tissue by using relatively mild procedures. So probably they are not very tightly bound to membranous structure in the cell. Calcium binding has been established by using equilibrium dialysis, and these proteins have dissociation constants of about 10^{-5} M.

These proteins often form higher molecular-weight complexes. The protein S-100, for example, exists in multiple forms in the presence of Ca^{2+} [8], and it undergoes conformational changes to a relatively unfolded structure. Although all of these proteins bind Ca^{2+}, their roles in the cell remain uncertain. They may simply act to localize calcium on membrane surfaces or they may interact with membranes to increase permeability or to act as carriers in transport systems.

Table 6-1. Calcium-Binding Proteins

Source and/or Name	Author and Reference	Molecular Weight	Dissociation Constant	Amino-Acid Composition
Brain, adrenal medulla, squid axon	Wolff & Siegel [4] Alema et al. [5]	12,000–15,000	1.5-2.5×10^{-5} M	18–21% Glut, 13–16% Asp
Brain, adrenal phosphoprotein	Brooks & Siegel [6]	11,500	2.5×10^{-5} M	21% Glut, 14.4% Asp
Sarcoplasmic reticulum (calsequestrin)	Ostwald et al. [7]	46,000	5-7×10^{-5} M ($-$KCl) 7-8×10^{-4} M (KCl)	37% Glut + Asp
Brain (S-100)	Calissano et al. [8]	6,300	10^{-3} M	44% Glut, 15.5% Asp
Intestine & brain	Wasserman et al. [9] Taylor [10]	28,000	2.6×10^{-5} M	16% Glut, 14% Asp

Phospho Transferase System
Roseman and his colleagues have elucidated the details of a sugar-transport system in microorganisms [11]. This is a group translocation system, in which the transported molecule is modified chemically during its passage through the membrane. Roseman and his colleagues have related the transport of nine sugars to a phospho transferase system (Fig. 6-2). The net process

sugar + phosphoenolpyruvate (PEP) → sugar-phosphate + pyruvate

involves three proteins and results in release of the phosphorylated derivative inside the cell. The sugar-phosphate cannot pass back through the membrane, thereby achieving an inward movement of the sugar. The energy donor for this system is PEP, making the phospho transferase system different from that of other known sugar kinases. A cytoplasmic enzyme (Enzyme I) mediates phosphorylation of a cytoplasmic energy-donating protein (HPr, a protein of molecular weight 9400) on a histidine residue to form phospho-HPr:

$$\text{PEP} + \text{HPr} \underset{\substack{\text{Mg}^{2+} \\ \text{(cytoplasm)}}}{\overset{\text{Enzyme I}}{\rightleftarrows}} \text{phospho-HPr} + \text{pyruvate}$$

This is followed by transfer of phosphate to the sugar, thereby regenerating dephospho-HPr:

$$\text{phospho-HPr} + \text{sugar} \xrightarrow[\text{Enzyme II}]{\text{(Membrane)}} \text{sugar-P} + \text{HPr}$$

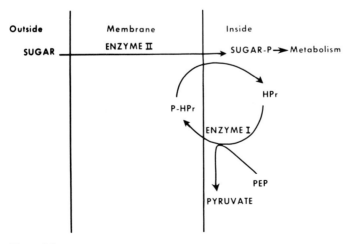

Figure 6-2
Sugar transport by the phospho transferase system.

The specificity of the system resides in the membrane-bound Enzyme II. Apparently, a specific Enzyme II exists for each sugar. Bacterial mutants that lack HPr or Enzyme I, which eliminates phospho-HPr as an energy donor, are not able to assimilate sugars. Also, there is little passive transport, suggesting that energy is required at the translocation step in this system.

Macrocyclic Compounds as Carrier Models

A large number of organic compounds that possess the ability to transport cations across membranes are now known. The earliest described were macrocyclic compounds synthesized by microorganisms and possessing antibiotic activity. More recently, structurally simpler cyclic polyethers have also been shown to exhibit transport properties. These substances have certain common features: they are low molecular-weight compounds (mol wt $<2,000$) with covalently linked or potential cyclic structures with lipophilic external surfaces. These compounds will sequester cations selectively. They are often able to bring about ion transport into mitochondria in vitro, including alkali-metal cation movement against concentration gradients by an energy-requiring process. The accumulation in mitochondria is presumed to be a form of active transport, as interrupting the energy supply reverses the direction of ion movement.

These active compounds also have demonstrable electrical effects on artificial bilayers. The factors underlying the ion selectivity of these compounds are discussed here briefly because the capacity of biological membranes to distinguish sodium and potassium ions underlies such important fundamental processes as excitation and conduction in the nervous system and active sodium and potassium transport in general. The detailed structures of these compounds are known, so the study of systems employing them as ion carriers offers an important alternate approach (as opposed to isolating carriers from the membrane) to the study of the molecular basis of membrane transport.

Table 6-2 lists the classes of ion carriers that have been studied. The valinomycin and cyclic polyether groups form positively charged ion pairs. The nigericin group forms neutral ion pairs. Alamethicin, a pure peptide containing one free carboxyl group, can form neutral or positively charged ion pairs. These compounds contain either covalently bonded closed rings or linear chains, which can assume a ring configuration by cyclizing noncovalently.

Many of these active compounds show a dramatic ability to discriminate between monovalent cations. For example, when added to mitochondrial systems, valinomycin has an apparent complexing preference of K^+ over Na^+ as high as 10,000 to 1. Complex formation in these compounds occurs by induced ion-dipole interactions with oxygen atoms of the ring. Polar groups face inward and hydrophobic groups face toward the exterior of the molecule. The number of ring atoms varies for the active polycyclic compounds. Those with fewer than 18 atoms, however, will not select for K^+. This is especially clear in the case of the cyclic polyethers, where there is excellent correlation between the ionic diameters of Li (1.20 Å), Na (1.90 Å), and K (2.66 Å) and the "holes" of the cyclic polyethers. The 4-oxygen, 5-oxygen, and 6-oxygen polyethers (Table 6-2) have holes approximately 1.8, 2.7, and 4.0 Å

Table 6-2. Macrocyclic Carrier Models

Active Compounds	No. Ring Atoms	Ion Selectivity
Valinomycin group (positively charged ion pairs formed)		
Nonactin, monactin, dinactin (macrotetrolides)	32	Potassium[a,b]
Enniatins A and B (depsipeptides)	18	Potassium[b]
Gramicidins A, B, and C (neutral linear polypeptides)	—	Potassium[b,c,d]
Valinomycin (depsipeptide)	36	Potassium[a,b,d,e]
Nigericin group (neutral ion pairs formed)		
Nigericin (macrocyclic monocarboxylic acid)	—	Potassium[a,d]
Monensin (macrocyclic monocarboxylic acid)	—	Sodium[a]
X537A (macrocyclic carboxylic acid)	—	Divalent cations[a]
A23187 (uncharacterized)	—	Divalent cations[a]
Cyclic polyethers (positively charged ion pairs formed)		
4-Oxygen polyethers (e.g., bis-butylcyclohexyl 14-crown-4)	14	Lithium[a]
5-Oxygen polyethers (e.g., butylcyclohexyl 15-crown-5)	15	Sodium[a,c]
6-Oxygen polyethers (e.g., dicyclohexyl 18-crown-6)	18	Potassium[a,d,e]
Alamethicin (neutral or positively charged ion pairs formed) (cyclic polypeptide)	53	Poor[a]

[a] Based on ion complexing power.
[b] Based on transport through artificial membranes.
[c] Based on cation permeability of red cells.
[d] Based on transport in mitochondria.
[e] Based on ion permeability of phospholipid micelles.

in diameter to accommodate ions, which probably accounts for their indicated ion preferences. Complexes are not formed if the ion is too large to lie in the hole of the polyether ring.

Interaction between the active compounds and the transported ions involves the unhydrated cation at the center of the active compound interacting with oxygen portions of the molecule. Hydrogen bonding is important in the formation of stable complexes in such open-chain compounds as monensin. In the cyclic polyethers, an increase in the number of oxygen atoms increases the stability of the complex, provided the oxygen atoms are coplanar and symmetrically distributed in the polyether ring.

Alamethicin, an extracellular cyclopeptide from *Trichoderma viride* (Table 6-2), is able to induce energy-linked K^+ accumulation by mitochondria [12] but shows little ion selectivity, probably due to the large ring size, which should be flexible enough to assume many ion-selective conformations. This compound differs from the other compounds discussed above in that unselective cation transport can be induced as a function of time and membrane current in experimental bilayers. When such basic proteins as histone, spermine, or protamine are present in the bilayer system, however, alamethicin can induce action potentials that are sensitive to the ionic gradient or applied voltage. In this system, therefore, conditions can be varied to permit electrical activity that resembles an action potential in nerve. These membrane characteristics can also be induced by a still-uncharacterized material, "excitability inducing material" (EIM), which is very likely a peptide or protein and which was isolated from *Enterobacter cloacae* by Mueller and Rudin [13]. These substances work, although with varying sensitivities, in membrane systems made from mixed lipids, phospholipids, or "oxidized" cholesterol. It has been proposed from electrokinetic and chemical data that six or more molecules form channels through the synthetic membrane, thereby permitting ion flow; assembly and disassembly of these aggregates by voltage and chemical parameters would regulate membrane conduction [14].

Ionophores that can transport divalent cations have also been identified. One of these compounds, A23187, is an antibiotic with a molecular weight of 523. It contains a carboxylic-acid moiety and probably transports divalent cations through biological membranes by an electroneutral process in which two molecules bind one free metal ion [15]. The relative affinities of this compound for divalent cations are reported to be $Ca^{2+} > Mg^{2+} > Sr^{2+} > Ba^{2+}$. Ionophore X537A, another carboxylic acid–containing compound, recently has been characterized [16] and shown to have low affinity for Na^+ and K^+ and higher affinities for $Ba^{2+} \gg Sr^{2+} > Ca^{2+} > Mg^{2+}$. These relative affinities indicate complexing power (Table 6-2). Interestingly, the ability of X537A to carry divalent ion current across lipid bilayers follows the order $Mg^{2+} > Ca^{2+} > Sr^{2+} > Ba^{2+}$. This difference may arise because ions with high relative affinity, e.g., Ba^{2+}, are not released; rather, they are retained by the carrier in membranes. The carrier is thereby inhibited by substrates that have too great an affinity for optimal net transport [16].

Carrier Mechanisms

These studies have given rise to two main schemes for molecular mechanisms by which cations might be transported across membranes through the mediation of macrocyclic compounds. In the first, channel formation, several active molecules would stack together and form a channel through the membrane (Fig. 6-3A). The tunnel or channel would be hydrophilic in nature so that anhydrous or partially hydrated ions could pass through; intermediary complexes could be formed with the active molecules that form the channel. In the process of channel formation, alamethicin molecules, for example, would stack together to form a channel lined with oxygen atoms. In this situation, water molecules would form a bridge between the K^+ and carbonyl group so that the partially hydrated ion could transit the membrane. Ion selectivity is presumed to arise from differences in hydrated ion size. Mueller and Rudin [14] favor channel formation in explaining the action of alamethicin and EIM. This is based on observations that the maximum change in steady-state

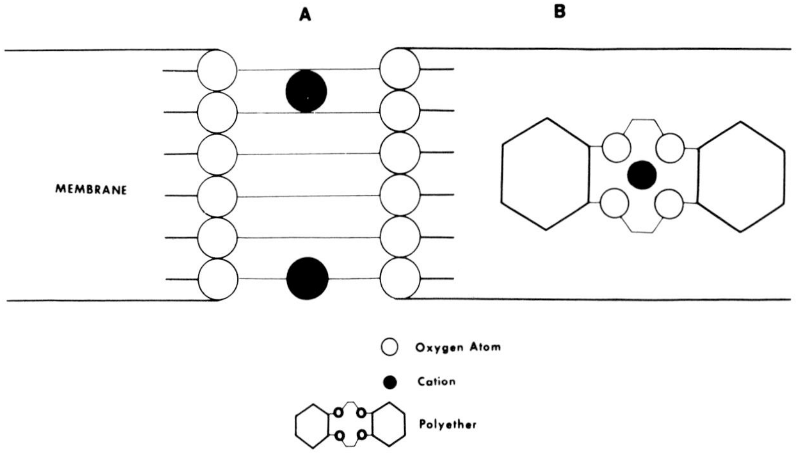

Figure 6-3
Possible mechanisms for movement of cations through biological membranes in the presence of
a macrocyclic compound. (A) Channel formation; (B) mobile carrier. (Adapted from [12].)

conductance in synthetic membranes varies exponentially with the sixth power of
the potential for a wide range of alamethicin concentrations. Also, as in nerve, the
apparent kinetic order of the reaction is six. The highly nonlinear cooperative effects
shown by alamethicin and EIM would appear to exclude a monovalent carrier process
and would favor a carrier or channel composed of six or more active molecules.

The alternative to a channel is a *mobile carrier.* In this case, the active complexing
molecule would displace the water of hydration from the cation and form a lipid-
soluble complex that could diffuse through the membrane and release the cation at
the opposite interface of the membrane (Fig. 6-3B). In such a mechanism, ion specificity can best be explained by a combination of unhydrated ion with the active molecule to form the mobile lipid-soluble complex. This mechanism is favored by evidence
gathered from x-ray crystallographic and nuclear magnetic resonance (NMR) analyses
for valinomycin, monactin, and the enniatins. For example, NMR data [17] indicate
that transfer of cations between valinomycin molecules in nonpolar media is unlikely,
thus favoring a mobile carrier mechanism.

ACTIVE TRANSPORT OF Na$^+$ AND K$^+$

Sodium and potassium ions are actively transported in a large number of cell types.
For example, Na$^+$ is transported from the cytoplasm of cells against an electrochemical gradient, and cells are able to maintain low internal Na$^+$ and high internal K$^+$ concentrations. The active transport of Na$^+$ depends on the presence of K$^+$ in the
extracellular fluid, indicating a coupling between the active outward transport of
Na$^+$ and inward transport of K$^+$. Na$^+$ extrusion and K$^+$ assimilation are greatly
reduced by 2,4-dinitrophenol, azide, and cyanide, and thus require metabolic energy.
The relationship between the adenosine triphosphate (ATP) supply and transport

was clearly demonstrated by Caldwell et al. [18]. They injected ATP, arginine phosphate, or phosphoenolpyruvate into axons of *Loligo* poisoned with 2,4-dinitrophenol or cyanide. Na^+ efflux was restored and the number of Na^+ ions extruded was proportional to the number of high-energy phosphate molecules injected. In other experiments, Hoffman [19] depleted human red cells of metabolic energy reserves and labeled them internally with ^{24}Na. After transfer to sodium-free media, only adenosine (and not inosine) stimulated ATP formation and sodium transport.

In contrast to the energy-requiring active transport of Na^+ and K^+, the shifts of these cations that occur during conduction of a nerve impulse take place without addition of an energy source as long as the necessary cation gradients are present (see Chap. 5 and [20]). Active transport is necessary for the maintenance of the cation gradients and hence for the membrane potential.

Electrogenic Sodium Pumping

Development of the membrane potential depends largely upon the relatively greater permeability of the neuronal membrane to K^+ than to Na^+ [21]. It is now believed that a component of the resting potential in neuronal and muscle cells is brought about by movement by the cation pump of unequal quantities of Na^+ and K^+ [22, 23]. Differing ratios of Na^+ out to K^+ in are reported; the most frequent ratios are 3:2 or 2:1. In each instance, however, more Na^+ ions than K^+ ions are transported actively.

Evidence for a contribution of the Na^+ pump to the membrane potential includes:

1. The membrane potential does not equal the potential predicted by a modified constant field equation that is based only on Na^+ and K^+ concentrations and permeabilities.
2. Ouabain produces depolarization in some invertebrate neurons.
3. Increasing internal Na^+ increases the potential in some cells.

The contribution to the membrane potential by the electrogenic pump has been estimated to be 1.8 millivolts (mv) for the squid giant axon, 5 to 15 mv for molluscan nerve cell bodies, and 10 to 16 mv for striated muscle.

$(Na^+ + K^+)$-Stimulated Adenosine Triphosphatase

Because the direct source of energy for extrusion of Na^+ ions from the interior of a cell and the coupled accumulation of K^+ appeared to be ATP, a system which consumes ATP (an adenosine triphosphatase) and which has properties differentiating it from other ATP-consuming systems was sought in membrane-derived subcellular fractions. In 1948 adenosine triphosphatase (ATPase) activity had been demonstrated in the sheath of the giant axon of the squid *Loligo pealii*. The activity of this bound enzyme was 19 to 100 times greater than the activity in the axoplasm. Appreciable magnesium-activated ATPase activity was also found in membrane-derived subcellular fractions prepared from rat peripheral nerve. Particular interest was generated by studies of Skou [24], which demonstrated an ATPase activity in microsomal membranes from crab nerve that was markedly stimulated by the presence of Na^+ plus K^+ in the incubation medium. Either cation alone was much less effective than both together. The requirement for K^+ was less specific than that for Na^+ because Rb^+ could substitute for K^+. Since this first detailed study, a great deal of circumstantial evidence has suggested involvement of this enzyme system in active Na^+ and K^+ transport.

The properties of the $(Na^+ + K^+)$-ATPase system that suggest its role in active Na^+ and K^+ transport are:

1. It is located in the cell membrane.
2. Cation requirements are similar to those of the active transport system: there is a higher affinity for Na^+ than K^+ on the intracellular side of cell membrane and opposite affinity on the outside of cell membrane.
3. Substrate requirements are similar to those of the active transport system: ATP is utilized as the energy source for movement of cations.
4. The enzyme system hydrolyzes ATP at a rate dependent on concentration of Na^+ inside and K^+ outside the cell.
5. The system is found in all cells that have active, linked transport of Na^+ and K^+.
6. The system is inhibited by cardiac glycosides.
7. There is good correlation between enzyme activity and rates of cation flux in different tissues.

Tissue Localization
The enzyme is found in very high activity in electrically excitable tissues such as eel electric organ, brain, nerve, and muscle. Appreciable activity also is present in secretory organs, including kidney, choroid plexus, and the ciliary body. Bonting and Caravaggio [25] correlated $(Na^+ + K^+)$-ATPase activities with rates of active cation fluxes (Na^+ efflux, K^+ influx, or their average) in six tissues in which flux data were available. Recognizing that the conditions of enzyme assay and those of flux determinations cannot exactly coincide, the ratios of enzyme and flux activity remain remarkably constant over a 22,000-fold range in magnitude (Table 6-3).

In every tissue which has been examined, $(Na^+ + K^+)$-ATPase activity has been found in subcellular fractions that contain particulate material derived from cell membranes. In dispersions prepared in isotonic sucrose from mammalian brain, the highest specific activity is found in the primary microsomal fraction, consisting morphologically of membrane fragments, and in the synaptosomal fraction that is prepared by centrifuging the primary "mitochondrial" fraction on a sucrose density gradient [26]. Lowest specific activities are found in the supernatant and purified mitochondrial fractions. When synaptosomes are disrupted by dispersion in water, $(Na^+ + K^+)$-ATPase activity is absent from the synaptic vesicle subfraction and is concentrated in fractions that contain membrane fragments derived from external synaptosomal membranes.

Vectorial Stimulation of $(Na^+ + K^+)$-ATPase
Comparisons between the activity of an enzyme system and the physiological function of that system are necessarily limited by the conditions under which each system is assayed. Most enzyme assays are carried out on purified or partially

Table 6-3. $(Na^+ + K^+)$-ATPase Activity and Cation Flux in Tissues

Tissue	Temp. (°C)	A Cation Flux $(10^{-14}$ moles/ sq cm/sec)	B $(Na^+ + K^+)$-ATPase[a] $(10^{-14}$ moles/ sq cm/sec)	(A)/(B)
Human erythrocytes	37	3.87	1.38 ± 0.36	2.80
Frog toe muscle	17	985	530 ± 94	1.86
Squid giant axon	19	1,200	400 ± 79	3.00
Frog skin	20	19,700	6,640 ± 1,100	2.97
Toad bladder	27	43,700	17,600 ± 1,640	2.48
Electric eel, noninnervated membranes of Sachs' organ	23	86,100	38,800 ± 4160	2.22
Average	....	....	....	2.56 ± 0.19

[a]ATPase activities were determined by standard methods on portions of lyophilized and re-constituted tissue homogenates or microdissected samples of frozen-dried tissue. (Data from Bonting and Caravaggio [25].)

purified material that has been separated, as far as possible, from other constituents of the tissue from which it is derived. In contrast, measurement of cation transport requires tissue that is as intact as possible in order to allow measurement of cation movements across membranes. Therefore, it is rather remarkable that, in tissues in which comparisons have been made, close correlations are found in the require-ments for the two processes. The erythrocyte has been particularly useful for such comparisons, and Post et al. [27] carefully documented the most important features of both systems. Both systems utilize adenosine triphosphate rather than inosine triphosphate as substrate. Both require Na^+ and K^+ together, rather than either cation alone, and NH_4^+ can substitute for K^+ but not for Na^+ in each system. In both systems, high concentrations of Na^+ competitively inhibit K^+ activation. In both systems, ouabain is inhibitory. The concentrations at which Na^+, K^+, ouabain, and NH_4^+ show half of their maximal effects are the same in both systems.

A technique has been developed with erythrocytes that makes possible the pro-duction of "ghosts," whose internal composition can be reconstituted with known concentrations of cations and other substances. In one series of experiments car-ried out by Whittam and Ager [28], the procedure used was the following:

Cells were hemolyzed in 5 to 60 volumes of a hypotonic solution containing 0.25 to 4 mM $MgCl_2$ and ATP. To the suspension, enough NaCl or other salt was added to raise the osmotic pressure to about 0.3 osmolar. The suspension was then incu-bated at 37° C for 30 min, during which time the cells regained their low permeability to Na$^+$ and K$^+$ and to ATP. The reconstituted ghosts were washed with 0.15 M NaCl or choline chloride, and they contained the cation in high concentration that had been added to the hemolysate. Determination of the amount of inorganic phosphate produced upon further incubation was used as a measure of ATP hydrolysis.

When cells were incubated in a medium containing 10mM KCl, the rate of ATP hydroly was greatest when the internal cation was Na$^+$. In addition, incubation in K$^+$-free medium or in medium containing ouabain decreased ATP hydrolysis. In contrast, external Na$^+$ failed to activate ATP hydrolysis. These and similar experiments have suggested that (Na$^+$ + K$^+$)-ATPase is stimulated by internal Na$^+$ and by external K$^+$ and that the ortho-phosphate produced from ATP hydrolysis is liberated inside the cell.

Lipid Requirements of the (Na$^+$ + K$^+$)-ATPase System

This enzyme system has been found to be bound particularly tightly to other mem-brane constituents. For optimal activity, the enzyme appears to require the presence of lipid. Removal or alteration of membrane phospholipids with detergents or phos-pholipases leads to loss of enzyme activity. Some studies have shown that only phos-phatidyl serine or acidic phospholipids are specifically required for reactivation of the enzyme [29], suggesting that the polar end groups of the phospholipids are very im-portant in maintaining the functional state of the (Na$^+$ + K$^+$)-ATPase system, perhaps by providing a negative charge at the active site. Others have found a less specific phospholipid requirement for enzyme reactivation [30, 31] and have stressed the *general* need for hydrophobic fatty acids and hydrophilic portions of the phospholipids for enzyme activity.

The role of cholesterol in the (Na$^+$ + K$^+$)-ATPase system remains uncertain. Removal of cholesterol from membranes does not inactivate the enzyme [32], although subsequent extraction of phospholipids leads to loss of enzyme activity. Addition of cholesterol alone restores at least a portion of this lost activity, perhaps through reorganization of the protein-lipid framework of the membrane. Studies with lipid monolayers have shown that cholesterol forms strong compact associations with long-chain lipids. Cholesterol's role in the enzyme system may be through such interactions.

Arrhenius plots for (Na$^+$ + K$^+$)-ATPase activity in isolated membranes are nonlinear with an apparent transition at about 20° C. Below this temperature the apparent activation energy is approximately 30 kcal per mol; above that point it is about 15 kcal per mol. Delipidation of the membrane matrix with phospholipases yields preparations of lower specific enzyme activity and with only the higher activation energy (no transi-tion point). Both the transition and higher specific activity of the enzyme may be restored by addition of exogenous phospholipids. The change in activation energy is likely caused by a lipid-phase transition since the lipids are more fluid in nature above 20° C. These and other studies with isolated spin-labeled lipids [33] suggest that, to function properly, the (Na$^+$ + K$^+$)-ATPase must be in a "fluid" environment with the matrix of the membrane.

Component Parts of the (Na$^+$ + K$^+$)-ATPase Reaction

Three reactions have been suggested to represent partial reactions of the (Na$^+$ + K$^+$)-ATPase system: (1) Na$^+$-stimulated incorporation of ^{32}P from ATP into membrane phosphoprotein; (2) Na^+-dependent ADP-ATP exchange reaction; and (3) K^+-stimulated phosphatase activity.

Incorporation of ^{32}P from ATP. Membrane preparations from kidney, brain, electrophorus electric organ, and erythrocytes incorporate labeled phosphate from γ-^{32}P-ATP after brief incubation. Incorporation is increased in the presence of Na^+ and is diminished by adding K^+ or NH_4^+. Cardiac glycosides slow the turnover of the "phosphoenzyme" and diminish its sensitivity to K^+ [34]. The rate of Na^+-dependent incorporation is consistent with its presumed role as part of the $(Na^+ + K^+)$-ATPase system. Peptic digests of membrane fractions have contained labeled peptides, and thus the labeled "intermediate" is probably bound to protein as an acyl phosphate. Recent work has shown that the labeled site is an aspartyl phosphate [35a, b].

ADP-ATP Exchange Reaction. Formation of labeled ATP occurs when membrane preparations are incubated in the presence of radioactively labeled adenosine diphosphate (ADP). Much of this ADP-ATP exchange activity is easily separated without loss of $(Na^+ + K^+)$-ATPase activity and is thus due to nonrelated reactions. A component that remains associated with $(Na^+ + K^+)$-ATPase activity is stimulated by Na^+ under carefully specified conditions [36, 37]. This Na^+-stimulated exchange reaction has substrate requirements similar to those of the overall $(Na^+ + K^+)$-ATPase activity and requires incubation at very low Mg^{2+} concentrations for its demonstration.

K^+-Stimulated Phosphatase Activity. Using either *p*-nitrophenyl phosphate or acetyl phosphate as substrate, $(Na^+ + K^+)$-ATPase preparations will form inorganic phosphate. This reaction is stimulated by K^+ and inhibited by ouabain, and it has therefore been regarded as resembling the dephosphorylation component of the overall $(Na^+ + K^+)$-ATPase reaction [29].

Interaction with Cardiac Glycosides

One of the chief pieces of evidence linking sodium and potassium transport and $(Na^+ + K^+)$-ATPase has been the specific nature of inhibition of both by such cardiac glycosides as ouabain. The asymmetry of the transport ATPase within plasma membranes was established in part through experiments which demonstrated that ouabain inhibits transport and ATPase activity when present on the outside, but not on the inside, of cells [28]. In microsomal preparations, these glycosides generally inhibit the net $(Na^+ + K^+)$-ATPase reaction by 50% in the range of 10^{-7} to 10^{-6} M. Structural features essential for interaction and inhibition of the enzyme have been reasonably well established [29] and may help to clarify the nature of the binding site. Cardiac glycosides interact slowly and noncovalently with the enzyme in the absence of physiological ligands, and the complex dissociates at an even slower rate. The affinity of the enzyme for cardiac glycosides is conformation dependent, and conditions that favor phosphorylation also favor binding; however, phosphorylation of the enzyme may not be necessary [34, 38].

Purification of $(Na^+ + K^+)$-ATPase

Separation of enzyme activity from other membrane components has been very difficult to achieve in spite of a great deal of effort. Material with high specific activity has been obtained by treating membrane fractions with a concentrated solution of sodium iodide.

Detergent treatment can also convert the enzyme system into a form that is not sedimented by high-speed centrifugation [39]. The "solubilized" enzyme system still contains considerable lipid and probably represents membrane fragments that are incorporated into detergent micelles. When electrophoresed, such material contains many protein components [40]. The purification procedure of Kyte [41], which utilizes canine renal medulla, contains the following steps: homogenization and differential centrifugation to obtain microsomes; treatment with deoxycholate; extraction with potassium iodide (a "chaotropic agent"); treatment with a low concentration of deoxycholate and salt; centrifugation and filtering; and fractionation on a Sepharose column. Gel electrophoresis of this and similar preparations in the presence of sodium dodecylsulfate has yielded two polypeptide fragments: a heavy chain of 84,000 to 139,000 daltons (most estimates are 90,000 to 100,000) and a light chain of 35,000 to 57,000, which is a glycoprotein [29]. Because the estimated molecular weight of the whole enzyme is 190,000 to 560,000 daltons, the functional enzyme must contain several subunits, but the exact number is uncertain.

The heavy chain appears to be specifically phosphorylated during turnover of the enzyme, and therefore this chain is now often referred to as the catalytic subunit. The heavy chain also appears to have binding sites for cardiac glycosides [42], sodium, potassium, and ATP. The role of the lighter glycoprotein chain remains unknown. The molar ratio of the catalytic subunit to the glycoprotein also remains uncertain, but ratios of 1:1 and 2:1 have been reported [35b]. The catalytic subunit appears to have the same amino-acid composition in all species studied, which suggests no fundamental evolutionary changes. Similarly, the glycoprotein shows little interspecies differences [35b].

Relationship of (Na⁺ + K⁺)-ATPase to Active Cation Transport

Relationship of $(Na^+ + K^+)$-ATPase to Active Cation Transport
Precise definition of the nature of the relationship of the enzyme to cation transport is not complete. Models are revised as new facts become known. Two such models are illustrated in Figure 6-4. One of the earliest proposals for coupling an enzyme reaction to transport was the Shaw model, in which a single carrier alternately transports sodium outward and potassium inward (quoted in [43]). The chemical free energy provided by the enzyme reaction is used to change the orientation and cation affinity of the ionophore.

The transport model by Albers, Koval, and Siegel (Fig. 6-4A) incorporates details about the partial reactions of the $(Na^+ + K^+)$-ATPase system and about the way in which inhibitors interfere with enzyme activity [44]. The reaction is pictured as taking place in several steps:

$$E_1 + MgATP \overset{Na^+}{\rightleftharpoons} E_1 \sim P + MgADP \tag{1, 6}$$

$$E_1 \sim P \xrightarrow{Mg^{2+}} E_2 - P \tag{2}$$

$$E_2 - P \xrightarrow{K^+} E_2 + P \tag{3, 4}$$

$$E_2 \rightleftharpoons E_1 \tag{5}$$

A

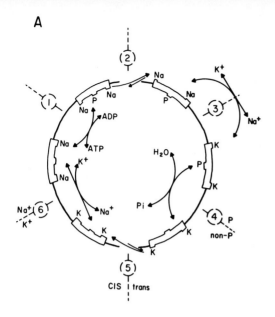

B

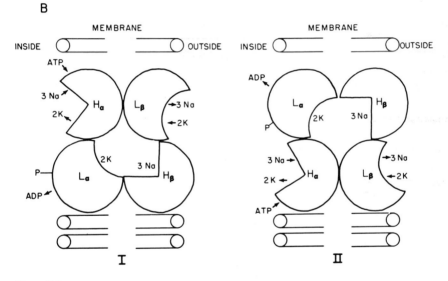

I

II

Figure 6-4
Two proposed mechanisms for active Na$^+$ and K$^+$ transport. (A) The (Na$^+$ + K$^+$)-ATPase brings about active transport through allosteric transition. (Adapted from [44].) (B) Tetramer model (adapted from [46]). The enzyme exhibits half-of-sites reactivity, in which only one of two identical subunits is expressed in the reaction. This is caused by negative cooperativity, in which binding of substrate by one subunit of the catalytic dimer reduces the affinity at the second subunit. In the model, the catalytic (α) units face the cell interior, and each subunit may exist in either high- or low-affinity states for Na$^+$ binding. The conformations I and II are energetically and functionally equivalent, and flipping from one conformation to the other brings about active transport. Stein et al. [46] postulate that in the first conformation (I), Na$^+$ binds to the H$_\alpha$ subunit and K$^+$ binds to L$_\beta$. As phosphorylation takes place, the tetramer flips to II, and ions brought into the internal cavity by the upper subunits redistribute so that Na$^+$ and K$^+$ bind at the H$_\beta$ and L$_\alpha$ subunits. With K$^+$ present, the upper subunit dephosphorylates and the tetramer returns to conformation I with expulsion of Na$^+$ to the exterior and K$^+$ to the interior of the cell. Cation movements on the lower two subunits are equivalent, but 180° out of phase.

Reaction (1), which requires Na^+, phosphorylates the enzyme reversibly and can be measured as an ADP-ATP exchange reaction and as a Na^+-dependent membrane phosphorylation. In reaction (2), the phosphorylated enzyme is converted to a second (trans) form of the enzyme, which can now react with K^+ (reaction 3) to cleave orthophosphate from the phosphorylated enzyme (reaction 4). In this scheme, inwardly directed sites have high affinity for Na^+, and outwardly directed sites have high affinity for K^+. In reactions (2) and (5), the orientation of the enzyme is altered by allosteric conversions and provides the vectorial force of transport. The cycle is completed when Na^+ displaces K^+ from the cis enzyme (reaction 6). The model may incorporate K^+-ATP interactions, in which K^+ must dissociate before ATP and Na^+ bind [38].

The foregoing model requires that a given subunit must interact sequentially with Na^+ and K^+ with enzyme phosphorylation intervening. There might be a single iono-phore with alternating affinities or separate ionophores for Na^+ and K^+. Other kinetic experiments [45] support the idea that Na^+ and K^+ may act at separate coexisting sites. Recently Stein et al. [46] and Repke and Schön [47] have suggested a "flip-flop" mechanism with "half-of-sites" reactivity. Their model visualizes two identical subunits which cooperate through reciprocating conformational coupling to transport Na^+ and K^+ simultaneously.

CALCIUM TRANSPORT

The concentration of ionized Ca^{2+} within cells has been estimated to be about 10^{-5} M, whereas a concentration of 1 M would be expected if this cation were distributed according to a Donnan ratio. Calcium moves into cells by passive transport, probably involving facilitated diffusion (Fig. 6-5). Perhaps one or more of the calcium-binding

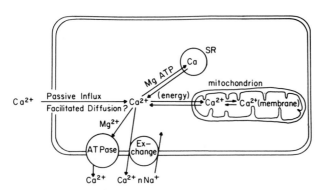

Figure 6-5
Calcium-transport processes. Uptake of calcium by cells is a passive process, and intracellular concentrations may be regulated: by Ca^{2+}-Na^+ exchange or by a $(Mg^{2+} + Ca^{2+})$-ATPase at the plasma membrane; by active transport into mitochondrion; and in muscle by the $(Mg^{2+} + Ca^{2+})$-ATPase, which permits sequestration of calcium by the sarcoplasmic reticulum (SR). The Ca^{2+}-Na^+ exchange is also energy dependent, because the $(Na^+ + K^+)$-ATPase maintains the sodium gradient across the plasma membrane.

proteins (Table 6-1) are involved in this process. Once inside the cell, the mitochondria actively accumulate Ca^{2+} [48]. However, in various tissues, several processes seem to exist for regulating intracellular Ca^{2+}. In muscle, sarcoplasmic reticulum contains an ATP-dependent mechanism [$(Mg^{2+} + Ca^{2+})$-ATPase] that sequesters Ca^{2+} and plays an important role in coupling between excitation of the muscle membrane and contraction of myofibrils. Erythrocyte membranes contain a ruthenium red-sensitive $(Mg^{2+} + Ca^{2+})$-ATPase, which appears to extrude Ca^{2+} actively. This enzyme has also been demonstrated in synaptosomes and, based on its specific activity, could be involved in calcium extrusion (Fig. 6-5) [49]. In excitable tissues, a calcium-sodium exchange process may be responsible for calcium transport at the level of the plasma membrane [50–52].

In neural tissue, detailed studies with squid axon [50] and brain slices [51] have shown that efflux of ^{45}Ca is reduced when external Na^+ is replaced by Li^+, choline, or dextrose. In brain slices, Ca^{2+} uptake is increased by ouabain [52, 53], which also inhibits active Na^+ and K^+ transport, as well as by omission of glucose or by addition of metabolic inhibitors [53]. These results, as well as similar findings with brain synaptosomes [54], have been interpreted in terms of a Ca^{2+}-Na^+ exchange process in brain plasma membranes (Fig. 6-5).

One may assume that each of these processes may play a role in regulating the concentration of cytoplasmic Ca^{2+} in brain cells. Under conditions of low Ca^{2+} influx at the plasma membrane, efflux may well be controlled by a Ca^{2+}-Na^+ exchange process coupled with the existing sodium gradient across the membrane or by a $(Mg^{2+} + Ca^{2+})$-ATPase. The capacity of this system could be exceeded during periods of greater ion influx, i.e., in depolarized cells, and excess calcium would be accumulated actively by mitochondria for later removal by the cell. Based on the available surface area of mitochondrial membranes and the rate constants for Ca^{2+} influxes in mitochondria, it has been concluded that mitochondria are potentially far more efficient in controlling cytoplasmic Ca^{2+} than are processes at the level of the plasma membrane [48]. Ultimately, however, the accumulated Ca^{2+} must be extruded from mitochondria to be transported from the cell, and little is known at present about mitochondrial mechanisms governing this extrusion process.

The ($Mg^{2+} + Ca^{2+}$) ATPase of Sarcoplasmic Reticulum

The mechanism of calcium transport has been studied extensively using fragmented sarcoplasmic reticulum (SR) from skeletal muscle. Isolated SR vesicles contain a $(Mg^{2+} + Ca^{2+})$ ATPase and take up calcium actively. A phosphorylated intermediate in the ATPase reaction was suggested by the discovery of an ADP-ATP exchange reaction in SR and, subsequently, Ca^{2+}-dependent formation of a phosphorylated protein (EP) in the presence of $AT^{32}P$ was found. In SR, two moles of Ca^{2+} are actively transported per mole of ATP hydrolyzed, and an efflux of Mg^{2+} and K^+ may be coupled to Ca^{2+} uptake. The Ca^{2+} and MgATP participating in EP formation bind on the outside of the membrane [55], and isolated SR vesicles are capable of reducing external calcium concentrations to less than 1 μmole [7]:

$$E + MgATP \underset{Ca_i^{2+}}{\overset{Ca_o^{2+}}{\rightleftarrows}} EP + ADP + Mg^{2+}$$

where the subscripts o and i refer to outside and inside of SR, respectively. This reaction has certain similarities to $(Na^+ + K^+)$-ATPase, and the phosphorylated intermediate is also believed to be labeled on an aspartyl residue [56]. The $(Mg^{2+} + Ca^{2+})$-ATPase in SR is reversible, but the relationship between uptake and release mechanisms remains to be clarified.

The $(Mg^{2+} + Ca^{2+})$-ATPase has been separated (about 95% pure) from such other membrane components as calsequestrin and has a molecular weight of about 100,000. Earlier work showed that digestion of SR vesicles with phospholipases removed 60% to 70% of the phospholipids and led to loss of $(Mg^{2+} + Ca^{2+})$-ATPase activity and Ca^{2+} uptake by the vesicles. As with the $(Na^+ + K^+)$-ATPase, these processes could be restored by addition of exogenous phospholipids to SR vesicles. Recent experiments [57] with the purified enzyme also showed that Ca^{2+} uptake could be demonstrated by insertion of the soluble enzyme into soybean phospholipid vesicles; Ca^{2+} uptake was sensitive to the calcium ionophores X537A and A23187, but resistant to monovalent ionophores such as valinomycin and nigericin. The transport enzyme has likewise been reconstituted in a system of defined exogenous phospholipids (dioleoyl lecithin). This approach, using model membranes of well-defined composition, promises to clarify how ATPases achieve vectorial movement of ions.

REFERENCES

*1. Stein, W. D. *The Movement of Molecules Across Cell Membranes.* New York: Academic, 1967.

2. Post, R. L., and Jolly, P. C. The linkage of sodium, potassium, and ammonium active transport across the human erythrocyte membrane. *Biochim. Biophys. Acta* 25:118, 1957.

3. Pardee, A. B. Membrane transport proteins. *Science* 162:632, 1968.

4. Wolff, D. J., and Siegel, F. L. Purification of a calcium binding protein from pig brain. *Arch. Biochem. Biophys.* 150:578, 1972.

5. Alema, S., Calissano, P., Rusca, G., and Guiditta, A. Identification of a calcium binding, brain specific protein in the axoplasm of squid giant axons. *J. Neurochem.* 20:681, 1973.

6. Brooks, J. C., and Siegel, F. L. Purification of calcium-binding phosphoprotein from beef adrenal medulla. *J. Biol. Chem.* 248:4189, 1973.

7. Ostwald, T. J., MacLennan, D. H., and Darrington, K. J. Effects of cation binding on the conformation of calsequestrin and the high affinity calcium binding protein of sarcoplasmic reticulum. *J. Biol. Chem.* 249:5867, 1974.

8. Calissano, P., Moore, B. W., and Firiesen, A. Effect of calcium ion on S-100, a protein of the nervous system. *Biochemistry* 8:4318, 1969.

9. Wasserman, R. H., Carradino, R. A., and Taylor, A. N. Binding Protein from Animals with Possible Transport Function. In *Membrane Proteins,* Proceedings of Symposium of N. Y. Heart Assn. Boston: Little, Brown, 1969, p. 114.

10. Taylor, A. N. Chick brain calcium binding protein: Comparison with intestinal vitamin D-induced calcium binding protein. *Arch. Biochem. Biophys.* 161:100, 1974.

*11. Roseman, S. The Transport of Carbohydrates by a Bacterial Phosphotransferase System in Membrane Proteins. In *Membrane Proteins,* Proceedings of Symposium of N. Y. Heart Assn. Boston: Little, Brown, 1969, p. 138.

*Asterisks denote key references.

*12. Eisenman, G. (Chairman). Symposium on biological and artificial membranes. *Fed. Proc.* 27:1249, 1968.

13. Mueller, P., and Rudin, D. O. Action potential phenomena in experimental lipid membranes. *Nature* 213:603, 1967.

14. Mueller, P., and Rudin, D. O. Action potentials induced in bimolecular lipid membranes. *Nature* 217:713, 1968.

15. Case, G. D., Vanderkooi, J. M., and Scarpa, A. Physical properties of biological membranes determined by the fluorescence of the calcium ionophore A23187. *Arch. Biochem. Biophys.* 162:174, 1974.

*16. Pressman, B. C. Properties of ionophores with broad range cation selectivity. *Fed. Proc.* 32:1698, 1973.

17. Haynes, D. H., Kowalsky, A., and Pressman, B. C. Application of nuclear magnetic resonance to the conformational changes in valinomycin during complexation. *J. Biol. Chem.* 244:502, 1969.

18. Caldwell, P. D., Hodgkin, A. L., Keynes, R. D., and Shaw, T. I. The effects of injecting "energy-rich" phosphate compounds on the active transport of ions in the giant axons of *Loligo. J. Physiol.* (Lond.) 152:561, 1960.

*19. Hoffman, J. F. Molecular Mechanism of Active Cation Transport. In Shanes, A. M. (Ed.), *Biophysics of Physiological and Pharmacological Action.* Washington, D.C.: American Association for the Advancement of Science, #69, 1961.

20. Baker, P. F., Hodgkin, A. L., and Shaw, T. I. The effects of changes in internal ionic concentrations on the electrical properties of perfused giant axons. *J. Physiol.* (Lond.) 164:355, 1962.

*21. Woodbury, J. W. The Cell Membrane: Ionic and Potential Gradients and Active Transport. In Ruch, T. C., and Patton, H. D. (Eds.), *Physiology and Biophysics.* Philadelphia: Saunders, 1965.

*22. Thomas, R. C. Electrogenic sodium pump in nerve and muscle cells. *Physiol. Rev.* 42:563, 1972.

23. Gorman, A. L. F., and Marmor, M. F. Steady-state contribution of the sodium pump to the resting potential of a molluscan neurone. *J. Physiol.* (Lond.) 242:35, 1974.

24. Skou, J. C. The influence of some cations on an adenosine triphosphatase from peripheral nerves. *Biochim. Biophys. Acta* 23:394, 1957.

25. Bonting, S. L., and Caravaggio, L. L. Studies on sodium-potassium-activated adenosine triphosphatase. V. Correlation of enzyme activity with cation flux in six tissues. *Arch. Biochem. Biophys.* 101:37, 1963.

26. Hosie, R. J. A. The localization of adenosine triphosphatase in morphologically characterized subcellular fractions of guinea pig brain. *Biochem. J.* 96:404, 1965.

27. Post, R. L., Merritt, G. R., Kinsolving, C. R., and Albright, C. D. Membrane adenosinetriphosphatase as a participant in the active transport of sodium and potassium in the human erythrocyte. *J. Biol. Chem.* 235:1796, 1960.

28. Whittam, R., and Ager, M. E. Vectorial aspects of adenosine triphosphatase activity in erythrocyte membranes. *Biochem. J.* 93:337, 1964.

*29. Dahl, J. L., and Hokin, L. E. The sodium-potassium adenosinetriphosphatase. *Annu. Rev. Biochem.* 43:327, 1974.

30. Stahl, W. L. Role of phospholipids in the Na^+, K^+-stimulated adenosine triphosphatase system of brain microsomes. *Arch. Biochem. Biophys.* 154:56, 1973.

31. Tanaka, R., and Sakamoto, T. Molecular structure in phospholipid essential to activate $(Na^+ + K^+ + Mg^{2+})$-dependent ATPase and $(K^+ + Mg^{2+})$-dependent phosphatase of bovine cerebral cortex. *Biochim. Biophys. Acta* 193:384, 1969.

32. Järnefelt, J. Lipid requirements of functional membrane structures as indicated by the reversible inactivation of $(Na^+ + K^+)$-ATPase. *Biochim. Biophys. Acta* 266:91, 1972.

33. Grisham, C. M., and Barnett, R. E. The role of lipid-phase transitions in the regulation of the $(Na^+ + K^+)$ adenosine triphosphatase. *Biochemistry* 12:2635, 1973.

34. Sen, A. K., Tobin, T., and Post, R. L. A cycle for ouabain inhibition of sodium and potassium-dependent adenosine triphosphatase. *J. Biol. Chem.* 244:6596, 1969.

35a. Post, R. L., and Kume, S. Evidence for an aspartyl phosphate residue at the active site of sodium and potassium ion transport adenosine triphosphatase. *J. Biol. Chem.* 248:6993, 1973.

35b. Hokin, L. E. Purification and properties of the $(Na^+ + K^+)$-activated adenosine-triphosphatase and reconstitution of sodium transport. *Ann. N. Y. Acad. Sci.* 242:12, 1974.

36. Fahn, S., Koval, G. J., and Albers, R. W. Sodium-potassium-activated adenosine triphosphatase of *Electrophorus* electric organ. I. An associated sodium-activated transphosphorylation. *J. Biol. Chem.* 241:1882, 1966.

37. Stahl, W. L. Sodium stimulated ^{14}C-adenosine diphosphate-adenosine triphosphate exchange activity in brain microsomes. *J. Neurochem.* 15:511, 1968.

38. Siegel, G. J., Goodwin, B. B., and Hurley, M. J. Regulatory effects of potassium on $(Na^+ + K^+)$-activated adenosinetriphosphatase. *Ann. N. Y. Acad. Sci.* 242:220, 1974.

39. Swanson, P. D., Bradford, H. F., and McIlwain, H. Stimulation and solubilization of the sodium ion-activated adenosine triphosphatase of cerebral microsomes by surface-active agents, especially polyoxyethylene ethers: Actions of phospholipases and a neuraminidase. *Biochem. J.* 92:235, 1964.

40. Medzihradsky, F., Kline, M. H., and Hokin, L. E. Studies on the characterization of the sodium-potassium transport adenosine triphosphatase. I. Solubilization, stabilization and estimation of apparent molecular weight. *Arch. Biochem. Biophys.* 121:311, 1967.

41. Kyte, J. Purification of the sodium- and potassium-dependent adenosine triphosphatase from canine renal medulla. *J. Biol. Chem.* 246:4157, 1971.

42. Ruoho, A., and Kyte, J. Photoaffinity labeling of the ouabain-binding site on $(Na^+ + K^+)$-adenosinetriphosphatase. *Proc. Natl. Acad. Sci. U.S.A.* 71:2352, 1974.

43. Caldwell, E. C. Energy relationships and the active transport of ions. *Curr. Top. Bioenerg.* 3:251, 1969.

44. Albers, R. W., Koval, G. J., and Siegel, G. J. Studies on the interaction of ouabain and other cardioactive steroids with sodium-potassium-activated triphosphatase. *Mol. Pharmacol.* 4:324, 1968.

45. Robinson, J. D. Interactions between monovalent cations and the $(Na^+ + K^+)$-dependent adenosine triphosphatase. *Arch. Biochem. Biophys.* 139:17, 1970.

46. Stein, W. D., Eilam, Y., and Lieb, W. R. Active transport of cations across biological membranes. *Ann. N. Y. Acad. Sci.* 277:328, 1974.

47. Repke, K. R. H., and Schön, R. Flip-flop model of (NaK)-ATPase function. *Acta Biol. Med. Ger.* 31:K19, 1973.

48. Borle, A. B. Calcium metabolism at the cellular level. *Fed. Proc.* 32:1944, 1973.

49. Ohashi, T., Uchida, S., Nagai, K., and Yoshida, H. Studies on phosphate hydrolyzing activities in the synaptic membrane. *J. Biochem.* (Tokyo) 67:635, 1970.

50. Blaustein, M. B., and Hodgkin, A. L. The effect of cyanide on the efflux of calcium from squid axons. *J. Physiol.* (Lond.) 200:497, 1969.

51. Stahl, W. L., and Swanson, P. D. Calcium movements in brain slices in low sodium or calcium media. *J. Neurochem.* 19:2395, 1972.

52. Stahl, W. L., and Swanson, P. D. Movements of calcium and other cations in isolated cerebral tissues. *J. Neurochem.* 18:415, 1971.
53. Stahl, W. L., and Swanson, P. D. Uptake of calcium by subcellular fractions isolated from ouabain-treated cerebral tissues. *J. Neurochem.* 16:1553, 1969.
54. Swanson, P. D., Anderson, L., and Stahl, W. L. Uptake of calcium ions by synaptosomes from rat brain. *Biochim. Biophys. Acta* 356:174, 1974.
55. Tonomura, Y., and Yamada, S. Molecular Mechanisms of Ca^{2+} Transport through the Membrane of the Sarcoplasmic Reticulum. In Nakao, M., and Parker, L. (Eds.), *Organization of Energy Transducing Membranes.* Tokyo: University Park Press, 1973, p. 107.
56. Degani, C., and Boyer, P. D. A borohydride reduction method for characterization of the acyl phosphate linkage in proteins and its application to sarcoplasmic reticulum adenosine triphosphatase. *J. Biol. Chem.* 248:8222, 1973.
57. Racker, E. Reconstitution of calcium pump with phospholipids and a purified Ca^{2+}-ATPase from sarcoplasmic reticulum. *J. Biol. Chem.* 247:8198, 1972.

Chapter 7

Molecular Biology of the Visual Process

Hitoshi Shichi

PHYSIOLOGICAL BACKGROUND

The eye is an organ that detects light and transmits its signal to the brain, so we shall expect the presence of light-absorbing pigments within the organ. In the vertebrate eye, such pigments are indeed found in the visual cells of the retina. Each visual cell contains two principal parts, the metabolically active inner segment that contains nucleus, mitochondria, and other subcellular organelles and the outer segment, in which the visual pigment is exclusively localized (see Fig. 7-1). At their inner segment terminals, the visual cells make synapses with horizontal cells and bipolar cells, which in turn form junctions with ganglion and amacrine cells, as illustrated in Figure 7-1.

The visual cells are classified into two types, based on their morphology. Rod cells have elongated outer segments and contain rhodopsin, the visual pigment responsible for dim-light (black-and-white) vision. Cone cells, which possess cone-shaped outer segments, are photoreceptors for day-light (color) vision. Although morphologically indistinguishable, microspectroscopic techniques have revealed the presence of three types of cone cells. Each cell contains one of three pigments with abosrption maxima at 445, 535, and 570 nanometers (nm), respectively. One estimate shows that a human eye contains 120 million rod cells in the peripheral region of the retina and 6.5 million cone cells concentrated mainly in the central (foveal) region.

For dim-light vision, absorption of light by a single pigment molecule in rod cells triggers a series of events that lead to excitation of the plasma membrane of the visual cell. This, in turn, results in excitation of optic neurons and, eventually, of cerebral neurons. In rod vision, the magnitude of neural excitation is directly related to perception of the brightness of light. For color vision, absorption of light by at least two cone pigments with different absorption maxima is essential. The ratio of magnitudes of excitation thus induced determines the type of color perceived. Visual signals generated by the photoreceptor cells and programmed by optic neurons, principally bipolar cells, are transmitted to the lateral geniculate bodies via ganglion cell axons which make up the optic nerves and tracts. Impulses are conveyed by the geniculo-calcarine radiations to the visual cortex of the brain where the signals representing light intensity and wavelength are presumed to be decoded separately by different neurons. The visual process related to the coding and decoding of visual signals is an important area of electrophysiology but is not covered here [1, 2]. This chapter deals primarily with structural and functional aspects of photoreceptors and molecular events that take place after photon absorption by the visual cells. General references are found in [3–8].

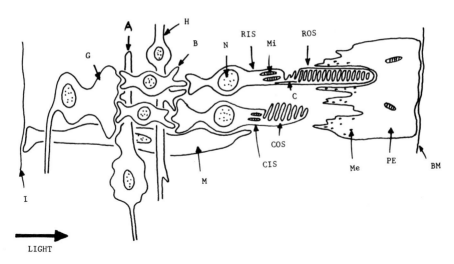

Figure 7-1
The vertebrate retina: (A) amacrine cell; (B) bipolar cell; (BM) Bruch's membrane; (C) cilium; (CIS) cone inner segment; (COS) cone outer segment; (G) ganglion cell; (H) horizontal cell; (I) inner limiting membrane; (M) Müller cell; (Me) melanin granule; (Mi) mitochondrion; (N) nucleus; (PE) pigment epithelium; (RIS) rod inner segment; (ROS) rod outer segment.

PHOTORECEPTOR MEMBRANES

Properties of Rod Membranes

Photoreceptor membranes are made of a continuous bilayer of phospholipids and the major membrane-protein rhodopsin [9]. The phospholipids contain high concentrations of polyunsaturated fatty acids. The outer segment of a rod visual cell is comprised of a stack of several hundreds of discs encased in a sack of the plasma membrane (Fig. 7-1). In 1963, Droz showed by means of autoradiography that, in animals injected with radioactive amino acids, labeled proteins migrate as a distinct band from the base toward the apex of a rod outer segment. He concluded that the proteins of rod outer-segment membranes undergo constant turnover. Young [10] showed that radioactive amino acids injected into animals are first incorporated into the proteins synthesized on the ribosomes in the inner segment, and a part of the synthesized protein, presumably opsin, moves to the Golgi apparatus and then to the junction between the inner and outer segments. The protein that reaches the outer segment through a narrow passage of the cilium is finally incorporated into the plasma membrane of the rod (Fig. 7-2). Incorporation of opsin into the membrane occurs as a part of disc formation, which is initiated by invagination of the plasma membrane in the basal region of the rod. The membranous infoldings thus formed become detached from the plasma membrane as they are displaced by the formation of newer discs. The discs that eventually reach the apical region of the rod are removed by the phagocytic action of the pigment epithelial cell. In this manner, the rod undergoes constant turnover with rates ranging from a few days to months depending upon species.

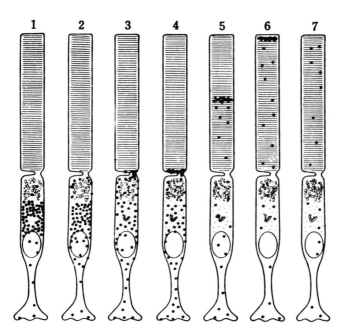

Figure 7-2
Protein turnover in the rod membrane of the frog. Radioactive amino acids are first incorporated
into proteins on the ribosomes and in the Golgi of the inner segment (1) and (2). Part of the
newly synthesized protein moves to the cilium region (3), then to the base of the outer segment
(4). The protein incorporated into discs reaches the apical region of the outer segment in two months
and eventually disappears (5) through (7). (From R. W. Young, Visual cells. Copyright © 1970 by
Scientific American, Inc. All rights reserved.)

This is a remarkably active membrane synthetic system. For example, the turnover
rate of a frog rod is 8.5 weeks. Assuming that a frog has 3 million rods (1,800 discs
per rod) per eye and a disc is 6×10^{-6} meter (m) in diameter, it is calculated that a
frog eye synthesizes 2.5 cm^2 of disc membrane every hour! Summarizing current
findings on the membrane synthetic system, Young, in [7], concluded that disc membranes are not capable of synthesizing membrane components and that such individual
membrane components as protein, carbohydrate, and lipid are all synthesized in the
inner segment, then transported to the outer segment, where membrane assembly
occurs.

In contrast to rod outer segments, the cone outer segments of animals injected with
radioactive amino acids do not show a distinct radioactive band of newly synthesized
protein.

In rats maintained on fat-free diets, the turnover of rod membranes is apparently
arrested. This is probably because a deficiency of polyunsaturated fatty acids prevents the assembly of lipid and protein for rod-membrane formation [11]. In rats
with inherited retinal dystrophy, outer segments grow abnormally long and accumulate as lamellar bundles in the extracellular space between the visual cell and the pigment epithelium. This overproduction is the result of a failure of the rod membranes
to be phagocytized by the pigment epithelium. In another type of hereditary retinal

Figure 7-3
Structure of docosahexaenoic acid.

dystrophy, the outer segment of mouse retina fails to develop to full maturation. This is undoubtedly caused by a biochemical abnormality of rod-membrane function. For instance, a cyclic nucleotide phosphodiesterase isozyme that is found in the normal retina is absent in the photoreceptor of dystrophic mice, and the elevated levels of cyclic guanosine monophosphate (c-GMP) resulting from the enzyme deficiency are suggested to be responsible for photoreceptor degeneration [12].

Rod outer segment membranes (>95 percent disc membranes, <5 percent plasma membrane) consist of 60 percent protein and 40 percent phospholipid. In vertebrate photoreceptors, about 80 percent of phospholipid is accounted for by phosphatidylethanolamine and phosphatidylcholine. The most abundant polyunsaturated fatty acid is docosahexaenoic acid, exclusively linked to the middle carbon of the glycerol moiety of phospholipids (Fig. 7-3).

Due to the high content of polyunsaturated fatty acids, rod membranes are highly fluid at physiological temperature and demonstrate various properties predicted by the fluid mosaic model of biomembranes [13]. In disc membranes, the retinal chromophore of rhodopsin is oriented in a plane perpendicular to the long axis of the rod. Does the rhodopsin molecule have rotational freedom in the membrane? If the chromophores are not flexible in the plane, the rods will demonstrate dichroic properties. That is because one of the linearly polarized light components that propagates parallel to the long axis is preferentially absorbed. In studies on fresh frog rods, the dichroism was not clearly demonstrated unless the tissue was fixed with glutaraldehyde [14]. By means of rapid-recording spectroscopy, however, a transient dichroism (life time = 20 μsec) was detected in unfixed rods. These results indicate that the chromophore of rhodopsin, i.e., rhodopsin itself, has rotational freedom along the long axis as well as translational freedom in the disc plane. Rhodopsin is a glycoprotein, and the carbohydrate moiety is believed to stick out on the surface of membrane. If the molecule rotated perpendicularly to the disc membrane, the carbohydrate moiety would have to go through the hydrophilic and hydrophobic layers of the membrane. This is not thermodynamically feasible, so the pigment is assumed not to tumble from one side of the membrane to the other.

Translational freedom of the molecule was recently demonstrated in an elegant fashion. When one side of the rod is bleached by light, the absorbance at 530 nm of the irradiated side is decreased. When the rod remains in the dark for a few seconds, the absorbance of the irradiated side rises, with a concomitant decrease in the absorbance of the unirradiated side (see Fig. 7-4). The rate of equilibration becomes greater as temperature is raised. Glutaraldehyde fixation of the tissue inhibits the equilibration process.

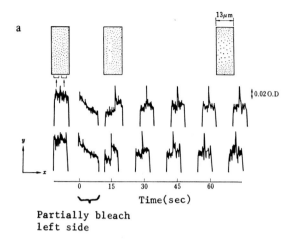

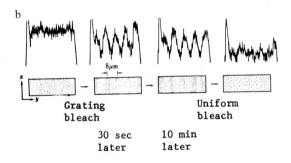

Figure 7-4
Lateral diffusion of rhodopsin in the disc membrane. (a) Rhodopsin diffusion after partial
bleaching of the left-hand side of the rod. (b) Absence of rhodopsin diffusion along the rod
axis. See Poo and Cone in [7].

If the rod is bleached with a grating pattern parallel to its long axis, the absorbance of the bleached region remains unchanged during the subsequent dark incubation. From these results, Poo and Cone in [7, 15] concluded that the rhodopsin
molecule can translate freely in the disc membrane but cannot migrate from one
disc to another. From the relaxation time for diffusion of rhodopsin, the viscosity
of the disc membrane was estimated to be about 2 poise (approximately equal to
the fluidity of olive oil).

Structure of Disc Membranes
Although the regular stacking of the discs in outer segments of vertebrate rod had
been known for some time, the two-membrane architecture of the disc was not
recognized until 1960. Electron microscopic studies indicated that the disc membrane is symmetrical in cross section, with the hydrocarbon region in the center
zone of the membrane, and that dense particles, presumably rhodopsin, are present

in the membrane. Studies on freeze-etched rod membranes indicate that the dense particles are associated with more than one layer of the membrane. The location of rhodopsin was also investigated by means of x-ray diffraction and chemical labeling of the functional groups of the membrane. Low-angle x-ray diffraction studies on outer segments in frog eyes showed that discs are stacked up with an interdiscal distance of 300 Å and contain a regular array of particles 40 Å in diameter.

Current models of the disc membrane deduced from the Fourier synthesis of diffraction patterns [7, 9] are summarized in Figure 7-5. Model 1 is based on the symmetric Fourier pattern and locates rhodopsin molecules (40 Å in diameter) on both sides of the membrane. In model 2, which takes into account the asymmetric pattern of the electron-dense peaks, rhodopsin is predicted to be located on the inner side of the disc membrane. The electron-dense regions in the x-ray diffraction pattern are often identified with protein. The technique does not distinguish true globular proteins from globular proteins with a peptide tail that extends through the membrane. Membrane proteins, such as cytochrome b_5 and glycophorine, are known to contain a peptide tail. The possibility cannot be excluded, therefore, that the two electron-dense peaks in the diffraction pattern represent a single pigment molecule with two separate but continuous core regions localized on opposite sides of the membrane. Such an elongated model is not inconsistent with the model suggested by experiments with fluorescent probes. The size of the pigment molecule

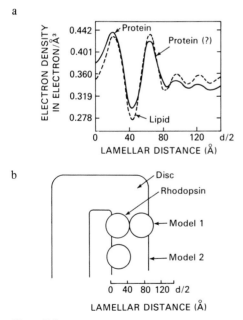

Figure 7-5
Structure of disc membrane. (a) Fourier series representations. (b) Localization of rhodopsin. Model 1, from the symmetric profile of electron-density distribution — dotted line in (a) — assumes rhodopsin molecules on both sides of membrane. From the asymmetric profile — solid line in (a) — rhodopsin is localized only on the inner layer of membrane in model 2.

determined by this method is at least 75 Å in length [16]. If this is true, the protein possibly spans the membrane. Hydrophilic reagents, such as glutaraldehyde, prevent lateral diffusion of the pigment in the membrane, as described above. Concanavalin A added to the disc membrane forms a 1:1 complex with rhodopsin. These results indicate that a part of the pigment molecule is exposed on the surface of the membrane.

PHOTOCHEMISTRY OF VISUAL PIGMENTS

General Structural and Spectral Properties

Rod pigments and cone pigments are both composed of protein (opsin), retinal, and associated phospholipids. Retinal$_1$, which has one unsaturated bond in the ionone ring, is the chromophore of visual pigments of terrestrial vertebrates. On the other hand, the chromophore of marine vertebrate visual pigments is retinal$_2$, which has two double bonds in the ionine ring. Whether a pigment takes retinal$_1$ or retinal$_2$ for its chromophore is determined by the wavelength of the sunlight that reaches the environment in which the animal lives. The major visual pigment of the tadpole has retinal$_2$ because the blue component of sunlight is absorbed by water before it reaches the eye. After metamorphosis of the tadpole, the visual pigment of frog contains retinal$_1$. Both types of retinal pigments are present in the rods and are called rhodopsin (λ_{max} = 498 nm) and porphyropsin (λ_{max} = 520 nm), respectively. In rhodopsin, the aldehyde group of retinal is linked to the ϵ-amino group of a lysine residue of opsin through Schiff base formation. Protonated Schiff-base complexes between retinal and a variety of amino acids show absorption maxima around 440 nm. Denaturation of the protein portion of visual pigments with guanidine hydrochloride and trichloroacetic acid gives rise to a pigment with λ_{max} = 440 nm. These findings suggest that the chromophore of visual pigments is primarily a protonated Schiff base (λ_{max} = 440 nm), and an additional shift toward the longer wavelength side is caused by interaction of the chromophore with opsin conformation.

As shown in Figure 7-6, rhodopsin has maximum absorption bands at 498 nm (α band, ϵ_M = 41,000) and 340 nm (β band, ϵ_M = 10,000) at room temperature. The α and β bands of porphyropsin are found at 523 nm and 360 nm. The conformation of the retinal in visual pigments was predicted to be in 11-*cis*, 12-*S-cis* form. X-ray crystallography of the so-called 11-*cis* retinal reveals that this isomer in crystalline form is indeed in the predicted conformation. In solution, however, retinal exists as an equilibrium mixture of low-energy *S-cis* and *S-trans* conformers [17]. Both

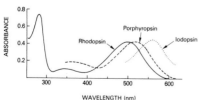

Figure 7-6
Absorption spectra of visual pigments.

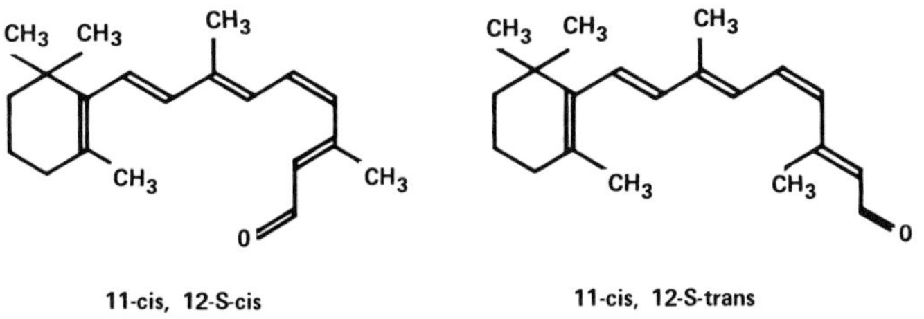

11-cis, 12-S-cis   11-cis, 12-S-trans

Figure 7-7
Formulas of the 11-*cis*, 12-*S-cis* and 11-*cis*, 12-*S-trans* conformers of retinal.

rhodopsin and porphyropsin show positive circular dichroic bands in wavelength regions corresponding to the α and β absorption maxima. The optical asymmetry of the chromophore, to which the circular dichroic properties are ascribed, may be partly due to preferential use of the 11-*cis*, 12-*S-cis* over the 11-*cis*, 12-*S-trans* isomer (Fig. 7-7).

Photon Absorption
Early studies established that photon absorption by rhodopsin results in the photic isomerization of the chromophore to the all-*trans* form. During the thermal process that follows the photochemical reaction, rhodopsin undergoes a series of spectrally distinct changes (Fig. 7-8) that can be identified by ordinary spectroscopy at different temperatures [8]. Upon irradiation of rhodopsin at $-268°$ C with light of 540 nm or longer wavelength, hypsorhodopsin is formed; this, in turn, is converted to batho-rhodopsin (previously called prelumirhodopsin). Honig and Karplus [18] reasoned that the 11-*trans*, 12-*S-cis* conformer is more planar than the 11-*cis*, 12-*S-cis* conformer and can account for the spectral shift to the red observed in the rhodopsin $\rightarrow$ bathorhodopsin conversion. According to their suggestion, the transformations: rhodopsin $\rightarrow$ bathorhodopsin $\rightarrow$ lumirhodopsin correspond to the conformational changes of retinal: 11-*cis*, 12-*S-cis* $\rightarrow$ 11-*trans*, 12-*S-cis* $\rightarrow$ 11-*trans*, 12-*S-trans*.

Kinetic parameters for the thermal process had been determined at low temperatures, because most of the bleaching intermediates are not stable at room temperature. By using a high-resolution flash photolysis apparatus, however, the rates of formation of bathorhodopsin and of metarhodopsin II from metarhodopsin I were recently determined at physiological temperatures [19, 20]. Metarhodopsin II formation from metarhodopsin I (3×10^{-4} sec), with a large negative entropy of activation, is believed to be related to the photocurrent generation ($\sim 10^{-4}$ sec) by the rod. Studies on detergent-extracted rhodopsin indicate that the conversion of metarhodopsin I to metarhodopsin II is accompanied by the appearance of SH groups and proton uptake. The equilibrium between metarhodopsin I and meta-rhodopsin II can be shifted toward metarhodopsin II under high pressures, 1,068 atmospheres (atm). Although these results suggest a conformational change of opsin protein during the conversion of metarhodopsin I to metarhodopsin II, circular dichroism measurements of membrane-associated rhodopsin in the far ultraviolet (UV) region show little change in opsin conformation produced by photic bleaching [7].

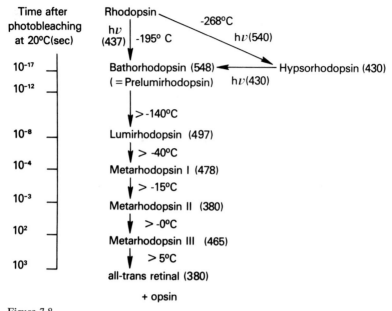

Figure 7-8
Intermediates detected spectrally after photic bleaching of vertebrate rhodopsin. The numbers in parentheses are wavelengths of light in nanometers used for irradiation or absorbed maximally by the individual intermediates.

Measurements of rod outer segments by use of ^{13}C nuclear magnetic resonance (NMR) have revealed no detectable change of the rod membrane structure in the light [21]. Any change of opsin conformation must be very small, therefore, and should not involve extensive unfolding or ordering of the peptide chain. A birefringence loss was observed in outer segments of frog rod exposed to light [22]. The birefringence change is related directly to the metarhodopsin I → metarhodopsin II conversion and is attributed to disorientation of one phospholipid molecule per mole of rhodopsin. As described above, the chromophore of rhodopsin is optically active. This property is retained to the level of metarhodopsin II. Metarhodopsin III is optically inactive. This suggests that, in going from metarhodopsin II to metarhodopsin III, retinal may be transferred from the original binding site to a different site (or different molecule). Studies involving chromophore fixation with borohydride at different bleaching stages support this possibility [23].

Color Vision

In contrast to what we know about rod pigments, very little is known about the chemistry of cone pigments. In 1937, Wald extracted a cone pigment named iodopsin (see Fig. 7-6) from the retinas of domestic fowl. After photic bleaching of iodopsin (λ_{max} = 562 nm), the following intermediates can be identified spectrally: bathoiodopsin (λ_{max} = 640 nm), lumiiodopsin (λ_{max} = 518 nm), metaiodopsin I

(λ_{max} = 495 nm), and metaiodopsin II (λ_{max} = 380 nm). Microspectroscopic measurements of single cones of isolated human retinas reveal the presence of three pigments with absorption maxima at 445 nm, 535 nm, and 570 nm. Examination of color-blind subjects by ophthalmoscopic densitometry has demonstrated that red blindness and green blindness are related to a deficiency of the 570-nm pigment and the 535-nm pigment, respectively [24].

REGENERATION OF VISUAL PIGMENTS

To maintain unimpaired vision, the visual pigments bleached by absorption of light must be regenerated. Regeneration of visual pigments takes place by two different mechanisms: (1) as a part of de novo synthesis of disc membrane; (2) reisomerization of all-*trans* retinal to 11-*cis* retinal and combination with opsin. As described already, opsin protein is synthesized in the inner segment and incorporated into the plasma membrane of outer segment. Thus there is little doubt that rhodopsin is formed (synthesized) by mechanism (1). It still remains unsettled, however, whether cone pigments are synthesized by a similar mechanism.

Pigment regeneration by mechanism (2) was first reported a century ago. Boll observed in 1876 that retinas excised from light-adapted animals were colorless, whereas the purple color was restored in retinas of animals that had been exposed to light and subsequently kept in the dark. Wald and associates established that the biochemical basis of pigment regeneration is a combination of opsin with a specific retinal isomer, 11-*cis* retinal [25]. How, then, is 11-*cis* retinal formed from all-*trans* retinal? Because rhodopsin regeneration took place in the dark, the isomerization of retinal was believed to be an enzymatic process, and attempts have been made to identify "retinal isomerase." However, all-*trans* retinal incubated in the dark with excised retinas or rod outer segments is converted consistently to 9-*cis* retinal, and the apparently nonenzymatic reaction is stimulated by nucleophilic reagents [26]. On the other hand, all-*trans* retinal is photochemically isomerized to 11-*cis* retinal in the presence of rod outer segments [27]. The activity is not destroyed by protein-denaturing agents and is fully accounted for by phosphatidylethanolamine catalysis [28]. Phosphatidylethanolamine forms a protonated Schiff base with retinal and, upon irradiation, effectively isomerizes all-*trans* retinal to the 11-*cis* form (Fig. 7-9). Transfer of the 11-*cis* isomer as a phospholipid complex to opsin explains rhodopsin regeneration in the dark. Photic isomerization of free retinal was long considered to be of little physiological importance because 380-nm light absorbed by free retinal is removed by the cornea and the lens before it reaches the retina and because little or no free retinal is found in the rod. These objections can be met, however, if in vivo isomerization of retinal occurs by absorption of the 460-nm light that the Schiff base complex absorbs maximally and the retinal that is released from bleached pigment forms immediately a complex with phosphatidylethanolamine.

Dowling [29] irradiated albino rats with intense light (1,000 foot-candles) and showed that the all-*trans* retinal released is reduced to retinol in the retina and stored in the pigment epithelium. He showed also that the retinol stored is sent back to the retina for pigment regeneration in the dark. From these results, a visual cycle involv-

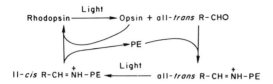

Figure 7-9
The in situ visual cycle in the rod outer segment of the vertebrate retina. The letters PE stand
for phosphatidylethanolamine.

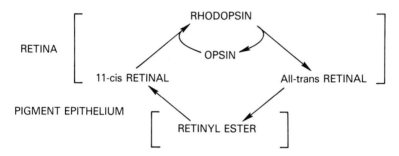

Figure 7-10
The visual cycle in the vertebrate eye in which both the retina and the pigment epithelium are
involved.

ing a shuttle of retinol between the retina and the pigment epithelium was proposed
(Fig. 7-10). Recent studies, however, suggest that this may not be the only mechanism
for regeneration of visual pigment. For example, rhodopsin regeneration in albino rats
takes place efficiently without the pigment epithelium if the retina is incubated in a
small sealed chamber [30]. If retinol is shuttled between the retina and the pigment
epithelium during adaptation cycles, the specific radioactivity of rhodopsin of animals
labeled with radioactive retinol would increase with repetitions of the adaptation
cycles. In both albino rats [31] and frogs [7], the specific radioactivity of the visual
pigment was not increased by alternate adaptations to moderate lighting and darkness.
These results suggest that, unless the rods are extensively bleached by intense light,
the retinol released from bleached pigment may not migrate to the pigment epithelium.
Intense light is known to cause disintegration of rod membranes. When this happens,
retinal is reduced to retinol and removed from the rod by the scavenger action of the
pigment epithelium. Under mild bleaching conditions, rhodopsin regeneration would
also occur by a retinal utilization mechanism in situ in the rod outer segment. A
scheme depicted in Figure 7-9 was suggested for such a mechanism [28]. However,
this mechanism has yet to be confirmed by in vivo observations, and there may be
other means for rhodopsin regeneration in vivo.

   In squid rhodopsin, all-*trans* retinal is not released from bleached pigment, and
pigment regeneration occurs by photic reisomerization of retinal at pH 10 and subsequent dark incubation [32]. Cone pigments are also regenerated by reisomerization

of all-*trans* retinal attached to opsin. In the regeneration of pigments from opsin and 11-*cis* retinal, the rate of iodopsin formation is considerably greater than that of rhodopsin regeneration.

VISUAL EXCITATION

Biophysical Model

The outer-segment membranes of vertebrate rods and cones show a resting potential of $-30 \sim -40$ millivolts (mv), inside negative. Vertebrate cells are hyperpolarized by photobleaching of the visual pigment, whereas invertebrate cells (e.g., in *Limulus*) are depolarized by light. In view of the highly ordered structure of photoreceptors that resemble macromolecular crystals, a mechanism involving electron- or hole-conduction was proposed for light-dependent modulation of membrane potential [33]. Accumulated evidence [34], however, is overwhelmingly in favor of the ionic mechanism and gives little support to the semiconduction mechanism.

According to the ionic mechanism, the potentials in the dark, as well as in the light, result from active transport of Na^+ across the photoreceptor membranes. This is supported by the observations that the transmembrane potentials are reduced by lowering Na^+ concentrations in medium and are inhibited by ouabain, the inhibitor of Na^+-K^+ ATPase. The extent of light-induced hyperpolarization of vertebrate photoreceptors — or depolarization, in the case of invertebrate photoreceptors — depends on wavelength and intensity of irradiating light. Rhodopsin is found in disc membranes that are detached from the plasma membrane of the rod. Bleaching of the visual pigment occurs in disc membranes, but it is the plasma membrane that is hyperpolarized. How can a signal generated by photon absorption by the disc membrane be transmitted to the plasma membrane? There is a growing evidence that Ca^{2+} can regulate the Na^+ current of photoreceptors. At 10^{-5} M Ca^{2+} or less, the inflow of Na^+ becomes maximum, whereas at Ca^{2+} concentrations as high as 20 mmol, the Na^+ current disappears in both the light and the dark. In other words, Ca^{2+} mimics the effect of light. From these results, Hagins [35] has proposed a model in which visual excitation of vertebrate rod and cone photoreceptors is modulated by Ca^{2+}.

According to this model (Fig. 7-11), in the dark, the outer-segment membrane shows high Na^+ permeability. As soon as rhodopsin molecules in the disc membrane are bleached by light, Ca^{2+} sequestered inside the disc is released and closes the Na^+ channels in the plasma membrane. The decreased Na^+ inflow results in hyperpolarization of the plasma membrane. The calculated number of Na^+ channels in the plasma membrane of a rod cell is about 1,000. Closing of a few percentages of the channels should suffice to induce the extent of hyperpolarization actually measured. Cone discs are continuous with the plasma membrane, and Ca^{2+} inside the disc and in the outside medium are not separated. The model suggests, therefore, that photobleaching of visual pigments in cone photoreceptors is coupled with rapid movement of external Ca^{2+} into the cytoplasm. In order for the model to function continuously, there must be a mechanism by which Ca^{2+} is accumulated within the discs in the dark. Experimental evidence for the existence of such a mechanism has yet to be presented. The structure of the infolded portion of the plasma membrane at the

a   Dark

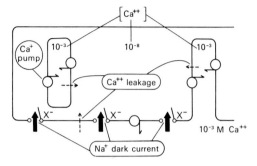

b   Light

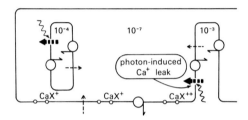

Figure 7-11
Modulation of Na^+ current by Ca^{++} for visual excitation in vertebrate rods and cones. See Hagins [35].

base of the rod is essentially identical with the infolded discs of a cone outer segment. Therefore, if one proves that light-induced uptake of external Ca^{2+} into the cytoplasmic space between the infolded membranes is of primary importance, visual excitation in rod and cone photoreceptors may be explained by a single model.

In addition to an inflow of Na^+, formation of a proton gradient across the photoreceptor membrane may also contribute to generation of the transmembrane potential. For example, the amplitude of electroretinograms evoked by a high-intensity light increases with a rise in the pH and decreases at lowered pH [36]. The pH of the medium in which retinas are bathed rises rapidly upon illumination [37]. Light-induced proton uptake in bovine outer segments accompanies changes in volumes of the discs [38]. In this connection, an interesting observation was made on bacteriorhodopsin, a rhodopsinlike pigment present in the purple membrane of halobacteria [39]. When the membrane is irradiated by light, a proton gradient is formed across the membrane and a transmembrane potential as high as 50 millivolts (mv) is generated [40].

Biochemical Model
Compared to the biophysical model described above, very little is known about biochemical reactions that occur immediately after photobleaching of visual pigments.

Some years ago Wald, in [7], postulated that rhodopsin is probably an enzyme and that photoactivation of the enzyme will lead to visual excitation through a cascade process somewhat similar to the blood-clotting process. This hypothesis, though interesting, is not of practical significance until an enzymic activity associated with rhodopsin is defined. Recently, rhodopsin was shown to be phosphorylated by ATP [41–43]. The reaction is strictly light dependent, and the action spectrum of the reaction is in good agreement with the absorption spectrum of rhodopsin. Peculiarly, only a minor fraction of rhodopsin is phosphorylated, and the phosphorylated rhodopsin molecule incorporates a considerable number of phosphate groups [44]. If further studies demonstrate that the phosphorylated rhodopsin species is localized in the plasma membrane, rather than in the disc, and is phosphorylated within a millisecond after irradiation, the component may become a candidate for the Ca^{2+} receptor at the Na^+ channel (see Fig. 7-11). Alternatively, the phosphorylation reaction may be related to a dark process, such as the pumping-out of Ca^{2+} across the plasma membrane. Regulation of membrane permeability and potential by cyclic nucleotides was reported in various membrane systems. In rod membranes, the phosphorylation of rhodopsin is not stimulated by cyclic nucleotides and cannot as yet be related to cyclic-nucleotide metabolism. Thus the significance of adenyl cyclase and guanyl cyclase, which may or may not be inhibited by light, and cyclic nucleotide phospho-diesterase, which is activated by light [45], is not known. The observation that AMP and ATP added to outer segments restore excitability (Na^+ permeability) in the dark [7] still remains to be confirmed.

REFERENCES

*1. Jung, R. (Ed.). *Handbook of Sensory Physiology*, Vol. 7/3. Part A, Central Processing of Visual Information; Part B, Visual Centers in the Brain. Heidelberg: Springer-Verlag, 1973.

*2. McIlwain, J. T. Central vision: Visual cortex and superior colliculus. *Annu. Rev. Physiol.* 34:291, 1972.

*3. Brindley, G. S. *Physiology of the Retina and Visual Pathway*. Baltimore: Williams & Wilkins, 1970.

*4. Dartnall, H. J. A. (Ed.). *Handbook of Sensory Physiology*, Vol. 7/1. Photochemistry of Vision. Heidelberg: Springer-Verlag, 1972.

*5. Honig, B., and Ebrey, T. G. The structure and spectra of the chromophore of the visual pigments. *Annu. Rev. Biophys. Bioeng.* 3:151, 1974.

*6. Fuortes, M. G. F. (Ed.). *Handbook of Sensory Physiology*, Vol. 7/2. Physiology of Photoreceptor Organs. Heidelberg: Springer-Verlag, 1972.

*7. Futterman, S., Kinoshita, J. H., and Shichi, H. (Ed.). Proceedings of the National Eye Institute Conference on "Rhodopsin, Its Chemistry and Function." *Exp. Eye Res.* Vol. 17/6 (1973); Vol. 18/1 (1974); Vol. 18/3 (1974).

*8. Langer, H. (Ed.). *Biochemistry and Physiology of Visual Pigments*. Heidelberg: Springer-Verlag, 1973.

*9. Wothington, C. R. Structure of photoreceptor membranes. *Annu. Rev. Biophys. Bioeng.* 3:53, 1974.

*10. Young, R. W. Visual cells. *Sci. Am.* 223:81, 1970.

*Asterisks denote key references.

11. Landis, D. J., Dudley, P. A., and Anderson, R. E. Alteration of disc formation in photoreceptors of rat retina. *Science* 182:1144, 1973.

12. Farber, D. B., and Lolley, R. N. Cyclic guanosine monophosphate: Evaluation in degenerating photoreceptor cells of the C3H mouse retina. *Science* 186:449, 1974.

13. Singer, S. J., and Nicolson, G. L. The fluid model of the structure of cell membranes. *Science* 175:720, 1972.

14. Brown, P. K. Rhodopsin rotates in the visual receptor membrane. *Nature* [*New Biol.*] 236:35, 1972.

15. Poo, M., and Cone, R. A. Lateral diffusion of rhodopsin in the photoreceptor membrane. *Nature* 247:438, 1974.

16. Wu, C. W., and Stryer, L. Proximity relationship in rhodopsin. *Proc. Natl. Acad. Sci. U.S.A.* 69:1104, 1972.

17. Rowan, R., Warshel, A., Sykes, B. D., and Karplus, M. Conformation of retinal isomers. *Biochemistry* 13:970, 1974.

18. Honig, B., and Karplus, M. Implications of torsional potential of retinal isomers for visual excitation. *Nature* 229:558, 1971.

19. Busch, G. E., Applebury, M. L., Lamola, A. A., and Rentzepis, P. M. Formation and decay of prelumirhodopsin at room temperatures. *Proc. Natl. Acad. Sci. U.S.A.* 69:2802, 1972.

20. Applebury, M. L., Zuckerman, D. M., Lamola, A. A., and Jovin, T. M. Rhodopsin. Purification and recombination with phospholipids associated by the metarhodopsin I—metarhodopsin II transition. *Biochemistry* 13:3448, 1974.

21. Millet, F., Hargrave, P. A., and Raftery, M. A. Natural abundance ^{13}C nuclear magnetic resonance spectra of the lipid in intact bovine retinal rod outer segment membranes. *Biochemistry* 12:3591, 1973.

22. Liebman, P. A., Jagger, W. S., Kaplan, M. W., and Bargoot, F. G. Membrane structure changes in rod outer segments associated with rhodopsin bleaching. *Nature* 251:31, 1974.

23. Rotmans, J. P., Daemen, F. J. M., and Bonting, S. L. Biochemical aspects of the visual process. XXVI. Binding site and migration of retinaldehyde during rhodopsin photolysis. *Biochim. Biophys. Acta* 357:151, 1974.

24. Rushton, W.A.H. A cone pigment in the protanope. *J. Physiol.* (Lond.) 168:345, 1963; A foveal pigment in the deuteranope. *J. Physiol.* (Lond.) 176:24, 1965.

25. Wald, G., and Brown, P. K. Synthesis and bleaching of rhodopsin. *Nature* 177:174, 1956.

26. Futterman, S., and Rollins, M. H. The catalytic isomerization of all-*trans* retinal to 9-*cis* retinal and 13-*cis* retinal. *J. Biol. Chem.* 248:7773, 1973.

27. Hubbard, R. Retinene isomerase. *J. Gen. Physiol.* 39:935, 1955.

28. Shichi, H., and Somers. R. L. Possible involvement of retinylidene phospholipid in photoisomerization of all-*trans* retinal to 11-*cis* retinal. *J. Biol. Chem.* 249:6570, 1974.

29. Dowling, J. E. Chemistry of visual adaptation in the rat. *Nature* 188:114, 1960.

30. Cone, R. A., and Brown, P. K. Spontaneous regeneration of rhodopsin in the isolated rat retina. *Nature* 221:818, 1969.

31. Bridges, C. D. B., and Yoshikami, S. Uptake of tritiated retinaldehyde by the visual pigment of dark-adapted rats. *Nature* 221:275, 1969.

32. Suzuki, T., Sugahara, M., and Kito, Y. An intermediate in the photoregeneration of squid rhodopsin. *Biochim. Biophys. Acta* 275:260, 1972.

33. Rosenberg, B. A physical approach to the visual receptor process. *Adv. Radiat. Biol.* 2:193, 1966.

*34. Tomita, T. Electrical activity of vertebrate photoreceptors. *Q. Rev. Biophys.* 3:179, 1970.

*35. Hagins, W. A. The visual process: Excitatory mechanisms in the primary receptor cells. *Annu. Rev. Biophys. Bioeng.* 1:131, 1972.

36. Winkler, B. S. The electroretinogram of the isolated rat retina. *Vision Res.* 12:1183, 1972.

37. Ostroy, S. E. Hydrogen ion changes of rhodopsin. pK changes and the thermal decay of metarhodopsin II_{380}. *Arch. Biochem. Biophys.* 164:275, 1974.

38. McConnell, D. G. Relationship of the light-induced proton uptake in bovine retinal outer segment fragments to Triton-induced membrane disruption and to volume changes. *J. Biol. Chem.* 250:1898, 1975.

39. Oesterhelt, D., and Stoeckenius, W. Rhodopsin-like protein from the purple membrane of *Halobacterium halobium*. *Nature [New Biol.]* 233:149, 1971.

40. Drachev, L. A., Kaulen, A. D., Ostroumov, S. A., and Skulachev, V. P. Electrogenesis by bacteriorhodopsin incorporated in a planar phospholipid membrane. *FEBS Lett.* 39:43, 1974.

41. Kuhn, H., Cook, J. H., and Dreyer, W. J. Phosphorylation of rhodopsin in bovine photoreceptor membranes. A dark reaction after illumination. *Biochemistry* 12:2495, 1973.

42. Bownds, D., Dawes, J., Miller, J., and Stahlman, M. Phosphorylation of frog photoreceptor membranes induced by light. *Nature [New Biol.]* 237:125, 1972.

43. Frank, R. N., Cavanaugh, H. D., and Kenyon, K. R. Light-stimulated phosphorylation of bovine visual pigments by adenosine triphosphate. *J. Biol. Chem.* 248:596, 1973.

44. Shichi, H., Somers, R. L., and O'Brien, P. J. Phosphorylation of rhodopsin: Most rhodopsin molecules are not phosphorylated. *Biochem. Biophys. Res. Commun.* 61:217, 1974.

45. Keirns, J. J., Miki, N., Bitensky, M. W., and Keirns, M. A link between rhodopsin and disc membrane cyclic nucleotide phosphodiesterase. Action spectrum and sensitivity to illumination. *Biochemistry* 14:2760, 1975.

Section III

Synaptic Function

Chapter 8

Overview of Chemical Transmission

Richard Hammerschlag
Eugene Roberts

A bacterium moving along a chemical gradient, a hydra initiating a stereotypic feeding behavior after detecting the presence of brine shrimp, the adrenal cortex producing steroids in response to blood-borne ACTH, and a muscle contracting after its motor nerve has been stimulated are all examples of a common process highly selected for in evolution: the ability of cells to respond to extracellular chemical stimuli. Chemical neurotransmission is an advanced development in this evolutionary sequence. In an organism with a nervous system, all normal or adaptive activity (information processing) is a result of the coordination of a dynamic interplay of excitation and inhibition within and between neuronal subsystems. Most communication that takes place between receptor cell and neuron, between neuron and neuron, and between neuron and effector cell probably occurs via the extracellular liberation of substances which interact with specialized regions of membranes of neurons or membranes of muscle or gland cells to produce either excitatory or inhibitory effects.

The region of functional contact between such cells is known as the *synapse.* Historically, the synapse was a morphological and physiological concept, proposed to counter a prevailing view of the nervous system as a single syncytial circuit of nerve cells. As presently conceived, the synaptic region includes a presynaptic nerve ending (a specialized region of axon), a synaptic cleft several hundred Å wide, and a sensitive region of postsynaptic membrane. Because most nerve cells can receive synaptic inputs over their entire surface, the postsynaptic element may be a membrane region of a dendrite, a soma, or an axon. The synapse appears to be insulated from adjacent synaptic sites by surrounding glial cells (see Chap. 1).

Typically, a wave of depolarization sweeps down the axon and triggers the release of a discrete amount of chemical transmitter from the presynaptic nerve ending into the cleft. Transmitter diffuses across the cleft and binds to specific receptors on the postsynaptic membrane. This binding is believed to cause conformational changes in molecules of the membrane in some unknown way; these, in turn, result in changes in ion permeabilities (see Chap. 3). The resultant ionic fluxes alter the membrane potential of the postsynaptic neuron and either bring the cell closer to firing (excitation) or keep it from firing (inhibition). Finally, transmitter action is thought to be terminated by removal of transmitter via active uptake processes in the neuronal and glial elements that surround the cleft and only in a few instances by its actual destruction in the cleft.

There is physiological and morphological evidence for a discontinuity between the pre- and postsynaptic (subsynaptic) regions of chemical synapses. The subsynaptic membrane regions are not considered to be electrically excitable. A delay is found between the time of application of a stimulus to a presynaptic cell and the recording of electrical activity from a postsynaptic cell. This delay is consistent with the calculated time required for diffusion of low-molecular-weight substances across the distance of approximately 200–300 Å which exists between the presynaptic terminal and the membrane of the postsynaptic cell.

Electrical stimulation of the postsynaptic cell, by contrast, has no detectable electrical effect on the presynaptic cell, demonstrating that chemical synapses are functionally unidirectional. The asymmetrical appearance of the synaptic region has suggested a basis for this one-way transmission. In the presynaptic cytoplasm adjacent to the terminal membrane are clusters of vesicles which "like chocolates come in a variety of shapes and sizes, and are filled with a variety of stuffings" [1]. Several lines of evidence suggest that the most important part of the stuffing consists of molecules of transmitter and that these vesicles are functionally important in both the storage and subsequent release of transmitter. Electron micrographs also reveal characteristic thickenings of the postsynaptic membrane that are restricted to those membrane regions in apposition to the presynaptic endings.

NEUROTRANSMITTER COMPOUNDS

Properties

At least a dozen compounds normally found in neural tissue are believed, with various degrees of certainty, to function as transmitters [2, 3]. Such compounds usually are considered to be transmitter candidates because they can excite or inhibit neurons and because they show uneven regional distributions in the nervous system. Virtually all putative transmitters are low-molecular-weight substances that are amino acids (γ-aminobutyrate, glycine, taurine, and glutamate) or amines (acetylcholine, dopamine, norepinephrine, and serotonin). Recently, a polypeptide, Substance P, has joined the ranks. Since all of the above substances have charged groups that are fully ionized at physiological pH, it is likely that electrostatic forces play a role in transmitter-receptor interaction. The same substances appear to be utilized in phyla at all levels of neural complexity, and many of the transmitter amines have been found to occur in plants and microorganisms, where their roles are still unknown. Transmitter substances are found in nonneural tissues as well as in nerve cells. Serotonin, for example, is found in blood platelets and intestinal mucosa [4], and high levels of acetylcholine (ACh) occur in mammalian placenta, a tissue that is not innervated [5].

The same transmitter may have different roles in different species and even at different synapses in the same species. ACh is the transmitter of motor neurons at vertebrate neuromuscular junctions, and probably is a transmitter at sensory endings in crustaceans and insects. Glutamate is a probable motor transmitter at many arthropod neuromuscular junctions and a possible sensory transmitter in vertebrates. GABA inhibits at numerous neuronal sites in vertebrates and invertebrates. Also it may have excitatory effects at synapses in vertebrates (see Chap. 11). ACh released at the vertebrate neuromuscular junction causes increased sodium permeability and excitation, whereas the same substance liberated in the heart from terminals of the vagus nerve produces increased potassium permeability and inhibition [6]. Thus the effects of transmitter substances are dependent on the nature of the postsynaptic receptors and do not simply follow from the chemical nature of the transmitters.

Criteria for Neurotransmitters

The two major requirements for establishing a compound as a transmitter are to collect it from a particular synapse after the presynaptic neuron has fired and to show that it can produce postsynaptic actions identical with those of the transmitter normally released at that synapse [7].

It can be shown at neuromuscular junctions and autonomic ganglia that the amount of material collected is related quantitatively both to the firing frequency and to the level of extracellular calcium ions. These relationships are presumed to hold in the central nervous system as well. These criteria distinguish transmitters from other substances (thiamine [8], and amino acids [9]) that may be released from nonsynaptic regions of conducting nerve cells. On the other hand, other materials may be released from synaptic endings together with the transmitter. For example, several proteins normally found within adrenergic vesicles are released together with norepinephrine during electrical stimulation of sympathetic nerves [10]. Thus the second criterion stated above, "identity of action," also must be satisfied. In order to become accepted as a transmitter, the transmitter candidate must mimic the excitatory or inhibitory action of the natural transmitter, bind to the same receptor, and cause permeability changes to the same ions(s).

Additional criteria often proposed for identifying transmitters include demonstrations at synapses of synthesis, storage, and inactivation of the transmitter candidate. These processes are clearly necessary but are not sufficient to prove transmitter function. However, determination of the distributions of putative transmitters and of their biosynthetic enzymes [11, 12], and autoradiographic studies that reveal differential cellular accumulation of labeled transmitter candidates [13], possibly reflecting uptake mechanisms involved in terminating transmitter action, have provided valuable information in the search for new transmitters as well as confirmatory evidence for transmitter assignments based on the more fundamental criteria. These neurochemical approaches, moreover, have produced potentially useful information about transmitter distribution in neural regions where the inaccessibility of the synapses has precluded monitoring of transmitter release or precise comparative measurements of postsynaptic events.

TRANSMITTER SYNTHESIS

Transmitter-forming enzymes, as well as other soluble proteins and even particulate structural materials, are believed to be loaded onto intracellular transport systems and transferred from their sites of synthesis in the cell bodies down the axons to their sites of utilization at the presynaptic endings (see Chap. 21). It follows that transmitters and their synthetic enzymes may be present in all regions of the neuron. It is not known as yet whether those molecules of transmitter substances that may be synthesized at cellular sites other than presynaptic axonal endings can find their way into nerve terminals, eventually to be available for release in informational transactions. Certainly it seems probable that most of the molecules of a substance used in neurotransmission are made very close to their site of release. Transmitters present in nonterminal regions of nerve cells may regulate intracellular ionic movements or modulate metabolic functions.

In order for there to be stability of communication at a synapse, a more or less constant pool of transmitter at the presynaptic ending must be maintained. How is this managed in the face of the widely fluctuating transient and long-range demands? Because the neuron must keep up with events that take place in seconds or fractions of seconds, adjustments to the transient demands must be made largely in the presynaptic endings themselves. A transmitter − norepinephrine, for instance − may have a reversible inhibitory effect on the enzyme that is rate-limiting in its own formation (Chap. 10). As its concentration builds up in the synaptic ending, the rate of transmitter formation is slowed. Thus, with little or no use of a synapse, the amount of transmitter remains at a certain maximal level. On the other hand, a greatly increased neuronal activity may cause a rapid depletion of transmitter from the ending. Under such circumstances, the rate-limiting step in its synthesis is disinhibited, and, given an adequate supply of precursor material, the concentration of a transmitter will rapidly be restored.

Of course, over a longer time scale, degradation of the transmitter-synthesizing enzymatic and transmitter-releasing machinery occurs. The shipment of replacement parts down the axon must balance the rate of breakdown. How this is regulated is not known, but it certainly must involve a two-way system by which the need for replacement is communicated via the axon in a retrograde fashion to the somal synthetic apparatus of the neuron. In turn, the neuron manufactures the needed parts and transports them down the axon to the presynaptic ending (see Chap. 21). It is known that the protein-synthetic capacity of the presynaptic ending is very limited (Chap. 16).

Sustained depletion of transmitter stores induced by chronic electrical stimulation can result in increased synthesis and subsequent axonal transport of additional transmitter-forming enzyme as well as in increased local synthesis of transmitter [14]. The consequences of major changes in functional demands usually have been attributed primarily to events that initially take place on the postsynaptic sites, the dendritic and somal membranes. The metabolic consequences of the postsynaptic actions of transmitters are believed to be mediated through effects on the levels of second messengers, the cyclic nucleotides, changes in the levels of which lead to initiation or interruption of cascades of intracellular reactions. Among the latter are those involved in the transcription of genetic information by the formation of messenger RNA and the translation of genetic messages carried in the RNA to form specific proteins. In response to increased activity, biosynthetic mechanisms are activated sufficiently to supply the accelerated needs for cell maintenance, and, in some instances, there actually could be enhancement of potency or size of the neuronal apparatus (hypertrophy), or both. In the extreme case where there is either prolonged partial or complete functional disuse or an actual interruption of the axonal link between the soma and presynaptic ending, reversible and irreversible degenerative changes might occur. Recent work suggests that chemical messages from postsynaptic back to presynaptic sites also are necessary for the maintenance of healthy synaptic relationships [15].

TRANSMITTER STORAGE AND RELEASE

A characteristic morphological feature of chemical synapses observed by electron microscopy is the presence of small vesicles in variable numbers and with various sizes and densities in the presynaptic terminals. The vesicles are bounded by a unit membrane and either appear clear or with a small, electron-dense core. At about the same time that synaptic vesicles first were observed, physiological measurements showed that ACh was released at the neuromuscular junction in vertebrate organisms in quantal packets [16]. The spontaneous release of ACh from nerve endings produced responses (called miniature end-plate potentials) with amplitudes that were discrete multiples of each other, but external application of varying amounts of ACh to postsynaptic membranes produced a smoothly graded response. This suggested that a characteristic feature of the release mechanism was responsible for the quantal response, and it seemed logical to propose that vesicles contained uniform packets of transmitter and released their contents in an all-or-none manner. At present, the phenomenon of quantal release has been confirmed in a large number of studies, but the vesicle hypothesis has not yet been proved conclusively. Studies employing techniques of subcellular fractionation and chemical and pharmacological assay have shown a significant association of ACh with a population of clear vesicles and of norepinephrine with dense-core vesicles [17]. It has not yet been possible, however, with similar techniques, to show such an association in the case of the amino acid transmitters, although clusters of vesicles are seen in electron micrographs of the synaptic terminals of neuronal sites which presumably use them as transmitters.

A MODEL OF VESICLE FUNCTION

A frankly speculative model of vesicle function, based in part on experimental findings, is presented here to serve as a framework for thinking about synaptic transmitter release and to illustrate some of the problems that must eventually be resolved experimentally. The vesicles in the internal volume of the presynaptic ending associate with transmitter-forming enzymes by ionic and hydrogen bonds or other adsorptive forces. Precursors in the cytoplasm bind to the active enzymatic sites and are converted to transmitter molecules, which, as a consequence of cooperative interactions of the enzyme and vesicular membrane, are moved or pumped into the internal volume of the vesicles. Whence come the substrates to furnish precursors for these enzyme pumps? Because the pathway down the axon is slow, the most likely proximate source is the extracellular environment in the region of the presynaptic ending, which in turn can get its molecular supply from the blood capillaries and the glial and postsynaptic elements in the vicinity. In the case of ACh (Chap. 9), the glucose required for mitochondrial generation of acetyl CoA enters the ending from extracellular sites. The choline moiety may consist of newly arrived molecules of choline from the bloodstream as well as choline recycled from ACh previously liberated into the cleft and hydrolyzed by acetylcholinesterase. Choline acetyltransferase associated with the vesicles then catalyzes the synthesis of ACh from acetyl CoA and choline. There seems to be little or no diffusional exchange of vesicle-contained ACh with free ACh.

In the case of GABA, it seems that glutamine coming from the extracellular environment and hydrolyzed internally to give glutamic acid may be an even more important source of the substrate, glutamic acid, than is its intraterminal synthesis by other routes [18]. It seems that the GABA in the vesicles in terminals of GABA neurons may be more readily exchangeable with nonvesicular GABA than ACh in cholinergic endings. Within vesicles there may be small or large molecules, or both, that may serve as "binders" for the transmitter molecules. Both ATP and specialized proteins serve this function in vesicles dealing with catecholamines [19]. In the latter instance, in addition to possibly being filled with transmitter by the enzyme-pump mechanism suggested above, the vesicles might serve as chemical vacuum cleaners, sweeping up from the synaptoplasm all molecules for which the binders have affinity, until the capacity of the vesicles is saturated. It might be for this reason that it is relatively easy to label vesicles in catecholamine-containing nerve endings with exogenously supplied transmitter and that it is possible to load such endings with false transmitters, substances that resemble the true transmitter chemically but which, themselves, have no transmitter function.

Of course, unattached enzyme molecules may produce transmitter molecules wherever they encounter substrate, or vesicle-contained molecules may leak out and a nonvesicular transmitter pool may arise in the presynaptic ending. Indeed, the data strongly suggest that intraterminal transmitter pools are not in simple diffusional equilibrium with each other [20]. Current results cannot be used with certainty in giving quantitative dimensions to estimating the extent to which enzymes associate with vesicles. Perhaps techniques currently being developed for the visualization of specific macromolecules at the electron-microscopic level eventually will give more definitive information.

The vesicles near the membrane surface are filled with transmitter to a capacity determined by the maximal extent to which their membranes can be stretched by the transmitter molecules and their associated water and binding materials. Thus vesicles may vary significantly in size. (This contrasts with the original version of the vesicle hypothesis, according to which all of the vesicles found in a particular nerve ending at a given time were believed to be filled with the same amount of transmitter.) As they approach the membrane, the vesicles enter collimators (seen with freeze-etching and regular electron-microscopic techniques) that serve to guide them to their membrane launching sites, while possibly keeping out those vesicles that are larger than a given predetermined size [21].

When a nerve is stimulated, depolarization of the presynaptic membrane by the invading action potential is assumed to result in an influx of calcium ions, causing vesicles adjacent to the presynaptic membrane to bind to it and fuse with it. A transient coordinated rupture of the membranes then takes place, during which the contents of the vesicles are extruded actively into the synaptic cleft, a process called *exocytosis.* Extrusion may be achieved by a contractile mechanism which may have much in common with processes of muscle contraction [22]. The pumping of calcium out of the presynaptic ending by a Ca^{2+}-activated ATPase may trigger the withdrawal of vesicles from the presynaptic membrane back into the cytoplasm. In any case, vesicles appear to be pinched off from their membrane release sites (via endocytosis) and to contain portions of the extracellular fluid with which their contents were transiently in equilibrium. Probably it is by this latter process that extracellular markers, such as horseradish peroxidase, enter the presynaptic endings during nerve stimulation [23].

Regardless of which transmitter is involved, the process of transmitter release appears to be similar at all chemical synapses in its dependence on calcium ions for stimulus-release coupling and in its quantal nature. The process is analogous to the mechanisms believed to be involved in the release of polypeptide hormones from posterior pituitary neurons [24] and of epinephrine from the chromaffin cells of the adrenal medulla [25]. Both types of substances are stored in granules, and calcium ions must be present in the extracellular medium for them to be released. In the above discussion, we have adhered to the assumption that vesicles and quantal release of transmitter are basic to chemical transmission by nerves. Alternative mechanisms, for which currently there is some circumstantial evidence, still are possible (see Chap. 9).

DALE'S LAW

One of the tenets by which several generations of neuroscientists have abided is Dale's law, which states that each neuron uses only one transmitter *in its informational transactions.* But this does not necessarily mean that only one type of potential transmitter molecule can be found anywhere in a given neuron. It is obvious that neurons that use ACh or catecholamines as transmitters will also contain the multipurpose amino acids glycine and glutamic acid, which are themselves prime transmitter candidates. There is an absolute necessity for the coexistence of L-glutamic acid, a probable excitatory transmitter, and GABA, a major inhibitory transmitter, in the same nerve ending, since L-glutamate is the immediate precursor of GABA. How could one envision that Dale's law would hold under the above circumstances, and how could one test it? If Dale's law were to hold, one would have to look for a specificity in storage, release, or postsynaptic action that would exclude the action of glutamate in GABAergic endings. It seems likely that such specialization would reside more economically in vesicular storage mechanisms. Thus a neuron in which GAD synthesis and axonal transport could take place also might synthesize a specific component in the vesicle membrane that would allow GABA, but not glutamate, to be stored in its vesicles and to be released subsequently for synaptic function.

Some evidence suggests that more than one transmitter or transmitter-forming enzyme, or both, may be present in a single nerve-cell body [26]; but there is no cogent evidence existing to date that a nerve *uses* more than one substance as a transmitter. Until such evidence becomes available, Dale's law will continue to be a guiding principle for the neuroscientist.

TRANSMITTER ACTION

Interaction of transmitter with a specific membrane receptor leads to one of two effects on the postsynaptic cell — excitation or inhibition. There are excitatory and inhibitory synapses on most neurons, and the balance of the two influences determines whether a neuron will increase or decrease its rate of discharge or generate an action potential at all. Typically, such effects are monitored electrophysiologically as changes in membrane potentials and conductances and as alterations in firing rates of neurons. Transmitters are presumed to induce cooperative conformational changes in receptor complexes that result in activation of particular ionophores (or ion channels) in the membrane regions in which they are located. In fact, current studies with the cholinergic receptor suggest that the ionophore site may be associated with the same protein molecule as is the receptor site [27].

In principle, the basic mechanisms for excitation and inhibition are similar, in that the action of the transmitters results in selective increases in permeability of the membrane for ions. Most commonly, excitation results from increased permeability to sodium ions. The normal gradient of these ions across the membrane is such that opening the ion channels results in a marked influx of ions. The electrochemical equilibrium reached by these ions has the effect of depolarizing the membrane potential, thus bringing it closer to its threshold level for firing off an action potential. Inhibition is generally associated with permeability changes to potassium or chloride ions or both. Since the normal electrochemical equilibria of both ions are near the resting level of membrane potential, the net effect is to stabilize the membrane potential near its resting level, thus keeping the neuron from firing.

While transmitters are often described as being excitatory or inhibitory, there are many examples of the same transmitter being able to exert either action, depending on the nature of the postsynaptic receptor and its associated ionophore. Receptors that modulate three distinct types of ionic channels have been described for ACh in vertebrate nervous systems; this transmitter mediates excitation at some sites and inhibition at others [6].

Norepinephrine and GABA, commonly thought of as inhibitory transmitters, also have been found to exert both excitatory and inhibitory effects. GABA acts on most membranes by increasing conductance to chloride, an ion to which permeability normally is low. The opening up of chloride channels makes the cell essentially a chloride electrode, and the intracellular levels of chloride determine whether the effect is hyperpolarizing with respect to the resting membrane potential (inhibitory) or depolarizing (excitatory). Several transmitters have been shown to exert three distinct conductance effects on a single cell in invertebrate ganglia, producing either excitation or inhibition [3]. Because of the diversity of the effects of individual transmitters, it is desirable to begin a systematic characterization of synapses based at least on the nature of the transmitter employed, the ionic mechanism, and the resultant excitation or inhibition [28].

INACTIVATION OF TRANSMITTER

There must be ways of removing transmitter rapidly from the synaptic cleft after its liberation from presynaptic endings, otherwise prolonged activation of receptors or their desensitization would lead to synaptic malfunctioning. ACh is the only known transmitter that has been found to be inactivated primarily by an enzyme. The major inactivation process for most neurotransmitters appears to be their removal from the synaptic cleft by sites located in pre- and postsynaptic membranes as well as in membranes of glial cells that surround the synapse [29, 30]. Such uptake systems show relatively high affinities for accumulating transmitter and have been found in many studies to be associated with the neural regions from which the transmitter is released. These high-affinity systems exist in addition to more widespread transport systems that possess considerably lower affinities for the same compounds. The latter systems may function to accumulate substrate for more general metabolic requirements.

Studies on the binding and intracellular accumulation of neurotransmitters, in which a variety of subcellular fractions and tissue slices have been used, have suggested a carrier-mediated model for transport. A reasonable picture of what happens to free extraneuronal transmitter that is liberated into the extracellular region of the synapse during activity of neurons is as follows: The membranes (presynaptic, postsynaptic, and possibly glial) contain highly mobile binding sites that have an absolute Na^+ requirement for their activity and need a high Na^+ concentration (0.1 M) for maximal activation. Thus transmitter binds to membranes to a greater extent in a high-Na^+ extracellular environment than in a low-Na^+ intracellular concentration. Transmitter that is bound on the outer surface of the membranes equilibrates rapidly with that in solution in the extracellular medium. The transmitter bound on the inside of the membranes equilibrates rapidly with transmitter in solution intracellularly. The binding sites are partially restricted to one or the other side of the membrane by a barrier. The frequency with which the binding sites traverse the barrier, cotransporting transmitter and Na^+, may depend upon various asymmetries on the two sides of the membranes, such as oxidation-reduction potential, degree of phosphorylation, and concentrations of particular ions or metabolites. In metabolizing tissue, because the intraneuronal concentrations of Na^+ are lower than the extraneuronal concentrations, transmitter tends to dissociate from the carrier on the intraneuronal side and to become available for mitochondrial metabolism. Thus the asymmetrical concentration of Na^+ sets up the conditions for rapid removal of transmitter from the extraneuronal synaptic environment into the intraneuronal, and possibly intraglial, environments and a rapid metabolism of the transmitter therein. These Na^+-requiring transport systems do not appear to have a direct requirement for metabolically generated ATP. The potential energy of the Na^+ gradient probably is a major source of energy for the carrier-mediated transport. ATP is, of course, required for operation of the Na^+ pump — the (Na^+, K^+) ATPase — that maintains the Na^+ gradient.

SYNAPTIC MODULATION

The finding that enzymes contain regulatory sites — at which metabolites other than substrates can bind and thereby modify enzyme-substrate interactions — was a major advance in understanding the control of enzymatic activity and the regulation of cellular metabolism. Accumulating evidence indicates that similar regulatory mechanisms exist at the synapse. Nontransmitter molecules, including substances released into the cleft along with transmitter, compounds released locally from glial cells or from nonsynaptic regions of neurons, or hormones released from distant tissue sites could serve as modulators of synaptic activity by interacting with regulatory sites on the postsynaptic membrane [31]. Such binding would induce conformational changes in adjacent receptor sites and thus modify transmitter-receptor interactions.

The concept of synaptic modulation can be invoked to suggest a nontransmitter function underlying the nonspecific depolarization of virtually all neurons by glutamate [32]. Varying background levels of this amino acid, as are released from nonsynaptic regions of conducting nerve bundles [9], could set the general level of neuronal excitability. The concept is also useful to explain the synaptic effects of various compounds that appear to have no independent action on postsynaptic membranes. Adenylic acid (5'-AMP) has been described as potentiating the postsynaptic effects of glutamate, the probable excitatory transmitter, at the crayfish neuromuscular junction [33]. These results may be of interest in light of other studies that show that adenine nucleotides are synaptically released together with transmitter in both adrenergic [25] and cholinergic [34] systems. Similarly, prostaglandin PGE_1 (see Chap. 13) has been found to block the inhibitory action of norepinephrine on cerebellar Purkinje cells, although it has no apparent postsynaptic effect when it is applied without norepinephrine [35]. Recent evidence also suggests that polypeptide hormones normally released from the hypothalamus to regulate secretion of anterior pituitary hormones may have additional modulatory effects on central neuronal activity [36].

The concept of synaptic modulation suggests a far more dynamic vision of a synapse than one limited to transmitter release, postsynaptic action, and clearance from the cleft.

MECHANISMS OF NEURAL INTEGRATION

The first level of neural integration resides in individual neurons. Excitatory and inhibitory influences coming from multiple synaptic inputs are integrated over the surfaces of dendrites and soma so that individual synaptic potentials seem to make the cell more or less excitable, increasing or decreasing, respectively, the readiness of the neuron to fire. Synaptic potentials, controlled by transmitters that affect ion permeabilities, are graded and usually decrementally conducted; they wax, then wane. On the other hand, the action potential, usually initiated in the region near the axon hillock when the integrated depolarization is sufficiently great, is an explosive phenomenon that propagates down the axon without decrement because the current flow produced by each membrane segment is more than sufficient to depolarize the one ahead of it, and so on. There can be hundreds, or even thousands, of excitatory and inhibitory synapses on the dendrites and somata of individual neurons, and there must be a multiplicity of transmitters of various presynaptic origins acting on a given cell at any time. When one deals with whole neuronal systems in intact organisms, the situation becomes even more complex, because the overall effect of inhibition at the synaptic level may be either activation or inhibition. Thus inhibition of inhibitory neurons may lead to disinhibition or excitation of the system as a whole; inhibition of excitatory neurons may lead to inhibition of the system; inhibition of excitatory neurons that act on inhibitory neurons may lead to disinhibition or excitation; and so forth. It is therefore understandable why so little definitive information is available about the nature of the transmitters that are operative at particular synapses in vertebrate nervous systems and why many neuroscientists have chosen to search for simpler invertebrate preparations in which to study basic synaptic properties.

Some of the goals in the study of chemical transmitters are: (1) to be able to identify the neurons that may employ the substances as transmitters; (2) to localize at an ultrastructural level those synaptic junctions at which these substances are released and act; (3) to elucidate the modes of action of the transmitters on neural membranes at a molecular level; and (4) to define the roles of such neurons in the neural systems of functioning organisms.

In the following several chapters data will be considered on various proved or putative transmitters that are pertinent to the attainment of these goals.

ACKNOWLEDGMENTS

This investigation was supported in part by the following grants: NS-09885 (R.H.) and NS-01615 (E.R.) from the National Institute of Neurological Diseases and Stroke, National Institutes of Health, and Grant MH-22438 (E.R.) from the National Institute of Mental Health.

REFERENCES

1. Palay, S. L. Principles of Cellular Organization in the Nervous System. In Quarton, G. C., Melnechuk, T., and Schmitt, F. O. (Eds.), *The Neurosciences: A Study Program.* New York: Rockefeller University Press, 1967, p. 24.

*2. Krnjević, K. Chemical nature of synaptic transmission in vertebrates. *Physiol. Rev.* 54:418, 1974.

*3. Gerschenfeld, H. M. Chemical transmission in invertebrate central nervous systems and neuromuscular junctions. *Physiol. Rev.* 53:1, 1973.

4. Erspamer, V. Occurrence of Indole-alkylamines in Nature. In Erspamer, V. (Ed.), *Handbuch der Experimentellen Pharmakologie,* Vol. 19. Berlin: Springer-Verlag, 1966, p. 132.

5. Hebb, C. O., and Krnjević, K. The Physiological Significance of Acetylcholine. In Elliott, K. A. C., Page, I. H., and Quastel, J. H. (Eds.), *Neurochemistry.* Springfield: Thomas, 1962, p. 452.

*6. Kravitz, E. A. Acetylcholine, γ-Aminobutyric Acid, and Glutamic Acid: Physiological and Chemical Studies Related to their Roles as Neurotransmitter Agents. In Quarton, G. C., Melnechuk, T., and Schmitt, F. O. (Eds.), *The Neurosciences: A Study Program.* New York: Rockefeller University Press, 1967, p. 433.

*7. Werman, R. CNS cellular level: Membranes. *Annu. Rev. Physiol.* 34:337, 1972.

8. Itokawa, Y., and Cooper, J. R. Ion movements and thiamine in nervous tissue. I. Intact nerve preparations. *Biochem. Pharmacol.* 19:985, 1970.

9. Weinreich, D., and Hammerschlag, R. Nerve impulse-enhanced release of amino acids from non-synaptic regions of peripheral and central nerve trunks of bullfrog. *Brain Res.* 84:137, 1975.

*Asterisks denote key references.

10. de Potter, W. P., de Schaepdryver, A. F., Moerman, E. J., and Smith, A. D. Evidence for the release of vesicle-proteins together with noradrenaline upon stimulation of the splenic nerve. *J. Physiol.* (Lond.) 204:102P, 1969.

11. Hildebrand, J. G., Barker, D. L., Herbert, E., and Kravitz, E. A. Screening for neurotransmitters: A rapid radiochemical procedure. *J. Neurobiol.* 2:231, 1971.

12. McCaman, R. E., and Dewhurst, S. A. Choline acetyltransferase in individual neurons of *Aplysia californica*. *J. Neurochem.* 17:1421, 1970.

13. Ehinger, B., and Falck, B. Autoradiography of some suspected transmitter substances: GABA, glycine, glutamic acid, histamine, dopamine, and L-DOPA. *Brain Res.* 33:157, 1971.

14. Thoenen, H. Neuronally mediated enzyme induction. *Biochem. J.* 128:69P, 1972.

*15. Harris, A. J. Inductive functions of the nervous system. *Annu. Rev. Physiol.* 36:251, 1974.

*16. Katz, B. *The Release of Neural Transmitter Substances.* Liverpool: Liverpool University Press, 1969.

*17. Whittaker, V. P. The Synaptosome. In Lajtha, A. (Ed.), *Handbook of Neurochemistry*, Vol. II. New York: Plenum, 1969, p. 327.

*18. Van den Berg, G. I. A Model of Glutamate Metabolism in Brain: A Biochemical Analysis of a Heterogeneous Structure. In Berl, S., Clarke, D. D., and Schneider, D. (Eds.), *Metabolic Compartmentation and Neurotransmission: Relation to Structure and Function in Brain.* New York: Plenum, 1975.

19. Geffen, L. B., and Livett, B. G. Synaptic vesicles in sympathetic neurons. *Physiol. Rev.* 51:98, 1971.

20. Birks, R., and MacIntosh, F. C. Acetylcholine metabolism of a sympathetic ganglion. *Can. J. Biochem. Physiol.* 39:787, 1961.

*21. Kornguth, S. E. The Synapse: A Perspective from *In Situ* and *In Vivo* Studies. In Ehrenpreis, S., and Kopin, I. (Eds.), *Reviews of Neurosciences*, Vol. I. New York: Raven, 1974, p. 63.

22. Berl, S., Puszkin, S., and Nicklas, W. J. Actomyosin-like protein in brain. *Science* 179:441, 1973.

23. Litchy, W. J. Uptake and retrograde transport of horseradish peroxidase in frog sartorius nerve *in vitro*. *Brain Res.* 56:377, 1973.

*24. Douglas, W. W. How Do Neurons Secrete Peptides? Exocytosis and Its Consequences, Including "Synaptic Vesicle" Formation, in the Hypothalamo-Neurohypophyseal System. In Zimmerman, E., Gispen, W. H., Marks, B. H., and deWied, D. (Eds.), *Drug Effects on Neuroendocrine Regulation.* Amsterdam: Elsevier, 1973, p. 21.

25. Douglas, W. W. Stimulus-secretion coupling: The concept and clues from chromaffin and other cells. *Br. J. Pharmacol.* 34:451, 1968.

26. Hanley, M. R., and Cottrell, G. A. Acetylcholine activity in an identified 5-hydroxytryptamine–containing neuron. *J. Pharm. Pharmacol.* 26:980, 1974.

27. Raftery, M. A., Vandlen, R., Michaelson, D., Bode, J., Moody, T., Chao, Y., Reed, K., Deutsch, J., and Duguid, J. The biochemistry of an acetylcholine receptor. *J. Supramol. Struct.* 2:582, 1974.

*28. Iversen, L. L. Neurotransmitters, Neurohormones, and Other Small Molecules in Neurons. In Schmitt, F. O. (Ed.), *The Neurosciences: Second Study Program.* New York: Rockefeller University Press, 1970, p. 768.

*29. Iversen, L. L. Neuronal Uptake Processes for Amines and Amino Acids. In Costa, E., and Giacobini, E. (Eds.), *Advances in Biochemical Psychopharmacology*, Vol. 2. New York: Raven, 1970, p. 109.

30. Snyder, S. H., Logan, W. J., Bennett, J. P., and Arregui, A. Amino acids as central nervous transmitters: Biochemical studies. *Neurosci. Res.* 5:131, 1973.
31. Florey, E. Neurotransmitters and modulators in the animal kingdom. *Fed. Proc.* 26:1164, 1967.
32. Crawford, J. M., and Curtis, D. R. The excitation and depression of mammalian cortical neurons by amino acids. *Br. J. Pharmacol.* 23:313, 1964.
33. Ozeki, M., and Sato, M. Potentiation of excitatory junctional potentials and glutamate-induced responses in crayfish muscle by 5'-ribonucleotides. *Comp. Biochem. Physiol.* 32:203, 1970.
34. Silinsky, E. M., and Hubbard, J. I. Release of ATP from rat motor nerve terminals. *Nature* 243:404, 1973.
35. Ramwell, P., and Shaw, J. E. Prostaglandins. *Ann. N. Y. Acad. Sci.* 180:1, 1971.
36. Renaud, L. P., Martin, J. B., and Brazeau, P. Depressant action of TRH, LH-RH, and somatostatin on activity of central neurones. *Nature* 225:233, 1975.

Chapter 9

Acetylcholine

F. C. MacIntosh

Acetylcholine (ACh) is a compound of considerable antiquity. We can be fairly sure that it arrived on the evolutionary scene before nervous systems, proved itself to be a versatile performer, and was tried out in a variety of parts before being assigned its major role as a synaptic transmitter in the higher animals. The evidence for this is that there are still many bacteria, fungi, protozoa, and plants that manufacture acetylcholine and store it. Even in mammals it can be found in high concentration at sites where it has no known function. The best known mammalian examples are the cornea, some ciliated epithelia, the spleens of certain ungulates, and the human placenta. As we have little information about the turnover or functional significance of ACh in any of these tissues, the rest of this chapter will be concerned with ACh in its role as a synaptic transmitter. In this role, ACh acts to open cation-selective channels in cell membranes, with consequences that depend both on the ionic species that can traverse the channels and on the nature of the ion-sensitive mechanisms inside the cell. If ACh itself ever acts by combining directly with intracellular receptors, that action has so far been overlooked.

CHEMISTRY OF ACETYLCHOLINE

The structure of ACh was deduced from crystallographic analyses by Pauling [1] and his colleagues. They pointed out that, though either end of the ACh ion (C3 − N − C4 − C5 or C5 − O1 − C6 − C7) has all four atoms coplanar, rotation can occur around the bonds C4 − C5 and C5 − O1.

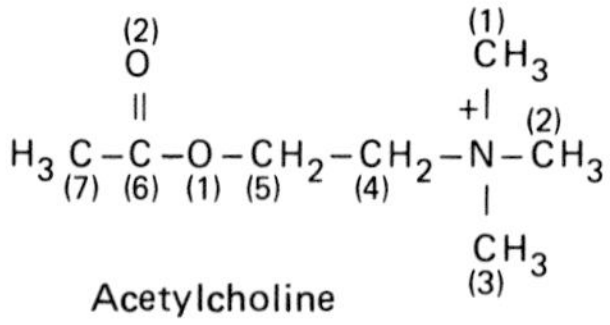

$$H_3\underset{(7)}{C}-\underset{(6)}{\overset{\overset{\displaystyle O}{\underset{(2)}{\|}}}{C}}-\underset{(1)}{O}-\underset{(5)}{CH_2}-\underset{(4)}{CH_2}-\overset{+}{\underset{\underset{\displaystyle CH_3}{\underset{(3)}{|}}}{N}}-\underset{(2)}{CH_3}$$

Acetylcholine

Because of this double hinge, the two planes are twisted somewhat away from each other, and further displacement can occur when the ion interacts with a protein receptor. As noted on page 184, there are two kinds of Ach receptors, nicotinic and muscarinic, mediating quick and slow responses respectively. Pauling's group

180

analyzed the structures of compounds that elicit, or block, one or another type of response, including a number of compounds, the molecules of which are more rigid than those of ACh. They concluded that as the ACh ion attaches to a receptor site, it has about the same conformation in each case but a different orientation: The *methyl side* of the ion has a high affinity for muscarinic receptors, and the *carbonyl side* has a high affinity for nicotinic receptors. Naturally, it is somewhat risky to suppose that the ACh ion when it is in solution, or close to a receptor, has the same shape as when it is in a crystal of an ACh salt; but the results of quantum-chemical analyses [2] seem to agree reasonably well with the work on crystals. There is no doubt that the onium head of ACh, with its positive charge spread over the methyl cluster, is an important structural feature for both muscarinic and nicotinic action, whereas the negatively charged carbonyl-O atom and the distance between it and the N atom are significant features on the nicotinic side.

Under ordinary laboratory conditions, or in vivo, the only part of the ACh molecule that is susceptible to chemical attack is the ester linkage. The hydrolysis of ACh at this site is accelerated by hydroxyl ions and, to a much smaller degree, by hydrions, as obtains with aliphatic esters generally. ACh is stable in solution at pH 4, and at pH 7.4 and $37°$ its half-life is many hours unless an esterase is present; but at pH 12 or higher it is destroyed within seconds: This is a useful test when it is suspected that a tissue extract owes its biological activity to ACh.

Both ACh and choline form salts of low water solubility with a number of inorganic and organic anions and may be separated in this way from other materials in tissue extracts. Chloroaurate, reineckate, dipicrylamate, and tetraphenylborate salts have been used for this purpose. Separation of ACh from choline can be achieved on the basis of the differential solubility of some of these salts in suitable organic solvents or by means of chromatography.

ASSAY OF ACETYLCHOLINE

Until the 1960s, anyone who wished to measure the ACh stored in, or released from, a tissue had to choose one of the bioassay methods: for references, see [3]. Among the favorite test objects have been the frog rectus, leech body wall, cat or rat blood-pressure, guinea-pig ileum, toad lung, and clam heart. These methods are tricky and tedious but sensitive (down to 1 to 10 pmol), and with experience they can be made to give results that are reliable within error limits of usually ±5% to 10%. During the last 10 years, a number of investigators have accepted the challenge of devising chemical assay methods of equal sensitivity, and as a result a dozen or more acceptable procedures are now available [3a].

CHOLINERGIC NEURONS

Definitive Constituents of Cholinergic Neurons

Cholinergic neurons are neurons that synthesize, store, and release ACh. Such neurons also synthesize choline acetyltransferase (ChAc), the enzyme that catalyzes the formation of ACh (1), and acetylcholinesterase (AChE), the most important of the enzymes that catalyze the hydrolysis of ACh (2):

$$\text{Choline} + \text{acetyl-CoA} \rightleftharpoons \text{ACh} + \text{CoA} \tag{9-1}$$

$$\text{ACh} + H_2O \rightleftharpoons \text{choline} + \text{acetic acid} \tag{9-2}$$

ChAc and AChE are made in the neuronal soma and are transported (see Chap. 21) down the axon to the synapses, where they are put to work. According to the current majority view, the presence of ACh in the axon is a by-product of the transport of ChAc: Acetylcholine functions only as a synaptic transmitter, and does not have a role in the conduction of impulses along the axon.

Choline acetyltransferase, and ACh itself, are present throughout the whole length of every cholinergic neuron. Their highest concentrations, however, are in the axon's terminal swellings, commonly referred to as *nerve endings*. (This term will be retained, although it is a misnomer in many cases. Usually the swellings are spaced out along the axon's distal course like beads on a string, but at some junctions, e.g., those between motor axons and the endplates of skeletal muscle fibers, all the presynaptic elements do have a terminal location.) Cell fractionation studies show that much of the ACh of the nerve endings is contained in the synaptic vesicles, but some of it is free in the cytosol. Most of the ChAc is also free in the cytosol, but some of it may adhere to the outside of the vesicles [4].

Acetylcholinesterase [5] has a somewhat wider distribution in the nervous system than have ACh or ChAc. Besides being a regular constituent of cholinergic neurons, it is made by some neurons, e.g., afferent neurons in vertebrates, that do not appear to be cholinergic. Whereas the presence of ChAc and ACh in a neuron is reliable evidence that the neuron is cholinergic, the presence of AChE is not, unless its concentration is very high. It has proved to be a useful marker, however, because it is much easier to detect histochemically than are ChAc or ACh. Except for such histochemical tests, there is no way of distinguishing cholinergic from noncholinergic neurons by their appearance. AChE is also made extraneuronally to a much greater extent than is ChAc. It is a constituent of erythrocyte membranes, and at some synapses, notably motor-axon/endplate synapses, most of the AChE is attached to the postsynaptic cell. As these examples indicate, AChE is characteristically a membrane-bound enzyme. At synapses, most of its active sites face outward, enabling the enzyme to hydrolyze ACh that has just been released; but cholinergic axons and nerve endings outside the central nervous system apparently contain a little inward-facing AChE as well.

ChAc and AChE have been studied intensively for decades: AChE has had special attention because a number of drugs, insecticides, and "nerve gases" act by inhibiting it. Neurochemical studies since 1970 have added two further items to the list of substances that seem to be specifically associated with cholinergic nerve endings: (1) a membrane carrier, presumed to be a protein but not yet isolated, which mediates the *high-affinity uptake of choline* for ACh synthesis [6]; and (2) vesiculin [7], an acidic protein of low molecular weight, which is present in the interior of the synaptic vesicles and may have a role in ACh storage.

Distribution of Cholinergic Neurons

Details of the distribution and functional significance of peripheral cholinergic pathways in mammals can be found in textbooks of physiology or pharmacology. Table 9-1 gives a summary and indicates also the nature of the postsynaptic receptors in each case. The cholinergic pathways in other vertebrates have about the same distribution as in mammals. The electric organ, present in some species of fish, is excellent material for neurochemical studies because of the density of its cholinergic innervation.

Any attempt to construct a similar table for the central cholinergic pathways would be grossly premature. At the time of writing, only two well-defined sets of synapses in the CNS have been proved to be cholinergic. First, the axons of motor nerves within the spinal cord send excitatory branches to a set of nearby neurons (Renshaw cells), whose discharge inhibits the same and other motor neurons. Because these axon collaterals arise from cholinergic neurons, they are themselves cholinergic, but the Renshaw cells are not. Second, a well-defined axonal tract originating from the septal region forms cholinergic synapses with neurons in the hippocampus: This pathway apparently accounts for most of the ACh and ChAc in the hippocampus.

Further neuronal mapping should soon identify other cholinergic pathways in the brain. At present, there is fairly strong evidence that the diffuse "arousal" projections passing to the neocortex from phylogenetically older parts of the brain (midbrain reticular formation, hypothalamus, striatum, and septum) are largely cholinergic. Indeed, the endings of these pathways may account for most of the ACh and ChAc in the cortex. On the other hand, it seems probable that the specific

Table 9-1. Cholinergic Pathways in the Peripheral Nervous System

Pathway	Neuronal Cell Bodies in:	Axonal Endings on:[a]	Receptors[f]	Action[g]
Somatic motor	CNS	Skeletal muscle fibers	N	
Preganglionic	CNS	Neurons in autonomic ganglia	N,M	E
		Interneurons in ganglia	M	E
		Adrenomedullary cells	N,M	E
		Neurons in gut plexuses	N	E
Postganglionic	Parasympathetic ganglia	Smooth muscle[b]	M	E
		Smooth muscle[c]	M	I
		Heart muscle fibers	M	I
		Gland cells[d]	M	E
Postganglionic	Sympathetic ganglia	Smooth muscle[e]	M	I
		Sweat gland cells	M	E
Gut wall, intrinsic	Gut plexuses	Gut smooth muscle	M	E

[a]Endings that supply smooth muscle are often distant from cells.
[b]Includes smooth muscle of iris (constrictor of pupil), ciliary body, bronchioles, bladder, etc.
[c]Includes vascular smooth muscle in brain, exocrine glands, genitalia, etc.
[d]Includes lachrymal, nasal, salivary, gastric glands, exocrine pancreas; also endocrine cells producing gastrin, insulin.
[e]Includes vascular smooth muscle in microcirculation of muscles in some animals.
[f]M, muscarinic; N, nicotinic; where both are listed, the first is more prominent.
[g]E, excitatory; I, inhibitory.

sensory projections from thalamus to cortex, the principal descending pathways including the pyramidal tracts, and the numerous neocortical and cerebellar inter-neurons, are nearly all noncholinergic. Several brain regions, especially the nuclei of the striatum, are richer in ACh and ChAc than are the neocortex or hippocampus and presumably have a high density of cholinergic synapses. The retina also should be mentioned in this context: It is not yet certain which kind of retinal neuron is cholinergic.

CHOLINERGIC RECEPTORS

Cells that respond to injected ACh, or to ACh released at nerve endings, are *cholino-ceptive*. Generally, their responses can be assigned to one or other of the two broad classes identified by Dale in 1914. *Muscarinic responses* are typically slow, lasting for seconds; they may be either excitatory or inhibitory; they can be evoked by the drug muscarine, a mushroom alkaloid, as well as by ACh; and they are blocked by atropine. *Nicotinic responses* are typically quick, lasting for milliseconds; they are always excitatory; they also can be evoked by nicotine; and they are blocked by drugs of the curare type or by an excess of either nicotine or ACh itself. These two responses result from the temporary combination of ACh with *muscarinic* or *nico-tinic receptors:* One or another kind of ACh receptor — occasionally both kinds — is present on the surface of all cholinoceptive cells.

The existence of these receptors was a convenient postulate for a long time before it was confirmed that they are specific proteins [8]. Like AChE, they are rather strongly membrane bound and are only solubilized by treatments that disrupt the membrane, e.g., treatment with a detergent. In addition to combining reversibly with ACh and ACh-like compounds, either type of ACh receptor also can combine selectively and irreversibly with certain toxic substances, especially alkylating agents that are structurally related to ACh and protein neurotoxins present in the venoms of several genera of snakes. The best known of the toxins is α-bungarotoxin from the krait, but several other venoms, such as that from *Naja naja*, also yield proteins with high affinity for nicotinic receptors. Since the discovery within the last decade of these remarkably specific blocking agents, the neurochemistry of cholinergic receptors in skeletal muscle and electric organ has become a lively topic [8a].
With the aid of radioactively labeled blocking agents, several teams of investigators have been able to estimate the number of receptor sites per unit area of postsynaptic membrane in muscle from human and other species and also to isolate purified receptor-blocker complexes. In the nicotinic receptor from the electric organ, it has been possible to dissociate the receptor protein from the complex by affinity chroma-tography [8]. The protein's amino acid composition is now known, and rapid further progress can be expected.

Nicotinic Receptors

Nicotinic agonists and antagonists [8] typically possess at least one strongly electro-positive group, usually centered on a quaternary N atom, so it has been believed for a long time that a nicotinic receptor has at least one anionic site, presumably carboxyl, as an essential feature. Otherwise our knowledge of the chemistry of the receptor sur-face is rudimentary. A disulfide bond may be present, for nicotinic responses are reversibly suppressed by the reducing agent dithiothreitol, which opens such bonds.

$(C_2 H_5)_4 N^+$

Tetraethylammonium

$(CH_3)_3 \overset{+}{N} -(CH_2)_6 -\overset{+}{N}(CH_3)_3$

Hexamethonium

$(CH_3)_3 \overset{+}{N} -(CH_2)_{10} -\overset{+}{N}(CH_3)_3$

Decamethonium

$O-CH_2 CH_2 -\overset{+}{N}(C_2 H_5)_3$
$O-CH_2 CH_2 -\overset{+}{N}(C_2 H_5)_3$
$O-CH_2 CH_2 -\overset{+}{N}(C_2 H_5)_3$

Gallamine

All nicotinic receptors do not behave alike. For example, tetraethylammonium and hexamethonium are classic competitive antagonists of ACh at ganglionic synapses, but they have little ability to block its action at neuromuscular synapses; gallamine has just the opposite specificity; decamethonium has a strong ACh-like action on muscle endplates in some situations, but little action on ganglion cells. Such findings have suggested that the spacing of the anionic sites may be regular, but different, in the two kinds of location.

The earliest attempts to harvest nicotinic receptors were based on a search for tissue macromolecules with a strong affinity for nicotinic antagonists of the curare type. These efforts were not very successful because cell surfaces are adorned with a variety of proteins and glycoproteins, most of which have anionic sites that have some tendency to combine with curare and similar bases. One such surface protein is AChE, and the hypothesis was current for some time that the receptor sites and the enzyme sites are closely linked and perhaps are parts of the same molecule. Now the hypothesis has been disproved. The proteins responsible for the two activities are distinct and can be separated from each other, but they are very similar chemically and may have a common evolutionary origin. They seem also to be about equally abundant (about 10^4 sites per μm^2) at postsynaptic surfaces, where together they may cover half or more of the total area.

The nicotinic [9] action of ACh is basically the same wherever it occurs, whether on muscle endplates, ganglion cells, or Renshaw cells: It makes the cell membrane more permeable to cations. Both G_{Na} and G_K (sodium and potassium conductances) are increased. In the absence of nicotinic action, G_{Na} is only a small fraction of G_K and rises to exceed it in response to nicotinic agonists. The opening of these ionic channels short-circuits the membrane potential, causing a local depolarization (endplate potential or excitatory postsynaptic potential). This local depolarization, on reaching a critical level, generates a conducted impulse (action potential or spike). Anion conductance is unaffected by nicotinic action, so it seems likely that the cation-selective channels opened by ACh are aqueous pores lined with fixed negative charges, which reject anions. Analysis of "endplate noise" [10] shows that the channels in muscle are open for a millisecond or so at a time, during which time they allow the passage of some 5×10^4 ions.

There has been a good deal of speculation about "receptor-ionophore coupling." On combining with the receptor, ACh presumably induces a conformational change in it; but how does that change open up cation-permeant channels through the membrane? One proposal, which seems attractive at present, is that the activated receptor is itself the ionophore with a central channel that extends right through the membrane.

Nicotinic responses often appear to be of brief duration, even when the nicotinic agonist remains in contact with the receptors. This may happen for either, or both, of two reasons. (1) The local depolarization, though initially sharply focused, becomes more diffuse as the adjacent membrane voltage is discharged by ion flow through the opened channels, and no longer generates spikes. In this situation, synaptic transmission is blocked. (2) In some cases, persistent contact with the nicotinic agent leads to "receptor desensitization," and the opened ionic channels gradually close; the mechanism is not known.

Muscarinic Receptors

Progress toward isolation and characterization of muscarinic receptors has been slower than for nicotinic receptors, perhaps partly because nature has not offered the neurochemist a reagent that can combine with them as specifically and irreversibly as α-bungarotoxin combines with nicotinic receptors. (Note that nicotinic blockade kills, whereas muscarinic blockade does not: a possible reason why venoms directed at muscarinic receptors have never evolved.) In 1966, however, a reagent with the necessary properties was discovered in the alkylating agent benzilylcholine mustard [11]. Although not completely specific, this compound and its analogs have a high affinity for muscarinic receptors in intestinal muscle and brain. At present it seems that these receptors are less densely packed on cholinoceptive cell surfaces than are nicotinic receptors on endplate and electroplaque membranes. In longitudinal gut muscle, for example, there are about 200 receptor sites per μm^2 [8], but muscarinic sites may often be scattered over the whole area of a cell, in contrast to nicotinic sites, which usually have a more restricted distribution. Tests on brain homogenates with muscarinic and nicotinic affinity labels (including α-bungarotoxin) have confirmed that muscarinic receptors are far more abundant in this tissue. As with nicotinic receptors, there are probably isoreceptors; apparently their reactivity to agonists differs more than their reactivity to antagonists.

The consequences of muscarinic receptor activation cannot be described simply: They differ from tissue to tissue [9]. In some cases, the unmasking of ionophores is a prominent, and perhaps the primary, effect: If both G_{Na} and G_K are increased, the response is depolarization, as in the excitation of intestinal smooth muscle; if only G_K is increased, the response is hyperpolarization, as in vagal inhibition of the heart. In other cases, the primary action may be to increase intracellular Ca, with reduction of K^+ efflux or activation of guanyl cyclase (see Chap. 12) as a secondary result. By comparison with nicotinic responses, all these muscarinic responses develop and fade slowly, even in brain.

SYNTHESIS OF ACETYLCHOLINE

Choline Acetyltransferase: Location

Choline acetyltransferase was recognized as a distinct enzyme by Nachmansohn and Machado in 1943; its second substrate, acetyl-CoA, was not isolated until a few years later. As already noted, its distribution is like that of ACh; both are present throughout the length of cholinergic neurons and are concentrated in the axonal endings. It is now agreed that most of the ChAc in brain synaptosomes is free in the cytoplasm, and this is probably true for all cholinergic nerve endings in situ. But it is recognized that ChAc, especially in some species, tends to adhere to membranes, and it has been suggested that part of the nerve-ending ChAc is attached during life to the outer surface of the synaptic vesicles. It is further proposed that ACh newly synthesized by ChAc in that location might be placed favorably for transport into the interior of the vesicles, a point that will be discussed later.

Choline Acetyltransferase: Properties

ChAc [5], extracted from the nervous tissue of different species and partially puri-fied, is a stable, relatively basic protein, the molecular weight of which, as determined by gel filtration, is about 65,000. There is some evidence that the native enzyme has a higher molecular weight and, during isolation, gives rise to two fragments that differ in size and activity. The positive charge on the molecule varies with the species, and indeed several isoenzymes with different isoelectric points may be present in a single tissue [4]. No doubt, these variations determine the proportion of enzyme that . becomes bound to vesicles, and that proportion, in turn, may influence the relative turnover rates of vesicular and extravesicular ACh.

Not much has been learned about the composition of ChAc except that basic amino acids predominate, and nothing has been learned about the chemistry of its active center. A plausible candidate for participation at that site is the imidazole ring of histidine. In fact, in the absence of any enzyme, free imidazole can catalyze transfer of acetyl groups from CoA to choline. ChAc has also been regarded for a long time as a sulfhydryl enzyme, but whether the SH groups lie close to the active center has not yet been decided.

The substrate specificity of ChAc is by no means extreme. It can acetylate many alcoholic bases besides choline, and it can accept acyl groups other than acetyl from CoA as donor. But normally its only product, at least in vertebrate nervous tissue, is ACh, because normally the only substrates that are presented to it are choline and acetyl-CoA. (Whittaker [3] has reviewed the occurrence of other choline esters in nonnervous tissue and in invertebrates.)

Choline Acetyltransferase: Kinetics

All investigators agree that the reaction catalyzed by ChAc is reversible, and that acetyl-CoA has a higher affinity for the enzyme than does choline. Estimated Michaelis constants are of the order of 1 mM for choline and 10 μM for acetyl-CoA. Kinetic analyses [12, 13] on the brain enzyme (the placental enzyme behaves differ-ently) are consistent with an ordered mechanism of the Theorell-Chance type, with the sequence of events shown in Figure 9-1.

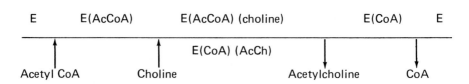

Figure 9-1
Reaction mechanism of choline acetyltransferase.

In life, the enzyme does not synthesize ACh at anything like the rate it can achieve under optimal conditions in free solution. After a lag at the outset, ACh synthesis in active nerve endings usually proceeds at a rate that roughly matches the rate of ACh release. In brain, this may be up to 25% of the enzyme's maximum rate in vitro [14] ; in peripheral tissues, the percentage is lower. Whether the limiting factor is substrate supply or product accumulation has been much discussed. No simple answer can yet be given.

Product inhibition by ACh, though it might seem teleologically suitable, can hardly be the major factor, for even isotonic ACh has been found to inhibit its own synthesis by less than 50%. Potter [13] suggests that both the rate of net synthesis of ACh and the level of free ACh in the cytosol depend on the equilibrium position of ChAc with respect to its substrates and products, i.e., on the mass-action relationship. This suggestion emphasizes that the ratio of CoA to acetyl-CoA in the vicinity of the enzyme may be at least as important as the ratio of ACh to choline. We have no reliable information, however, about what either ratio may be in any set of nerve endings during life.

Source of Choline

Neurons, unlike liver cells, cannot synthesize choline de novo. The ultimate source of choline for ACh synthesis is, therefore, the free choline of the plasma. This is homeostatically maintained at a remarkably constant level, 10 to 20 μM, by renal and other mechanisms. In vivo, the small pool of extracellular free choline undergoes constant turnover, exchanging with both intracellular free choline (probably 20 to 30 μM in brain [15]) and the much larger pool of choline covalently bound in tissue phospholipid. In vitro, phospholipid catabolism is likely to be dominant, and the free choline that diffuses out of a tissue placed in an initially choline-free artificial medium may soon attain a concentration that is adequate to support net ACh synthesis at an optimal rate. This fact sometimes has been ignored by neurochemists who apply radioactive choline to isolated tissues in order to study choline uptake or ACh turnover.

No convincing evidence exists as yet that nerve endings can make direct use of either endogenous or exogenous phospholipid as a source of choline for ACh synthesis. There is some evidence that lysophosphatidylcholine (and perhaps other choline esters) in the circulation can provide choline to make ACh in brain [16], but probably the bound choline only becomes available for uptake after it has been split off extraneurally. Concentrative uptake of radioactive choline has been demonstrated for a number of tissues. Many other quaternary bases, including ACh, are taken up by the same mechanism, though usually more slowly; hemicholinium, HC-3, is the most potent of many bases that have been found to block the uptake. In the first studies of this kind, the Michaelis constants for the uptake of choline appeared to be in the range 10 to 100 μM for most tissues, including brain. It seemed odd that the ChAc of nervous tissue should have to depend for its supply of substrate on a carrier that could not be efficiently loaded at a level of 10 μM, which is the plasma concentration of free choline, especially as experiments on perfused ganglia [17] had shown this level to be adequate to maintain ACh synthesis, even during maximal demand. In fact, according to the same experiments, active nerve endings have an extraordinary capacity for converting plasma choline into ACh: Some 20% of the choline in the plasma can be trapped in this way, though the plasma leaves the ganglion only one second after it enters.

This apparent paradox was resolved by the discovery of a transport system [6] in brain synaptosomes that has a much higher affinity for choline, with K_m values in the range 1 to 5 $\times$ 10^{-6} M. There is now convincing evidence that this high-affinity system is located in cholinergic nerve endings and that most of the choline it transports is converted into ACh. Noncholinergic cells and terminals possess only the low-affinity mechanism discovered earlier. The high-affinity mechanism is carrier mediated and Na dependent; it is inhibited by low concentrations of HC-3 (K_i values below 0.1 μM have been reported); and it does not significantly mediate the uptake of ACh. The chemistry and the precise location of the high-affinity carrier, and the manner in which it is coupled (as it seems to be) to both ACh release and ACh synthesis, are problems for future research. There is no evidence that the high-affinity transport is directly coupled to an energy transducer such as an ATPase, but the process seems to be more complex than facilitated diffusion across the nerve-terminal plasma membrane. It may turn out that the carrier is somehow activated or made accessible to choline by vesicular exocytosis or some other event associated with the arrival of the nerve impulse.

There is now much evidence [18] that cholinergic nerve endings recapture and reuse up to 60% of the choline formed by hydrolysis of the ACh they release. They do not recapture ACh, the transmitter itself; in this they differ from their adrenergic cousins and probably from most other kinds of chemically transmitting nerve endings. So much of the uptake of choline during activity is accounted for by the reuptake, that it seems likely that the high-affinity uptake mechanism is particularly active for a few milliseconds after the arrival of each nerve impulse. After any longer period, the choline just formed from released ACh would have diffused away from the synapses and could not compete for uptake with the choline supplied by the circulation.

Source of Acetyl-CoA

Experiments with labeled precursors leave little doubt that, in mammalian brain, the acetyl moiety of ACh is derived almost wholly from pyruvate, generated from glucose via the Embden-Meyerhof pathway [12, 19]. Rather curiously, acetate is a relatively poor source of acetyl-CoA for ACh synthesis in mammalian brain, though it is a good source of acetyl-CoA for lipid synthesis in that tissue and for ACh synthesis in *Torpedo* electroplaques, rabbit cornea, and lobster nerve. In all studied tissues, acetyl-CoA is primarily a product of the mitochondrial inner compartment and is supposed not to leak out, so the problem of how mitochondrially generated acetyl-CoA can give rise to cytoplasmic acetyl-CoA has been a continuing puzzle. Several ingenious explanations based on known metabolic pathways have been suggested, but at present none of them seems very likely, and one is tempted to guess that cholinergic nerve endings generate acetyl-CoA by some special route that bypasses the mitochondrial permeability barrier.

Inhibitors of Acetylcholine Synthesis

A number of reversible and irreversible inhibitors of ChAc have been synthesized and studied [20]. The best known are a series of styrylpyridines and some halogenated analogs of ACh. Unfortunately these compounds, though active in vitro,

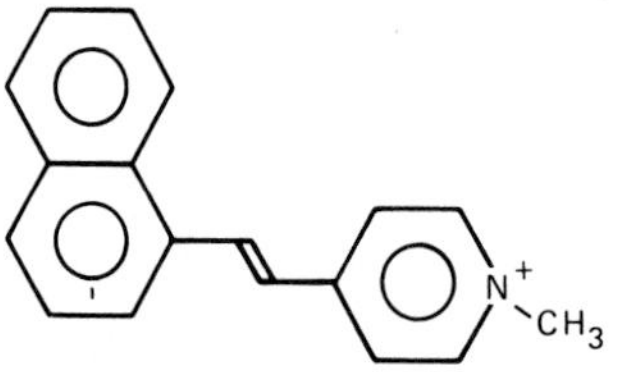

4 - (1- naphthylvinyl)- 1 - methylpyridine

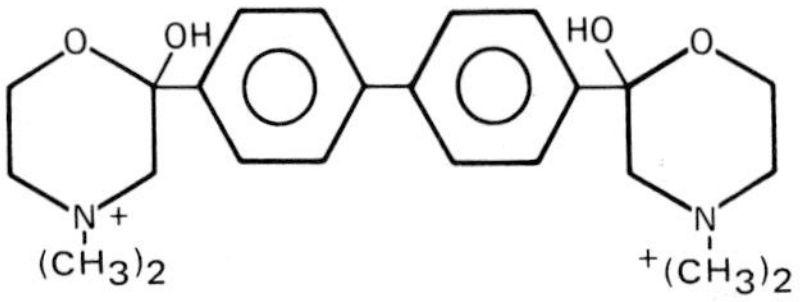

HC−3

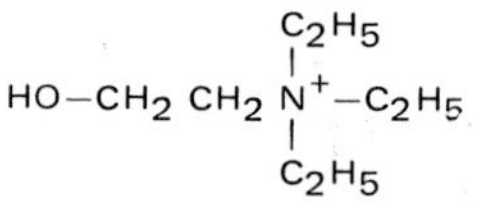

TEC

have little effect on ChAc in its natural setting within the nerve ending, probably because they do not reach the enzyme. At present the only well-established inhibitors of ACh synthesis are the compounds that compete with choline for the high-affinity carrier. They are all quaternary bases [20]. The best known, and perhaps still the most potent, is HC-3, but some of the other hemicholiniums are little, if any, less active. (The hemicholiniums are characterized by the presence of a choline moiety cyclized through hemiacetal formation into a 6-membered ring.) Another inhibitor of interest is the triethyl analog of choline, known as triethylcholine, TEC. In vivo, these bases are delayed-action poisons, effective in minute dosage. In such dosage, they have no obvious effect on cholinergic transmission until the synapses have been activated long enough, or at high enough frequency, to deplete stored ACh below a critical level. Because the site of the high-affinity transport system is still uncertain, the site at which its inhibitors act is also uncertain. Once it was thought that HC-3 and TEC, or their acetyl esters formed in vivo, might act as false transmitters, usurping ACh storage sites in vesicles. This has now been disproved, but it is still far from certain that the only significant action of the drugs is to prevent the entry of choline into the presynaptic cytosol. Inhibition of ACh transfer from cytosol to vesicles remains a distinct possibility.

Acetylcholine Uptake by Brain Tissue
Though peripheral cholinergic endings trap choline efficiently, they cannot trap ACh [21]. Brain tissue can do so, however, apparently by the same mechanism that mediates the low-affinity uptake of choline. This uptake of ACh can be observed when an organophosphate AChE inhibitor is used to protect the ACh; the uptake is blocked by eserine as well as by HC-3. The ACh that enters the tissue does not gain access to synaptic vesicles, and it is not released by stimulation. The phenomenon is thus an artifact, the principal significance of which is that it can mislead a neurochemical experimenter.

STORAGE OF ACETYLCHOLINE

Compartmentation of Neuronal Acetylcholine
ACh is presumed to be present with ChAc in the cell bodies and dendrites of vertebrate cholinergic neurons, but its concentration and location are not known. Axonal ACh appears to be free in the axoplasm, because when a nerve trunk is homogenized in an anticholinesterase-containing medium, nearly all of its ACh is found in the supernatant. The concentration of ACh in the axoplasm has been estimated as 0.3 mM. About 20% of the preformed ACh of brain tissue is released by simple homogenization: No doubt much of this "free ACh," as it is called, originates from cell bodies and axons.

ACh in nerve endings behaves differently [22]. Osmotic or mechanical lysis of synaptosomes releases no more than half of this ACh ("labile-bound ACh"). The rest of it remains sedimentable ("stable-bound ACh"): Much of this can be recovered in a well-defined fraction of very small particles, which, when examined with the electron microscope, cannot be distinguished from synaptic vesicles. The remainder is in vesicle-containing debris. It seems reasonable to conclude that stable-bound ACh represents ACh that was in the vesicles during life. But it is probably less safe to conclude that labile-bound ACh represents ACh that was in the terminal cytosol during life: Some of the vesicular ACh might have leaked back into the cytosol during the manipulations.

Reliable information about ACh levels in the presynaptic cytosol and vesicles of brain neurons would be of great interest. The information we have is not very reliable, unfortunately, partly because of the just-mentioned uncertainty about leakage, but even more because we do not know what proportion of the harvested synaptosomes are cholinergic. Whittaker's guess, 10% to 15%, seems as good as any. On that basis, he calculated (see [22]) that there would be some 2,000 ACh molecules in a vesicle, which would make the ACh concentration in the vesicle core about 0.2 M if all the ACh was in the form of free ions. Whittaker's estimate agrees well with one based on data for vesicles harvested from sympathetic ganglia, in which most of the synapses can be assumed to be cholinergic. After correcting for various biasing factors, Wilson and his colleagues [23] arrived at a figure of 1,630 molecules per vesicle. Taken at face value, these estimates for mammalian vesicles suggest that the vesicular fluid is approximately isotonic with its surroundings and that ACh is its principal cation.

Cholinergic vesicles from the electric organ can be obtained in greater purity and yield, and the data for these tend to support a similar conclusion. The concentration of ACh in *Torpedo* vesicles is estimated at 0.65M, which roughly matches the plasma electrolyte concentration. *Torpedo* vesicles should not, however, be regarded as identical little bags of isotonic ACh solution; first, they differ in size; second, some of the vesicular ACh is less firmly bound than the rest, either because it is in a different sort of vesicle or because it is attached to the vesicle membrane rather than free inside [24]. There is some evidence also that cholinergic vesicles from mammalian brain do not constitute a single population: The most recently synthesized ACh tends to be associated with vesicles that adhere to broken synaptosomal membranes [22]. Further research may show that differences between vesicles are related to their age, degree of filling, or nearness to release sites, but speculation on these points would be premature.

Binding of Vesicular Acetylcholine

Torpedo vesicles [7] contain ATP and vesiculin, an acidic protein of low molecular weight; the molecular ratio of ACh to each of these is about 12. No doubt they are the principal counter-ions to ACh, but there is no evidence yet that either vesiculin or a vesiculin-ATP complex can bind ACh. In life, the vesicles must take up and retain ACh against a fairly steep concentration gradient, which suggests that there is some sort of pump in the vesicular membrane. Its action could be to exchange Na or choline for ACh [10], or to create an electrical gradient favoring ACh accumulation, or simply to move ACh uphill, but none of these mechanisms has been detected. Indeed, isolated vesicles placed in an ACh-containing medium have little capacity for either net uptake or exchange of ACh.

Recent experiments [7] on the electric organ have shown that, during repetitive stimulation of its nerve supply, ACh, ATP, and vesiculin are all lost, but ACh is lost and recovers more rapidly than the other two. No doubt, some sort of diffusion barrier hinders the escape of the larger molecules, but its location is unknown.

Surplus Acetylcholine

When a peripheral tissue supplied by cholinergic axons is treated with a suitable anticholinesterase, its ACh content rises during an hour or so to about twice its initial value, even in the absence of nerve impulses [25, 26]. The ACh formed under these conditions has been called surplus ACh. It is clearly intracellular, and there are several reasons for thinking that it is presynaptic: The postsynaptic elements have little capability for making ACh or for taking it up from outside; a tissue deprived of its cholinergic supply loses its ability to form surplus ACh; and cholinergic axons whose

AChE is inactivated also accumulate additional ACh, and to about the same extent [27]. These findings, and others, suggest that some of the presynaptic AChE in peripheral tissues is inward-facing and that when it is inhibited the level of ACh in the nerve endings rises as a result of continuing synthesis to a higher-than-normal level. There is now conclusive evidence [26] that surplus ACh in ganglia and muscles cannot be released by nerve impulses (though in muscles it can slowly exchange with releasable ACh). It can, however, be released by K^+ and by nicotinic agonists, including ACh itself; these agents release no ACh from the normal synaptic store. Surplus ACh must therefore be held in some compartment other than the one that holds most of the normal store, which is 80% to 90% releasable. Certainly the simplest interpretation is to suppose that surplus ACh is in the presynaptic cytosol outside the vesicles, but decisive evidence is lacking. Surplus ACh is an experimental artifact, but one of considerable significance. Its rate of formation is a measure of intracellular ACh turnover at rest. If the location just suggested for it can be proved, the evidence that nerve impulses release ACh by vesicular exocytosis — evidence that is still rather weak — will be greatly strengthened.

Surplus ACh formation, as just discussed, is a phenomenon that has been described for peripheral synapses. The evidence that it occurs at central cholinergic synapses is weak or lacking. Perhaps only peripheral cholinergic endings possess inward-facing AChE and, therefore, normally have little ACh outside the vesicles.

RELEASE OF ACETYLCHOLINE

Quantal Transmission

At some, and probably at all, cholinergic synapses, ACh is released in multimolecular packets or quanta. When the nerve ending is at rest, the quanta are released singly and with nearly random timing. When the nerve ending is partly depolarized, e.g., by applying K^+ or a cathodal current, the quanta are released more often: the greater the depolarization, the higher the frequency. When the nerve ending is invaded by an impulse, and therefore fully depolarized (there is actually a reversal of its membrane potential), the frequency of release is greatly increased for a brief instant. In the special case of the neuromuscular junction, a few hundred quanta may be released almost synchronously from numerous sites on the motor axon's terminal expanse. At more typical nerve endings, an impulse probably releases no more than one or two quanta from any one varicosity.

The probability that an ACh quantum will emerge from a given release site at any selected instant depends, as has just been seen, on the presynaptic transmembrane potential at that site. It also depends on a second major factor, the concentration of Ca^{2+} in the extracellular medium, or — more accurately — at some critical site or sites in the nerve ending itself. As the nerve ending becomes depolarized, Ca^{2+} enters (along with Na^+) and forms a temporary complex with some nerve-ending constituent. The amount of complex formed determines the probability of quantal release. This probability is small in the absence of external Ca^{2+} and also if the external Mg^{2+} concentration is high; Mg^{2+} competes with Ca^{2+}, preventing formation of the complex.

The material with which Ca^{2+} combines has not been identified. One mathematical analysis of the biophysical data has suggested that four Ca ions must attach in order to activate a single release site.

An intriguing finding, the significance of which has not yet been evaluated, is the presence of a discrete Ca-binding site on each synaptic vesicle [28]. Of course, it is easy to theorize that the divalent Ca ion is able to attach itself simultaneously to presynaptic membrane and vesicles, and thus to promote vesicular exocytosis, but hard evidence in support of the theory has not yet been produced. It does seem safe to conclude, however, that the Ca complex has only a brief existence and that thereafter the Ca that has entered is trapped by other nerve-ending constituents, perhaps the mitochondria, before eventually being extruded. It is also probable that intracellular Na, here as in other kinds of cells, can displace trapped intracellular Ca and so make it available for promoting release. This may be the reason why repetitive synaptic activation, even though it tends to deplete the stock of releasable quanta, often increases the number of quanta released by each impulse. Metabolic poisons also potentiate release temporarily in many cases and can do so by interfering with both Na and Ca extrusion.

Quantum Size

All the influences discussed above act to determine the number of quanta released, not the size of the individual quanta. Quantum size, i.e., the amount of ACh per quantum, is generally uniform. So far the only agents known to decrease it are HC-3 and the other inhibitors of choline transport [29]. At neuromuscular synapses poisoned with HC-3 and repetitively stimulated, the normal number of quanta is released by each impulse, but the quanta become smaller and smaller until transmission fails. The spontaneously released quanta are reduced to the same extent as those released by nerve stimulation.

This effect of HC-3 on quantum size, though well established, may not be universal, for it has been reported, rather surprisingly, that in sympathetic ganglia it is the number of quanta per impulse that declines during stimulation in the presence of HC-3, not the size of the individual quanta.

Quanta Versus Vesicles

Biophysical data allow estimates of the number of quanta released but not of the amount of ACh in each quantum. Neurochemical data (including bioassay data) do not give quantum counts but do provide absolute measurements of released ACh. By combining the two kinds of data, it should be possible to estimate the number of ACh molecules in a quantum. According to the vesicle hypothesis in its usual form, the estimate should agree with estimates of the number of molecules in a vesicle. At present, disappointingly, the agreement seems to be poor, even when it is admitted that both sorts of estimates are rough. Mammalian vesicles, as already noted, are reckoned to hold some 2,000 ACh molecules apiece, but mammalian quanta are believed to be an order of magnitude larger, with estimates (summarized in [30]) ranging from 12,000 to 60,000 molecules. The figures at the upper end of this range are certainly hard to reconcile with the vesicle hypothesis as it is generally conceived. Potter [13] calculates that a mammalian vesicle 31 nm in diameter could hold 63,000 molecules of ACh chloride in crystal packing, but he points out that such vesicles would be heavier than the ones isolated by centrifuging on a density gradient. Presumably, too, some room must be found in the vesicles for ATP and vesiculin, if not for water. The best single comparison of vesicles and quanta is probably the one made by Nishi and his colleagues [31]. Working on frog ganglia, they measured quantum content electrophysiologically and ACh by bioassay and estimated quantum size at 8,000 to 12,000 molecules, just below the cited range for mammals. Unfortunately, no one has estimated the ACh content of a frog vesicle. After making every reasonable allowance for the unavoidable errors in all these measurements, one seems to be left with the conclusion that there is a real discrepancy between the estimates of ACh in a vesicle and ACh in a quantum. The discrepancy is awkward, but not necessarily fatal, for the vesicular-exocytosis hypothesis of ACh release [30]. (This hypothesis is further described in Chapter 8.)

Stimulation of adrenergic pathways releases the soluble proteins and ATP of the vesicles along with the norepinephrine. From this result it is naturally inferred that the release depends upon exocytosis. Similar experiments on cholinergic pathways have been performed but so far have not provided convincing evidence for exocytosis [30].

Some Suggested Release Mechanisms

The hypothesis that ACh release is *regenerative*, i.e., that released ACh, besides acting postsynaptically, also acts presynaptically to release more ACh, has not withstood critical testing [30]. Axonal and presynaptic ACh receptors do exist and have nicotinic properties, but they play little or no part in normal transmission. Moreover, activation of the presynaptic receptors appears to reduce, not increase, ACh release.

A role for *cyclic nucleotides* in ACh release has been postulated to account for the finding that both caffeine and dibutyryl cyclic AMP increase nerve-evoked ACh discharge at the neuromuscular junction. The numerous examples of mutual feedback between Ca transport and nucleotide-cyclase systems make such a mechanism plausible enough in principle, but there have been some negative findings [30], and a convincing case has not yet been made. This is further discussed in Chapter 12.

One of the *prostaglandins,* PGE$_1$, is known to be involved in the negative-feedback control of norepinephrine release (see Chap. 13), but it seems to have no significant action on ACh release [30]. The prostaglandins are more likely to be involved in ACh mobilization, the process by which ACh in the depots is made more readily available for release.

The evidence that *actomyosinlike complexes* (neurostenin) and *microfilaments* are specifically involved in ACh release is at present too fragmentary to require discussion. *Microtubules*, which certainly are important for axonal growth and transport, are probably unimportant for transmitter liberation per se. Nevertheless, it is possible that some process involving filamentous proteins and activation by Ca will be discovered to play an essential part in release.

REMOVAL OF ACETYLCHOLINE

Cholinesterases

Acetylcholinesterase is one of a number of cholinesterases. It is the one that splits ACh most rapidly and is the only one, so far as we know, that is functionally important at cholinergic synapses. The other cholinesterases are collectively referred to as acylcholinesterases, or sometimes as pseudocholinesterase, and the best known of them is a plasma enzyme, the preferred substrate of which is butyrylcholine: Its physiological role is ill-defined.

Acetylcholinesterase: Location

AChE is manufactured in cholinergic cell bodies, and its intracellular location there has been verified by histochemistry at the ultrastructural level. It is delivered to the synapses by axonal transport, mainly at the slower rate (Chap. 21). Somewhere along its route it becomes bound to the outer face of the axolemma. A minor fraction of it, however, moves at the faster rate and is probably attached to some sort of intraaxonal particle. Koelle and his colleagues [32] noted that only part of the AChE of brain behaves as if it were accessible to quaternary substrates and inhibitors. They supposed that this fraction of the enzyme ("functional AChE") is outward-facing and the remainder ("reserve AChE") is inward-facing and in transit. The ability of ganglia and muscles to form surplus ACh [25, 26] in the presence of an anticholinesterase suggests that at least a small part of the enzyme transported by peripheral axons is still in the "reserve" orientation as it nears the synapses; cerebral cortex, which apparently lacks that ability, may possess only "functional" AChE.

As has been noted, much of the body's AChE does not originate from neurons. In skeletal muscle, even the junctional AChE is largely postsynaptic and is synthesized by muscle rather than by nerve cells.

Acetylcholinesterase: Properties and Kinetics

Though AChE, as has been pointed out, is a membrane-bound enzyme, it can be made soluble with the aid of detergents or polar organic solvents and has been isolated in crystalline form from electric tissue. As so prepared, it has a molecular weight of 230,000 and appears to be composed of four subunits, each of which has a center that can bind and split ACh. Traditionally, the center has been described as incorporating "anionic" and "esteratic" sites, but recent descriptions, e.g. [33], are more complex. The so-called anionic site apparently can accommodate uncharged groups, and the esteratic site, where splitting occurs after attachment of the ACh carbonyl, incorporates a serine side chain and probably a histidine imidazole as well. The enzyme's substrates and inhibitors are varied chemically, but all of them combine with it in precisely the same way. The irreversible (or very slowly reversible) organophosphorus inhibitors, including the awesomely toxic "nerve gases," inactivate the enzyme by phosphorylating the serine hydroxyl; most of the other well-known inhibitors, including physostigmine (eserine) and neostigmine, are carbamates, and carbamylate the same group, but reversibly. Kinetic studies of the enzyme indicate a three-step reaction with ACh (Fig. 9-2). The final deacetylation is probably the rate-controlling step.

Acetylcholinesterase: Physiological Significance

The role of AChE at synapses can be studied by applying a drug such as physostigmine, the principal action of which is to inhibit the enzyme. The lifetime of

Figure 9-2
Reaction mechanism of acetylcholinesterase.

synaptically released ACh is then prolonged. Ongoing muscarinic actions are intensified. Ongoing nicotinic actions are also intensified initially, but nicotinic blockade may follow if the frequency of synaptic activation is high. The reason that the latter effect is not seen at lower activation frequencies is that simple diffusion can remove released ACh from the synaptic area within milliseconds. As was pointed out earlier, ACh, unlike norepinephrine and probably unlike most other transmitters, is not recaptured to any extent by the nerve endings that have released it; only the choline derived from its breakdown can be taken up and reused. Therefore AChE is essential for inactivation of ACh effects on the receptor.

Anticholinesterase poisoning, if not too severe, can be relieved by treatment with atropine, supplemented by a suitable oxime ($R \cdot CH \cdot NOH$), if the poisoning is the result of a long-acting organophosphorus inhibitor (Chap. 33). The oxime combines chemically with the phosphorus atom of inhibitor and so reactivates the enzyme. The outstanding example is pralidoxime (N-methylpyridinium-2-aldoxime chloride). Pharmacology texts give further information.

TURNOVER OF ACETYLCHOLINE

Resting Turnover
Two kinds of observation show that synaptic ACh is being synthesized and destroyed even in the absence of nerve impulses. The ACh store increases in size (surplus ACh formation) when the extracellular medium contains a penetrating anticholinesterase, and the store becomes labeled when the medium contains radioactive choline. Reported ACh turnover rates, measured in either of these ways, are in the range 0.5% to 1.5% per minute in ganglia, muscles, and salivary glands [30]. Most of this resting turnover is intracellular, at least in ganglia; no more than 10% to 20% of it can be accounted for by the ACh that escapes into the medium. ACh turnover at central synapses during rest appears to be faster than at these peripheral synapses. Rates of about 5% per minute can be calculated from measurements by the labeling method on anesthetized whole brain in situ [14] and on cortical slices treated with tetrodotoxin to eliminate nerve impulses. In brain, however, the ACh that is released probably accounts for most of the turnover, for brain tissue is less able to form surplus ACh than are peripheral tissues. Relatively high rates of resting ACh release and turnover also have been reported for neuroeffector synapses in intestine [30].

Turnover During Activity
The experimental data from ganglia [17] and skeletal muscles [25] are in good agreement. In these tissues, synaptic activity accelerates both synthesis and release of ACh,

and the increase in both is frequency-dependent up to a maximum, which is reached at about 16 Hz in the case of plasma-perfused ganglia subjected to prolonged stimulation. When driven at this frequency, either set of synapses can maintain an ACh turnover rate of 7% to 10% per minute, whereas their stored ACh remains at about the resting level.

These rates are impressive, but those for mammalian brain in situ are an order of magnitude higher. Several laboratories [34] have now reported mean turnover times of only 1 to 2 minutes for part or all of the brain ACh. The fastest of these turnover rates is about 25% of the rate at which the ChAc present could synthesize ACh under optimal conditions. There is some reason to think that an even higher rate might be reached in animals subjected to stress. The effects of anesthetics and other drugs suggest that ACh turnover in brain, as in peripheral tissues, depends on the frequency of synaptic activation, but no quantitative relationship has been established.

Pools, Compartments, and Mobilization

The preceding paragraphs have dealt with synaptic ACh turnover under steady-state conditions, but often such conditions do not exist. The commonest reason for departure from steady-state conditions is a change in the frequency of synaptic activation. When this occurs, the amounts of ACh released by successive nerve impulses are determined by the interaction of several processes that are poorly understood at present. Some of these processes increase and some decrease the release. According to the quantum hypothesis, which assumes that there is a store of preformed quanta in the nerve ending, all the processes influence one or both of two parameters: (1) the number of quanta that at a given moment are available for release, because they are in a particular state or location, and (2) the probability that any of these quanta will be released by the next impulse.

As to (1), many neurochemical and neurophysiological analyses support the idea that all the nerve-ending ACh is not equally releasable but is distributed between two or more *pools*. In the simplest model [17], there is a smaller "readily available" pool and a large "reserve" pool. The process by which the first pool is replenished from the second has been called *mobilization,* and there is no doubt that it is often rate-limiting for release. In an effort to accommodate all the experimental data, subdivisions of the two pools have sometimes been postulated. One further modification is certainly required: Experiments on several tissues whose ACh stores were partly replaced by radioactive ACh showed clearly that newly synthesized ACh is preferentially released [30]. Apparently the readily available pool is replenished largely by new synthesis, not just by mobilization of old ACh from the reserve pool.

As to (2), studies of peripheral synapses have shown that the major factor determining the probability of release is undoubtedly the level of Ca in the nerve ending [35]. This Ca includes both the Ca that enters during a given impulse and the Ca that entered previously and became trapped reversibly. The exact locations of the active and the trapped Ca are unknown, as was stated earlier.

It would be a forward step if the notion of pools could be discarded and, instead, ACh metabolism could be described in terms of morphologically identifiable *compartments.* At present, however, this would be a step into the dark or, at best, into an area of dimly lit pitfalls. The readily releasable pool has been plausibly conceived as representing the ACh in those vesicles that are close to the presynaptic release sites, or perhaps actually adherent to them. There is some morphological evidence for this concept, but so far it is weak, and alternative or modified schemes have been proposed. Reference has already been made to the hypotheses that there are two vesicle populations of different lability [22] or degree of filling [7], and two kinds of vesicular ACh, internal and surface-bound [24]. Both hypotheses deserve further examination. Recently the possibility that the cytosol is the immediate source of released ACh has had renewed attention [36]. On that hypothesis, the ACh quanta are not preformed but are generated only at the moment of their release, as free ACh leaks out through briefly opened membrane gates.

Rate-Limiting Factors

ACh synthesis in nerve endings depends on the availability of choline, glucose (or pyruvate), and O_2 from outside sources. Extracellular Na is also a requisite, supporting both choline uptake and ACh storage. In vivo, the levels of all four materials are well maintained by efficient homeostatic mechanisms. There is no evidence that the extracellular level of any of them is limiting for ACh synthesis, unless a disease process or a poison (e.g., HC-3) is present. Normally, synthesis is coupled to release, by mechanisms that are poorly understood, so that ACh depots tend to remain well stocked during increased activity. This coupling is highly effective in peripheral nerve endings, at least with activation frequencies up to about 20 Hz. In central synapses, where ACh turnover is faster, the coupling is much less effective, and ACh stores, at least in cortex, are always in a state of partial depletion, except perhaps during sleep. The higher the frequency of synaptic activation, the greater the depletion. The rate-limiting factor for ACh synthesis in brain is possibly the supply of acetyl-CoA, but more probably it is the fact that the coupling process, whatever it may be, is not instantaneous.

   ACh mobilization is the limiting process for ACh release at peripheral synapses that are being activated at high frequency. Under these conditions, ACh output per unit of time cannot be maintained at its initial maximum but declines to a steady level that is independent of the frequency [17]. The decline is not the result of failure of ACh synthesis, because there is little change in the level of stored ACh and there is no reason to suppose that release has become less efficient. By exclusion, release must be determined by the rate of mobilization, and neurochemical studies have revealed only that in ganglia it is promoted by an unidentified factor present in normal plasma [17]. Little is known about ACh mobilization at central synapses.

   ACh release at physiological rates of activation is controlled by Ca influx, as has been stressed. Some secondary factors that promote release may be mentioned: CO_2, which may increase intracellular Ca ionization; intraterminal Na, which competes with Ca for binding sites; presynaptic hyperpolarization, which results from repetitive activation and augments the nerve-ending spike; and numerous drugs, acting by a variety of mechanisms [30]. As yet, there is no biochemical explanation for the extraordinary ability of botulinum toxin (and, to a lesser degree, certain other toxic proteins made by bacteria and animal venom glands) to block ACh release irreversibly, while doing nothing to interfere with the functioning of noncholinergic synapses.

As indicated earlier (see also [38]), the major problem in deciding among models for ACh turnover is how to translate data about ACh pools into morphological terms. Three recent reviews [35, 37, 38] present in diagrammatic form the divergent concepts of their authors. In this writer's view, the original vesicular-exocytosis hypothesis of ACh release is as plausible, and as far from being proved, as it was 20 years ago.

REFERENCES

*1. Pauling, P. The Shapes of Cholinergic Molecules. In Waser, P. G. (Ed.), *Cholinergic Mechanisms.* New York: Raven, 1974

*2. Green, J. P., Johnson, C. L., and Kang, S. Application of quantum chemistry to drugs and their interactions. *Annu. Rev. Pharmacol.* 14:319, 1974.

*3. Whittaker, V. P. Identification of Acetylcholine and Related Choline Esters of Biological Origin. In Koelle, G. B. (Ed.), *Handbuch der experimentellen Pharmakologie: Cholinesterases and Anticholinesterase Agents.* Berlin: Springer-Verlag, 1963.

*3a. Hanin, I. (Ed.). *Choline and Acetylcholine: Handbook of Chemical Assay Methods.* New York: Raven, 1974.

*4. Fonnum, F. Molecular Aspects of Compartmentation of Choline Acetyltransferase. In Balazs, R., and Cremer, J. E. (Eds.), *Molecular Compartmentation in the Brain.* New York: Wiley, 1973.

*5. Potter, L. T. Acetylcholine, Choline Acetyltransferase and Acetylcholinesterase. In Lajtha, A. (Ed.), *Handbook of Neurochemistry.* Vol. 4. New York: Plenum, 1970.

6. Yamamura, H. I., and Snyder, S. H. Affinity transport of choline into synaptosomes of rat brain. *J. Neurochem.* 21:1355, 1973.

7. Dowdall, M. J., and Zimmermann, H. Evidence for heterogeneous pools of acetylcholine in isolated synaptic vesicles. *Brain Res.* 71:160, 1974.

*8. Rang, H. P. Acetylcholine receptors. *Q. Rev. Biophys.* 7:283, 1974.

*8a. 1975 Myasthenia Gravis Symposium. *Ann. N. Y. Acad. Sci.* In press, 1976.

*9. Krnjević, K. Chemical nature of synaptic transmission in vertebrates. *Physiol. Rev.* 54:418, 1974.

10. Katz, B., and Miledi, R. The statistical nature of the acetylcholine potential and its molecular components. *J. Physiol.* (Lond.) 224:665, 1972.

11. Gill, E. W., and Rang, H. P. An alkylating derivative of benzilylcholine with specific and long-lasting parasympatholytic activity. *Mol. Pharmacol.* 2:284, 1966.

*12. Hebb, C. Biosynthesis of acetylcholine in nervous tissue. *Physiol. Rev.* 52:918, 1972.

*13. Potter, L. T. Synthesis, Storage and Release of Acetylcholine from Nerve Terminals. In Bourne, G. H. (Ed.), *The Structure and Function of Nervous Tissue.* Vol. IV. New York: Academic, 1972.

*14. Sparf, B. On the turnover of acetylcholine in brain. *Acta Physiol. Scand.* [Suppl. 397], 1973.

15. Eade, I., Hebb, C., and Mann, S. P. Free choline levels in the rat brain. *J. Neurochem.* 20:1499, 1973.

*Asterisks denote key references.

16. Illingworth, D. R., and Portman, D. W. The uptake and metabolism of plasma lysophosphatidylcholine in vivo by the brain of squirrel monkeys. *Biochem. J.* 130:557, 1972.

17. Birks, R., and MacIntosh, F. C. Acetylcholine metabolism of a sympathetic ganglion. *Can. J. Biochem. Pharmacol.* 39:787, 1961.

18. Collier, B., and Katz, H. S. Acetylcholine synthesis from recaptured choline by a sympathetic ganglion. *J. Physiol.* (Lond.) 238:639, 1974.

19. Tucek, S., and Cheng, S. C. Provenance of acetylcholine and compartmentation of acetyl-CoA and Krebs cycle intermediates in the brain in vivo. *J. Neurochem.* 22:893, 1974.

*20. Bowman, W. C., and Marshall, I. G. Inhibitors of Acetylcholine Synthesis. In Cheymol, J. (Ed.), *International Encyclopedia of Pharmacology and Therapeutics, Section 14, Vol. I: Neuromuscular Blocking and Stimulating Agents.* Oxford: Pergamon, 1972.

21. Katz, H. S., Salehmoghaddam, S., and Collier, B. The accumulation of radioactive acetylcholine by a sympathetic ganglion and by brain: Failure to label endogenous stores. *J. Neurochem.* 20:569 (1973).

*22. Barker, L. A., Dowdall, M. H., Essman, W. B., and Whittaker, V. P. The Compartmentation of Acetylcholine in Cholinergic Nerve Terminals. In Heilbronn, E., and Winter, A. (Eds.), *Drugs and Cholinergic Mechanisms in the CNS.* Stockholm: Försvarets Forskningsanstalt, 1970.

23. Wilson, W. S., Schulz, R. A., and Cooper, J. R. The isolation of cholinergic synaptic vesicles from bovine superior cervical ganglion and estimation of their acetylcholine content. *J. Neurochem.* 20:659, 1973.

*24. Marchbanks, R. M. Problems Concerning the Compartmentation of Acetylcholine in the Synaptic Region. In Balazs, R., and Cremer, J. E. (Eds.), *Molecular Compartmentation in the Brain.* New York: Wiley, 1973.

25. Potter, L. T. Synthesis, storage and release of ^{14}C acetylcholine in isolated rat diaphragm muscles. *J. Physiol.* (Lond.) 206:145, 1970.

26. Collier, B., and Katz, H. S. The synthesis, turnover and release of surplus acetylcholine in a sympathetic ganglion. *Br. J. Pharmacol.* 39:428, 1970.

27. Evans, C. A. L., and Saunders, N. R. An outflow of acetylcholine from normal and regenerating ventral roots of the cat. *J. Physiol.* (Lond.) 240:15, 1974.

28. Politoff, A. L., Rose, S., and Pappas, G. D. The calcium-binding sites of synaptic vesicles of the frog neuromuscular junction. *J. Cell Biol.* 61:818, 1974.

29. Elmqvist, D., and Quastel, D. M. J. Presynaptic action of hemicholinium at the neuromuscular junction. *J. Physiol.* (Lond.) 177:463, 1965.

*30. MacIntosh, F. C., and Collier, B. The Neurochemistry of Cholinergic Terminals. In Zaimis, E., and MacLagan, J. (Eds.), *Handbook of Experimental Pharmacology: Organization, Function and Pharmacology of the Neuromuscular Junction.* Berlin: Springer-Verlag, 1976.

31. Nishi, S., Noeda, H., and Koketsu, K. Release of acetylcholine from sympathetic preganglionic nerve terminals. *J. Neurophysiol.* 30:114, 1967.

*32. Koelle, G. B. Cytological Distributions and Physiological Functions of Cholinesterases. In Koelle, G. B. (Ed.), *Handbuch der experimentellen Pharmakologie: Cholinesterases and Anticholinesterase Agents.* Berlin: Springer-Verlag, 1963.

33. O'Brien, R. D. The Design of Organophosphate and Carbamate Inhibitors of Cholinesterases. In Ariens, E. J. (Ed.), *Drug Design.* New York: Academic, 1971.

34. Jenden, D. J., Choi, L., Silverman, R. W., Steinborn, J. A., Roch, M., and Booth, R. A. Acetylcholine turnover estimation in brain by gas chromatography/mass spectrometry. *Life Sci.* 14:55, 1974.

*35. Hubbard, J. I. Neuromuscular Transmission: Presynaptic Factors. In Hubbard, J. I. (Ed.), *The Peripheral Nervous System.* New York: Plenum, 1974.

36. Birks, R I. The relationship of transmitter release to fine structure in a sympathetic ganglion. *J. Neurocytol.* 3:133, 1974.

*37. Fonnum, F. Review of Recent Progress in the Synthesis, Storage and Release of Acetylcholine. In Waser, P. G. (Ed.), *Cholinergic Mechanisms.* New York: Raven, 1975.

*38. Dunant, Y., and Israel, M. Acetylcholine Turnover in the Course of Stimulation. In Waser, P. G. (Ed.), *Cholinergic Mechanisms.* New York: Raven, 1975.

Chapter 10

Catecholamines, Serotonin, and Histamine

Solomon H. Snyder

The catecholamines, norepinephrine and dopamine, as well as the indoleamine, serotonin, are putative neurotransmitters in certain neuronal tracts in the brain. They occupy a uniquely important place in neurobiology because they are the only putative neurotransmitters whose localization in particular brain tracts has been established and whose relationship to specific animal or human behaviors has, at least in part, been worked out. Hence, a discussion of the neurochemistry of these compounds can integrate findings derived from electron microscopy, histochemistry, neurophysiology, enzymology, pharmacology, psychology, and even clinical psychiatry.

A great deal of information about catecholamines and serotonin had accumulated prior to knowledge of which neuronal tracts contained them or even whether they were localized in neurons at all. Nonetheless, for clarity's sake, it might be best to begin by presenting the histochemical observations that have delineated the neuronal systems which contain these compounds in the brain.

HISTOCHEMICAL MAPPING OF MONOAMINE NEURONS IN THE BRAIN

The histochemical fluorescence method for the identification of catecholamines and serotonin, developed by a group of Swedish workers [1, 2, 3], incorporates the condensation of these amines in tissue sections with formaldehyde in a humid environment to form intensely fluorescent isoquinolines. In this way, cell bodies, axons, and nerve terminals of the monoamine-containing neurons have been mapped throughout the brain. Then, by making selective brain lesions and following the anterograde loss or retrograde build-up of amine, the course of the tracts has been clarified.

In the histochemical fluorescence method, serotonin can be distinguished from catecholamines by the wavelength of fluorescence, which varies in such a way that serotonin appears bright yellow and the catecholamines appear bright green. Although both dopamine and norepinephrine fluoresce green, they can be differentiated by their response to drugs.

What are the pathways of the monoamine neuronal systems? The cell bodies of all of them are in one or another location within the brain stem. Axons ascend or descend throughout the brain and into the spinal cord. For each of the amines there are several separate and distinct tracts (Fig. 10-1).

Norepinephrine

The cell bodies of the norepinephrinergic neurons occur in ten or more clusters in the medulla oblongata, pons, and midbrain, mostly scattered throughout the reticular formation. In the locus coeruleus, however, norepinephrinergic cell bodies are so

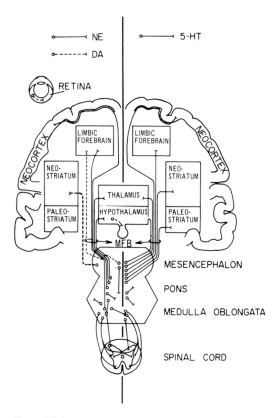

Figure 10-1
Pathways of the monoamine tracts in the brain. (From Anden et al. [3].)

densely packed that they may be the only ones in that nucleus. Axons from some of the norepinephrinergic cells in the medulla oblongata descend in the lateral sympathetic columns of the spinal cord and terminate at various levels. Others ascend primarily in the medial forebrain bundle, giving off terminals in all areas of the brain; the highest density is in the hypothalamus, a lesser density is in the limbic system, and the lowest frequency is in the cerebral cortex and cerebellum.

There are two major groups of norepinephrine pathways, the dorsal and ventral bundles, respectively [1]. Cell bodies for the dorsal pathway are localized to the locus coeruleus with terminals in the cerebellum, hippocampus, and cerebral cortex. The ventral bundle derives from less circumscribed cell groups in the brain stem and projects primarily to the hypothalamus and some limbic structures. Thus a discrete lesion of the locus coeruleus depletes cerebral cortical norepinephrine without altering hypothalamic levels.

Serotonin
The cell bodies of the serotonin-containing neurons are localized in a series of nuclei in the lower midbrain and upper pons that are called the *raphe* nuclei. These are a

group of old structures, phylogenetically, whose function, until the discovery that they contained serotonin, was obscure. As with norepinephrine and the locus coeruleus, serotonin is so highly concentrated in the raphe nuclei that probably all of its cells are serotoninergic. Serotoninergic axons, like those containing norepinephrine, ascend primarily in the medial forebrain bundle and also give off terminals in all brain regions, with the major proportion in the hypothalamus and the least in the cerebellum and cerebral cortex. Despite the low levels of serotonin in the cerebral cortex, terminals are so uniformly distributed that almost all cells probably receive serotoninergic innervation.

Dopamine

Dopaminergic neuronal tracts are more circumscribed than those of norepinephrine or serotonin. The largest dopaminergic tract originates in the zona compacta of the substantia nigra (stereotaxic designation A9) with terminals in the caudate nucleus and putamen of the corpus striatum. In Parkinson's disease, this nigrostriatal tract degenerates, with an attendant depletion of brain dopamine. That the dopamine deficiency accounts for the symptoms of this disease is attested to by the therapeutic efficacy of L-DOPA, the amino acid precursor of dopamine, in treating Parkinson's disease (see also Chap. 28).

Another dopamine tract arises from cell bodies in the midbrain near the interpeduncular nucleus (A10) and terminates principally in the nucleus accumbens of the septal region, olfactory tubercle, and the central nucleus of the amygdala. Yet a third dopamine system begins in the arcuate and anterior periventricular nuclei of the hypothalamus and passes downward to terminate in the primary capillary plexus that supplies the pituitary gland. It is via this capillary plexus that hypothalamic releasing factors reach the pituitary gland to regulate discharge of the anterior pituitary tropic hormones (see Chap. 22). Dopamine can stimulate the release of some of these hormones. The retina also contains isolated dopamine fibers, the function of which is obscure.

Biochemical studies [4] and modifications of the histochemical technique [5] have recently permitted identification of a prominent dopamine innervation of the frontal and cingulate areas of the cerebral cortex, not depicted in Figure 10-1. Cell bodies are closely allied to the A9 and A10 groups, which project to the corpus striatum and limbic system, respectively, so that it has not yet been possible to produce lesions in these pathways selectively.

CATECHOLAMINES

Biosynthesis

While histochemical studies have delineated the tracts containing biogenic amines, investigation of their metabolism and turnover and the effects of drugs has contributed greatly to an understanding of their function.

Tyrosine is the dietary amino acid precursor of dopamine and norepinephrine (Fig. 10-2). The rate-limiting step in the biosynthesis of both these catecholamines

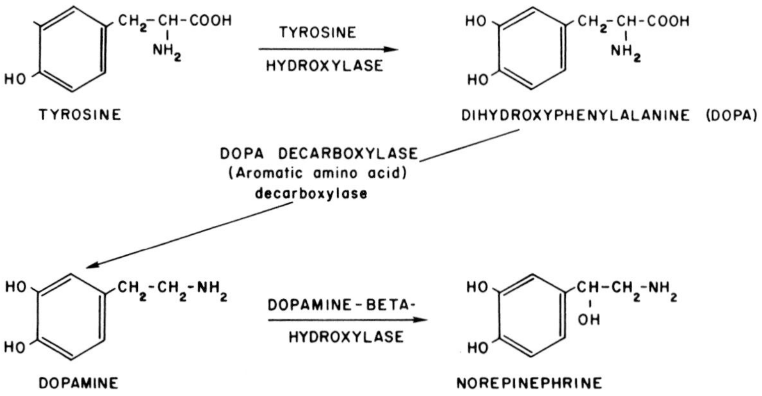

Figure 10-2
Pathways of catecholamine synthesis. (From Snyder [22].)

is the hydroxylation of tyrosine to form DOPA by the enzyme tyrosine hydroxylase. The activity of this enzyme is relatively feeble, so it can be detected only radioiso-topically. It requires a biopterin as a cofactor; biopterin is a pteridine compound that also enhances the activity of several other hydroxylating enzymes [6, 7].

Researchers have been eager to define the localization of tyrosine hydroxylase within neurons. For instance, if this enzyme, whose activity determines the rate of catecholamine synthesis, were primarily confined to neuronal cell bodies, it would support the concept that catecholamines are primarily synthesized there and then transported down the axon to the nerve terminal. If the enzyme were mostly in the synaptic vesicles, one would conclude that those vesicles are factories for synthesizing, as well as for storing and releasing, the amines. Although these questions are not altogether resolved, it appears likely that, while some tyrosine hydroxylase can be located in cell bodies, the majority is confined to catecholamine nerve terminals. Within the nerve terminals, the enzyme is largely cytoplasmic, so that at least the initial step of catecholamine synthesis probably does not take place within the synaptic vesicles.

Several tyrosine analogs, such as α-methyltyrosine, can inhibit tyrosine hydroxylase. These drugs deplete dopamine and norepinephrine in the brain. When the synthesis of a neurotransmitter is inhibited, its levels can be expected to fall at a rate determined by the rate at which the neurons fire and by the subsequent catabolism of the transmitter molecule. Thus, the rate of decline of brain catecholamine in animals that have been treated with α-methyltyrosine provides a reflection of the turnover rate of catecholamines [8]. Catecholamine turnover measured in this way corresponds well with measurements of the rate of catecholamine synthesis from tyrosine, and the two types of measures are affected similarly by drugs or altered physiological conditions.

Biosynthesis and Neuronal Activity
Tyrosine hydroxylase activity also can be inhibited by the catecholamines, suggesting that normally they may exert a type of "feedback" inhibition [6]. This notion would

imply that catecholamine release produced by nerve stimulation or drugs would induce enhanced catecholamine synthesis (to replace that which was discharged) by stimulating the hydroxylation of tyrosine without actually increasing levels of the tyrosine hydroxylase itself. In support of this are findings that stimulation of peripheral sympathetic nerves (which seems to regulate norepinephrine metabolism as it does in the brain) produces a rapid acceleration of norepinephrine synthesis from tyrosine [9]. In such a situation, the formation of norepinephrine from administered DOPA is not enhanced, showing that hydroxylation of tyrosine was the "stimulated" step in the pathway. Cold stress, as well as a variety of drugs, increases tyrosine hydroxylation in the intact brain without increasing the activity of this enzyme when subsequently the tissues are assayed in vitro. If catecholaminergic nerve activity is enhanced for several days, the levels of tyrosine hydroxylase become elevated. Thus, after brief stresses, tyrosine hydroxylation can be stimulated by apparent release from feedback inhibition without an increase in enzyme levels; prolonged stimuli provoke the synthesis of new tyrosine hydroxylase molecules (see Chap. 16).

Variations in the activity of norepinephrine synapses alter the synthesis of this amine in a predictable way. For instance, drugs that block adrenergic receptors accelerate the formation of norepinephrine from tyrosine. Presumably a message from the postsynaptic neuron, saying "I am not getting enough norepinephrine; send me more," is transmitted via neuronal feedback to the presynaptic neurons, which then speed both their firing rates and the synthesis of norepinephrine. Drugs such as haloperidol, which block dopamine receptors, and those such as apomorphine, which stimulate the same receptors, respectively enhance and slow the synthesis of dopamine.

The mechanism whereby accelerated neuronal discharge enhances catecholamine synthesis involves the influx of calcium into the nerve terminal upon depolarization. Calcium increases the affinity of tyrosine hydroxylase for both tyrosine and biopterin [10, 11]. These drug interactions have implications for the pathophysiology of schizophrenia [12].

Formation of Dopamine and Norepinephrine

DOPA is decarboxylated to dopamine by an enzyme, DOPA decarboxylase. The same enzyme can decarboxylate 5-hydroxytryptophan, the amino acid precursor of serotonin, as well as several other aromatic amino acids. Accordingly, it has also been called "aromatic amino acid decarboxylase." Like tyrosine hydroxylase, DOPA decarboxylase seems to be contained in the soluble portion of catecholamine nerve terminals in the brain. The activity of DOPA decarboxylase is not rate limiting in the synthesis of the catecholamines and hence is not a regulating factor in their formation. Normally, there seems to be a great excess of DOPA decarboxylase present in the brain, so that drugs that inhibit its activity by as much as 95 percent fail to lower brain levels of the catecholamines.

The hydroxylation of dopamine at the beta carbon to form norepinephrine is mediated by the enzyme dopamine β-hydroxylase. This enzyme requires copper for its optimal functioning. The enzyme is difficult to assay in crude tissue preparations because of the presence of unidentified inhibitory substances. In the adrenal medulla, where it participates in the synthesis of epinephrine, dopamine β-hydroxylase is localized, at least in part, in the wall of the epinephrine-storing granules. It is not released by the adrenal gland at times that epinephrine is released, and this has been taken as evidence that the secretion of epinephrine involved exocytosis, a process by which the granule or storage vesicle fuses with the cell membrane, discharges its contents, and returns to the cell interior. One would expect that only soluble constituents of the vesicles are released during exocytosis, and that chemicals in the vesicle wall would stay inside. Although there is no direct evidence, it is reasonable to suppose that the process of catecholamine release in the brain might be similar to or the same as that in the adrenal gland.

As might be expected, dopamine β-hydroxylase is lacking in those parts of the brain, such as the caudate nucleus, in which dopamine is the predominant catecholamine and which have only negligible concentrations of norepinephrine.

Catecholamine Catabolism
Two enzymes are primarily responsible for the degradation of catecholamines. One of these is monoamine oxidase (MAO) (see also Chap. 32), which oxidatively deaminates dopamine or norepinephrine to the corresponding aldehydes (Fig. 10-3). These, in turn, can be converted by aldehyde dehydrogenase to analogous acids. The aldehydes may also be reduced to form alcohols. Dietary factors, such as the ingestion of

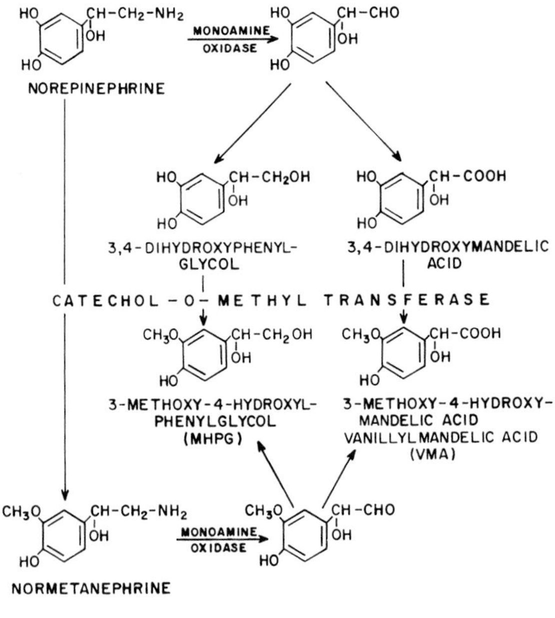

Figure 10-3
Pathways of catecholamine degradation.

ethanol, can determine the relative amounts of catechol acids or alcohols formed from the catecholamines because ethanol also competes for aldehyde dehydrogenase.

Catecholamines can be methylated by the enzyme catechol-O-methyl transferase (COMT), which transfers the methyl group of S-adenosylmethionine to the meta (3 position) hydroxyl of the catecholamines [13]. To a very limited extent, COMT can also methylate the para hydroxyl grouping. COMT will act on any catechol compound, including the aldehydes and acids formed from the action of MAO on the catecholamines. When norepinephrine is methylated by this enzyme, the product is called normetanephrine. There is no corresponding name for the methylated derivative of dopamine, which is simply referred to as 3-O-methyldopamine.

Like COMT, MAO is relatively nonspecific and will act on any monoamine, including normetanephrine and 3-O-methyldopamine, converting these first into their respective aldehydes and then into acids or alcohols. The formation of the acid is probably mediated by aldehyde oxidase and the alcohol by alcohol dehydrogenase. Thus an O-methylated alcohol or acid will result from the combined actions of MAO and COMT. Measurement of the levels in tissues or body fluids of O-methylated acids or alcohols, of course, will give no indication of which enzyme acted first: To find this out, one must measure the O-methylated amines or the catechol acids. Determining if released catecholamines were first acted on by MAO or by COMT conveys useful information about neuronal function. In the peripheral sympathetic nervous system, the extent of oxidation of the aldehyde product exceeds the reductive pathway so that the O-methylated acid product of norepinephrine degradation, called vanillylmandelic acid (VMA), is the final breakdown product of norepinephrine. Quantitatively, VMA is the major metabolite of norepinephrine in the periphery and is readily detectable in the urine, so its levels are measured in clinical laboratories as an index of sympathetic nervous function and to diagnose tumors that produce norepinephrine or epinephrine, such as pheochromocytomas and neuroblastomas.

In the brain, reduction of the aldehyde formed from the action of MAO on norepinephrine or normetanephrine predominates, so that the major norepinephrine metabolite in the brain is an alcohol derivative called 3-methoxy-4-hydroxylphenylglycol (MHPG). The MHPG formed in the brain is conjugated to sulfate. Because MHPG can diffuse from brain to the general circulation, estimates of its levels in the urine might be thought to reflect directly the activity of norepinephrinergic neurons in the brain. However, although MHPG is proportionately only a minor metabolite of norepinephrine in the peripheral sympathetic nervous system, quantitatively most of it in the urine still derives from the periphery and to only a minor extent (about 30 percent) from the brain.

Whether the aldehyde formed from dopamine or 3-O-methyl dopamine is primarily oxidized or reduced is not altogether certain. Most evidence suggests that it is largely oxidized to an acid, 4-hydroxy-3-methoxy-phenylacetic acid, more commonly known as homovanilic acid (HVA). HVA levels are often measured in the brain or cerebrospinal fluid, and those levels are taken as reflections of the activity of dopaminergic neurons, especially of those with endings in the caudate nucleus and putamen. HVA levels are markedly lower in the spinal fluid of parkinsonian patients (see also Chap. 32).

Epinephrine differs chemically from norepinephrine only in the addition of a methyl group to the amine nitrogen. It is the predominant catecholamine in the adrenal glands of most species. Only extremely small amounts of epinephrine are found in sympathetic nerves or in the brain. The synthesis of epinephrine occurs by the same steps as that of norepinephrine, with the addition of an *N*-methylating enzyme which transforms norepinephrine to epinephrine. That enzyme is designated phenyl-ethanolamine-*N*-methyl transferase (PNMT) [13]. PNMT is almost wholly confined to the adrenal gland, although small amounts are in the brain. The activity of this enzyme and, accordingly, the synthesis of epinephrine are regulated by hormones of the adrenal cortex. Thus, removal of the pituitary gland results in a profound lowering of adrenal PNMT, which can be restored to normal levels either by ACTH or by large doses of adrenal glucocorticoids. This provides a means whereby the adrenal cortex can regulate the functions of the adrenal medulla. Epinephrine is catabolized by the same enzymes that act on the other catecholamines, and its major end-metabolite is VMA.

The subcellular localization of MAO and COMT in brain can provide important information about how these enzymes function in vivo. MAO is present in the outer membrane of mitochondria in almost every tissue of the body and, in the brain, in glia as well as in neurons. However, the portion of MAO that is primarily concerned with the deamination of norepinephrine and dopamine is localized in mitochondria within the nerve terminals of the catecholaminergic neurons, where it deaminates the surplus catecholamines that leak out of synaptic vesicles [13]. Accordingly, levels of catechol acids, which arise when MAO is the first enzyme to act on these compounds, reflect catecholamines that leak out of vesicles within nerve terminals and are metabolized before leaving them. Hence, these catecholamines cannot reach the synaptic cleft or act on postsynaptic receptors.

There are several isozymes (distinct proteins that catalyze the same chemical reaction) of MAO in the brain. In the caudate nucleus, the area of the brain richest in dopamine, one of the MAO isozymes is extraordinarily active in deaminating dopamine, with much less effect on other monoamines. This suggests that different classes of aminergic neurons may have their own "tailor-made" MAO isozymes.

COMT also occurs in a wide variety of tissues. The COMT moiety that has first access to catecholamines in the brain, however, seems to be located largely outside the catecholamine nerve terminals. Hence, levels of normetanephrine or 3-*O*-methyl-dopamine reflect catecholamines that are released outside the nerve terminals into the synaptic cleft with access to postsynaptic receptors before being metabolized by COMT [13].

Reuptake Inactivation of Synaptically Released Catecholamines

After discharge at synapses, acetylcholine is inactivated via hydrolysis by the enzyme acetylcholinesterase. Two enzymes can degrade the catecholamines. Surprisingly, neither of them is the primary mode of catecholamine inactivation. Instead, synaptically discharged catecholamines are inactivated by reuptake into the nerve terminals that had released them (Fig. 10-4) [13]. Proof that this mechanism terminates the actions of released catecholamines was obtained largely from the peripheral sympathetic

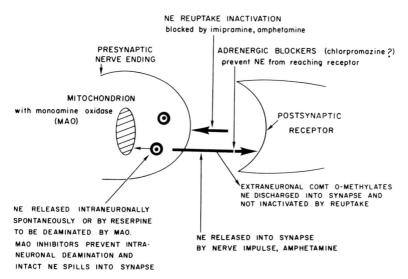

Figure 10-4
Model of postulated events at catecholamine synapses. (From Snyder [22].)

nervous system, but the process is probably the same in the brain. Reuptake mechanisms are described more fully in Chapter 8.

Catecholamine reuptake was discovered by Axelrod and his co-workers [13]. They found that when radioactive norepinephrine is injected intravenously, it accumulates primarily in tissues that have a dense sympathetic innervation. Cutting the sympathetic nerves to these organs abolishes the accumulation of the radioactive amine, so its uptake must be into sympathetic nerves. When these tissues are homogenized and centrifuged on sucrose density gradients, the accumulated norepinephrine is localized in a fraction of particles about the size of microsomes and with sedimentation properties similar to those of synaptic vesicles (see Chap. 2). Drugs that inhibit this uptake process increase the quantity of norepinephrine released after sympathetic nerves are stimulated and also potentiate the effects of sympathetic nerve stimulation. Inhibitors of MAO or of COMT do not potentiate the effects of sympathetic nerve stimulation or injected catecholamines.

There are no simple ways to measure the effects of nerve stimulation in the brain, so the physiological role of catecholamine reuptake cannot be readily assessed. However, brain tissue is useful for characterizing the uptake process itself. Catecholamine uptake in the brain can be measured after radioactive amine has been injected intraventricularly or if brain slices or even synaptosomes (pinched-off nerve terminals, see Chap. 2) are incubated with the catecholamine. The uptake process is energy dependent, since it is depressed at $0°C$ and by a variety of metabolic inhibitors. It is saturable, obeys Michaelis-Menten kinetics, and requires sodium and potassium, functioning optimally at the physiological concentrations of these ions. Catecholamine uptake is inhibited by ouabain, an inhibitor of (Na^+, K^+)-ATPase. Its relationship to the cell's sodium pump probably reflects a dependence of catecholamine uptake on the relative concentrations of sodium within and without the neuron. The norepinephrine-containing neurons in the brain can accumulate dopamine and the dopamine neurons can take up norepinephrine. Interestingly, while the norepinephrine nerve terminals have four times greater affinity for the physiologically occurring *l*-norepinephrine than for *d*-norepinephrine, the dopaminergic neurons of the caudate nucleus have equal affinities for the two isomers. Because dopamine is optically symmetrical and has no stereoisomers, perhaps its neuronal membrane has no physiological need to distinguish optical isomers of the catecholamines [14].

Catecholamine uptake can take place at the neuronal membrane of cell bodies, axons, and nerve terminals. It can be distinguished from the process whereby catecholamines are retained within synaptic vesicles, or storage granules inside the nerve terminal. The granular storage process is disrupted by the drug reserpine, which does not affect the neuronal membrane transport system [15]. Other drugs, such as the tricyclic antidepressant imipramine, are selective inhibitors of the neuronal membrane transport [15].

SEROTONIN

Metabolism

Tryptophan, the dietary amino acid precursor of serotonin, is hydroxylated by the enzyme tryptophan hydroxylase to form 5-hydroxytryptophan (Fig. 10-5). The enzymology of tryptophan hydroxylase is not worked out as well as is that of tyrosine hydroxylase, and it is not yet clear whether tryptophan hydroxylase is "inducible" when the activity of nerves that contain serotonin is altered. There are some conditions under which the conversion of tryptophan to serotonin in the intact animal is changed, but whether this is accompanied by variations in tryptophan hydroxylase activity is not known. *d*-Lysergic acid diethylamide (LSD) slows the turnover rate of serotonin by retarding its formation from tryptophan [16]. When iontophoresed onto raphe cells, LSD decreases the firing of cells by acting on the serotonin receptors

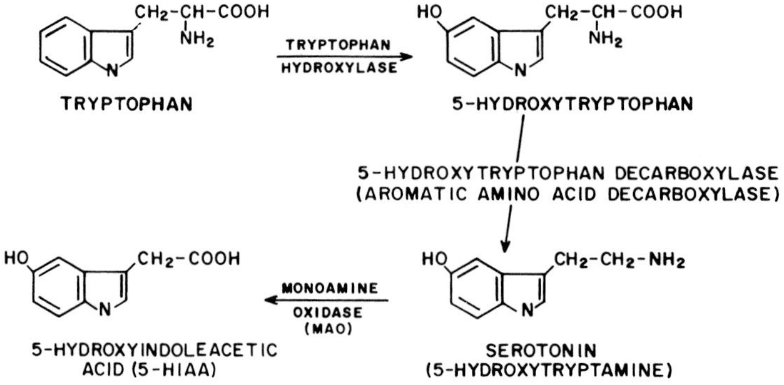

Figure 10-5
Metabolism of serotonin.

on cell body membranes and nerve terminals. After its synaptic release, serotonin normally acts on these presynaptic receptors to inhibit further release. The specificity of pre- and postsynaptic receptors differs, as LSD is potent at pre- but not at post-synaptic receptors, whereas the psychedelically inactive 2-bromo-LSD is active at post- but not at presynaptic receptors [17, 18]. Presumably the slowed discharge rate of the neuron results in a decrease in the synthesis of new serotonin, as there is less need for the synthetic machinery to keep up with the loss of serotonin by neuronal discharge.

5-Hydroxytryptophan is decarboxylated to serotonin by 5-hydroxytryptophan decarboxylase, an enzyme whose range of substrate preference is the same as DOPA decarboxylase, so that it also can be termed an aromatic amino acid decarboxylase. Serotonin itself is metabolized by only one enzyme, MAO (Fig. 10-5).

Serotonin Transport Into Neurons
Serotonin is not established as a neurotransmitter in any major system of peripheral nerves, so there is no simple way to determine what mechanism accounts for its synap-tic inactivation. As with norepinephrine, serotonin can be accumulated by serotoniner-gic neurons via a highly specific uptake system operating at the level of the neuronal membrane. Like the catecholamine process, the serotonin uptake system is energy dependent, requires sodium, is saturable, obeys Michaelis-Menten kinetics, and can be demonstrated both by intraventricular injection of radioactive serotonin and by incu-bation of the radioactive amine with brain slices or synaptosomes. Serotonin also appears to be stored in vesicles within the nerve endings by a process which can be disrupted by reserpine. Although both the serotonin and the catecholamine uptake systems can be inhibited by some of the same drugs, especially the tricyclic antidepres-sants, there are differences in the relative affinities of various antidepressants for the neurons that contain serotonin and catecholamine. It is commonly thought that the mechanism of action of the tricyclic antidepressants (see Chap. 34) involves inhibition of norepinephrine uptake, with resultant potentiation of amine released into the synaptic cleft. These drugs might also owe their therapeutic efficacy to a similar action on the neuronal uptake of serotonin. Thus, by analogy, a good case can be made for a major role of neuronal reuptake in terminating the effects of synaptically released serotonin in the brain.

The notion that neurotransmitter actions can be terminated by reuptake into the nerve terminals that had released them was first thought to be unique for the catechol-amines. As other putative neurotransmitters have been investigated, it begins to appear that reuptake may be a universal means of terminating transmitter action, and that enzymatic degradation, as with acetylcholine, may be the exception rather than the rule. Besides norepinephrine, dopamine, and serotonin, there is strong evidence for a neuronal uptake system that accumulates γ-aminobutyric acid (GABA) and glycine. (See also Chap. 8 for discussion on transmitter uptake.)

Physiological Role of Serotonin in the Brain
Histochemical identification of the raphe nuclei as a repository of cell bodies of serotoninergic neurons opened the way for studies that have elucidated, in part,

the function of these neurons. Jouvet, the French neurophysiologist, destroyed the raphe nuclei of cats and examined their subsequent sleeping patterns [19]. Serotoninergic neurons degenerate when their cell bodies are destroyed, so serotonin levels in the brain are depleted proportionally to the extent of the lesion. With 90 percent destruction of the raphe nuclei, cats became totally insomniac. Jouvet then administered p-chlorophenylalanine to cats. This drug profoundly inhibits tryptophan hydroxylase activity in intact animals, although in test-tube systems it is only a weak inhibitor of the enzyme. After drug treatment, the serotonin content of the brain is depleted, and in Jouvet's experiments the cats became markedly insomniac. Of course, p-chlorophenylalanine, an amino acid analog, might have many effects besides inhibition of serotonin synthesis.

To establish the specificity of its action, Jouvet administered 5-hydroxytryptophan, which can be converted into serotonin even when tryptophan hydroxylase is inhibited because it bypasses the enzymatic block. Treatment with 5-hydroxytryptophan reversed the p-chlorophenylalanine-induced insomnia and put the cats to sleep. This pharmacological paradigm has been repeated in other animal species.

Serotonin and Its Metabolites in the Pineal Gland

The pineal gland, although situated in the middle of the head, does not appear to be part of the brain in most species. In the rat, in which the best neuroanatomical studies have been performed, the pineal gland is attached to the brain only by some connective tissue and receives its major or sole innervation from postganglionic sympathetic nerve fibers, the cell bodies of which originate in the superior cervical ganglion [20]. The pineal gland contains the highest serotonin concentration of any known animal tissue. In the rat, serotonin concentration is 200 times greater in the pineal gland than in the brain. The levels of serotonin in the pineal gland undergo marked diurnal fluctuations; at noon, levels are ten times higher than they are at midnight. This biological rhythm persists in the absence of light-dark changes in illumination and in blinded animals. It is not synchronized within the pineal gland but receives regulatory information from the central nervous system via the sympathetic nerves. Almost every chemical that has been examined in the pineal gland exhibits rhythmicity, although sometimes with altered phases.

The pineal gland also has unique pathways for the metabolism of serotonin (Fig. 10-6). The aldehyde formed from the action of MAO can be oxidized to 5-hydroxyindoleacetic acid or reduced to 5-hydroxytryptophol. 5-Hydroxytryptophol can then be methylated on the 5-hydroxyl group of the indole ring to form 5-methoxytryptophol. This reaction is mediated by an enzyme, hydroxyindole-O-methyl transferase (HIOMT), which is found only in the pineal gland. Serotonin can be acetylated by N-acetyl transferase to form N-acetylserotonin, which can be acted on by HIOMT to form a compound known as melatonin. Both melatonin and 5-methoxytryptophol are produced solely in the pineal gland.

These pineal-specific metabolic pathways may have important physiological significance [20]. Melatonin and 5-methoxytryptophol are potent inhibitors of gonadal activity. The influences of light on gonadal activity are paralleled by effects on the biosynthesis of these two inhibitors. In addition, surgical denervation of the pineal

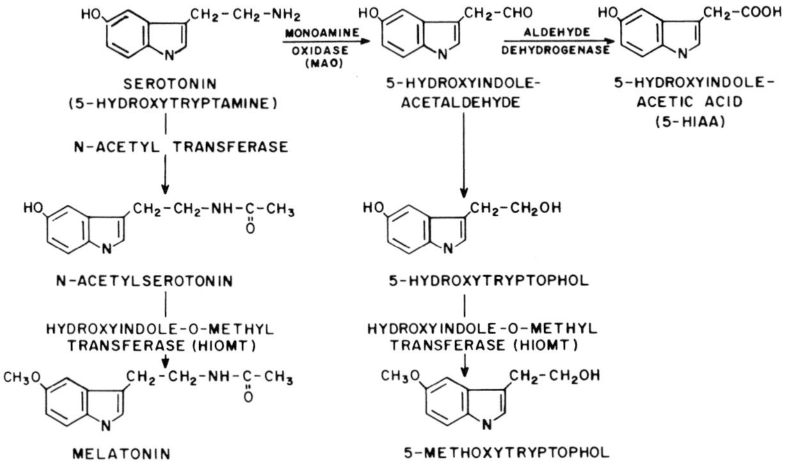

Figure 10-6
Unique pathways of serotonin metabolism in the pineal gland.

gland abolishes the effects of light on the gonads. Melatonin synthesis is regulated by the activity of the *N*-acetyltransferase, which acts upon serotonin. This enzyme is dramatically inducible by β-adrenergic stimulation, which is, therefore, the mechanism that mediates the effects of light [21].

HISTAMINE

The evidence for histamine as a central neurotransmitter is not as impressive as it is for the other biogenic amines. Essentially no synaptic neurophysiology shows specific actions of histamine that mimic a natural transmitter and are blocked by drugs that impair natural transmission. Moreover, no histochemical technique is available to visualize brain histamine. Nonetheless, an abundance of biochemical information supports the role of histamine as a central transmitter [22, 23].

Histamine is distributed in a nonuniform fashion throughout the brain, with highest levels in the hypothalamus. There are marked variations in the histamine concentration among different nuclei within the hypothalamus. The turnover rate of histamine is more rapid than for any other biogenic amine in the brain except, perhaps, acetylcholine. Histamine turnover is accelerated by stressful stimuli. Histamine biosynthesis can be regulated by drugs that inhibit the activity of its synthetic enzyme, histidine decarboxylase. These drugs partially deplete brain levels of histamine by interfering with its formation. Depolarization of brain slices results in a calcium-dependent release of endogenous histamine. In subcellular fractionation experiments, histamine is localized to pinched-off nerve terminals (synaptosomes), and, when these are lysed by hypotonic shock, a substantial amount of histamine is contained in those fractions enriched with synaptic vesicles.

In sum, histamine satisfies several criteria for a neurotransmitter. It is localized within nerve terminals and possibly in synaptic vesicles. It is released selectively upon neuronal depolarization. Its heterogeneous distribution throughout the brain is consistent with unique histaminergic neuronal pathways. There is even evidence to suggest the pathways of histamine neurons. Lesions of the medial forebrain bundle elicit a marked reduction of histidine decarboxylase activity in the forebrain. This suggests that axons of histaminergic neurons, like those for other aminergic neurons, ascend in the medial forebrain bundle.

ACKNOWLEDGMENTS

The author is a recipient of National Institute of Mental Health Research Scientist Development Award, K3-MH-33128.

REFERENCES

*1. Ungerstedt, U. Stereotaxic mapping of the monamine pathways in the rat brain. *Acta Physiol. Scand.* 367 [Suppl.]: 1, 1971.
2. Hillarp, N. A., Fuxe, K., and Dahlström, A. Demonstration and mapping of central neurons containing dopamine, noradrenaline and 5-hydroxytryptamine and their reactions to psychopharmaca. *Pharmacol. Rev.* 18:727, 1966.
3. Anden, N. E., Dahlström, A., Fuxe, K., Larsson, K., Olson, L., and Ungerstedt, U. Ascending monoamine neurons to the telencephalon and diencephalon. *Acta Physiol. Scand.* 67:313, 1966.
4. Thierry, A. M., Stinus, L., Blanc, G., and Glowinski, J. Some evidence for the existence of dopaminergic neurons in the rat cortex. *Brain Res.* 50:230, 1973.
5. Lindvall, O., Bjorklund, A., Moore, R. Y., and Stenevi, U. Mesencephalic dopamine neurons projecting to neocortex. *Brain Res.* 81:325, 1974.
6. Nagatsu, T., Levitt, M., and Udenfriend, S. Tyrosine hydroxylase. The initial step in norepinephrine biosynthesis. *J. Biol. Chem.* 239:2910, 1964.
7. Shiman, R., Akino, M., and Kaufman, S. Solubilization and partial purification of tyrosine hydroxylase from bovine adrenal medulla. *J. Biol. Chem.* 246:1330, 1971.
8. Costa, E., and Neff, N. H. Estimation of Turnover Rates to Study the Metabolic Regulation of the Steady-State Level of Neuronal Monoamines in the Central Nervous System. In Lajtha, A. (Ed.), *Handbook of Neurochemistry,* Vol. 4. New York: Plenum, 1970, p. 45.
9. Alousi, A., and Weiner, N. The regulation of norepinephrine synthesis in sympathetic nerves: Effects of nerve stimulation, cocaine and catecholamine releasing agents, *Proc. Natl. Acad. Sci. U.S.A.* 56:1491, 1966.
10. Morgenroth, V. H., Boadle-Biber, M., and Roth, R. H. Tyrosine hydroxylase: Activation by nerve stimulation. *Proc. Natl. Acad. Sci. U.S.A.* 71:4283, 1974.
11. Zivkovic, B., Guidotti, A., and Costa, E. Effects of neuroleptics on striatal tyrosine hydroxylase: Changes in affinity for the pteridine cofactor. *Mol. Pharmacol.* 10:727, 1974.
*12. Snyder, S. H., Banerjee, S. P., Yamamura, H. I., and Greenberg, D. Drugs, neurotransmitters and schizophrenia. *Science* 184:1243, 1974.
*13. Axelrod, J. The metabolism, storage, and release of catecholamines. *Recent Prog. Horm. Res.* 21:597, 1965.
14. Coyle, J. T., and Snyder, S. H. Catecholamine uptake by synaptosomes in homogenates of rat brain: Stereospecificity in different areas. *J. Pharmacol. Exp. Ther.* 170:221, 1969.

*Asterisks denote key references.

*15. Iversen, L. L. *The Uptake and Storage of Noradrenaline in Sympathetic Nerves.*
      London: Cambridge University Press, 1967.
 16. Schubert, J., Nyback, H., and Sedvall, G. Accumulation and disappearance of
      ^{3}H-5-hydroxytryptamine formed from ^{3}H-tryptophan in mouse brain: Effect of
      LSD-25. *Eur. J. Pharmacol.* 10:215, 1970.
 17. Aghajanian, G. K., Haigler, H. J., and Bloom, F. E. Lysergic acid diethylamide
      and serotonin: Direct actions on serotonin containing neurons. *Life Sci.*
      17:615, 1972.
 18. Haigler, H. J., and Aghajanian, G. K. Lysergic acid diethylamide and serotonin:
      A comparison of effects on serotonergic neurons and neurons receiving a sero-
      tonergic input. *J. Pharmacol. Exp. Ther.* 188:688, 1974.
 19. Jouvet, M. Biogenic amines and the states of sleep. *Science* 163:32, 1969.
 20. Wurtman, R. J., Axelrod, J., and Kelly, D. E. *The Pineal.* New York: Academic,
      1968.
 21. Deguchi, T., and Axelrod, J. Supersensitivity and subsensitivity of the β-adrener-
      gic receptor in pineal gland regulated by catecholamine transmitter. *Proc. Natl.
      Acad. Sci. U.S.A.* 70:2411, 1973.
 22. Snyder, S. H., and Taylor, K. M. Histamine in the Brain: A Neurotransmitter?
      In Snyder, S. H. (Ed.), *Perspectives in Neuropharmacology — A Tribute to Julius
      Axelrod.* New York: Oxford University Press, 1972, p. 43.
 23. Garbarg, M., Barbin, G., Feger, J., and Schwartz, J. C. Evidence for a new
      aminergic pathway in rat brain: Reduction in histamine and histidine decar-
      boxylase in telencephalon after lesions in the lateral hypothalamic area. *Science*
      186:833, 1974.

Chapter 11
Amino Acid Transmitters

Eugene Roberts
Richard Hammerschlag

GAMMA-AMINOBUTYRIC ACID

γ-Aminobutyric acid (GABA) was the first amino acid shown conclusively to function as a neurotransmitter in both vertebrate and invertebrate nervous systems. Much of the key evidence for this role of GABA has been obtained at a few relatively accessible synaptic sites: the crayfish stretch-receptor sensory neuron; the mammalian cerebellar Purkinje cell; and the neuromuscular junction of the lobster, where release of GABA related to nerve stimulation has been detected. Whether released from neurons or externally applied, GABA most often exerts inhibitory or hyperpolarizing effects. Recent data, however, suggest a greater versatility for GABA, as it also appears to have excitatory or depolarizing actions on selected neurons [1–8].

The widespread neural distribution of both GABA and the major enzyme for its biosynthesis, glutamic acid decarboxylase (GAD), has been established in numerous biochemical and immunocytochemical studies. Such studies, coupled with extensive physiological data, suggest that GABA-releasing neurons play a key role at virtually all levels of nervous-system function [8a].

Metabolism of GABA in the Nervous System

An outline of the principal reactions of GABA is shown in Figure 11-1. GABA is formed in the nervous systems of vertebrates and invertebrates mainly by the single step α-decarboxylation of glutamic acid.* The reaction is catalyzed by an L-glutamic acid decarboxylase (GAD), L-glutamate-1-carboxy-lyase (EC 4.1.1.15). The enzyme, as purified from a synaptosomal fraction of mouse brain, has a coenzyme requirement for pyridoxal phosphate and shows a high degree of substrate specificity for L-glutamate [8b]. A nonneuronal GAD, present in glia as well as in blood vessels and several peripheral tissues, has been highly purified from beef-heart muscle and has been found to be chemically and immunologically distinct from the brain enzyme [12]. The major catabolic route for GABA is the reversible transamination with α-ketoglutarate catalyzed by an aminotransferase (GABA-T), 4-aminobutyrate:2-oxoglutarate aminotransferase (EC 2.6.1.19). The succinic semialdehyde formed in this reaction is rapidly oxidized to succinate by succinic semialdehyde-NAD$^+$-oxido-reductase (EC 1.2.1.16), a dehydrogenase. The enzyme has a high affinity for its substrate and is present at higher levels than GABA-T. Thus, succinic semialdehyde in nervous tissue is below the level of detection. GABA-T, as purified and characterized

---

*Surprisingly, recent data indicate that the GABA in the small precursor pool for the formation of homocarnosine (γ-aminobutyryl-L-histidine) in brain probably arises largely from putrescine and that the latter pool is only in slow equilibrium with the pool or pools that comprise most of the brain GABA [9]. The metabolic pathway in mouse brain from putrescine to GABA is believed to take place via the following reactions: putrescine → monoacetyl putrescine → N-acetyl-γ-amino-butyraldehyde → N-acetyl GABA → GABA [10]. The occurrence and metabolism of homocarnosine, guanidinobutyric acid, and other possible metabolites of GABA are reviewed elsewhere [11].

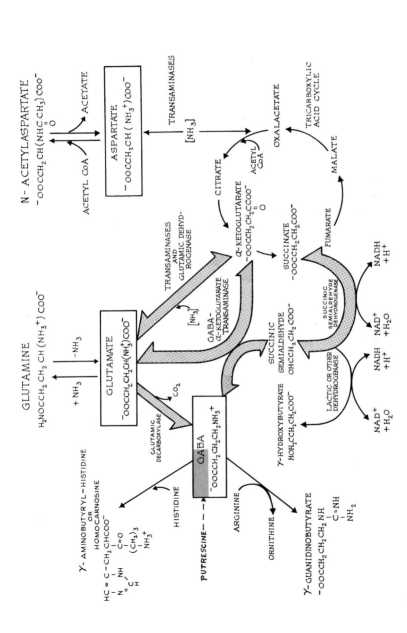

Figure 11-1
Principal known reactions of GABA, glutamate, and aspartate in the nervous system. The reactions pertinent to GABA metabolism are emphasized by the large arrows.

from mouse brain, is a B_6-requiring enzyme with an absolute specificity for α-keto-glutarate as amino group acceptor. However, β-aminoisobutyric acid and β-alanine, the homologue with one carbon less than GABA, as well as δ-aminovaleric acid, the compound longer by one carbon, also are effective amino donors. GABA-T also is found in nonneural tissues, so it may be important in several aspects of metabolism of ω-amino acids and not related just to the GABA system in neural tissue [8b, 12].

Of particular importance to the functional significance of GABA metabolism is that the synthesis and breakdown of this amino acid appears to occur largely in structurally distinct cellular regions. Subcellular distribution studies indicate that, although both enzymes are associated with particulate fractions, GAD is enriched in the presynaptic nerve-ending fraction whereas GABA-T is found chiefly in the mito-chondrial fraction. In addition, the mitochondrial fraction (comprising a mixture from neuronal perikarya and dendrites and from glial and endothelial cells) is rela-tively much richer in GABA-T than are mitochondria isolated from lysed presynaptic nerve endings. Thus it appears that GABA, after being released from the synaptic cleft and then removed by active uptake mechanisms (see subsequent sections), is metabolized mainly in the mitochondria of postsynaptic neurons and perisynaptic extraneuronal cells [3, 5, 11].

Consistent with this concept are findings that the steady-state concentrations of GABA, as studied in various brain regions, are not normally regulated by GABA-T but by GAD, presumably at presynaptic sites [3]. Estimates of GABA content in nerve endings of Purkinje cells show that strikingly high concentrations, 90 to 140 millimolars (mM), of this amino acid are maintained with corresponding GAD activities of 0.5 to 1.0 mmol/hr/gm [8c]. Similarly high levels of GABA have been found in peripheral inhibitory axons in the lobster [2].

Pharmacological studies also suggest that GAD and GABA-T may not be present in the same location in the vertebrate central nervous system (CNS) [3, 5]. When hydroxylamine or aminooxyacetic acid, substances that are potent inhibitors of both GAD and GABA-T in vitro, are administered to animals, marked elevations of GABA content are observed in brain and only the GABA-T appears to be inhibited. One of the simplest explanations for these observations is that the two enzymes are present in different cell types or in different intracellular sites in the same cells and that the inhibitors penetrate to the regions containing the GABA-T but not to those in which the GAD is located. Different sites of the two enzymes of GABA metabolism within neurons have been shown most directly by using antibodies produced against GAD and GABA-T purified from mouse brain. (See Fig. 11-4 and relevant discussion.) This differential localization of the two GABA enzymes is important also in inter-preting the markedly increased levels of GABA found in various pathological or experimental conditions. One of the substrates for GABA-T (α-ketoglutarate) may be limiting in hypoglycemia; oxidized coenzyme (NAD^+) for succinic semialdehyde dehydrogenase may be depleted in anoxia; or GABA-T may be inhibited directly, as with aminooxyacetic acid. Thus some, or even all, of the increase in GABA content observed in some abnormal metabolic states or after treatment with drugs may reflect increases in nonterminal sites and may not necessarily be associated with an increased functional capacity of GABA neurons.

The spatial separation of GABA synthesis and breakdown is relevant to the concept of the so-called GABA shunt pathway. In addition to the usual tricarboxylic acid cycle, which contains the two-step conversion of α-ketoglutarate to succinate, the unique presence of GAD and GABA in neural tissue allows α-ketoglutarate to be metabolized to succinate via glutamate, GABA, and succinic semialdehyde. The latter pathway would give three moles of ATP for each mole of α-ketoglutarate metabolized, while metabolism via α-ketoglutarate oxidase gives three moles of ATP and one mole of GTP.

Caution should be exercised in correlating measurements of GABA levels in whole brain or in grossly dissected neural regions with such global phenomena as convulsive seizures, EEG changes, or various behavioral paradigms: Such changes in total content of any given transmitter compound may be attributable to many factors not relevant to its use in informational transactions at synapses. Measurements of turnover rates of GABA in selected neural regions have potentially much greater functional significance. Studies of turnover rates of specific transmitter pools of GABA in specific regions of brain or spinal cord await the development of adequate strategies and microtechniques. Techniques that are analogous to those employed to study turnover of catecholamines or serotonin are not applicable to GABA because it is formed from ubiquitously occurring glutamate and is metabolized to CO_2 and H_2O extremely rapidly in many tissues.

Presynaptic Release of GABA

The clearest demonstration of GABA release has been achieved after several nerves that are inhibitory to various muscles of the lobster claw have been stimulated. Reduction of calcium levels in the extracellular medium, a condition that generally is found to block transmitter release from all synapses where it has been tested, prevents this impulse-related release of GABA. Stimulation of excitatory nerves that synapse on the same lobster muscles does not lead to GABA release [2, 8m].

Data showing the liberation of GABA after stimulation of specific inhibitory neurons in the vertebrate nervous system are extremely difficult to obtain, and we must content ourselves at the present time with successive approaches to this problem. More GABA and less glutamic acid are liberated from the perforated pial surface of the cortex of cats during sleep than in the aroused state [13], and GABA has been shown also to be released specifically from the surface of the posterior lateral gyri of cats during several conditions that produce cortical inhibition [14]. But the neuronal source of the GABA released in these experiments is uncertain. The vertebrate neuron most clearly identified by chemical, physiological, and immunochemical criteria as "GABAergic" is the cerebellar Purkinje cell. In cats pretreated with aminooxyacetic acid to block GABA metabolism, the rate of GABA release in a perfusate of the fourth ventricle increased threefold over the control level during stimulation of the cerebellar cortex [15]. It is likely that the GABA in these perfusates was released from Purkinje cells because these neurons form abundant inhibitory synapses in the deep cerebellar nuclei that face the fourth ventricle.

Whereas GABA does not readily pass the blood-brain barrier (see Chap. 20), ^{3}H-GABA is rapidly taken up into slices or synaptosomes prepared from brain tissue and labels a pool of the amino acid that is releasable by either electrical stimulation or by increased extracellular K^+ [16]. A drawback of the tissue-slice preparation for such studies is that both high K^+ and electrical-field stimulation are used as non-specific stimuli so that the neurons being stimulated, and hence the synapses from which GABA is released, cannot be identified. An additional complicating factor for slice preparations is the finding that glial cells in dorsal-root ganglia take up ^{3}H-GABA and release it in response to high K^+ by a calcium-dependent mechanism [17]. Synaptosomes appear to be a more promising model system in which to study GABA release. Such release has been demonstrated after ^{3}H-GABA-labeled synaptosomes have been exposed briefly to high K^+ [18].

Ionic Basis of GABA Action

Experiments carried out at the crustacean neuromuscular junction have compared the ionic basis of the postsynaptic inhibitory action of GABA with the inhibitory effects produced by nerve stimulation. Both GABA and the natural transmitter produce an increase in membrane permeability to Cl^- ions, measured as an increase in membrane conductance. Chloride then tends to distribute across the membrane according to its equilibrium potential, which is similar to the resting potential of the cell, but often is more negative. The resultant effect is to clamp the membrane potential near its resting level and thereby to decrease the sensitivity of the membrane to ongoing or subsequent depolarizing stimuli. GABA and the natural inhibitory transmitter both are ineffective in Cl^--free solutions. In addition, both show the same reversal potential — the level of experimentally determined membrane potential at which any given substance has neither a hyperpolarizing nor depolarizing effect — and both are blocked by picrotoxin [4].

Axons that mediate a postsynaptic inhibition of crayfish muscle also send collaterals to exert presynaptic inhibition on the terminals of axons that cause muscle excitation. GABA mimics the action of the natural presynaptic inhibitory transmitter by increasing permeability to Cl^-. Inhibition in this case, however, results from a *depolarization* of the excitatory nerve endings, presumably because those endings contain a relatively high concentration of intracellular Cl^-. This results in a depolarized, less negative, membrane potential and in a decreased probability of quantal release of excitatory transmitter from the excitatory nerve endings.

The actions of GABA observed in the vertebrate CNS appear essentially similar to those in the crayfish. Striking similarities exist in the mammalian cerebral cortex between intracellularly recorded inhibition produced by electrical stimulation and that produced by direct application of GABA from micropipettes (iontophoresis). Both produce hyperpolarization and a marked increase in membrane permeability, as shown by a large rise in electrical conductance; and both show closely similar reversal potentials. The mechanism of the inhibition is again most likely through an increase in permeability by the Cl^- ion of the postsynaptic membrane [6, 7]. The action of GABA or of natural inhibition differs greatly from that of glycine or acetylcholine, both of which have some depressant effects on cortical neurons. Studies correlating physiology and morphology suggest that inhibition in the cortex may be attributable almost exclusively to intracortical interneurons, and pharmacologica studies seem to demonstrate that at least some of these are GABA neurons. Careful investigations of the effects of GABA on neurons in other regions of the vertebrate CNS, e.g., Deiters' nucleus and spinal cord, have shown that the inhibitory action of GABA takes place by the same ionic mechanism.

Recent evidence suggests that GABA also may be a transmitter mediating presynaptic inhibition in the spinal cord and other regions in the vertebrate CNS and that it acts by producing depolarization of primary afferent terminals. GABA has been reported also to have a depolarizing action on cell bodies of primary afferent neurons located in sensory ganglia. This may well result from the cell body's having similar receptors to the synaptic endings, a phenomenon observed in several invertebrate ganglia. It seems that in all of the latter instances GABA also acts by increasing permeability of membranes to Cl^-. The decreases, rather than increases, in membrane potential produced by GABA action in these instances, similar to its presynaptic actions in crayfish, probably are attributable to relatively high intracellular Cl^- concentrations [6, 7].

The GABA Receptor Problem

Results from several invertebrate preparations suggest that activation of the GABA receptor (as measured by conductance changes) occurs only after more than one molecule of GABA binds to the active site. A combination of two molecules of GABA appears to be required for receptor activation at the crayfish neuromuscular junction [8d], and three molecules seem to be required at a receptor site on locust muscle [19]. It is possible that several receptor sites must be activated by GABA prior to their cooperative assembly into the ionophore. Similar analyses of GABA receptors in vertebrate systems have not been attempted, largely because of the relative inaccessibility of the GABA synapses and the difficulties of performing accurate quantitative measurements.

There is no hint to date about the biophysical mechanisms by which the interaction of GABA with membranes produces increases in Cl^- conductance. From work that has been done on the nicotinic cholinergic receptor, it is apparent that only the isolation and characterization of a GABA receptor-ionophore complex will give the kind of information necessary to begin to achieve an understanding of the molecular biology of this process. At present there are no high affinity ligands for the GABA receptor analogous to α-bungarotoxin from snake venom for the cholinergic receptor. An additional problem is that the GABA transport system has membrane binding sites for GABA that are extremely difficult to distinguish unequivocally from those that may be involved in GABA receptor function. Certainly, all crude membrane fractions prepared either from invertebrate or vertebrate preparations would be likely to contain both types of sites. In addition, there are GABA-metabolizing enzymes (GABA-T, homocarnosine synthetase, GABA transaminidase, etc.) that have affinities for GABA. Therefore, numerous problems must be faced when labeled GABA itself is used as ligand in studies of the physiological GABA receptor. Nevertheless, several laboratories have begun studies of GABA receptors, employing as ligand [14C]- or [3H]-labeled GABA in cell-free preparations from shrimp [20], crayfish [8e], rat brain [8f], and rat cerebellar cortex [8g].

Active Transport of GABA into Intracellular Sites
The major degradative enzyme for GABA is GABA-T, predominantly a mitochondrial enzyme not associated with neuronal membranes. Compounds that inhibit GABA-T, such as hydroxylamine and aminooxyacetic acid, fail to potentiate either synaptic inhibition or the inhibitory effects of iontophoretically applied GABA on cortical neurons. It appears that active uptake mechanisms, such as those described in Chapter 8, rather than enzymatic destruction or slow diffusion processes, account for the rapid cessation of the postsynaptic effects of GABA [3, 8h]. The membrane site for GABA transport appears to be distinct from the receptor that is required for GABA to exert its synaptic effects on membrane potential. Imipramine and desmethyl-imipramine inhibit the binding of GABA to membranous subcellular particles in brain, but these substances do not antagonize the inhibitory action of iontophoretically applied GABA on the synaptic responses of spinal neurons. Similarly, the Na^+-dependent binding of GABA at the crustacean neuromuscular junction is prevented by substituting choline for Na^+, whereas the typical inhibitory actions of GABA are unaffected by this ionic change. It seems clear that different strategies must be designed for studying the mechanisms and molecular entities involved in the binding and uptake of GABA and those involved in the effects of GABA on ion movement.

Tracking Down the GABA Neuron
Until very recently, the localization of GABA neurons has been inferred by correlating microchemical, electrophysiological, and iontophoretic studies with the known cytoarchitecture of specific regions of brain and spinal cord [1–8]. The purification of GAD and GABA-T from mouse brain now has provided a means for the direct visualization of GABA neurons at the light- and electron-microscopic levels through the use of immunocytochemical procedures.

Briefly, the technique involves treating suitably prepared neural-tissue sections with rabbit serum containing antibodies produced against purified enzyme. An antibody sandwich then is created by treating the sections with goat antibodies to which horse-radish peroxidase has been conjugated and which are directed against the rabbit γ-globulin. The endogenous GAD- or GABA-T-containing neuronal sites finally are visualized by allowing the peroxidase to oxidize diaminobenzidine in the presence of H_2O_2 to a brown product that can be seen in the light microscope and that also is electron dense and thus can be detected by the electron microscope.

The neural region to which the immunocytochemical approach was first applied is the cerebellum, the most favorable site for investigating structural and functional properties of inhibitory neurons [21, 22]. The overall function of the cerebellar cortex probably is entirely inhibitory. Cells that lie entirely in the cerebellar cortex — the basket, stellate, and Golgi type-II cells — are believed to play inhibitory roles within the cerebellum (Fig. 11-2). The basket cells make numerous powerful inhibitory synapses on the lower region of the somata of the Purkinje cells and on their basal processes, or "preaxons." The superficial stellate cells form inhibitory synapses on the dendrites of Purkinje cells. The Golgi cells make inhibitory synapses on the

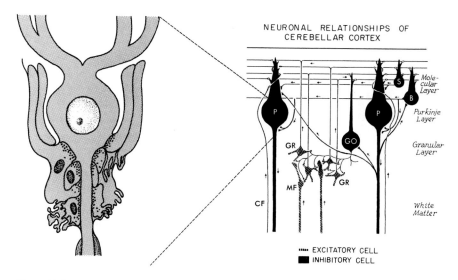

Figure 11-2
Neuronal relationships in the mammalian cerebellar cortex (after Fig. 120 of [21]). (B) Basket
cell; (CF) climbing fiber; (GO) Golgi cell; (GR) granule cell; (MF) mossy fiber; (P) Purkinje cell;
(S) stellate cell.

dendrites of the granule cells. Afferent excitatory inputs reach the cerebellum via
the climbing and mossy fibers, which excite the dendrites of the Purkinje and granule
cells, respectively. The granule cells are believed to be the only cells that lie entirely
within the cerebellum that have an excitatory function. An afferent inhibitory noradrenergic input (not shown in the figures) also is believed to reach the Purkinje cells
from cells in the locus coeruleus [23]. The Purkinje cells, which are the only output
cells of the cerebellar cortex, inhibit monosynaptically in Deiters' and intracerebellar
nuclei (Fig. 11-3). They also probably inhibit other cells in the cerebellar cortex
through axon collaterals.

Biochemical laminar analyses of the GABA system have suggested the possibility
that all of the inhibitory cells of the cerebellum (Purkinje, basket, stellate, and Golgi)
might use GABA as transmitter [3]. Direct evidence for the above supposition now
has come from the immunocytochemical localization of GAD in rat cerebellum [24,
25]. Figure 11-4 illustrates the differential localization of GAD and GABA-T. At
the light microscopic level (Fig. 11-4, A and E) GAD was visualized in boutonlike
structures; at the electron microscopic level, the enzyme was seen to be highly localized in synaptic terminals in close association with the membranes of synaptic vesicles
and mitochondria, but not within these organelles. In contrast to the finding of GAD-
positive boutons *on* cell somata and dendrites, GABA-T (Fig. 11-4C) has been visualized
inside Purkinje cell bodies and dendrites and *in* the somata of stellate cells and of
neurons in the nucleus interpositus. Also, it is associated with structures which may
belong to the glia of the cerebellar cortex. All the results are consistent with the
concept that GAD resides largely in presynaptic terminals and that GABA-T is mainly
in neuronal sites that are postsynaptic to these terminals or in glia in their vicinity.

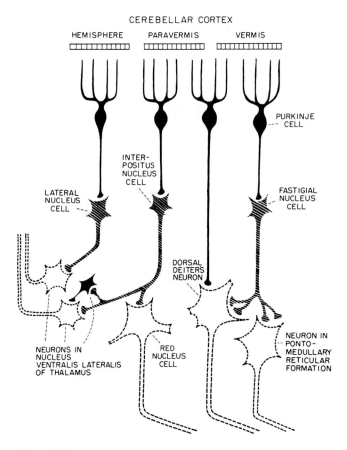

Figure 11-3
Neuronal relationships in the cerebellar efferent system. (After Fig. 124 of [21]). The black
cells are inhibitory, the others excitatory. Climbing and mossy fibers are not shown in this
diagram.

The immunocytochemical technique has also been applied to developmental
problems in the cerebellum. GAD was found to be present in the terminals of growing neurites prior to the time at which synaptic contact is first initiated [26]. The
data suggest that the initial signal for GAD synthesis predates the establishment of
contacts between pre- and postsynaptic elements of a developing synapse. The possibility then exists that the passage of transmitter molecules to receptors on potential
postsynaptic membranes or the release of the transmitter-forming enzyme itself, or
both, might be part of a recognition process that results in synaptic contact between
specific cell types.

The GABA system has been studied in almost every region of the vertebrate CNS,
with various degrees of thoroughness. Immunocytochemical studies have confirmed
the presence of GABA neurons among the inhibitory amacrine cells in the inner
synaptic region of the retina, and the presence of GAD in the indigenous inhibitory

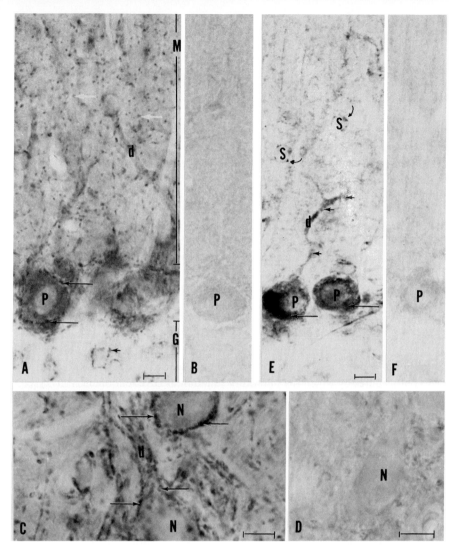

Figure 11-4
Immunocytochemical visualization of GAD and GABA-T in the rat cerebellum at the light microscopic level.

(A) Photomicrograph of a cerebellar folium from a preparation treated with anti-GAD serum and
stained by the procedure described in the text. Punctate GAD-positive boutons, which correspond to the
expected distribution of basket-cell endings (long arrows), surround the Purkinje cell soma (P). In the
granule-cell layer (G), GAD-positive boutons (short arrows) are in a pattern consistent with the distribution of Golgi II endings upon dendrites of the granule cells. In the molecular layer (M), GAD-positive
boutons (white arrows) are not associated with obvious structures and may be stellate endings on small
dendrites. Some GAD-positive boutons are seen on the larger Purkinje cell dendrites (d). Bar represents 10 μ.

(B) Control section for A treated with nonimmune rabbit serum and carried through the staining
procedure. Purkinje cell soma (P).

(C) The nucleus interpositus visualized in a preparation treated with anti-GAD serum. Boutonlike GAD-positive structures (arrows) are seen studding the cell bodies (N) and dendrites (d) of the
neurons in this deep cerebellar nucleus. This conforms to the known distribution of axon terminals
of Purkinje cells in these regions. Bar represents 10 μ.

(D) Control section for C, treated with normal rabbit serum. Cell body (N). The bar represents 10 μ.

(E) Cerebellar folium from a preparation treated with anti-GABA-T serum. The GABA-T-positive
product is found in the cytoplasm (long arrows) of the Purkinje cell somata (P) and in their dendrites
(d), short arrows. The cell bodies (S) contain small punctate deposits of GABA-T-positive product at
points around the cell body (curved arrows). The bar represents 10 μ.

(F) Control section for E, treated with normal serum. Purkinje cell soma (P).

interneurons of the hippocampus, i.e., the basket cells [8i, 8j] . The role of the GABA system in the spinal cord is reviewed in a following section.

Pharmacology of the GABA System

Pharmacological manipulation of the GABA system could occur at least at four sites: GABA biosynthesis (GAD) and release; the GABA receptor; the GABA uptake system; and GABA-T. Much of the present complexity in this field lies in the difficulty of finding or designing drugs that are selective antagonists or mimetics for any one of these sites. As an example, the first pharmacologically active compounds that were tested in relation to the GABA system were the convulsant hydrazides. Studies initially focused on the GABA system because these compounds were shown to inhibit GAD. Correlations between decreased GABA levels and occurrence of seizures produced by the hydrazides, however, have become increasingly difficult to interpret because it was found that these highly reactive substances can also inhibit GABA-T and a host of other B_6-requiring enzymes. In addition, they can react with carbonyl groups present in many tissue components to cause a variety of physiological and pathological effects [27] . A truly specific inhibitor of GAD still remains to be found.

From the point of view of therapy, it would be worthwhile to establish pharmacological procedures by which the inhibitory function of GABA in the nervous system could be altered. A variety of clinical conditions, such as Huntington's disease, epilepsy, and schizophrenia, may involve defects in key inhibitory elements either in the whole brain or in specific regions [8a] . A strategy should be sought by which the effect of GABA liberated from nerve endings that normally use GABA as a transmitter could be modulated at pertinent synapses. Substances that can alter the rate of uptake or release of GABA, while they themselves have no GABA-mimetic or GABA-antagonist properties, should be sought. Although hundreds of compounds have been tested as inhibitors of GABA uptake in test systems employing brain slices or subcellular particles, none has yet been found with an affinity which approaches that required for such an agent.

Newly discovered relationships among GABA, cyclic GMP, and the benzodiazepines suggest that the latter may facilitate GABA release [8k] . The excitatory transmitter liberated by the climbing fibers at their synapses on Purkinje cells may activate guanylcyclase in these cells, resulting in an increase in their content of cyclic GMP. The action of GABAergic neurons (e.g., basket cells) upon the Purkinje cells counteracts the effects of the climbing-fiber input on cyclic GMP. Diazepam and other benzodiazepines exert a GABAlike action in preventing those increases in cyclic GMP in the cerebellum that appears to be closely correlated with their anxiolytic potency in humans. In the spinal cord these drugs increase presynaptic inhibition, and their effects are blocked by picrotoxin and bicuculline, but not by general anesthetics. Here, too, the action of these substances is remarkably similar to that of presynaptically released GABA, and they may be presumed to act by a GABAergic mechanism.

Drugs that interact specifically with GABA receptors would also be valuable. The most extensively studied compounds that antagonize the effects of GABA are the convulsants, bicuculline and picrotoxin [8 1]. Because these substances have been shown to have antagonistic effects to GABA on firing rates of neurons in various physiological preparations [5–7], it has become commonplace to infer the demonstration of GABA-mediated synapses in both vertebrate and invertebrate nervous systems on the basis of such effects. Obviously, when effects on firing rates of neurons are observed, nonsynaptic as well as synaptic actions must be considered. Although there often is a close relationship between the generator potential and frequency of firing, it is well known that a variety of chemical and physical changes can produce a dissociation, or uncoupling, between the two. When the more direct conductance measurements were made, it became obvious that neither bicuculline nor picrotoxin is a competitive antagonist of GABA action on postsynaptic membranes in invertebrate preparations. To our knowledge, there have been no quantitative intracellular studies of GABA-bicuculline and -picrotoxin antagonism on conductance changes in vertebrate neurons. Actually, it appears more likely from current data that picrotoxin, and possibly bicuculline, might react with the ionophore rather than with the GABA receptor itself. It is possible either that GABA receptors and associated ionophores are closely coupled, but not necessarily identical, entities or that the ionophores are made up of associations of GABA-receptor protein complexes.

The results from a large number of studies with picrotoxin and bicuculline are complex and seem unlikely to be interpretable on the basis of single mechanisms of action [5, 7]. At best, it might be said that actions of these substances counter to those of applied GABA at a given neural site at least suggest that GABA might be the transmitter, but proof must be established by more specific means. The search is still very much on for a potent and specific competitive antagonist of GABA at postsynaptic sites.

AMINO-ACID TRANSMITTERS IN SPINAL CORD

GABA is believed to function as a transmitter in all regions of the vertebrate CNS, but evidence for other amino acids as transmitter candidates in vertebrates has come in large part from studies on spinal cord. Before reviewing in detail the data for glycine, glutamate, aspartate, and Substance P as transmitters, the functional interrelations of spinal neurons that have been proposed to traffic in these compounds as well as in GABA will be considered. The spinal cord is one of the few regions of the vertebrate nervous system in which transmitter substances have been tentatively assigned to as many as a half-dozen functionally distinct types of synapses. Although much of the evidence must be described as suggestive, a working model showing sites of transmitter action is presented in Figure 11-5. Representative analytical data on which this scheme is based are compiled in Table 11-1.

Most of the sensory fibers that enter the spinal cord terminate on interneurons in the dorsal regions of the cord, forming the first synapses of polysynaptic pathways that eventually end on ventral motor neurons. Only a relatively small proportion of these afferent fibers form direct monosynaptic contacts with motor neurons. Therefore, the content of an excitatory transmitter stored in primary afferent fibers would be expected to be higher in dorsal roots and dorsal-cord gray matter than in corresponding ventral regions. Two compounds that show such a distribution in concentration and, in addition, exert strong depolarizing actions on spinal neurons are glutamic acid and Substance P, a polypeptide. Both of these and other amino acids will be discussed in the following sections. At the present time there is no compelling

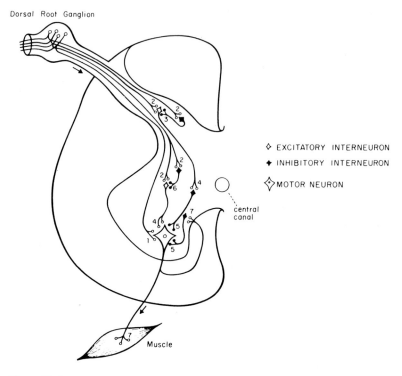

Figure 11-5
Suggested loci of action of putative postsynaptic transmitter substances in spinal cord: (1, 2) glutamic acid or Substance P; (3) GABA; (4) aspartic acid; (5, 6) glycine or GABA; (7) acetyl-choline. For the sake of simplicity presynaptic inhibitory connections have not been indicated.

reason to think that there is only one transmitter employed among all of the dorsal root fibers in the cat.

The dorsal portion of the spinal gray matter contains laminar arrays of many small interneurons that play important roles in information processing at the point at which somatic sensory information enters the cord. Probably most of these interneurons are inhibitory, producing postsynaptic inhibition of other interneurons or motor neurons and presynaptic inhibition of primary afferent nerve endings. Studies of the regional distribution of GAD activity and GABA concentration correlate well with both electrophysiological data and immunocytochemical localization of GAD-containing synaptic endings. They suggest that the capacity to form GABA in the cord is present almost entirely in interneurons which have a higher distribution in the dorsal than ventral gray matter.

Recently GAD has been visualized by light (Fig. 11-6A) and electron microscopy (Fig. 11-6D) in the rat lumbosacral spinal cord through the immunocytochemical procedures described previously [8i, 8j]. The light-microscopic level showed a heavy reaction for GAD in laminae I to III. Moderately heavy reaction product also was seen in the deeper laminae, IV to VI, the medial aspect of the intermediate gray (lamina VII), and the region around the central canal (lamina X). Although a

Table 11-1. Amino Acids and Related Enzymes in Spinal Cord[a]

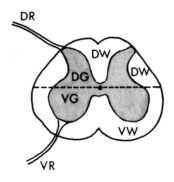

Spinal Cord Areas

Amino Acid-Enzyme System	Dorsal Roots (DR)	Dorsal Gray (DG)	Ventral Gray (VG)	Ventral Roots (VR)	Dorsal White (DW)	Ventral White (VW)
GABA system						
GABA	0.06	2.23(2.12)	1.07(1.33)	0.08	0.43(0.43)	0.44(0.32)
GAD	4	13	7	3	0	0
GABA-T	1	14	13	1	3	3
Glycine system						
Glycine	0.64	5.65(4.05)[b]	7.08(4.95)[b]	0.64	3.04(2.25)	4.39(3.32)[b]
D-Amino acid oxidase (glycine oxidase)	–	2	2	–	0.8	0.4
Glycine amino transferase	–	10	10	–	7	7
Glutamic-aspartic system						
Glutamate	4.61	6.48(4.72)[b]	5.39(4.48)[b]	3.14	4.80(4.18)[c]	3.89(3.41)[c]
Glutamine	1.88	5.30(5.09)	5.35(5.01)	1.90	3.59(3.60)	3.81(3.71)
Glutaminase	12	397	331	13	67	66
Glutamine synthetase	7	67	67	4	20	19
Glutamate dehydrogenase	17	446	448	16	149	139
Aspartate amino transferase	545	5412	5418	537	2454	2480
Aspartate	1.02	2.05(1.54)[c]	3.05(2.41)[c]	1.27	1.11(1.19)	1.29(1.27)

[a]Amino acids: μmoles/g fresh wt tissue; enzymes; μmoles/g fresh wt/hr.
[b]Significant at $P < 0.005$.
[c]Significant at $P < 0.05$.
Sources: See [28, 30, 42, 43] for pertinent data. The values in parentheses were obtained 11 to 35 days after aortic occlusion.

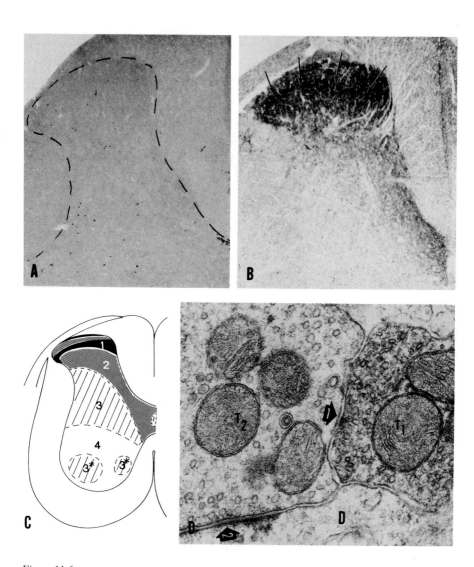

Figure 11-6
Immunocytochemical visualization of GAD at the light- and electron-microscopic levels in the rat spinal cord.

(A) Photomicrograph of control section of the dorsal horn of rat lumbar spinal cord treated with nonimmune rabbit serum. Dashed line outlines the dorsal horn. ($\times$ 70.) Dense bodies are red blood cells.

(B) Section of the dorsal horn of rat lumbar spinal cord treated with anti-GAD rabbit serum. Arrows point to intense band of punctate GAD-positive product in the substantia gelatinosa. ($\times$ 70.)

(C) The diagram shows different regions of spinal cord, numbered 1 to 4 to represent differing relative intensities of GAD-positive product in different areas.

(D) Axo-axonal synapse in the substantia gelatinosa, in which a GAD-positive terminal (T_1) is presynaptic (arrow 1) to another terminal (T_2), which is not GAD positive and is, itself, presynaptic (arrow 2) to a dendrite (D). ($\times$ 60,000.) This is taken to be evidence in support of the idea that GABA neurons are, indeed, mediators of presynaptic inhibition in the dorsal cord.

moderately light reaction for GAD was observed in the ventral horn, numerous punctate deposits (boutons) were observed on bodies of motor neurons. At the electron-microscopic level, many more GAD-positive terminals were observed in laminae I to III than in IV to VI. These terminals were presynaptic to dendrites and somata and also to other presynaptic endings (axo-axonal synapses believed to mediate presynaptic inhibition), the latter types of endings being more numerous in laminae II and III than elsewhere. The overall data are compatible with the interpretation that in the spinal cord GABA neurons may mediate postsynaptic inhibition on dendrites and somata of interneurons and motor neurons and that GABA probably is also the transmitter that mediates presynaptic inhibition of primary afferent terminals, whether they end on cells in the dorsal horn or on motor neurons in the ventral horn. In some instances, however, GABA may exert depolarizing effects at postsynaptic as well as at presynaptic sites, so the possibility remains that in some parts of the spinal cord GABA endings may transmit excitatory signals. Obviously, definitive information will depend on much more extensive neurophysiological data than hitherto have been available. At this time it cannot be said that all GABA endings in the cord are inhibitory.

Excitatory, as well as inhibitory, interneurons are found on both dorsal and ventral regions of the cord in complex polysynaptic circuitry. Amino acid analyses show that, in contrast to the distribution of glutamate and GABA, the amounts of glycine and aspartic acid in cat spinal cord are higher in ventral than in dorsal regions. The distribution of glycine, taken together with its hyperpolarizing action on motor neurons, suggests that this amino acid may be a major inhibitory transmitter in the ventral cord (Fig. 11-5; Table 11-1). Aspartate shares with structurally similar glutamate the property of depolarizing motor neurons. Moreover, its distribution in the cord conforms to what would be expected for an excitatory interneuronal transmitter in polysynaptic pathways. Consistent with the synaptic assignments for glutamate and aspartate are the findings that decreases in aspartate, but not glutamate, correlate well with the decrease in the number of interneurons that follows a period of anoxia produced by aortic occlusion. Similar anoxic insult also leads to a decrease in glycine, but the GABA levels, paradoxically, do not decrease significantly. [28]. It remains possible that GABA is in interneurons that are less sensitive to anoxia. At present, GABA and glycine-releasing interneurons cannot be assigned different functions with certainty, but it appears that a major role of GABA neurons, not shared by glycine neurons, is in mediating presynaptic inhibition of primary afferent nerve endings. Immunocytochemical localization of GAD-containing nerve endings that show direct GABA synapses on somata of motor neurons [8i, 8j] suggests that GABA-ergic neurons also share with glycinergic ones the task of postsynaptic inhibition in the ventral cord.

The scheme of transmitter assignments in the spinal cord can be completed by noting the probable sites of action of several transmitters that are not amino acids. It generally is accepted that acetylcholine functions as an excitatory transmitter released from motor neurons at peripheral neuromuscular junctions, as well as at central synapses via motor axon collaterals that terminate on inhibitory interneurons called Renshaw cells. These axon collaterals can function as part of negative feedback loops by exciting Renshaw cells, which in turn inhibit either the motor neurons that excited them or neighboring motor neurons. Finally, there is evidence to suggest that both serotonin and norepinephrine may function as inhibitory transmitters released from cells whose perikarya lie within the brain stem and whose descending axons terminate at various levels of the cord [8a]. Future studies will doubtless clarify this general scheme as well as point up its inadequacies to explain the richness of synaptic connections in the spinal cord.

GLYCINE
Neurochemically, glycine would seem an unlikely candidate for a transmitter role. It has a relatively simple structure, numerous metabolic functions (as precursor of other amino acids, purines, and proteins), and is synthesized and broken down by several different enzymatic routes. A neurochemical study that showed the uneven distribution of glycine in brain and spinal cord suggested, however, that the earlier findings of a hyperpolarizing action of glycine on spinal neurons might be of physiological significance and not just a pharmacological curiosity [28]. At present, a large body of electrophysiological data and pharmacological evidence strongly supports a role for glycine as an inhibitory transmitter in the spinal cord and brain stem [6, 7, 28].

Metabolism of Glycine in Nervous Tissue
The central problem in seeking biochemical evidence to support a transmitter role for glycine is to identify the rate-limiting metabolic process that makes this amino acid available for synaptic function. This could involve entry of extracellular glycine into neurons, a specific presynaptic localization of a biosynthetic enzyme, or the sequestering of a portion of neuronal glycine into a unique metabolic or structural compartment.

Experiments in intact animals suggest that blood-borne glycine or its precursor, serine, are not important sources for maintaining the glycine content in the CNS [28]. However, within the CNS, either serine formed from glucose or glyoxalate formed from isocitrate can serve as a direct precursor of glycine (Fig. 11-7). Serine is converted to glycine by the enzyme L-serine hydroxymethyltransferase (SHMT; EC 2.1.2.1.), which requires tetrahydrofolate as cofactor. Serine is formed from glucose by alternate pathways, with 3-phosphoglycerate serving as the branch point. In the route via 3-phosphoserine, the rate-limiting enzyme may be 3-phosphoglycerate dehydrogenase; in the alternate route via glycerate, the key enzyme may be D-glycerate dehydrogenase. When glycine levels, as well as activities of both of these enzymes, were measured in various regions of brain and spinal cord, the D-glycerate dehydrogenase activities showed a correlation coefficient with glycine levels of approximately 0.8, whereas the corresponding value for 3-phosphoglycerate dehydrogenase was only 0.07 [29]. The latter results, together with the observation that glycine is a relatively potent noncompetitive inhibitor of D-glycerate dehydrogenase, suggest that the activity of this enzyme may be rate limiting in the formation of a glycine precursor pool of serine and that glycine exerts feedback inhibition of its own synthesis at this step. These alternate pathways have a common final step in the conversion of serine to glycine via SHMT. The regional distribution of glycine levels correlates well with activity levels

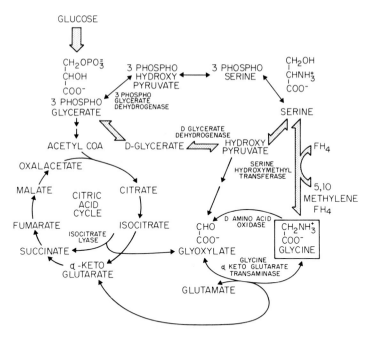

Figure 11-7
Principal known reactions of glycine that may be operative in nervous tissue.

of SHMT in spinal cord, with both enzyme and product present at highest levels in ventral gray matter [30].

Another possible biosynthetic pathway for glycine is via the glyoxylate, which is formed by cleavage of isocitrate, and subsequent transamination with glutamate serving as amino donor (Fig. 11-7). The activity of glycine aminotransferase is higher in spinal gray matter than in white matter, but there are no significant dorsoventral differences (Table 11-1). Also, no apparent correlations between activity of the latter enzyme and glycine content have been found in various areas of the rat nervous system. At present, it seems likely that serine is the major metabolic precursor of glycine in the CNS and that D-glycerate dehydrogenase and serine hydroxymethyl-transferase are the most probable candidates for the rate-limiting enzymes in biosynthesis of transmitter glycine.

Routes of glycine catabolism include conversion back to serine via SHMT and back to glyoxylate via transamination, or a less-likely direct oxidation via D-amino acid oxidase. The relative importance of these reactions for metabolism of transmitter glycine is not yet understood.

Independent data on the routes of glycine metabolism have been obtained by utilizing labeled precursors [31]. Studies with a variety of natural metabolites have shown a consistently higher specific activity for serine than for glycine, suggesting that there is no major pathway of glycine synthesis other than via serine. In studying glycine biosynthesis from glucose and other precursors, the general concept of multiple metabolic pools for citric-acid-cycle intermediates and related amino acids in CNS tissue must also be considered. As an example, the specific radioactivity of glycine formed from glucose labeled in carbons 3 and 4 is relatively higher than that of glycine formed from glucose labeled in position 1. The question of biosynthetic routes for glycine must be approached at a finer level than regional measurements of total activity of specific enzymes or total content of glycine.

Presynaptic Release
Several attempts have been made to show a specific release of glycine from spinal cord during nerve stimulation. Preliminary results, which suggested that labeled glycine was released from isolated toad hemicord after stimulation of the dorsal root [32], have proved difficult to reproduce. Recent evidence for release of endogenous glycine from a perfused in vivo preparation of rat spinal cord appears promising [33].

Action on Postsynaptic Membranes
The effects of glycine are best observed in iontophoretic studies in spinal cord and brain stem. In higher regions of the CNS, such as cerebellum and cerebral cortex, neurons are virtually insensitive to glycine. When tested on spinal motor neurons, the hyperpolarizing action of glycine is identical with that of the naturally released inhibitory transmitter. More precisely, both effects appear mediated by increases in chloride permeability, and the equilibrium potentials for both processes are the same before and after the internal anionic concentration of motor neurons is altered by intracellular injection of anions that can move through the membrane [6, 7, 28].

The inhibitory mechanisms underlying the action of glycine on spinal neurons appear identical to those mediating GABA action, but an antagonism to glycine action is seen with strychnine, a convulsant substance with no effect on GABAergic synapses. Strychnine, therefore, is useful for identifying presumptive sites of glycine action, although it has multiple pharmacological actions, including blocking depressant effects of certain amines and competitively inhibiting acetylcholinesterase.

Glycine Receptor
Studies of the binding of ^{3}H-strychnine to a membrane fraction from rat spinal cord membranes have suggested that glycine and strychnine bind to distinct sites which interact in a cooperative fashion [34]. The ability of a series of anions to inhibit ^{3}H-strychnine binding correlated closely with their capacity to invert inhibitory postsynaptic potentials when injected iontophoretically into spinal motor neurons [35]; thus strychnine binding may be associated closely with the ionic conductance mechanism for Cl^- in the glycine receptor. Although the glycine receptor has not yet been solubilized and more definitive work remains to be done, the above experiments represent the initial attempts to study the receptor-ionophore complex.

Active Uptake
The postsynaptic action of glycine is believed to be terminated by its active transport into pre- and postsynaptic neuronal sites and into glia. A high-affinity, sodium-dependent uptake mechanism for glycine has been demonstrated in spinal cord but

not in higher regions of the CNS, consistent with the proposed distribution of glycine-releasing inhibitory neurons. No substance has been described that is a potent blocker of glycine uptake, but γ-hydroxymercuribenzoate is partially effective both in inhibiting uptake and in potentiating the postsynaptic action of glycine [36].

GLUTAMATE AND ASPARTATE

The collected data on the synaptic properties of L-glutamate at vertebrate central synapses suggest that this amino acid may have a major role as an excitatory transmitter (for reviews, see [6, 7, 37]. An excitatory transmitter function has also been proposed for L-aspartate, the structurally similar amino acid, but the evidence is limited to the spinal cord.

Metabolism

Despite the ubiquitous presence of high levels of glutamate in all areas of the nervous system, its unequal regional distribution in spinal cord was a main factor in proposing glutamate as an excitatory transmitter released at the first sensory synapses in the cord [38]. Higher levels of glutamate are present in dorsal roots (the primary afferent axon bundles) and in dorsal regions of the cord (sites of most primary afferent nerve endings) than in corresponding ventral regions. In addition, higher levels of glutamate are present in the central than in the peripheral branches of the primary afferent axons as they bifurcate and leave the ganglion, suggesting a preferential shunting of glutamate toward the nerve endings [39]. Similar differences do not exist for any of the other amino acids, including aspartate and glutamine. These regional differences in glutamate concentration apparently are not maintained by any of the glutamate-synthesizing enzymes studied to date, as similar activities were found in dorsal and ventral roots and in corresponding regions of cord (Table 11-1). The higher levels of glutamate that may reflect a transmitter pool in primary afferent neurons could be supplied to presynaptic endings from their cell bodies by axoplasmic transport. This seems unlikely, however, because sectioning of dorsal roots did not significantly reduce the levels of glutamate in dorsal regions of the cord [40].

A unique approach to providing neurochemical evidence for glutamate as a transmitter in other CNS regions has been the use of a virus that selectively destroys granule cells in the cerebellum [41]. With greater than 95 percent of this cell type depleted after infection, a marked decrease in cerebellar glutamate was detected. No decrease in other amino acids, including aspartate, was observed. The high-affinity uptake system for glutamate also was reduced significantly.

Release

Increased levels of glutamate have been demonstrated in solutions that perfuse exposed surfaces of various brain and spinal cord regions after specific fiber tracts to these areas have been stimulated. In particular, the stimulated release of glutamate from brain-stem nuclei rich in primary afferent nerve endings showed a marked dependence on the presence of extracellular calcium, indicative of release from presynaptic sites [42]. Interpretations of the release of glutamate enhanced by nerve impulses, however, should consider that conducting nerve trunks appear to release glutamate from *nonsynaptic* regions, as well, by a mechanism that is not calcium dependent [43]. This suggests that, if glutamate release is being studied in a poly-synaptic pathway, a decreased release of glutamate in a calcium-free medium may not necessarily constitute evidence for glutamate as a transmitter. Reduced levels of calcium will block synaptic transmission − whatever the nature of the trans-mitter − so fewer nerves will conduct action potentials and correspondingly less "nonsynaptic" glutamate will be released.

Postsynaptic Action

Small amounts of L-glutamate or L-aspartate exert a rapid and potent depolarization of neurons in virtually all areas of the vertebrate CNS. At one time, this relatively nonspecific action was considered to be evidence against a specific transmitter role for either glutamate or aspartate. Recent studies, however, have demonstrated a differential sensitivity to the two amino acids between physiologically distinct types of spinal interneurons [44], results which are consistent with neurochemical evidence for possibly separate transmitter roles for glutamate and aspartate in spinal cord. The depolarizing action of glutamate seems mainly dependent on an increase in mem-brane permeability to sodium ions. An increased influx of sodium ions also is a requisite step in the conduction of action potentials, but different sodium channels are involved in the two processes, as tetrodotoxin blocks sodium-mediated conduc-tion but not amino acid-induced depolarization [45].

A large number of structural analogs of glutamate have been screened for their potency as antagonists, as well as agonists, of glutamate action. Synaptic effects in spinal cord and in brain stem produced by primary afferent nerve stimulation or by applied glutamate are blocked by a cyclic compound, 1-hydroxy-3-aminopyrrolidone-2 [46]. Several glutamate-mimetic compounds have been found to be more effective as depolarizing agents than glutamate. The most potent of these are kainic acid, a conformationally restricted analog of glutamate isolated from a Japanese seaweed, and *N*-methyl-D-aspartate, a synthetic analog [47]. Glutamate appears to function only as a depolarizing, excitatory transmitter in the vertebrate and arthropod nervous systems, but it is of interest that inhibitory, as well as excitatory, actions of glutamate have been described in molluscan ganglia [48].

Inactivation

Pretreatment of spinal neurons with a variety of substances that inhibit glutamate metabolism was found to have no effect on the latency of glutamate-induced excita-tion. These results, together with the similarities in the time course of action of L- and D-glutamate as neural excitants, suggest that enzymatic inactivation is not an important means of terminating the synaptic action of glutamate. An alternate mechanism for rapid removal of glutamate from the synaptic cleft is via the high-affinity uptake system present in glia [49] and in synaptosomes isolated from brain and spinal cord [50].

Both glia and nerve endings also possess a low-affinity uptake system for glutamate but, similarly to the two uptake systems for other transmitter candidates, only the high-affinity system has an absolute requirement for sodium ions. Glutamate, like the other amino-acid transmitters and unlike the amine transmitters, has not been found to be selectively concentrated in isolated synaptosomes. As discussed earlier, this may be because only a small percentage of total neuronal glutamate may serve as a transmitter. However, glutamate has been shown to accumulate selectively in a specific subpopulation of synaptosomes prepared from either cerebral cortex or spinal cord. In addition, the uptake of relatively low levels of glutamate is dependent on the presence of physiological concentrations of sodium ions, suggesting that the high-affinity, sodium-requiring uptake system for glutamate is associated with the unique subpopulation of nerve endings that accumulate this amino acid. Glutamate uptake is competitively inhibited by aspartate, and both glutamate and aspartate are taken up by the same characteristic population of nerve endings. Thus, it appears that both amino acids bind to the same membrane site prior to being transported into cells. By contrast, it appears that the postsynaptic membrane receptors that mediate excitation can distinguish to some extent between glutamate and aspartate [44]. The receptor sites subserving excitation and uptake probably are conformationally distinct, because compounds such as DL-homocysteate and N-methyl-DL-aspartate, more potent excitants than glutamate, show negligible inhibitory effects on the high affinity uptake of glutamate [51].

The question of whether glutamate is accumulated either by neuronal or glial elements or by both has not yet been settled, but the case for a predominantly glial localization of the high-affinity uptake system is fairly strong [49]. Autoradiographic studies of retina and dorsal root ganglia show that grains from labeled glutamate are highly restricted to the glia. Isolated glia, as well as cultured glioma cells, show high- and low-affinity uptake systems similar to that of synaptosomes. Proposals based on data from metabolic compartmentation studies also favor glia as a major site of glutamate uptake; glia appear to convert glutamate to glutamine, which then is transferred to neurons and metabolized back to glutamate [52].

Evidence for Glutamate as an Excitatory Neuromuscular Transmitter in Arthropods

The probable role of glutamate as an excitatory transmitter released from motor nerves of crustacea and insects [4, 48, 53] is an example of how the same physiological function may be served by different transmitters in different species, because neuromuscular excitation is mediated by acetylcholine in vertebrates.

The depolarizing action of glutamate on crustacean muscle was first observed during screening of various extracts of mammalian brain tissue. Later studies showed that glutamate, of the various excitatory substances isolated from the CNS of lobster, could best mimic the actions of the natural excitatory transmitter released at the neuromuscular junction in this species. The best evidence for a release of glutamate during electrical stimulation of motor nerves is from studies with locust. Although other amino acids are also released at rest, only the release of glutamate is increased during stimulation, shows a degree of proportionality to stimulus frequency, and is increased by increasing extracellular calcium levels during stimulation.

The specificity of the muscle receptor that mediates the depolarizing action of L-glutamate is more stringent than that of vertebrate glutamate receptors because neither D-glutamate nor L-aspartate shows appreciable excitatory effects. In crustacea and insects, the action of iontophoretically applied glutamate has been shown to be localized to specific patches of postsynaptic muscle membrane and to those patches that correspond to the particular membrane regions from which excitatory junctional potentials can be recorded after nerve stimulation. Glutamate also mimics the natural excitatory transmitter in that the depolarizing action of both substances results in an increased permeability to sodium ions and is depressed by inhibitory nerve stimulation. In addition, prolonged bath administration of glutamate leads to desensitization of the muscle to iontophoretically applied glutamate and to motor nerve stimulation, without affecting muscle sensitivity to the hyperpolarizing effects of GABA, the probable inhibitory nerve transmitter.

Glutamate also appears to exert a presynaptic effect in several arthropod nerve-muscle preparations, increasing the frequency of miniature end-plate potentials, thus implying an amplification effect of glutamate on its own release. The possibility that glutamate could be acting presynaptically to cause the release of a natural transmitter other than glutamate seems unlikely, as glutamate causes depolarization of denervated insect muscles. Denervation experiments in locust have also demonstrated a resulting spread of glutamate-sensitive regions of muscle, similar to the extension of acetyl-choline-sensitive regions from the specific end-plate sites to the total muscle surface after vertebrate preparations have been denervated.

As with putative glutamate synapses in vertebrates, there appears to be no evidence for an enzymatic termination of the junctional effects of glutamate in arthropods. A high-affinity, sodium-dependent uptake system for glutamate has been demonstrated in a lobster nerve-muscle preparation. The technique of electron-microscope autoradiography has been utilized in parallel with chemical procedures to demonstrate glutamate uptake from neuromuscular junctions of cockroach. The uptake appears greater at junctional than at nonjunctional regions and is seen to increase after nerve stimulation.

In summary, while the case for glutamate as the excitatory transmitter at arthropod neuromuscular junctions could be strengthened — particularly by the development of specific drugs that selectively block the postsynaptic action and the uptake of glutamate — this amino acid, as Kravitz et al. [53] point out for the lobster, "remains not only the leading candidate for the excitatory transmitter compound but also the only one."

Evidence for Other Amino Acids and Polypeptides as Transmitters

Among an increasing number of naturally occurring amino acids being studied in the search for new transmitters, the most promising evidence has been assembled for the sulfonic amino acid taurine [6]. The ability of taurine to depress spontaneous neuronal firing, as well as its uptake and metabolism, has been demonstrated in many areas of the CNS, the most extensive studies being carried out in retina [54]. Both labeled and endogenous taurine are released from perfused preparations of retina after electrical stimulation or flashing light. The membrane interactions underlying taurine action appear to be similar to those mediating the action of glycine, because in retina, as well as in spinal cord and brain stem, the depressant effects are blocked by strychnine rather than by GABA-blocking drugs. A high-affinity transport system has been demonstrated for taurine, and autoradiographic data show labeled taurine to be taken up mainly into amacrine cells and glial elements in the retina. Thus, taurine has been proposed as the postsynaptic transmitter that mediates the lateral inhibitory function of some amacrine cells in the retina. Taurine is synthesized in neural tissue from cysteine sulfinic acid via a decarboxylase (present in nerve-ending fractions of brain) and a subsequent oxidative step.

The last few years have seen an acceleration of interest in the possible role of polypeptides as neurotransmitters. Most of this research has been focused on the undecapeptide, Substance P, that has been postulated to function as a transmitter at the first sensory synapse in spinal cord [55–57]. Substance P (Arg-Pro-Lys-Pro-Gln-Gln-Phe-Phe-Gly-Leu-Met-NH$_2$) was first detected in 1931 as the active principle in extracts of brain and intestine that stimulated the contraction of smooth muscle. Its purification from hypothalamus has been achieved recently, after the chance finding in this tissue of a sialogogic peptide that proved to be identical in chemical and biological properties to Substance P. The hypothalamic peptide also appears identical to a Substance P-like compound that is preferentially concentrated in dorsal over ventral spinal roots. Substance P is also highly concentrated in the most dorsal parts of cat spinal cord, where its level drops significantly after dorsal roots have been ligated or sectioned. These data, coupled with results of subcellular fractionation studies, suggest that Substance P is synthesized in the perikarya of primary afferent neurons and is actively transported toward the axon terminals, where it is stored. Demonstration of the release of Substance P from spinal cord during nerve stimulation has not yet been achieved.

The excitatory effects of Substance P, after it has been applied to spinal motor neurons iontophoretically, is several-hundredfold greater on a molar basis than that of L-glutamate although a more rapid uptake of glutamate than of Substance P may underlie a part of this difference. This potent depolarizing action of Substance P has also been observed on neurons in the cerebellar cortex and caudate nucleus. Although Substance P has been proposed as a neurotransmitter liberated from sensory nerve endings, some physiological data suggest that its action may be considerably slower than that of the natural transmitter released from such endings [7].

Evidence for other polypeptides as transmitters is only very preliminary, but it seems likely that peptide-releasing neurons may exist at various sites in the nervous system in addition to the hypothalamus, where their neuroendocrine function is well documented. It is also possible that polypeptides, like the prostaglandins, may serve as important modulators of synaptic activity in ways other than by functioning as transmitters.

ACKNOWLEDGMENTS

This investigation was supported in part by the following grants: NS-01615 (E.R.) and NS-09885 (R.H.) from the National Institute of Neurological Diseases and Stroke, National Institutes of Health, and grant MH-22438 (E.R.) from the National Institute of Mental Health. The authors would like to thank Mr. Robert P. Barber for the preparation of Figures 11-4 and 11-6.

REFERENCES

*1. Roberts, E., and Eidelberg, E. Metabolic and neurophysiological roles of γ-aminobutyric acid. *Int. Rev. Neurobiol.* 2:279, 1960.

*2. Kravitz, E. A. Acetylcholine, γ-Aminobutyric Acid, and Glutamic Acid: Physiological and Chemical Studies Related to Their Roles as Neurotransmitter Agents. In Quarton, G. C., Melnechuk, T., and Schmitt, F. O. (Eds.), *The Neurosciences.* New York: Rockefeller University Press, 1967, p. 433.

*3. Roberts, E., and Kuriyama, K. Biochemical-physiological correlations in studies of the γ-aminobutyric acid system. *Brain Res.* 8:1, 1968.

*4. Takeuchi, A., and Takeuchi, N. Actions of transmitter substances on the neuromuscular junctions of vertebrates and invertebrates. *Adv. Biophys.* 3:45, 1972.

*5. Roberts, E. γ-Aminobutyric acid and nervous system function − a perspective. *Biochem. Pharmacol.* 23:2637, 1974.

*6. Curtis, D. R., and Johnston, G. A. R. Amino acid transmitters in the mammalian central nervous system. *Ergeb. Physiol.* 69:97, 1974.

*7. Krnjević, K. Chemical nature of synaptic transmission in vertebrates. *Physiol. Rev.* 54:418, 1974.

*8. Roberts, E., Chase, T. N., and Tower, D. B. (Eds.). *GABA in Nervous System Function.* New York: Raven, 1976. See in particular:

  8a. Roberts, E. Disinhibition as an Organizing Principle in the Nervous System: Role of the GABA System; Application to Neurological and Psychiatric Disorders. P. 515.

  8b. Wu, J.-Y. Purification, Characterization, and Kinetic Studies of GAD and GABA-T from Mouse Brain. P. 7.

  8c. Storm-Mathisen, J. Distribution of the Components of the GABA System in Neuronal Tissue: Cerebellum and Hippocampus-Effects of Axotomy. P. 49.

  8d. Takeuchi, A. Studies of Inhibitory Effects of GABA in Invertebrate Nervous Systems. P. 255.

  8e. Olsen, R. W. Approaches to Study of GABA Receptors.

  8f. Young, A. B., Enna, S. J., Zukin, S. R., and Snyder, S. H. The Synaptic GABA Receptor in the Mammalian CNS. P. 305.

  8g. Peck, E. J., Jr., Schaeffer, J. M., and Clark, J. H. In Pursuit of the GABA Receptor. P. 319.

  8h. Martin, D. L. Carrier-Mediated Transport and Removal of GABA from Synaptic Regions. P. 347.

  8i. Barber, R., and Saito, K. Light Microscopic Visualization of L-Glutamate Decarboxylase and GABA-Transaminase in Immunocytochemical Preparations of Rodent CNS. P. 113.

  8j. Wood, J. G., McLaughlin, B. J., and Vaughn, J. E. Immunocytochemical Localization of GAD in Electron Microscopic Preparations of Rodent CNS. P. 133.

  8k. Costa, E., Guidotti, A., and Mao, C. C. A GABA Hypothesis for the Action of Benzodiazepines. P. 413.

  8 l. Johnston, G. A. R. Physiological Pharmacology of GABA and its Antagonists in the Vertebrate Nervous System. P. 395.

  8m. Otsuka, M. GABA in the Crustacean Nervous System: A Historical Review. P. 245.

9. Nakajima, T., Kakimoto, Y., Tsiyi, M., Knoishi, H., and Sano, Y. Metabolism of polyamines in mammals: Metabolic pathways of putrescine. *Bull. Jpn. Neurochem. Soc.* 13:174, 1974.

*Asterisks denote key references.

10. Seiler, N., and Al-Therib, M. J. Putrescine catabolism in mammalian brain. *Biochem. J.* 144:29, 1974.

*11. Baxter, C. F. The Nature of γ-Aminobutyric Acid: In Lajtha, A. (Ed.), *Handbook of Neurochemistry.* Vol. 3. New York: Plenum, 1970, P. 289.

12. Wu, J.-Y. Purification and characterization of L-glutamate decarboxylase from bovine heart. *Proceedings of the 5th International Meeting of the Int. Soc. Neurochem.* (Barcelona), Sept. 2–6, 1975.

13. Jasper, H. H., Khan, R. T., and Elliott, K. A. C. Amino acids released from the cerebral cortex in relation to its state of activation. *Science* 147:1448, 1965.

14. Mitchell, J. F., and Srinivasan, V. Release of ^{3}H-γ-aminobutyric acid from the brain during synaptic inhibition. *Nature* 224:663, 1969.

15. Obata, K., and Takeda, K. Release of γ-aminobutyric acid into the fourth ventricle induced by stimulation of the cat's cerebellum. *J. Neurochem.* 16:1043, 1969.

16. Srinivasan, V., Neal, M. J., and Mitchell, J. F. The effect of electrical stimulation and high potassium concentrations on the efflux of [^{3}H] γ-aminobutyric acid from brain slices. *J. Neurochem.* 16:1235, 1969.

17. Kelly, J. S., Iversen, L. L., Minchin, M., and Schon, F. Inactivation of amino acid transmitters. *Proceedings of the 4th International Meeting of the Int. Soc. Neurochem.* (Tokyo) 4:27, 1973.

18. Levy, W. B., Redburn, D. A., and Cotman, C. W. Stimulus-coupled secretion of GABA from synaptosomes. *Science* 181:676, 1973.

19. Brookes, N., and Werman, R. The cooperativity of γ-aminobutyric acid action on the membrane of locust muscle fibers. *Mol. Pharmacol.* 9:571, 1973.

20. DeRobertis, E., and Fiszer de Plazas, S. Isolation of hydrophobic proteins binding neurotransmitter amino acids: γ-Aminobutyric acid receptor of the shrimp muscle. *J. Neurochem.* 23:1121, 1974.

*21. Eccles, J. C., Ito, M., and Szentagothai, J. *The Cerebellum as a Neuronal Machine.* Berlin: Springer-Verlag, 1967.

*22. Llinás, R. (Ed.). *Neurobiology of Cerebellar Evolution and Development.* Chicago: American Medical Association, 1969.

23. Bloom, F. E., Hoffer, B. J., and Siggins, G. R. Studies on norepinephrine-containing afferents to Purkinje cells of rat cerebellum. I. Localization of the fibers and their synapses. *Brain Res.* 25:501, 1971.

24. Saito, K., Barber, R., Wu, J.-Y., Matsuda, T., Roberts, E., and Vaughn, J. E. Immunohistochemical localization of glutamate decarboxylase in rat cerebellum. *Proc. Natl. Acad. Sci. U.S.A.* 71:269, 1974.

25. McLaughlin, B. J., Wood, J. G., Saito, K., Barber, R., Vaughn, J. E., Roberts, E., and Wu, J.-Y. The fine structural localization of glutamate decarboxylase in synaptic terminals of rodent cerebellum. *Brain Res.* 76:377, 1974.

26. McLaughlin, B. J., Wood, J. G., Saito, K., Roberts, E., and Wu, J.-Y. The fine structural localization of glutamate decarboxylase in developing axonal processes and presynaptic terminals of rodent cerebellum. *Brain Res.* 85:355, 1975.

*27. Roberts, E., Wein, J., and Simonsen, D. G. γ-Aminobutyric acid (γABA), vitamin B_6, and neuronal function – a speculative synthesis. *Vitam. Horm.* 22:503, 1964.

*28. Aprison, M. H., Davidoff, R. A., and Werman, R. Glycine: Its Metabolic and Possible Transmitter Roles in Nervous Tissue. In Lajtha, A. (Ed.), *Handbook of Neurochemistry.* Vol. 3. New York: Plenum, 1970, p. 381.

29. Uhr, M. L., and Sneddon, M. K. The regional distribution of D-glycerate dehydrogenase and 3-phosphoglycerate dehydrogenase in the cat central nervous system: Correlation with glycine levels. *J. Neurochem.* 19:1495, 1972.

30. Daly, E. C., and Aprison, M. H. Distribution of serine hydroxymethyltransferase and glycine transaminase in several areas of the central nervous system of the rat. *J. Neurochem.* 22:877, 1974.

31. Shank, R. P., Aprison, M. H., and Baxter, C. F. Precursors of glycine in the nervous system: Comparison of specific activities in glycine and other amino acids after administration of [U-^{14}C]glucose, [3,4-^{14}C]glucose, [1-^{14}C]glucose, [U-^{14}C]serine or [1,5-^{14}C]citrate to the rat. *Brain Res.* 52:301, 1973.

32. Aprison, M. H. Studies on the release of glycine in the isolated spinal cord of the toad. *Trans. Am. Soc. Neurochem.* 1:25, 1970.

33. Cutler, R. W. P. Glycine release from rat spinal cord in vivo. *Proc. Am. Soc. Neurochem.* (Mexico City) 6:191, 1975.

34. Young, A. B., and Snyder, S. H. Strychnine binding in rat spinal cord membranes associated with the synaptic glycine receptor: Cooperativity of glycine interactions. *Mol. Pharmacol.* 10:790, 1974.

35. Young, A. B., and Snyder, S. H. The glycine synaptic receptor: Evidence that strychnine binding is associated with the ionic conductance mechanism. *Proc. Natl. Acad. Sci. U.S.A.* 71:4002, 1974.

36. Curtis, D. R., Duggan, A. W., and Johnston, G. A. R. The inactivation of extracellularly administered amino acids in the feline spinal cord. *Exp. Brain Res.* 10:447, 1970.

*37. Hammerschlag, R., and Weinreich, D. Glutamic acid and primary afferent transmission. *Adv. Biochem. Psychopharmacol.* 6:165, 1972.

38. Graham, L. T., Jr., Shank, R. P., Werman, R., and Aprison, M. H. Distribution of some synaptic transmitter suspects in cat spinal cord. *J. Neurochem.* 14:465, 1967.

39. Duggan, A. W., and Johnston, G. A. R. Glutamate and related amino acids in cat spinal roots, dorsal root ganglia and peripheral nerves. *J. Neurochem.* 17:1205, 1970.

40. Roberts, P. J., and Keen, P. Effect of dorsal root section on amino acids of rat spinal cord. *Brain Res.* 74:333, 1974.

41. Young, A. B., Oster-Granite, M. L., Herndon, R. M., and Snyder, S. H. Glutamic acid: Selective depletion by viral induced granule cell loss in hamster cerebellum. *Brain Res.* 73:1, 1974.

42. Roberts, P. J. The release of amino acids with proposed neurotransmitter function from the cuneate and gracile nuclei of the rat in vivo. *Brain Res.* 67:419, 1974.

43. Weinreich, D., and Hammerschlag, R. Nerve impulse-enhanced release of amino acids from non-synaptic regions of peripheral and central nerve trunks of bullfrog. *Brain Res.* 84:137, 1975.

44. Duggan, A. W. The differential sensitivity to L-glutamate and L-aspartate of spinal interneurones and Renshaw cells. *Exp. Brain Res.* 19:522, 1974.

45. Curtis, D. R., Duggan, A. W., Felix, D., Johnston, G. A. R., Tebecis, A. K., and Watkins, J. C. Excitation of mammalian central neurons by acidic amino acids. *Brain Res.* 41:283, 1972.

46. Curtis, D. R., Johnston, G. A. R., Game, C. J. A., and McCulloch, R. M. Antagonism of neuronal excitation by 1-hydroxy 3-aminopyrrolidone-2. *Brain Res.* 49:467, 1973.

47. Johnston, G. A. R., Curtis, D. R., Davies, J., and McCulloch, R. M. Spinal interneurone excitation by conformationally restricted analogues of L-glutamic acid. *Nature* 248:804, 1974.

*48. Gerschenfeld, H. M. Chemical transmission in invertebrate central nervous systems and neuromuscular junctions. *Physiol. Rev.* 53:1, 1973.

49. Henn, F. A., Goldstein, M. N., and Hamberger, A. Uptake of the neurotransmitter candidate glutamate by glia. *Nature* 249:663, 1974.

*50. Snyder, S. H., Logan, W. J. Bennett, J. P., and Arregui, A. Amino acids as central nervous transmitters: Biochemical studies. *Neurosci. Res.* 5:131, 1973.

 51. Roberts, P. J., and Watkins, J. C. Structural requirements for the inhibition of L-glutamate uptake by glia and nerve endings. *Brain Res.* 85:120, 1975.

 52. Benjamin, A. M., and Quastel, J. H. Locations of amino acids in brain slices from the rat: Tetrodotoxin-sensitive release of amino acids. *Biochem. J.* 128:631, 1972.

*53. Kravitz, E. A., Slater, C. R., Takahashi, K., Bownds, M. D., and Grossfeld, R. M. Excitatory Transmission in Invertebrates – Glutamate as a Potential Neuromuscular Transmitter Compound. In Andersen, P., and Jansen, J. K. S. (Eds.), *Excitatory Synaptic Mechanisms.* Oslo: Universitets-forlaget, 1970, pp. 85–93.

 54. Pasantes-Morales, H., Klethi, J., Urban, P. F., and Mandel, P. The effects of electrical stimulation, light and glutamic acid on the efflux of [^{35}S] taurine from retina of the domestic fowl. *Exp. Brain Res.* 19:131, 1974.

 55. Takahashi, T., and Otsuka, M. Regional distribution of Substance P in the spinal cord and nerve roots of the cat and the effect of dorsal root section. *Brain Res.* 87:1, 1975.

 56. Konishi, S., and Otsuka, M. Excitatory action of hypothalamic Substance P on spinal motoneurones of newborn rats. *Nature* 252:734, 1974.

*57. Leeman, S. E., and Mroz, E. A. Substance P. *Life Sci.* 15:2033, 1974.

Chapter 12
Cyclic Nucleotides and Synaptic Transmission

James A. Nathanson
Paul Greengard

Many of the biochemical and physiological responses of target cells to circulating hormones depend on adenosine $3':5'$-monophosphate (cyclic AMP, cAMP). Thus cAMP is viewed as a "second messenger" that translates extracellular messages into an intracellular response. Because the receptors for neurotransmitters show many similarities to those for systemic hormones, the role of cyclic AMP in the nervous system may be analogous to its role in nonneuronal tissues — it may translate certain synaptic chemical messages into responses of receptor cells.

The cyclic AMP molecule is part of a complex system of cellular regulation and represents one link in a chain of biochemical reactions, each chain having its own substrates, enzymes, and control mechanisms. The variation in one or more of the components of this system allows cyclic AMP to exert diverse effects on the cellular functions of different tissues. A less extensive body of evidence suggests that another cyclic nucleotide, guanosine $3':5'$-monophosphate (cyclic GMP, cGMP), may also be involved in translating certain hormonal and neurotransmitter signals into appropriate cellular responses. In this chapter we try to explain the mechanism of action of these cyclic nucleotide systems and their relationship to neuronal function.

THE ROLE OF CYCLIC AMP IN HORMONE ACTION

Relationship Between Hormone Receptors and Adenylate Cyclase

The actions of many hormones on their target cells are mediated through an increase in the intracellular concentration of cAMP [1, 2]. It is thought that the membrane receptors for this class of hormones may be intimately associated with the regulatory subunit of adenylate cyclase, the membrane-bound enzyme which catalyzes the formation of cAMP from adenosine triphosphate (ATP) [3]. Thus, when the proper hormone binds with the regulatory subunit of the enzyme, the catalytic subunit, thought to be located on the inner surface of the cell membrane, is stimulated to form cyclic AMP at an increased rate. It is the resulting increase in cAMP that mediates the intracellular effects of the hormone. Cyclic AMP is broken down in the cell by another enzyme, cyclic nucleotide phosphodiesterase, which hydrolyzes cyclic AMP to $5'$-AMP [4, 5].

Mechanism of Action of Cyclic AMP

In many cells, the intracellular "receptor" for the action of cyclic AMP may be a cAMP-dependent protein kinase [6–8]. This enzyme catalyzes the phosphorylation, by ATP, of protein substrates, according to the reaction:

$$\text{ATP} + \text{protein} \longrightarrow \text{ADP} + \text{phosphoprotein}$$

Cyclic AMP stimulates this enzymatic reaction. Protein kinases that are cAMP dependent are composed of a regulatory (inhibitory) subunit and a catalytic subunit. By binding to the regulatory subunit, cAMP leads to a dissociation of the enzyme, resulting in activation of the catalytic subunit and an increase in the rate of phosphorylation of the appropriate substrate. Phosphorylation usually occurs on serine residues, but occasionally on threonine residues, of the substrate protein. This phosphorylation of the substrate protein leads to an alteration of intracellular physiology. Phosphate is removed from the substrate protein by another enzyme, phosphoprotein phosphatase, which cleaves phosphate from the serine or threonine residues. Recent evidence suggests that, in a limited number of instances, cAMP may inhibit protein kinase activity or stimulate protein phosphatase activity. In any case, whether by stimulation or inhibition of protein kinase activity or by stimulation of protein phosphatase activity, the intracellular action mechanism of cAMP appears, in almost all instances studied so far, to be through a change in the level of phosphorylation of substrate proteins.

The nature of the proteins phosphorylated in cAMP-dependent reactions varies considerably. The protein may be a known enzyme, such as phosphorylase b kinase (Chap. 14); a nuclear protein, such as histone; or a membrane-bound protein, such as those found in synaptic-membrane fractions. This last type of substrate may be important in relation to the possible role of cAMP in synaptic transmission, for it provides a means through which cAMP could regulate the ion permeability of membranes (see Mechanism of Action of Cyclic AMP in Synaptic Transmission, p. 256).

Determinants of Specificity of Hormone Action

Viewed overall, specificity of hormone action resides at several levels. First, the nature of the membrane receptor (regulatory subunit of adenylate cyclase) will determine if the hormone will increase cAMP levels. Even though a cell has an adenylate cyclase, the enzyme will not respond to a particular hormone unless it has a receptor site that can bind and be activated by that hormone. Second, the nature of the cAMP-dependent protein kinases (or phosphoprotein phosphatases) present in the cell will determine the pattern of protein substrates phosphorylated. Third, the nature of the substrate(s) present will determine the ultimate physiological response. For instance, a liver cell does not have the synaptic-membrane proteins that are present in a nerve cell, so it cannot be subject to the same phosphorylation reaction that might affect membrane permeability in the neuron. Figure 12-1 summarizes this concept as it may apply to the role of cyclic AMP in mediating synaptic transmission in neurons.

CYCLIC AMP AND NEURONAL FUNCTION

Enzymatic Machinery in Nervous Tissue

Part of the evidence that cyclic AMP may mediate the actions of certain neurotransmitters is the presence, in excitable tissue, of the enzymes and substrates necessary for cAMP-mediated reactions [9—12].

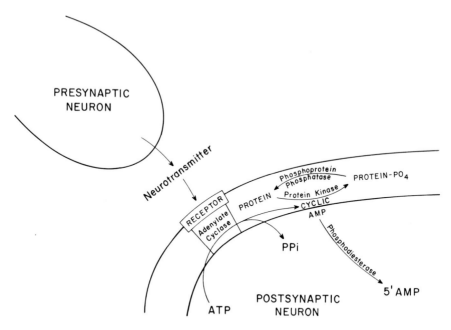

Figure 12-1
Model of a proposed mechanism by which cyclic AMP may mediate synaptic transmission at
certain types of synapses.

Adenylate Cyclase

The adenylate cyclase activity of neural tissue is almost entirely particulate. The
enzyme utilizes ATP as the preferred substrate (K_m 0.3–1.5 mM) and needs a
divalent cation, such as magnesium, manganese, or cobalt. Certain additional cofactors, such as guanosine triphosphate (GTP) or phospholipid fractions, may also be
necessary for optimal enzyme activity.

Relevant to its possible role in neuronal function is the fact that adenylate
cyclase activity is higher in brain than in most other tissues. The exact distribution
of activity among the various regions of the brain varies in different species, although
in all species it is greater in gray than in white matter. Subcellular distribution studies
have shown that the highest levels of adenylate cyclase are found in those fractions
rich in synaptic membranes, strengthening the belief that cyclic AMP found in the
brain may have relevance to neurons and to synaptic function.

Brain adenylate cyclase is stimulated by micromolar concentrations of calcium
and by millimolar concentrations of fluoride. Although the mechanism of calcium
stimulation is unclear, it is generally believed that fluoride ion acts directly on the
catalytic subunit of the enzyme, bypassing the hormonal receptor. The adenylate
cyclase of neural tissue also has been shown to be sensitive to a number of putative
neurotransmitters, including norepinephrine, dopamine, serotonin, histamine, and
octopamine. A number of similarities have been found between the physiological

receptor for a particular neurotransmitter and the regulatory subunit of the adenylate cyclase with which it interacts. Such similarities constitute important evidence that a particular neurotransmitter-sensitive adenylate cyclase may mediate the physiological action of that neurotransmitter. This is discussed in detail later in this chapter.

Phosphodiesterase

Cyclic-nucleotide phosphodiesterase converts cAMP to $5'$-AMP, and thus serves to terminate reactions initiated by the cyclic nucleotide [4, 5]. Like adenylate cyclase, magnesium or manganese ion is required for activity. Phosphodiesterase is stimulated by imidazole, and is inhibited by the methyl xanthines, including caffeine, theophylline, and theobromine. Inhibition of phosphodiesterase has the effect of augmenting and prolonging certain cAMP-mediated reactions. There appear to be multiple forms of phosphodiesterase, each with different properties, including substrate specificities. In rat cerebrum and cerebellum, for example, there are at least four separable forms, some of which can be stimulated and inhibited differentially. Certain of the phosphodiesterases from brain can be activated by an endogenous, heat-stable, calcium-binding protein.

Phosphodiesterase activity is greater in brain than in most other tissues. Subcellular distribution studies show a high concentration of the enzyme in synaptic-membrane fractions. Histochemical studies have revealed intense phosphodiesterase activity at the postjunctional thickening of synapses in the cerebral cortex of the rat. This localization tends to support a role for cyclic AMP in neurotransmission: a phosphodiesterase at the postjunctional site would have access to, and thus be able to degrade, cyclic AMP that had been synthesized as a result of stimulation of adenylate cyclase by a neurotransmitter.

Protein Kinase and Phosphoprotein Phosphatase

Protein kinase activity dependent on cyclic AMP has been found in all excitable and nonexcitable tissues so far examined. Under appropriate conditions, the activity of this enzyme is stimulated as much as twentyfold in the presence of cAMP [6–8]. The level of cAMP normally found in brain is about 0.3 μM — the same concentration that is needed for half-maximal stimulation of protein kinase of brain. This means that small changes in the usual brain concentrations of cAMP may regulate the activity of the enzyme. In subcellular distribution studies, the highest specific activity of the enzyme is found in synaptic-membrane fractions, as is that of adenylate cyclase and phosphodiesterase.

A phosphoprotein phosphatase also has been demonstrated in brain tissue. This enzyme is able to hydrolyze the phosphate from phosphoserine residues of proteins that have previously been phosphorylated in cAMP-dependent reactions. In this way, the enzyme may serve to terminate the action of cAMP. Under certain conditions, the rate at which endogenous protein phosphatase can remove phosphate from substrate proteins in the synaptic-membrane fraction can be stimulated by cAMP, thereby hastening the return of the protein to its dephosphorylated form. Like protein kinase, the highest concentration of phosphoprotein phosphatase is found in synaptic-membrane fractions.

Substrates for cAMP-Dependent Phosphorylation Reactions
Ultimately, the identity of the phosphorylated substrates determines the outcome of
cAMP-dependent reactions. A number of studies have shown that synaptic-membrane
fractions from brain are excellent substrates for endogenous cAMP-dependent protein
kinases and phosphoprotein phosphatases. In fact, the subcellular distribution of
substrates for the protein kinases and phosphatases parallels that of the protein
kinases and phosphatases themselves. The phosphorylation of synaptic-membrane
proteins by endogenous protein kinase(s) exhibits considerable specificity with respect
to the individual substrate proteins that are phosphorylated. In rat cerebral cortex,
for example, separation of phosphorylated synaptic membrane proteins by SDS-polyacrylamide gel electrophoresis reveals that, out of several dozen protein bands, the
phosphorylation of only two is markedly affected by cAMP. These experiments also
suggest that the phosphorylated proteins may be situated in the immediate vicinity
of a cyclic nucleotide-dependent protein kinase in the synaptic membrane.

Certain protein components in brain other than those in synaptic membranes also
can be phosphorylated in cAMP-dependent reactions. For example, synaptic vesicles,
neurotubules, myelin membranes, and ribosomes all contain proteins that serve as
effective substrates for intrinsic cAMP-dependent protein kinases.

Studies of Cyclic AMP Levels in Whole Brain and Brain Slices

Whole Brain
The levels of cyclic AMP in intact vertebrate brain are higher than those in almost
any other tissue. In most unstimulated, nonneural tissues, the cAMP concentration
is on the order of 0.5 to 2.5 picomoles (pmoles) per milligram of protein. In mammalian brain, the range is from 10 to 20 pmoles per mg protein. The hypothalamus
and striatum have the highest levels; those in the cerebellum and hippocampus are
somewhat lower, although the range is not great. Immunohistochemical techniques
for localizing cAMP have shown that in the cerebellum the nucleotide is present
mostly in the granule and Purkinje cells.

Endogenous cAMP levels are markedly affected by the method of tissue fixation
used. After decapitation of the rabbit, for example, cAMP levels in rabbit brain rise
as much as eightfold within 90 seconds. The mechanism of this postdecapitation
rise is not understood, although it may be related to anoxia or release of some endogenous activator of cAMP or both.

Brain Slices
Effect of Putative Neurotransmitters. Many laboratories have studied the effects
of putative neurotransmitters on respiring slices of brain tissue because such slices
demonstrate responsiveness to hormonal stimulation [12–14]. The use of such
brain slices has the advantage, over broken-cell preparations, of more nearly approximating the condition of the intact brain. It also has some disadvantages for the study
of adenylate cyclase receptors. First, levels of cAMP in brain slices represent the net
difference between synthesis by adenylate cyclase and degradation by phosphodiesterase. If there is an increase in cAMP level in response to a hormone, it is not

always clear whether the increase is a result of activation of adenylate cyclase, inhibition of phosphodiesterase, or some indirect effect such as a change in the ATP content of cells. Second, it is difficult to ascertain if the putative neurotransmitter being tested is the immediate cause of the increase in cAMP content observed. For example, it may be that the putative neurotransmitter that is applied depolarizes certain neurons, which then release a second, endogenous (and unidentified) neurotransmitter — the one that actually interacts with the adenylate cyclase receptor responsible for the observed change in cAMP. These studies with brain slices have produced important, but indirect, evidence for the presence of hormone-sensitive adenylate cyclase in neuronal tissue.

Norepinephrine, epinephrine, dopamine, serotonin, histamine, adenosine, glutamate, aspartate, and a variety of analogs of these suspected neurotransmitters have been shown to increase the cAMP content of cortical brain slices from two- to tenfold or more. Interestingly, there is a variation among different species. Histamine, for example, causes large increases of cyclic AMP in slices from rabbit cerebral cortex, cerebellum, or striatum, but has little effect in the rat. Regional differences also exist. Some of these studies are summarized in Table 12-1, and the reader is referred to several recent reviews for further details [10–15]. In general, these studies support the concept that neurotransmitters have a profound effect on cAMP metabolism. However, because of the problems described above, it is difficult to interpret those results rigorously in terms of adenylate cyclase receptors.

Effect of Electrical Pulses and Depolarizing Agents. The cAMP content of brain slices can be elevated severalfold either by stimulating the slices electrically or by applying depolarizing agents, such as high extracellular potassium, ouabain, batrachotoxin, or veratridine. The increase in cyclic AMP brought about by these procedures is blocked by low-calcium, high-magnesium, membrane-stabilizing agents such as tetrodotoxin or local anesthetics, and (paradoxically) by theophylline. It is thought that electrical stimulation and depolarizing agents may cause release of endogenous substances, which then stimulate adenylate cyclase receptors. Specifically, it has been proposed that adenosine may be the mediator of some depolarization-induced increases in cyclic AMP. It has been shown that endogenous adenosine is released during depolarization and that exogenous adenosine is potent in increasing cAMP levels in brain slices. In addition, the effect of adenosine in elevating cAMP levels is blocked by theophylline. It is uncertain whether adenosine can be considered a neurotransmitter, but current evidence certainly does suggest that it may have a role in cAMP metabolism.

Neurotransmitter-Sensitive Adenylate Cyclases
and Their Possible Role in Neuronal Function

If cyclic AMP does mediate the postsynaptic action of a particular neurotransmitter, the adenylate cyclase and the physiological receptor should share properties in common that are specific for that neurotransmitter. That is, agents that activate the adenylate cyclase should also activate the physiological receptor and, conversely, agents that antagonize the activation of the enzyme should antagonize the activation of the receptor. Furthermore, the test agents used should affect the two systems at similar

Table 12-1. Stimulation by Various Putative Neurotransmitters of Cyclic AMP Accumulation in Respiring Slices of Nerve Tissue[a]

	Cerebral Cortex		Cerebellum		Striatum		Ganglion	
Norepinephrine								
Rat	2–12X	(10–100)	1–10X	(10–100)	2–5X	(10)	6X	(1,000)[b]
Guinea pig	1.5–3X	(10–100)	8–20X	(100)	—	—	—	—
Rabbit	1.5–4X	(100)	3–11X	(100)	1–4X	(100)	—	—
Monkey	2–6X	(100)	1–2X	(100)	1.2X	(100)	—	—
Human	3–30X	(100)	2–4X	(100)	—	—	—	—
Epinephrine								
Rat	4X	(100)	—	—	—	—	16X	(100)[b]
Guinea pig	1.75X	(10)	—	—	—	—	—	—
Human	30–50X	(100–500)	—	—	—	—	—	—
Dopamine								
Aplysia	—	—	—	—	—	—	6X	(200)[c]
Cockroach	—	—	—	—	—	—	3.5X	(250)[d]
Rat	1–1.3X	(10–100)	1X	(10)	2X	(100)	2X	(10,000)[b]
Rabbit	—	—	—	—	—	—	1.2X	(100)[b]
Cow	—	—	—	—	—	—	7X	(100)[b]
Octopamine								
Aplysia	—	—	—	—	—	—	3–5X	(100)[c]
Cockroach	—	—	—	—	—	—	7X	(100)[d]
Histamine								
Rat	1–2X	(10–100)	1X	(10)	1–1.8X	(10–100)	—	—
Guinea pig	3–15X	(100)	—	—	—	—	—	—
Rabbit	16–28X	(100)	2.5–10X	(100)	4–30X	(100)	—	—
Monkey	1–1.5X	(100)	1X	(100)	1X	(100)	—	—
Human	1–7X	(100–1,000)	1.2–1.8X	(100)	—	—	—	—

Serotonin										
Aplysia	—	—	—	—	—	—	—	—	—	—
Cockroach	—	—	—	—	—	—	—	—	—	—
Rat	1–1.5X	(10–100)	—	—	1X	—	—	—	8X	(200)c
Monkey	1.25–1.75X	(100)	—	—	1X	(10)	1X	(10)	1.5X	(250)d
Human	1.0–2.1X	(100–1,000)	—	—	1X	(100)	—	—	—	—
Glutamate										
Rat	68X	(10,000)	—	—	—	—	—	—	—	—
Guinea pig	10–30X	(1,000–10,000)	—	—	—	—	—	—	—	—
Aspartate										
Rat	6–8X	(10,000)	—	—	—	—	—	—	—	—
Guinea pig	10–30X	(1,000–10,000)	—	—	—	—	—	—	—	—
Adenosine										
Rat	4X	(100)	1.2X	(100)	—	—	—	—	—	—
Guinea pig	8–25X	(100)	50X	(100)	—	—	—	—	—	—
Monkey	10–15X	(100)	—	—	—	—	—	—	—	—
Human	4–7X	(200)	—	—	—	—	—	—	—	—

aValues represent the fold-stimulation over control (1X = no stimulation): the concentration in micromolars required to cause this stimulation is given in parentheses. In cases in which several values have been published, the range of values is given. Results only for selected species and areas of the nervous system are tabulated here. The values are taken from the published reports of a number of laboratories, including those of J. Daly, G. Krishna, G. Palmer, J. Perkins, T. Rall, G. Robison, and H. Shimizu. Refer to [10–15] for original references.

bSuperior cervical.

cAbdominal.

dThoracic.

concentrations. It appears that at least four neurotransmitters — dopamine, nor-epinephrine, serotonin, and octopamine — fulfill these criteria, suggesting that some of their actions may be mediated through cyclic AMP [16, 17].

Dopamine

A dopamine-sensitive adenylate cyclase has been identified in several mammalian tissues, including the sympathetic ganglion, retina (see Chap. 7), caudate nucleus, and limbic areas of the brain, as well as in invertebrate ganglia. This enzyme has its highest specific activity in synaptic membrane fractions of the mammalian brain, is activated by dopamine at concentrations of 1 μM or less, and is inhibited by agents known to block the dopamine receptor.

A number of observations of the mammalian sympathetic ganglion suggest that dopamine may hyperpolarize postganglionic neurons and thereby modulate cholinergic neurotransmission. Small chromaffinlike interneurons within these ganglia are thought to release dopamine in response to preganglionic stimulation. Preganglionic stimulation, which results in synaptic transmission, also increases cAMP content in the ganglion. Dopamine, too, increases cAMP levels in the ganglion, and histochemical techniques have localized this increase to the postganglionic neurons. Furthermore, the hyperpolarization of postganglionic neurons, seen in response to preganglionic stimulation or the application of dopamine, is mimicked by cAMP, potentiated by theophylline (which augments the stimulation-induced increase in cAMP levels), and is blocked by prostaglandin E_1 or by alpha-adrenergic antagonists (which block the stimulation-induced increases in cAMP levels). This and other evidence suggest that dopamine, released from interneurons by preganglionic stimulation, activates the dopamine receptor — a dopamine-sensitive adenylate cyclase — which increases the cAMP content in the postganglionic neurons. This increase in cAMP may then cause a change in membrane permeability (see p. 256), which results in hyperpolarization of the postganglionic neuron.

A variety of evidence suggests that dopamine may also function as a neurotransmitter in the caudate nucleus and limbic areas of the mammalian brain. Hypoactivity of the dopaminergic pathway to the caudate nucleus is thought to be involved in the etiology of Parkinson's disease, and hyperactivity of the dopaminergic pathway to the limbic forebrain has been implicated as a possible cause of schizophrenia. In the treatment of Parkinson's disease (see Chap. 32), L-dopa, which is converted to dopamine in the brain, is used. Antipsychotic agents such as the phenothiazines, which are known to block dopamine receptors, are used to treat schizophrenia. A dopamine-sensitive adenylate cyclase has been identified in the caudate and limbic areas of the brain. This enzyme has properties very similar to those of the dopamine receptor in brain. The enzyme is stimulated by low concentrations of dopamine and apomorphine, and the activation by dopamine is blocked by a variety of antipsychotic drugs in a fashion correlated with the known clinical potency of these agents. Such antipsychotic agents as fluphenazine competitively inhibit activation of the enzyme by dopamine at extremely low concentrations (K_i = 5–8 nM), similar to those present in brain after therapeutic doses. Compounds structurally related to fluphenazine, but lacking antipsychotic effectiveness, show only a slight ability to block enzyme activation. These results, together with other studies, suggest that both the therapeutic effects and the extrapyramidal side effects of certain antipsychotic agents may be caused, at least in part, by their ability to block the activation by dopamine of certain dopamine-sensitive adenylate cyclases in the brain. This same concept of interference with adenylate cyclase receptors may also apply to the mechanism of action of other psychoactive drugs, such as LSD (see below under Serotonin).

Norepinephrine
In the cerebellum, a norepinephrine-sensitive adenylate cyclase has been identified which may mediate the inhibitory effect of norepinephrine on Purkinje cells [11]. This enzyme is stimulated by low concentrations of norepinephrine, and such stimulation is antagonized by beta-adrenergic blocking agents. Electrophysiological studies have demonstrated that norepinephrine applied iontophoretically results in a hyperpolarization and slowing in the firing rate of Purkinje cells. This effect is potentiated by phosphodiesterase inhibitors, and blocked by prostaglandin E_1 (see Chap. 13). The same effect can be produced by iontophoretically applied cAMP. Furthermore, the ability of various cAMP analogs to mimic the effect of norepinephrine correlates with the ability of these analogs to activate cAMP-dependent protein kinase from brain: those analogs which are the most effective activators of the kinase likewise cause the greatest slowing of firing rate.

Electrical stimulation of the locus coeruleus also causes slowing in the firing rate of Purkinje cells through norepinephrine-containing projections, which terminate on Purkinje cell dendrites. The effect of such locus coeruleus stimulation is potentiated by phosphodiesterase inhibitors and blocked by prostaglandin E_1. Furthermore, histochemical localization studies have shown that locus coeruleus stimulation results in a dramatic increase in the number of Purkinje cells that show cAMP staining. Evidence similar to that just described supports a role for cAMP in mediating the effects of norepinephrine on the firing rate of hippocampal and cerebral cortical pyramidal cells.

Serotonin
There is much evidence that serotonin may function as a neurotransmitter in various invertebrate ganglia, as well as in the central and peripheral nervous system of mammals. The actions of serotonin are blocked by various antiserotonergic agents, including such hallucinogens as LSD. In insect thoracic ganglia, an adenylate cyclase sensitive to low (<1 μM) concentrations of serotonin has been identified. This enzyme has characteristics similar to those of known invertebrate serotonin receptors. Activation of the enzyme by serotonin is blocked in a competitive manner by extremely low concentrations of LSD (K_i = 5 nM), 2-bromo-LSD (K_i = 5 nM), and the structurally unrelated serotonin blocker, cyproheptadine (K_i = 0.25 μM). The effect of these blockers on the serotonin-sensitive adenylate cyclase is specific, as evidenced by the fact that a dopamine-sensitive adenylate cyclase and an octopamine-sensitive adenylate cyclase present in the same ganglion are blocked only poorly by the same antiserotonergic agents. Serotonin is known to increase cAMP levels in other invertebrate ganglia as well, and a serotonin-sensitive adenylate cyclase has been identified in several areas of immature rat brain.

Octopamine
Octopamine, found in both vertebrates and invertebrates, is a phenethylamine that lacks the catechol structure of norepinephrine and dopamine. Recent studies have shown the existence, in arthropod and molluscan ganglia, of neuronal receptors for octopamine that are distinct from those for dopamine and norepinephrine. An octopamine-sensitive adenylate cyclase, distinct from adenylate cyclases activated by dopamine or serotonin, has been identified in insect thoracic ganglia, supporting the possibility that octopamine may be a neurotransmitter. This enzyme is activated by low concentrations ($<0.1 \mu M$) of octopamine. The activation is blocked by alpha- but not by beta-adrenergic antagonists. As yet, an octopamine-sensitive adenylate cyclase has not been identified in mammalian nervous tissue.

Mechanism of Action of Cyclic AMP in Synaptic Transmission

cAMP-Regulated Phosphorylation of Specific Membrane Proteins
Phosphorylation of membrane proteins that is cAMP dependent presents a possible mechanism for producing the ion-permeability changes associated with activation of certain neurotransmitter-sensitive adenylate cyclases. The endogenous phosphorylation of two synaptic-membrane proteins has been found to be markedly affected by cAMP in the rat cerebrum. Phosphorylation of both proteins is rapid, reaching maximal levels in less than five seconds, the shortest interval studied. One of these proteins has been observed to be present only in neural tissues that contain synapses. For example, this protein has been found in cerebral cortex, cerebellum, and caudate nucleus, but not in peripheral lingual nerve, liver, lung, kidney, or spleen. In contrast, the second protein has been found in both particulate and soluble fractions of all vertebrate tissues examined. Under certain conditions, it can undergo cAMP-stimulated dephosphorylation, and apparently it is similar to one in the amphibian bladder that may be involved in hormone-mediated sodium transport.

In addition to occurring in broken-cell preparations, phosphorylation of specific proteins has been demonstrated in intact neural tissue. In slices of the caudate nucleus, for example, the phosphorylation of two proteins (which may be identical to those described above) is selectively stimulated by both phosphodiesterase inhibitors and 8-bromo-cyclic AMP, a potent cyclic AMP analog. In intact *Aplysia* ganglia, both serotonin and octopamine cause a delayed (22-hr) phosphorylation of a specific membrane protein. In slices of guinea pig cerebral cortex, norepinephrine has been reported to stimulate the phosphorylation of total protein of neuronal, but not of glial, fractions.

Correlation Between Membrane Permeability and the
State of Phosphorylation of Membrane Proteins
Cyclic AMP is known to cause a slow hyperpolarization of neurons in several regions of the nervous system. These areas include the cerebellum, caudate nucleus, hippocampus, cerebral cortex, and sympathetic ganglion [11]. In the cerebellum, where the hyperpolarization of Purkinje cells has been studied by intracellular electrode techniques, it appears that cAMP-induced hyperpolarization is consistently associated

with an increase in transmembrane resistance (decrease in conductance). This is different from the classical mechanism of inhibitory postsynaptic potentials, in which the hyperpolarization is associated with an increased conductance to potassium or chloride ions. The increased membrane resistance observed with cAMP may possibly be the result of a decrease in the resting conductance to sodium or calcium ions.

If protein phosphorylation is responsible for changes in ionic permeability, a correlation should exist between the state of phosphorylation of a particular membrane protein(s) and the state of ion permeability of the membrane. In the nervous system, the short duration of permeability changes in membrane has made it difficult to establish such a correlation. However, in at least three vertebrate tissues (avian erythrocytes, amphibian bladder epithelium, and mammalian heart muscle), where the duration of hormone-induced permeability changes is much longer, it appears that a correlation does exist. In the avian erythrocyte, for example, catecholamines have been shown to increase passive sodium and potassium transport, apparently through a mechanism involving cAMP. Activation of beta-adrenergic receptors is known to cause an increase in the cAMP content of the erythrocytes, and exogenous cAMP mimics the effect of the catecholamines in increasing cation transport. Both beta-adrenergic agonists and cAMP have been shown to stimulate the rapid phosphorylation of a single, high molecular-weight protein located in the erythrocyte plasma membrane. The time course and the dose-response relationship of the catecholamine-stimulated phosphorylation of this protein agree well with the time course and dose-response relationship for the catecholamine-induced increase in cation transport. In addition, certain agents that block the catecholamine-induced increase in phosphorylation also block the catecholamine-induced increase in sodium transport. Thus it is clearly possible that the phosphorylation of this membrane protein may be associated with the increased permeability of the erythrocyte membrane to cations.

Other Possible Actions of Cyclic AMP in Neural Tissue
The above sections describe the possible involvement of cyclic AMP in certain types of synaptic transmission. Much evidence has accumulated to support such a role. This nucleotide may also have other important functions in the nervous system. Because evidence to support these postulated functions is not yet as complete as that for the involvement of cAMP in synaptic transmission, they will be mentioned only briefly.

Cell Growth and Differentiation. Cyclic AMP and its synthetic derivatives slow the division of nerve and glial cell lines in culture and cause them to express differentiated morphology (see Chap. 18).

Microtubule Function. Cyclic AMP stimulates neurite outgrowth and elongation in cultured sensory ganglia through a mechanism apparently unrelated to that used by nerve growth factor. Moreover, when highly purified preparations of neurotubules are used, a protein component of high molecular weight is capable of undergoing phosphorylation catalyzed by an endogenous cAMP-dependent protein kinase.

Glial Cell Function. Catecholamines, histamine, and adenosine cause large increases in cAMP levels in various glial tumor cell lines. The function of these increases is not yet known.

Transmitter Release. The catecholamines enhance the release of acetylcholine at the neuromuscular junction under certain conditions, and this effect is mimicked by cAMP.

Carbohydrate Metabolism. Cyclic AMP may be involved in the activation of brain phosphorylase (see Chap. 14).

Enzyme Induction and Enzyme Activation. Cyclic AMP causes the induction of tyrosine hydroxylase in adrenal medulla, sympathetic ganglia, and certain neuroblastoma lines. In the rat brain, soluble tyrosine hydroxylase is activated both by cAMP and by cAMP-dependent protein kinase. As mentioned above, cerebral ribosomal proteins serve as substrates for cAMP-dependent phosphorylation. In the pineal gland, cAMP is involved in the regulation of melatonin synthesis (see Chap. 10).

Drug Effects. In addition to the antipsychotics and hallucinogens mentioned above, certain other drugs that act on the central nervous system also affect cAMP metabolism. Among these are lithium, morphine, and ethanol, all of which appear to inhibit adenylate cyclase activity in brain.

Vision. A light-dependent activation of cyclic nucleotide phosphodiesterase has been observed that may regulate the sensitivity of photoreceptors to incident illumination (see Chap. 7).

CYCLIC GMP AND NEURONAL FUNCTION

Evidence gathered in recent years has suggested that a second cyclic nucleotide, cyclic GMP (cGMP), may be important in regulating cellular metabolism and function [18]. Although, in general, much less is known about the physiological role of cyclic GMP than that of cyclic AMP, it appears that cGMP may be involved in nervous system function. Specifically, a number of reports suggest that cGMP may mediate the muscarinic effect of acetylcholine and, possibly, of certain other neurotransmitters.

Cyclic GMP and Associated Enzymes in Nervous Tissue

In general, in both the brain and the peripheral tissues, levels of cGMP are only 1/10 to 1/50 those of cAMP. These cGMP levels in brain are similar to those in several nonneural tissues. In the cerebellum, however, concentrations of cGMP are ten times higher and nearly equal to those of cAMP.

Guanylate cyclase, which catalyzes the formation of cGMP from GTP, is found to a substantial degree in the soluble cellular fraction (in contrast to adenylate cyclase, which is almost entirely particulate). Brain guanylate cyclase isolated from the soluble fraction of osmotically shocked synaptosomes has been reported to have an especially high specific activity. Particulate fractions do have significant enzyme activity, nevertheless, and it has been shown that this activity can be increased substantially by nonionic detergents. A major hindrance to the study of cGMP function in the nervous system has been the great difficulty in demonstrating hormonal stimulation of guanylate-cyclase activity in broken-cell preparations.

Cyclic GMP is hydrolyzed to 5'-GMP by cyclic nucleotide phosphodiesterase. Although the several identified molecular species of phosphodiesterase will hydrolyze both cAMP and cGMP to a significant degree, some forms of the enzyme are more active against one nucleotide and some against the other.

Specific cGMP-dependent protein kinases have also been identified. These enzymes, which have been found in both vertebrate and invertebrate neural and nonneural tissue, are selectively stimulated by low concentrations of cGMP but not of cAMP. Endogenous substrates for cGMP-dependent protein kinase have been found in membrane fractions from each of several types of mammalian smooth muscle that have muscarinic cholinergic receptors, but such substrates have not as yet been found in neural tissue.

Taken together, the above evidence suggests that cellular regulation by cGMP may involve a set of enzyme reactions generally similar to those of the cAMP system. Thus, the presence of guanylate cyclase, cGMP phosphodiesterase, cGMP-dependent protein kinase, and specific substrates for cGMP-dependent phosphorylation provides the necessary elements for a chain of intracellular reactions that may mediate the physiological effects of cGMP.

Cyclic GMP Studies in Intact Preparations

Central Nervous System

In contrast to the difficulty of demonstrating a hormonal stimulation of cGMP formation in broken-cell systems, investigators have reported neurotransmitter-induced increases in cGMP in intact nerve tissue [10–12]. Several reports have suggested that cGMP may be involved in the action of acetylcholine in the central nervous system. In mice, systemic administration of oxytremorine, a cholinergic agent, increases the cGMP content of both cerebral cortex and cerebellum, an effect that is blocked by pretreatment with atropine. In slices of rabbit cerebral cortex or cerebellum, low concentrations of acetylcholine and such other muscarinic agonists as bethanechol increase cGMP content; this effect is prevented by the muscarinic antagonist, atropine, but not by the nicotinic antagonist, hexamethonium. The acetylcholine-induced increase in cGMP requires the presence of calcium ions in the extracellular medium. It has been suggested that an increase in intracellular calcium, which occurs in response to the applied acetylcholine, stimulates guanylate cyclase activity, and that this causes the observed increase in cGMP level.

Depolarizing agents, such as veratridine, ouabain, and potassium, also increase cGMP levels in slices. This rise in cGMP also requires the presence of calcium. The effect is not blocked by atropine, indicating that acetylcholine is probably not involved in this phenomenon. The amino acids glutamate and glycine increase cerebellar cGMP when applied to slices or when given by intraventricular injection. Gamma-aminobutyric acid (GABA), on the other hand, causes a decrease in cerebellar cGMP when given intraventricularly.

Electrophysiological studies of neurons in the rat cerebral cortex have shown that iontophoretically applied cGMP, like acetylcholine, increases the rate of firing of certain pyramidal cells. Cortical pyramidal cells which are unaffected by acetyl-choline show no response to cGMP. These latter cells often display a decrease in firing rate in response to norepinephrine, an effect mimicked by cAMP.

Peripheral Nervous System

In both frog and rabbit sympathetic ganglia, stimulation of the preganglionic fibers causes a slow excitatory (depolarizing) postsynaptic potential, which is blocked by atropine. The same effect can be produced by applying acetylcholine or bethanechol. Substantial evidence suggests that cGMP may mediate the muscarinic effects of acetyl-choline in these peripheral sympathetic ganglia. In the frog ganglion, brief presynaptic stimulation at physiological rates causes a twofold increase in cGMP content. This increase in cGMP is blocked by the muscarinic antagonist atropine. When the release of neurotransmitter is prevented by a high-magnesium, low-calcium Ringer solution, the same preganglionic stimulation no longer raises cGMP levels.

In slices of bovine superior cervical ganglion, both acetylcholine and bethanechol increase cGMP content. This effect is blocked by atropine, but not by the nicotinic blocking agent hexamethonium. Histochemical evidence also shows that the increase in cGMP caused by acetylcholine takes place in the postganglionic neurons.

Electrophysiological studies in the rabbit have shown that the dibutyryl derivative of cGMP mimics electrical stimulation and acetylcholine in causing a slow depolariza-tion of the sympathetic ganglion, whereas cAMP causes a slow hyperpolarization.

Muscarinic cholinergic end organs, such as intestine, heart, and ductus deferens, all display increases in cGMP when exposed to acetylcholine, as does the sympathetic ganglion. Furthermore, these increases are blocked specifically by muscarinic antag-onists. Electrical stimulation of the vagus nerve has been shown to increase cGMP in stomach mucosa, and this effect, too, is abolished by atropine. A cGMP-dependent protein kinase, found in intestine, ductus deferens, and uterus, causes the phosphoryl-ation of specific membrane proteins. It seems reasonable to speculate that the mem-brane-permeability changes that occur in response to activation of muscarinic receptors in cholinergically innervated tissues may be mediated, in part, through an acetylcholine-induced increase in cGMP. This increase, in turn, causes an activation of a cGMP-dependent protein kinase and, thereby, a phosphorylation of specific membrane proteins.

CONCLUSION

The available evidence supports a role for cAMP as mediator of the postsynaptic effects of a number of neurotransmitters. Neurotransmitter diffusing across the synaptic cleft and contacting the postsynaptic membrane is thought to activate a specific adenylate cyclase located in that membrane. The resulting increase in cyclic AMP in the receptor cell activates a cAMP-dependent protein kinase (or phospho-protein phosphatase), which changes the level of phosphorylation of a particular membrane-bound protein (or proteins). It is postulated that the phosphorylated protein then causes a change in ion permeability of the nerve membrane. Finally, the altered membrane permeability results in a change in membrane potential and nerve excitability. In some cases, this change in nerve excitability may be relatively long lasting, and it is possible that, under certain circumstances, such a mechanism could form the basis for plastic changes in the nervous system. Less extensive data suggest an analogous role for cGMP as mediator of the postsynaptic effects of some other neurotransmitters.

The full extent of the involvement of cAMP and cGMP in neuronal function is not yet known. It seems likely that these nucleotides are involved in several aspects of the functioning of the nervous system. The nature and extent of this involvement represent important areas for future study.

ACKNOWLEDGMENT

Preparation of this chapter was supported in part by H. E. W. grants MH-17387 and NS-08440.

REFERENCES

*1. Robison, G. A., Butcher, R. W., and Sutherland, E. W. *Cyclic AMP.* New York: Academic, 1971.
*2. Bitensky, M. W., and Gorman, R. E. Cellular Responses to Cyclic AMP. In Butler, J. A. V., and Noble, O. (Eds.), *Progress in Biophysics and Molecular Biology.* Vol. 26. New York: Pergamon, 1973, p. 409.
*3. Perkins, J. D. Adenyl cyclase. *Adv. Cyclic Nucleotide Res.* 3:1, 1973.
*4. Appleman, M. M., Thompson, W. J., and Russell, T. R. Cyclic nucleotide phosphodiesterases. *Adv. Cyclic Nucleotide Res.* 3:65, 1973.
*5. Amer, M. S., and Kreighbaum, W. E. Cyclic nucleotide phosphodiesterases: Properties, activators, inhibitors, structure-activity relationships, and possible role in drug development. *J. Pharm. Sci.* 64:1, 1975.
*6. Greengard, P., and Kuo, J. F. On the mechanism of action of cyclic AMP. *Adv. Biochem. Psychopharmacol.* 3:287, 1970.
*7. Krebs, E. G. Protein kinases. *Curr. Top. Cell. Regul.* 5:99, 1972.
*8. Langan, T. A. Protein kinases and protein kinase substrates. *Adv. Cyclic Nucleotide Res.* 3:99, 1973.
*9. Drummond, G. I., Greengard, P., and Robison, G. A. (Eds.), Second International Conference on Cyclic AMP. *Adv. Cyclic Nucleotide Res.* 5, 1975 (several articles).
*10. Drummond, G. I., and Ma, Y. Metabolism and Function of Cyclic AMP in Nerve. In Kerkut, G. A., and Phillis, J. W. (Eds.), *Progress in Neurobiology.* Vol. 2. New York: Pergamon, 1973, p. 119.
*11. Bloom, F. E. The Role of Cyclic Nucleotides in Central Synaptic Function. In *Reviews of Physiology, Biochemistry, and Experimental Pharmacology.* New York: Springer-Verlag, 1976.
*12. Nathanson, J. A., and Greengard, P. Cyclic nucleotides and the nervous system. *Physiol. Rev.* In press.
*13. Daly, J. W. The Role of Cyclic Nucleotides in the Nervous System. In Iversen, L. L., Iversen, S. D., and Snyder, S. H. (Eds.), *Handbook of Psychopharmacology.* New York: Plenum, 1975, p. 47.
*14. Rall, T. W. Role of adenosine $3',5'$-monophosphate (cyclic AMP) in actions of catecholamines. *Pharmacol. Rev.* 24:399, 1972.

*Asterisks denote key references.

*15. McIlwain, H. Cyclic AMP and Tissues of The Brain. In Rabin, B. R., and Freedman, R. B. (Eds.), *Effects of Drugs on Cellular Control Mechanisms.* New York: Macmillan, 1972, p. 281.

*16. Greengard, P., Nathanson, J. A., and Kebabian, J. W. Dopamine- , Octopamine- , and Serotonin-Sensitive Adenylate Cyclase: Possible Receptors in Aminergic Neurotransmission. In Usdin, E., and Snyder, S. (Eds.), *Frontiers in Catecholamine Research.* New York: Pergamon, 1973, p. 377.

*17. Von Hungen, K., and Roberts, S. Neurotransmitter-Sensitive Adenylate Cyclase Systems in the Brain. In Ehrenpreis, S., and Kopin, I. J. (Eds.), *Reviews of Neuroscience.* Vol. 1. New York: Raven, 1974, p. 231.

*18. Goldberg, N. D., O'Dea, R. F., and Haddox, M. K. Cyclic GMP. *Adv. Cyclic Nucleotide Res.* 3:155, 1973.

*Key references.

Chapter 13

Prostaglandins and Synaptic Transmission

Leonhard S. Wolfe

The discovery of the prostaglandins dates back to 1933, when von Euler and Gold-blatt described active principles in human seminal fluid and in extracts of seminal vesicular glands of sheep that stimulated isolated intestinal and uterine muscle and lowered the blood pressure in intact animals. It was shown that these activities were due to highly potent acidic lipids, and von Euler proposed the name "prostaglandin" because initially they were thought to derive from the prostate gland. Subsequently, this proved to be wrong; however, the name has persisted.

Intensive work on the purification and elucidation of the structure of these active principles was carried out by Bergström and his team in Sweden over the years 1956–1963 [1]. Two classes of prostaglandins were separated by solvent partition between ether and phosphate buffer. The compounds more soluble in ether were called prostaglandin E and those more soluble in the phosphate buffer were called prosta-glandin F (after *Fosfat*). Elegant chemical degradation studies and the powerful techniques of gas liquid chromatography combined with mass spectrometry, which also were being developed in Sweden at the same time, resulted in the elucidation of the structures of this new group of compounds, even though only a few milligrams of the purified materials were then available. The prostaglandins were found to be a family of twenty-carbon cyclopentane carboxylic acids in which individual members differ in the positions of oxygenation and unsaturation.

Immediately after the structures were established, their origin from the essential fatty acids was demonstrated by two groups of scientists — Bergström in Sweden and van Dorp in Holland — who showed that microsomal fractions from sheep vesicular glands bioconverted $\Delta^{5,8,11,14}$-all *cis*-eicosatetraenoic acid (arachidonic acid) into PGE_2. Subsequently, it was shown that $PGF_{2\alpha}$ was also formed from arachidonic acid; that $\Delta^{8,11,14}$ eicosatrienoic acid (homo-γ-linolenic acid) was the precursor of PGE_1 and $PGF_{1\alpha}$; and that $\Delta^{5,8,11,14,17}$ eicosapentaenoic acid was the precursor of PGE_3 and $PGF_{3\alpha}$ [1]. These transformations are illustrated in Figure 13-1.

The capacity for biosynthesis of prostaglandins is widely distributed in the animal kingdom. In most mammalian tissues (including nervous tissue, particularly in the human), arachidonic acid is quantitatively the most important precursor for local endogenous biosynthesis of prostaglandins.

In the older literature, one finds many names given independently to the biological activity of lipid extracts of tissues (slow-reacting substance C, Darmstoff, intestinal stimulant acid lipids, irin, menstrual stimulant, vasodepressor lipid, medullin, unsat-urated hydroxy fatty-acid fractions from brain), which are now known to owe much of their biological activity to the presence of one or more of the prostaglandins, prostaglandin endoperoxides, and lipoperoxides [1, 2].

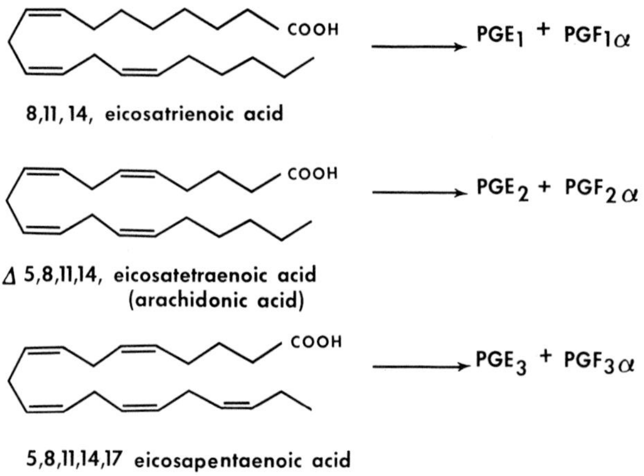

8,11,14, eicosatrienoic acid $\longrightarrow$ PGE$_1$ + PGF$_{1\alpha}$

Δ 5,8,11,14, eicosatetraenoic acid (arachidonic acid) $\longrightarrow$ PGE$_2$ + PGF$_{2\alpha}$

5,8,11,14,17 eicosapentaenoic acid $\longrightarrow$ PGE$_3$ + PGF$_{3\alpha}$

Figure 13-1
Formation of prostaglandins from polyunsaturated fatty acids.

The pharmacological actions of the prostaglandins are exceedingly diverse and differ according to type, tissue, species, and whether tested in vivo or in vitro. The physiological functions subserved by them are still far from clear. There is growing appreciation, however, that this new class of molecules is formed and acts locally in tissues in response to a wide range of stimuli, both physiological and pathophysiological. They can be regarded as mediators, modulators, or regulators of many processes coupled with stimulus secretion. They are involved in the action of many hormones on their target cells, in the translation of membrane responses into intracellular processes, and in the communication between cells, especially in defensive reactions induced by trauma or stress. Prostaglandins do not appear essential for life, but their absence or excessive formation may profoundly modify homeostasis at the cellular level [3, 4].

NAMES AND TYPES

The chemical names of all prostaglandins are based at present on prostanoic acid, a hypothesized twenty-carbon cyclopentane carboxylic acid. The type designations for the abbreviated names of the primary prostaglandins are based on the functionality in the cyclopentane ring, i.e., β-hydroxyketone for E-type; 1,3-diol for F-type; α,β unsaturated ketone for A-type; and so on. Class designations within each type are indicated as subscripts one to three, which refer to the number of carbon-carbon double bonds. The primary prostaglandins from mammalian sources have a single stereochemical form; the carboxyhexyl side chain on the cyclopentane ring is in the *trans* configuration to the hydroxyoctyl side chain; the hydroxyl groups in the cyclopentane ring are in the α-configuration; and the hydroxyl group is in the S-configuration at carbon-15. Figure 13-2 lists the structures and names of natural prostaglandins derived from arachidonic acid.

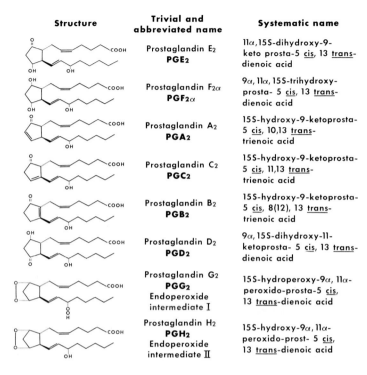

Structure	Trivial and abbreviated name	Systematic name
	Prostaglandin E₂ **PGE₂**	11α,15S-dihydroxy-9-keto prosta-5 <u>cis</u>, 13 <u>trans</u>-dienoic acid
	Prostaglandin F₂α **PGF₂α**	9α,11α,15S-trihydroxy-prosta- 5 <u>cis</u>, 13 <u>trans</u>-dienoic acid
	Prostaglandin A₂ **PGA₂**	15S-hydroxy-9-ketoprosta-5 <u>cis</u>, 10,13 <u>trans</u>-trienoic acid
	Prostaglandin C₂ **PGC₂**	15S-hydroxy-9-ketoprosta-5 <u>cis</u>, 11,13 <u>trans</u>-trienoic acid
	Prostaglandin B₂ **PGB₂**	15S-hydroxy-9-ketoprosta-5 <u>cis</u>, 8(12), 13 <u>trans</u>-trienoic acid
	Prostaglandin D₂ **PGD₂**	9α, 15S-dihydroxy-11-ketoprosta- 5 <u>cis</u>, 13 <u>trans</u>-dienoic acid
	Prostaglandin G₂ **PGG₂** Endoperoxide intermediate I	15S-hydroperoxy-9α, 11α-peroxido-prosta-5 <u>cis</u>, 13 <u>trans</u>-dienoic acid
	Prostaglandin H₂ **PGH₂** Endoperoxide intermediate II	15S-hydroxy-9α, 11α-peroxido-prost- 5 <u>cis</u>, 13 <u>trans</u>-dienoic acid

Figure 13-2
Structures and names of prostaglandins formed from arachidonic acid. The endoperoxide prostaglandins have considerably higher biological potency on certain smooth-muscle preparations and on human platelets than do PGE_2 or $PGF_{2\alpha}$.

MECHANISM OF BIOSYNTHESIS

The multienzyme system required for the biosynthesis of prostaglandins, termed *prostaglandin synthetase,* is tightly bound to membrane elements, and to date attempts to solubilize and purify enzyme components of the system have only been partly successful. Knowledge of the overall mechanism for prostaglandin biosynthesis is largely due to the investigations of Samuelsson and his co-workers in Sweden [5, 6, 7]. The initial step is the removal, by a lipoxygenaselike reaction, of a prochiral hydrogen in the *S*-configuration at C-13 of the polyunsaturated fatty acid substrate. The Δ^{11} double bond isomerizes into the Δ^{12} position, and molecular oxygen is inserted at carbon-11 to form 11-peroxy eicosaenoic acids. These intermediates are then converted by dioxygenase reactions, recently termed the *fatty acid cyclooxygenases,* into 15-hydroperoxy prostaglandin endoperoxides through cyclization and formation of a further double bond at Δ^{13}. Reduction of the 15-hydroperoxy group by a peroxidase produces the 15-hydroxy prostaglandin endoperoxides. Endoperoxide isomerase converts the endoperoxide intermediates into E-type prostaglandins, and a reductase (or reducing cofactors) converts them into F-type prostaglandins. Further transformations of the E-type prostaglandins into the A, C, or B types may occur, depending on the particular tissue or animal species. A schematic outline of the formation of PGE_2 and $PGF_{2\alpha}$ from arachidonic acid with certain side reactions is shown in Figure 13-3.

Two endoperoxide intermediates in the biosynthesis of prostaglandins from arachidonic acids, i.e., PGG_2 and PGH_2, have been isolated from several tissues (seminal vesicles, platelets, lung) and have a half-life in aqueous solutions of about 5 min. What is very interesting is that their biological activity in several systems (contraction of the rabbit aorta and isolated tracheal smooth muscle, aggregation of human platelets, and increase in airway resistance in the guinea pig) is many times greater than that of the final products, PGE_2 or $PGF_{2\alpha}$, or both. The endoperoxides may have considerable importance in physiological processes and also in such pathophysiological mechanisms as anaphylaxis and thrombosis [2, 4, 8–11].

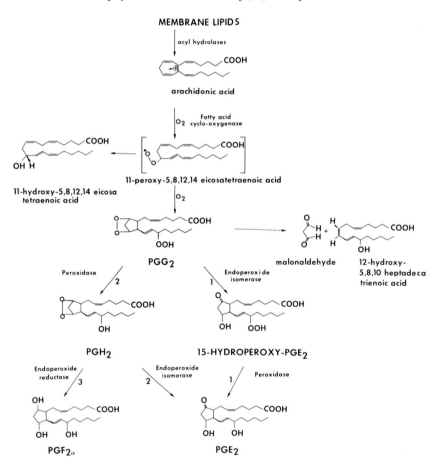

Figure 13-3
Pathways in the biosynthesis of prostaglandin endoperoxides, PGE_2 and $PGF_{2\alpha}$ from arachidonic acid. The rate of endogenous biosynthesis is controlled by the activity of acyl hydrolases, which release free arachidonic acid from complex lipids, particularly phospholipids. Side reactions are indicated by dotted arrows. The favored pathway for the biosynthesis of PGE_2 is $PGG_2 \rightarrow$ 15-hydroperoxy-$PGE_2 \rightarrow PGE_2$.

The prostaglandins or their endoperoxides cannot be formed from precursor fatty acids that are esterified to triglycerides or phospholipids. Unesterified precursor fatty acids, such as arachidonic acid, are virtually undetectable in most tissues in vivo, so a controlled release from complex membrane lipids by acyl hydrolases to the multienzyme complex of the prostaglandin synthetase is thought to be the rate-limiting step for endogenous biosynthesis. Stimulation of phospholipase A_2 activity within the cell membrane is the most likely reaction, but this is by no means proved. The delivery of specific substrate fatty acids to the fatty-acid cyclooxygenases may be controlled and compartmentalized. In some tissues, the biosynthetic sequence may stop at the endoperoxide stage, and in such cases endoperoxides would accumulate and be released; in other tissues, the reaction would form the prostaglandins.

It has been known for some time that the biosynthesis of prostaglandins can be inhibited by several fatty acids that are not substrates for the cyclooxygenases (i.e., γ-linolenic acid, eicosatetraynoic acid). In 1971, Vane and his co-workers found that nonsteroidal, antiinflammatory drugs, such as aspirin, indomethacin, and the fenamates, were potent irreversible inhibitors of the prostaglandin synthetase. This discovery has been of great importance in determining physiological and pathophysiological processes that involve the mediation of the prostaglandins [4, 8]. A specific inhibitor of prostaglandin action has not yet been discovered.

CATABOLISM

Studies with tritium-labeled prostaglandins have shown that both the E and F types are converted extremely rapidly into inactive metabolites when injected into the systemic circulation. These metabolites are excreted principally in the urine [6, 12–14]. Four main types of transformation take place:

1. Oxidation of the allylic alcohol group at carbon-15 to the 15-ketoprostaglandins by a specific NADH-dependent 15-hydroxyprostaglandin dehydrogenase. The highest activities of this enzyme are found in the lung, spleen, and kidney cortex and the lowest in brain.
2. Reduction of the Δ^{13} double bond to form dihydro compounds. There is much variability in the activity of the reductase in different tissues, among animal species, and with stage of development.
3. β-Oxidation occurs principally in the liver to form *dinor* or *tetranor* prostaglandins.
4. ω-Oxidation by microsomal enzymes transforms the metabolites further into ω-1, ω-2 hydroxylated compounds and then to dicarboxylic acids.

The major urinary metabolites in man from $PGF_{1\alpha}$ and $PGF_{2\alpha}$ are 5α,7α-dihydroxy-11-keto-tetranor-prostane-1,16-dioic acid. The analogous metabolite from PGE_1 and PGE_2 is 7α-hydroxy-5,11-diketo-tetranor-prostane-1,16-dioic acid. Figure 13-4 illustrates these interconversions. There are many other urinary metabolites; these vary with the animal species. A 9-keto reductase activity that would lead to the conversion

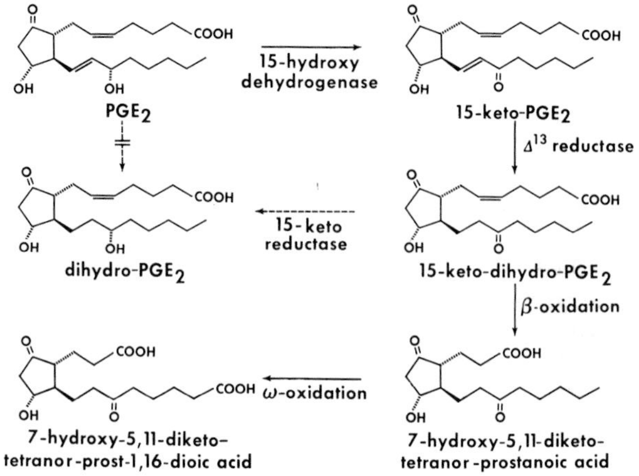

Figure 13-4
Major pathways in the metabolism of PGE$_2$ in man. The tetranor dicarboxylic acids are found in the urine.

of E to F prostaglandins has been reported recently in a number of tissues and species. The details can be obtained from the literature. An interesting finding in man is that each day males excrete in the urine considerably more prostaglandin metabolites than do females, indicating a greater total body synthesis [13].

FORMATION AND RELEASE FROM THE CENTRAL NERVOUS SYSTEM

The spontaneous release of prostaglandins into fluids that superfuse the cerebral cortex, cerebellum, spinal cord, and the cerebral ventricles is now well documented. Direct electrical stimulation of the cerebral cortex or evoked responses through stimulation of peripheral nerves increase the amounts of prostaglandins released severalfold. This accelerated release during stimulation does not derive from pre-formed stores but from acceleration of de novo synthesis. It seems clear that prosta-glandin formation and release from brain can be affected by many different stimuli — neural, hormonal, drugs, and trauma [1, 15–17].

The capacity for prostaglandin biosynthesis in tissue slices, homogenates, or microsomal fractions is often measured by determining the conversion to prosta-glandins of precursor fatty acids added to the incubation medium. By these meth-ods, brain tissue shows an exceedingly low capacity to form prostaglandins from exogenously added substrates. Considerable amounts are formed, however, during the incubation from endogenous precursors. Measurement of endogenous synthetic capacity is a more valid procedure by which to study prostaglandin biosynthesis in brain [16]. Table 13-1 illustrates this point for rat and cat cerebral tissues. The endogenous formation of PGF$_{2\alpha}$ and PGE$_2$ by slices of rat cerebral cortex is linear, with time up to 60 min of incubation. This contrasts with incubated homogenates,

Table 13-1. Endogenous Biosynthesis of Prostaglandins by Slices of Rat Cerebral Cortex

Tissue	$PGF_{2\alpha}$	PGE_2	$F_{2\alpha}/E_2$
	ng/100 mg tissue[a]		
Immediately frozen	1.1	–	–
Slices not incubated	8.6	4.2	2.0
60-min incubations[b]			
Slices	59.3	17.6	3.4
Slices + norepinephrine 1 mM	125.1	24.5	5.1
Slices + indomethacin 1×10^{-5} M	11.7	7.1	1.6
Homogenate	75.4	28.9	2.7
Homogenate + norepinephrine 1 mM	248.2	40.1	6.2
Homogenate + indomethacin 1×10^{-4} M	15.6	11.1	1.4

[a]The prostaglandins were measured by gas chromatography–mass spectrometry, using deuterated prostaglandins as carriers and internal standards. Values are means of 4 to 6 determinations.
[b]Incubations in ringer-bicarbonate-glucose (RBG) at pH 7.4.
Source: Wolfe and Pappius (in preparation and ref. [16]).

in which the most rapid synthesis is in the first 5–10 min. During incubation in the intact tissue, arachidonate is liberated from a complex membrane-lipid compartment, and its rate of release limits the rate of formation of prostaglandins. Biogenic amines (serotonin, dopamine, norepinephrine) stimulate greatly the endogenous synthesis. This is due, at least in part, to a stimulation of the liberation of unesterified precursor fatty acids from membrane phospholipids, thus making more substrate available.

The capacity of brain tissue to degrade prostaglandins is low, compared with most nonneural tissues. The prostaglandins formed during normal neuronal activity, or in response to other stimuli, diffuse into the cerebrospinal fluid and are transported out to the circulation by the choroid plexus and are inactivated in extraneural tissues. Thus, measurement of prostaglandin levels in cerebrospinal fluid can give a good indication of conditions causing increased prostaglandin formation in vivo. Normally, human CSF contains less than 100 picograms per milliliter (pg/ml) of $PGF_{2\alpha}$. Greatly increased levels are found in patients with epileptic seizures, following cerebral infarction, meningoencephalitis, or surgical brain trauma [16]. Prostaglandin levels are also significantly elevated in ventricular CSF of cats made febrile when bacterial pyrogens are administered.

RELEASE FROM AUTONOMIC NERVE SYNAPSES
AND REGULATION OF NEUROTRANSMISSION

Stimulation of sympathetic or parasympathetic nerves is associated with release of prostaglandins, principally PGE_2 and $PGF_{2\alpha}$. The release rate is, in general, related to the stimulus frequency and decreases to spontaneous levels when stimulation ceases. The postsynaptic effector-cell membrane is most likely the site of synthesis because receptor blockade by drugs (hyoscine, atropine, phenoxybenzamine, dibenzyline) inhibits prostaglandin release. Addition of the neurotransmitter also stimulates prostaglandin release. The many investigations of Hedqvist and co-workers on stimulation of sympathetic nerves to various organs (heart, oviduct, spleen, vas deferens) have shown that the effector responses are inhibited by PGE_1 and PGE_2 but not by the PGF series of prostaglandins [12, 15, 16, 18, 19]. The inhibition can be produced by picogram and nanogram amounts well within the range of physiological concentrations.

Hedqvist put forward the hypothesis that endogenous PGE_2 formed and released from the postsynaptic effector membrane during stimulation inhibits the release of norepinephrine from presynaptic terminals. Good evidence for this hypothesis of negative-feedback control was obtained by using specific inhibitors of endogenous prostaglandin synthesis, namely 5,8,11,14-eicosatetraynoic acid and indomethacin [8, 19]. In the presence of these compounds, the outflow of norepinephrine from sympathetic nerve terminals is increased with a corresponding increase in the effector responses. The influx of calcium into nerve terminals is an essential part of stimulus-coupled exocytosis of norepinephrine along with adenine nucleotides and dopamine-β-hydroxylase from sympathetic nerve terminals (see Chaps. 8 and 10), and the E-type prostaglandins inhibit this process. The Hedqvist hypothesis is diagrammatically represented in Figure 13-5.

Release of PGE_2 from adrenergic pathways in the central nervous system may also be involved in the regulation of neurotransmitter release. The stimulated release of norepinephrine and dopamine from rat cerebral cortex and neostriatum in vitro is reduced by PGE_2. The disappearance time of dopamine histofluorescence in the neostriatum of rats that have been pretreated with a tyrosine hydroxylase inhibitor is decreased when PGE_2 is infused into the carotid artery.

Although much evidence favors a role for E-type prostaglandins in the control of autonomic neurotransmission, evidence is growing that the F-type prostaglandins also are involved, but appear to have a facilitatory action [16, 18]. Prostaglandin $F_{2\alpha}$ is known to potentiate the action of norepinephrine on vascular systems. Thus, the two major types of prostaglandins may have competitive interactions at adrenergic

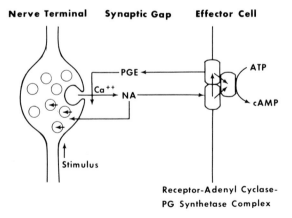

Figure 13-5
Diagrammatic representation of the feedback control of the release of norepinephrine by E-type prostaglandins (Hedqvist's hypothesis).

synapses. There is no evidence that prostaglandins have any action on the release of acetylcholine or on the effector responses at nicotinic cholinoreceptors in sympathetic ganglia or at the neuromuscular junction.

EFFECTS ON CYCLIC NUCLEOTIDES

There is much evidence that prostaglandins of the E type stimulate cyclic adenosine monophosphate (cAMP) formation in many tissues, including cerebral cortex slices in vitro and in cultured neural tissues [12, 20–22]. Which brain-cell types – neurons, or glial cells – are specifically affected is not certain. Synergism of action on adenylate cyclase occurs with combinations of hormones and prostaglandins in intact brain tissue, and the meaning of these complex interactions is far from clear (see also Chap. 12). Homogenization destroys this hormonal responsiveness. There is growing appreciation, however, that high concentrations of intracellular cAMP are associated with inhibited cell growth of various cell lines in culture. Mouse neuroblastoma cells in culture, for example, respond to dibutyryl cAMP and prostaglandins E_1 and E_2 with a striking morphological differentiation and the development of axonlike processes. Is this solely a pharmacological response, or does it indicate a physiological implication of E-type prostaglandins in the cellular differentiation process? This deserves further investigation.

Another aspect of the relationship between cAMP and the E-type prostaglandins that is of particular relevance to the central nervous system is found in the report by Collier and Roy [23] that morphine and related drugs inhibit the stimulation of cAMP formation by E-type prostaglandins. These studies raise the possibility that the pharmacological effects of opiates are related to inhibition of the action of endogenously produced prostaglandins on cAMP levels in brain.

High activities of guanylate cyclase are present in brain. A surprising and interesting finding is that cholinergic agonists, serotonin and $PGF_{2\alpha}$, elevate tissue levels of cGMP. It has been speculated that $PGF_{2\alpha}$ facilitates excitation of cholinergic and serotoninergic pathways in brain by activating guanylcyclase, but as yet no solid evidence exists.

EFFECTS ON CNS FUNCTION

Direct intraventricular injections of E-type prostaglandins in experimental animals has a marked sedative effect, which may progress to a stuporous and catatoniclike state. The effect of barbiturates is potentiated and animals are protected to some degree against the convulsions produced by pentylenetetrazol or electroshock. The effects of injected prostaglandins are prolonged and are present long after the injected prostaglandins have been removed through the choroid plexus. The mechanism of the sedative effects is unclear but is likely caused by a slowly reversible depression of neurotransmission in certain central pathways through the reduction of neurotransmitter release, as has been demonstrated in peripheral adrenergic pathways.

Prostaglandins of both E and F types have complex actions on spinal-cord reflexes and brain-stem cardioregulatory and respiratory centers. Inhibitory and excitatory effects have been found, the results varying with the animal species and type of prostaglandin studied. The interested reader should consult the literature for details of these pharmacological actions and their relationships to neurophysiological mechanisms [24–26]. Clearly, both excitatory and inhibitory actions are produced, and the effects are prolonged. Such prolonged changes are inconsistent with a transmitter-like role of prostaglandins and favor a modulatory action on synaptic transmission.

FEVER AND TEMPERATURE REGULATION

In recent years, convincing evidence implicates stimulated endogenous biosynthesis of prostaglandins by the anterior hypothalamus in the genesis of fever caused by bacterial pyrogens and endotoxins [4, 8, 12, 15, 16]. This is based on the following results:

1. Prostaglandins of both E and F types are synthesized by hypothalamic tissue in vitro.
2. The E-type prostaglandins, injected intraventricularly, raise the temperature in all animal species examined so far in a dose-dependent fashion. PGE_1 is the most powerful pyretic agent known when injected directly into the anterior hypothalamus.
3. Fever is produced in unanesthetized cats by microinjections of as little as 2 ng of PGE_1 directly into the anterior hypothalamus. This fever is unaffected by paracetamol (4-acetamidophenol) or aspirinlike drugs.
4. Pyrexia produced by intracerebral injections of pyrogens is inhibited by potent prostaglandin synthetase inhibitors, e.g., indomethacin.
5. The amounts of PGE_2 in cerebrospinal fluid are greatly increased during fever produced by bacterial pyrogens and endotoxins.

The current hypothesis is that pyrogens accelerate endogenous prostaglandin biosynthesis and that the endoperoxide intermediates or prostaglandins liberated locally alter the balance of neurotransmitter release in the hypothalamus. According to recent theory, temperature regulation is explained in terms of an antagonistic relationship (thermogenic or thermolytic) between norepinephrine and 5-hydroxytryptamine. Excessive production of prostaglandins would change the balanced release and actions of these two neurotransmitters and so mediate the temperature rise. It should be clearly understood, however, that a mediator role of prostaglandins is invoked only in fevers sensitive to those antipyretic drugs that inhibit prostaglandin biosynthesis, and there is little direct evidence to implicate prostaglandins in the regulation of normal body temperature.

RELEASE OF ANTERIOR PITUITARY HORMONES

Another area of much current interest and in which prostaglandins may be involved is the regulation of the synthesis and secretion of hormones by the anterior pituitary through peptide neurohormones (releasing factors) that are synthesized and released by hypothalamic neurons into the hypophyseal portal blood vessels [15, 16]. Inhibitors of prostaglandin synthesis are known to interfere with pituitary gonadotropin secretion. When prostaglandin E_2 and, to a lesser extent, PGE_1 are given intraventricularly to rats, the plasma level of luteinizing hormone increases greatly. The site of action is at the hypothalamic level since direct injections into the pituitary gland have absolutely no effect. Microinjections of aspirin into the hypothalamus of the rat inhibit ovulation but, when given simultaneously with $PGF_{2\alpha}$ or dopamine, ovulation is restored. Dopamine alone had no effect. Direct application of both E- and F-type prostaglandins to basal hypothalamic neurons in doses in the nanogram range are effective in stimulating ACTH secretion, an effect which is abolished by morphine. These results, together with a variety of other studies, strongly suggest that prostaglandins can regulate the release and action of neurotransmitters, such as norepinephrine and dopamine, in the hypothalamus. These, in turn, can alter the release of the hypophysiotropic hormones from the hypothalamus (Chap. 22). Specific neuronal pathways are possible for the production and release of each hypophysiotropic hormone, and these pathways may respond differently to the E- or F-type prostaglandin. Thus we come back to a recurring theme on the physiological role of prostaglandins in nervous tissue: that the local formation of prostaglandins in the nervous system, when triggered by specific stimuli, can interact with and change monoaminergic release mechanisms.

CEREBRAL BLOOD FLOW

Much recent interest centers on the contributions that prostaglandins might make to both normal and pathophysiological responses of cerebral blood vessels. Agents that constrict cerebral blood vessels are present in platelet-rich plasma. Serotonin is one of these, but its vasoconstrictive action is short lived. Of much greater importance is the release of prostaglandin endoperoxides derived from arachidonic acid and release of PGE_2 and $PGF_{2\alpha}$ [9–11]. Both PGE_2 and $PGF_{2\alpha}$ have vasoconstrictive actions on cerebral arterioles and reduce cerebral blood flow [16]. Furthermore, the vasoconstriction is prolonged, indeed, showing many similarities to the vasospasm observed in clinical situations after subarachnoid hemorrhage. Prostaglandin endoperoxides are known to form in considerable amounts during platelet aggregation and have an exceedingly potent contracting action on vascular smooth muscle. Thus, in cerebral thromboembolism (stroke), the release of prostaglandin endoperoxides with their potent vasoconstrictive actions could greatly contribute to the development of local ischemia and expansion of the damage to the brain. In addition, the formation and release of prostaglandins, particularly $PGF_{2\alpha}$, is increased in ischemic brain and could contribute to the expansion of the brain lesion through restriction of blood flow. Evidence accumulating from clinical trials shows that aspirin is of real value in the prevention of focal cerebral ischemia due to platelet emboli. This protective action appears directly related to inhibition of the biosynthesis of prostaglandins and the endoperoxide intermediates by platelets.

SUMMARY

Prostaglandins are formed and released from central and peripheral neural pathways, and the rate of biosynthesis can be increased by both physiological and pathological stimuli. Prostaglandins (1) have direct effects on autonomic and catecholaminergic neurons; (2) cause tranquilization; (3) affect neurons of the cerebellum, brain stem, and spinal cord; (4) affect the hypothalamus, leading to alterations in body temperature, feeding behavior, and release of pituitary hormones; and (5) alter cerebral blood flow. The wide variety of neuropharmacological effects that the prostaglandins exert suggests that they may regulate many neurophysiological mechanisms. At the present state of knowledge, the basis of prostaglandin action in nervous tissue appears to be alteration of neurotransmitter release mechanisms and stimulation of cyclic nucleotide synthesis [27].

REFERENCES

*1. Bergström, S., Carlson, L. A., and Weeks, J. R. The prostaglandins: A family of biologically active lipids. *Pharmacol. Rev.* 20:1, 1968.

*2. Vargaftig, B. B. Search for Common Mechanisms Underlying the Various Effects of Putative Inflammatory Mediators. In Ramwell, P. W. (Ed.), *The Prostaglandins.* Vol. 2. New York: Plenum, 1974, p. 205.

*3. Andersen, N. H., and Ramwell, P. W. Biological aspects of prostaglandins. *Arch. Intern. Med.* 133:30, 1974.

*4. Ferreira, S. H., and Vane, J. R. Aspirin and Prostaglandins. In Ramwell, P. W. (Ed.), *The Prostaglandins.* Vol. 2. New York: Plenum, 1974, p. 1.

*5. Hamberg, M., Samuelsson, B., Bjorkhem, I., and Danielsson, H. Oxygenases in Fatty Acid and Steroid Metabolism. In Hayaishi, O. (Ed.), *Molecular Mechanisms of Oxygen Activation.* New York: Academic, 1974.

*6. Samuelsson, B. Structures, Biosynthesis and Metabolism of Prostaglandins. In Wakil, S. J. (Ed.), *Lipid Metabolism.* New York: Academic, 1970.

*7. Samuelsson, B. Biosynthesis of prostaglandins. *Fed. Proc.* 31:1442, 1972.

*8. Robinson, H. J., and Vane, J. R. (Eds.). *Prostaglandin Synthetase Inhibitors.* New York: Raven, 1974.

9. Hamberg, M., Svensson, J., Wakabayashi, T., and Samuelsson, B. Isolation and structure of two prostaglandin endoperoxides that cause platelet aggregation. *Proc. Natl. Acad. Sci. U.S.A.* 71:345, 1974.

10. Willis, A. L., Vane, R. M., Kuhn, D. C., Scott, C. G., and Petrin, M. An endoperoxide aggregator (LASS), formed in platelets in response to thrombotic stimuli. *Prostaglandins* 8:453, 1974.

11. Hamberg, M., and Samuelsson, B. Prostaglandin endoperoxides. Novel transformations of arachidonic acid in human platelets. *Proc. Natl. Acad. Sci. U.S.A.* 9:3400, 1974.

*12. Bergström, S. (Ed.). *International Conference on Prostaglandins, Vienna, September 25 to 28, 1972.* New York: Pergamon, 1973.

13. Samuelsson, B. Quantitative Aspects on Prostaglandin Synthesis in Man. In [12], p. 7.

*14. Marrazzi, M. A., and Anderson, N. H. Prostaglandin Dehydrogenase. In Ramwell, P. W. (Ed.), *The Prostaglandins.* Vol. 2. New York: Plenum, 1974, p. 99.

*15. Coceani, F. Prostaglandins and the central nervous system. *Arch. Intern. Med.* 133:119, 1974.

*Asterisks denote key references.

*16. Wolfe, L. S. Possible Roles of Prostaglandins in the Nervous System. In Agranoff, B. W., and Aprison, M. H. (Eds.), *Advances in Neurochemistry.* Vol. 1. New York: Plenum, 1975, p. 1.

*17. Ramwell, P. W., and Shaw, J. E. (Eds.). *Prostaglandins. Ann. N.Y. Acad. Sci.* 180:5, 1971.

*18. Brody, M. J., and Kadowitz, P. J. Prostaglandins as modulators of the autonomic nervous system. *Fed. Proc.* 33:48, 1974.

*19. Hedqvist, P. Autonomic Neurotransmission. In Ramwell, P. W. (Ed.), *The Prostaglandins.* Vol. 1. New York: Plenum, 1973, p. 101.

20. Bergström, S. Prostaglandins: Members of a new hormonal system. *Science* 157:1, 1967.

*21. Greengard, P., and Robison, G. A. (Eds.). *Advances in Cyclic Nucleotide Research.* Vols. 1 and 3. New York: Raven, 1972 and 1973.

*22. Kahn, R. H., and Lands, W. E. M. (Eds.). *Prostaglandins and Cyclic AMP.* New York: Academic, 1973.

23. Collier, H. O. J., and Roy, A. C. Morphine-like drugs inhibit the stimulation by E prostaglandins of cAMP formation by rat brain homogenate. *Nature* 248:24, 1974.

*24. Horton, E. W. Prostaglandins. *Monographs on Endocrinology.* Vol. 7. New York: Springer-Verlag, 1972.

*25. Horton, E. W. The Prostaglandins. In Goodwin, T. W. (Ed.), *Biochemistry of Lipids.* MTP International Review of Science. Baltimore: University Park Press, 1974.

*26. Potts, W. J., East, P. F., and Mueller, R. A. Behavioral Effects. In Ramwell, P. W. (Ed.), *The Prostaglandins.* Vol. 2. New York: Plenum, 1974, p. 157.

27. Pike, J. E., and Weeks, J. R. (Eds.). *The Prostaglandins, Bibliography.* Kalamazoo: Upjohn, 1974. (A complete, updated source of the prostaglandin literature.)

*Key references.

Section IV
Metabolism

Chapter 14

Intermediary Metabolism of Carbohydrates and Amino Acids

Howard S. Maker
Donald D. Clarke
Abel L. Lajtha

This chapter outlines aspects of carbohydrate and amino-acid metabolism that are important for brain. Alterations in energy and carbohydrate metabolism, which are produced by varying functional demands of the nervous system and by pathological conditions, are discussed in Chapters 18, 19, 24, 25, and 31.

Present methods limit our ability to interpret carbohydrate metabolism. Measurements of blood gases and metabolites give a picture of overall metabolism, but they do not indicate a precise localization of intermediate metabolic steps. Studies in vitro are limited by tissue disjunction and damage, as well as by alteration of oxygenation and metabolite supply to the tissue. Maximal velocities of brain enzymes can be estimated from various tissue preparations in vitro, but, in the living brain, enzyme and substrate compartmentation, as well as steady-state conditions, produce significant departures from assay conditions.

The search for metabolic reactions or pathways that might be unique to brain has met with little success. In glycolysis and the operations of the Krebs cycle, however, there are marked quantitative differences between brain and such other tissues as liver. Two compounds that are present in high concentrations in brain, but are barely detectable in other tissues, are γ-amino butyrate and N-acetyl-aspartate. Because these compounds are metabolized actively in other tissues when added exogenously, the enzymes concerned with their degradation are certainly not unique to the brain, although the enzymes involved in their biosynthesis are barely detectable in tissues other than brain. Glutamic acid is another amino acid that has attracted the interest of neurochemists because of its unusual concentration in brain (roughly five times as high in brain as in liver).

The metabolic response in brain that probably is most different from those of other organs is its reaction to ammonia. Glutamine levels increase markedly in brain, whereas excess ammonia is detoxified in liver as urea. This is not caused by the absence in brain of the enzymes of the urea cycle.

The high levels in brain of such amino acids as glutamate, aspartate, and glycine, all of which may be derived from glucose, do not appear to be related to their general function as substrates for protein synthesis, but rather to their special function as neurotransmitters (see Chap. 11). A further characteristic of the metabolism of these amino acids is that high concentrations may be toxic to brain, and they seem to be actively excluded from the brain by the blood-brain barrier (see Chap. 20). In immature animals, in which the blood-brain barrier to glutamate is not fully developed, high levels of glutamate in the diet lead to brain damage [1]. Another example of glutamate toxicity is "Chinese restaurant syndrome," in which certain individuals show a marked sensitivity to large amounts of monosodium glutamate in a meal.

279

SOURCE OF AMINO ACIDS FOR BRAIN

It is clear that brain cannot make carbon skeletons of the amino acids that are essential in the diet and must depend on transport of these amino acids from the blood. On the other hand, the amino acids that are made in the animal body also appear to be synthesized in brain. There is no convincing evidence that amino acids made in the liver or other organs are transported via the blood to the brain in significant quantities. Ammonia is taken up actively by brain if circulating levels are increased, either artificially or when liver function is impaired. The uptake of ammonia from normal circulatory ammonia levels probably is sufficient to maintain the brain in nitrogen balance, although the portion of nitrogen supplied as ammonia vs. that available as preformed amino acids is not well established. Glucose metabolism appears to be adequate to replenish the carbon supply for the skeletons of nonessential amino acids.

ENERGY METABOLISM

Oxidative steps of carbohydrate metabolism normally contribute 36 of the 38 high-energy phosphate bonds ($\sim$P) generated during the aerobic metabolism of a single glucose molecule. Approximately 15 percent of brain glucose is converted to lactate and does not enter the citric acid cycle. There are indications, however, that this might be matched by a corresponding uptake of ketone bodies. The total net gain of $\sim$P is 33 equivalents per mole of glucose utilized. The steady-state level of ATP is high and represents the sum of a very rapid synthesis and utilization. Half of the terminal phosphate groups of ATP turn over in about three seconds, on the average, and probably in considerably shorter periods in certain regions [2]. The level of $\sim$P is kept constant by regulation of ADP phosphorylation in relation to ATP hydrolysis. The active adenylate kinase reaction, which forms equal amounts of ATP and AMP from ADP, prevents any great accumulation of ADP. Only a small amount of AMP is present under steady-state conditions; consequently, a relatively small percentage decrease in ATP may lead to a relatively large percentage increase in AMP. AMP is a positive modulator of several reactions that lead to increased ATP synthesis, so such an amplification factor provides a sensitive control for maintenance of ATP levels [3].

The level of creatine phosphate in brain is even higher than that of ATP, and creatine phosphokinase is extremely active. The creatine phosphate level is exquisitely sensitive to changes in oxygenation, providing $\sim$P for ADP phosphorylation, thus maintaining ATP levels. The creatine phosphokinase system may also function in regulating mitochondrial activity. The isoenzyme of creatine phosphokinase in the soluble fraction of brain differs from that in muscle; however, there appears to be more than one isoenzyme present in the particulate fraction of brain [4–9].

Currently, it is common to relate metabolism to the "energy charge" of a tissue, which is equal to one-half the average number of anhydride-bound phosphate groups per adenosine moiety:

$$\frac{(ATP) + \frac{1}{2}(ADP)}{(ATP) + (ADP) + (AMP)}$$

Calculated in this way, the steady-state energy charge of most tissues is approximately the same and declines rapidly when energy demand exceeds supply (see Chap. 31). It is a convenient shorthand for expressing relative energy states. It ignores nonadenylate energy stores, however, and its application to brain, which possesses finely tuned mechanisms for regulation and coupling of energy supply to utilization, can be misleading. Because brain metabolism actually responds in various ways to changes in the levels of individual nucleotide cofactors, metabolites, and ion fluxes, the calculated energy charge is an oversimplification that may cause overconfidence in one's knowledge of the true energy state and metabolic activity level of the tissue.

Glucose Transport

Under ordinary conditions, the basic substrate for brain metabolism is glucose. The brain depends on glucose for energy and as the major carbon source for a wide variety of simple and complex molecules. A transient decline in the oxidative metabolism of glucose may lead to an abrupt disruption of brain function. Despite this dependence on glucose, the brain at rest extracts only about 10 percent of the glucose from the blood flowing through it — about 5 mg (28 μmol) glucose per 100 g per minute. If the flow of blood is slowed, a relatively greater fraction of both the oxygen and the glucose of the blood is taken up by the brain (see Chap. 19). The entry of most water-soluble substances into brain from blood is restricted. Even small molecules, the size of glucose or fructose, do not simply diffuse across this anatomico-physiological barrier. For selected metabolites important to brain metabolism, however, specific mechanisms exist to carry these molecules across the barrier (see Chap. 20). Although many glycolytic metabolites and substances that can be transformed into these metabolites can sustain brain metabolism in vitro, they fail to do so in the intact animal simply because they cannot penetrate into brain at sufficient rates. Glucose crosses the barrier by a carrier mechanism specific to D-glucose. Certain analogs, such as 2-deoxyglucose, can be used to study this transport process because they are not metabolized beyond the hexokinase step [10]. Despite such active transport, the concentration of glucose in brain is lower than it is in structures lacking a barrier, e.g., sympathetic ganglia or peripheral nerve [11]. Although mannose may sustain brain metabolism in vivo, most other sugars, such as fructose, cannot be taken up rapidly enough, and brain metabolism cannot continue at normal rates when dependent on fructose alone. The capacity of fructose to sustain metabolism in the immature brain may be related to an incomplete barrier or to a lower metabolic demand before the brain matures [12].

Glucose uptake can be related to the affinity of the carrier system for glucose and may be expressed as a K_m, analogous to enzyme activity (see Chap. 6). The K_m of glucose uptake in mammalian brains is 7 to 8 mM, approximately the level of plasma glucose. With a substrate level at or below the K_m of the system, small changes in plasma glucose cause significant changes in the amount transported. Thus, within limits, glucose and glycogen levels in brain vary directly with blood glucose concentrations. The rate of influx of glucose increases linearly up to about 7 mM plasma glucose and at a lesser rate to about 20 mM. Energy is not supplied to the transport system, so no pumping mechanism is possible to facilitate glucose entry against a concentration gradient. Brain glucose is always lower than blood glucose. There is, then, no safety device to supplement such a small carbohydrate reserve during hypoglycemia. Beyond the barrier, brain tissue itself takes up glucose much more avidly. Glucose can be transported from the cell body into its processes or can be taken up into certain subcellular regions, such as synaptosomes, by carrier processes that possess an affinity for glucose 30 times higher than that of the blood-brain barrier transport system [10]. These regions may be able to function despite low overall brain-glucose levels. Brain tissue beyond the barrier is apparently sensitive to insulin, and an increase in glucose uptake and glycogen storage can be demonstrated in vitro. However, despite many studies, no effect of insulin on glucose entry into whole brain via the barrier transport mechanism has ever been demonstrated satisfactorily. This may not be due simply to the exclusion of insulin from the brain, as insulin has been detected in CSF, although at present there is no report of a definite effect of circulating insulin on the intact brain [9, 13].

Glycogen

Although present in relatively low concentration in brain (3.3 μmol per kg, rat), glycogen is the single largest energy reserve and requires no energy (ATP) for initiation of its metabolism. As with brain glucose, glycogen levels in brain appear to vary with plasma-glucose concentrations. Glycogen is stored in cytoplasmic granules, which can be seen in electron micrographs. If the anesthetized animal has been perfused arterially with cold fixative, these granules may be seen in neurons, neuropil, and glia, but only if ischemic changes have not allowed glycogen metabolism during fixation. Associated with the granules are enzymes concerned with glycogen synthesis and, perhaps, degradation. The increased glycogen found in areas of brain injury may be due to glial changes or to decreased utilization during tissue preparation. The accepted role of glycogen is that of a carbohydrate reserve utilized when glucose falls below need. There is, however, a rapid continual breakdown and synthesis of glycogen (17 μmol/kg/min, in rat) [14]. This is approximately 2 percent of the normal glycolytic flux in brain and is subject to elaborate control mechanisms. This suggests that, even under steady-state conditions, local carbohydrate reserves are important for brain function. If glycogen were the sole energy supply, however, the normal glycolytic flux in brain would be maintained for less than 25 minutes.

The enzyme systems that synthesize and catabolize glycogen in other tissues also are found in brain, but their kinetic properties and modes of regulation appear to differ [15]. Glycogen levels in liver are regulated to reduce blood glucose below the renal threshold after a carbohydrate meal (via insulin) and to maintain blood glucose levels (via decreased insulin, glucagon, and epinephrine) at other times. Resting aerobic muscles utilize glucose and fatty acids for energy, but a large store of glycogen is utilized under the relatively hypoxic conditions present during strong rapid contractions. Glycogenolysis in muscle is linked to contraction by sensitivity to the same changes in ionized calcium that couple muscle-membrane excitation to contraction. Circulating epinephrine also will increase muscle glycogenolysis controlled by the cyclic AMP (cAMP) system. Liver glycogen is stored for use by the entire body, and the regulation of muscle glycogen reflects its use during brief periods of intense activity. Glycogen metabolism in brain, however, is controlled locally. Evidently because of the blood-brain barrier, this metabolism is isolated from the tumult of systemic activity. Although glucocorticoid hormones that penetrate the brain will increase glycogen turnover, circulating protein hormones and biogenic amines are without effect [14]. Beyond the barrier, cells are sensitive to local amine levels, so drugs that penetrate the barrier and modify amine levels or membrane receptors cause metabolic changes [16].

Separate systems for the synthesis and degradation of glycogen provide a greater degree of control than would be the case if glycogen were degraded simply by reversing the steps in its synthesis (see Fig. 14-1). The level of glucose 6-phosphate, the initial synthetic substrate, usually varies inversely with the rate of brain glycolysis because of a greater facilitation of the phosphofructokinase step relative to glucose transport and phosphorylation. Thus, the decline in glucose 6-phosphate at times of energy need decreases glycogen formation.

The glucosyl group of uridine diphosphate (UDP) is transferred to the terminal glucose of the nonreducing end of an amylose chain in an α-1,4-glycosidic linkage (Fig. 14-1). This reaction is catalyzed by glycogen synthetase, and is the rate controlling reaction in glycogen synthesis [15]. In brain, as in other tissues, glycogen synthetase occurs in both a phosphorylated (D) form, which is dependent for activity on glucose 6-phosphate as a positive modulator, and as a dephosphorylated, independent (I) form sensitive to, but not dependent on, the modulator. Although in

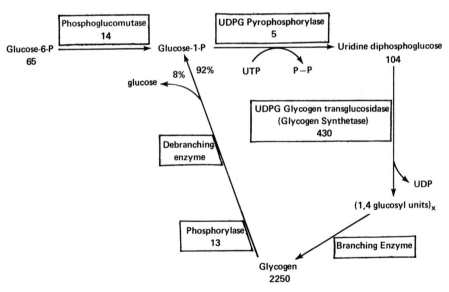

Figure 14-1
Glycogen metabolism in brain. Enzyme data from mouse brain homogenates. The figures under each enzyme represent V_{max} at 38° C in mmol per kilogram wet weight per minute. Metabolite levels from quick-frozen adult mouse brain are indicated in micromoles per kilogram wet weight. (Metabolite data from Passonneau et al., *J. Biol. Chem.* 244:902, 1969. Enzyme data from Breckenridge and Gatfield, *J. Neurochem.* 3:234, 1961.)

brain the independent form of the glycogen synthetase requires no stimulator, it has a relatively low affinity for UDP-glucose. At times of increased energy demand, there is a change not only from the D to the I form, but an I form with an even lower affinity for the substrate develops. The inhibition of glycogen synthesis is enhanced, and this increases the availability of glucose 6-phosphate for energy needs. Goldberg and O'Toole [15] hypothesize that, in brain, the I form is associated with inhibition of glycogen synthesis under conditions of energy demand, whereas the D form is responsible for a relatively small regulated synthesis under resting conditions. The regulation of the D form may be responsible for reducing the rate of glycogen formation in brain to about 5 percent of its potential rate. In liver, where large amounts of glycogen are synthesized and degraded, the I form of the synthetase is associated with glycogen formation. At the present time, it appears that the two tissues use the same biochemical apparatus in different ways in relation to differences in overall metabolic patterns.

Under steady-state conditions, it is probable that less than 10 percent of phosphorylase in brain (Fig. 14-1) is in the unphosphorylated *b* form (requiring AMP), which is inactive at the very low AMP concentrations present normally. When the steady state is disturbed, there may be an extremely rapid conversion of the enzyme to the *a* form, which is active at low AMP levels. Brain phosphorylase *b* kinase is indirectly activated by cAMP and by the micromolar levels of ionized calcium released during neuronal excitation (see Chap. 12). Brain endoplasmic reticulum, like that in muscle, is capable of taking up this calcium to terminate its stimulatory effect. These reactions provide energy from glycogen during excitation and when cAMP-forming systems are activated. It has not been possible to confirm directly, however, that the conversion from phosphorylase *b* to *a* is a control point of glycogenolysis in vivo. The hydrolysis of the α-1,4-glycoside linkages leaves a limit dextrin that turns over at only half the rate of the outer chains [14]. The debrancher enzyme that hydrolyzes the α-1,6-glycoside linkages may be rate limiting if the entire glycogen granule is to be utilized. Because one product of this enzyme is free glucose, approximately one glucose molecule for every eleven of glucose 6-phosphate is released if the entire glycogen molecule is degraded (Fig. 14-1). α-Glucosidase (acid maltase) is a lysosomal enzyme whose precise function in glycogen metabolism is not known. In Pompe's disease (the hereditary absence of the enzyme), glycogen accumulates in brain lysosomes as well as in those elsewhere (Chaps. 23 and 25). The steady-state level of glycogen is regulated precisely by the coordination of synthetic and degradative processes through enzymatic regulation at several metabolic steps [6, 9].

Glycolysis

The terms aerobic and anaerobic glycolysis may be misleading. Since the time when metabolism began to be studied in the Warburg apparatus, workers usually have referred to the production of lactate under conditions of "adequate" oxygen as *aerobic glycolysis* and that under anoxia as *anaerobic glycolysis*. Aerobic glycolysis, thus defined, measures only a small portion of total glycolysis [17]. Recently it has become common to refer to the Embden-Meyerhoff glycolytic sequence from glucose (or glycogen glucosyl) to pyruvate by the term glycolysis. Changes in the activity of the sequence under various oxygen levels have been termed aerobic or anaerobic glycolysis. Failure to differentiate between these definitions may lead to confusion, particularly if one is examining data pertinent to neurochemistry derived from more recent methods, such as cell and tissue culture. *Glycolytic flux* is defined indirectly; it is the rate at which glucose must be utilized to produce the observed rates of ADP phosphorylation.

Figure 14-2 outlines the flow of glycolytic substrates in brain. Glycolysis first involves phosphorylation by hexokinase. The reaction is essentially irreversible and is a key point in the regulation of carbohydrate metabolism in brain. The electrophoretically slow-moving (type 1) isoenzyme of hexokinase is characteristic of brain. Hexokinase (in most tissues) may exist in the cytosol (soluble), or it may be firmly attached to mitochondria [18]. Under conditions in which no special effort is made to stop metabolism while isolating mitochondria, 80 to 90 percent of brain hexokinase is bound. In the live steady state, however (i.e., when availability of substrates keeps up with metabolic demand and end products are removed), an equilibrium exists between the soluble and the bound enzyme. Binding changes the kinetic properties of hexokinase and its inhibition by glucose 6-phosphate, so that the bound

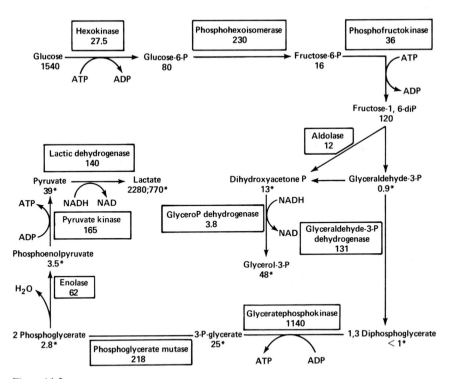

Figure 14-2
Glycolysis in brain. Enzyme and metabolite data expressed as in Figure 14-1. (Data from McIlwain, *Biochemistry and the Central Nervous System.* Boston: Little, Brown, 1966, pp. 1–26.) Asterisk indicates 10-day-old mouse brain [8].

enzyme on mitochondria is more active. The extent of binding is inversely related to the ATP/ADP ratio, so conditions in which energy utilization exceeds supply shift the solubilization equilibrium to the bound form and produce a greater potential capacity for initiating glycolysis to meet the energy demand. This mechanism allows ATP to function both as a substrate of the enzyme and, at another site, as a regulator to increase ATP production through its influence on enzyme binding. It also confers preference on glucose in the competition for the $MgATP^{2-}$ generated by mitochondrial oxidative phosphorylation. Thus a process that will sustain ATP production continues at the expense of other uses of energy. Because energy reserves are rapidly exhausted postmortem, it is not surprising that brain hexokinase is found to be almost entirely bound.

The significance of reversible binding of other enzymes to mitochondria is not clear. The measured glycolytic flux, when compared to the maximal velocity of hexokinase, indicates that, in the steady state, the hexokinase reaction is 97 percent inhibited. Brain hexokinase is inhibited by its product, glucose 6-phosphate, to a lesser extent by ADP, and allosterically by 3-phosphoglycerate and several nucleoside phosphates, including cAMP and free ATP^{4-}. The ratio of ATP to magnesium may also have a regulatory action. In addition to acting on enzyme kinetics, glucose 6-phosphate solubilizes hexokinase, thus reducing the enzyme's efficiency when the reaction product accumulates. The sum total of these mechanisms is a fine tuning of the activity of the initial enzyme of glycolysis, in response to changes in the cellular environment, to increase glycolytic flow when needed and suppress it when demands for its eventual products (energy and glycolytic metabolites) are less. Glucokinase (low-affinity hexokinase), a major component of liver hexokinase, has not been found in brain.

Glucose 6-phosphate represents a branch point in metabolism because it is a common substrate for enzymes involved in glycolytic, pentose-phosphate shunt, and glycogen-forming pathways. There is also slight, but detectable, glucose 6-phosphatase activity in brain, the significance of which is not clear. The liver requires this enzyme to convert glycogen to glucose. The differences between the liver and brain hexokinase and the differences between the modes of glycogen metabolism of these two tissues can be related to the function of liver as a carbohydrate storehouse for the body, whereas brain metabolism is adapted for rapid carbohydrate utilization for energy needs.

In glycolysis, glucose 6-phosphate is the substrate of phosphohexose isomerase. This is a reversible reaction (small free-energy change), whose five-to-one equilibrium ratio in brain favors glucose 6-phosphate.

Fructose 6-phosphate is the substrate of phosphofructokinase, a key regulatory enzyme controlling glycolysis [3]. The other substrate is $MgATP^{2-}$. Like other regulatory reactions, it is essentially irreversible. It is modulated by a large number of metabolites and cofactors, whose concentrations under different metabolic conditions have great effects on glycolytic flux. Prominent among these are availability of $\sim$P and citrate levels. Brain phosphofructokinase is inhibited by ATP, magnesium, and citrate and is stimulated by NH_4^+, K^+, Pi, 5'-AMP, cyclic 3',5'-AMP, ADP, and fructose 1,6-diphosphate. Possibly its inhibition by reduced pyridine nucleotides can be overcome by an increase in pH. The effects of NH_4^+, inorganic phosphate, and AMP are additive, and there is evidence for several different allosteric sites on this complex enzyme. In general, activity is inhibited by ATP and enhanced by substances that accumulate under conditions in which $\sim$P reserves are diminished. Activity is also increased by $3',5'$-cAMP, the presence of which leads ultimately to increased energy use. Inhibition by citrate would amount to an end-product inhibition if acetate, its precursor, is considered the final product of glycolysis. The concentration of NH_4^+, a product of amino acid and amine metabolism, is controlled by transamination reactions coupled to citric acid cycle intermediates, as well as by glutamic dehydrogenase and glutamine synthetase. Increase of glycolysis, by raising pyruvate and α-ketoglutarate levels, would tend to reduce NH_4^+ levels. The stimulation by NH_4^+ is unaffected by high ATP levels, allowing acceleration of glycolysis to increase the availability of α-ketoglutarate and pyruvate for transamination even when $\sim$P levels are adequate.

When oxygen is admitted to cells metabolizing anaerobically, utilization of O_2 increases and utilization of glucose and lactate production drops (Pasteur effect). Modulation of the phosphofructokinase reaction can account directly for the Pasteur effect. In the steady state, ATP and citrate levels in brain apparently are sufficient to keep phosphofructokinase relatively inhibited as long as the level of positive modulators (or disinhibitors) is low. When the steady state is disturbed, activation of this enzyme produces an increase in glycolytic flux that takes place almost as fast as the events changing the internal milieu.

Fructose 1,6-diphosphate is split by brain aldolase to glyceraldehyde 3-phosphate and dihydroxyacetone phosphate. Dihydroxyacetone phosphate is the common substrate for both glycerophosphate dehydrogenase, an enzyme active in NADH oxidation and lipid pathways (see Chap. 15), and triose phosphate isomerase, which maintains an equilibrium between dihydroxyacetone phosphate and glyceraldehyde 3-phosphate; the equilibrium strongly favors accumulation of dihydroxyacetone phosphate.

After the reaction with glyceraldehyde phosphate dehydrogenase, glycolysis in the brain proceeds through the usual steps. It is of interest that brain phosphoenolpyruvate kinase controls an essentially irreversible reaction, which requires not only Mg^{2+} (as do several other glycolytic enzymes) but also K^+ or Na^+. This step also may be regulatory.

Brain tissue, even when it is at rest and well oxygenated, produces a small amount of lactate, which is removed in the venous blood. This amount is derived from 13 percent of the pyruvate produced by glycolysis. The measured lactate level in brain depends on the success of halting brain metabolism rapidly during tissue processing. Five isoenzymes of lactate dehydrogenase are present in adult brain. The one that moves electrophoretically most rapidly toward the anode (band 1) predominates. This isoenzyme is generally higher in those tissues that are more dependent on aerobic processes for energy; the slower-moving isoenzymes are relatively higher in such tissue as white skeletal muscle, which is better adapted to function at suboptimal oxygen levels (see Chap. 23). The activities of lactate dehydrogenases and their distribution in various brain regions, layers of the retina, brain neoplasms, brain tissue cultures, and during development indicate that the synthesis of these isoenzymes might be controlled by oxygen levels in the tissues. Lactate dehydrogenase functions in the cytoplasm as a means of oxidizing NADH, which accumulates as the result of the activity of glyceraldehyde phosphate dehydrogenase in glycolysis. It thus permits glycolytic ATP production to continue under anaerobic conditions. Lactate dehydrogenase also functions under aerobic conditions because NADH cannot easily penetrate the mitochondrial membrane. The oxidation of NADH in the cytoplasm depends on this reaction and on the activity of "shuttle mechanisms" that transfer reducing equivalents to the mitochondria.

Glycerol phosphate dehydrogenase is another enzyme indirectly associated with glycolysis that participates in the cytoplasmic oxidation of NADH. This enzyme reduces dihydroxyacetone phosphate to glycerol 3-phosphate, oxidizing NADH in the process. Under hypoxic conditions, the levels of α-glycerophosphate and lactate increase initially at comparable rates, although the amount of lactate produced greatly exceeds that of α-glycerophosphate. The relative levels of the oxidized and reduced substrates of these reactions indicate much higher local levels of NADH in brain than are found by gross measurements. In fact, the relative proportions of oxidized and reduced substrates of the reactions that are linked to the pyridine nucleotides may be a better indicator of local oxidation-reduction states (NAD^+/ NADH) in brain than is provided by the direct measurement of the pyridine nucleotides themselves [5, 6, 8, 9, 13, and Chap. 31].

An aspect of glucose metabolism which has led to much confusion among neurochemists is the observation that labeled glucose appears in carbon dioxide much more slowly than might be suggested from a cursory examination of the glycolytic pathway plus the citric-acid cycle [19]. Glucose flux is ~0.5 to 1.0 μmol/min/g wet weight of brain in a variety of species. The level of glycolytic plus tricarboxylate cyclic intermediates is ~2 μmol per g. Hence these intermediates might be predicted to turn over every 2 to 4 min, and $^{14}CO_2$ production might be predicted to reach a steady state in 5 to 10 min. This is not observed experimentally. In addition, large quantities of radioactivity are trapped in amino acids related to the Krebs (TCA) cycle (70 to 80 percent) from ten minutes to half an hour after a glucose injection. This is due to the high activity of transaminase in comparison with the flux through the citric-acid cycle, and the amino acids developed by transamination behave as if they were a part of the Krebs cycle. When the pools of these amino acids (approx. 20 μmol per g) are added to the levels of TCA cycle components plus glycolytic intermediates, the time for $^{14}CO_2$ evolution is increased by a factor of ten, and this agrees with the values observed experimentally.

In contrast, in tissues such as the liver, the amino acids related to the Krebs cycle are present at much lower steady-state values, and 20 percent of the radioactivity from administered glucose is trapped in these amino acids at short times after injection. As a result, ignoring the radioactivity trapped in amino acids has a relatively small effect on estimates of glycolytic fluxes in liver but makes an enormous difference in brain. Immature brains more nearly resemble liver in this respect. The relationship of the Krebs cycle to glycolysis undergoes a sharp change during development, coincident with the development of the metabolic compartmentation of amino-acid metabolism that is characteristic of adult brain.

Citric Acid Cycle

The energy output and oxygen consumption in adult brain are associated with high levels of enzyme activity in the citric acid cycle [20]. The actual flux through the citric acid cycle (Krebs cycle) depends on glycolysis and active acetate production, which can "push" the cycle, the possible control at several enzymatic steps of the cycle, and the local ADP level, which is known to be a prime activator of the mitochondrial respiration to which the citric-acid cycle is linked. The steady-state level of citrate in brain is about one-fifth that of glucose. This is relatively high as compared with levels of glycolytic intermediates or with that of isocitrate.

As in other tissues, there are two isocitrate dehydrogenases in brain. One is active primarily in the cytoplasm and requires NADP as cofactor; the other, bound to mitochondria and requiring NAD, is the enzyme that participates in the citric acid cycle. The NAD-linked enzyme catalyzes an essentially irreversible reaction, has allosteric properties, is inhibited by ATP and NADH, and may be stimulated by ADP. The function of cytoplasmic NADP-isocitrate dehydrogenase is uncertain, but it has been postulated that it supplies the reduced NADPH that is necessary for many reductive synthetic reactions. The relatively high activity of this enzyme in immature brain and white matter is consistent with such a role. α-Ketoglutarate dehydrogenase, which oxidatively decarboxylates α-ketoglutarate, requires the same cofactors as does the pyruvate decarboxylation step (Chap. 11).

In adult mouse brain, the enzyme succinate dehydrogenase, which catalyzes the oxidation of succinate to fumarate, is tightly bound to mitochondrial membrane. In brain, succinate dehydrogenase may also have a regulatory role when the steady state is disturbed. Isocitrate and succinate levels in brain are little affected by changes in the flux of the citric acid cycle, as long as an adequate glucose supply is available. The highly unfavorable free-energy change of the malate dehydrogenase reaction is overcome by the rapid removal of oxaloacetate, which is maintained at low concentrations under steady-state conditions by the condensation reaction with acetyl-CoA [4–6, 9, 13].

Malic dehydrogenase is one of several enzymes in the citric acid cycle that is present in both the cytoplasm and mitochondria. The function of the cytoplasmic components of these enzyme activities is not known, but they may function in the transfer of hydrogen from the cytoplasm into mitochondria.

Another source of much misunderstanding in the interpretation of the citric-acid cycle is the distinction between its function as an oxidative scheme for energy production and its function as a biosynthetic process for the production of amino acids, e.g., glutamate, glutamine, GABA, aspartate, and asparagine, as well as other amino acids needed for protein synthesis. In order to export net amounts of α-ketoglutarate or oxaloacetate from the Krebs cycle, the supply of dicarboxylic acids must be replenished. The major route for this seems to be the fixation of CO_2 to pyruvate or another substrate at this three-carbon level. Thus, the rate of CO_2 fixation sets the upper limit at which biosynthetic reactions can occur. This has been estimated as approximately 0.15 μmol per g wet weight of brain per minute in studies of acute ammonia toxicity in cats, or approximately 10 percent of the flux through the citric acid cycle. Liver, on the other hand, appears to have 10 times the capacity of brain for CO_2 fixation, as is appropriate for an organ geared to making large quantities of protein for export.

The activity of pyruvate dehydrogenase seems to be the rate-limiting step for the entry of pyruvate from glycolysis into the Krebs cycle.

Intracellular Messengers

The action of many hormones and some neural transmitters is indirect. These substances act at membranes of specific receptor cells to cause the intracellular release of an active molecule ("second messenger"), which in turn modifies cellular metabolism and function. Ionic calcium in muscle is such a substance; $3',5'$-cAMP is another (see Chap. 12).

In sensitive cells, epinephrine activates the membrane-associated adenylate cyclase system that converts ATP into $3',5'$-cAMP. This substance, unlike $5'$-AMP, retains a high-energy phosphate bond and can activate various protein kinases, among which is phosphorylase b kinase kinase. This enzyme catalyzes the phosphorylation of phosphorylase b kinase, which in turn stimulates the conversion of phosphorylase b to the a form; this facilitates the breakdown of glycogen. Also $3',5'$-cAMP activates the kinase that converts glycogen synthetase to the regulated D form. Relatively high levels of $3',5'$-cAMP are found in brain and may function postsynaptically in interneural transmission, particularly at adrenergic synapses. Preganglionic sympathetic stimulation increases cAMP in the postsynaptic cell in proportion to the rate of stimulation.

Pentose Phosphate Shunt

Controversy exists concerning the proportion of brain glucose that is metabolized along the pentose phosphate shunt. The most common technique for determining that proportion has been to compare $^{14}CO_2$ release or ^{14}C appearing in triose phosphates, lactate, or glycogen from glucose $6\text{-}^{14}C$ to that from glucose $1\text{-}^{14}C$. The C-1 carbon of glucose is lost to CO_2 during flow along the shunt, so equal yields of ^{14}C from glucose labeled with C-1 and C-6 may indicate that glucose metabolism occurs primarily by the glycolytic route, whereas a high relative yield of $^{14}CO_2$ from glucose labeled with C-1 suggests a greater flow along the shunt. Such procedures are known to be approximations that are subject to a variety of errors that have produced estimates of pentose-phosphate flow ranging from zero to 21 percent of glycolysis. The shunt pathway is definitely active in brain, and it is probable that, under basal conditions in the adult, at least 5 to 8 percent of brain glucose is metabolized through the shunt [21]. The pathway has relatively high activity in developing brain, reaching a peak during myelination. Its main contribution probably produces the NADPH required for the reductive reactions necessary for lipid synthesis (Chap. 15). The shunt pathway also provides pentose for nucleotide synthesis; however, only a small fraction of the activity of this pathway would be required to meet such a need (see in Chap. 17). As with glycogen synthesis, turnover in the pentose-phosphate pathway decreases under conditions of increased energy need, e.g., during and after high rates of stimulation [22]. Pentose phosphate flux apparently is regulated by the concentrations of glucose 6-phosphate,

NADP, glyceraldehyde 3-phosphate, and fructose 6-phosphate [22]. Transketolase, one of the enzymes in this pathway, requires thiamine pyrophosphate as a cofactor. Poor myelin maintenance in thiamine deficiency may be due to the failure of the pathway to provide sufficient NADPH for lipid synthesis ([5, 6, 9,] and Chap. 29).

Metabolic Compartmentation
Clearly, protein synthesis has a major requirement for amino acids in all tissues, and brain is no exception. However, there is increasing evidence that many of the amino acids derived from the Krebs cycle also function as neurotransmitters (Chap. 11). If one were to assume that the difference in glutamate levels in brain and other tissue (an estimated 8 μmol per g) is related to the special function of glutamate as a neurotransmitter, it might be expected that glutaminergic nerve endings would predominate in certain brain areas, particularly cortex. (This estimate of the quantity of glutamate that functions as a neurotransmitter is probably an upper limit.) Because GABA and glutamate are intimately related, not only metabolically but also as opposing neurotransmitters (inhibitory vs. excitatory), another estimate of the quantity of glutamate that functions as a neurotransmitter is similar to the level of GABA, i.e., approximately 2 μmol per g. Of course, this would vary according to the particular brain area and is probably a lower-limit estimate. The recent work of Snyder et al. does in fact suggest that glutamate and GABA may be among the major neurotransmitters in the CNS [23].

The compartmentalization in brain of glutamate into separate pools that equilibrate only slowly is a vital factor in regulating separately such special functions of glutamate as neurotransmission and such general functions as protein biosynthesis [14]. This does not mean that glutamate metabolism may not be compartmented in other tissues, but that the characteristics of such pools may be quite different.

Not only is glutamate metabolism in brain characterized by the existence of at least two distinct pools; the Krebs cycle intermediates associated with these pools are also distinctly compartmented [24–26]. A mathematical model to fit data from radiotracer experiments that require separate Krebs cycles to satisfy the hypotheses of compartmentation has been developed [24, 27] and further refined [28]. A key assumption of the current models is that GABA is metabolized at a site different from that at which it is synthesized. The best fit of the kinetic data is obtained when glutamate, from a small pool that is actively converted to glutamine, flows back to the large pool that is converted to GABA. An enzymatic basis for such predictions is now developing; it has been shown that GAD (E.C.3.1.1.15) is localized at or near nerve terminals, whereas GABA-transaminase, the major degradative enzyme, is mitochondrial [29].

Increasing evidence points to the small pool of glutamate as probably glial. Thus it seems that glutamate released from nerve endings is taken up by glia (and by pre- and postsynaptic terminals, or both), converted to glutamine, and recycled to glutamate and GABA. Various estimates of the proportion of glucose carbon that flows via the GABA shunt (Chap. 11) have been published, but the most definitive experiments show the value to be approximately 10 percent [13] of the total glycolytic flux. Although this may seem small, it should be understood that the portion of the

Krebs cycle flux that is used for energy production (ATP synthesis, maintenance of ionic gradients) does not require CO_2 fixation but that the portion used for biosynthesis of amino acids does. By recycling the carbon skeleton of glutamate via glutamine and GABA to succinate, the need for dicarboxylic acids to replenish intermediates of the Krebs cycle is diminished when export of α-ketoglutarate takes place.

It is difficult to get good estimates of the extent of CO_2 fixation in brain, but estimates of the maximum capability, obtained under conditions of ammonia stress, when glutamine levels increase rapidly, suggest that CO_2 fixation occurs at 0.15 μmol/g/min (in cat) and 0.33 μmol/g/min (in rat), i.e., at about the same rate as for the GABA shunt.

For comparison, it should be pointed out that only about 2 percent of the glucose flux in whole brain goes toward lipid synthesis and approximately 0.3 percent is used for protein synthesis. Thus the turnover of neurotransmitter amino acids is a major biosynthetic effort in brain.

Metabolic compartmentation of glutamate is usually observed when ketogenic substrates are administered to animals. Interestingly enough, acetoacetate and β-hydroxybutyrate do not show this effect. Apparently this is because ketone bodies are a normal substrate for brain and are taken up into all types of cells. Acetate and similar substances, which are not taken up into brain efficiently, appear to be more readily taken up or activated in glia, or both. This may lead to the abnormal glutamate/glutamine ratio that is observed. Similarly, metabolic inhibitors, such as fluoroacetate, appear to act selectively in glia and produce their neurotoxic action without marked inhibition of the overall Krebs cycle flux in brain.

The occurrence of a nonuniform distribution of various compounds in living systems up to a certain level is widely accepted. Steady-state levels of GABA are well documented as varying over a fivefold range in discrete brain regions (2 to 10 mM), and it has been estimated that GABA may be as high as 50 mM in nerve terminals. Observations in brain indicate the existence of pools with half-lives of many hours for mixing, which is most unusual. The discovery of a subcellular morphological compartmentation, i.e., that there are different populations of mitochondria in cerebral cortex that have distinctive enzyme complements, may provide a somewhat better perspective by which to visualize such separation of metabolic function. We still do not know if different mitochondrial populations are present within single cells or are characteristic of different cell populations. The heterogeneity of brain tissue fosters the latter as the simpler explanation, but there is no positive evidence for either assumption.

Tapia [30] has proposed that, in addition to the phasic release of both excitatory and inhibitory transmitter, there is a continuous tonic release of GABA, dependent only on the activity of the enzyme responsible for its synthesis and independent of the depolarization of the presynaptic membrane. Such inhibitory neurons could act tonically by constantly maintaining an elevated threshold in the excitatory neurons so that the latter would start firing when a decrease occurs in the continuous release of GABA acting upon them. This is consistent with a good correlation between the inhibition of GAD and the appearance of convulsions after certain drug treatments. GABA levels have been observed to be depleted by some convulsant drugs and elevated by others. Wood et al. [31] introduced the concept of a critical GABA factor that combines the activity of GAD and the level of GABA because the degradation of GABA is not the rate-limiting step in the overall process that determines the availability of GABA at the synapse; rather, it is the activity of GAD. To describe the relative excitability of the CNS, this formula uses the change in GAD activity combined with an empirical factor (0.4) times the square root of the change in the GABA concentration. Such ideas fit well with the observation that the so-called GABA shunt involves a significant part (ca. 10 percent) of the total glycolytic flux in brain. The role of amino acids as neurohumoral agents and transmitters is more fully described in Chapter 11.

Mitochondria
As in other tissues, brain mitochondria apparently are self-replicating bodies, although most of their active enzymes depend on the cell's chromosomal-ribosomal apparatus for synthesis. As noted above, not all mitochondria are identical in enzyme complement or function. Because current methods of separating mitochondria from other cell components depend on sedimentation from homogenates (Chap. 2), differences related to function and intracellular provenance are difficult to define. Furthermore, mitochondrial size and function do change during maturation. Therefore, functional heterogeneity among brain mitochondria may be related to their location in perikarya, synaptosomes, or various glial cells.

The mitochondrial membranes and matrix form metabolic compartments separate from the cytosol so that the entry and egress of metabolites and ions are selective. This allows a degree of metabolic control that is not possible otherwise. For example, mitochondrial membranes are not freely permeable to pyridine nucleotides. The brain contains several enzymes that could function in postulated shuttle systems that transfer hydrogen generated in the cytosol (e.g., NADH) to the mitochondria to be oxidized by the electron-transport system. These enzymes include mitochondrial and cytoplasmic malate dehydrogenases (NAD) as well as cytoplasmic (NAD) and mitochondrial (flavoprotein) glycerol phosphate dehydrogenases.

The sine qua non of brain metabolism is its high rate of respiration. In the coupled, controlled state, the level of mitochondrial function depends on local concentrations of ADP. The entry of ADP into mitochondria is restricted insofar as it must exchange with intramitochondrial ATP. The high steady-state mitochondrial respiration in brain is related to local availability of substrates and the ratio of ADP to ATP. This may not be reflected in the average ratio of ADP to ATP in whole brain, which is quite low. Brain mitochondria also differ from those in other tissues because they contain higher concentrations of certain "nonmitochondrial" enzymes. Hexokinase, creatine kinase, and perhaps lactic dehydrogenase are partially mitochondrial. Hexokinase and creatine kinase may function to maintain local levels of ADP by transferring ~P from ATP to creatine or glucose [4–6, 9].

Relation of Carbohydrates to Lipid Metabolism
The principal source of lipid carbon in brain is blood glucose. Carbohydrate intermediates and related metabolites — such as acetate (fatty acid and cholesterol),

dihydroxyacetone phosphate (glycerol phosphate), mannosamine and pyruvate (neuraminic acid), glucose 6-phosphate (inositol), galactose, and glucosamine — supply the building blocks of the complex lipids (Chap. 15). NADPH is also necessary for the reductive synthesis of lipids. When a few polyunsaturated fatty acids and sulfate are supplied, immature brain slices readily form all the lipids of myelin and cell membranes, utilizing glucose as the only substrate. Energy supply as ATP via the carbohydrate pathways is also required to supply nucleoside phosphates for lipid assembly [6, 22].

Carbohydrates in Neuronal Function

The high functional requirement of nervous tissue for $\sim$P is probably related to energy demands for transmitter synthesis, packaging, secretion, uptake, and sequestration; for ion pumping to maintain ionic gradients; for intracellular transport; and for synthesis of complex lipids and macromolecules in both neurons and glial cells. The manner in which metabolism is coupled to function is conceptualized easily from the discussion of the regulatory mechanisms that control carbohydrate metabolism, and can be illustrated by (Na^+, K^+)-ATPase, which functions in the Na^+, K^+ exchange reaction so essential for maintaining electrolyte gradients. (See also Chap. 6.) This enzyme is particularly active in regions with high concentrations of synaptic membranes, e.g., gray matter and synaptosomes, the subcellular fractions that contain nerve endings. This membrane-associated, topographically oriented enzyme is stimulated by extracellular K^+ and intracellular Na^+, so its activity is increased by the ionic changes that accompany depolarization (Chap. 5). The ADP that is released intracellularly is an activating modulator for mitochondria and several rate-limiting reactions: glycolytic (hexokinase, phosphofructokinase), Krebs cycle (NAD-dependent isocitric dehydrogenase), and glycogenolytic (phosphorylase). The changes accompanying the consequently increased metabolic flux lead to inhibition of other pathways, such as glycogen synthesis and the pentose phosphate shunt, whereas the lowered $\sim$P levels and NADPH levels inhibit various synthetic reactions. Because the products of reactions that use $\sim$P are accelerators of reactions leading to $\sim$P formation, energy supply as ATP is regulated by utilization. The set points of this regulation may be altered in certain toxic-metabolic states and physiologic changes in behavior [32].

Regional Differences

It is assumed that lipid represents a relatively inert metabolic compartment (although many enzymes are membrane associated, and even lipid dependent). Therefore, comparisons of different areas of brain are often based on lipid-free dry weight. In this way, one hopes to gain a better estimate of the actively metabolizing compartments in regions of widely varying lipid content. In general, those regions of brain with higher metabolic requirements have higher activity of enzymes in the glycolytic series and the citric acid cycle as well as higher levels of respiration. Several glycolytic and mitochondrial enzymes are more active in regions with large numbers of synaptic endings (neuropil) than in areas rich in neuronal cell bodies. On the other hand, the activity of glucose 6-phosphate dehydrogenase (a rate-limiting enzyme of the pentose phosphate pathway) is high in myelinated fibers and tends to vary with the degree of myelination. Phosphofructokinase and phosphorylase are distributed in a relatively constant ratio in different brain regions, suggesting a relationship between glycolysis and glycogenolysis. Hexokinase distribution is more closely linked to mitochondrial enzymes than to glycolytic enzymes.

Blood flow to gray matter is greater than to white, and many in vitro studies indicate that, even when corrected for lipid content, the metabolic activity of white matter is less than that of gray. However, ~P and glycolytic flux may be as high in white matter as it is in gray. The depressing effect of anesthetic agents on these factors of metabolism may also be greater in white than in gray matter [2]. The neuropil-rich molecular layer of cerebellum has only slightly higher ~P flux than has white matter. High metabolic rates in white matter might be related to axonal transport mechanisms and the maintenance of myelin. The oligodendroglial cell that maintains myelin in the CNS probably ranks, along with the neuron and its processes, as one of the cells with the highest known metabolic requirements [7, 8].

Metabolism of the Retina

The rabbit retina is avascular and depends almost entirely on diffusion from choroidal capillaries. The primate retina, however, is vascularized from the vitreal surface as far as the bipolar cell layer. The rabbit retina shows high rates of glucose and oxygen consumption and also of lactate formation. The high rate of aerobic lactate formation might be due to segregation of glycolytic and oxidative processes as well as to the adaptation of the poorly vascularized inner layers to a relatively anoxic existence. The rod inner segment contains packed mitochondria and has high levels of all mitochondrial enzymes and hexokinase. This is the region closest to the choroidal nutrient supply.

In the vascularized inner layers of the monkey retina, hexokinase activity is almost twice that in the homologous layers of the rabbit retina. Several glycolytic enzymes, including phosphofructokinase and glyceraldehyde phosphate dehydrogenase, as well as lactic dehydrogenase and glycogen phosphorylase, tend to be higher toward the vitreous surface in rabbit than in monkey retina. The total energy reserves (especially high glycogen levels) and lactate levels increase as the avascular vitreous surface of the rabbit retina is approached. Malate dehydrogenase and NAD-dependent isocitrate dehydrogenase, the citric acid cycle enzymes, vary with the relative density of mitochondria, which is high in the layer of rod inner segments and in the synaptic layers. Data such as these suggest that adaptive changes occur in carbohydrate enzymes and metabolism and that these changes are dependent on the local availability of substrates and oxygen. As in brain, however, continued electrical responsiveness of the retina depends upon oxidative metabolism. The response to light is dependent on metabolic processes and control mechanisms that are similar to those described for the maintenance of electrical activity in brain [8 and Chap. 7].

Peripheral Nerve

In mice, the metabolic rate of peripheral nerve is about 7 percent of that in brain. As in brain, carbohydrate reserves are low. In the sympathetic system, as in brain, glucose is the major metabolite of both nerve and ganglion. Unlike brain, however, there is no apparent barrier to glucose uptake in ganglion, and glucose levels are close to those in blood. Glucose levels in peripheral nerve are intermediate between those of brain and the ganglia. The patterns of substrate distribution and utilization of stimulated nerve differ from those of nerve at rest. Depletion of energy reserves leads to the failure of synaptic transmission before failure of conduction. Transmission in sympathetic ganglia can fail after carbohydrate reserves (glucose and glycogen) are depleted despite maintenance of high levels of ~P. This situation is similar to that in hypoglycemic brain (see in Chap. 25), and similar mechanisms may be involved [11].

Glial Tumors

Glial tumors might serve as models for glial metabolism. The metabolic rates in these tumors (with the possible exception of oligodendrogliomas) are relatively low; the reason, at least in part, may be due to poor vascularization of large areas of the tumors. Both the architecture and the blood supply of tumors are complex; the overall metabolism of tumors may represent adaptation to their environment, rather than the intrinsic metabolic capacity of the parent (glial) tissue or the character of the neoplastic change. As would be expected, metabolic fluxes are higher in areas well supplied with nutrients than in areas more distant from the blood supply, and cellular proliferation is more active in well-supplied regions [7]. Some neuronal and glial neoplastic cell lines will, under certain conditions, manifest morphological, biochemical, and physiological properties that resemble those of neurons or glial cells. Although they are useful for studying differentiation and cell interaction, metabolic investigations of these cell lines are valid only for the particular set of microenvironmental conditions under which the cells are studied. For instance, the rate of glucose and oxygen consumption, and the proportion of glucose metabolized by the citric acid cycle, will vary with pH, a factor that may not be well controlled by current methods.

THE COMPOSITION OF THE FREE AMINO ACID POOLS

The Level of Amino Acids and Related Compounds

Most of the amino acids in brain, as in other tissues, are found as components of enzymes, membranes, and other structures. Small amounts of amino acids, which are the synthetic intermediates of proteins, are present in the soluble extracts of the tissue and constitute the so-called free pool of amino acids. Neurochemists have long been interested in the observation that brain contains relatively higher concentrations of certain amino acids of the glutamate family than do other tissues. Within brain, these compounds comprise more than half of the total amino acid content.

The ability of glutamate to support respiration of brain-tissue slices was suggested as a provision for a buffer system to maintain oxygen uptake and energy production under hypoglycemic conditions. This idea has led to many trials of glutamate feeding as a treatment for mental illnesses. In early studies, it was demonstrated that such feeding experiments were unlikely to have any effect on glutamate levels in brain because of a marked blood-brain barrier to the entry of this amino acid. Studies with radioactively labeled glutamate, which did not raise circulating levels of the amino acid, have shown that the label enters the brain slowly. This suggests that the high endogenous levels of glutamate in brain arise from local synthesis. This has been confirmed in other studies, especially those related to acute ammonia toxicity, in which levels of brain glutamine increase rapidly without sufficient uptake of material from the circulation to account for those increases.

The content of free amino acids in the brain is maintained at fairly constant levels under most conditions and is a characteristic of that organ. In general, three groups can be distinguished: (1) the essential amino acids, which are present at fairly low levels — close to those in plasma; (2) the nonessential amino acids, which are present at concentrations several times higher than the essential ones; and (3) compounds that are specifically present in brain, such as GABA and acetylaspartic acid [33].

The composition of the free amino acid pool as shown in Table 14-1 is similar in most species [34]. The amino acid level in the spinal fluid is much lower than

Table 14-1. Free Amino Acid Levels[a]

| | Brain | | | Human | | |
	Cat	Rat	Carp	Brain	CSF	Plasma
Glutamic acid	790	1160	550	660	0.8	2
Taurine	230	660	480	120	0.6	6
N-acetylaspartic acid	600	560	80	490		0
Glutamine	280	450	770	580	50	60
Aspartic acid	170	260	350	96	0.02	0.2
γ-Aminobutyric acid	140	230		42		0
Glycine	78	68	62	40	0.7	22
Alanine	48	65	66	25	2.6	35
Serine	48	98	33	44	2.5	11
Threonine	17	66	36	27	2.5	14
Lysine	8	21	34	12	2.1	19
Arginine		11	14	10	1.8	8
Histidine	2	5	36	9	1.3	9
Valine	6	7	15	13	1.6	23
Leucine	7	5	22	7	1.1	12
Isoleucine	3	2	13	3	0.4	6
Phenylalanine	2	5	13	5	0.8	5
Tyrosine	3	7	9	6	0.8	5
Proline	3	8	12	10		18
Methionine	2	4		3	0.3	2
Ornithine	1	2	7	3	0.6	5
P-ethanolamine	120	200	62	110		0.2
Cystathionine	14	2	32	200		
Homocarnosine	0.4	6		23	0.3	0
Glutathione	49	260		200		

[a]The values presented, which are averages from many publications, are expressed as μmol amino acid per 100 g of brain or 100 ml of CSF or plasma.

in brain. CSF amino acid levels are also lower than those in plasma. The amino acid composition of spinal fluid is not parallel to that of brain; for example, note the high glutamine and low glutamate in the CSF. The free pool in brain also does not reflect the amino acid composition of the cerebral proteins. The amino acids are present at much higher levels in the protein-bound than in the free form. The concentration ratio of protein-bound to free amino acids in brain varies from 10 for glutamic acid to 1,800 for isoleucine. The amino acids present at high level (glutamate, taurine, GABA, and glycine) are the most active physiologically; a number of peptides are also of physiological interest. The physiological activity of these amino acids and peptides, their regional distributions, and compartmentalization are further discussed in Chapter 11.

The composition of the protein-free amino acid pool in peripheral nerve is different from that in the brain; most amino acids in mammalian nerve are lower than in brain. GABA is nearly absent in vertebrate peripheral nerve. In some crustacean species, a few specific compounds are at very high level (for example, aspartate, glycine, and alanine in some). In other species, the levels (for example, taurine) are 10 to 100 times higher in peripheral nerve than in brain.

Much less is known about the distribution of peptides in the organism, principally because methods are not as well developed for separation and detection of these compounds, which usually are present at very low levels. Only a few peptides, such as glutathione, are at high levels in the brain. Interest in this class of compounds is great because of their high physiological activity. There are indications that a number of peptides are present exclusively in brain. Many γ-glutamyl peptides are found, including γ-glutamyl derivatives of glutamate, glutamine, glycine, alanine, β-aminoisobutyric acid, serine, and valine. These peptides are present at levels of 10 to 700 μg per g of fresh tissue, and they may be formed by transpeptidation reactions from glutathione. In fairly large amounts, N-acetyl-α-aspartylglutamic acid is present (about 10 to 30 mg per 100 g of fresh brain). No function has been attributed to this peptide, but it is known that it forms the N-terminal sequence of the actin molecule in muscle, and it may play a similar role in some brain protein. Homocarnosine and homoanserine, two peptides of histidine, are also unique to brain. Homocarnosine is γ-aminobutyrylhistidine, the homolog of the long-known muscle constituent, β-alanylhistidine (carnosine). The γ-aminobutyryl derivative of anserine (β-alanyl-l-methylhistide) is also present. These histidine peptides are at much higher levels in this tissue than are their more widely distributed relatives, carnosine and anserine.

Developmental Changes

The levels of most components, including the components of the free amino acid pool, undergo changes during the maturation of the brain. The changes are complex; different compounds change in different developmental periods, and a few compounds show several changes — increase followed by decrease, then increase again, for example. Many of the essential amino acids show some decreases, while some nonessential amino acids increase. Some changes are illustrated in Table 14-2.

Quantitatively, the greatest changes are a decrease in taurine and an increase in glutamic acid. These changes are gradual, compared with the rapid decrease in alanine

Table 14-2. Changes in Amino Acid Levels During Development[a]

	Fetal 15 day	Newborn 1 day	Adult
Taurine	14	16	8.0
Glutamic acid	7.5	5.0	12
Aspartic acid	2.4	2.3	3.8
Threonine	4.3	0.90	0.56
Proline	0.89	0.57	0.15
Glycine	2.26	2.30	1.27
Alanine	5.08	0.80	0.56
Leucine	0.53	0.18	0.06
Tyrosine	0.24	0.20	0.08
Phenylalanine	0.24	0.13	0.07
γ-Aminobutyric acid	0.50	1.62	2.37
Arginine	0.45	0.14	0.11

[a]Values are from mouse brain, expressed as μmol amino acid per gram brain tissue.

around birth. Although such changes in amino acid levels indicate developmental changes in metabolism (in the relative rates of various metabolic pathways), the connection between substrate levels and metabolism is not clear. It is tempting to theorize that the decrease in essential amino acids parallels the decrease in the rate of protein turnover, for example, but a decrease in amino acids does not necessarily cause a decrease in protein turnover. In spite of such difficulties, developmental studies are helpful toward understanding the function of amino acids in the nervous system [35].

Alterations of Amino Acid Levels
The composition of the free amino acid pool remains constant under most conditions. This stability of composition in brain is not the result of "binding": The amino acids are in a dynamic equilibrium with many that are undergoing rapid metabolism, and there is also a rapid exchange between plasma and free amino acids. Changes in metabolism and transport therefore can alter the level of cerebral amino acids [36].

Although changes in amino acid levels in plasma do not produce comparable changes in brain, nutritional and pathological changes in plasma can affect amino acids in brain. The daily rhythm of increase in plasma tryptophan after meals is not reflected in brain, because both tryptophan and other related amino acids also increase in plasma, competing for the same transport system and jointly inhibiting each other's uptake. A selective increase of tryptophan, especially when accompanied by a decrease in the level of other competing amino acids, increases the cerebral levels of tryptophan and of its metabolic products, such as serotonin [37]. Such nutritional variations in the level of essential amino acids can be expected but, because the penetration of nonessential amino acids into brain is much less, plasma-level alterations are not likely to affect brain concentrations.

In severe malnutrition, the changes of the brain amino acid pools are rather specific: There is a large increase in histidine and homocarnosine [38] whereas some amino acids decrease (valine, serine, aspartate). Among endocrine influences, the effects of insulin in particular have been studied ever since insulin was first used in the treatment of depressive states. In insulin hypoglycemia, the major changes are a decrease of nonessential amino acids; most likely this reflects the changes in the activity of the citric-acid cycle. Hyperthyroidism increases most of the nonessential amino acids. In general, relatively smaller changes in the reaction rates involved in carbohydrate or energy metabolism do not affect the levels of cerebral amino acids, but major rate changes do shift the levels of nonessential amino acids, primarily those that react (via amino transfer) with the intermediates of energy metabolism (glutamate, aspartate, alanine) and GABA. For example, in ischemia, hypoxia, and hypothermia, glutamate and aspartate decrease and GABA increases. Similar changes may be found in hibernating animals; upon arousal, levels return to normal [39].

The changes are somewhat different in convulsions. In human epileptogenic brain tissue, the most consistent changes reported were decreases in glutamic acid and taurine and an increase in glycine; these changes were localized in areas of pathological changes; in surrounding areas, aspartate and GABA also were lowered. Such changes could be reproduced in experimentally induced convulsive states. Upon the administration of taurine, amino acid levels tended to return to normal [40]. In induced convulsions, such changes also depend on the convulsant used: Pentamethylenetetarole causes an increase in alanine (interference with the entry of pyruvate into the citric acid cycle); inhibitors of glutamic acid decarboxylase result in increased levels of glutamate.

Drugs, especially at high dosages, also have been reported to affect the levels of nonessential amino acids; the relation of such changes to any of the pharmacological effects of the drugs has not been established. Chlorpromazine lowers glutamate, aspartate, and GABA; drugs altering catecholamine metabolism (reserpine, 6-hydrodopamine) have similar effects; ethanol was reported to lower GABA levels [41].

TRANSPORT OF AMINO ACIDS

When slices from brain are incubated, they accumulate amino acids with tissue levels several times higher than those in the surrounding medium. This uptake against a concentration gradient that is higher in slices from brain than in slices from most other organs illustrates the active-transport processes in brain (see Chaps. 6 and 20). In most cases, such uptake processes are better controlled, more selective, and more rapid than diffusion: The present concept is that the distribution of most physiological metabolites in the brain is governed by mediated transport, rather than by passive diffusion [42].

Mechanisms

Mediated transport implies the presence of specific membrane components that have an affinity to the transported substrate and that possess sufficient asymmetry near the inner and outer membrane surface to permit net unidirectional movement of the substrate (see also Chap. 20). The specificity of the transport, described below, indicates that there is a participation of carrier proteins. The net transport is not solely in the direction of uptake; specific mediated exit of metabolites from the brain also occurs; that is, exit of a compound from the brain is observed even when the level of this compound is higher in plasma than in brain. The properties of uptake and exit are not identical; factors that influence transport do not necessarily affect uptake and exit similarly, and the sites at which uptake is most active differ from those of exit (see Regional Differences, page 294).

Although specific amino acid "binding proteins" have been isolated from bacterial membranes, the precise mechanism of transport of metabolites through membranes is not well understood. There are indications that the heterogeneous structure and function of the nervous system are reflected in the heterogeneity of its membrane transport systems; important differences exist in transport systems of capillaries, neurons, glia, choroid plexus, nerve endings, peripheral nerve, vesicles, mitochondria, and ganglia.

Recently, an enzymatic mechanism for amino acid transport was proposed, involving γ-glutamyl peptide formation of the transported amino acid with glutathione by γ-glutamyl transpeptidase, with subsequent hydrolysis of the peptide and resynthesis of glutathione [43]. The enzyme γ-glutamyl transpeptidase is low in brain but high in brain capillaries and in choroid plexus. Under most conditions, simultaneous uptake and exit occur.

A third aspect of transport (exchange) that differs somewhat from uptake and exit can be recognized. Exchange may be important. If heteroexchange occurs, the uptake of one compound drives the countertransport of a structurally related compound. Heteroexchange has been observed in isolated systems from brain, and it may be quantitatively significant in some pathological cases, such as the aminoacidurias. The half-life (50 percent exchanged) for most amino acids via exchange of brain and plasma free pools is in minutes. Active protein metabolism of the brain results in a very high rate of exchange between the free and the protein-bound forms. The half-life via exchange of free amino acids to protein-bound forms in brain is usually a few hours.

Specific Systems

When uptake of one amino acid was studied in the presence or absence of another, inhibition by related compounds was found. This shows that related compounds are transported by the same system and can compete for carrier sites. Eight such transport classes may be identified in slices, according to the structure of the amino acids: small neutral, large neutral, acidic, small basic, and large basic amino acids, and three additional classes represented by GABA, proline, and taurine [42]. In vivo, the transport by brain capillaries appears to utilize only three classes, large neutral, large basic, and acidic; the others are absent or too weak to detect in measurements of capillary transport rates. Transport classes exist for many other metabolites as well. They have been identified for hexoses, pentoses, carboxylic acids, mono-, di- and polyamines, and nucleotides. The high rate of cellular transport for the nonessential amino acids, combined with low capillary transport of the same compounds, is specific for brain.

There is some overlap among the classes because some amino acids have an affinity to another "carrier" in addition to their own. In substrate specificity, the transport classes in brain are similar to, but not identical with, those described in other systems. In bacteria, classes with narrower specificity are found; in other organisms, there are multiple classes. For example, there are three systems for the large basic amino acids: one specific for lysine alone; another for lysine, arginine, and ornithine; still another for basics and some neutrals.

More recently, additional amino acid transport classes with high affinity and high substrate specificity have been described in brain. These were found primarily in synaptosomal preparations. It was proposed that the low-affinity (K_m approx. 10^{-3} M), more generally distributed transport system serves metabolic functions; the high-affinity (K_m approx. 10^{-5} M) system removes the physiologically active (neurotransmitter) amino acids (see Chap. 11). This high-affinity transport was suggested as another criterion for assignment of neurotransmitter function. It was found for a number of amino acids in the brain (glutamate, aspartate, GABA, glycine, proline, tryptophan, and taurine), each of which is also a substrate for low-affinity transport. The participation of exchange in these studies must be determined because exchange can simulate high-affinity transport but cannot remove neurotransmitters [44].

The requirements of structure are fairly strict; decarboxylated amino acids (amines) are not transported by the amino acid transport systems. The stereospecificity is not absolute: D-amino acids are transported in most cases, but to considerably lower levels than are the L isomers. Despite this, some D-amino acids can penetrate the brain because, although their uptake is slower, so is their exit, which allows slow accumulation. This, again, illustrates that uptake and exit both influence the level of compounds. Some compounds, related to amino acids although not normal components of biological fluids (such as synthetic amino acid analogs), have affinity for the various carriers; nonmetabolizable analogs are useful for studying transport independent of metabolism.

Requirements

Transport against a concentration gradient requires energy. Metabolite transport in the brain, as in most other systems, is energy dependent. The primary source of the energy that fuels amino acid transport is not known. Inhibitors of metabolic energy also inhibit amino acid uptake, and in many cases, but not all, the decrease in ATP levels is accompanied by a decrease in uptake. Such inhibition, however, may be indirect.

Amino acid uptake in brain, as in other systems, is dependent on Na^+ (Chap. 11). Inhibition of the Na^+ pump or the absence of Na^+ abolishes most (but not all) of the amino acid transport. These and other indications strongly support the idea that Na^+ gradients (potential gradients) may be the driving force for transport, although altering Na^+ levels can affect metabolite distribution indirectly in many systems. Not all compounds show the same dependence on Na^+: diamine uptake is independent and basic amino acid uptake only partially dependent on Na^+ levels. Thus, lowering Na^+ does not affect all amino acids to the same degree. The Na^+ dependence of high-affinity uptake appears to be greater than that of low-affinity uptake. These and other results indicate that alterations in Na^+ levels or Na^+ gradients may influence various amino acids to different degrees. Although other ions (probably K^+, possibly Ca^{2+}) also may influence transport, Na^+ seems to be a primary factor [45].

Changes in Development

The composition of the brain, including the free amino acid pool, undergoes large changes during development (Table 14-2). Brain permeability for most compounds, and not only for amino acids, is greater in young than in adult brain. Although developmental changes in the free pool and in permeability to amino acids have been studied in detail, changes in amino acid transport are not as well known.

The elevation of an amino acid in plasma causes a greater elevation of levels in brain in young, as compared with adult, animals, as has been shown for most amino acids. Despite this, barriers and transport processes are not absent in the immature brain. For example, amino acid levels in fetal brain differ from those in fetal blood. Also, although the lability of the amino acid pool in young brain is greater, most experiments show that amino acid levels in brains of young animals do not equilibrate with plasma levels. As in adults, the barrier to nonessential amino acid penetration into the brain is stronger in young animals than is the barrier to essential amino acids. Transport properties, such as apparent affinity (K_m), show no developmental alterations, with GABA an exception to this rule. The amounts of carriers increase without any change in properties. Important metabolic changes affect amino acid movements during development: protein turnover is higher in young brain (both protein synthesis and protein breakdown); during growth, the net protein deposition utilizes in a few hours as much amino acid as is equivalent to the contents of the free pool (Chap. 18).

Regional Aspects of Transport

Distribution, especially of nonessential amino acids, is heterogeneous in the various areas of the brain. Only gross distribution has been studied, but there are some indications that further analytical refinements will show more extensive heterogeneity and compartmentation. There are indications that the amino acid pool in neurons is different from that in glia and that additional differences exist between nuclear and mitochondrial compartments. Lysosomes (in which protein degradation takes place) and the nerve-ending region (where release and removal of neurotransmitter amino acids occur) also represent special compartments.

Structural and metabolic heterogeneity is most likely to be paralleled by regional and structural heterogeneity of transport mechanisms, e.g., the distribution of "carriers" is not the same in all membranes and structures. The active cellular transport of glycine and proline seems to be absent in capillaries; glutamate uptake from capillaries is also low when compared with that of brain slices. This indicates a difference between the capillaries and the cellular membranes in the quantitative distribution of amino acid transport systems.

The choroid plexus, a structure with still different transport properties, influences the composition of spinal fluid and brain (Chap. 20). In vivo and in vitro transport have been observed in choroid plexus, which may contain additional and specialized transport mechanisms, one for organic acids and one for bases. These are absent in brain but present in kidney [46].

Perhaps the most heterogeneous distribution of transport systems is represented by the high-affinity transport in synaptosomes. Synaptosomal subpopulations, each containing a specific transport system, were separated. That is, synaptosomes containing the high-affinity glutamate system were separated from those containing the GABA system, and synaptosomes from spinal cord, but not those from brain, contained the glycine system. It is not known if such regional heterogeneity exists in glial high-affinity transport systems [23]. Similarly, some differences probably exist between mitochondrial and nuclear membranes and among mitochondria, and differences can be expected among various mitochondrial subpopulations.

Influences of Transport

Normal levels of amino acids in brain are fairly constant, but a number of alterations have been found in physiological and pathological conditions. Some of these changes may be mediated by alterations of transport.

The best-studied alterations of transport occur when a plasma amino acid is increased. Increase of one member of a transport class inhibits the uptake by the brain of the other members of that class. Under pathological conditions (aminoacidurias), the situation is more complex, since the elevation of a particular amino acid in plasma also causes an increase of that amino acid in the brain (see Chap. 24). The increased tissue level, by inhibiting exit or stimulating heteroexchange, may partially counteract the inhibition of uptake by the same increased amino acid in plasma. It was proposed that in phenylketonuria the increased plasma phenylalanine lowers several amino acids in brain. Among these is tryptophan; this results in a lowered brain serotonin, and may play a role in the development of mental retardation. The lowering of an amino acid such as tryptophan may also be the cause of a permanently decreased brain protein level if malnutrition persists throughout brain development. In contrast, in adult protein deficiency, brain protein content is maintained despite decreasing proteins of most other organs. This is thought to be the consequence of the more active amino acid transport in the adult brain that maintains the free amino acid pool to a greater degree than in other organs [33]. Pathological changes in protein metabolism could alter the free amino acid pool because the major portion of amino acids is protein bound; e.g., a net 1 percent breakdown of proteins would increase most amino acids severalfold. An important, but yet undecided, question is what effect an altered amino acid pool has on brain function. In particular, the effects of an increased phenylalanine concentration have been studied with regard to cerebral protein, lipid, and energy metabolism. At present, there are indications that the changes are specific, rather than general. Not all proteins are affected, and changes in some myelin components have been observed ([47, 48,] and Chap. 24).

REFERENCES

1. Olney, J. W. Brain lesions, obesity, and other disturbances in mice treated with monosodium glutamate. *Science* 164:719, 1969.
2. Gatfield, P. D., Lowry, O. H., Schulz, D. W., and Passonneau, J. V. Regional energy reserves in mouse brain and changes with ischaemia and anaesthesia. *J. Neurochem.* 13:185, 1966.
3. Lowry, O. H., and Passoneau, J. V. The relationships between substrates and enzymes of glycolysis in brain. *J. Biol. Chem.* 239:31, 1964.

*4. Abood, L. G. Brain Mitochondria. In Lajtha, A. (Ed.), *Handbook of Neurochemistry*. Vol. 2. New York: Plenum, 1970.

*5. Bradford, H. F. Carbohydrate and Energy Metabolism. In Davison, A. N., and Dobbing, J. (Eds.), *Applied Neurochemistry*. Philadelphia: Davis, 1968.

*6. Lehninger, A. L. *Biochemistry*. New York: Worth, 1970, pp. 267–564.

*7. Maker, H. S., and Lehrer, G. M. Effect of Ischemia. In Lajtha, A. (Ed.), *Handbook of Neurochemistry*. Vol. 6. New York: Plenum, 1971.

*8. Matschinsky, F. M. Energy Metabolism of the Microscopic Structures of the Cochlea, the Retina, and the Cerebellum. In Costa, E., and Giacobini, E. (Eds.), *Advances in Biochemical Psychopharmacology*. Vol. 2. New York: Raven, 1970.

*9. McIlwain, H., and Bachelard, H. S. *Biochemistry and the Central Nervous System*. Baltimore: Williams & Wilkins, 1971, pp. 1–170.

10. Diamond, I. A., and Fishman, R. A. High affinity transport and phosphorylation of 2-deoxyglucose in synaptosomes. *J. Neurochem.* 20:1533, 1973.

11. Stewart, M. A., and Moonsammy, G. I. Substrate changes in peripheral nerve recovering from anoxia. *J. Neurochem.* 13:1433, 1966.

12. Seta, K., Sershen, H., and Lajtha, A. Cerebral amino acid uptake in vivo in newborn mice. *Brain Res.* 47:415, 1972.

*13. Balazs, R. Carbohydrate Metabolism. In Lajtha, A. (Ed.), *Handbook of Neurochemistry*. Vol. 3. New York: Plenum, 1970.

14. Watanabe, H., and Passonneau, J. V. Factors affecting the turnover of cerebral glycogen and limit dextrin in vivo. *J. Neurochem.* 20:1543, 1973.

15. Goldberg, N. D., and O'Toole, A. G. The properties of glycogen synthetase and regulation of glycogen biosynthesis in rat brain. *J. Biol. Chem.* 244:3053, 1969.

16. Estler, C. J., and Ammon, H. P. T. Antagonistic effects of dopa and propranolol on brain glycogen. *J. Pharm. Pharmacol.* 22:146, 1970.

*17. van Eys, E. Regulatory Mechanisms in Energy Metabolism. In Bonner, D. M. (Ed.), *Control Mechanisms in Cellular Processes*. New York: Ronald, 1961.

18. Knull, H. R., Taylor, W. F., and Wells, W. W. Effects of energy metabolism on in vivo distribution of hexokinase in brain. *J. Biol. Chem.* 248:5415, 1973.

19. Sacks, W. Phenylalanine metabolism in control subjects, mental patients, and phenylketonurics. *J. Appl. Physiol.* 17(6):985, 1962.

20. Goldberg, N. D., Passonneau, J. V., and Lowry, O. H. Effects of changes in brain metabolism on the levels of citric acid cycle intermediates. *J. Biol. Chem.* 24:3997, 1966.

21. Hostetler, K. Y., Landau, B. R., White, R. J., Albon, M. S., and Yashor, D. Contribution of the pentose cycle to the metabolism of glucose in the isolated perfused brain of the monkey. *J. Neurochem.* 17:33, 1970.

22. Kauffmann, F. C., Brown, J. B., Passonneau, J. V., and Lowry, O. H. Effect of changes in brain metabolism on levels of pentose phosphate pathway intermediates. *J. Biol. Chem.* 244:3547, 1969.

*23. Snyder, S. H., Young, A. B., Bennett, J. P., and Mulder, A. H. Synaptic biochemistry of amino acids. *Fed. Proc.* 32:2039, 1973.

24. van den Berg, C. J., and Garfinkel, D. A simulation study of brain compartments. *Biochem. J.* 123:211, 1971.

25. Gaitonde, M. J., Dahl, D. R., and Elliott, K. A. C. Entry of glucose carbon into amino acids of rat brain and liver in vivo after injection of uniformly ^{14}C-labeled glucose. *Biochem. J.* 94:345, 1965.

26. Waelsch, H., Berl, S., Rossi, C. A., Clarke, D. D., and Purpura, D. Quantitative aspects of CO_2 fixation in mammalian brain in vivo. *J. Neurochem.* 11:717, 1964.

*Asterisks denote key references.

27. Garfinkel, D. A simulation study of the metabolism and compartmentation in brain of glutamate, aspartate, the Krebs cycle, and related metabolites. *J. Biol. Chem.* 241:3918, 1966.

28. Clarke, D., and Garfinkel, D. To be published.

29. Wu. J. Y., and Roberts, E. Properties of brain L-glutamate decarboxylase: Inhibition studies. *J. Neurochem.* 23:759, 1974.

*30. Tapia, R. The Role of γ-Aminobutyric Acid Metabolism in the Regulation of Cerebral Excitability. In Myers, R. D., and Drucker-Colin, R. R. (Eds.), *Neurohumoral Coding of Brain Function.* New York: Plenum, 1974.

31. Wood, J. D., and Peesker, S. J. Development of an expression which relates the excitability of the brain to the level of GAD activity and GABA content, with particular reference to the action of hydrazine and its derivatives. *J. Neurochem.* 23:703, 1974.

32. Van der Noort, S., Eckel, R. E., Brine, K. L., and Hrdicka, J. Brain metabolism in experimental uremia. *Arch. Intern. Med.* 126:831, 1970.

*33. Gaull, G. E., Tallan, H. H., Lajtha, A., and Rassin, D. K. Pathogenesis of Brain Dysfunction in Inborn Errors of Amino Acid Metabolism. In Gaull, G. E. (Ed.), *Biology of Brain Dysfunction.* Vol. 3. New York: Plenum, 1974.

*34. Himwich, W. A., and Agrawal, H. C., Amino Acids. In Lajtha, A. (Ed.), *Handbook of Neurochemistry.* Vol. 1. New York: Plenum, 1969.

*35. Himwich, W. A. (Ed.). *Biochemistry of the Developing Brain.* Vol. 1. New York: Dekker, 1973.

*36. Lajtha, A. Alterations in the Amino Acid Content of the Brain. In Towers, D. B., and Brady, R. O. (Eds.), *The Nervous System.* Vol. 1. New York: Raven, 1975.

37. Fernstrom, J. D., Madras, B. K., Munro, H. N., and Wurtman, R. J. Nutritional Control of the Synthesis of 5-Hydroxytryptamine in the Brain. In *Aromatic Amino Acids in the Brain, CIBA Foundation Symposium 22.* New York: American Elsevier, 1974.

38. Enwonwu, C. O., and Worthington, B. S. Regional distribution of homocarnosine and other ninhydrin-positive substances in brains of malnourished monkeys. *J. Neurochem.* 22:1045, 1974.

39. Palladin, A. V., Belik, Ya. V., and Polyakova, N. M. *Protein Metabolism of the Brain.* New York: Plenum, 1976.

40. van Gelder, N. M., Sherwin, A. L., and Rasmussen, T. Amino acid content of epileptogenic human brain: Focal versus surrounding regions. *Brain Res.* 40:385, 1972.

41. Himwich, W. A., and Davis, J. M. Free amino acids in the developing brain as affected by drugs. *Adv. Behav. Biol.* 8:231, 1974.

*42. Lajtha, A. Amino Acid Transport in the Brain in Vivo and in Vitro. In *Aromatic Amino Acids in the Brain, CIBA Foundation Symposium 22.* New York: American Elsevier, 1974.

43. Meister, A. Glutathionine; metabolism and function via the γ-glutamyl cycle. *Life Sci.* 15:177, 1974.

*44. Iversen, L. L. Neuronal Uptake Processes. In Callingham, B. A. (Ed.), *Drugs and Transport Processes.* Baltimore: University Park Press, 1974.

45. Banay-Schwartz, M., Teller, D. N., and Lajtha, A. Energy Supply and Amino Acid Transport in Brain Preparations. In Levi, G., Battistin, L., and Lajtha, A. (Eds.), *Advances in Experimental Biology and Medicine.* New York: Plenum, 1976.

46. Franklin, G. M., Dudzinski, D. S., and Cutler, R. W. P. Amino acid transport into the cerebrospinal fluid of the rat. *J. Neurochem.* 24:367, 1975.

*47. Agrawal, H., and Davison, A. N. Myelination and Amino Acid Imbalance in the Developing Brain. In Himwich, W. A. (Ed.), *Biochemistry of the Developing Brain.* Vol. 1. New York: Dekker, 1973.

*48. Roberts, S. Effects of Amino Acid Imbalance on Amino Acid Utilization, Protein Synthesis, and Polyribosome Function in Cerebral Cortex. In *Aromatic Amino Acids in the Brain, CIBA Foundation Symposium 22.* New York: American Elsevier, 1974.

Chapter 15

Chemistry and Metabolism of Brain Lipids

Kunihiko Suzuki

Studies of lipids in the nervous system form a significant part of neurochemical investigations. There are several obvious reasons for the importance of brain lipids, both as structural constituents and as participants in the functional activity of the brain. Among various body organs, the brain is one of the richest in lipids. It contains a unique structure, the myelin sheath, which has the highest concentration of lipid among any normal tissue or subcellular components, except for adipose tissue, and which has been the subject of intensive and extensive studies in recent years. Another important aspect is the existence of a number of genetically determined metabolic disorders involving brain lipids. Identification of abnormally stored lipids and the search for underlying enzymatic defects have been giving strong impetus not only to the investigations of these pathological conditions but also to the study of chemistry and metabolism of brain lipids in general. There is increasing evidence, furthermore, that lipid molecules play important functional roles within the membrane. Some of the postulated physiological functions of membrane lipid include the site for the cell-to-cell recognition process, specific cell-surface antigens, and specific receptors for toxins or other physiological compounds.

This chapter is designed to provide basic reference knowledge regarding brain lipids. The chapter, therefore, will cover only the most basic and elementary aspects of brain lipid, its chemistry, the lipid composition of normal brain, the peculiarities of brain lipids, and the major metabolic pathways. Many important aspects of biochemistry of brain lipids are covered elsewhere in this volume, such as the lipid and its metabolism in myelin (Chap. 4), developmental changes (Chap. 18), demyelinating diseases (Chap. 28), or inborn errors of sphingolipid metabolism (Chap. 26). There are also several excellent and reasonably up-to-date review articles on the general subject of brain lipids. They are given at the beginning of the reference list for this chapter and are recommended for more details on brain lipids and as the source of additional references [1–8].

CHEMISTRY OF MAJOR BRAIN LIPIDS

The lipid composition of the brain is unique not only in the high total lipid concentration but also in the types of lipids present. Three major categories include almost all of the lipids of normal brain: cholesterol, sphingolipids, and glycerophospholipids.

Cholesterol

Cholesterol (Formula 15-A) is the only sterol present in normal adult brains in significant amounts. The alcohol group at position 3 may be esterified with a long-chain fatty acid. Esterified cholesterol is present in normal brain only at very low

concentrations. Desmosterol, which is the immediate metabolic precursor to cholesterol and has an additional double bond at C-24 (Fig. 15-5), is present in substantial
amounts in normal developing brain. Careful examination with highly sensitive
analytical methodologies indicates the presence in normal brain of other sterols and
their precursors, such as squalene. The amounts of these other compounds found in
normal brain, however, are minute.

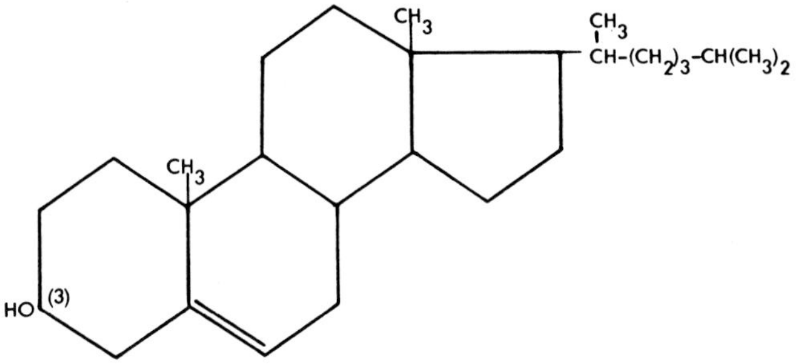

15-A   Cholesterol

Sphingolipids

Sphingosine
The basic building block of all sphingolipids is sphingosine, which is a long-chain
amino diol with one unsaturated bond (Formula 15-B). The major sphingosine in
the brain is C_{18}-sphingosine, but smaller amounts of C_{16}-, C_{20}-, and C_{22}-sphingo-
sines are known to occur. Also, a small portion exists in the saturated form as
dihydrosphingosine. Psychosine is sphingosine with additional galactose at the
primary alcohol group at C-1 (galactosylsphingosine). Although it is a potentially
important metabolic intermediate (see below), only trace amounts are present in
normal brain.

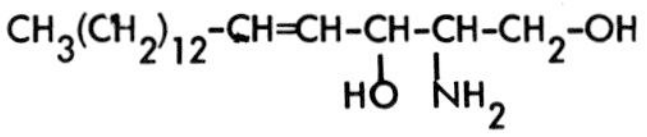

15-B   Sphingosine

Ceramide
The amino group of sphingosine is almost always acylated with a long-chain fatty
acid, ranging from C_{14} to C_{26}. N-Acylsphingosine is generically called ceramide
(Formula 15-C).

$$CH_3(CH_2)_{12}-CH=CH-CH-CH-CH_2-OH$$

$$HO \quad NH$$

$$C=O$$

$$R$$

$$R = -(CH_2)_n CH_3$$

15-C   Ceramide (*N*-acylsphingosine)

A variety of compounds can be substituted for the C-1 alcohol group of ceramide to form different sphingolipids. Ceramide-A is any sphingolipid characterized by a substituent, A, at the terminal hydroxyl group of ceramide. We can now define individual sphingolipids in the brain by defining the substituent A.

Sphingomyelin
This is the only phospholipid in the brain that is also a sphingolipid; in this case, A is phosphorylcholine, so sphingomyelin may be defined as ceramide-phosphorylcholine.

Cerebroside and Sulfatide
Cerebroside is a generic term for monohexosylceramide, i.e., the substituent A is a hexose. Ceramide-galactose is also known as galactocerebroside, and ceramide-glucose as glucocerebroside.

All the cerebroside in normal adult brain is galactocerebroside. Glucocerebroside occurs in the brain in small amounts in certain pathological conditions, as well as in the immature brain. Sulfatide is galactocerebroside with an additional sulfate group at C-3 of the galactose moiety, i.e., sulfatide is ceramide-galactose-SO_4^-.

A few glycosphingolipids structurally related to cerebroside are known to occur in small amounts in the brain. One is cerebroside plasmalogen in which the normally unsubstituted hydroxy group at C-3 of sphingosine is substituted with a fatty aldehyde with the α,β-unsaturated ether linkage. The other two are acylated galactocerebrosides, one at C-3, and the other at C-6 of galactose.

Ceramide Oligohexosides (2–4 Sugars)
A series of sugar-containing sphingolipids is known. Unlike cerebroside, all ceramide oligohexosides in the brain have a glucose moiety linked to ceramide. Although these compounds are virtually absent in normal adult brains, they are present in measurable amounts in immature brains and in some pathological conditions. They are important in relation to the metabolism of brain ganglioside. Some of these substances are ceramide–Glc–Gal, or ceramide dihexoside, also called ceramide lactoside; ceramide–Glc–Gal–GalNAc, or ceramide trihexoside; and ceramide–Glc–Gal–GalNAc–Gal, or ceramide tetrahexoside (Glc = glucose; Gal = galactose; GalNAc = *N*-acetylgalactosamine).

Gangliosides
Gangliosides are defined as sphingoglycolipids that contain sialic acid. Sialic acid is a generic name for *N*-acylneuraminic acid, and the acyl group of sialic acid

in gangliosides of the brain is always the acetyl form. *N*-Acetylneuraminic acid is commonly abbreviated as NeuNAc (Formula 15-D).

OH
|
C-COOH
|
H C H
|
H C OH
|
O   Ac-NH-C H   Ac = CH_3CO-
|
C H
|
H C OH
|
H C OH
|
CH_2OH

15-D   *N*-Acetylneuraminic acid (NeuNAc)

The series of gangliosides in the brain has the above ceramide oligohexosides as the backbone, with one or more NeuNAc moieties attached. Major gangliosides of the brain are depicted in Figure 15-1. Several other minor gangliosides have been identified in the nervous system. They include G_{D3}, which is the disialo-form of G_{M3}, G_{D2} which is the disialo-form of G_{M2}, and sialyl-galactosylceramide. The last is unusual in that the hexose residue on ceramide is galactose rather than glucose. It is therefore sialylated galactocerebroside. This compound is attracting attention as a specific component of myelin, particularly in human brain. In the peripheral nerve, a substantial portion of G_{M1}-ganglioside contains *N*-acetylglucosamine in place of *N*-acetylgalactosamine.

The monosaccharide units of the series of glycosphingolipids described above are linked together with the glycosidic linkage, which is defined as a covalent linkage of the reducing aldehyde group of a sugar to a hydroxy group. Because the configuration around C-1 of hexose defines the anomeric forms, these monosaccharide units are linked together with fixed anomeric configurations. Essentially all glycosidic linkages known to occur in the brain glycolipids have β-configuration. This consideration becomes important in relation to genetic disorders of sphingolipid metabolism (Chap. 26), because hydrolytic enzymes are specific with respect to the anomeric configuration.

Glycerophospholipids

Glycerophospholipids with two acyl ester linkages are termed phosphatidyl compounds (Formula 15-E). In Formula 15-E, R_1 and R_2 represent long-chain fatty acid moieties which form ester linkages to two alcohol groups of the parent glycerol molecule. The third alcohol group of the glycerol molecule is linked to a phosphate group. Like sphingolipids, one of the hydroxyl groups of the phosphate moiety is substituted with a substituent, A, to form various phosphatidyl compounds. When substituent A is hydrogen, the formula is that of phosphatidic acid. Other groups that may be substituted for A are listed in Formula 15-E(a).

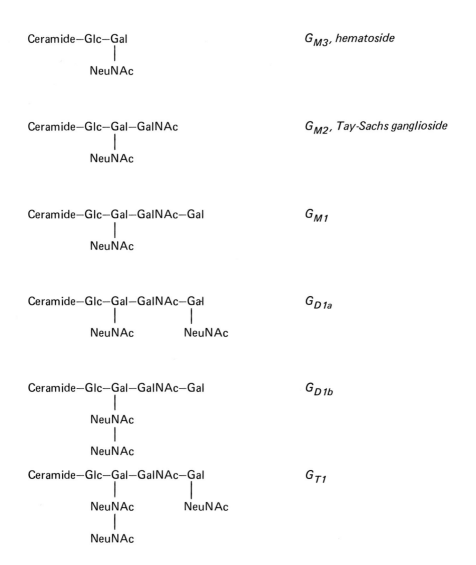

Figure 15-1
Major gangliosides of the brain. The nomenclature of gangliosides is that of Svennerholm [49], probably most widely used. For other nomenclatures, refer to Ledeen [50].

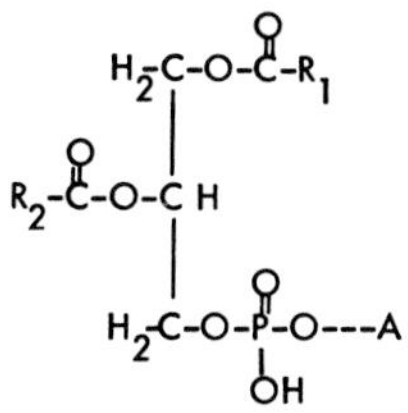

$$H_2C-O-\overset{\overset{O}{\|}}{C}-R_1$$

$$R_2-\overset{\overset{O}{\|}}{C}-O-\overset{|}{C}H$$

$$H_2C-O-\overset{\overset{O}{\|}}{\underset{OH}{P}}-O---A$$

15-E   Phosphatidyl compounds
(Phosphatidic acid when A = H)

Phosphatidylethanolamine:

$$A = -CH_2-CH_2-\overset{+}{N}H_3$$

Phosphatidylcholine (lecithin):

$$A = -CH_2-CH_2-\overset{+}{N}(CH_3)_3$$

Phosphatidylserine:

$$A = -CH_2-CH-(\overset{+}{N}H_3)-COOH$$

Phosphatidylinositol (monophosphoinositide):

A =

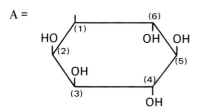

15-E(a)   Additional phosphate groups may be present at
position 4, or at positions 4 and 5. Then the
compounds are called phosphatidylinositol-4-
phosphate (diphosphoinositide) and phosphatidyl-
inositol-4,5-diphosphate (triphosphoinositide).

Another group of glycerophosphatides of potential importance is characterized
by the presence of an α,β-unsaturated ether linkage, which replaces the acyl ester
at C-1 of phosphatidyl compounds. These are termed *phosphatidal compounds* or
plasmalogens (Formula 15-F).

Although substituent A may be any one of those described for phosphatidyl
compounds, almost all of the phosphatidal lipid in the brain is phosphatidalethanolamine, with much smaller amounts of phosphatidalcholine and phosphatidalserine.

15-F   Phosphatidal compounds

In addition to the major glycerophospholipids above, many glycerolipids are known to be present in small amounts in the brain. Mono-, di-, and triglycerides, which are major lipids in many systemic organs, are present in the brain only in small amounts. Monogalactosyldiglyceride, although present as a minor constituent, appears to be highly localized in the myelin sheath. Other glycerolipids known to occur in the brain include cardiolipin (diphosphatidylglycerol), phosphatidic acid, phosphatidylglycerol, and phosphatidylglycerophosphate. Very small amounts of lysophosphatidyl compounds, which result from removal of one of the two acyl groups from glycerophosphatides, also appear to occur as normal constituents of the brain.

LIPID COMPOSITION OF NORMAL ADULT HUMAN BRAIN

General Analytical Scheme

Although there are numerous minor modifications, the method of Folch, Lees, and Sloane-Stanley [10] is almost always the starting point for any investigation of lipids in the nervous system. The basic analytical scheme now being used in our laboratory for general analysis of brain lipid is depicted in Figure 15-2.

Polyphosphoinositides are not extracted by this procedure unless the insoluble residue is reextracted with an acidic mixture of chloroform and methanol. In order to extract ganglioside completely, it is necessary to reextract the chloroform-methanol insoluble residue with a mixture of chloroform-methanol of a reversed ratio (1:2, v/v) containing 5 percent water [9]. Otherwise the procedure is sufficiently quantitative for most purposes, so the final chloroform-methanol soluble fraction can be used for detailed lipid analysis and the retentate of the upper phase for ganglioside analysis. When analysis of specific lipid types is intended, the methodology for general survey analysis may be inadequate. For example, quantitative determination of nonpolar gangliosides requires column chromatographic separation of the total lipid extract because the solvent partition procedure does not recover nonpolar gangliosides quantitatively [11].

The proteolipid protein-free lipid fraction and the ganglioside fraction are then ready for detailed analysis of individual lipids. It is not the purpose of this chapter to describe the analytical methods for individual lipids in any detail. Colorimetry and column and thin-layer chromatography are utilized extensively. Thin-layer chromatography has become established in the analytical investigations of lipids as one of the most convenient and rapid procedures. Figures 15-3 and 15-4 show

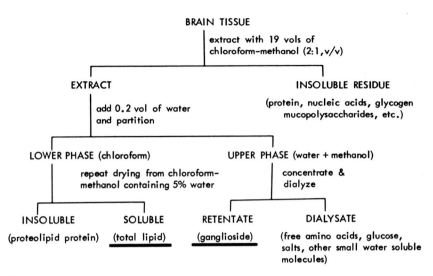

Figure 15-2
A typical scheme of brain lipid analysis.

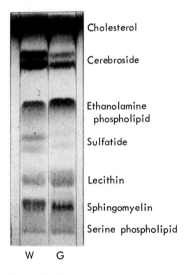

W   G

Figure 15-3
Thin-layer chromatogram of the total lipid fractions of normal gray (G) and white (W) matter.
Approximately 250 μg of total lipid was spotted on a silica gel G plate 250 μ in thickness.
Solvent system: chloroform-methanol-water (70:30:4, by volume). Spots were visualized by
spraying with 50% sulfuric acid and heating. Note the greater amounts of cerebroside and
sulfatide and lesser amounts of phospholipids in white matter. Serine phospholipid streaks
from the origin to the area of sphingomyelin in this solvent.

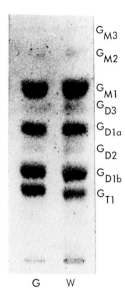

G_{M3}
G_{M2}
G_{M1}
G_{D3}
G_{D1a}
G_{D2}
G_{D1b}
G_{T1}

G   W

Figure 15-4
Thin-layer chromatogram of the ganglioside fraction of normal gray (*G*) and white (*W*) matter.
The dialyzed upper phase containing approximately 30 μg of NeuNAc was spotted on a silica
gel G plate 250 μ in thickness. Chromatography was run in two solvent systems successively:
the first solvent was chloroform-methanol-2.5 N ammonia (60:40:9, by volume), and the second
solvent, *n*-propanol and water (7:3 by volume). Spots were visualized by 50% sulfuric acid spray
and heating. The four major gangliosides are clearly visible, but other minor gangliosides also
are present in very small amounts in normal brain.

separations of total lipid and gangliosides in gray and white matter. The compounds
have been separated by thin-layer chromatography in some of the routine solvent
systems.

Gas-liquid chromatography is now used extensively for the analysis of almost
every constituent moiety of brain lipids. The combination of gas-liquid chromatography with mass spectrometry promises to be a useful tool for simultaneous
definitive identification and quantitative determination of such compounds.

Lipid Composition

Typical analytical results of normal adult human brain are given in Tables 15-1 and
15-2. Because the lipid compositions of gray and white matter differ both in total
concentration and in the distribution of individual lipids, it is usually important to
treat them separately. Depending on the purpose of investigation, selection of
specific regions of the nervous system could become of fundamental importance.
Equally important is the selection of the basis of reference, such as whole brain,
wet weight, dry weight, protein, DNA, or, in cases of lipid studies, total lipid. No
single basis of reference is satisfactory or proper in all situations. In Table 15-1
values are given according to three bases of reference commonly used in analytical
studies of brain lipids: fresh weight, dry weight, and lipid weight. Fresh weight is

Table 15-1. Typical Lipid Composition of Normal Adult Human Brain

	Gray Matter			White Matter		
Constituents	% fresh wt.	% dry wt.	% lipid	% fresh wt.	% dry wt.	% lipid
Water	81.9	–	–	71.6	–	–
Chloroform-methanol insoluble residue	9.5	52.6	–	8.7	30.6	–
Proteolipid protein	0.5	2.7	–	2.4	8.4	–
Total lipid	5.9	32.7	100	15.6	54.9	100
Upper phase solids	2.2	12.1	–	1.7	6.0	–
Cholesterol	1.3	7.2	22.0	4.3	15.1	27.5
Phospholipid, total	4.1	22.7	69.5	7.2	25.2	45.9
Ethanolamine phospholipid	1.3	7.2	22.7	2.3	8.2	14.9
Lecithin	1.6	8.7	26.7	2.0	7.0	12.8
Sphingomyelin	0.4	2.3	6.9	1.2	4.2	7.7
Monophosphoinositide	0.16	0.9	2.7	0.14	0.5	0.9
Serine phospholipid	0.5	2.8	8.7	1.2	4.3	7.9
Plasmalogen	0.7	4.1	8.8	1.8	6.4	11.2
Galactolipid, total	0.4	2.4	7.3	4.1	14.5	26.4
Cerebroside	0.3	1.8	5.4	3.1	10.9	19.8
Sulfatide	0.1	0.6	1.7	0.9	3.0	5.4
Ganglioside, total[a]	0.3	1.7	–	0.05	0.18	–

[a]The amounts of ganglioside were calculated from the total sialic acid, on the assumption that sialic acid constitutes 30% of ganglioside weight in a typical ganglioside mixture of normal brain. Note: Polyphosphoinositides are not included in this table.

Table 15-2. Composition of Major Gangliosides in Adult Human Brain[a]

	Gray Matter		White Matter	
Constituents	Average	Range	Average	Range
Total NeuNAc (μg/g wet wt)	812	744–918	110	80–180
G_0	3.9	3.2–4.8	4.8	2.8–6.1
T_{T1}	19.7	15.8–25.7	19.1	14.1–21.2
G_{D1b}	16.7	14.3–19.9	14.8	12.2–18.1
G_{D2}	3.0	1.2–4.2	1.6	1.2–3.1
G_{D1a}	38.0	29.1–43.7	36.2	30.0–38.2
G_{D3}	2.0	1.0–2.8	3.2	1.2–5.0
G_{M1}	14.2	13.0–15.6	18.8	14.6–21.2
G_{M2}	1.7	1.5–2.0	1.0	0.6–2.0
G_{M3}	<1	–	<1	–

[a]Values are expressed as percent of total NeuNAc in each ganglioside, except for total NeuNAc. G_0 represents all sialic acid which has mobility slower than G_{T1}. G_{D2} is Tay-Sachs ganglioside (G_{M2}) with an additional NeuNAc. G_{D3} is hematoside (G_{M3}) with an additional NeuNAc.

often the most proper reference in pathological conditions involving lipids because lipids are major constituents of the brain and often decrease drastically, thus substantially altering dry weight. When the water content of the tissue is in question, as is often the case in casually stored frozen specimens, dry weight may be the reference of choice. To compare relationships among various lipids, total lipid weight often provides the clearest picture.

White matter contains less water and much more lipid and proteolipid than does gray matter. On the basis of wet weight, there is almost three times as much lipid in white matter as in gray matter. There is almost twice as much on a dry-weight basis. As a consequence, there is more of each of the individual lower-phase lipids in white than in gray matter on the basis of wet weight, with the possible exception of monophosphoinositide. In contrast, gangliosides are characteristically gray matter lipids. Gray matter is ten times richer in gangliosides than is white matter, on dry-weight basis.

When the lipid compositions of gray and white matter are compared, the most conspicuous difference is that white matter is relatively much richer in galactolipids and relatively poor in phospholipids. Galactolipids constitute 25 to 30 percent of the lipids in white matter, whereas they are only 5 to 10 percent of those in gray matter. Phospholipids account for two-thirds of the total lipids in gray matter, but less than half of those in white matter. Plasmalogen constitutes 75 to 80 percent of ethanolamine phospholipid in white matter, but less than half of this phospholipid in gray matter. There are no measurable amounts of sphingoglycolipids, other than galactocerebroside and sulfatide, in the lower-phase lipid of normal adult human brain. These glycolipids are primarily localized in white matter. In contrast, gangliosides are highly localized in neuronal membranes and consequently in gray matter, although the molecular distribution pattern is similar in gray and white matter (see Table 15-2).

Except for the enrichment of ganglioside in gray matter, all other differences in the lipid composition between gray and white matter appear to be due mostly to the presence of myelin in the latter. Myelin constitutes half or more of the total dry weight of white matter, and it is poor in water content and rich in proteolipid protein and galactolipids. Readers are referred to Chapter 4 for details of composition and metabolism of myelin lipids.

The lipid composition of the brain is relatively stable throughout adult life, and the above data may serve as a basis for judging conditions that are altered pathologically. It is essential to keep in mind, however, that substantial changes in lipid composition take place during early development, particularly during the period of active myelination. Before myelination, both gray and white matter have a similar lipid composition, which resembles adult gray matter composition. During active myelination, the brain loses water, predominantly in white matter, the lipid content increases rapidly, and the differences between gray and white matter become more apparent. Characteristically, galactocerebroside is virtually nonexistent before myelination begins and increases concomitantly with the amount of myelin formed [12–15]. Ganglioside is also known to undergo substantial developmental changes [16]. For more details of developmental changes, the reader should refer to Chapters 4 and 18.

In addition to myelin, the lipid composition of specific subcellular fractions, such as nuclei, mitochondria, synaptic elements, or microsomes, have been investigated [17–20]. With the recent development of procedures to isolate relatively intact neurons and glial cells (Chap. 2) the lipid compositions of different cell types are also being investigated actively [21].

Characteristics of Brain Lipids

The lipid of the brain possesses several unique characteristics. As mentioned earlier, the brain is one of the richest portions of the body in total lipid content; approximately half of the dry weight of the entire brain is lipid. Although by no means constant, the lipid composition of the brain tends to remain relatively unaffected by various external factors, such as the nature of dietary intake, malnutrition, and other conditions that would alter drastically the lipid composition of systemic organs or plasma.

The brain is unique in its lack of certain lipids that are abundantly present elsewhere in the body. Triglycerides and free fatty acids constitute, at the most, only a few percent of the total lipid of the brain. Some of these are probably contributed by blood and blood vessels, rather than by neural tissues. Esterified cholesterol is always much less than 1 percent of the total cholesterol in normal brain, although it is present in slightly higher amounts just prior to active myelination. In pathological conditions in which massive myelin breakdown occurs, esterified cholesterol is often present in high concentrations. Because of this characteristic, the sum of cholesterol, phospholipid, and glycolipid comprises nearly the total lipid, an often useful criterion for judging analyses of brain lipid. If the sum of the three major classes of lipids is substantially less than total lipid weight (e.g., less than 85 percent), the analysis is technically faulty, unless it can be explained otherwise.

Glycolipids of the brain are unique in many ways. Normal adult brain contains only galactocerebroside, whereas almost all systemic organs contain glucocerebroside primarily and very little galactocerebroside. The kidney is the only systemic organ known to contain approximately equal amounts of galactocerebroside and glucocerebroside, but even in this organ the actual amount is almost negligible compared to that in the brain. Brain galactolipids contain high proportions of long-chain fatty acids, predominantly C_{24} to C_{26}, which are rarely found in systemic organs. Also, there are high proportions of α-hydroxy fatty acids in brain galactolipids: approximately two-thirds of cerebroside and one-third of sulfatide. α-Hydroxy fatty acids are very unusual in systemic organs.

Brain gangliosides are also unique in their high concentration and molecular distribution. The level of total ganglioside in the brain is rarely approached by any systemic organs; hematoside (G_{M3}) is usually the only ganglioside present in a significant amount in most systemic organs.

BIOSYNTHESIS AND CATABOLISM OF MAJOR LIPIDS

The biosynthetic and catabolic pathways of brain lipids are generally similar to those in systemic organs. Some of the general references at the beginning of the reference list contain excellent chapters on details of the metabolic pathways of individual lipids. Only the basic outline will be reviewed here.

Fatty Acids

Although free fatty acids are very minor constituents in brain, they are important components of all complex brain lipids. The mechanism of fatty acid biosynthesis in the brain appears to be essentially identical with that in other organs, such as liver [22]. De novo synthesis of palmitic acid ($C_{16:0}$) involves acetyl-CoA and malonyl-CoA, and further elongation takes place in mitochondria. Although the brain is capable of synthesizing most fatty acids, it cannot synthesize certain of them, which therefore must be provided from dietary sources and be transported into the brain [23]. These are known as essential fatty acids and include linoleic acid (ω-6$C_{18:2}$) and linolenic acid (ω-3$C_{18:3}$). In the nomenclature of fatty acids, the number following ω indicates the position of the first double bond, counting the carbon atoms from the methyl end; the number before the colon is the number of carbon atoms; and the number after the colon, the number of double bonds. Since the brain contains relatively large amounts of polyunsaturated fatty acids, the essential fatty acids are metabolically significant. Desaturation of fatty acids appears to occur at the stage of acyl-CoA. In systemic organs, α-hydroxylation of fatty acids is an intermediate step in C-1 degradation of fatty acids through dehydrogenation and decarboxylation. In the brain, α-hydroxy fatty acids are found as such in galacto-cerebroside and sulfatide. Prostaglandins are C-20 hydroxylated cyclic fatty acids and derive metabolically from arachidonic acid (ω-6$C_{20:4}$) (Chap. 13). Phytanic acid is a C-20 branched fatty acid (3,7,11,15-tetramethyl-hexadecanoic acid) which derives only from chlorophyll in dietary sources. Genetic inability to degrade this fatty acid results in Refsum's disease (Chap. 26).

Cholesterol Metabolism

Although not every step of the biosynthetic pathway of cholesterol known to occur in other tissues has been demonstrated in brain, it is safe to assume that cholesterol synthesis takes place in brain through the same pathway as in systemic organs. Acetate and its precursors are transformed through mevalonic acid to cholesterol (Fig. 15-5).

Desmosterol, the immediate precursor of cholesterol, is known to be present in brain in measurable amounts just prior to myelination [24, 25] and also in the myelin sheath itself in the early stage of myelination [26]. Biosynthesis of cholesterol in brain is most rapid during the period of active myelination, but adult brain retains the capacity to synthesize cholesterol when precursors such as acetate or mevalonate are available. Although most of the cholesterol in brain appears to be synthesized from endogenous precursors, experimental evidence indicates that a small amount of systemically injected cholesterol can be taken up intact and that the rate of uptake is greatest when the rate of cholesterol deposition in brain is most rapid, i.e., during active myelination [27]. Once deposited in brain, cholesterol, particularly that incorporated into myelin, is relatively inert metabolically [28, 29]. When radioactive cholesterol is injected into newly hatched chicks, considerable

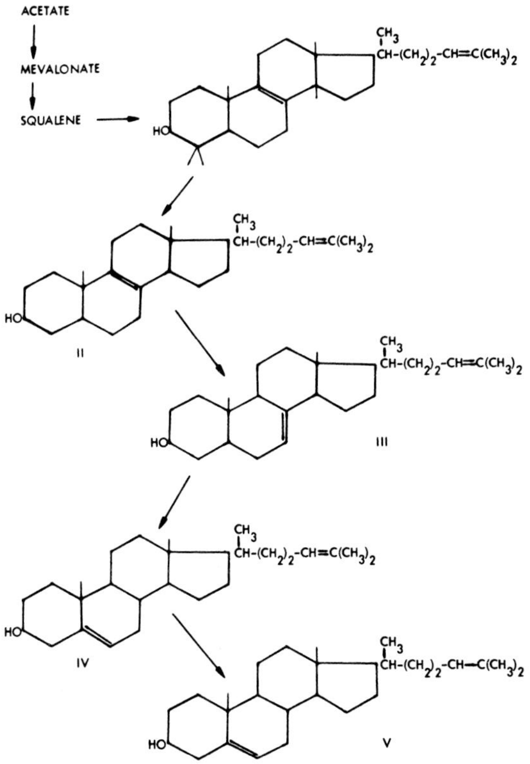

Figure 15-5
Outline of cholesterol biosynthesis. I = lanosterol; II = zymosterol; III = cholesta-7,24-dienol; IV = desmosterol (24-dehydrocholesterol); V = cholesterol.

radioactivity remains in cholesterol of brain, whereas that of liver and plasma virtually disappears within 3 to 8 weeks. This finding is consistent with the apparent lack in brain of an enzyme system for cholesterol degradation.

Phospholipid Metabolism

Metabolic pathways of phospholipid in brain are similar to those in systemic organs [30]. Figure 15-6 depicts the main pathways involving major brain diacylglycerophospholipids (phosphatidylphospholipid) and sphingomyelin.

There are two major synthetic pathways: one involves diglyceride and CDP-choline or CDP-ethanolamine for the formation of phosphatidylethanolamine, lecithin, and sphingomyelin; the other passes through CDP-diglyceride to form phosphoinositide and possibly phosphatidylserine (I and II in Fig. 15-6). The formation of phosphatidic acid is an important preliminary pathway common to both the major pathways (III in Fig. 15-6). Choline or ethanolamine is first phosphorylated and is then converted to an active form, CDP-choline or CDP-ethanolamine, by a reaction with cytidine triphosphate. This activated form of choline

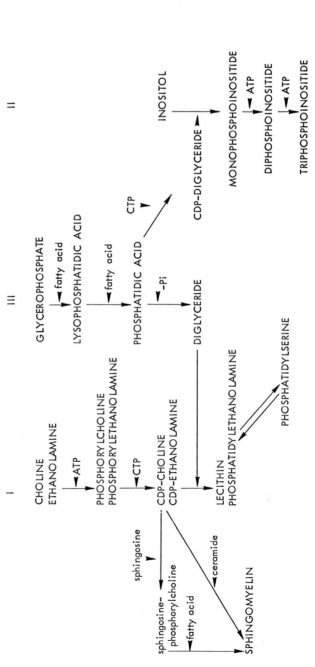

Figure 15-6
Major pathways of phospholipid biosynthesis.

or ethanolamine then forms phosphatidylcholine (lecithin) or phosphatidylethanol-amine by a reaction with diglyceride derived from phosphatidic acid. The direct conversion of phosphatidylethanolamine to lecithin by methylation apparently does not occur in the brain, although it has been demonstrated in some systemic organs. CDP-choline also reacts with ceramide to form sphingomyelin (ceramide-phosphoryl-choline) [31]. At least in the brain, an alternate pathway appears to exist in which the first reaction is the formation of sphingosine phosphorylcholine, followed by its acylation [32]. Direct transfer of phosphorylcholine from lecithin to ceramide has recently been demonstrated, and the importance of this pathway for synthesis of brain sphingomyelin must also be kept in mind.

Phosphoinositides are synthesized through a different mechanism (II in Fig. 15-6). Instead of forming CDP-compounds of inositol, CTP reacts with phosphatidic acid to form CDP-diglyceride, which, in turn, reacts with free inositol to form phospha-tidylinositol. Polyphosphoinositides are synthesized by stepwise phosphorylation of monophosphoinositide. Interconversion of phosphatidylserine and phosphatidyl-ethanolamine is known to take place in brain [33–35].

There are probably several hydrolytic enzymes in the brain which deacylate phos-phatidyl compounds to lysophosphatidyl compounds. Phospholipase A deacylates the β position of phosphatidylethanolamine, phosphatidylserine, or lecithin. A spe-cific enzyme to deacylate the α' position of these compounds has also been reported [36]. On the other hand, the lyso compounds can be reacylated by acyl-CoA to the original phosphatidyl compounds. This mechanism makes it possible for fatty acids of these phospholipids to turn over independently of the whole phospholipid molecule.

The first step of sphingomyelin degradation is catalyzed by sphingomyelinase, which cleaves sphingomyelin into ceramide and phosphorylcholine. The lack of this enzyme is the cause of at least some forms of Niemann-Pick disease (see Chap. 26).

The high metabolic activity of phosphoinositides is well documented and at pres-ent is the focus of intensive investigations regarding their physiological role in brain function.

Biosynthesis of plasmalogen in the brain has not been completely elucidated, but it appears that the pathway is similar to that for the diacyl form of glycerophospho-lipids [37]. However, plasmalogens might also be formed by reduction of phospha-tidyl compounds [38].

Cerebroside and Sulfatide

As stated earlier, the normal adult brain contains only galactocerebroside, and alternative pathways have been proposed for its biosynthesis. One is through psycho-sine, which is formed from sphingosine and UDP-galactose (I of Fig. 15-7). Psycho-sine, in turn, may be acylated by acyl-CoA to form galactocerebroside [39]. More recent experimental data indicate, however, that galactocerebroside can be formed through ceramide and UDP-galactose, and the investigators were unable to confirm the acylation of psychosine [40, 41]. At present it appears more likely that the main biosynthetic pathway of brain galactocerebroside is pathway II of Figure 15-7. Biosynthesis of sulfatide occurs through cerebroside with the "active sulfate," 3'-phosphoadenosine 5'-phosphosulfate (PAPS), as the sulfate donor.

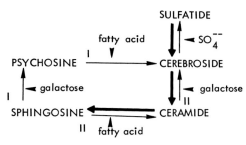

Figure 15-7
Outline of galactocerebroside biosynthesis.

The initial step of sulfatide degradation is removal of the sulfate group and its conversion back to galactocerebroside. The reaction is catalyzed by cerebroside sulfate sulfatase. The lack of this enzyme characterizes metachromatic leukodystrophy, in which excess sulfatide accumulation occurs (Chap. 26). Then galactocerebroside is degraded to ceramide and galactose by galactocerebroside β-galactosidase, the genetic lack of which causes Krabbe's globoid cell leukodystrophy (Chap. 26). Ceramide is further degraded to sphingosine and fatty acid by ceramidase. Metabolic pathways of neutral glycosphingolipids have recently been reviewed in some detail [7].

As mentioned earlier, galactocerebroside is almost nonexistent in the brain before myelination. The rate of biosynthesis parallels that of myelin deposition and declines in adult brains. Once deposited, most of the brain cerebroside undergoes only slow turnover, as compared to many other phospholipids.

Ganglioside Metabolism
Biosynthesis of gangliosides appears to take place by sequential addition of monosaccharides or NeuNAc, catalyzed by glycosyl transferases specific for each step [42] (Fig. 15-8). The active forms of all monosaccharides are UDP compounds and that of NeuNAc is CMP-NeuNAc. The synthetic steps appear to occur in the microsomal fraction.

Degradation of brain gangliosides also proceeds by sequential removal of monosaccharides and NeuNAc by glycosidases and neuraminidase [8]. These hydrolytic enzymes are localized in the lysosomal fraction. Although details of some of the pathways have not been completely established, the steps indicated by bold arrows in Figure 15-8 appear to be the main degradative pathway. Brain ganglioside metabolism is of fundamental importance to the understanding of several of the most important sphingolipid storage disorders, including four enzymatically distinct G_{M2}-gangliosidoses, G_{M1}-gangliosidosis (possibly two forms), and Gaucher's disease. The topic is covered elsewhere in this volume (Chap. 26).

Because of the high concentration in neuronal membranes, particularly in synaptic membranes, and because of the unusual chemical structure of both hydrophilic and lipophilic chains, the functional roles of brain gangliosides — such as their static role as a membrane constituent or their more dynamic roles in ion transport or

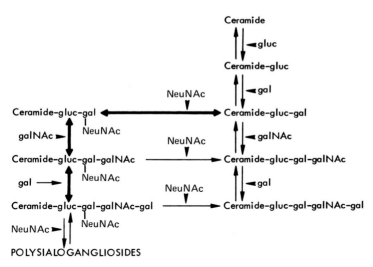

Figure 15-8
Simplified outline of ganglioside metabolism. In the direction of synthesis, hexose intermediates occur in the form of UDP esters. NeuNAc is incorporated via CMP-NeuNAc. Free sugars are released in the degradative pathways.

nerve transmission — have often been implicated but have not been experimentally substantiated.

Some gangliosides appear to act as the specific receptor for certain toxins. Tetanus toxin binds specifically with those gangliosides, with NeuNAc residues located at both the internal and terminal galactose residues [43]. Thus, among the major gangliosides, only G_{D1a} and G_{T1} react with tetanus toxin. On the other hand, the major monosialoganglioside, G_{M1}, appears to be the specific receptor for cholera toxin [44–47]. Gangliosides also bind botulinum toxin, but the binding appears to be less specific than that for the other two toxins [48].

REFERENCES

*1. Davison, A. N. Lipid Metabolism of Nervous Tissue. In Davison, A. N., and Dobbing, J. (Eds.), *Applied Neurochemistry*. Philadelphia: Davis, 1968, p. 178.

*2. Brady, R. O. Sphingolipid Metabolism in Neural Tissues. In Ehrenpreis, S., and Solnitzky, O. C. (Eds.), *Neurosciences Research*. Vol. 2. New York: Academic, 1969, p. 301.

*3. Eichberg, J., Hauser, G., and Karnovsky, M. L. Lipids of Nervous Tissue. In Bourne, G. H. (Ed.), *The Structure and Function of Nervous Tissue*. Vol. 3. New York: Academic, 1969, p. 185.

*4. Lajtha, A. (Ed.). *Handbook of Neurochemistry*. New York: Plenum. See in particular:
D'Adamo, A. F., Jr. Fatty Acids. Vol. 3, 1970, p. 525.
Davison, A. N. Cholesterol Metabolism. Vol. 3, 1970, p. 547.

*Asterisks denote key references.

Hawthorne, J. N., and Kai, M. Metabolism of Phosphoinositides. Vol. 3, 1970, p. 491.

Paoletti, R., Grossi-Paoletti, E., and Fumagalli, R. Sterols. Vol. 1, 1969, p. 195.

Radin, N. S. Cerebrosides and Sulfatides. Vol. 3, 1970, p. 415.

Rapport, M. M. Lipid Haptens. Vol. 3, 1970, p. 509.

Rosenberg, A. Sphingomyelin: Enzymatic Reactions. Vol. 3, 1970, p. 453.

Rossiter, R. J., and Strickland, K. P. Metabolism of Phosphoglycerides. Vol. 3, 1970, p. 467.

Rouser, G., and Yamamoto, A. Lipids. Vol. 1, 1969, p. 121.

Svennerholm, L. Gangliosides. Vol. 3, 1970, p. 425.

*5. Ansell, G. B., Hawthorne, J. N., and Dawson, R. M. C. *Form and Function of Phospholipids.* Amsterdam: Elsevier, 1972.

*6. Stanbury, J. B., Wyngaarden, J. B., and Frederickson, D. S. (Eds.). *The Metabolic Basis of Inherited Disease* (3rd ed.). New York: McGraw-Hill, 1972. (A standard reference book for biochemistry of inherited metabolic diseases, including those involving brain lipids.) The 4th edition is in preparation.

*7. Morell, P., and Braun, P. Sphingolipid metabolism: Biosynthesis and metabolic degradation of sphingolipids not containing sialic acid. *J. Lipid Res.* 13:293, 1972.

*8. Ledeen, R., and Yu, R. Structure and Enzymatic Degradation of Sphingolipids. In Hers, H. G., and van Hoof, F. (Eds.), *Lysosomes and Storage Diseases.* New York: Academic, 1973, p. 105.

 9. Suzuki, K. The pattern of mammalian brain gangliosides. II. Evaluation of the extraction procedures; post-mortem changes and the effect of formalin preservation. *J. Neurochem.* 12:629, 1965.

10. Folch-Pi, J., Lees, M., and Sloane-Stanley, G. H. A simple method for the isolation and purification of total lipides from animal tissues. *J. Biol. Chem.* 226:497, 1957.

11. Ledeen, R. W., Yu, R. K., and Eng, L. F. Gangliosides of human myelin: Sialosylgalactosylceramide (G_7) as a major component. *J. Neurochem.* 21:829, 1973.

12. Wells, M. A., and Dittmer, J. C. A comprehensive study of the postnatal changes in the concentration of the lipids of developing rat brain. *Biochemistry* 6:3169, 1967.

13. Cuzner, M. L., and Davison, A. N. The lipid composition of rat brain myelin and subcellular fractions during development. *Biochem. J.* 106:29, 1968.

14. Norton, W. T., and Poduslo, S. E. Myelination in rat brain: Changes in myelin composition during brain maturation. *J. Neurochem.* 21:759, 1973.

*15. Norton, W. T. The Myelin Sheath. In Shy, G. M., Goldensohn, E. S., and Appel, S. H. (Eds.), *The Cellular and Molecular Basis of Neurologic Diseases.* Philadelphia: Lea & Febiger, 1976.

16. Suzuki, K. The pattern of mammalian brain gangliosides. III. Regional and developmental differences. *J. Neurochem.* 12:969, 1965.

17. Eichberg, J., Whittaker, V. P., and Dawson, R. M. C. Distribution of lipids in subcellular particles of guinea-pig brain. *Biochem. J.* 92:91, 1964.

18. Seminario, L. M., Hren, H., and Gomez, C. J. Lipid distribution in subcellular fractions of the rat brain. *J. Neurochem.* 11:197, 1964.

19. Lapetina, E. G., Soto, E. F., and DeRobertis, E. Lipids and proteolipids in isolated subcellular membranes of rat brain cortex. *J. Neurochem.* 15:437, 1968.

20. Breckenridge, W. C., Gombos, G., and Morgan, I. G. The lipid composition of adult rat brain synaptosomal plasma membranes. *Biochim. Biophys. Acta* 266:695, 1972.

*21. Norton, W. T., Abe, T., Poduslo, S. E., and DeVries, G. H. The lipid composition of isolated brain cells and axons. *J. Neurosci. Res.* 1:57, 1975.

*22. Olson, J. A. Lipid metabolism. *Annu. Rev. Biochem.* 35:559, 1966.

*23. Guarnier, M., and Johnson, R. M. The essential fatty acids. *Adv. Lipid Res.* 8:115, 1970.

24. Kritchevsky, D., Tepper, S. A., DiTullio, N. W., and Holms, W. L. Desmosterol in developing rat brain. *J. Am. Oil Chem. Soc.* 42:1024, 1965.

25. Paoletti, R., Fumagalli, R., and Grossi, E. Studies in brain sterols in normal and pathological conditions. *J. Am. Oil Chem. Soc.* 42:400, 1965.

26. Smith, M. E., Fumagalli, R., and Paoletti, R. The occurrence of desmosterol in myelin of developing rats. *Life Sci.* 6:1085, 1967.

27. Dobbing, J. The entry of cholesterol into rat brain during development. *J. Neurochem.* 10:739, 1963.

28. Davison, A. N., Dobbing, J., Morgan, R. S., and Payling Wright, G. The deposition of [4-^{14}C] cholesterol in the brain of growing chickens. *J. Neurochem.* 3:89, 1958.

29. Khan, A. A., and Folch-Pi, J. Cholesterol turnover in brain subcellular particles. *J. Neurochem.* 14:1099, 1967.

*30. Rossiter, R. J. Biosynthesis of Phospholipids and Sphingolipids in the Nervous System. In Rodahl, K., and Issekutz, B. (Eds.), *Nerve as a Tissue.* New York: Harper & Row, 1966, p. 175.

31. Sribney, M., and Kennedy, E. P. The enzymatic synthesis of sphingomyelin. *J. Biol. Chem.* 233:1315, 1958.

32. Brady, R. O., Bradley, R. M., Young, O. M., and Kaller, H. An alternative pathway for the enzymatic synthesis of sphingomyelin. *J. Biol. Chem.* 240:PC 3693, 1965.

33. Borkenhagen, L. F., Kennedy, E. P., and Fielding, L. Enzymatic formation and decarboxylation of phosphatidylserine. *J. Biol. Chem.* 236:PC 28, 1961.

34. Ansell, G. B., and Spanner, S. The incorporation of the radioactivity of [3-^{14}C] serine into brain glycerophospholipids. *Biochem. J.* 84:12P, 1962.

35. McMurray, W. C. Metabolism of phosphatides in developing rat brain. I. Incorporation of radioactive precursors. *J. Neurochem.* 11:287, 1964.

36. Gatt, S. Purification and properties of phospholipase A$_1$ from rat and calf brain. *Biochim. Biophys. Acta* 159:304, 1968.

37. McMurray, W. C. Metabolism of phosphatides in developing rat brain. II. Labeling of plasmalogens and other alkali-stable lipids from radioactive cytosine nucleotides. *J. Neurochem.* 11:315, 1964.

38. Ansell, G. B., and Spanner, S. The metabolism of labeled ethanolamine in the brain of the rat in vivo. *J. Neurochem.* 14:873, 1966.

39. Brady, R. O. Studies on the total enzymatic synthesis of cerebrosides. *J. Biol. Chem.* 237:PC 2416, 1962.

40. Morell, P., and Radin, N. S. Synthesis of cerebroside by brain from uridine diphosphate galactose and ceramide containing hydroxy fatty acid. *Biochemistry* 8:506. 1969.

41. Morell, P., Costantino-Ceccarini, E., and Radin, N. S. The biosynthesis by brain microsomes of cerebrosides containing nonhydroxy fatty acids. *Arch. Biochem. Biophys.* 147:738, 1970.

42. Kaufman, B., Basu, S., and Roseman, S. Studies on the Biosynthesis of Gangliosides. In Volk, B. W., and Aronson, S. M. (Eds.), *Inborn Disorders of Sphingolipid Metabolism.* Oxford: Pergamon, 1967, p. 193.

43. van Heyningen, W. E., and Mellanby, J. The effect of cerebroside and other lipids on the fixation of tetanus toxin by gangliosides. *J. Gen. Microbiol.* 52:447, 1968.

44. van Heyningen, W. E., Carpenter, W. B., Pierce, N. F., and Greenough, W. B. Deactivation of cholera toxin by ganglioside. *J. Infect. Dis.* 124:415, 1971.

45. King, C. A., and van Heyningen, W. E. Deactivation of cholera toxin by a sialidase-resistant monosialoganglioside. *J. Infect. Dis.* 127:639, 1973.

46. Holmgren, J., Lönnroth, I., and Svennerholm, L. Tissue receptor for cholera toxin: Postulated structure for studies with G_{M1}-ganglioside and related glycolipids. *Infect. Immun.* 8:208, 1973.

47. Cuatrecasas, P. Vibrio cholerae choleragenoid. Mechanism of inhibition of cholera toxin action. *Biochemistry* 12:3577, 1973.

48. Simpson, L. L., and Rapport, M. M. The binding of botulinum toxin to membrane lipids: Sphingolipids, steroids and fatty acids. *J. Neurochem.* 18:1751, 1971.

49. Svennerholm, L. Chromatographic separation of human brain gangliosides. *J. Neurochem.* 10:613, 1963.

*50. Ledeen, R. The chemistry of gangliosides: A review. *J. Am. Oil Chem. Soc.* 43:57, 1966.

Chapter 16

Protein Metabolism in the Regulation of Nervous System Function

Samuel H. Barondes

Proteins have the widest diversity of functions of any class of biological compounds. This is because of the vast number of structures that can be generated by combining 20 different amino acids into long polymers. The resultant polypeptides assume unique conformations, largely determined by the interactions of their constituent amino acids. They then may aggregate with identical or different polypeptide chains to form a wide variety of functional units. The completed aggregated proteins can function in many ways: as enzymes that catalyze biological reactions; as materials that maintain or control cell structure (e.g., tubulin, actin, myosin); as receptors for neurotransmitters, hormones, or other regulatory substances; and as regulators of specific messenger-RNA synthesis (e.g., repressor proteins).

Because of the enormous importance of maintaining precise levels of these specialized substances, an elaborate apparatus has evolved to regulate their biosynthesis and functional state. The details of these processes continue to be under intensive investigation. Studies of protein metabolism in the nervous system indicate that these mechanisms are shared with other eukaryotic tissues and resemble those in prokaryotes, about which more detailed information is available.

The major purpose of this chapter is to indicate the application of general knowledge about protein metabolism to an analysis of regulatory processes in the nervous system. To understand brain protein metabolism, it is necessary to understand the general principles that have been elucidated in other tissues. Chapters in textbooks of general biochemistry or molecular biology, therefore, should be consulted to supplement the broad and sometimes oversimplified view given here. References have been restricted either to review articles [1−12] or to original articles that illustrate an approach or technique directly relevant to studies of the nervous system.

BIOSYNTHESIS AND REGULATION OF FUNCTIONAL PROTEIN LEVELS

General Considerations

The general scheme of protein synthesis has been well defined. The genetic material, DNA, directs the synthesis of specific messenger RNA, the polynucleotide sequence of which directs the insertion of individual amino acids into a growing polypeptide chain. Synthesis occurs on ribosomes and involves interaction of messenger RNA with transfer RNAs. Each type of transfer RNA combines with a single specific amino acid and carries it into the protein-synthesizing complex under the direction of a specific sequence of three nucleotides in messenger RNA (see also Chap. 17). A number of other macromolecules, including initiation factors, elongation factors, and release factors, mediate the translation of the polynucleotide sequence of the messenger RNA into the amino acid sequence of the polypeptide chain. Upon release from the protein-synthesizing complex, the polypeptide twists and folds to take on its three-dimensional structure. It may then interact with other polypeptides [3], either identical or different, to form a functional unit such as an enzyme.

The functional amount of a given protein in a cell or tissue is regulated in three general ways: (1) by alterations in rate of synthesis of its polypeptide chains; (2) by structural modification of the polypeptide chains after synthesis; (3) by regulation of its rate of degradation. These mechanisms have been studied in greatest detail with enzyme proteins because enzymatic activity is easier to assess quantitatively than are other aspects of protein function. The same regulatory processes, however, apply to proteins with nonenzymatic functions.

Regulation of Synthesis of a Specific Protein

Protein synthesis is usually limited by the rate of synthesis of specific messenger RNAs. This has been studied in detail in bacteria. In *E. coli*, synthesis of the enzyme β-galactosidase is induced by introducing lactose or related compounds into the medium. The inducer (usually a compound related to lactose that cannot be metabolized) has been shown to combine with the repressor protein that blocks the transcription of the messenger RNA that codes the synthesis of β-galactosidase. Another protein also plays a role in regulating transcription, and its action is modulated by binding cyclic AMP (Chap. 12). The increased messenger RNA is then translated by the protein-synthesizing system, and increased β-galactosidase is made within a few minutes after the inducer has been introduced. In eukaryotes, there is also increased transcription of specific messenger RNAs in response to specific inducers. In addition, there is evidence that the regulation of the rate of translation of messenger RNA already present in the cell can also alter levels of a specific protein [6].

The regulatory role of increased synthesis of specific proteins in the nervous system is presently under investigation (see Chap. 18). Many studies have shown that there are critical periods in the development of the nervous system during which there is markedly increased synthesis of specific proteins such as enzymes and other proteins (e.g., myelin proteins). In the adult nervous system, regulation of synthesis of enzymes involved in catecholamine metabolism has been demonstrated [9]. Many of these studies have been done with adrenal medulla, but others have been done with nervous system tissue as well. It has been shown that increased cholinergic input to cells that synthesize catecholamines leads to an increased synthesis of both tyrosine hydroxylase and dopamine β-hydroxylase [9, 13, 14]. Available evidence indicates that the increased synthesis of tyrosine hydroxylase is mediated by increased synthesis of specific messenger RNA. There is also evidence, as mentioned in Chapter 10, for similar regulation of tyrosine hydroxylase levels in brain [15], although it remains to be determined if this is due to increased synthesis of this protein or to a structural modification.

Structural Modification of Proteins

The functional level of a protein can be regulated both by changing the number of protein molecules and by changing the conformational state (see Chap. 3) of the protein [2]. Both physical and chemical modifications of proteins can change their functional state. Chemical modifications are produced when a specific group is attached to or removed from a protein by a formation or cleavage of a covalent bond. For example, when a phosphate group is covalently attached to the enzyme phosphorylase *b* in liver, it is converted to phosphorylase *a,* the active form which degrades glycogen to glucose 1-phosphate. The activated enzyme is converted back to the inactive form by the action of a phosphatase that removes the covalently bound phosphate group. Covalent addition of the phosphate group to phosphorylase *b* by transfer from ATP is mediated by another enzyme, called phosphorylase *b* kinase. The latter is a protein kinase, i.e., an enzyme that catalyzes phosphorylation of a protein. *Phosphorylase b kinase* is, in turn, activated through phosphorylation catalyzed by another protein kinase which is often called *phosphorylase b kinase kinase.* The activity of this latter protein kinase is regulated by $3',5'$-cyclic AMP, as mentioned below.

There are many other examples of both physical and chemical modifications of proteins. Chemical modification by phosphorylation under the control of cAMP (see Chap. 12) has been shown both in homogenates of brain tissue [16] and in intact invertebrate ganglia maintained in culture [17]. Chemical modification may involve groups other than phosphate [2]. For example, xanthine oxidase of rat liver changes its electron acceptor from NAD to molecular oxygen if its sulfhydryl groups are oxidized. Reversal by reduction to the sulfhydryl form reverses the functional preference of the protein. If an ADP residue is added covalently to an enzyme called translocase, its enzyme activity is changed.

Physical modifications are also employed extensively. In some cases it is clear that a regulatory compound binds (noncovalently) to a regulatory site that differs from the active site of the protein. This modifies the conformation of the protein so that its enzymatic activity is altered. This regulatory site is often called the allosteric site. In other cases the regulatory compound may bind to specific subunits of a protein made up of more than one subunit type. The binding may dissociate the two types of subunits, thereby activating one of them. For example, interaction of cAMP with subunits associated with a protein kinase may dissociate these subunits, thereby promoting the protein kinase activity of the freed subunits.

Both physical and chemical modifications may be rapidly reversible in that the protein may revert quickly from an active to an inactive form. With physical modification, reversal may be affected by rapid dissociation of the regulatory compound from the protein (which could occur in seconds or less). Chemical modifications might also be reversed quickly. Chemical reversal, however, requires the breaking of a covalent bond (e.g., removal of a phosphate from a protein by a phosphatase). There is evidence that reversal might sometimes be quite slow. For example, phosphorylation of a specific protein controlled by cAMP may last for hours [17]. The regulatory consequences of chemical modifications may, therefore, be more durable than physical modifications.

Regulation of Degradation of a Specific Protein

The proteins in all eukaryotic cells are being degraded continuously and replaced by the synthesis of new protein molecules [1, 4, 5]. The time that it takes a tissue to degrade half the number of molecules of a specific protein is called the half-life of the protein. Half-lives of proteins in mammalian cells vary from ten minutes to many months. Most proteins in brain have half-lives in the range of days. Studies of liver suggest that the enzymes that are present in rate-limiting amounts for a specific bio-synthetic pathway tend to turn over more rapidly than do enzymes normally present in great excess. This presumably allows for regulation of metabolic pathways through alterations in the levels of the rate-limiting enzymes. There is now considerable evidence that the functional level of a specific protein may be regulated by control of its rate of degradation. The rate of degradation of liver tryptophan oxygenase, for example, is reduced if tissue levels of tryptophan increase [1]. Presumably, similar types of process are operative in nervous system as well.

Regulation of Protein Metabolism Not Always Completely Selective

It should be apparent from the general discussion above that: (1) the synthesis, degradation, and modification of a specific protein in a given cell may be regulated independently of the other proteins in that cell; (2) metabolism of the same protein may be under different controls (e.g., specific hormones) in different cells. This latter consideration is particularly important in the nervous system, which contains a number of morphologically and biochemically distinct glial and neuronal cell types.

Because of these considerations, it is generally improper to investigate protein metabolism in the nervous system *as a whole* (except in certain cases, e.g., determining the efficacy of a drug that inhibits protein synthesis). Rather, in most cases, attention should be directed to metabolism of a *specific* protein in a *specific* cell type. Regulation may not be completely specific, however, for a given protein or cell type in that: (1) a group of proteins within a cell may be influenced by a similar regulatory compound (e.g., glucocorticoids affect the metabolism of many liver proteins); (2) a certain regulatory process may have the same influence on the same protein in different cell types (e.g., cholinergic stimulation appears to have the same effects on tyrosine hydroxylase in adrenal medulla and adrenergic neurons); (3) in certain situations there may be marked changes in the overall protein metabolism of a cell. For example, growing cells in differentiating brain synthesize proteins more actively than do mature neurons. There appear to be differentiative programs which markedly affect the metabolism of large groups of proteins at specific stages in development. Furthermore, trophic effects of neurons on muscle or activation of muscle (or both) may affect the metabolism of many different muscle proteins [4] (also see Chap. 23). There might be similar regulation of neuronal protein metabolism. In certain cases, therefore, changes in protein metabolism may be so general that an effect on overall protein metabolism in a specific cell (or homogeneous populations of cells, such as muscle) may be detectable.

TECHNIQUES FOR STUDYING PROTEIN METABOLISM IN THE NERVOUS SYSTEM

Determination of Rates of Protein Synthesis Using Radioisotopes

The rate of synthesis of a protein can be estimated (subject to major qualifications as discussed below) by determining the rate of incorporation of a radioactive amino acid into this product. Such studies are generally done in vivo. This is because synthesis of proteins in tissue slices or homogenates may not reflect the in vivo situation. Ganglia maintained in organ culture, however, might provide a reasonable representation of the in vivo state. In vivo studies are generally done with a single injection of a radioactive amino acid, which is administered either subcutaneously or into the cranium. There is a potential advantage to each method. In the former, the precursor is distributed relatively uniformly in the brain; the latter results in relatively high concentrations of the precursor in brain. This increases the amount of radioactivity in the product. At a number of times after injection, measurements are made of the amount of radioactive amino acid incorporated into protein. This is done by determining the amount of radioactivity incorporated into substances that are precipitable by trichloracetic acid, which precipitates proteins but not amino acids. Using this approach, the rate of incorporation of radioactivity into whole brain protein can be estimated. By appropriate fractionation, attention can be focused on a given brain region, a fractionated cell type, a subcellular fraction, or an individual protein.

A major assumption made in these studies is that the rate of incorporation of the administered radioactive amino acid directly reflects the rate of protein synthesis. This would be true if the administered radioactive amino acid equilibrated with the endogenous amino acid precursor pool that provides the amino acids used for protein synthesis. It is never possible, however, to determine this with certainty. Awareness of that assumption is of critical importance for interpreting studies in which protein synthesis is compared under experimental and control conditions. Thus, one may administer a radioactive amino acid to animals and observe a change in the amount of labeled amino acid incorporated into a brain protein as a result of some specific treatment. One must, in addition, determine whether the treatment influences the specific activity of the precursor pool. If a given treatment increases the amount of radioactive amino acid incorporated into a brain protein, it could be the result either of an increase in synthesis of this protein or of increased access of the *labeled* amino acid to the protein synthesizing system.

Increased access might occur in any one of several ways. For example, the treatment being studied could so change blood flow to the brain that a larger proportion of the total amount of subcutaneously administered radioactive amino acid is taken up by the treated brain. This might produce a particularly high specific activity of the precursor pool of the administered amino acid; that is, there might be a higher ratio of labeled to unlabeled amino acid molecules. In that case, incorporation of more radioactivity into protein would reflect this higher specific activity of precursor and not an increased rate of protein synthesis. To evaluate the possible contribution of such access, one could measure both the amount of unlabeled amino acid and the amount of labeled amino acid present in the brains of both treated and untreated animals so that the specific activities of the precursor pools (radioactivity per mole of amino acid) in the brain are known. Ideally, this should be determined in different groups of animals at a number of times in the interval during which amino acid incorporation is being studied. If the specific activity differs in the treated and untreated conditions, appropriate corrections can then be made to improve the estimate of the influence of the treatment on protein synthesis per se.

Even these corrections, however, do not allow an absolute determination of the rate of synthesis of a protein because the pool of amino acids in tissues is segregated so that some of each amino acid is employed for protein synthesis and some is not. Studies with other tissues indicate that, in some cases, amino acids derived from ongoing protein degradation are preferentially utilized for protein synthesis [18]. Other studies have shown preferential utilization of exogenous amino acid applied in the medium [19]. Therefore, unless one knows the specific activity of the true precursor pool (that is, the radioactivity per mole of amino acid that is actually being used by the protein-synthesizing apparatus) one cannot get an absolute estimate of the rate of synthesis of a protein. Lack of awareness of these various considerations has led to many claims in the literature that may not be valid.

In spite of these difficulties, it is possible to estimate *relative* rates of synthesis of specific proteins in specific nerve cells. This is best done if the metabolism of a single cell is observed, which is possible with large identified neurons in a molluscan central nervous system. If one isolates the protein from the single cell and makes the reasonable assumption that all the proteins of this cell are synthesized from a precursor pool with the same specific activity, it is possible to compare the *relative* rates of synthesis of specific proteins within the cell as well as the relative rates between cells treated in different ways. A number of studies have shown that there are differences in the relative amounts of specific proteins made by different identified large cells obtained from ganglia of *Aplysia californica* and other mollusks [20–23]. In these studies, the constituent proteins are resolved by polyacrylamide gel electrophoresis after the protein has been made soluble in sodium dodecyl sulfate (SDS), an ionic detergent. Electrophoresis in sodium dodecyl sulfate fractionates proteins on the basis of their subunit molecular weight. The gels are calibrated by using standard proteins of known molecular weight. After electrophoresis of the radioactive protein, the gel is sliced into a number of fractions, each of which is counted to determine protein radioactivity. A profile of radioactive proteins is thereby obtained, and the amount of radioactivity incorporated into protein in each molecular weight class can be determined. A number of different identified cells in molluscan ganglia show different profiles [20–22]. Furthermore, by comparing the relative height of a specific peak of the profile to the heights of the other peaks in experimental and control conditions, the relative rates of synthesis of different classes of proteins in the experimental and control conditions can be determined. With this method it was demonstrated that neuronal inputs can change the relative rates of synthesis of a specific protein in an identified neuron in *Aplysia californica* [23]

A similar strategy can be applied to a heterogeneous cell mixture such as that in a piece of brain tissue. It should be recognized, however, that interpretation is more difficult because the specific activities of the precursor pools in the various cell types may be different, and therefore the relative rates at which the protein in question is synthesized by the various cell types may not be immediately apparent. Overall protein synthesis by the different cell types could be estimated by radioautographic studies of the tissue and quantitative counting of the numbers of incorporated grains of radioactive protein, but this is tedious and is generally avoided. One way around this is to study the metabolism of a specific protein that is unique to a specific subpopulation within the bit of tissue if one assumes that the specific activity of the precursor pool will be fairly uniform throughout that subpopulation. If this protein is abundant enough and unique in some properties, such as molecular weight on SDS gels, polyacrylamide gel electrophoresis could be the technique used for fractionation. The protein could also be separated by migration in other electrophoretic systems that separate on the basis of either size and charge or isoelectric point [22]. Another method is to prepare an antibody to the specific protein and then to use that antibody to specifically precipitate the radioactive protein in question from the radioactive protein mixture [1]. Conventional protein purification techniques or specific techniques, such as affinity chromatography, could also be used to estimate the amount of radioactivity in a given protein.

Determination of Rates of Protein Degradation Using Radioisotopes

As pointed out above, the rate of degradation of a protein may be an important regulatory device for controlling its functional levels in a tissue [1, 4, 5]. Rates of protein degradation can be determined after incorporation of radioactive amino acids, as described above. Sequential determinations of the specific activity of the protein are made at a number of times after incorporation of radioactive amino acid is completed. The specific activity of the protein should fall exponentially as the protein is degraded. The major problem in interpreting degradation studies lies in the reutilization of radioactive amino acid for resynthesis. As radioactive protein is degraded, the radioactive amino acids are reutilized to synthesize protein, often quite efficiently. Therefore, the true rate of degradation could be substantially greater than the rate actually observed, which could be obscured by resynthesis of radioactive molecules of the protein in question. Attempts to inhibit reutilization with inhibitors of protein synthesis (discussed in the next section) are complicated by the fact that these drugs may change degradation rates [1]. Double-isotope techniques to determine relative rates of degradation have been described [1].

Drugs for Studying Brain Protein Metabolism

Many studies attempt to determine if a specific adaptive response in the nervous system is accompanied by a change in functional activity of a specific protein. If the protein is an enzyme, its enzymatic activity can be determined in a crude tissue homogenate since enzymatic activity (amount of substrate utilized per mole of enzyme) is generally similar in a crude homogenate and in a pure state, although there are exceptions. Therefore the protein purification required to study aspects of protein metabolism with the radioisotope technique may not be necessary if one measures the specific enzymatic activity (or some other specific activity such as simple binding to another molecule, e.g., colchicine binding of tubulin).

As indicated above, a change in enzyme activity in a tissue can be caused by a change in its rate of synthesis or degradation or by structural modifications. One way to determine if an experimental variable produces a change in enzyme synthesis is to block protein synthesis during the experimental period with an appropriate drug. By measuring the enzymatic activity in the presence and absence of the drug, it is possible to determine if the change in enzyme activity is the result of a change in its rate of synthesis. Drugs used for this purpose include cycloheximide, acetoxycycloheximide, and anisomycin, all of which can inhibit nervous system protein synthesis almost completely. The only eukaryotic protein-synthesizing system that is resistant to these drugs is in mitochondria, which synthesize only a tiny fraction of total nervous system protein. Chloramphenicol inhibits mitochondrial protein synthesis but does not affect the remainder of the protein-synthesizing system in the nervous system. Puromycin affects both types of protein-synthesizing systems.

Of the drugs mentioned, the most practical for studies of protein metabolism in nervous system tissue is cycloheximide. This is because it is commercially available (in contrast with anisomycin and acetoxycycloheximide) and is inexpensive (in contrast with puromycin). It is useful for studying protein metabolism in the central nervous system of mammals because it crosses the blood-brain barrier and can therefore be administered systemically by either intravenous or subcutaneous injection. Much smaller total doses can be administered intracerebrally or intraventricularly to inhibit brain protein synthesis, but the distribution of the drug is much less uniform; it may take hours for it to diffuse throughout the brain, during which time it is also leaving the brain by passage into the systemic circulation.

Cycloheximide, administered subcutaneously, has been used to inhibit brain protein synthesis in many studies with mice (for example, see [24]). When administered subcutaneously (120 mg per kg), brain protein synthesis is rapidly inhibited. Within the first 10 min after injection of this dose, brain protein synthesis is inhibited more than 90% and more than 95% in the second 10 min. Inhibition is maintained at this level for about an hour and then falls. Examples of the time course of inhibition with several doses in one strain of mice and a simple technique for estimating degree of inhibition have been published [24]. Strains of mice differ somewhat, however, in both the magnitude and duration of inhibition. If reasonably precise estimates of inhibition are desired for a specific experiment, therefore, it is necessary to make these measurements in the specific strain actually in use. Cycloheximide has also been used with rats, but, for reasons that are not known, they cannot tolerate nearly as much cycloheximide as can mice. The very high and relatively sustained levels of inhibition of activity readily achieved with mice by using large doses of cycloheximide are difficult to achieve with rats since large doses are much more lethal in that species.

As indicated above, specific protein synthesis is generally limited by specific messenger RNA synthesis, although there may be exceptions [6]. If inhibition of protein synthesis blocks an increase in enzyme activity, it is generally inferred that the latter was due to increased synthesis of the enzyme protein, directed by increased synthesis of the specific messenger RNA. One way to test this is to administer drugs that block messenger RNA synthesis. If the increased enzyme synthesis is blocked by these drugs, it is taken as confirmation that increased messenger RNA synthesis is the primary event. Two drugs, actinomycin-D and α-amanitin, have been used to block RNA synthesis in the nervous system. Neither drug is very effective if administered subcutaneously so that intracranial injections have been used. Therefore, it is not possible to obtain a uniform distribution of the drug throughout brain tissue. Another major limitation of the use of these drugs is their great toxicity. When administered in doses large enough to inhibit about 95% of RNA synthesis in brain, actinomycin-D [25] invariably produces death in mice within a few days. α-Amanitin [26] is also very toxic. Thus, cycloheximide is the preferred drug for studies of this type. The work of Mueller et al. [14] provides an example of the use of cycloheximide and actinomycin-D to evaluate the role of protein and RNA synthesis in transsynaptic induction of tyrosine hydroxylase.

SPECIAL ASPECTS OF PROTEIN METABOLISM IN THE NERVOUS SYSTEM

Sites of Synthesis of Neuronal Proteins

A unique characteristic of nerve cells is that they may have extremely long axonal processes that extend for long distances from the cell body. The ribosomes in neurons are clustered in the nerve cell body, and these clusters are visible by light microscopy as the so-called Nissl substance [Chap. 1]. This high concentration of ribosomes supports a high rate of protein synthesis in the nerve cell bodies. In contrast, there are no identifiable ribosomes in the mature axon except in the initial segment. Thus the mature axon contains no morphologically identifiable protein-synthesizing system except the very limited system present in intraaxonal mitochondria. The nerve terminals are also devoid of identifiable ribosomes, but they contain a high concentration of mitochondria. It has been shown that mitochondria in other tissues make only a small number of proteins, probably about a dozen, and these are believed to be used exclusively for mitochondrial function. Cycloheximide, which does not affect mitochondrial protein synthesis, can inhibit about 98% of radioactive amino acid incorporation into brain protein in vivo, so mitochondrial protein synthesis in brain comprises at most 2% of total brain protein synthesis and probably much less.

   Are the proteins of the axons in nerve endings metabolically active, and where do the newly synthesized proteins come from? These have been major questions in studies of protein metabolism in the nervous system. A large body of work, reviewed in Chapter 21, indicates that proteins are transported from the nerve cell body to the axon and the nerve terminal. Because nerve terminals in brain are often close to their cell bodies, there may be rapid regulation of nerve-ending function by changes in protein synthesis in the cell body. Given rates of fast axoplasmic transport in the range of 0.3 mm per min, transport to nerve terminals 1 mm from the cell body can occur within about 3 min. Electron-microscope studies with autroradiography show that labeled protein is readily found in nerve endings of mouse brain within 15 min after a radioactive amino acid precursor has been injected [27].

Other sources of axonal and nerve ending protein have been identified. Some axonal proteins may be taken into the axon after having been synthesized in surrounding glia [28]. Other proteins, such as nerve growth factor [29], may be taken up by nerve endings and transported along the axon to the cell body. Only a small amount of neuronal protein may be derived in this way, but it could be very important for regulation of neuronal function. In addition, some studies suggest that protein is synthesized in nerve endings not only in mitochondria but also at other sites. These studies show that subcellular fractions (synaptosome fractions) enriched in nerve endings incorporate radioactive amino acid into protein. These fractions, however, are always contaminated to some extent with other particles that also can incorporate radioactive amino acid into protein. On the basis of a critical review of this evidence, it was concluded that there is presently no definitive proof for nonmitochondrial protein synthesis at nerve endings [7].

Structural Modification of Proteins at Nerve Endings

Although nerve endings are dependent on the cell body for regulating the amounts of protein that control their function, physical and chemical modifications of protein permit local functional alterations. Such regulation would be more prompt than that requiring transport of proteins from the nerve cell body. As mentioned above, norepinephrine inhibits tyrosine hydroxylase (which is active at noradrenergic nerve endings) by a physical modification. Proteins at nerve endings may be chemically modified by addition of carbohydrate residues [7, 11]. Rapid regulation of synaptic function through proteins occurs, therefore, even in the absence of protein-synthesizing systems in nerve endings. Structural modifications of proteins in the postsynaptic membrane (e.g., phosphorylation [12]) may also have important consequences.

Special Nervous System Proteins

Most nervous system proteins such as the enzymes that are involved in glucose metabolism are common to all the other cells in the organism. Regulation of common proteins like these may produce specific effects in the nervous system in which specific functions may be regulated by common metabolic pathways. Although some proteins attract particular attention because they are found only in the nervous system, common proteins may be no less important for regulating nervous system function.

Most of the proteins known to be specific to the nervous system are enzymes. These include, for example, the enzymes involved in the biosynthesis of neurotransmitters. The biosynthesis and metabolism of these enzymes are of great interest to those concerned with the general problem of protein metabolism in brain and are discussed in other sections of this book. Receptor proteins for specific neurotransmitters are also considered elsewhere. In addition to these nervous system proteins, several others have received considerable attention. One (tubulin) has been popular, in part, because it is the most abundant soluble protein in brain. Because of its abundance and because it can be partially purified by precipitation with vinblastine, it lends itself readily to investigation. This protein, composed of two different polypeptides, aggregates to form microtubules. It has a half-life in the range of four days.

It appears to be the major protein transported along the axon with the slow component of axoplasmic transport [8]. Its role in axoplasmic transport is considered in Chapter 21.

Several other proteins have received considerable attention because they appear to be specific to the nervous system and are present in large enough quantities to be detectable by simple fractionation procedures such as ammonium sulfate fractionation or gel electrophoresis. Examples are S-100 protein [10], 14-3-2 protein [10], and glial fibrillary acidic protein [30]. Each has been purified and used as an antigen to generate specific antibodies. The antibodies have been useful in determining the cellular localization of these proteins and as a tool for studying their metabolism. Their functions remain to be determined.

Summary

The functional levels of nervous system proteins are controlled by the same general processes that are employed in other eukaryotic tissues. What complicates research on nervous system protein metabolism is the marked cellular heterogeneity of this organ. Different cell types may contain different proteins that are regulated in special ways. It is difficult to estimate *overall* protein metabolism in the nervous system or brain, and it is extremely difficult to interpret results. Of major concern is an understanding of the regulation of specific proteins in specific cells and the consequences of this regulation for establishing and maintaining the critical intercellular interactions that characterize nervous system function. An understanding of the mechanisms that regulate specific functional proteins will ultimately play a crucial role in explaining the complex adaptive and regulatory processes of the nervous system.

REFERENCES

*1. Schmike, R. T. Principles Underlying the Regulation of Synthesis and Degradation of Protein in Animal Tissues. In Schmitt, F. O., and Worden, F. G. (Eds.), *The Neurosciences: Third Study Program.* Cambridge, Mass.: MIT Press, 1974, p. 813.

*2. Holzer, H., and Duntze, W. Metabolic regulation by chemical modification of enzymes. *Annu. Rev. Biochem.* 40:345, 1971.

*3. Frieden, C. Protein-protein interaction and enzymatic activity. *Annu. Rev. Biochem.* 40:653, 1971.

*4. Goldberg, A. L. Regulation and Importance of Intracellular Protein Degradation. In Schmitt, F. O., and Worden, F. G. (Eds.), *The Neurosciences: Third Study Program.* Cambridge, Mass.: MIT Press, 1974, p. 827.

*5. Goldberg, A. L., and Dice, J. F. Intracellular protein degradation in mammalian and bacterial cells. *Annu. Rev. Biochem.* 43:835, 1974.

*6. Tomkins, G. M., Gelehrter, T. D., Granner, D., Martin, D., Samuels, H. H., and Thompson, E. B. Control of specific gene expression in higher organisms. *Science* 166:1474, 1969.

*7. Barondes, S. H. Synaptic macromolecules: Identification and metabolism. *Annu. Rev. Biochem.* 43:147, 1974.

*Asterisks denote key references.

*8. Barondes, S. H. Neuronal Proteins: Synthesis, Transport and Neuronal Regulation. In Schneider, D. J., Angeletti, R. H., Bradshaw, R. A., Grasso, S., and Moore, B. W. (Eds.), *Proteins of the Nervous System.* New York: Raven, 1973, p. 215.

*9. Axelrod, J. Regulation of the Neurotransmitter Norepinephrine. In Schmitt, F. O., and Worden, F. G. (Eds.), *The Neurosciences: Third Study Program.* Cambridge, Mass.: MIT Press, 1974, p. 863.

*10. Moore, B. W. Brain Specific Proteins. In Schneider, D. J., Angeletti, R. H., Bradshaw, R. A., Grasso, S., and Moore, B. W. (Eds.), *Proteins of the Nervous System.* New York: Raven, 1973, p. 1.

*11. Barondes, S. H. Brain Glycomacromolecules and Interneuronal Recognition. In Schmitt, F. O. (Ed.), *The Neurosciences: Second Study Program.* New York: Rockefeller University Press, 1970, p. 747.

*12. Greengard, P., McAfee, D. A., and Kababian, J. W. On the mechanism of action of cyclic AMP and its role in synaptic transmission. *Adv. Cyclic Nucleotide Res.* 1:337, 1972.

13. Molinoff, P. B., Brimijoin, S., and Axelrod, J. Induction of dopamine β-hydroxylase and tyrosine hydroxylase in rat hearts and sympathetic ganglia. *J. Pharmacol. Exp. Ther.* 182:116, 1972.

14. Mueller, R. A., Thoenen, H., and Axelrod, J. Inhibition of trans-synaptically increased tyrosine hydroxylase activity by cycloheximide and actinomycin-D. *Mol. Pharmacol.* 5:463, 1969.

15. Segal, D. S., Sullivan, J. L., III, Kuczenski, R. T., and Mandell, A. J. Effects of long-term reserpine treatment on brain tyrosine hydroxylase and behavioral activity. *Science* 173:847, 1971.

16. Ueda, T., Maeno, H., and Greengard, P. Regulation of endogenous phosphorylation of specific proteins in synaptic membrane fraction from rat brain by adenosine 3':5' monophosphate. *J. Biol. Chem.* 248:8295, 1973.

17. Levitan, I. B., and Barondes, S. H. Octopamine- and serotonin-stimulated phosphorylation of specific protein in the abdominal ganglion of *Aplysia californic Proc. Natl. Acad. Sci. U.S.A.* 71:1145, 1974.

18. Righetti, P., Little, E. P., and Wolf, G. Reutilization of amino acids in protein synthesis in HeLa cells. *J. Biol. Chem.* 246:5724, 1971.

19. Hider, R. C., Fern, E. G., and London, D. R. Relationship between intracellular amino acids and protein synthesis in the extensor digitorum longus muscle of rats. *Biochem. J.* 114:171, 1969.

20. Wilson, D. L. Molecular weight distribution of proteins synthesized in single identified neurons of *Aplysia*. *J. Gen. Physiol.* 59:26, 1971.

21. Loh, Y. P., and Peterson, R. P. Protein synthesis in phenotypically different single neurons of *Aplysia*. *Brain Res.* 78:83, 1974.

22. Gainer, H., and Wollberg, Z. Specific protein metabolism in identifiable neurons of *Aplysia californica*. *J. Neurobiol.* 5:243, 1974.

23. Gainer, H., and Barker, J. L. Synaptic regulation of specific protein synthesis in an identified neuron. *Brain Res.* 78:314, 1974.

24. Barondes, S. H. Is the Amnesic Effect of Cycloheximide Due to Specific Interference with a Process in Memory Storage? In Lajtha, A. (Ed.), *Protein Metabolism of the Nervous System.* New York: Plenum, 1970, p. 371.

25. Cohen, H. D., and Barondes, S. H. Further studies of learning and memory after intracerebral actinomycin-D. *J. Neurochem.* 13:207, 1966.

26. Montanaro, N., Novello, F., and Stirpe, F. Effect of α-amanitin on ribonucleic acid polymerase II of rat brain nuclei and on retention of avoidance conditioning. *Biochem. J.* 125:1087, 1971.

27. Droz, B., and Barondes, S. H. Nerve endings: Rapid appearance of labelled protein shown by electron microscopic radioautography. *Science* 165:1131, 1969.
28. Lasek, R. J., Gainer, H., and Przybylski, R. J. Transfer of newly synthesized proteins from Schwann cells to the squid giant axon. *Proc. Natl. Acad. Sci. U.S.A.* 71:1188, 1974.
29. Hendry, I. A., Stach, R., and Herrup, K. Characteristics of the retrograde axonal transport system for nerve growth factor in the sympathetic nervous system. *Brain Res.* 82:117, 1974.
30. Dahl, D., and Bignami, A. Glial fibrillary acidic protein from normal human brain: Purification and properties. *Brain Res.* 57:343, 1970.

Chapter 17

Nucleic Acid Metabolism

Henry R. Mahler

NUCLEIC ACIDS AND BRAIN FUNCTION

Any attempt at formulating neurobiological phenomena in molecular terms must take cognizance of the nature, expression, and regulation of the genetic capabilities of the nervous system and its component parts. At one level, this entails systematic studies of neurological and behavioral mutants and how alterations in the genome affect defined functions of the nervous system; such studies can be of the utmost importance in providing clues concerning the contributions made by single, defined, gene products to complex and integrated brain function and behavior (see Part One, Section 3). At another level, one deals with emergent properties as a function of time as the independent variable. These will include qualitative or quantitative changes in the pattern of the proteins present within a given cell, in a population of cells linked to one another structurally and functionally, or in even more complex systems (Chap. 18). By their intervention as enzymes and membrane constituents, these proteins in turn influence all other constituents of the affected cells.

But such changes in protein patterns are themselves caused by shifts in gene expression and regulation. This is the nature of the problem we need to solve. Our concern is not just with those hypothetical or actual changes in response to electrical activity, sensory inputs, and behavioral challenges that produce or reflect specific products in, and alter the properties of, certain families of neurons and the synapses between them [1–3], itself a formidable task. We must also be concerned with the events responsible for the ontogeny of neuronal formation and localization in the course of development. In biochemical terms, the questions posed constitute an inquiry into the parameters that govern the biosynthesis and degradation of DNA and RNA in brain and nerve, and that is the topic of this chapter.

DNA VS. RNA METABOLISM

DNA synthesis and turnover in all somatic cells usually reflect the cell's ability to undergo division and to effect the excision and repair of certain damages. As a first approximation, these events appear directly correlated with the mitotic index of the cells, i.e., their rate of division. Although there is now good evidence that certain brain cells (interneurons, astrocytes, glia) continue to divide into infancy or even adulthood (see p. 348), the ordinary long-axoned neurons (macroneurons) of mammals cease dividing and proliferating shortly after birth or in early infancy (see Chap. 18). For instance, in rat cerebral cortex there is no net increase in DNA beyond the 18th day after birth. Furthermore, the amount of DNA per diploid nucleus and the genetic information residing therein are believed to be constant for all the cells of any one individual. In consequence, most of the current literature has tended to neglect DNA and instead has been concerned with the synthesis, degradation, and functional involvement of various species of RNA found in the nervous system [4, 5].

The coverage in this chapter reflects this state of the art. It finds its justification in the compelling evidence in favor of the continuity, stability, and identity of the genetic constitution of all cells of a metazoan organism in contrast to their structural and functional diversity, which, in turn, is subject to alteration as a function of time. These alterations, in turn, are caused by changes in the expression of the genetic information and its regulation. In biochemical terms, this involves the transcription of DNA base sequences into RNA and the translation of some of the latter into the amino acid sequences of polypeptides. Although we might expect patterns of RNA synthesis and nucleic acid metabolism in brain to be similar to those of other tissues in their broad and general outline, quantitative differences should exist. In particular, we might anticipate that the existence of the blood-brain barrier would place serious and preferential constraints on the entry of certain precursors for macromolecular biosynthesis. Additional difficulties arise in consequence of the inordinate cellular and regional complexity of brain tissue.

METABOLISM OF NUCLEIC ACID PRECURSORS

Synthesis of Purine and Pyrimidine Nucleotides

De Novo Synthesis
Qualitatively, the pathways established for the de novo biosynthesis of the purine and pyrimidine nucleotides in other tissues appear to be operative also in nerve and brain. This assertion rests on the demonstration of (1) the conversion of labeled precursors to the expected intermediates in the correct time course, (2) the presence of certain critical enzymes characteristic of and exclusive to the two pathways, and (3) the proper blocks by specific and characteristic inhibitors of key steps, such as the glutamine antagonists azaserine (O-diazoacetyl-L-serine) and DON (6-diazo-5-oxo-L-norleucine) for purine, and azauridine for pyrimidine biosynthesis.

Quantitatively, most of the nucleic acid purines probably arise de novo [6], although the operation of other routes (see below) appears to be of great regulatory significance. For pyrimidines, the situation may be different [7]: the rate of flux through the small pool of orotate actually present is probably insufficient to satisfy the demand for pyrimidines for nucleic acid and coenzyme synthesis, and the block by azauridine can be effectively circumvented by raising the level of available uridine. Furthermore, uridine, but not orotate, can serve as an adequate precursor of nucleic acid pyrimidines with brain preparations in vitro.

Salvage Pathways and Orotic Acidurias
These observations direct attention to the possibility that a substantial portion of the bases do not arise de novo in the brain, but instead originate in the liver and are then transported by the circulation in the form of nucleosides, mainly uridine in the case of pyrimidines. This, the so-called salvage pathway, appears to be of great potential significance in both normal function and certain pathological states of the brain [8].

Administration of uridine and cytidine protect a perfused cat brain from degradation and allow it to exhibit normal carbohydrate metabolism and electrical activity for periods of up to 4 hours. Repeated injections of azauridine block the de novo formation of pyrimidine nucleotides [9] ; a similar, but permanent defect is characteristic of the autosomally inherited, recessive genetic lesion in humans called orotic aciduria. Azauridine, after conversion to the nucleotide, specifically inhibits orotidylate decarboxylase; in orotic aciduria Type I, both this enzyme and orotidylate pyrophosphorylase — the enzyme that catalyzes the preceding step in the de novo pathway — drop to undetectable levels. (In Type II, only the decarboxylase is missing.) Severe neurological and electrophysiological disturbances are observed in both instances. These effects can be circumvented or reversed by the administration not only of uridine monophosphate (UMP), the absent direct product, but also of uridine, and this observation permits the inference that the salvage pathway is operative.

These hereditary orotic acidurias, although quite rare, represent exceedingly interesting genetic diseases for the following reasons [8] : (1) they constitute the only described instances of blocks in pyrimidine biosynthesis, not just in man but in higher organisms in general; (2) because their effects can be overcome by administration of pyrimidines, they also constitute the clearest evidence of auxotrophism in higher organisms; (3) Type II may be the result of a lesion in a structural gene, Type I of a lesion in a regulatory gene; (4) these defects result in a breakdown of the regulatory machinery of the cells affected, which leads both to overproduction of orotic acid and to a pattern of "pyrimidine starvation" previously observed in pyrimidine auxotrophs in microorganisms; (5) their consequence — and alleviation — can be mimicked by a synthetic pharmacological agent.

Lesch-Nyhan Syndrome

Even more striking are the manifestations of another genetic lesion in male children that affects hypoxanthine-guanine phosphoribosyl transferase, one of the enzymes of the salvage pathway for purines, responsible for the conversion of hypoxanthine and guanine into their respective nucleotides. Individuals who carry this defect on their X-chromosomes exhibit the Lesch-Nyhan syndrome [10] , characterized by an elevated level of uric acid in blood and urine (hyperuricemia), spastic cerebral palsy (choreoathetosis), aggressive and destructive behavior leading to self-mutilation, and mental retardation. Although it is not yet certain to what extent these manifestations are consequences of the primary enzymatic defect rather than a response to the attendant deranged purine metabolism in general, the whole picture makes clear just how finely poised and closely interacting are the various pathways for purine biosynthesis and degradation and points toward a crucial, perhaps regulatory, function of the salvage pathway.

Nucleotides and Behavior

Finally, it has been claimed that distinct behavioral effects can be elicited in animals and humans by feeding heterologous RNA, an entity that is almost certainly degraded to nucleotides and nucleosides either prior to or in the course of administration.

Because the most striking results were the correction of a memory defect in aged subjects, it is not impossible that upon stress or aging the demand on the salvage pathway becomes great enough so that it can no longer be satisfied wholly by endogenous sources, and dietary supplementation might indeed prove beneficial.

Nature of Precursor Pools

For all practical purposes, the only kind of nucleic acid synthesis with which we need to concern ourselves takes place in the nucleus. Thus the immediate precursor pools involved are constituted of the various nuclear (not the total cellular) nucleoside triphosphates, and the information that we require is their concentration both in the steady state and in its more-or-less transient response to a particular stimulus. Unfortunately, this information is not yet available. There are indications that the nuclear pools can be refilled from cytoplasmic nucleoside 5'-phosphates, but whether the diphosphates and triphosphates can also be transported across the nuclear membrane or whether intranuclear phosphotransferases have to intervene are unknown factors, as are the levels and rates of the various enzymatic reactions involved. The relative total concentrations in whole rat brain for the four triphosphates are $ATP \gg GTP \approx UTP \gg CTP$ at all ages, but whereas the concentrations of the first three remain relatively invariant during maturation (at about 200, 30, and 30 μmol per 100 g wet weight, respectively), the last drops to very low levels (less than 1 μmol per 100 g) [11].

Functional Changes

General Considerations in Tracer Experiments

The usual, and frequently the only practical, method for investigating the effect of perturbations on the rate and extent of synthesis of cellular, cytoplasmic, or nuclear RNA — either in its totality or in any of its subpopulations — consists in measuring the difference in uptake of some radioactive precursor, e.g., uridine (Urd). The scheme shown in Formula 17-1, though simplified, depicts the flow of this precursor into various species of RNA.

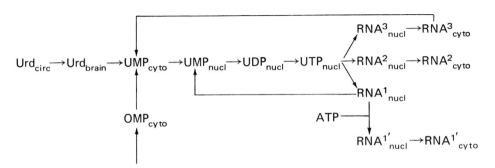

Formula 17-1
Flow of uridine into various species of RNA.

Clearly, what needs to be done is (1) to establish the complete kinetics of the flux of label in the unperturbed state and its relevance to RNA synthesis, assuring that what is measured is indeed synthesis rather than transfer or the filling of one or more of the precursor pools, and (2) to extend this study to the much more difficult problem posed by the introduction of the perturbation. Very few of the studies that will be discussed meet these stringent criteria. What is commonly done is to take one or, at the most, a selected few time points subsequent to the administration of label and to compare the appearance of the label in some RNA (and in one or more of the intermediates in the better investigations) in appropriately selected experimental and control groups. The only permissible inference from studies of this type is the rather trivial one that any differences observed indicate differences in the net rate of appearance of label in the RNA, but they cannot be interpreted in terms of differences in the rate of synthesis of this species. In fact, more often than not the differences may be due to differences in the interconversion of precursors, in transport phenomena, or in rates of degradation of short-lived intermediates or products.

Electrical Stimulation

Electrical stimulation — either briefly at high intensity during electroshock in vivo, or at low intensity, but for a sustained period, of slices in vitro — elicits profound effects on the metabolism of the total nucleotide population in these systems. The most striking results of electric shock in vivo are relatively persistent alterations in the concentrations of UTP and GTP, as well as in their rates of turnover. In the latter, investigations have been restricted so far to measurements of the conversion of uridine to its phosphorylated derivatives. A diminution of some 40 percent coincident with maintenance of the total uridine-containing pool was found, and this was ascribed to effects on the cation pump that resulted in disturbances in the rate of influx of Na^+ into, and efflux of K^+ from, the cells.

To what extent such phenomena are relevant for a control of nuclear RNA synthesis in vivo is uncertain. In general, one might expect that the RNA polymerases would respond only if some ribonucleoside triphosphates either dropped to or rose from (1) extremely low levels (less than 10 percent of the K_m), in which case the regulation would be of the "on/off" variety, or (2) the region around its K_m, in which case the rate would be directly proportional to concentration. In view of the results already cited, it might prove fruitful to explore the response of nuclear GTP levels to stimulation in some well-defined system. In any event, any such regulation by availability of a critical substrate may be expected to be nonselective and to affect to the same extent the transcription of all active genes by any one polymerase.

SYNTHESIS AND TURNOVER OF DNA

Localization and Species

The only well-authenticated sites at which DNA is found in the cells of all animals, including brain and nerves, are the nucleus (this includes the nucleolus) and the mitochondria. No hard evidence for cellular or tissue specificity has yet been adduced for either of these entities, in contrast to the compelling indications of their species specificity. To the best of our knowledge, then, all the somatic cells of a metazoan animal contain, encoded in their DNA, the identical genetic information. Possible exceptions are somatic mutations in individual structural genes of cells engaged in the formation of immunoglobulins and quantitative differences in the number of certain highly reiterated base sequences [12].

The DNA content of diploid mammalian cells, including neurons and other brain cells, is of the order of 6.5 picograms (pg) per cell (or about 1 μg per milligram [mg] wet brain weight). Virtually all of it is nuclear. This corresponds to a particle weight of 3.8×10^{12} daltons for the total chromosomal complement of such cells. The numbers are lower and subject to much greater variation for other animals. The base composition of all animal DNA is close to 60 percent guanine plus cytosine (G + C). This DNA in nondividing (interphase) nuclei is present as a complex or aggregate called chromatin, which consists of approximately 30 percent DNA, 10 percent RNA, and 60 percent protein by weight. The proteins are composed of both basic species (histones account for about 80 percent of the total) and acidic species. The DNA of even a particular chromosome is not homogeneous but varies with respect to size, base composition, and redundancy of its constituent or subunit segments.

Nucleolar DNA is much more homogeneous with respect to all these criteria: Its base composition is also different (about 70 percent G + C), all reflecting its principal function of serving as the depository for the information that encodes several hundred copies of the 45S (4×10^6 daltons) precursor of ribosomal RNA.

Mitochondrial DNA accounts for less than 0.1 percent of total cellular DNA. Its base composition is close to that of nuclear DNA, but it is much smaller, more uniform in size (5 μ in length or a particle weight of 10×10^6 daltons), and highly homogeneous with regard to base composition and lack of redundancy [13]. Mitochondrial DNA is present in many copies per cell (corresponding, on the average, to several molecules per mitochondrion) and is not tightly complexed with proteins. It can be isolated as a covalently closed, twisted circle (superhelix), and this configuration probably corresponds to that actually existing inside the cell.

Turnover

As already mentioned, the nuclear DNA of mature macroneurons is not subject to metabolic turnover: Replication is absent and repair synthesis, if it occurs at all, has so far escaped detection. In general, these findings can be correlated with the presence of DNA polymerases in developing brain and the absence of activities of this kind in adult brain in vitro. Such an activity has recently been demonstrated, however, in rat brain preparations [14]. What remains to be established is its origin — neuronal or glial. Whether there is any further modification of nucleotides (e.g., by methylation) in the intact polymer of adult animals has not been determined. In contrast, mitochondrial DNA is subject to turnover [15]: Its half-life in brain has been estimated at 31 days, as compared to 7.5 days in the liver of the same species.

DNA and Neurogenesis

Autoradiographic Studies

Altman and his collaborators [16] have used autoradiography of cell nuclei labeled with ^{3}H-thymidine to determine the nature and extent of postnatal neurogenesis in rodents. By this technique, they were able to demonstrate that considerable DNA synthesis (or at least turnover) and cell proliferation, migration, and differentiation take place in the brain of infant rats (see Chap. 18) and that these events appear to be initiated by the cells of the outer granular layers of the cerebellum, hippocampus, olfactory bulb, and ventral cochlear nucleus. The formation and differentiation of cells, classified as microneurons, continues during infancy, although at declining rates in most structures. Thus, in the cerebellum, almost all the cells of the granular and molecular layers that form the precursors of granule, stellate, and basket cells are formed and completed during·infancy, the last-named earlier than the first two types, but most of them completing their differentiation and migration by 21 days post-natally. In contrast, almost all Golgi cells have already been formed at birth. In the wall of the olfactory ventricle cell multiplication continues at very high rates for 6 days and declines dramatically by 13 days, but it is still detectable even in the young adult. In the hippocampus, the bulk of the cell population is formed postnatally and the neurogenesis of granule and stellate cells in the dentate gyrus may continue over a prolonged span of time.

From these observations, Altman has concluded that the long-axoned or macro-neurons of the brain are formed prenatally, while many of the short-axoned or axon-less microneurons are formed postnatally. He further proposes that this continued genesis of microneurons may be required for the modulation and feedback regulation of (1) the transmission of sensory information in primary afferent relay areas, (2) the execution of motor action in the cerebellar cortex, and (3) the motivational efficacy of need-catering (appetitive and consumatory activities) in the hippocampus and limbic system in general.

DNA Content and Ploidy

Perhaps related to some of these considerations is the fact that the chromosome number (ploidy), and hence the DNA content, of certain neuronal types in the higher integrative centers is about twice (i.e., 13 pg DNA per cell) that of the ordinary diploid brain cells. The nuclei of these unusual cells also are larger than their standard counterparts: Among them are the Purkinje cells of the cerebellum and the pyramidal cells of the hippocampus (but not their granule, interneuron, and glial cells) [17, 18] .

Effects of Growth Hormone

The weight of the brain, its content of total DNA (i.e., its cell number), as well as the relative proportions of its neurons and glia in newborn or infant rats can be significantly increased if the mother is injected subcutaneously or intravenously with purified growth hormone (from bovine pituitary glands) during the period between the seventh and twentieth days of pregnancy. The significance of these findings and their generality as to strain and species remain to be explored.

Inhibitors
Arabinosylcytosine (cytosine arabinoside), an analog in which arabinose has taken
the place of ribose or deoxyribose, is an effective and selective inhibitor of DNA
synthesis in animals in vivo and in their cells in culture. This inhibition is probably
elicited after its conversion to the triphosphate. Because it blocks the uptake of
radioactive thymidine into various areas of goldfish brain, it appears to function in
the expected manner in this system also. The block is, however, without any effect
on certain behavioral parameters that are affected by blocks of protein synthesis [8].
Thus neurogenesis or other concomitants of DNA synthesis do not appear involved,
at least in these particular behavioral tasks.

The alkylating agents, cyclophosphamide and methylazoxymethanol, act as inhibitors of mitosis of stem cells in the fetal or newborn animal and block subsequent
neurogenesis and development of specific brain areas, depending on the time of their
administration (i.e., at 14 to 16 days for cerebellum) [9].

BIOSYNTHESIS OF RNA

Localization
In every cell not infected by an RNA virus, all RNA synthesis is absolutely dependent
on the presence of a DNA template and is catalyzed by a family of enzymes called
DNA-dependent RNA polymerases. From our considerations so far, we would expect
RNA synthesis to be localized at only two intracellular sites in all cells of the nervous
system: in the nucleus (including the nucleolus) and in the mitochondria, with the
former providing for the bulk of this material in both variety and amounts. All the
evidence so far available bears out these expectations.

Properties
Isolated nuclei of brain cells, particularly those of neurons, are capable of effective
incorporation of radioactive precursors into RNA at rates at least equal to those
observed in the nuclei of cells with a much higher mitotic index [20]. These and
other properties of the incorporation reaction, as observed with intact and disrupted
nuclei as well as with different enzyme preparations at various stages of solubilization
and purification, suggest a strong analogy (if not an identity) between the responsible
enzymes in neuronal tissue [21, 22] and those in such other tissues as liver. The
enzymes responsible are now known to consist of at least three species that differ in
localization, requirements, inhibition pattern, type of product synthesized, and,
presumably, function [23]. Type I (or A) polymerases are localized in the nucleolus,
Type II (or B) and Type III (or C) polymerases, in the nucleoplasm. The former are
active at low ionic strength and can utilize either Mg^{2+} or Mn^{2+} as a divalent cation
cofactor; the second exhibit optimal activity only at higher salt concentrations,
0.1 to 0.2 M $(NH_4)_2SO_4$ being usually employed, and Mn^{2+} is the preferred cofactor.
The reactions of all three enzymes are inhibited by actinomycin D (with enzymes of
Type I the most sensitive [<1 μg per ml]), and they are resistant to the action of
antibiotics of the rifamicin, streptovaricin, and streptolydigin groups as well as to
intercalating dyes (ethidium bromide, acridines) at low concentrations. Only

polymerase II is inhibited (with a $K_i \approx 10^{-8}$ M) by α-amanitin, a bicyclic octapeptide isolated from the poisonous mushroom *Amanita phalloides.* The product of polymerase I appears to be the large, guanine-rich, unmethylated precursor of ribosomal RNA (rRNA) (sedimentation coefficient equal to or less than 45S) in most animals. The product of polymerase II is much more heterogeneous, in agreement with its probable function as the principal enzyme responsible for the transcription of active genes in chromatin. The function of polymerase III is not yet certain, but it may be responsible for the synthesis of tRNA.

At present, there is no information concerning mitochondrial RNA synthesis in brain. By analogy with the data obtained for other cells, one might expect it to be highly susceptible to inhibition by ethidium bromide and resistant to α-amanitin and, in intact particles, to actinomycin D; its stable products would then be the mitochondrial RNAs with sedimentation coefficients of about 12S and 18S and messenger RNAs for the small number of polypeptides ($\leqslant$10) specified by this DNA [24].

Transport of RNA

Nucleus to Cytoplasm
Most of the RNA of neurons is localized in the perikaryon and there, as is true of all cells, the majority forms an integral constituent of ribonucleoprotein particles, which are themselves aggregated into polymeric structures (polyribosomes, polysomes). These structures exist either free in the cytosol or attached to membranes of the endoplasmic reticulum. They are held together by messenger RNA (mRNA) and, in conjunction with the soluble transfer RNA (tRNA) of the cytosol, constitute the protein-synthesizing machinery of the cell. After administration of a radioactive precursor, nuclear and nucleolar RNA are the first to become labeled, as might be expected. The latter provides the precursor of ribosomal RNA, the former of mRNA, in the form of an exceedingly polydisperse group of molecules known as heterogeneous nuclear RNA (HnRNA). Most of these never leave the nucleus and are degraded there, but a minority, after polyadenylation at their 3'-terminus and, perhaps, additional processing, enter the cytoplasm within a few minutes, probably in association with some proteins. This flux of mRNA species is followed within 30 minutes to 1 hour by the stable ribosomal RNAs already integrated into ribonucleoprotein particles.

Nucleus to Axons
In addition to their perikarya, certain neurons may contain some RNA also in their axons, including the presynaptic thickening (nerve ending, bouton). In peripheral nerve, particularly in the giant Mauthner neurons of fish or the stretch receptor of the lobster, this axonal RNA may contribute some 0.5 mg per g of wet weight, although in most other instances values are considerably lower [25]. The origin and significance of this axonal RNA are open to question. Because in addition to 4S RNA it consists of the usual ribosomal 18S and 28S species and because its synthesis is blocked by actinomycin D, a mitochondrial origin of these stable species is unlikely.

Similar considerations apply to the labeled RNA that makes its appearance in and down the axon shortly after a labeled precursor has been administered to the cell bodies of neurons that have been the object of investigation of axoplasmic flow, such as the sciatic and other peripheral nerves of mammals or the optic nerve of fish and amphibians (see also Chap. 21). It remains uncertain to what extent this labeled RNA represents *macromolecules* synthesized in the cell body and then transported down the axon, as contrasted to RNA synthesized in the immediate vicinity by supporting cells (glia) from low molecular weight *precursors* that have been so transported [25, 25a, b]. Even more controversial is whether RNA in or on axons and nerve endings participates in protein synthesis [5]. Although it is generally accepted that these structures do not contain any free ribosomes and polysomes, such ribonucleoprotein particles present in low concentration and tightly integrated into the axonal or presynaptic membrane might have escaped detection.

Changes in RNA Synthesis Related to Function

Electrical Stimulation in Vivo
The sea hare *Aplysia* provides a particularly useful system for correlating electrophysiological and biochemical findings: Certain giant neurons are readily identifiable in its abdominal ganglia; these structures or individual neurons can be removed, stimulated, and monitored electrophysiologically and, at the same time, be exposed to radioactive precursors of RNA. The products formed from them can be isolated subsequently and the extent and the nature of the incorporation explored. Prolonged stimulation (about 90 min) of the ganglion, sufficient to elicit postsynaptic spikes in cell R2, results in a significant increase in the incorporation of labeled nucleosides into RNA as compared with unstimulated controls. The newly formed material is found in both the nucleus and the cytoplasm; both ribosomal and heterodisperse RNA (presumably mRNA) appear to be affected [26, 27].

Changes in the specific activity of the precursor pool or in the rate of degradation might serve as alternate explanations for the results obtained. Their interpretation in terms of an increase in the rate of synthesis, however, is consistent with many reports on net increases of stable RNA as a result of stimulation in many species, including earthworm neurons, rat Purkinje cells, frog retinal ganglion, and cat sympathetic ganglion cells.

Electrical Stimulation in Vitro
Contrary to the results obtained in vivo, electrical stimulation of slices of rat or cat cerebral cortex in vitro [28, 29] decreases the rate of incorporation of labeled uridine into RNA by about 50 percent. This effect is blocked by 5 μM tetrodotoxin, a neurotoxin known to inhibit Na^+ influx. An inhibition of the conversion of labeled UMP into UDP plus UTP can be demonstrated to accompany the inhibition of RNA labeling, so the most likely explanation, consistent also with a number of additional experiments, is that electrical activity affects the Na^+ pump, and a disturbance in the balance between Na^+ and K^+ is responsible for affecting the rate of phosphorylation and thereby the availability of labeled nucleoside triphosphates for RNA synthesis. Similar effects may also account for some of the reported decreases in RNA and its synthesis that result from gross stimulation of the cortex in vivo, particularly during electroshock (see p. 346).

Visual Stimulation

When the incorporation of orotate into the RNA in the visual cortex of a totally blind strain of rats is compared with that of normal rats [30], both a delay and a decrease in the rate of labeling of ribosomal and of a heterogeneous 5S to 8S RNA are observed. There are no differences in the labeling of these RNAs in the motor cortex, while differences in the visual cortex are accentuated by using a flashing light rather than continuous illumination for the stimulation of the sighted rats. Thus, in this instance, there appears to be a positive correlation between neuronal activity and the kinetics of RNA labeling. Since the visual system of the blind animals might be expected to have suffered atrophy long before the start of the experiment, the ability of the system to synthesize certain enzymes required for the incorporation reaction also may have been impaired. In fact, it is known that blinded animals or those kept in the dark from birth show deficits in RNA in various cells of the visual system. The dynamics of RNA and protein synthesis have also been reported [3] for the visual cortex of young rats upon first exposure to light after rearing for 7 weeks in the dark from birth.

Visual imprinting of day-old chicks by flashing light is also reported to coincide with a relatively profound and rapid stimulation of RNA synthesis (as well as protein synthesis), particularly in the forebrain roof, as compared with slower increases elicited by continuous illumination, which does not result in imprinting [31].

Olfaction

Catfish brains in vivo or the experimental half of split-brain preparations in vitro have been shown to respond in a specific manner to odorants such as morpholine by affecting the amount as well as the base ratios of nuclear RNA, but not of cytoplasmic RNA.

Drugs

1,3-Dicyanoaminopropene is reported to increase the amount of RNA and its rate of synthesis in the nervous system. Because it also affects a variety of growth processes, can act as a general mild stimulant of the CNS, and is completely inert insofar as affecting RNA synthesis by isolated brain nuclei, its mode of action is probably not specific for RNA but instead is more general, perhaps metabolic. The claims that magnesium pemoline, a combination of 2-amino-5-phenyl-4-oxazolidinone and magnesium hydroxide, affects incorporation of some precursors into nuclear RNA, but not into total brain RNA, have not been substantiated and probably can be discounted.

Development

During the early stages of development, rapid cellular proliferation is dependent on high levels of the nucleolar polymerase I and extensive synthesis of rRNA. As maturation, accompanied by myelination, progresses this pattern is replaced by one geared to the production of HnRNA and mRNA requiring high levels of polymerase II [32, 33]. Complementary patterns are also observed for tRNA methyltransferase. Possible alterations in the levels of the polyamines putrescine, spermidine, and spermine (and in those of the enzymes responsible for their synthesis and degradation), which can act as specific activators for the polymerases [21], must also be considered.

Nerve Growth Factor (NGF)

When NGF is added to embryonic sensory or sympathetic nerve fibers in vivo or in vitro, their growth and proliferation are enhanced but remain largely restricted to axonal processes that are free of myelin sheaths (see Chap. 18). Under these conditions, autoradiography shows rapid incorporation of labeled uridine into the RNA of both the perikaryon and the axon, with a time course consistent with the uridine being the precursor of the RNA [34].

Training

Carefully controlled experiments seem to indicate that in mice [4] the acquisition of a new behavioral pattern, at least in response to certain tasks, or of a specific emotional response elicited by or associated with this performance, results in an increased flux of labeled pyrimidines into nucleotide pools and thence into RNA. These increases are restricted to the nuclei of the neurons of the limbic system, while decreases are observed in those of the outer layer of the neocortex. The increases in radioactivity do not appear to be restricted to any one type of RNA and do not involve any changes in base ratio. Similar results have also been reported for rats, including actual increases in the amount (not just the rate of labeling) in the RNA of hippocampal nuclei [35]. In contrast, acquisition of a new swimming skill by goldfish is not accompanied by any change in tRNA amino acid acceptor activity or methylation. (Additional discussion can be found in Chap. 36.)

MOLECULAR SPECIES OF RNA AND THEIR TURNOVER

If the internal or external milieu can bring about more-or-less selective effects on RNA synthesis, this will generate a qualitative or quantitative alteration of the profile of the various RNA species in the population. Some of the effects observed have already been mentioned in earlier sections: Here we address ourselves to this problem in a more explicit fashion.

Hybridization

With prokaryotic organisms (bacteria and viruses), it is possible, at least in principle, to determine by appropriate DNA-RNA hybridization techniques the nature and the length of a specific section of the total genome represented by any given RNA transcribed either in vivo or in vitro. The enormous complexity and the large number of partially overlapping (reiterated or repetitive) long sequences of the nuclear DNA of metazoan animals, and especially of mammals, impose severe constraints on similar quantitative hybridization experiments, as well as formidable hazards in the interpretation of this type of experiment in general [27].

As ordinarily performed (i.e., at high RNA/DNA ratios), the restrictions mentioned virtually assure that what are measured are not locus specific, unique sequences corresponding to what would be expected for mRNAs of structural genes. Instead, they are (a) sequences present in many copies, some of the most abundant ones with multiple tandem reiterations, or (b) the extent of overlap between sequences that are partially homologous structurally but unrelated functionally. These reservations do not hold for properly performed determinations on ribosomal RNAs or their precursors, precisely because they are specified by clusters containing more than 100 genes for the various species. The values of the particular parameters obtained for brain agree reasonably well with those for other tissues.

In addition, properly controlled experiments designed to assess the extent of transcription of unique sequences have now been reported for both mouse and human brain [36, 37] with the following results: (1) Approximately 10 percent of such sequences could be shown to be present in the RNA isolated from the brain of either species, while the values for all other mouse tissues tested (liver, kidney, spleen) were much lower (≤5 percent); (2) the number of such sequences appeared to increase with increasing age in both mouse and man (12 percent for fetal vs. 24 percent for adult); (3) there appeared to be regional and positional variations, with the highest values (close to 50 percent of theoretical) in RNA from left frontal cortex, one-half to one-third that in RNA from right and left parietal and temporal cortex as well as right frontal cortex, and lowest in brain stem; (4) similar low values were obtained with RNA prepared from temporal cortex of a patient with severe cerebral atrophy. Interpretation of these important findings is rendered difficult by two sets of facts: (a) what is measured is mainly nuclear RNA, which consists of a majority of HnRNA and a minority of functional mRNA, and (b) the hybridization plateaus observed are contributed to not by a single but a multiple — and in certain of the experiments a variable — population of cell types.

Rapidly Labeled RNA

As discussed earlier, at short times after administration of label to the brain of an animal or to even an isolated neuron, labeled (i.e., newly synthesized) RNA is restricted to the nucleus and nucleolus. As in other tissues, the RNA appears to be mainly of two kinds: ribosomal precursors that first appear in the nucleolus, with sedimentation coefficients of 45S, 35S, and 32S; and a polydisperse (HnRNA) species, with sedimentation coefficients varying from about 50S to about 5S and a base composition close to that of DNA. This is also the species capable of effective hybridization with DNA. Most of this RNA decays without ever leaving the nucleus and, for this reason as well as because of its other properties, the bulk of HnRNA is not the direct precursor of the mRNA found associated with the polysomes of the cytoplasm. It may, however, play a significant part in the selection of the kind and amount of mRNA that does find its way into this compartment [38].

Cytoplasmic mRNA makes its first appearance shortly (several minutes) after extensive labeling of nuclear RNA has already been completed. The molecules leaving the nucleus and entering the cytoplasm exhibit the following properties: They are heterodisperse, reflecting the variation in length of the polypeptides to be translated; they have been posttranscriptionally modified by the addition of polyriboadenylate tracts to their 3'-termini; and they are associated with ribonucleoprotein particles (informosomes or informofers). These mRNA particles become associated with 80S ribosomes to form polysomes in order to discharge their function in translation. The labeling rate of the ribosomal RNA proper exhibits a lag and, although slower, proceeds for a longer time, relative to mRNA, before reaching an isotopic steady state. The explanation of this phenomenon is probably that it is due to gross differences in the kinetics of interconversion and degradation of various macromolecular precursors, as well as of the final products themselves.

Polysome Turnover and State of Aggregation

In both structure and function, brain and neuronal polysomes exhibit properties similar to those of other tissues and cells. Maintenance of their proper state of aggregation and configuration, and hence their activity, appears to be particularly finely poised and responsive to changes in environment: These parameters appear to be modulated in vitro by the level of Mg^{2+}, the presence and proper balance of K^+ and Na^+, and the concentrations of various amino acids and certain transmitters (e.g., γ-aminobutyrate).

Such effects may also be operative in vivo. In addition, we might anticipate from studies on other systems that changes in the functional state of a cell might affect the size and turnover profiles of polysomal populations of selected cell populations or brain regions if they influence (1) the rate of translation of any mRNA (as, for instance, in the presence of cycloheximide and its derivatives or other inhibitors of protein synthesis), (2) the rate or extent of its formation (by activation or inhibition of specific genes, as well as of transcription in general by agents such as actinomycin or amanitin), or (3) the rate or extent of its degradation (for example, by activation of specific nucleases or by affecting accessibility). So far, studies have been restricted to whole brain or cerebral cortex. Extensions of these studies to more localized areas and small groups of homogeneous neurons are badly needed.

In cortex we find polysomes in both glial and neuronal perikarya. In the latter, they appear both free and attached to endoplasmic reticulum; both types are capable of and responsible for effective protein synthesis. In adult rats the turnover rate of both the RNA and the proteins of the various constituent ribosomes is 12 days. Thus, the particles are subject to turnover as a unit. It is known that protein synthesis is most extensive in the brain and in brain preparations of young rats and mice shortly after birth (it reaches a maximum at about 10 weeks for rats) and declines as the animals mature. This decline in activity is correlated with ATP levels in cortex slices; it is also correlated, at least in part, with a decrease in polysomal RNA, a change in the size distribution in favor of smaller aggregates, and decreased stability of isolated polysomes.

Coincident with these changes is a change in the base composition [39] as reflected in the ratio of guanine + cytosine to adenine + uracil $[(G + C)/(A + U)]$: This ratio has been reported to rise during maturation from 1.30 to 1.50 in whole rat brain and from 1.41 to 1.66 in single pyramidal cells of the hippocampus. In the latter, the amount of RNA per cell is reported to increase from 24 pg to 110 pg for mature rats and to drop to 53 pg per cell in very old rats. In the latter, the ratio $(G + C)/(A + U)$ rises to 1.95.

The proportion of total ribosomal RNA found in polysomes has also been reported to increase as a function of environmental or behavioral stimulation. For instance, when adult rats were kept first in the dark for 3 days and then exposed to the lights and sounds of the laboratory for 15 min, the polysome-to-monomer ratio increased by 83 percent in cerebral cortex but was unaffected in the liver. On the other hand, electroshock had the opposite effect: Polysomes were dissociated as a result.

Disaggregation of whole brain polysomes has also been observed in response to the administration to rats of the "unnatural" amino acids L-dihydroxyphenylalanine (L-DOPA) and L-5-hydroxytryptophan, mediated by their conversion to the cate-cholamine transmitters dopamine and serotonin, respectively [40].

Behavior and Changes in Base Ratios

The literature contains several reports concerning changes in base ratios resulting from or coincident with the acquisition of new behavioral patterns (see [4] and Chap. 36). In general, the experiments have been of two types and both are subject to serious reservations. In the first, total RNA of small cell populations is analyzed by micro-techniques and profound changes are observed, such as formation of RNA with $(G + C)/(A + U)$ ratios in the range of 0.70 to 0.90. These, then, must have been the result of alterations in ribosomal RNA since it is the only species present in amounts large enough for its base composition to affect cellular base composition. (Messenger RNA accounts for about 5 percent of the total cellular RNA, and, in its base composition, it must reflect the average of a large multiplicity of different base sequences. Even under hyperinduction in prokaryotes, a single species never accounts for more than 20 percent of the total population. Thus, total base composition could be altered by 1 percent in the most extreme case.) No such fluctuations in ribosomal RNA have ever been observed in any other system and are highly unlikely in view of the fact that these RNAs are, as we have mentioned, specified by a cluster of highly homogeneous genes.

In the second type of experiment, newly synthesized, rapidly labeled RNA is ana-lyzed by determining the ratio of radioactivity in the uridylate and cytidylate frac-tions of RNA hydrolysates formed from labeled orotate. However, in the absence of data for nuclear UTP and CTP, the direct precursors, the results are not amenable to any ready interpretations. It is of interest to note in this context that the increased synthesis of RNA, which was discussed previously in regard to the effects of behav-ioral training on RNA synthesis (see above under Training), is not accompanied by any changes in base ratio.

EFFECTS OF ANTIBIOTICS AND OTHER DRUGS

Actinomycin D

As outlined earlier, this antibiotic, at least under the conditions usually employed in whole-animal experiments, is a general inhibitor for DNA-dependent RNA synthesis. As a result of this interference, it might be assumed that it will also block protein synthesis, but that is not necessarily true. (These differential effects are not observed in prokaryotic cells, in which transcription and translation are tightly coupled and mRNA turnover is much more uniform.) Synthesis supported by inherently unstable mRNAs, or on polysomes in the perikaryon close to the site of the block, will be affected much more immediately than will synthesis supported by relatively stable mRNAs or under conditions in which some of the nucleic acid may still be in transit to its site of utilization somewhere along the axon. This reasoning may explain some of the results on protein synthesis obtained with actinomycin D. Brief exposure to the drug, even at levels that inhibit RNA synthesis almost completely, elicits no effects on resting or action potentials, on acquisition of new tasks, or on short-term memory.

Interpretations of the retrograde amnesic effects (or their absence) obtained with the drug are complicated by its production of necroses and electrophysiological abnormalities as found, for instance, in the hippocampus of rodents. These reactions may occur either in addition to or, perhaps, sometimes in the absence of effects on RNA synthesis.

8-Azaguanine

Intracisternal injection of 8-azaguanine leads to the incorporation of this analog of guanine into the RNA of rat brain, but the qualitative and quantitative consequences of this intervention have not been explored. It is reported to exert behavioral effects, which, however, may be a secondary consequence of a generalized depression of motor activity.

6-Azauracil

The effects produced by repeated administration of this uracil analog have been described above under Salvage Pathways and Orotic Acidurias. The available evidence suggests that, although 6-azauracil profoundly affects the levels of pyrimidine precursors available de novo, the operation and the quantitative importance of the salvage pathway insure their adequate supply, so the rate and extent of RNA formation becomes relatively insensitive to the inhibitor even when it is present in high concentration. Any neurophysiological and behavioral effects produced are therefore probably referrable to disturbances of metabolism rather than to the biosynthesis of RNA or protein.

5-Fluorouracil and 5-Fluoroorotate

Unlike the substitution of nitrogen for carbon in 6-azauracil, the substitution of fluoride for hydrogen forming fluorouracil results in a molecule that is sufficiently similar to its parent to permit its participation in all reactions, including the conversion to the nucleoside triphosphate and the incorporation of the resultant nucleotide into RNA at nearly normal rates and to the same extent. This phenomenon, established in systems other than nervous tissue, also appears to occur in cat spinal cord, the only nervous tissue for which intensive studies are available. The most interesting result of exposure to 5-fluorouracil is its well-authenticated ability to produce (or correct) miscoding, either at the transcriptional or the translational level, by virtue of "faulty" mRNAs or tRNAs containing 5-fluorouracil capable of forming base pairs with guanine rather than adenine. This, in turn, might result in the production of specific faulty proteins; however, this possible reaction has not yet been explored. The possible subtle behavioral consequences arising from such an alteration of protein synthesis also have not yet been investigated.

When injected into cats at a relatively high dose (about 10 mg), fluoroorotate or fluorouridylate (but not fluorouridine) produces severe neuroanatomical alterations in both the central and the peripheral nervous systems.

Uric Acid

Because it is an end product rather than a precursor of purine metabolism, uric acid probably cannot affect nucleic acid biosynthesis directly. Nevertheless it may serve as an indicator of disturbances in purine or nucleic acid catabolism.

ENZYME INDUCTION

The most precise and direct way of determining alterations in gene expression is to measure their effect on the formation of defined final products, i.e., specific proteins such as enzymes. Whether any changes in observed enzyme activity are due to specific de novo synthesis (induction proper), to activation of precursors, to relatively nonspecific stimulation of the synthesis of several proteins, or to the inhibition of their degradation is, however, a problem that must be solved in each particular instance. Two examples relevant to the nervous system are discussed here.

The activity of glutamine synthetase in the retina of embryonic chicks follows a characteristic developmental pattern typical for this tissue and is temporally and spatially correlated with other aspects of embryonic development. A characteristic rise in enzyme activity occurs which can be induced precociously with 11-β-hydroxy-corticosteroids, either in the intact animal or in isolated retina in culture, without any generalized cell proliferation but within the context of other well-defined changes. The effects observed are therefore specific and characteristic [41–43]. They involve the de novo synthesis of the enzyme and its accumulation. Whether they are due to an increase in the rate of synthesis rather than to a decrease in the rate of degradation remains to be established; what has been ascertained by the use of actinomycin D is that the process also requires de novo synthesis of some species of RNA (but not protein) during the initial phases, although the process later becomes partially independent of it (neither high nor low concentrations of actinomycin D have an effect). These observations are consistent with analogous patterns obtained in the induction of a wide variety of enzymes by glucocorticoids and other hormones in many other systems.

Nadler et al. [44] removed one side of the entorhinal cortex of 11-day-old rats and studied the resulting formation of new cholinergic septo-hippocampal synapses in the ipsilateral dentate gyrus — with the contralateral side as control — at the structural, cytochemical, and biochemical levels. They found that within 5 days the absolute and specific activities of choline acetyltransferase and acetylcholinesterase became significantly elevated in the outer part of the molecular layer on the operated, relative to the unoperated, side. These differences decreased to extinction before the animals reached maturity.

ACKNOWLEDGMENTS

I would like to express my appreciation to Dr. M. Maguire and A. L. de Blas for their invaluable help in surveying the voluminous recent literature on these topics and to Drs. E. Shooter and B. W. Agranoff for providing me with many stimulating suggestions for the first edition. I would also like to acknowledge receipt of Research Career Award No. GM 05060 from the National Institutes of Health.

REFERENCES

*1. Ansell, G. B., and Bradley, P. B. (Eds.). *Macromolecules and Behavior.* London: Macmillan, 1973.
*2. Bourne, B. H. (Ed.). *Structure and Function of Nervous Tissue.* New York: Academic, 1971.
*3. Horn, G., Rose, S. P. R., and Bateson, P. P. G. Experience and plasticity in the central nervous system. *Science* 181:506, 1973.
*4. Glassman, E. Biochemistry of learning: An evaluation of the role of RNA and proteins. *Annu. Rev. Biochem.* 38:605, 1969.
*5. Barondes, S. H. Synaptic macromolecules: Identification and metabolism. *Annu. Rev. Biochem.* 43:147, 1974.
 6. Held, I., and Wells, W. Observations on purine metabolism in rat brain. *J. Neurochem.* 16:529, 1969.
 7. Appel, S. H., and Silberberg, P. H. Pyrimidine synthesis in tissue culture. *J. Neurochem.* 15:1437, 1968.
*8. Smith, L. H., Jr., Huguley, C. M., Jr., and Bain, J. A. Orotic Acidurias. In Stanbury, J. B., Wyngaarden, J. B., and Fredrickson, D. S. (Eds.), *The Metabolic Basis of Inherited Disease.* New York: McGraw-Hill, 1972, p. 1003.
*9. Koenig, H. Neurobiological action of some pyrimidine analogs. *Int. Rev. Neurobiol.* 10:199, 1967.
*10. Kelley, W. N., and Wyngaarden, J. B The Lesch-Nyhan Syndrome. In Stanbury, J. B., Wyngaarden, J. B., and Fredrickson, D. S. (Eds.), *The Metabolic Basis of Inherited Disease.* New York: McGraw-Hill, 1972, p. 969.
11. Mandel, P., and Jacob, M. Regulation of Transcription in Nervous Cells. In Lajtha, A. (Ed.), *Protein Metabolism of the Nervous System.* New York: Plenum, 1970.
12. Ranjekar, P. K., and Murthy, M. R. V. Rapidly reassociating DNA of rat brain. *J. Neurochem.* 20:1257, 1973.
*13. Borst, P. Mitochondrial nucleic acids. *Annu. Rev. Biochem.* 41:333, 1972.
14. Chin, J. F., and Sung, S. C. Particulate form of DNA polymerase in rat brain. *Nature [New Biol.]* 239:176, 1972.
15. Gross, N. J., Getz, G. S., and Rabinowitz, M. Apparent turnover of mitochondrial deoxyribonucleic acid and mitochondrial phospholipids in the tissues of rat. *J. Biol. Chem.* 244:1552, 1969.
*16. Altman, J. DNA Metabolism and Cell Proliferation. In Lajtha, A. (Ed.), *Handbook of Neurochemistry.* Vol. 2. New York: Plenum, 1969.
17. Herman, C. J., and Lapham, L. W. DNA content of neurons of the cat hippocampus. *Science* 160:537, 1968.
18. Casola, L., Lim, R., Davis, R. E., and Agranoff, B. W. Behavioral and biochemical effects of intracranial injection of cytosine arabinoside in goldfish. *Proc. Natl. Acad. Sci. U.S.A.* 60:1389, 1968.

*Asterisks denote key references.

19. Chanda, R., Woodward, D. J., and Griffin, S. Cerebellar development in the rat after early postnatal damage by methylazoxymethanol: DNA, RNA and protein during recovery. *J. Neurochem.* 21:547, 1973.

20. Monotamaro, N., Novello, F., and Stirpe, F. Effect of α-amanitin on ribonucleic acid polymerase II of rat brain nuclei and on retention of avoidance conditioning. *Biochem. J.* 125:1087, 1971.

21. Singh, V. K., and Sung, S. C. Effect of spermidine on DNA-dependent RNA polymerases from brain cell nuclei. *J. Neurochem.* 19:2885, 1972.

22. Thompson, R. Studies on RNA synthesis in two populations of nuclei for the mammalian cerebral cortex. *J. Neurochem.* 21:19, 1973.

*23. Chambon, P. Eucaryotic RNA Polymerases. In Boyer, P. D. (Ed.), *The Enzymes,* 3rd ed. Vol. 10. New York: Academic, 1974, p. 261.

*24. Attardi, G., Constantino, P., England, J., Lynch, D., Murphy, W., Ojala, D., Posakony, J., and Storrie, B. The biogenesis of mitochondria in HeLa cells: A molecular and cellular study. In Birkey, C. W., Jr., Perlman, P. S., and Byers, T. J. (Eds.), *College of Biological Sciences Colloquium. Genetics and Biogenesis of Mitochondria and Chloroplasts.* Columbus, Ohio: Ohio State University Press, 1975.

*25. Koenig. E. Nucleic Acid and Protein Metabolism of the Axon. In Lajtha, A. (Ed.), *Handbook of Neurochemistry.* Vol. 2. New York: Plenum, 1969.

25a. Gambetti, P., Autilio-Gambetti, L., Shafer, B., and Pfaff, L. Quantitative autoradiographic study of labeled RNA in rabbit optic nerve after intraocular injection of [^{3}H] uridine. *J. Cell Biol.* 59:677, 1973.

25b. Ingoglia, N. A., Grafstein, B., and McEwen, B. S. Effect of actinomycin-D on labelled material in the retina and optic tectum of goldfish after intraocular injection of tritiated RNA precursors. *J. Neurochem.* 23:681, 1974.

26. Berry, R. W. Ribonucleic acid metabolism of a single neuron: Correlation with electrical activity. *Science* 166:1021, 1970.

27. Peterson, R. P. RNA in single identified neurons of *Aplysia. J. Neurochem.* 17:325, 1970; Peterson, R. P., and Kernell, D. Effects of nerve stimulation on the metabolism of ribonucleic acid in a molluscan giant neuron. *J. Neurochem.* 17:1075, 1970.

28. Orrego, F. Synthesis of RNA in normal and electrically stimulated brain cortex slices in vitro. *J. Neurochem.* 14:851, 1967.

29. Prives, C., and Quastel, J. H. Effects of cerebral stimulation in biosynthesis of nucleotides and RNA in brain slices in vitro. *Biochim. Biophys. Acta* 182:285, 1969.

30. Dewar, A. J., and Reading, H. W. Nervous activity and RNA metabolism in the visual cortex of rat brain. *Nature* (Lond.) 225:869, 1970.

31. Rose, S. P. G., Bateson, P. P. G., Horn, A. L. D., and Horn, G. Effects of an imprinting procedure on regional incorporation of tritiated uracil into chick brain RNA. *Nature* (Lond.) 225:650, 1970.

32. Banks, S. P., and Johnson, T. C. Maturation-dependent events related to DNA dependent RNA synthesis in intact mouse brain nuclei. *Brain Res.* 41:155, 1972; Developmental alterations in RNA synthesis in isolated mouse brain nucleoli. *Biochim. Biophys. Acta* 294:450, 1973; Synthesis of RNA polyadenylic acid by isolated brain nuclei. *Science* 181:1064, 1973.

33. Berthold, W., Lin, L., and Davison, A. N. Nucleic acid metabolism in the developing rat brain. *Biochem. Soc. Trans.* 2:244, 1974.

34. Amaldi, P., and Rusca, G. Autoradiographic study of RNA in nerve fibers of embryonic sensory ganglia cultured in vitro under NGF stimulation. *J. Neurochem.* 17:767, 1970.

35. Bowman, R. E., and Strobel, D. A. RNA metabolism in the rat brain during learning. *J. Comp. Physiol. Psychol.* 67:448, 1969.
36. Grouse, L., Chilton, M. D., and McCarthy, B. J. Hybridization of ribonucleic acid with unique sequences of mouse deoxyribonucleic acid. *Biochemistry* 11:798, 1972.
37. Grouse, L., Omenn, G. A., and McCarthy, B. J. Studies by DNA-RNA hybridization of transcriptional diversity in human brain. *J. Neurochem.* 20:1063, 1973.
*38. Brawerman, G. Eukaryotic messenger RNA. *Annu. Rev. Biochem.* 43:621, 1974.
39. Yamagami, S., and Mori, K. Changes in polysomes of developing rat brain. *J. Neurochem.* 17:721, 1970.
40. Weiss, B. F., Wurtman, R. J., and Munro, H. N. Disaggregation of brain polysomes by L-5-hydroxytryptophan: Mediation by serotonin. *Life Sci.* 13:411, 1973.
41. Sarkar, P. K., and Moscona, A. A. Glutamine synthetase induction in embryonic neural retina: Immunochemical identification of polysomes involved in enzyme synthesis. *Proc. Natl. Acad. Sci. U.S.A.* 70:1667, 1973.
42. Sarkar, P. K., and Moscona, A. A. Changes in macromolecular synthesis associated with the induction of glutamine synthetase in embryonic retina. *Proc. Natl. Acad. Sci. U.S.A.* 68:2308, 1971.
43. Moscona, A. A., Moscona, M., and Saenz, N. Enzyme induction in embryonic retina: The role of transcription and translation. *Proc. Natl. Acad. Sci. U.S.A.* 61:160, 1968.
44. Nadler, J. V., Cotman, C. W., and Lynch, G. S. Altered distribution of choline acetyltransferase and acetylcholinesterase activities in the developing rat dentate gyrus following entorhinal lesion. *Brain Res.* 63:215, 1973.

Section V

Physiological Integration

Chapter 18
Development, Regeneration, and Aging
Joyce A. Benjamins
Guy M. McKhann

The precise cellular and subcellular mechanisms that regulate development have not been clearly defined in any mammalian system, certainly not in one so complex as the nervous system. At present, we know a great deal about biochemical changes that occur in whole brain or nerve during development [1–6; for historical review see ref. 7]. Some progress has been made in dissecting the origin and development of various cell types in the nervous system and the factors that control their maturation [8].

The first section of this chapter discusses overall changes in composition and metabolism of the developing nervous system and relates them to cellular changes, where possible. In the second section, the maturation of the two major cell populations in the nervous system, neurons and glia, is described, with emphasis on systems in which the biochemical events underlying the morphological and functional changes have been studied.

The events of nervous system development often have been divided into stages for ease of discussion and organization. These stages are artificial, at best, particularly when applied to a dynamic system. In addition, the nervous system is heterogeneous from region to region in the timing of development of cell types and complexity of interaction among those types. The following general scheme [1] serves to indicate major changes applicable to all species:

1. Organogenesis and neuronal multiplication
2. The brain-growth spurt
   a. Axonal and dendritic growth, glial multiplication, and myelination
   b. Growth in size
3. Mature, adult size
4. Aging

BIOCHEMICAL CHANGES IN DEVELOPMENT
The most obvious index of brain growth is weight. The most rapid growth may occur before, after, or at the time of birth, depending on the species (Fig. 18-1). In general, brain weight reaches its maximum before body weight, as illustrated for rat brain in Figure 18-2. The period of maximal rate of weight increase takes place at various times in different regions of the nervous system, in the following sequence: peripheral nerves, spinal cord, cerebrum, cerebellum. The cerebellum shows the sharpest rate of growth, the spinal cord the most gradual. In general, the increase in weight of a region of the brain corresponds to the proliferation of neuronal processes and myelination. The increase in membranous structures is accompanied by an increase in brain solids and a decrease in brain water. For

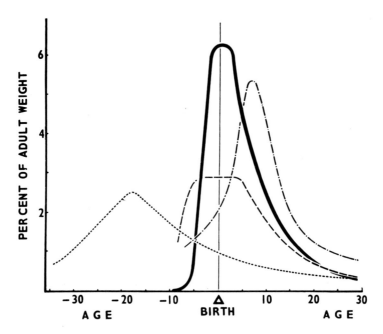

Figure 18-1
Rate of brain growth (wet weight) in relation to birth in different species. Increments were calculated as percentage of adult weight gained in a given interval. Human _____, in months; guinea pig _____, in days; pig _ _ _ _, in weeks; rat _._._._._., in days. (From Dobbing [95]. Reprinted by permission; courtesy of W. B. Saunders Co.)

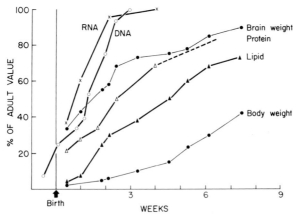

Figure 18-2
Increases in body weight and in several brain constituents during development in rat. Values are expressed as percentages of maximal (adult) values. Results for body weight, brain weight, and brain lipid have been calculated from the data of Kishimoto, Davies, and Radin [9], DNA values from the data of Fish and Winick [10], RNA values from the data of Mandel [11], and protein values from the data of Himwich [12, 13].

example, in rat brain, water is 90 percent of the total brain weight at birth and 83 percent at maturity [12].

The developmental changes in four major constituents of brain – DNA, RNA, protein, and lipid – are summarized for the forebrain of rat in Figure 18-2, with values expressed as percentages of adult levels. In Figure 18-3, changes in brain weight, DNA, and cholesterol are expressed in terms of rate of change, emphasizing the time of most rapid increase. The sequential changes in these components illustrate the events that take place in the developing nervous system. Early proliferation of cells is indicated by synthesis of DNA. As differentiation occurs, DNA replication is followed by increased transcription of DNA to RNA, then translation of RNA to protein. With the appearance of cell-specific enzymes and structural proteins, each cell type acquires its unique metabolism and morphology. Maturation of such elements as neuronal processes, synaptic endings, and myelin involves deposition of lipids into those membranous structures. DNA content of brain is considered to be a reliable indicator of cell number, and the ratio of protein to DNA indicates cell size. This ratio increases after neuronal division ends, reflecting, in part, the arborization of neuronal processes. Increasing lipid content indicates compartmentation and membrane formation, particularly of synaptic endings and myelin membranes.

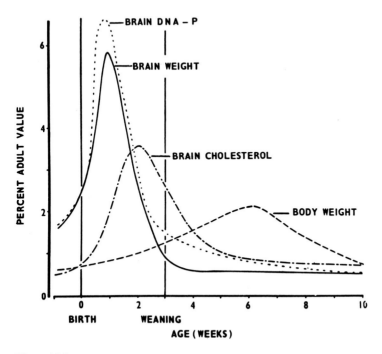

Figure 18-3
Rates of change in wet weight, DNA, and cholesterol of developing rat brain, and their relationship to the later bodily growth spurt. Values were calculated as increments (percentage of adult values) at two-day intervals. (From Dobbing [95]. Reprinted by permission; courtesy of W. B. Saunders Co.)

These developmental events occur at different ages in various species (Table 18-1). It is impossible to give exact times for the onset or end of these events, but the ages in Table 18-1 are intended to serve as approximate guidelines. Cortex was chosen because more complete data are available for this brain region than for others.

Nucleic Acids

The first increases in brain DNA reflect proliferation of neuronal precursors. Later increases are due primarily to the appearance of glial precursors. In human brain, two major periods of cell proliferation have been detected by measuring DNA levels [95]. The first period begins at 15 to 20 weeks of gestation and corresponds to neuroblast proliferation. The second begins at 25 weeks and continues into the second year of postnatal life. This latter period corresponds to multiplication of glial cells and includes a second wave of neuronogenesis, restricted to microneurons in certain regions of the brain. (In the rat, these regions are primarily the granular layer of the cerebellar cortex, the olfactory lobe, and hippocampus [22].)

In smaller species, DNA in brain appears to accumulate in a linear manner until the adult level is reached. Perhaps two phases of DNA accumulation have not been detected because of a lack of sufficient data during the period of neuronal division. In rat, neuronal division is largely prenatal and glial division, postnatal. The values in Figure 18-2 suggest two phases of DNA accumulation in rat brain, and one study has demonstrated that the rate of increase in DNA and the activity of DNA polymerase are high at 6 days before birth, are lower at birth, then are higher again, reaching a peak between 6 and 10 days [23].

As with weight increase, the period of most rapid DNA synthesis and accumulation is earliest in spinal cord, followed by cerebrum, then cerebellum. For example, at birth, rat cerebrum has 50 percent of its adult DNA content, whereas cerebellum has only 3 percent [10]. The rate of increase in cell number is most rapid in cerebellum, slower in cerebrum, and most gradual in cord. Thus in postnatal rat brain DNA content increases five times more rapidly in cerebellum than in cerebrum.

Table 18-1. Comparison of Developmental Milestones in Cortex of Several Species[a]

Species	Rapid Increase in Maturity and Number of Nerve Processes	Histological Appearance of Myelin	DNA at Maximal Level	Protein at Maximal Level
Rat	6–24 days [14, 17]	10–40 days [18]	20 days [10]	99 days [10]
Mouse	2–17 days [15]	9–30 days [19]	16 days [13]	80–90 days [13]
Guinea pig	(41–45 days) [16]	(63 days) to 7 days [16]	7 days [16]	55 days [13]
Man	(6 months) to 3 months [13, 60]	(7 months) to 4 years [20]	6–8 months [21]	2 years [21]

[a]Values in parentheses represent days or months of gestation, the other values the time after birth. Pertinent references are in brackets.

Incorporation of radioactive thymidine [22] and activity of DNA polymerase [23] increase during periods of cell proliferation in a given region, then decrease as cell division ends; thus these factors give a reasonable approximation of mitotic activity.

As DNA accumulation ends, the ratio of RNA to DNA increases. On a cellular level [8], most of the newly synthesized RNA remains in the nucleus until cell division has ended. Upon differentiation, the nucleolus matures, with increased RNA synthesis, especially of ribosomal RNA, and transfer of RNA into the cytoplasm. In neurons, Nissl substance (ribosomal RNA) first becomes prominent at this stage. The rate of RNA synthesis decreases with maturation, together with the activity of RNA polymerase in nuclei of both neurons and glia [24].

Characterization of the RNA in rat brain shows that a larger proportion of RNA synthesis results in messenger RNA (mRNA) synthesis in brain than in liver [25]. In developing brain, RNA characteristics are compatible with the rapid rate of protein synthesis observed. Compared with adult brain, (a) the mRNA turns over more rapidly; (b) more of the ribosomal RNA is membrane bound (polysomal); and (c) these polysomes are more stable [26]. The base ratios of nucleotides in RNA show some changes in development in various regions of brain, but the contributions of individual cell types to these changes are not yet known [27].

The individual ribosomal monomers of rat brain may acquire structural and functional differences by interaction with mRNA and the endoplasmic reticulum [25]. Any factor affecting the relative proportions of free and membrane-bound ribosomes may change the rate of protein synthesis. There may be qualitative differences as well; ribosomes attached to the membrane may synthesize protein for secretion (or axoplasmic flow, in the case of neurons) whereas free ribosomes may synthesize proteins for more local intracellular use.

Regional differences in protein synthesis may be related to the content and degree of aggregation of the ribosomes in various cell types. In large neurons, ribosomes are concentrated near the nucleus, with others scattered throughout the cytoplasm. Some are seen in proximal dendrites, but not in the distal axons. Many of the ribosomes are in monomeric form or in small aggregates. In glial cells and microneurons, the ribosomes appear to be distributed more evenly through the cytoplasm, and a larger proportion is attached to the endoplasmic reticulum than is found in macroneurons.

Proteins

In developing rat brain, protein increases more rapidly than does wet weight. As previously mentioned, the ratio of protein to DNA indicates cell size, and this ratio is elevated during development. The time at which the protein content of brain reaches its maximal level in various species is summarized in Table 18-1. As with most other constituents, proteins increase most rapidly in rat cortex during the first two to three weeks of postnatal life. In this same period, there is a shift from water-soluble to membrane-bound proteins. Changes in both membrane and soluble proteins during development have been detected by acrylamide gel electrophoresis, but the functional significance of these changes remains to be determined [28].

Among the brain-specific proteins appearing during development is one localized in astroglial processes [29, 93]. Three brain-specific proteins appear about the time myelin formation begins: the acidic S-100 protein, which is primarily glial, and the basic protein(s) and proteolipid protein(s) of myelin. The subcellular fraction studied most extensively during development in brain is the myelin membrane (see Chap. 4). Changes in the protein composition of isolated myelin fractions have been demonstrated with increasing proportions of basic and proteolipid proteins relative to proteins of higher molecular weight.

Developmental changes in a number of enzymes have been observed, both histochemically [30] and biochemically [2, 4, 5, 31]. The sequence of appearance of proteins with enzymatic activity determines the metabolic properties of a given cell. (Metabolic changes occurring in brain during development are discussed later.) Isozyme patterns change for several enzymes, including fructose 1,6-diphosphate aldolase and lactic dehydrogenase; the latter changes probably are associated with increasing rates of respiration in maturing brain [27]. Mechanisms of enzyme induction during development are discussed in Chapter 17.

In the rat, incorporation of amino acids into brain proteins is most rapid during the fetal and newborn periods, as measured both in vivo and in vitro [5]. The rate of protein synthesis in brain slices or isolated ribosomal systems decreases sharply at the end of the first week; the incorporation of radioactive valine into protein in brain slices at day 16 is only one-sixth the rate found at day 7. Determination of the rate of protein synthesis in vivo with radioactive precursors is complicated by a number of factors [32] including changes in the available amino acid pools during development (see Chaps. 14 and 16). If correction is made for the specific activity of the labeled amino acid in the brain, there appear to be two peaks of protein synthesis in the weanling rat, at 5 to 7 days and at 19 days [33]. In addition to decreasing rates of protein synthesis with development, the average half-lives of proteins become greater, indicating decreasing rates of breakdown. In parallel, brain proteinases increase in activity just after birth, then decrease [34].

Lipids

The most rapid increase in lipid content of brain begins after the periods of greatest increase of DNA and protein. Changes in specific lipids have been well documented [35, 36]. Changes in fatty acid chain length and degree of saturation have been determined for several groups of lipids [35, 37].

The rapid increase in lipid content is closely related to the onset of myelination. About 50 percent of all lipid in white matter, or 30 percent of total brain lipid, has been estimated to belong to myelin sheaths in rat brain (see Chap. 4).

Comprehensive analysis of lipids in developing rat brain [36] indicates that they may be divided into groups on the basis of their period of most rapid change with respect to myelination. Sterol esters and gangliosides undergo marked shifts before myelination begins; other lipids show marked, moderate, or small increases during myelination. In rat brain, total levels of sterol ester increase from birth to 40 days [38]. Unlike other lipids, sterol ester decreases in concentration during this period, except for a transient increase apparently related to onset of myelination [38].

The gangliosides are 27 percent of their adult concentration on day 3, and increase rapidly to 90 percent by day 24. The pattern of gangliosides (see Chap. 15 for nomenclature) in rat brain also changes [17]. G_{M1} and G_{T1} comprise the major proportion at birth and then fall; G_{D1a} increases to become the predominant ganglioside in the adult brain. Similar changes take place in human cortex, although G_{M1} remains a major ganglioside after an initial drop. Increases in gangliosides and in the activity of the enzyme that synthesizes their precursor, glucocerebroside, are related temporally to the outgrowth of axons and dendrites in agreement with the possible localization of much of the ganglioside in nerve processes.

A second category of lipids is characterized by their low concentration at 3 days (less than 10 percent of the adult level), followed by dramatic increases between 12 and 18 days. This category includes galactocerebroside, cerebroside sulfate, sphingomyelin, and triphosphoinositide. In keeping with their high degree of localization in the myelin membrane is the low concentration of these lipids at birth and their subsequent increase at the time myelination starts. The activity of the enzymes that synthesize galactocerebroside and cerebroside sulfate also parallels myelination, with low levels at birth and marked peaks of activity at about 17 to 20 days [39, 40].

Lipids in a third category occur at 12 to 34 percent of their adult value at birth and increase during myelination, although the increase is less than in those in the second category. Ethanolamine and choline plasmalogens, cholesterol, and phosphatidylserine are in this group. These lipids are associated with nonmyelin membranes in the brain of newborn rat, and the increases seen during the period of myelination are probably due to elaboration of both myelin and nonmyelin membranes.

Three of the major phospholipids — phosphatidylethanolamine, phosphatidylinositol, and phosphatidylcholine — comprise the fourth category. These lipids are 40 to 50 percent of their adult value at 3 days and increase only moderately in concentration during development. They are ubiquitous components of most membrane structures; their temporal pattern of development is not related to any specific morphological change but reflects increasing synthesis of membranes by brain cells.

Maturation is accompanied by an increase in both hydroxy and long-chain saturated fatty acids, due in large part to the abundance of these fatty acids in the lipids of the myelin membrane. Fatty acid aldehydes are found primarily in ethanolamine plasmalogen, which is also enriched in myelin, so fatty acid aldehydes of brain increase with the long-chain saturated fatty acids and the hydroxy fatty acids.

In summary, elevation of the galactolipid and plasmalogen content of brain correlates well with the morphological appearance of myelin, and increases in gangliosides appear to be related to the increasing arborization of neurons. In the rat, the maximal rate of change in ganglioside content takes place at about day 10, and the maximal rate of change in content of galactocerebroside plus sulfatide occurs about day 20 (Fig. 18-4). These times correspond, respectively, to the middle of the periods characterized by the outgrowth of neuronal processes and by the formation of myelin.

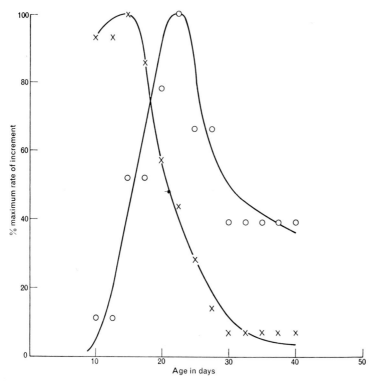

Figure 18-4
Rates of change in gangliosides and in cerebrosides plus sulfatides of the developing rat brain [1].
Results have been calculated from the data of Kishimoto, Davies, and Radin [9], and are expressed
as (O) mg of galactolipid (cerebroside and sulfatide) or as (x) mg of ganglioside stearate per 2.5
days in whole rat brain. Stearate was found to be the major ganglioside fatty acid (83% to 91%
of the total fatty acids) [9]. (From Davison and Dobbing [1]. Reprinted by permission; courtesy of Blackwell Scientific Publications, Ltd.)

Metabolic Changes

Many of the metabolic changes that take place during development can be placed in
one of three categories that emphasize their relation to morphological or functional
events: (a) energy availability and substrate utilization; (b) metabolism of compounds
related to neural transmission; and (c) myelin formation. These areas are covered
elsewhere in this book (Chaps. 1, 4, 9, 10, 11, and 14).

REGULATORY MECHANISMS IN DEVELOPMENT

Perhaps the best example of a substance known to have a direct and specific effect on
nervous system maturation is nerve growth factor (NGF). This protein is required
for survival and maturation of peripheral sensory and sympathetic neurons [41] and
promotes regeneration of axons from lesioned adrenergic neurons in the central nervous system [42]. NGF stimulates fiber outgrowth from embryonic ganglia in

culture. RNA synthesis is not required for this outgrowth, and cyclic AMP does not appear to be involved. A possible role of cyclic GMP is under investigation. Antibodies to NGF have a devastating effect on immature sympathetic neurons in vivo and in culture. NGF has been extensively purified and characterized [41, 43]. Under certain conditions of isolation, the factor consists of three different proteins, designated alpha, beta, and gamma, which interact in specific ways. The beta protein has the biological activity; the roles of alpha and gamma proteins are not clearly defined. Alpha protein is known to possess esterase activity and gamma protein protects ganglion cells during dissociation [92]. The active beta protein is similar to insulin in a number of ways: they share many common amino acid sequences, both have a general anabolic effect, and both apparently act by binding to receptors on the cell surface. The receptors for NGF are different from those for insulin.

During specific critical periods of development, hormones may act directly on the central nervous system to produce permanent changes. The same hormone may have different actions at different times. Most hormones produce changes in the rat brain only when administered in the prenatal or immediate postnatal period [45]. Different regions or cell types may be affected preferentially. For example, testosterone administered to newborn rats apparently acts on the hypothalamus to cause male sexual behavior at maturity; its absence leads to female sexual behavior [45]. See Chapter 35 for additional discussion.

A number of hormones have widespread effects on the developing nervous system; the effects of corticosteroids and thyroid hormone have been the most extensively studied [44, 45]. Administration of either hormone to newborn rats leads to decreased brain weight and cell proliferation (presumably glial cells). Corticosteroids appear to have selective effects on the timing of brain maturation; they reduce dendritic branching transiently and probably decrease myelin formation [46]. The onset of maturation of glucose metabolism and metabolic compartmentation is not affected, but several enzymes can be induced by administration of corticosteroids [45].

The decrease in cell number found after thyroid hormone has been administered may be secondary to an acceleration of maturation which causes cell proliferation to end prematurely. There is an acceleration of dendritic branching, of the onset of myelination, and of biochemical maturation, as indicated by changes in glucose metabolism and metabolic compartmentation. In contrast, hypothyroidism leads to a retardation of brain development. There is a reduction in brain weight, total cell number, the number and size of neurons, density of axons and synapses, and amount of myelin. Further, there is delay in the onset of dendritic branching and myelination. Behavioral changes accompanying these altered hormonal states also have been documented.

One caution in interpreting experiments on, for example, the effects of thyroid deficiency is that any changes observed may be due to nutritional deprivation rather than to a specific effect of the lack of hormone. The effects of undernutrition and other deficiencies on the developing brain are discussed in greater detail in Chapter 29.

Experiments in vivo, in brain slices, and in culture have demonstrated the appearance during development of specific receptors for a number of neurohormones [47]. The action of several of these hormones is apparently mediated through cAMP.

Thus, addition of prostaglandins, norepinephrine, isoproteronol, or 5-hydroxytrypt-amine to brain slices causes rapid accumulation of cAMP. This response appears with development: norepinephrine will not stimulate cAMP accumulation in brain slices from newborn rats, but the response appears at day 4 and is maximal at day 10 [48], apparently due to the appearance of β-adrenergic receptors on the cell surface.

Induction of several enzymes in brain appears to be controlled by specific factors during development. For example, the developmental increase of cytoplasmic glycerol phosphate dehydrogenase depends on the presence of glucocorticoids, both in vivo and in culture, and levels of lactic dehydrogenase can be regulated specifically by catecholamines and cAMP [49]. Indirect evidence and studies in glioblastomas suggest that these responses, and the stimulation of cAMP levels by norepinephrine, may reside primarily in glial cells rather than in neurons [47, 49].

CHANGES WITH AGING

Brain growth in rodents continues slowly long into adulthood, whereas in primates brain weight and other factors show a plateau. Studies on senescent·changes in brain are complicated by these differences in growth patterns, the large individual variation in aging, and the scarcity of samples. Little change in brain weight or water content accompanies aging in rats or mice. In primates, including man, brain weight decreases and water content increases. Species differences have also been noted in changes in DNA content and neuronal number. Little change in either occurs in aging rodent brain [44]. One study in cerebral cortex of rat showed that neuronal and microglial densities were constant between 100 and 700 days, but densities of astroglia and oligodendroglia increased [50]. In primate and human brains, concentration of DNA and numbers of neurons in cerebral cortex decrease significantly with aging [51, 52], with an accompanying increase in glial density [50]; little change is found in other regions, including brain stem.

CELLULAR DEVELOPMENT OF THE NERVOUS SYSTEM

During embryogenesis in mammals, the neural tube gives rise to the precursors of neurons in the central nervous system, whereas the neurons of the peripheral nervous system derive primarily from the neural crest [8, 53]. Epithelial cells that line the neural tube give rise to migrating neuroblasts, some of which then form the multi-polar long-axoned macroneurons of the central nervous system (brain and spinal cord). Another pool of migrating cells gives rise to microneurons (short-axoned modulating interneurons). Migrating neuroblasts from the neural crest develop into the multipolar neurons of the sympathetic and parasympathetic ganglia of the peripheral nervous system. Similarly, bipolar neuroblasts from the neural crest become the unipolar and bipolar neurons of the sensory ganglia of the peripheral nervous system.

In the CNS, the astroglia and oligodendroglia originate in the neural-tube epithelium; the Schwann cells and satellite cells of the PNS arise from both the neural tube and the neural crest. In most brain regions that have been studied, large neurons arise first, followed by small neurons, then glia. Phylogenetically older regions of brain develop earlier [8].

Because the cerebral hemispheres of the rat are so commonly used for neurochemical studies of the developing nervous system, we will emphasize what occurs in that brain region and species.

Neurons

Proliferation and Migration

Neuronogenesis in the rat cerebrum extends from about day 11 or 12 postconception to birth. The cerebral vesicle forms in the rat about day 10 of gestation. Shortly thereafter, the ventricular zone of pseudostratified neuroepithelial cells appears. This zone reaches its maximum thickness at about day 14 of gestation. In this zone, the proliferating neuroepithelial cells first give rise to differentiating neurons and then to differentiating glia. Gliogenesis commences, however, when neuronogenesis is still continuing [14].

Once the primitive neuron (neuroblast) leaves the ventricular zone, it is no longer capable of further division. This is in contrast to glial cells and neurons of neural-crest origin, which can divide after they have migrated away from their original area of proliferation.

The time of origin and subsequent migration of neurons have been mapped by using tritiated thymidine. After it has been injected into the mother, thymidine is available for DNA synthesis for about 60 minutes [22]. Those cells undergoing DNA replication at the time of injection will incorporate [^{3}H] thymidine. Those cells continuing to divide will dilute the label with each subsequent division. The neurons undergoing their final cell division at the time of injection will remain heavily labeled throughout the life of the animal. These techniques have been used to study both normal neuronal formation and abnormal formation in genetically determined and exogenously induced (x-ray, nutritional deprivation, viral infection) defects of the nervous system.

It is not known why neuroblasts lose their ability to divide. The neuroblast may irreversibly lose some of its DNA either by deletion or by permanent repression [54]. Two lines of study indicate that a cytoplasmic factor may be responsible for DNA stability after neurons have differentiated. Nuclei from cells of adult amphibian brain transplanted to oöcytes begin to divide, although at a somewhat slower rate than do nuclei from less mature stages [55]. Fusion of adult neurons with an established fibroblast line results in reactivation of DNA synthesis in the neuronal nuclei [56].

The migration of neurons is a carefully programmed process. The cerebral cortex is formed from the inside out, that is, the early migrating cells form the deeper layers of cortex. The later migrating cells pass through these formed layers to constitute more superficial layers.

In other regions of the nervous system, such as the anterior horn of the spinal cord, this pattern of migration may not take place. Here first-formed cells migrate distally, and later-forming cells position themselves more centrally. In spinal cord, large neurons arise first. In cortex, the smaller neurons of layers VI and IV clearly arise before the large neurons of layer III.

The events underlying neuronal migration are not clear. Some authors suggest that there is only an intracytoplasmic migration of the nucleus, the cell extending from lumen to pia. This migration might be related to the proliferation of microtubules and microfilaments in migrating neurons. Other theories relate to the role of the axon in directing migration or the guidance of migrating neurons by contact with already-formed glial processes.

Maturation

Other than its morphology, the characteristics that distinguish a mature neuron from other cells are its abilities to propagate an action potential and to make specialized contacts, via the synapses, with other neurons.

It is not known precisely when the mature membrane properties of a neuron first appear, but maturation may begin shortly after mitosis or after a long period of growth. Two extreme examples are (1) the motor neurons of chick spinal cord, in which axonal growth begins while the cell is migrating out of the germinal layer, and (2) the retinal ganglion cells, in which axonal outgrowth does not take place until the neuron is in its adult position and has grown in size [8].

On a cellular level, the maturing neuron exhibits (a) decreasing numbers of microtubules and increasing numbers of microfilaments in the cytoplasm; (b) shift from multiple nucleoli to one large nucleolus, accompanied by an increase in nucleolar RNA; (c) decreasing amounts of free ribosomes and increasing amounts of rough endoplasmic reticulum and the smooth membrane-bounded cisternae of the Golgi apparatus; and (d) growth in size [8, 57, 58, 96].

Axonal outgrowth accompanies these changes in the neuronal cell body and involves the amoeboid movement of the growth cone, with new molecules apparently inserted into the membrane near the growing tip. Synapse formation begins when the axon reaches its destination; increased numbers of mitochondria and vesicles appear, followed by thickening of the synapse region [59].

Biochemical maturation of synapses accompanies these morphological changes, as illustrated by the development of noradrenergic endings [60]. Activity of synthetic enzymes, levels of norepinephrine, and a specific high-affinity transport system for norepinephrine all increase markedly during the period of synapse formation, first appearing about a week before birth and increasing to adult levels by three weeks after birth (Fig. 18-5). Norepinephrine and the synthetic enzymes are increasingly localized in synaptic endings. The appearance of the membrane system for uptake of norepinephrine precedes by a day or two the appearance of the synaptic vesicles which store norepinephrine in the endings.

Studies of Neuronal Characteristics

The diversity of neuronal types in cerebrum prohibits definitive study of the biochemical events accompanying their maturation and interaction with other cells. Studies in simpler systems have allowed more detailed correlation of morphological

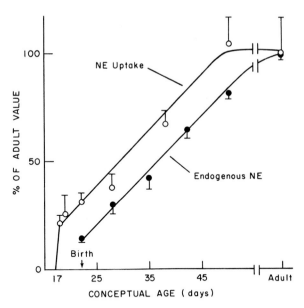

Figure 18-5
Correlation between the development of the high-affinity uptake of norepinephrine and the level of endogenous norepinephrine in rat brain at various stages of development [60]. Values are expressed as percentages of the adult value. Initial uptake is expressed in terms of the percentages of the adult V_{max} of transport. Adult norepinephrine: 0.47 ± 0.01 μg per g. Adult V_{max} of uptake: 2.216 ± 0.423 nmol per g in 5 min. The upper curve represents uptake of norepinephrine by whole-brain homogenates (open circles); the lower curve the content of norepinephrine in whole brain (closed circles). (From Coyle and Axelrod [60]. Reprinted by permission; courtesy of Pergamon Press.)

and biochemical changes in developing neurons. Nervous tissue in culture has been used successfully to study neuronal maturation [61].

1. *Cell Lines in Culture.* Clonal cell lines derived from the mouse neuroblastoma C 1300 or from chemically induced brain tumors exhibit various combinations of neuronal properties, including electrically excitable membranes, high levels of enzymes involved in transmitter metabolism, receptors for transmitters, and the neuronal specific protein 14-3-2 [62–64]. These cell lines offer the potential of dissecting the control and interaction of the various genes that give rise to neuron-specific properties [63].

When cultured neuroblastoma cells attach to a solid substratum, these neuronal characteristics may become more pronounced, and the cells extend "neurites," long processes initially containing mitochondria, ribosomes, and dense core vesicles and, later, microtubules and filaments [65]. This differentiation can be stimulated by a variety of conditions such as decreasing the serum content in the growth medium or exposing cultures to dibutyryl cAMP, acetyl choline, prostaglandins, or bromodeoxyuridine. Increased glycosylation of one protein and exposure of another protein on the cell surface accompany this differentiation [66].

A number of individual clonal lines, both neuronal and glial in origin, have been derived from rat brain by chemical induction of tumors [64]. In the neuronal lines, low serum content or the presence of dibutyryl cAMP in the culture medium induces process extension, and bromodeoxyuridine causes only cell flattening. Protein synthesis is not required for process extension in either the neuroblastoma C 1300 or induced tumor lines. Colchicine, an alkaloid which dissociates microtubule subunits, causes very rapid retraction of the processes in both systems.

Low serum concentrations and addition of substances of low molecular weight may induce differentiation, in part, by enhancing interaction between the cell surface and the culture dish. Some conditions that induce neurite formation also stop cell division, but several lines of evidence indicate there is no causal relationship between these two events [64].

2. *Reaggregating Systems.* Cells dissociated from fetal or developing brain and maintained in rotating flasks aggregate in specific and highly organized patterns [47, 67], and formation of both synapses and myelin has been observed. Abnormal patterns of reaggregation occur with cells dissociated from the brain of the Reeler mutant mouse, a mutant characterized by defective orientation and layering of neurons in both cerebral and cerebellar cortex [68].

Under appropriate conditions, some components exhibit biochemical differentiation similar to that in developing brain [47]. For example, three neuronal-enriched enzymes — choline acetyltransferase, acetylcholinesterase, and glutamate decarboxylase — show a pattern of increase similar to that seen in vivo. Receptors for norepinephrine appear in culture, as in vivo, and a shift toward more oxidative metabolism also occurs.

3. *PNS Ganglia.* Superior cervical ganglia (sympathetic) and dorsal root ganglia (sensory) are useful systems for examining neuronal maturation both in vivo and in culture. Morphological and biochemical events during maturation are well documented [69–71], and the effects of preganglionic and postganglionic manipulation on neuronal maturation and maintenance can be observed readily in a relatively homogeneous system. For example, in superior cervical ganglia of mouse or rat, formation of preganglionic cholinergic synapses correlates well with a fiftyfold increase in activity of choline acetyltransferase in the ganglia between birth and 30 days. Maturation of the adrenergic cell body is accompanied by increases in tyrosine hydroxylase and dopamine-β-hydroxylase. Preganglionic blockade prevents the increase in tyrosine hydroxylase activity and morphological maturation of the neurons; administration of nerve growth factor at the proper time stimulates maturation [70, 71]. Administration of antiserum to NGF to newborn mice results in destruction of undifferentiated ganglionic neurons, with the earliest lesions occurring in the nucleus. Administration of 6-hydroxydopamine also destroys these young neurons, but apparently by a different mechanism, as the earliest changes are cytoplasmic and NGF cannot reverse the effects. In the adult animal, the differentiated neurons show less severe reversible changes with both insults. In the adults as well, antiserum to NGF produces nuclear changes and some decrease in cell volume, while 6-hydroxydopamine acts primarily at the adrenergic terminals [72].

4. *Retinal Neurons: Recognition and Locus Specificity.* Development of specific synaptic connections has been studied in greatest detail in the visual system of frogs and chicks. Each retinal ganglion cell sends out an axon to a specific location in the optic tectum, so that "the spatial pattern of individual retinal axon terminals in the tectum reduplicates the spatial relationship of the corresponding retinal ganglion cells" [73]. With electrophysiological techniques, Jacobson [73] has mapped the relationship of retinal cells to tectal cells in embryonic frog brain and has found that specification of the gradients that determine the connections takes place after DNA synthesis has ceased and precedes fiber outgrowth in retinal neurons. Some preliminary evidence for recognition and interaction between retinal and tectal cells has come from experiments with dissociated and reaggregating tissue [74]; surface proteins or glycoproteins have been implicated in these interactions. Refined electrophysiological and culture techniques promise to define these interactions in greater detail.

Neuronal Responses in Regeneration and Aging

Once differentiated, neurons cannot divide, so their response to injury may be either death or degeneration of damaged processes, followed by regeneration [75, 96]. If a population of neurons dies or is removed, other neurons may be able to reinnervate the same field if the injury occurs early enough in development [8, 96]. Cell death occurs normally during development of the nervous system. Experimentally, removing the field into which a group of neurons is sending processes will lead to the death of that group, suggesting a trophic function of the target tissue [8].

With increasing age, multiple morphological changes are seen. Many of these changes have been observed in pathological situations in less mature brains, as well. The most prominent neuronal changes are appearance of senile plaques (areas of degenerating neuronal processes, reactive nonneuronal cells, and amyloid), increasing deposits of lipofuscin, and areas of neurofibrillary tangles (twisted tubules) [76]. The deposition of lipofuscin may be related to increased oxidation of unsaturated fatty acids; whether its accumulation leads to cell death or decreased function is not known [52].

Axon severance leads to injury or death of the neuronal perikarya. The sensitivity of the neuron decreases with age and with the distance of the cut from the cell body [77]. Axotomy provides a model system for observing degenerative and regenerative changes in the neuron; the events in regeneration recapitulate to some extent those occurring during development. Axotomy in very young neurons causes nucleolar loss but no chromatolysis; in mature neurons, chromatolysis occurs without nucleolar loss.

Glia

Proliferation, Migration, and Turnover of Oligodendroglia and Astrocytes
Macroglia, which are composed of oligodendrocytes and astrocytes, have a neuroectodermal origin. Their initial site of proliferation is the ventricular zone, but once active cell division ceases in this zone, the glioblasts migrate into the surrounding nervous tissue and continue to divide in situ [8]. It is generally accepted that formation of

neurons and neuroglia in a specific region of the CNS may overlap, but this takes place only toward the end of neuronal proliferation. The appearance of astrocytes usually precedes that of oligodendrocytes, with both cell types forming concurrently at later developmental stages [78].

Little is known about the cells responsible for the formation of astrocytes and oligodendrocytes. There are essentially two possibilities. First, there may be a multipotential glial cell [79] that can form both astrocytes and oligodendrocytes. Second, there may be two distinct cell lines, one astrocytic and one oligodendrocytic, that are determined genetically at very early developmental stages.

It is uncertain whether there is a turnover of oligodendroglia in mature CNS. Electron-microscopic quantitative estimates of neuroglia in cerebral cortex at different ages show no increase in the number of oligodendrocytes in aging rats [80]. It is possible, however, that some oligodendrocytes die and are replaced by new cells.

In developing optic nerve and corpus callosum [81, 82] and in optic nerve undergoing Wallerian degeneration, electron-microscopic autoradiographic studies indicate that astrocytes and oligodendrocytes that have differentiated to some degree can still incorporate tritiated thymidine. Immature oligodendroglia that have initiated myelin formation and fibrous astrocytes will still incorporate [^{3}H]-thymidine. These data suggest that the long-held tenet that cell division and differentiation are two separate stages of development may not be universally correct.

Maturation of Oligodendroglia

Appearance of oligodendroglia and, subsequently, myelin generally proceeds in a caudocranial direction, with spinal cord first and frontal cortex last. In the corpus callosum and the pyramidal tract of the rat, oligodendroglial proliferation, or "myelination gliosis," begins shortly before birth and reaches a maximum about 4 days after birth [83]. As the rate of cell proliferation decreases, RNA content and protein synthesis increase, and lipid droplets appear in the oligodendroglia. Myelin sheaths first appear in these regions at days 10 to 12. Some cell proliferation continues, though at a lower rate, during the period of myelination.

Interaction Between Axons and Myelin-Forming Cells

In the central nervous system, oligodendroglia migrate toward the unmyelinated axon. Presumably there is some form of chemotaxis between the immature axon and the oligodendrocyte. Little is known about the initiation of myelination other than that processes of oligodendroglia myelinate axons of a larger diameter first. There have been a number of speculations about the interrelationship between axons and myelin formation, including the suggestion that precursors or formed compounds can be transported from the axon to the forming myelin sheath.

In the peripheral nervous system, Schwann cells myelinate axons. The cell of origin is in the neural crest, and the maturing cell migrates to the axon.

The elegant model of Wood and Bunge [84] has been used to analyze the relationship between axons and developing Schwann cells. By using explants of dorsal root ganglia, it has been shown that contact of naked neurites with quiescent

Schwann cells stimulates the mitogenesis of Schwann cells. This system may provide the mechanism for the direct study of interaction between Schwann cells and axons and, if the model can be adapted to explants of central nervous system, of interaction between oligodendroglia and axons.

Myelination

The end point of differentiation of the oligodendrocyte is the formation of myelin. The biochemical events involved in myelin formation are outlined in Chapter 4.

Oligodendroglia may be separated from other cell types in brain [85], but separation of actively myelinating from myelin-maintaining oligodendroglia is not yet possible. Isolated mature oligodendroglia have a composition different from that of astrocytes or neurons isolated under similar conditions. The oligodendroglia fractions are enriched in myelin components [86]. It is possible to develop the concept that, during myelination, the plasma membrane of the oligodendrocyte changes, and components are added. The composition of isolated oligodendroglial plasma membranes supports this hypothesis [87].

Oligodendroglial Responses in Regeneration

The regenerative response of oligodendroglia is remyelination, with or without preceding proliferation of oligodendroglia. There have been a number of experimental models in which demyelination was followed by remyelination. These have included, in the cat, removal and replacement of cerebrospinal fluid, transient compression of the spinal cord, and cyanide intoxication; in the rabbit, experimental allergic encephalomyelitis; and, in mice, intoxication with cuprizone [88].

In these studies, the criteria of remyelination have been morphological and include altered ratios of myelin-sheath thickness to axon diameter and reduction of internodal length.

Recently a viral mode of demyelination and subsequent remyelination in the mouse has been described [89]. Remyelination occurred, along with formation of new oligodendroglia, as judged by incorporation of [3H] thymidine and increased incorporation of [35S] sulfate into sulfatide.

Whether remyelination takes place in the human central nervous system is less certain. Around the edges of a demyelinative plaque, the so-called shadow plaque in multiple sclerosis, axons with short internodes of myelin are seen. It is not clear if this represents remyelination or partial demyelination.

Role of Astrocytes in Development

Investigators have suggested a role for "glia" in a number of developmental systems. Often it is not clear what cell types are being described by this term. Presumably the cells are not oligodendroglia, but mixtures of primitive stem cells and astrocytes.

As previously mentioned, it has been suggested that glial processes form a trellis along which neurons migrate during development. It is not clear what cells would provide this trellis. Rakic and Sidman [90] have suggested that Bergmann astrocytes might play such a role for granular cells migrating from the external granular cell zone to the internal granular cell zone of the cerebellum and that abnormalities of fiber outgrowth would lead to abnormalities of neuronal migration.

Other possible roles for "glia," which could be important during development, are regulation of extracellular concentrations of sodium and potassium, inactivation of transmitters (Chap. 8), and secretion of substances similar to nerve growth factor.

Such hypotheses are difficult to substantiate in the intact nervous system. In tissue-culture systems, however, neuronal-glial interactions can be evaluated. Neuroblasts from chicken cerebrum attach quickly to a monolayer of astroblasts (primitive glial cells from a 13-day chick embryo) and differentiate more rapidly than do those cultured on plastic or on layers of fibroblasts or meningeal cells [91]. In the peripheral nervous system, a similar phenomenon has been observed, in that dissociated neurons from newborn mouse dorsal root ganglia will attach and grow on a layer of nonneuronal cells from the ganglion. Further, those neurons that have attached to ganglionic nonneuronal cells have a decreased requirement for nerve growth factor. It has been suggested that, in this system, the growth-promoting effects of nonneuronal ganglionic cells and NGF have a common mechanism [92].

Astrocytes contain a specific protein, glial fibrillary acidic protein (GFA) [93]. The protein, or a closely related group of proteins, is intracellular and probably is related to glial filaments within fibrous processes. Immunofluorescence, using antisera to GFA, has been used to detect astrocytic development in cerebellum, cerebral cortex, and spinal cord.

Because glial processes are readily detected, antisera to GFA have been useful in studying the development of Bergmann astrocytes in mouse mutants with cerebellar abnormalities [94]. In all instances, fibers from Bergmann glia appear at the normal time in homozygously affected animals. The subsequent development of fibers, however, is quite different in the particular mutants.

Responses of Astrocytes

It is difficult to comment about changes in astrocytes specific for age because these cells are so sensitive to various stimuli. For example, when one uses the term *gliosis* in a pathological sense, one is referring to the proliferation of astrocytes and microglia. Such responses are seen in a variety of situations such as trauma, infection, or anoxia. The factors that induce astrocytic proliferation are unknown.

ACKNOWLEDGMENTS

This report represents activities supported by funds from The John A. Hartford Foundation, Inc., the National Institute of Neurological and Communicative Disorders and Stroke, and the National Multiple Sclerosis Society.

The authors thank Dr. Robert P. Skoff and Dr. Mark E. Molliver for their assistance in reviewing our comments about morphological aspects of development.

REFERENCES

*1. Davison, A. N., and Dobbing, J. (Eds.). *Applied Neurochemistry*. Philadelphia: Davis, 1968.

*2. Ford, D. E. (Ed.). *Neurobiological Aspects of Maturation and Aging*. Vol. 40. Progress in Brain Research series. New York: Elsevier, 1973.

*3. Himwich, W. A. (Ed.). *Developmental Neurobiology*. Springfield, Ill.: Thomas, 1970.

*4. Himwich, W. A. (Ed.). *Biochemistry of the Developing Brain*. Vols. 1 and 2. New York: Dekker, 1973, 1974.

*5. McIlwain, H., and Bachelard, S. Chemical and Enzymic Make-up of the Brain During Development. In McIlwain, H., and Bachelard, S. (Eds.), *Biochemistry and the Central Nervous System*. London: Churchill Livingstone, 1971, p. 406.

*6. Waelsch, H. (Ed.). *Biochemistry of the Developing Nervous System*. New York: Academic, 1955.

*7. Himwich, H. E. Early Studies of the Developing Brain. In Himwich, W. A. (Ed.), *Biochemistry of the Developing Brain*. Vol. I. New York: Dekker, 1973, p. 1.

*8. Jacobson, M. *Developmental Neurobiology*. New York: Holt, Rinehart and Winston, 1970.

9. Kishimoto, Y., Davies, W. E., and Radin, N. S. Developing rat brain: Changes in cholesterol, galactolipids, and the individual fatty acids of gangliosides and glycerophosphatides. *J. Lipid Res.* 6:532, 1965.

10. Fish, I., and Winick, M. Effect of malnutrition on regional growth of the developing rat brain. *Exp. Neurol.* 25:534, 1969.

*11. Mandel, P., Rein, H., Harth-Edel, S., and Mardell, R. Distribution and Metabolism of Ribonucleic Acid in the Vertebrate Central Nervous System. In Richter, D. (Ed.), *Comparative Neurochemistry*. New York: Macmillan, 1964, p. 149.

12. Himwich, W. A. Appendix. In Lajtha, A. (Ed.), *Handbook of Neurochemistry*. Vol. 1. New York: Plenum, 1969, p. 469.

*13. Himwich, W. A. Biochemistry and Neurophysiology of the Brain in the Neonatal Period. In Pfeiffer, C. C., and Smythies, J. R. (Eds.), *International Review of Neurobiology*. Vol. 4. New York: Academic, 1962, p. 117.

14. Eayrs, J. R., and Goodhead, B. Postnatal development of the cerebral cortex in the rat. *J. Anat.* 93:385, 1959.

15. Kobayashi, T., Inman, O. R., Buno, W., and Himwich, H. E. Neurohistological studies of developing mouse brain. *Prog. Brain Res.* 9:87, 1954.

16. Dobbing, J., and Sands, J. Growth and development of the brain and spinal cord of the guinea pig. *Brain Res.* 17:115, 1970.

17. Vanier, M. T., Holm, M., Ohman, R., and Svennerholm, L. Developmental profiles of gangliosides in human and rat brain. *J. Neurochem.* 18:581, 1971.

18. Jacobson, M. Sequence of myelinization in the brain of the albino rat. A. Cerebral cortex, thalamus and related structures. *J. Comp. Neurol.* 121:5, 1963.

19. Uzman, L. L., and Rumley, M. K. Changes in the composition of developing mouse brain during early myelination. *J. Neurochem.* 3:171, 1958.

20. Yakovlev, P. I., and Lecours, A. R. The Myelogenetic Cycles of Regional Maturation of the Brain. In Minkowski, A. (Ed.), *Regional Development of the Brain*. Oxford: Blackwell, 1967, p. 3.

21. Winick, M., Rosso, P., and Waterlow, J. Cellular growth of the cerebrum, cerebellum and brain stem in normal and marasmic children. *Exp. Neurol.* 26:393, 1970.

*Asterisks denote key references.

22. Altman, J. DNA Metabolism and Cell Proliferation. In Lajtha, A. (Ed.), *Handbook of Neurochemistry.* Vol. 2. New York: Plenum, 1969, p. 133.
23. Brasel, J. A., Ehrenkranz, R. A., and Winick, M. DNA polymerase activity in rat brain during ontogeny. *Dev. Biol.* 23:424, 1970.
24. Guiffrida, A. M., Cox, D., and Mathias, A. P. RNA polymerase activity in various classes of nuclei from different regions of rat brain during development. *J. Neurochem.* 24:749, 1975.
25. Murthy, M. R. V. Membrane-Bound and Free Ribosomes in the Developing Rat Brain. In Lajtha, A. (Ed.), *Protein Metabolism of the Nervous System.* New York: Plenum, 1970, p. 109.
26. Rappoport, D. A., Fritz, R. R., and Myers, J. L. Nucleic Acids. In Lajtha, A. (Ed.), *Handbook of Neurochemistry.* Vol. 1. New York: Plenum, 1969, p. 101.
27. Ford, D. H. Selected Maturational Changes Observed in the Postnatal Rat Brain. In Ford, D. E. (Ed.), *Neurobiological Aspects of Maturation and Aging.* Vol. 40. Progress in Brain Research Series. New York: Elsevier, 1973, p. 1.
28. Shooter, E. M. Some Aspects of Gene Expression in the Nervous System. In Schmitt, F. O. (Ed.), *The Neurosciences: Second Study Program.* New York: Rockefeller University Press, 1970, p. 812.
29. Bignami, A., and Dahl, D. Astrocyte-specific protein and neuroglial differentiation. An immunofluorescence study with antibodies to the glial fibrillary acidic protein. *J. Comp. Neurol.* 153:27, 1974.
*30. Adams, C. W. M. *Neurohistochemistry.* New York: Elsevier, 1965.
*31. Van den Berg, D. J. Enzymes in Developing Brain. In Himwich, W. A. (Ed.), *Biochemistry of the Developing Brain.* Vol. 2. New York: Dekker, 1973, p. 149.
32. Oja, S. S. Comments on the measurement of protein synthesis in the brain. *Int. J. Neurosci.* 5:31, 1973.
33. Miller, S. A. Protein Metabolism During Growth and Development. In Munro, H. N. (Ed.), *Mammalian Protein Synthesis.* Vol. 3. New York, Academic, 1969.
34. Marks, M., and Lajtha, A. Developmental Changes in Peptide-Bond Hydrolases. In Lajtha, A. (Ed.), *Protein Metabolism of the Nervous System.* New York: Plenum, 1970, p. 39.
*35. Rouser, G., and Yamamoto, A. Lipids. In Lajtha, A. (Ed.), *Handbook of Neurochemistry.* Vol. 1. New York: Plenum, 1969, p. 121.
36. Wells, M. A., and Dittmer, J. C. A comprehensive study of the postnatal changes in the concentration of the lipids of developing rat brain. *Biochemistry* 6:3169, 1967.
*37. O'Brien, J. S. Lipids and Myelination. In Himwich, W. A. (Ed.), *Developmental Neurobiology.* Springfield, Ill.: Thomas, 1970, p. 262.
38. Eto, Y., and Suzuki, K. Cholesterol esters in developing rat brain: Concentration and fatty acid composition. *J. Neurochem.* 19:109, 1972.
39. Costantino-Ceccarini, E., and Morrell, P. Quaking mouse: In vitro studies on brain sphingolipid biosynthesis. *Brain Res.* 29:75, 1971.
40. McKhann, G. M., and Ho, W. The in vivo and in vitro synthesis of sulfatides during development. *J. Neurochem.* 14:717, 1967.
*41. Angeletti, P. U., Angeletti, R. H., Frazier, W. A., and Bradshaw, R. A. Nerve Growth Factor. In Schneider, D. J. (Ed.), *Proteins of the Nervous System.* New York: Raven, 1973, p. 133.
42. Bjerre, B., Bjorklund, A., and Stenevi, U. Stimulation of growth of new axonal sprouts of some lesioned monoamine neurons in adult rat brain by nerve growth factor. *Brain Res.* 60:161, 1973.

*43. Herrup, K., Stickgold, R., and Shooter, E. M. The role of the nerve growth factor in the development of sensory and sympathetic ganglia. *Ann. N.Y. Acad. Sci.* 228:381, 1974.

44. Howard, E. Hormonal Effects on the Growth and DNA Content of the Developing Brain. In Himwich, W. A. (Ed.), *Biochemistry of the Developing Brain.* Vol. 2. New York: Dekker, 1974, p. 1.

45. Balazs, R., and Richter, D. Effects of Hormones on the Biochemical Maturation of the Brain. In Himwich, W. A. (Ed.), *Biochemistry of the Developing Brain.* Vol. I. New York: Dekker, 1973, p. 253.

46. Howard, E., and Benjamins, J. A. DNA, ganglioside and sulfatide in brains of rats given corticosterone in infancy, with an estimate of cell loss during development. *Brain Res.* 92:73, 1975.

47. Seeds, N. W. Differentiation of Aggregating Brain Cell Cultures. In Sato, G. (Ed.), *Tissue Culture of the Nervous System.* New York: Plenum, 1973, p. 35.

48. Schmidt, M. H., Palmer, E. C., Dettbarn, W. D., and Robison, G. A. Cyclic AMP and adenyl cyclase in the developing rat brain. *Dev. Psychobiol.* 3:53, 1970.

49. DeVellis, J., and Brooker, G. Induction of Enzymes by Glucocorticoids and Catecholamines in a Rat Glial Cell Line. In Sato, G. (Ed.), *Tissue Culture of the Nervous System.* New York: Plenum, 1973, p. 231.

50. Brizzee, K. R. Quantitative Histological Studies on Aging Changes in Cerebral Cortex of Rhesus Monkeys and Albino Rats with Notes on Effects of Prolonged Low-Dose Ionizing Irradiation in the Rat. In Ford, D. E. (Ed.), *Neurobiological Aspects of Maturation and Aging.* Vol. 40. Progress in Brain Research series. New York: Elsevier, 1973, p. 141.

51. Samorajski, T., and Rolsten, C. Age and Regional Differences in the Chemical Composition of Brains of Mice, Monkeys, and Humans. In Ford, D. E. (Ed.), *Neurobiological Aspects of Maturation and Aging.* Vol. 40. Progress in Brain Research series. New York: Elsevier, 1973, p. 253.

52. Brody, H. Aging of the Vertebrate Brain. In Rochstein, M. (Ed.), *Development and Aging in the Nervous System.* New York: Academic, 1973, p. 121.

53. Crelin, E. S. Development of the Nervous System. *Ciba Clin. Symp.* Vol. 26, No. 2, 1974.

*54. Ebert, J. D., and Kaighn, M. E. The Keys to Change: Factors Regulating Differentiation. In Locke, M. (Ed.), *Major Problems in Developmental Biology.* New York: Academic, 1966, p. 29.

55. Gurdon, J. B. Transplanted nuclei and cell differentiation. *Sci. Am.* 161:24, 1968; Changes in somatic cell nuclei inserted into growing and maturing amphibian oöcytes. *J. Embryol. Exp. Morphol.* 20:401, 1968.

56. Jacobson, C.-O. Reactivation of DNA synthesis in mammalian neuron nuclei after fusion with cells of an undifferentiated fibroblast line. *Exp. Cell Res.* 53:316, 1968.

57. LaVelle, A., and LaVelle, F. W. Cytodifferentiation in the Neuron. In Himwich, W. A. (Ed.), *Developmental Neurobiology.* Springfield, Ill.: Thomas, 1970, p. 117.

58. Caley, D. W., and Maxwell, D. S. An electronmicroscopic study of the neuroglia during postnatal development of the rat cerebrum. *J. Comp. Neurol.* 133:17, 1968.

*59. Cotman, C. W., and Banker, G. A. The Making of a Synapse. In Ehrenpreis, S., and Kopin, I. (Eds.), *Reviews of Neuroscience.* Vol. I. New York: Raven, 1974, p. 1.

60. Coyle, J. T., and Axelrod, J. Development of the uptake and storage of L-[^{3}H] norepinephrine in the rat brain. *J. Neurochem.* 18:2061, 1971.

*61. Sato, G. (Ed.). *Tissue Culture of the Nervous System.* New York: Plenum, 1973.

62. Amano, T., Richelson, E., and Nirenberg, M. Neurotransmitter synthesis by neuroblastoma clones. *Proc. Natl. Acad. Sci. U.S.A.* 61:258, 1972.

63. Minna, J. D. Genetic Analysis of the Mammalian Nervous System Using Somatic Cell Culture Techniques. In Sato, G. (Ed.), *Tissue Culture of the Nervous System.* New York: Plenum, 1973, p. 161.

64. Schubert, D. Induced differentiation of clonal rat nerve and glial cells. *Neurobiology* 4:376, 1974.

65. Schubert, D., Harris, A. J., Heinemann, D., Kidkoro, Y., Patrick, J., and Steinbach, J. H. Differentiation and Interaction of Clonal Cell Lines of Nerve and Muscle. In Sato, G. (Ed.), *Tissue Culture of the Nervous System.* New York: Plenum, 1973, p. 55.

66. Truding, R., Shelanski, M. L., Davids, M. P., and Morrell, P. Comparison of surface membranes isolated from cultures of murine neuroblastoma cells in the differentiated or undifferentiated state. *J. Biol. Chem.* 249:3973, 1974.

67. Garber, B. G., and Moscona, A. A. I. Aggregation patterns of cells dissociated from different regions of the developing brain. *Dev. Biol.* 27:217, 1972.

68. DeLong, G. R., and Sidman, R. Alignment defect of reaggregating cells in cultures of developing brains of Reeler mutant mice. *Dev. Biol.* 22:584, 1970.

69. Eranko, L. Ultrastructure of the developing nerve cell and the storage of catecholamines. *Brain Res.* 46:159, 1972.

70. Black, I. B., Joh, T. H., and Reis, D. J. Accumulation of tyrosine hydroxylase molecules during growth and development of the superior cervical ganglion. *Brain Res.* 75:133, 1974.

71. Thoenen, H., Kettler, R., and Sauer, A. Time course of development of enzymes involved in the synthesis of norepinephrine in the superior cervical ganglion of the rat from birth to adult life. *Brain Res.* 40:459, 1972.

72. Angeletti, P. V., Levi-Montalcini, R., and Caramia, F. Analysis of the effects of the antiserum to the nerve growth factor in adult mice. *Brain Res.* 27:343, 1971.

*73. Jacobson, M. Genesis of Neuronal Locus Specificity. In Rochstein, M. (Ed.), *Development and Aging in the Nervous System.* New York: Academic, 1973, p. 105.

74. Barbera, A. J., Marchase, R. B., and Roth, F. Adhesive recognition and retinotectal specificity. *Proc. Natl. Acad. Sci. U.S.A.* 70:2482, 1973.

*75. Ramon y Cajal, S. *Degeneration and Regeneration in the Nervous System.* London: Oxford University Press, 1928.

76. Wisniewski, H. M., and Terry, R. D. Morphology of the Aging Brain, Human and Animal. In Ford, D. H. (Ed.), *Neurobiological Aspects of Maturation and Aging.* Vol. 40. Progress in Brain Research series. New York: Elsevier, 1973, p. 167

77. Lavelle, A. Levels of Maturation and Reactions to Injury During Neuronal Development. In Ford, D. H. (Ed.), *Neurobiological Aspects of Maturation and Aging.* Vol. 40. Progress in Brain Research series. New York: Elsevier, 1973, p. 161.

78. Vaughn, J. E. An electron microscopic analysis of gliogenesis in rat optic nerve. *Z. Zellforsch.* 94:293, 1969.

79. Vaughn, J. E., Hinds, P. L., and Skoff, R. P. Electron microscopic studies of Wallerian degeneration in rat optic nerve. I. The multipotential glia. *J. Comp. Neurol.* 140:175, 1970.

80. Vaughan, D. W., and Peters, A. Neuroglia cells in the cerebral cortex of rats from young adulthood to old age: An electron microscopic study. *J. Neurocytol.* 3:405, 1974.

81. Mori, S., and Leblond, C. P. Electron microscopic features and proliferation of astrocytes in the corpus callosum of the rat. *J. Comp. Neurol.* 137:197, 1969.

82. Skoff, R. P. The fine structure of pulse labelled (^{3}H thymidine cells) in degenerating rat optic nerve. *J. Comp. Neurol.* 161:595, 1975.

83. Schonbach, J., Hu, K. H., and Friede, R. L. Cellular and chemical changes during myelination: Histologic, autoradiographic, histochemical and biochemical data on myelination in the pyramidal tract and corpus callosum of rat. *J. Comp. Neurol.* 134:21, 1968.

84. Wood, P. M., and Bunge, R. P. Evidence that sensory axons are mitogenic for Schwann cells. *Nature.* 256:662, 1975.

85. Poduslo, S. E., and Norton, W. T. The Isolation of Specific Brain Cells. In Colowick, S. P., and Kaplan, N. O. (Eds.), *Methods in Enzymology.* Vol. 35. New York: Academic, 1975, p. 561.

86. Poduslo, S. E., and Norton, W. T. Isolation and some chemical properties of oligodendroglia from calf brain. *J. Neurochem.* 19:727, 1972.

87. Poduslo, S. E. The isolation and characterization of a plasma membrane and a myelin fraction derived from oligodendroglia of calf brain. *J. Neurochem.* 24:647, 1975.

88. McDonald, W. I. Remyelination in relation to clinical lesions of the central nervous system. *Br. Med. Bull.* 30:186, 1974.

89. Herndon, R. M., Griffin, D. E., McCormick, U., and Weiner, L. P. Mouse hepatitis virus-induced recurrent demyelination. *Arch. Neurol.* 32:32, 1975.

90. Rakic, P., and Sidman, R. L. Sequence of developmental abnormalities leading to granule cell deficit in cerebellar cortex of Weaver mutant mice. *J. Comp. Neurol.* 152:103, 1973.

91. Sensenbrenner, M., and Mandel, P. Behavior of neuroblasts in the presence of glial cells, fibroblasts, and meningeal cells in culture. *Exp. Cell Res.* 87:159, 1974.

92. Varon, S. Nerve growth factor and its mode of action. *Exp. Neurol.* 48 (3) Part 2:75, 1975.

93. Eng, L. F., and Kosek, J. C. Light and electron microscopic localization of the glial fibrillary acidic protein and S-100 protein by immunoenzymatic techniques. *Trans. Am. Soc. Neurochem.* 5:160, 1974.

94. Bignami, A., and Dahl, D. The development of Bergmann glia in mutant mice with cerebellar malformations: Reeler, Staggerer and Weaver. Immunofluorescence study with antibodies to the glial fibrillary acidic protein. *J. Comp. Neurol.* 155:219, 1974.

95. Dobbing, J. The Later Development of The Brain and Its Vulnerability. In Davis, J. A., and Dobbing, J. (Eds.), *Scientific Foundations of Paediatrics.* Philadelphia: Saunders, 1974, p. 565.

*96. Growth and Regeneration in The Nervous System. A Report by the Ad Hoc Subcommittee of the Advisory Council of the NINCDS of the NIH. *Exp. Neurol.* 48 (3) Part 2:1, 1975.

Chapter 19

Circulation and Energy Metabolism of the Brain

Louis Sokoloff

The biochemical pathways of energy metabolism in the brain are, in most respects, similar to those of other tissues (see Chap. 14). Special conditions that are peculiar to the central nervous system in vivo limit quantitative expression of its biochemical potentialities. In no tissue are the discrepancies between in situ and in vitro results greater, or the extrapolations from in vitro data to conclusions about in vivo metabolic functions more hazardous. Valid identification of the normally used substrates and the products of cerebral energy metabolism, as well as reliable estimation of their rates of utilization and production, can be obtained only in the intact animal; in vitro studies serve to identify pathways of intermediary metabolism, mechanisms, and potential rather than actual performance.

In addition to the usual causes of differences between in vitro and in vivo studies that pertain to all tissues, there are two unique conditions that pertain only to the central nervous system.

First, in contrast to cells of other tissues, nerve cells do not function autonomously. They are generally so incorporated into a complex neural network that their functional activity is integrated with that of various other parts of the central nervous system and, indeed, with most somatic tissues as well. It is obvious, then, that any procedure which interrupts the structural and functional integrity of the network or isolates the tissue from its normal functional interrelationships would inevitably grossly alter — at least quantitatively and, perhaps, even qualitatively — its normal metabolic behavior.

The second is the phenomenon of the blood-brain barrier (described in Chap. 20), which selectively limits the rates of transfer of soluble substances between blood and the brain. This barrier, which probably developed phylogenetically to protect the brain against noxious substances, serves also to discriminate among various potential substrates for cerebral metabolism. The substrate function is confined to those compounds in the blood which not only are suitable substrates for cerebral enzymes but which also can penetrate from blood to the brain at rates adequate to support the brain's considerable energy demands. Substances which can be readily oxidized by brain slices, minces, or homogenates in vitro and which are effectively utilized in vivo when formed endogenously within the brain are often incapable of supporting cerebral energy metabolism and function when present in the blood because of restricted passage through the blood-brain barrier. The in vitro techniques establish only the existence and potential capacity of the enzyme systems required for the utilization of a given substrate; they do not define the extent to which such a pathway is actually utilized in vivo. This can be done only by studies in the intact animal, and it is this aspect of cerebral metabolism with which this chapter is concerned.

STUDIES OF CEREBRAL METABOLISM IN VIVO

A variety of suitable methods have been used to study the metabolism of the brain in vivo; these vary in complexity and in the degree to which they yield quantitative results. Some require such minimal operative procedures on the experimental animal that no anesthesia is required, and there is no interference with the tissue except for the effects of the particular experimental condition being studied. Some of these techniques are applicable to normal, conscious human subjects, and consecutive and comparative studies can be made repeatedly in the same subject. Other methods are more traumatic and either require killing the animal or involve such extensive surgical intervention and tissue damage that the experiments approach an in vitro experiment carried out in situ. All, however, are capable of providing specific and relevant information.

Behavioral Physiological and Chemical Correlations

The simplest way to study the metabolism of the central nervous system in vivo is to correlate spontaneous or experimentally produced alterations in the chemical composition of the blood, spinal fluid, or both, with changes in cerebral physiological functions or gross CNS-mediated behavior. The level of consciousness, reflex behavior, or the electroencephalogram is generally used to monitor the effects of the chemical changes on the functional and metabolic activities of the brain. For example, such methods first demonstrated the need for glucose as a substrate for cerebral energy metabolism; hypoglycemia produced by insulin or other means altered various parameters of cerebral function that could not be restored to normal by the administration of substances other than glucose.

The chief virtue of these methods is their simplicity, but they are gross and nonspecific and do not distinguish between direct effects of the agent on cerebral metabolism and those secondary to changes produced initially in somatic tissues. Also, negative results are often inconclusive, for there always remain questions of insufficient dosage, inadequate cerebral circulation and delivery to the tissues, or impermeability of the blood-brain barrier.

Biochemical Analyses of Tissues

The availability of analytical chemical techniques makes it possible to measure specific metabolites and enzyme activities in brain tissue at selected times during or after exposure of the animal to an experimental condition. (For details of these extensively used techniques, the reader should consult the first edition of *Basic Neurochemistry*, Boston: Little, Brown, 1972, Chap. 20.) This approach has been very useful in studies of the intermediary metabolism of the brain. It has permitted the estimation of the rates of flux through the various steps of established metabolic pathways and the identification of control points in the pathways where regulation may be exerted. Such studies have helped to define more precisely the changes in energy metabolism associated with altered cerebral functions produced, for example, by anesthesia, convulsions, or hypoglycemia. Although these methods require killing the animal and analyzing tissue samples, they are in vivo methods in effect since they attempt to describe the state of the tissue while it is still in the animal at the moment of sacrifice. These methods have encountered their most serious problems regarding this point. Postmortem changes in brain are extremely rapid and are not always completely retarded by the most rapid freezing techniques available. These methods have proved to be very valuable, nevertheless, particularly in the area of energy metabolism. (A detailed discussion of results obtained with these methods is presented in Chap. 14.)

Radioisotope Incorporation

The technique of administering radioactive precursors followed by the chemical separation and assay of products in the tissue has added greatly to the armamentarium for studying cerebral metabolism in vivo. Labeled precursors are administered by any one of a variety of routes; at selected later times the brain is removed and the precursor and the various products of interest are isolated. The radioactivity and quantity of the compounds in question are assayed. Such techniques facilitate the identification of metabolic routes and the rates of flux through various steps of the pathway. In some cases the comparison of the specific activities of the products and precursors has led to the surprising finding of higher specific activities in the products than in the precursors. This is conclusive evidence of the presence of compartmentation [1]. These methods have been used effectively in studies of amine and neurotransmitter synthesis and metabolism, lipid metabolism, protein synthesis, amino acid metabolism, and the distribution of glucose carbon through the various biochemical pathways present in the brain.

The methods are particularly valuable for studies of intermediary metabolism that generally are not feasible by most other in vivo techniques. They are without equal for the qualitative identification of the pathways and routes of metabolism. They suffer, however, from a disadvantage; only one set of measurements per animal is possible because the animal must be killed. Quantitative interpretations are often confounded by the problems of compartmentation. Also, they are all too frequently misused; unfortunately, quantitative conclusions are often drawn on the basis of radioactivity data without appropriate consideration of the specific activities of the precursor pools.

Polarographic Techniques

The oxygen electrode has been employed for measuring the amount of oxygen consumed locally in the cerebral cortex in vivo [2]. The electrode is applied to the surface of the exposed cortex, and the local PO_2 is measured continuously before and during occlusion of the blood flow to the local area. During occlusion the PO_2 falls linearly as the oxygen is consumed by the tissue metabolism, and the rate of fall is a measure of the rate of oxygen consumption locally in the cortex. Repeated measurements can be made successively in the animal, and the technique has been used to demonstrate the increased oxygen consumption of the cerebral cortex and the relation between the changes in the EEG and the metabolic rate during convulsions [2]. The technique is limited to the measurements in the cortex and, of course, to oxygen utilization.

Arteriovenous Differences

The primary function of the circulation is to replenish the nutrients consumed by the tissues and to remove the products of their metabolism. This function is reflected in the composition of the blood traversing the tissue. Substances taken up by the tissue from the blood are higher in concentration in the arterial inflow than in the venous outflow, and the converse is true for substances released by the tissue. The convention is to subtract the venous concentration from the arterial concentration so that a positive arteriovenous difference represents net uptake and a negative difference means net release. In nonsteady states there may be transient, but significant, arteriovenous differences which reflect reequilibration of the tissue with the blood. In steady states, in which it is presumed that the tissue concentration remains constant, positive and negative arteriovenous differences mean net consumption or production of the substance by the tissue. Zero arteriovenous differences indicate neither. This method is useful for all substances in blood that can be assayed with enough accuracy, precision, and sensitivity to enable the detection of arteriovenous differences. The method is useful only for the tissues from which mixed representative venous blood can be sampled. Arterial blood has essentially the same composition throughout and can be sampled from any artery. On the other hand, venous blood is specific for each tissue, and to establish valid arteriovenous differences the venous blood must represent the total outflow or the flow-weighted average of all the venous outflows from the tissue under study, uncontaminated by blood from any other tissue. It is not possible to fulfill this condition for many tissues.

The method is fully applicable to the brain, particularly in man, in whom the anatomy of venous drainage is favorable for such studies. Representative cerebral venous blood, with no more than about 3 percent contamination with extracerebral blood, is readily obtained from the superior bulb of the internal jugular vein. The venipuncture can be made percutaneously under local anesthesia [3, 4] so the measurements can be made during the conscious state, undistorted by the effects of general anesthesia. The monkey is similar, although the vein must be surgically exposed before puncture. Other common laboratory animals are less suitable because extensive communications between cerebral and extracerebral venous beds are present, and uncontaminated, representative venous blood is difficult to obtain from the cerebrum without major surgical intervention. In these cases, one can obtain relatively uncontaminated venous blood from the confluence of the sinuses even though it does not contain representative blood from the brain stem and some of the lower portions of the brain.

The chief advantages of these methods are their simplicity and applicability to unanesthetized man. They permit the qualitative identification of the ultimate substrates and products of cerebral metabolism. They have no applicability, however, to those intermediates which are formed and consumed entirely in the brain without being exchanged with blood, or to those substances which are exchanged between brain and blood with no net flux in either direction. Furthermore, they provide no quantitation of the rates of utilization or production because arteriovenous differences depend not only on the rates of consumption or production by the tissue but also on the blood flow (see below). Blood flow affects all the arteriovenous differences proportionately, however, and comparison of the arteriovenous differences of various substances obtained from the same samples of blood reflects their relative rates of utilization or production.

Combination of Cerebral Blood Flow and Arteriovenous Differences

In a steady state, the tissue concentration of any substance utilized or produced by the brain is presumed to remain constant. When a substance is exchanged between brain and blood, the difference in its rates of delivery to the brain in the arterial blood and removal in the venous blood must be equal to the net rate of its utilization or production by the brain. This relation can be expressed as follows:

$$CMR = CBF\,(A - V)$$

where $(A - V)$ is the difference in concentration in arterial and cerebral venous blood, CBF is the rate of cerebral blood flow in volume of blood per unit time, and CMR (cerebral metabolic rate) is the steady-state rate of utilization or production of the substance by the brain.

If both the rate of cerebral blood flow and the arteriovenous difference are known, therefore, the net rate of utilization or production of the substance by the brain can be calculated. This has been the basis of most quantitative studies of the cerebral metabolism in vivo [5–8].

There are a number of methods for determining the cerebral blood flow, many of them of questionable value or validity. One type of technique involves isolating the cerebral circulation by surgical means; the brain is then perfused at a fixed rate with blood or perfusate, or the rate of flow in the cerebral vascular bed is measured by any one of a variety of flowmeter devices [7, 9]. Such methods are, of course, limited to animals. They can be carried out usually only under general anesthesia and require extensive surgical procedures that make truly normal preparations impossible.

A thorough discussion of the methods for measuring cerebral blood flow is beyond the scope of this chapter, but the subject has been comprehensively reviewed in recent years [6, 7, 10–12].

The most reliable method for determining cerebral blood flow is the inert gas method of Kety and Schmidt [4]. It was originally designed for use in studies of conscious, unanesthetized man, and it has been most widely employed for this purpose, but it also has been adapted for use in animals [13]. The method is based on the Fick principle, or the law of conservation of matter, and it utilizes low concentrations of a freely diffusible, chemically inert gas as a tracer substance. The original gas was nitrous oxide, but subsequent modifications have substituted other gases, such as ^{85}Kr, ^{79}Kr, or hydrogen [10, 11, 14–16], which can be measured more conveniently in blood. During inhalation of 15 percent N_2O in air, for example, arterial and cerebral venous blood samples are withdrawn at intervals and analyzed for N_2O content. The cerebral blood flow (in milliliters per 100 g of brain tissue per minute) can be calculated from the equation:

$$CBF = 100\lambda V_{10} \Big/ \int_{0}^{10} (A - V)\,dt$$

where A and V = arterial and cerebral venous blood concentrations of N_2O

$\qquad V_{10}$ = concentration of N_2O in venous blood at 10 min

$\qquad \lambda$ = partition coefficient for N_2O between brain tissue and blood

$\qquad t$ = time of inhalation in minutes

$\int_0^{10}(A-V)\,dt$ = integrated arteriovenous difference in N_2O concentration over 10 min of inhalation.

The partition coefficient for N_2O is approximately one when equilibrium has been achieved between blood and brain tissue. It has been found that at least 10 minutes of inhalation are required to establish equilibrium. At the end of this interval, the N_2O concentration in brain tissue is about equal to the cerebral venous blood concentration.

Because the method requires sampling of both arterial and cerebral venous blood, it lends itself readily to the simultaneous measurement of arteriovenous differences of substances involved in cerebral metabolism. This method and its modifications have provided most of our knowledge of the rates of substrate utilization or product formation by the brain in vivo.

REGULATION OF CEREBRAL METABOLIC RATE

Normal Consumption of Oxygen by the Brain

The brain is metabolically one of the most active of all the organs of the body. This is reflected in its relatively enormous rate of oxygen consumption, which provides the energy required for its intense physicochemical activity. The most reliable data on cerebral metabolic rate have been obtained in man, although the rates in lower mammals appear to be comparable on a unit-weight basis. Cerebral oxygen consumption in normal, conscious, young men is about 3.5 ml/100 g brain/minute (Table 19-1); the rate is similar in young women. The rate of oxygen consumption by an entire brain of average weight (1,400 g) is then about 49 ml O_2 per minute.

Table 19-1. Cerebral Blood Flow and Metabolic Rate in Normal Young Adult Man

	Rate	
Function	Per 100 g Brain Tissue	Per Whole Brain[a]
Cerebral blood flow (ml/min)	57	798
Cerebral O_2 consumption (ml/min)	3.5	49
Cerebral glucose utilization (mg/min)	5.5	77

[a]Brain weight assumed to be 1,400 g.
Source: Based on data derived from literature [8].

The magnitude of this rate can be more fully appreciated when it is compared with the metabolic rate of the body as a whole. The average man weighs 70 kg and consumes about 250 ml O_2 per minute in the basal state. Therefore, the brain alone, which represents only about 2 percent of total body weight, accounts for 20 percent of the resting total body oxygen consumption. In children the brain takes an even larger fraction [17].

Oxygen is utilized in the brain almost entirely for the oxidation of carbohydrate [8, 18]. The energy equivalent of the total cerebral metabolic rate is, therefore, approximately 20 watts or 0.25 kcal per minute. If it is assumed that this energy is utilized mainly for the synthesis of high-energy phosphate bonds, that the efficiency of the energy conservation is about 20 percent, and that the free energy of hydrolysis of the terminal phosphate of ATP is about 7 kcal per mole, this energy expenditure then can be estimated to support the steady turnover of close to 7 mmol or about 4×10^{21} molecules of ATP per minute in the entire brain.

The brain normally has no respite from this enormous energy demand. The cerebral oxygen consumption continues unabated day and night. Even during sleep there is no reduction in cerebral metabolic rate; indeed, it may even be markedly increased in rapid eye movement (REM) sleep (see p. 406).

Energy-Demanding Functions

There has been considerable speculation about the nature of the functions responsible for the brain's energy demands. The brain does not do mechanical work, like that of cardiac and skeletal muscle. It does not do osmotic work, as does the kidney in concentrating urine. It does not have the complex energy-consuming metabolic functions of liver, nor, despite the synthesis of a few hormones and neurotransmitters, is it noted for its biosynthetic activities. Recently, considerable emphasis has been placed on the extent of macromolecular synthesis in the central nervous system, an interest stimulated by the recognition that there are some proteins with short half-lives in the brain. These represent, however, relatively small numbers of molecules, and, in fact, the average protein turnover and the rate of protein synthesis in the mature brain are slower than in most other tissues except, perhaps, muscle [19]. Clearly, the functions of nervous tissues are mainly excitation and conduction, and these are reflected in the unceasing electrical activity of the brain. The electrical energy is ultimately derived from chemical processes, and it is likely that most of the brain's energy consumption is utilized for active transport of ions to sustain and restore the membrane potentials discharged during the process of excitation and conduction (see Chaps. 5, 6, and 31).

Not all of the oxygen consumption of the brain is used for energy metabolism. The brain contains a variety of oxidases and hydroxylases which function in the synthesis and metabolism of a number of neurotransmitters (see Chap. 10). For example, tyrosine hydroxylase is a mixed-function oxidase which hydroxylates tyrosine to 3,4-dihydroxyphenylalanine (DOPA), and dopamine-β-hydroxylase hydroxylates dopamine to form norepinephrine. Similarly, tryptophane hydroxylase hydroxylates tryptophane to form 5-hydroxytryptophane in the pathway of serotonin synthesis. These enzymes are oxygenases which utilize molecular oxygen and incorporate it into the hydroxyl group of the hydroxylated products. Oxygen also is consumed in the metabolism of these monamine neurotransmitters, which are oxidatively deaminated to their respective aldehydes by monamine oxidase. All of these enzymes are present in brain, and the reactions catalyzed by them utilize oxygen. When, however, the total turnover rates of the neurotransmitters and the sum total of the maximal velocities of all the oxidases involved in their synthesis and degradation are considered, it is clear that the oxygen consumed in the turnover of the neurotransmitters can account for only a very small, possibly immeasurable, fraction of the total oxygen consumption of the brain.

Role of Cerebral Circulation

Not only does the brain utilize oxygen at a very rapid rate, it is absolutely dependent on continuously uninterrupted oxidative metabolism for maintenance of its functional and structural integrity. There is a large Pasteur effect in brain tissue [20], but, even at its maximum rate, anaerobic glycolysis is unable to provide sufficient energy to meet the brain's demands. Since the oxygen stores of the brain are extremely small compared with its rate of utilization, the brain requires the continuous replenishment of its oxygen by the circulation. If the cerebral blood flow is completely interrupted, consciousness is lost within less than 10 sec, or the amount of time required to consume the oxygen contained within the brain and its blood content [21]. Loss of consciousness as a result of anoxemia, caused by anoxia or asphyxia, takes only a little longer because of the additional oxygen present in the lungs and still-circulating blood. There is evidence that the critical level of oxygen tension in the cerebral tissues, below which consciousness and the normal EEG pattern are invariably lost, lies between 15 and 20 mm Hg [7]. This appears to be so whether the tissue anoxia is achieved by lowering the cerebral blood flow or the arterial oxygen content.

Cessation of cerebral blood flow is followed within a few minutes by irreversible pathological changes within the brain, readily demonstrated by microscopic anatomical techniques. It is well known, of course, that in medical crises, such as cardiac arrest, damage to the brain occurs earliest and is most decisive in determining the degree of recovery.

The cerebral blood flow must be able to maintain the brain's avaricious appetite for oxygen. The average rate of blood flow in the brain as a whole is about 57 ml/ 100 g tissue/min (see Table 19-1). For the whole brain this amounts to almost 800 ml per minute or approximately 15 percent of the total basal cardiac output. This level must be maintained within relatively narrow limits, for the brain cannot tolerate any major drop in its perfusion. A fall in cerebral blood flow to half of its normal rate is sufficient to cause loss of consciousness in normal, healthy young men [6, 7, 22]. There are, fortunately, numerous reflexes and other physiological mechanisms to sustain adequate levels of arterial blood pressure at the head level and to maintain the cerebral blood flow, even when arterial pressure falls in times of stress [6, 7]. There are also mechanisms to adjust the cerebral blood flow to changes in cerebral metabolic demand.

Regulation of the cerebral blood flow is achieved mainly by control of the tone or the degree of constriction or dilatation of the cerebral vessels. This, in turn, is controlled mainly by local chemical factors, such as PCO_2, PO_2, and pH. High PCO_2, low PO_2, and low pH — products of metabolic activity — tend to dilate the blood vessels and increase cerebral blood flow; changes in the opposite direction constrict the vessels and decrease blood flow [6, 7, 23]. The cerebral blood flow is regulated through such mechanisms to maintain homeostasis of these chemical factors in the local tissue. Their rates of production depend on the rates of metabolism, so cerebral blood flow is, therefore, also adjusted to the cerebral metabolic rate [6, 7, 23].

Local Cerebral Blood Flow and Metabolism
The rates of blood flow and metabolism presented in Table 19-1 and discussed above represent the average values in the brain as a whole. The brain is not a homogeneous organ, however, but is composed of a variety of tissues and discrete structures which often function independently or even inversely with respect to one another. There is little reason to expect that their perfusion and metabolic rates would be similar. Indeed, experimental evidence clearly indicates that they are not (see Chap. 14). Local cerebral blood flow in laboratory animals has been determined from the local tissue concentrations, measured by a quantitative autoradiographic technique, and from the total history of the arterial concentration of a freely diffusible, chemically inert, radioactive tracer introduced into the circulation [24, 25]. The results reveal that blood-flow rates vary widely throughout the brain with average values in gray matter approximately four to five times those of white matter [24–26].

Recently a method has been devised to measure glucose consumption in the discrete functional and structural components of the brain in intact conscious laboratory animals [27, 28]. This method also employs quantitative autoradiography to measure local tissue concentrations but utilizes ^{14}C-deoxyglucose as the tracer. The local tissue accumulation of ^{14}C-deoxyglucose as ^{14}C-deoxyglucose 6-phosphate in a given interval of time is related to the amount of glucose that has been phosphorylated by hexokinase over the same interval, and the rate of glucose consumption can be determined from the ^{14}C-deoxyglucose 6-phosphate concentration by appropriate consideration of (1) the relative concentrations of ^{14}C-deoxyglucose and glucose in the plasma; (2) their rate constants for transport between plasma and brain tissue; and (3) the kinetic constants of hexokinase for deoxyglucose and glucose. This method has demonstrated that local consumption of cerebral glucose varies as widely as blood flow throughout the brain. Indeed, in the normal animal there is a remarkably close correlation between local cerebral blood flow and glucose consumption [29]. Changes in functional activity produced by physiological stimulation, anesthesia, or deafferentation result in corresponding changes in blood flow [26] and glucose consumption [30] in the structures involved in the functional change. The ^{14}C-deoxyglucose method for the measurement of local glucose utilization has been used recently to map the functional visual pathways and to identify the locus of the visual cortical representation of the retinal "blind spot" in the brain of the rhesus monkey [30]. These results establish that local energy metabolism in brain is coupled to local functional activity and also confirm the long-held belief that local cerebral blood flow is adjusted to metabolic demand in local tissue.

SUBSTRATES OF CEREBRAL METABOLISM

Normal Substrates and Products

In contrast to most other tissues, which exhibit considerable flexibility in regard to the nature of the foodstuffs absorbed and consumed from the blood, the normal brain is restricted almost exclusively to glucose as the substrate for its energy metabolism. Despite long and intensive efforts, the only incontrovertible and consistently positive arteriovenous differences demonstrated for the human brain under normal conditions have been for glucose and oxygen [8, 18]. One report [31] of glutamate uptake and equivalent release of glutamine has never been substantiated and must at present be considered doubtful. Negative arteriovenous differences, significantly different from zero, have been found consistently only for carbon dioxide, although it is likely that water, which has never been measured, is also produced [8]. Pyruvate and lactate production have been observed occasionally, certainly in aged subjects with cerebral vascular insufficiency [20], but irregularly in subjects with normal oxygenation of the brain [8].

It appears, then, that in the normal in vivo state glucose is the only significant substrate for the brain's energy metabolism. Under normal circumstances no other potential energy-yielding substance has been found to be extracted from the blood in more than trivial amounts. The stoichiometry of glucose utilization and oxygen consumption is summarized in Table 19-2. The normal, conscious human brain consumes oxygen at a rate of 156 μmol/100 g tissue/min. Carbon dioxide production is the same, leading to a respiratory quotient of 1.0, further evidence that carbohydrate is the ultimate substrate for oxidative metabolism. The O_2 consumption and CO_2 production are equivalent to a rate of glucose utilization of 26 μmol glucose/100 g tissue/min, assuming 6 μmol of O_2 consumed and CO_2 produced for each μmol of glucose completely oxidized to CO_2 and H_2O. The glucose

Table 19-2. Relationship Between Cerebral Oxygen Consumption and Glucose Utilization in Normal Young Adult Man

Function	Value
O_2 consumption (μmol/100 g brain tissue/min)	156
Glucose utilization (μmol/100 g brain tissue/min)	31
O_2/glucose ratio (mole/mole)	5.5
Glucose equivalent of O_2 consumption (μmol glucose/100 g brain tissue/min)	26[a]
CO_2 production (μmol/100 g brain tissue/min)	156
Cerebral respiratory quotient	0.97

[a]Calculated on the basis of 6 moles of O_2 required for complete oxidation of 1 mole of glucose.
Source: Values are the median of the values reported in the literature [8].

utilization actually measured is, however, 31 μmol/100 g/min, which indicates that glucose consumption is not only sufficient to account for the total O_2 consumption but is in excess by 5 μmol/100 g/min. For the complete oxidation of glucose, the theoretical ratio of oxygen-to-glucose utilization is 6.0; the excess glucose utilization is responsible for a measured ratio of only 5.5 μmol of O_2 per μmol of glucose. The fate of the excess glucose is unknown, but it is probably distributed in part in lactate, pyruvate, and other intermediates of carbohydrate metabolism that are released from the brain into the blood, each in insufficient amount to be detectable in significant arteriovenous differences. Probably some of the glucose is also utilized not for the production of energy but for the synthesis of other chemical constituents of the brain.

Some oxygen is known to be utilized for the oxidation of substances not derived from glucose, as, for example, in oxygenase reactions involved in the synthesis of monamine neurotransmitters, as mentioned above. The amount of oxygen utilized for these processes is, however, extremely small and is undetectable in the presence of the enormous oxygen consumption used for carbohydrate oxidation. Also, the amount of amino acids taken up for these purposes is so small as to be undetected in the arteriovenous differences.

The combination of a cerebral respiratory quotient of unity, an almost stoichiometric relationship between oxygen uptake and glucose consumption, and the absence of any significant arteriovenous difference for any other energy-rich substrate is strong evidence that the brain normally derives its energy from the oxidation of glucose. In this respect, cerebral metabolism is unique because no other tissue, with the possible exception of the testis [32], has been found to rely only on carbohydrate for energy. This does not imply that the pathways of glucose metabolism in the brain lead, like combustion, directly and exclusively to oxidation. Various chemical and energy transformations occur between the uptake of the primary substrates, glucose and oxygen, and the liberation of the end products, carbon dioxide and water. Various compounds derived from glucose or produced through the energy made available from glucose catabolism are intermediates in the process. Glucose carbon is incorporated, for example, into amino acids, protein, lipids, glycogen, and so on. These are turned over and act as intermediates in the overall pathway from glucose to carbon dioxide and water. There is clear evidence from studies with ^{14}C-glucose that the glucose is not entirely oxidized directly, and that at any given moment some of the carbon dioxide being produced is derived from sources other than the glucose that enters the brain at the same moment or just prior to that moment [9, 33]. That oxygen and glucose consumption and carbon dioxide production are essentially in stoichiometric balance and no other energy-laden substrate is taken from the blood means, however, that the net energy made available to the brain must ultimately be derived from the oxidation of glucose. It should be noted that this is the situation in the normal state; as will be discussed later, other substrates may be used in special circumstances or in abnormal states.

Obligatory Utilization of Glucose
The brain normally derives almost all its energy from the aerobic oxidation of glucose, but this does not distinguish between preferential and obligatory utilization of glucose.

Most tissues are largely facultative in their choice of substrates and can use them interchangeably more or less in proportion to their availability. This does not appear to be so in brain. The present evidence indicates that, except in some unusual and very special circumstances, only the aerobic utilization of glucose is capable of providing the brain with enough energy to maintain normal function and structure. The brain appears to have almost no flexibility in its choice of substrates in vivo. This conclusion is derived from the following evidence.

Effects of Glucose Deprivation

It is well known clinically that a drop in blood glucose content, if of sufficient degree, is rapidly followed by aberrations of cerebral function. Hypoglycemia, produced by excessive insulin or occurring spontaneously in hepatic insufficiency, is associated with changes in mental state ranging from mild, subjective sensory disturbances to coma, the severity depending on both the degree and the duration of the hypoglycemia. The behavioral effects are paralleled by abnormalities in EEG patterns and cerebral metabolic rate. The EEG pattern exhibits increased prominence of slow, high-voltage delta rhythms, and the rate of cerebral oxygen consumption falls. In studies of the effects of insulin hypoglycemia in man [34], it was observed that, when the arterial glucose concentration fell from a normal level of 70 to 100 mg per 100 ml to an average level of 19 mg per 100 ml, the subjects became confused and their cerebral oxygen consumption fell to 2.6 ml/100 g/min or 79 percent of the normal level. When the arterial glucose level fell to 8 mg per 100 ml, a deep coma ensued and the cerebral oxygen consumption decreased even further to 1.9 ml/100 g/min.

These changes are not caused by insufficient cerebral blood flow, which actually increases slightly during the coma. In the depths of the coma, when the blood glucose content is very low, there is almost no measurable cerebral uptake of glucose from the blood. Cerebral oxygen consumption, although reduced, is still far from negligible, and there is no longer any stoichiometric relationship between glucose and oxygen uptakes by the brain — evidence that the oxygen is utilized for the oxidation of other substances. The cerebral respiratory quotient remains approximately one, however, indicating that these other substrates are still carbohydrate, presumably derived from the brain's endogenous carbohydrate stores. The effects are clearly the result of hypoglycemia and not some other direct effect of insulin in the brain. In all cases, the behavioral, functional, and cerebral metabolic abnormalities associated with insulin hypoglycemia are rapidly and completely reversed by the administration of glucose. The severity of the effects is correlated with the degree of hypoglycemia and not the insulin dosage, and the effects of the insulin can be completely prevented by the simultaneous administration of glucose with the insulin.

Similar effects are observed in hypoglycemia produced by other means, such as hepatectomy. The inhibition of glucose utilization at the phosphohexoseisomerase step with 2-deoxyglucose also produces all the cerebral effects of hypoglycemia despite an associated elevation in blood glucose content [8]. It appears, then, that when the brain is deprived of its glucose supply in an otherwise normal individual, no other substance present in the blood can satisfactorily substitute for it as the substrate for the brain's energy metabolism.

Other Substrates in Hypoglycemia

The hypoglycemic state provides convenient test conditions to determine whether a substance is capable of substituting for glucose as a substrate of cerebral energy metabolism. If it can, its administration during hypoglycemic shock should restore consciousness and normal cerebral electrical activity without raising the blood glucose level. Numerous potential substrates have been tested in man and animals. Very few can restore normal cerebral functions in hypoglycemia, and of these all but one appear to operate through a variety of mechanisms to raise the blood glucose level rather than by serving as a substrate directly (Table 19-3).

Mannose appears to be the only substance that can be utilized by the brain directly and rapidly enough to restore or maintain normal function in the absence of glucose [34]. It traverses the blood-brain barrier and is converted to mannose 6-phosphate. This reaction is catalyzed by hexokinase as effectively as the phosphorylation of glucose. The mannose 6-phosphate is then converted to fructose 6-phosphate by

Table 19-3. Effectiveness of Various Substances in Preventing or Reversing the Effects of Hypoglycemia or Glucose Deprivation on Cerebral Function and Metabolism

Effectiveness	Substance	Comments
Effective	Epinephrine	Raises blood glucose concentration
	Maltose	Converted to glucose and raises blood glucose level
	Mannose	Directly metabolized and enters glycolytic pathway
Partially or occasionally effective	Glutamate Arginine Glycine p-Aminobenzoate Succinate	Occasionally effective by raising blood glucose level
Ineffective	Glycerol Ethanol Lactate Pyruvate Glyceraldehyde Hexosediphosphates Fumarate Acetate β-Hydroxybutyrate Galactose Lactose Inulin	Some of these substances can be metabolized to varying extent by brain tissue and could conceivably be effective if it were not for the blood-brain barrier

Source: Summarized from literature [8].

phosphomannoseisomerase, which is active in brain tissue [34]. Through these reactions mannose can enter directly into the glycolytic pathway and replace glucose.

Maltose also has been found to be effective occasionally in restoring normal behavior and EEG activity in hypoglycemia, but only by raising the blood glucose level through its conversion to glucose by maltase activity in blood and other tissues [8, 35].

Epinephrine is effective in producing arousal from insulin coma, but this is achieved through its well-known stimulation of glycogenolysis and the elevation of blood glucose concentration. Glutamate, arginine, glycine, p-aminobenzoate, and succinate also are effective occasionally, but they probably act through adrenergic effects which raise the epinephrine level and consequently the glucose concentrations of the blood [8].

It is clear, then, that no substance normally present in the blood can replace glucose as the substrate for the brain's energy metabolism. Thus far, the one substance found to do so — mannose — is not normally present in blood in significant amounts and is, therefore, of no physiological significance. It should be noted, however, that failure to restore normal cerebral function in hypoglycemia is not synonymous with an inability of the brain to utilize the substance. Many of the substances that have been tested and found ineffective are compounds normally formed and utilized within the brain and are normal intermediates in its intermediary metabolism. Lactate, pyruvate, fructose, 1,6-diphosphate, acetate, β-hydroxybutyrate, and acetoacetate are such examples. These can all be utilized by brain slices, homogenates, or cell-free fractions, and the enzymes for their metabolism are present in the brain. Enzymes for the metabolism of glycerol [36] or ethanol [37], for example, may not be present in sufficient amounts. For other substrates, for example, β-hydroxybutyrate and acetoacetate [38–40], the enzymes are adequate, but the substrate is not available to the brain because of inadequate blood level or restricted transport through the blood-brain barrier.

Nevertheless, the functioning of the nervous system in the intact animal depends on substrates supplied by the blood, and no satisfactory, normal, endogenous substitute for glucose has been found. Glucose must, therefore, be considered essential for normal physiological behavior of the central nervous system.

Cerebral Utilization of Ketones in Ketosis

Recent developments indicate that the brain may be somewhat more flexible in its choice of nutrients than was once believed and that, in special circumstances, it may fulfill its nutritional needs partly, although not completely, with substrates other than glucose. Normally there are no significant cerebral arteriovenous differences for D-β-hydroxybutyrate and acetoacetate [8, 18], which are "ketone bodies" formed in the course of the catabolism of fat. Owen et al. [41] observed, however, that when human patients were treated for severe obesity by complete fasting for several weeks, there was considerable uptake of both substances by the brain. Indeed, if one assumed that the substances were completely oxidized, their rates of utilization would have accounted for more than 50 percent of the total cerebral oxygen consumption — more than that accounted for by the glucose uptake. The uptake of D-β-hydroxybutyrate was several times greater than that of acetoacetate, a reflection of its higher concentration in the blood. The enzymes responsible for their metabolism,

D-β-hydroxybutyric dehydrogenase, acetoacetate-succinyl-CoA transferase, and acetoacetyl thiolase, have been demonstrated to be present in brain tissue in sufficient amounts to convert them into acetyl-CoA and to feed them into the tricarboxylic cycle at a sufficient rate to satisfy the brain's metabolic demands [38–40].

Under normal circumstances, when there is ample glucose and the levels of ketone bodies in the blood are very low, the brain apparently does not resort to their use in any significant amounts. In prolonged starvation, the carbohydrate stores of the body are exhausted, and the rate of gluconeogenesis is insufficient to provide glucose fast enough to meet the requirements of the brain; blood ketone levels rise as a result of the rapid fat catabolism. The brain then apparently turns to the ketone bodies as the source of its energy supply.

Cerebral utilization of ketone bodies appears to follow passively their levels in arterial blood [42, 43]. In normal adults, ketone levels are very low in blood, and cerebral utilization of ketones is neglible. In ketotic states resulting from starvation, fat-feeding or ketogenic diets, diabetes, or any other condition which accelerates the mobilization and catabolism of fat, cerebral utilization of ketones is increased more or less in direct proportion to the degree of ketosis [42, 43]. Significant utilization of ketone bodies by brain is, however, normal in the neonatal period. The newborn infant tends to be hypoglycemic but becomes ketotic when it begins to nurse because of the high fat content of mother's milk [42, 43]. When weaned onto the normal relatively high carbohydrate diet, the ketosis and cerebral ketone utilization disappear [42, 43]. The studies in infancy have been carried out mainly in the rat, but there is evidence that the situation is similar in the human infant [42]. The first two enzymes in the pathway of ketone utilization are D-β-hydroxybutyrate dehydrogenase and acetoacetyl-succinyl-CoA transferase. These exhibit a postnatal pattern of development in brain that is well adapted to the nutritional demands of the brain. At birth, the activities of these enzymes in brain are low; they rise rapidly with the ketosis that develops with the onset of suckling; reach their peaks just before weaning; and then gradually decline after weaning to normal adult levels of about one-third to one-fourth the maximum levels attained [39, 40, 42–44].

It should be noted that D-β-hydroxybutyrate is incapable of maintaining or restoring normal cerebral function in the absence of glucose in the blood [45]. This suggests that, although it can partially replace glucose, it cannot fully satisfy the cerebral energy needs in the absence of some glucose consumption. One possible explanation may be that the first product of D-β-hydroxybutyrate oxidation, acetoacetate, is further metabolized by its displacement of the succinyl moiety of succinyl-CoA to form acetoacetyl-CoA. A certain level of glucose utilization may be essential to drive the tricarboxylic cycle and provide enough succinyl-CoA to permit the further oxidation of acetoacetate and hence pull along the oxidation of D-β-hydroxybutyrate.

INFLUENCE OF AGE AND DEVELOPMENT ON
CEREBRAL ENERGY METABOLISM

The energy metabolism of the brain and the blood flow which sustains it vary considerably from birth to old age. Data on the cerebral metabolic rate obtained directly in vivo are lacking for the early postnatal period, but the results of in vitro measurements in animal brain preparations [46] and inferences drawn from cerebral blood flow measurements in intact animals [47] suggest that the cerebral oxygen consumption is low at birth, rises rapidly during the period of cerebral growth and development, and reaches a maximal level at about the time maturation is completed (Chap. 18). This rise is consistent with the progressive increase in the levels of a number of enzymes of oxidative metabolism in the brain [44]. The rates of blood flow in different structures of the brain reach peak levels at different times, depending on the maturation rate of the particular structure [47]. In the structures that consist predominantly of white matter, the peaks coincide roughly with the times of maximal rates of myelination [47]. From these peaks, blood flow and, probably, cerebral metabolic rate decline to the levels characteristic of adulthood.

Reliable quantitative data on the changes in cerebral circulation and metabolism in man from the middle of the first decade of life to old age are summarized in Table 19-4. By 6 years of age, the cerebral blood flow and oxygen consumption have already attained their high levels, and they decline thereafter to the levels of normal young adulthood [17]. A cerebral oxygen consumption of 5.2 ml/100 g brain tissue/min in a 5- to 6-year-old child corresponds to a total oxygen consumption by the brain of approximately 60 ml per minute, or more than 50 percent of the total body basal oxygen consumption, a proportion markedly greater than that occurring in adulthood. The reasons for the extraordinarily high cerebral metabolic rates in children are unknown, but presumably they reflect the extra energy requirements for biosynthetic processes associated with growth and development.

Despite reports to the contrary [48], cerebral blood flow and oxygen consumption normally remain essentially unchanged between young adulthood and old age.

Table 19-4. Cerebral Blood Flow and Oxygen Consumption in Man from Childhood to Old Age and Senility

Life Period and Condition	Age (years)	Cerebral Blood Flow (ml/100 g/min)	Cerebral O_2 Consumption (ml/100 g/min)	Cerebral Venous O_2 Tension (mm Hg)
Childhood	6[a]	106[a]	5.2[a]	—
Normal young adulthood	21	62	3.5	38
Aged:				
Normal elderly	71[a]	58	3.3	36
Elderly with minimal arteriosclerosis	73[a]	48[a]	3.2	33[a, b]
Elderly with senile psychosis	72[a]	48[a, b]	2.7[a, b]	33[a, b]

[a]Statistically significant difference from normal young adults ($P < 0.05$).
[b]Statistically significant difference from normal elderly subjects ($P < 0.05$).
Source: From [17] and [49].

In a population of normal elderly men in their eighth decade of life — who were carefully selected for good health and freedom from all disease, including vascular — both blood flow and oxygen consumption were not significantly different from those of normal young men 50 years younger (see Table 19-4) [49]. In a comparable group of elderly subjects, who differed only by the presence of objective evidence of minimal arteriosclerosis, cerebral blood flow was significantly lower. It had reached a point at which the oxygen tension of the cerebral venous blood declined, which is an indication of cerebral hypoxia. Cerebral oxygen consumption, however, was still maintained at normal levels through removal of larger-than-normal proportions of the arterial blood oxygen. In senile psychotic patients with arteriosclerosis, cerebral blood flow was no lower, but cerebral oxygen consumption had declined. These data suggest that aging per se need not lower the cerebral oxygen consumption and blood flow, but that when blood flow is reduced, it probably is secondary to arteriosclerosis, which produces cerebral vascular insufficiency and chronic hypoxia in the brain. Because arteriosclerosis is so prevalent in the aged population, most individuals probably follow the latter pattern.

CEREBRAL METABOLIC RATE IN VARIOUS PHYSIOLOGICAL STATES

Cerebral Metabolic Rate and Functional Activity

In organs such as the heart or skeletal muscles, which perform mechanical work, increased functional activity clearly is associated with increased metabolic rate. In nervous tissues outside the central nervous system, the electrical activity is an almost quantitative indicator of the degree of functional activity, and in structures such as sympathetic ganglia and postganglionic axons [8, 50] increased electrical activity produced by electrical stimulation is definitely associated with increased utilization of oxygen. Within the central nervous system, local changes in metabolism can be correlated with changes in relatively discrete functions of particular brain regions, as described above (Local Cerebral Blood Flow and Metabolism). On the other hand, more complex brain functions involve the electrical activities of heterogeneous units which are integrated into a composite record; EEG data cannot always be interpreted readily in terms of total functional activity, and the relationship between total brain functional activity and metabolic rate may be obscured.

Convulsive activity, induced or spontaneous, has often been employed as a method of increasing electrical activity of the brain (see Chap. 31). Davies and Rémond [2] used the oxygen electrode technique in the cerebral cortex of cat and found increases in oxygen consumption during electrically induced or drug-induced convulsions. Because the increased oxygen consumption either coincided with or followed the onset of convulsions, it was concluded that the elevation in metabolic rate was the consequence of the increased functional activity produced by the convulsive state. Similar results have been obtained in the perfused cat brain in which cerebral oxygen consumption was determined from the combined measurements of blood flow and arteriovenous oxygen differences [9].

In the lightly anesthetized monkey, Dumke and Schmidt [51] found an excellent correlation between cerebral oxygen consumption and cerebral functional activity; the latter was judged by muscular movements, ocular reflexes, character of respiration, and level of arterial pressure. Changes in functional activity either occurred spontaneously or were caused (1) by altering cerebral blood flow by means of hemorrhage, transfusions, or epinephrine infusion, or (2) by producing convulsions with analeptic drugs. Cerebral oxygen consumption during the convulsions was double that of the preconvulsive state and fell to half the resting level during the postconvulsive state of depression. Similar increases in cerebral oxygen consumption have been observed in human epileptic patients during seizures [52].

Convincing correlations between cerebral metabolic rate and mental activity have been obtained in man in a variety of pathological states of altered consciousness [5, 8, 53]. Regardless of the cause of the disorder, graded reductions in cerebral oxygen consumption were accompanied by parallel graded reductions in the degree of mental alertness, all the way to profound coma (Table 19-5). This subject is discussed further in Chapter 31.

Mental Work
It is difficult to define or even to conceive of the physical equivalent of mental work. A common view equates concentrated mental effort with mental work, and it is also fashionable to attribute a high demand for mental effort to the process of problem solving in mathematics. Nevertheless, there appears to be no increased energy utilization by the brain during such processes. From the resting levels, total cerebral blood flow and oxygen consumption remain unchanged during the exertion of the

Table 19-5. Relationship Between Level of Consciousness and Cerebral Metabolic Rate

Level of Consciousness	Cerebral Blood Flow (ml/100 g/min)	Cerebral O_2 Consumption (ml/100 g/min)
Mentally alert: Normal young men	54	3.3
Mentally confused: Brain tumor Diabetic acidosis Insulin hypoglycemia Cerebral arteriosclerosis	48	2.8
Comatose: Brain tumor Diabetic coma Insulin coma Anesthesia	57	2.0

Source: From [5].

mental effort required to solve complex arithmetical problems (Table 19-6) [54].
It may be that the assumptions which relate mathematical reasoning to mental work
are erroneous, but it seems more likely that the areas which participate in the processes
of such reasoning represent too small a fraction of the brain for changes in their
functional and metabolic activities to be reflected in the energy metabolism of the
brain as a whole.

Sleep
Sleep is a naturally occurring, periodic, reversible state of unconsciousness, and the
EEG pattern in deep sleep is characterized by high-voltage, slow rhythms very similar to those often seen in pathological comatose states. In contrast to the depressed
cerebral energy metabolism seen in pathological coma, nevertheless, there is surprisingly no change in the oxygen-consumption of the brain as a whole in deep, slow-
wave sleep (see Table 19-6) [54, 55]. There are no comparable data in man for the
state of "paradoxical" sleep (REM sleep), the stage characterized by rapid, low-voltage
frequencies in the EEG and believed to be associated with the dream state. However,
measurements of local blood flow in most areas of the brain of animals indicate an
enormous increase in blood flow throughout the brain in REM sleep [56]. The basis
for this increase is unknown. It cannot be accounted for by any of the known extracerebral factors which could change cerebral blood flow to this degree, and it must
be tentatively attributed to a comparable increase in cerebral metabolic rate.

CEREBRAL ENERGY METABOLISM IN PATHOLOGICAL STATES
The cerebral metabolic rate is normally relatively stable and varies little under
physiological conditions. There are, however, a number of pathological states of
the nervous system and other organs that affect the functions of the brain either
directly or indirectly, and some of these have profound effects on the cerebral
metabolism. Some of these disorders are also discussed in Chapter 31.

Psychiatric Disorders
In general, disorders which alter the quality of mentation but not the level of consciousness — for example, the functional neuroses, psychoses, and psychotomimetic

Table 19-6. Cerebral Blood Flow and Metabolic Rate in Normal Young Men During Sleep and
Mental Work

Condition	Cerebral Blood Flow (ml/100 g/min)		Cerebral O_2 Consumption (ml/100 g/min)	
	Control	Exptl.	Control	Exptl.
Deep (slow-wave) sleep	59	65[a]	3.5	3.4
Mental arithmetic	69	67	3.9	4.0

[a]Statistically significant difference from control level ($P < 0.05$).
Source: From [54] and [55].

states — have no apparent effects on the blood flow and oxygen consumption of the brain. Thus, no changes in either function are observed in schizophrenia [34, 54] or LSD intoxication (Table 19-7) [54]. There is still uncertainty about the effects of anxiety, mainly because of the difficulties in evaluating quantitatively the intensity of anxiety. It is generally believed that ordinary degrees of anxiety or "nervousness" do not affect the cerebral metabolic rate, but severe anxiety or "panic" may increase the cerebral oxygen consumption [5, 54]. This may be related to the level of epinephrine circulating in the blood. Small doses of epinephrine that raise heart rate and cause some anxiety do not alter cerebral blood flow and metabolism, but large doses that are sufficient to raise the arterial blood pressure cause significant increases in the levels of both [54].

Psychosis related to parenchymatous damage in the brain is usually associated with depression of the cerebral metabolic rate. Thus, although normal aging per se has no effect on cerebral blood flow and oxygen consumption, both are reduced in the psychoses of senility (see Table 19-4) [49, 54, 57]. Senile dementia is always the result of degenerative changes in the brain tissue. In most cases these are secondary to cerebral vascular disease and the ensuing chronic tissue hypoxia. In some cases there is primary parenchymatous degeneration which may even occur prematurely, as in Alzheimer's disease. In either case, cerebral blood flow and metabolic rate are reduced.

Convulsive Disorders

Convulsions are discussed above in regard to the relation of cerebral metabolic rate to functional activity. In most types of convulsions studied, the seizures were induced by electroshock or analeptic drugs, or they occurred spontaneously in epileptic disease. In all these, cerebral blood flow and oxygen consumption are significantly increased in the course of the seizure (see Chap. 31 and [52]). Increased energy utilization is not, however, an essential component of the convulsive process. Convulsions may also be induced by impairment of the energy-generating system. Oxygen or glucose deficiency and fluoroacetate poisoning, all associated with a deficient availability or utilization of essential substrates for the brain's metabolism, cause seizures without increasing cerebral metabolic rate. Hypoglycemic convulsions are, in fact, associated with lowered rates of oxygen and glucose utilization [34] with no apparent depletion of high-energy phosphate stores [52].

Table 19-7. Cerebral Blood Flow and Metabolic Rate in Schizophrenia and in Normal Young Men During LSD-induced Psychotomimetic State

Condition	Cerebral Blood Flow (ml/100 g/min)	Cerebral O_2 Consumption (ml/100 g/min)
Normal	67	3.9
LSD intoxication	68	3.9
Schizophrenia	72	4.0

Source: From [54].

Pyridoxine deficiency (see Chap. 29) produces a substrate-deficiency type of convulsion with reductions in cerebral blood flow and oxygen consumption [52]. The reason is unknown but, since pyridoxine deficiency leads to lowered γ-amino-butyric acid levels in the brain, it has been presumed that the lowered cerebral metabolic rate reflects impairment of the α-ketoglutarate-to-glutamate-to-γ-amino-butyric acid-to-succinic semialdehyde-to-succinate shunt pathway around the α-ketoglutarate-to-succinate step in the tricarboxylic acid cycle in the brain (see Chaps. 11 and 14).

Coma

Coma is correlated with depression of cerebral oxygen consumption; progressive reductions in the level of consciousness are paralleled by corresponding graded decreases in cerebral metabolic rate (see Table 19-5). There are almost innumerable derangements that can lead to depression of consciousness (Chap. 31). Table 19-8 includes only a few typical examples that have been studied by the same methods and by the same or related groups of investigators.

Table 19-8. Cerebral Blood Flow and Metabolic Rate in Humans with Various Disorders Affecting Mental State

Condition	Mental State	Cerebral Blood Flow (ml/100 g/min)	Cerebral O_2 Consumption (ml/100 g/min)
Normal	Alert	54	3.3
Increased intracranial pressure (brain tumor)	Coma	34[a]	2.5[a]
Insulin hypoglycemia: Arterial glucose level:			
74 mg/100 ml	Alert	58	3.4
19 mg/100 ml	Confused	61	2.6[a]
8 mg/100 ml	Coma	63	1.9[a]
Thiopental anesthesia	Coma	60[a]	2.1[a]
Postconvulsive state:			
Before convulsion	Alert	58	3.7
After convulsion	Confused	37[a]	3.1[a]
Diabetes:			
Acidosis	Confused	45[a]	2.7[a]
Coma	Coma	65[a]	1.7[a]
Hepatic insufficiency	Coma	33[a]	1.7[a]

[a]Denotes statistical significant difference from normal level ($P < 0.05$).
Source: All studies listed were carried out by Kety and/or his associates, employing the same methods. For references see [8].

Inadequate cerebral nutrient supply leads to decreases in the level of consciousness, ranging from confusional states to coma. The nutrition of the brain can be limited by lowering the oxygen or glucose levels of the arterial blood, as in anoxia or hypoglycemia, or by impairment of their distribution to the brain through lowering of the cerebral blood flow, as in brain tumors. Consciousness is then depressed, presumably because of inadequate supplies of substrate to support the energy metabolism necessary to sustain the appropriate functional activities of the brain.

In a number of conditions, the causes of depression of both consciousness and the cerebral metabolic rate are unknown and must, by exclusion, be attributed to intracellular defects in the brain. Anesthesia is one example. Cerebral oxygen consumption is always reduced in the anesthetized state regardless of the anesthetic agent used, whereas blood flow may or may not be decreased and may even be increased [58]. This reduction is the result of decreased energy demand and not an insufficient nutrient supply or a block of intracellular energy metabolism [53]. There is evidence that general anesthetics interfere with synaptic transmission, thus reducing neuronal interaction and functional activity and, consequently, metabolic demands [50, 53].

Several metabolic diseases with broad systemic manifestations are also associated with disturbances of cerebral functions. Diabetes mellitus, when allowed to progress to states of acidosis and ketosis, leads to mental confusion and, ultimately, to deep coma, with parallel proportionate decreases in cerebral oxygen consumption (see Table 19-8) [59]. The abnormalities are usually completely reversed by adequate insulin therapy. The cause of the coma or depressed cerebral metabolic rate is unknown. Deficiency of cerebral nutrition cannot be implicated because the blood glucose level is elevated and cerebral blood flow and oxygen supply are more than adequate. Neither is insulin deficiency, which is presumably the basis of the systemic manifestations of the disease, a likely cause of the cerebral abnormalities since no absolute requirement of insulin for cerebral glucose utilization or metabolism has been demonstrated [8]. Ketosis may be severe in this disease, and there is disputed evidence that a rise in the blood level of at least one of the ketone bodies, acetoacetate, can cause coma in animals [53]. In studies of human diabetic acidosis and coma, a significant correlation between the depression of cerebral metabolic rate and the degree of ketosis has been observed, but there is an equally good correlation with the degree of acidosis [59]. It is possible that ketosis, acidosis, or the combination of both may be responsible for the disturbances in cerebral function and metabolism.

Coma is occasionally associated with severe impairment of liver function, or hepatic insufficiency. In human patients in hepatic coma, cerebral metabolic rate is markedly depressed (see Table 19-8) [53]. Cerebral blood flow is also moderately depressed, but not sufficiently to lead to limiting supplies of glucose and oxygen. The blood ammonia level is usually elevated in hepatic coma, and significant cerebral uptake of ammonia from the blood is observed. Ammonia toxicity has, therefore, been suspected as the basis for cerebral dysfunction in hepatic coma. Because ammonia can, through glutamic dehydrogenase activity, convert α-ketoglutarate to glutamate by reductive amination, it has been suggested that ammonia might thereby deplete α-ketoglutarate and thus slow the Krebs cycle [60]. The correlation between the degree of coma and the blood ammonia level is far from convincing, however, and coma has, in fact, been observed in the absence of an increase in blood ammonia concentration [53]. Although ammonia may be involved in the mechanism of hepatic coma, the mechanism remains unclear, and other causal factors are probably involved [53].

Depression of mental functions and cerebral metabolic rate has been observed in association with kidney failure, i.e., uremic coma [53]. The chemical basis of the functional and metabolic disturbances in the brain in this condition also remains undetermined.

In the comatose states associated with these systemic metabolic diseases, there is depression of both the level of conscious mental activity and cerebral energy metabolism. From the available evidence, it is impossible to distinguish which, if either, is the primary change. It is more likely that the depressions of both functions, although well correlated with each other, are independent reflections of a more general impairment of neuronal processes by some unknown factors incident to the disease.

ACKNOWLEDGMENTS

This work was sponsored by the Intramural Research Program of the National Institute of Mental Health and is therefore not subject to copyright.

REFERENCES

*1. Berl, S., and Clarke, D. D. Compartmentation of Amino Acid Metabolism. In Lajtha, A. (Ed.), *Handbook of Neurochemistry*. Vol. 2. New York: Plenum, 1969.

2. Davies, P. W., and Rémond, A. Oxygen consumption of the cerebral cortex of the cat during metrazole convulsions. *Res. Publ. Assoc. Res. Nerv. Ment. Dis.* 26:205, 1946.

3. Myerson, A., Halloran, R. D., and Hirsch, H. L. Technique for obtaining blood from the internal jugular vein and internal carotid artery. *Arch. Neurol. Psychiatry* 17:807, 1927.

4. Kety, S. S., and Schmidt, C. F. The nitrous oxide method for the quantitative determination of cerebral blood flow in man: Theory, procedure, and normal values. *J. Clin. Invest.* 27:476, 1948.

*5. Kety, S. S. Circulation and metabolism of the human brain in health and disease. *Am. J. Med.* 8:205, 1950.

*6. Lassen, N. A. Cerebral blood flow and oxygen consumption in man. *Physiol. Rev.* 39:183, 1959.

*7. Sokoloff, L. The action of drugs on the cerebral circulation. *Pharmacol. Rev.* 11:1, 1959.

*8. Sokoloff, L. The Metabolism of the Central Nervous System in Vivo. In Field, J., Magoun, H. W., and Hall, V. E. (Eds.), *Handbook of Physiology-Neurophysiology*. Vol. 3. Washington, D. C.: American Physiological Society, 1960.

*Asterisks denote key references.

*9. Geiger, A. Correlation of brain metabolism and function by use of a brain perfusion method *in situ. Physiol. Rev.* 38:1, 1958.

*10. Kety, S. S. The Cerebral Circulation. In Field, J. (Ed.), *Handbook of Physiology-Neurophysiology.* Vol. 3. Washington, D. C.: American Physiological Society, 1960.

*11. Sokoloff, L. Quantitative Measurements of Cerebral Blood Flow in Man. In Bruner, H. D. (Ed.), *Methods in Medical Research.* Vol. 8. Chicago: Year Book, 1960.

*12. Betz, E. Cerebral blood flow: Its measurement and regulation. *Physiol. Rev.* 52(3):595, 1972.

13. Page, W. F., German, W. J., and Nims, L. F. The nitrous oxide method for measurement of cerebral blood flow and cerebral gaseous metabolism in dogs. *Yale J. Biol. Med.* 23:462, 1951.

14. Lassen, N. A., and Munck, O. The cerebral blood flow in man determined by the use of radioactive krypton. *Acta Physiol. Scand.* 33:30, 1955.

15. Lewis, B. M., Sokoloff, L., Wechsler, R. L., Wentz, W. B., and Kety, S. S. A method for the continuous measurement of cerebral blood flow in man by means of radioactive krypton (^{79}Kr). *J. Clin. Invest.* 39:707, 1960.

16. Gotoh, F., Meyer, J. S., and Tomita, M. Hydrogen method for determining cerebral blood flow in man. *Arch. Neurol.* 15:549, 1966.

17. Kennedy, C., and Sokoloff, L. An adaptation of the nitrous oxide method to the study of the cerebral circulation in children; normal values for cerebral blood flow and cerebral metabolic rate in childhood. *J. Clin. Invest.* 36:1130, 1957.

*18. Kety, S. S. The General Metabolism of the Brain in Vivo. In Richter, D. (Ed.), *The Metabolism of the Nervous System.* London: Pergamon, 1957.

*19. Waelsch, H., and Lajtha, A. Protein metabolism in the nervous system. *Physiol. Rev.* 41:709, 1961.

20. Meyer, J. S., Ryu, T., Toyoda, M., Shinohara, Y., Wiederholt, I., and Guiraud, B. Evidence for a Pasteur effect regulating cerebral oxygen and carbohydrate metabolism in man. *Neurology* 19:954, 1969.

21. Rossen, R., Kabat, H., and Anderson, J. P. Acute arrest of cerebral circulation in man. *Arch. Neurol. Psychiatry* 50:510, 1943.

22. Finnerty, F. A., Jr., Witkin, L., and Fazekas, J. F. Cerebral hemodynamics during cerebral ischemia induced by acute hypotension. *J. Clin. Invest.* 33:1227, 1954.

*23. Sokoloff, L., and Kety, S. S. Regulation of cerebral circulation. *Physiol. Rev.* 40(Suppl. 4):38, 1960.

24. Landau, W. H., Freygang, W. H., Rowland, L. P., Sokoloff, L., and Kety, S. S. The local circulation of the living brain; values in the unanesthetized and anesthetized cat. *Trans. Am. Neurol. Assoc.* 80:125, 1955.

25. Freygang, W. H., and Sokoloff, L. Quantitative measurements of regional circulation in the central nervous system by the use of radioactive inert gas. *Adv. Biol. Med. Phys.* 6:263, 1958.

26. Sokoloff, L. Local Cerebral Circulation at Rest and During Altered Cerebral Activity Induced by Anesthesia or Visual Stimulation. In Kety, S. S., and Elkes, J. (Eds.), *The Regional Chemistry, Physiology and Pharmacology of the Nervous System.* Oxford: Pergamon, 1961.

27. Sokoloff, L., Reivich, M., Patlak, C. S., Pettigrew, K. D., Des Rosiers, M., and Kennedy, C. The [^{14}C] deoxyglucose method for the quantitative determination of local cerebral glucose consumption. *Trans. Am. Soc. Neurochem.* 5:85, 1974.

28. Kennedy, C., Des Rosiers, M., Patlak, C. S., Pettigrew, K. D., Reivich, M., and Sokoloff, L. Local cerebral glucose consumption in conscious laboratory animals. *Trans. Am. Soc. Neurochem.* 5:86, 1974.

29. Des Rosiers, M. H., Kennedy, C., Patlak, C. S., Pettigrew, K. D., Sokoloff, L., and Reivich, M. Relationship between local cerebral blood flow and glucose utilization in the rat. *Neurology* 24:389, 1974.

30. Kennedy, C., Des Rosiers, M. H., Reivich, M., Sharpe, F., Jehle, J. W., and Sokoloff, L. Mapping of functional neural pathways by autoradiographic survey of local metabolic rate with [^{14}C] deoxyglucose. *Science* 187:850, 1975.

31. Adams, J. E., Harper, H. A., Gordon, G. S., Hutchin, M., and Bentinck, R. C. Cerebral metabolism of glutamic acid in multiple sclerosis. *Neurology* 5:100, 1955.

32. Himwich, H. E., and Nahum, L. H. The respiratory quotient of testicle. *Am. J. Physiol.* 88:680, 1929.

*33. Sacks, W. Cerebral Metabolism in Vivo. In Lajtha, A. (Ed.), *Handbook of Neurochemistry.* Vol. 1. New York: Plenum, 1969.

34. Kety, S. S., Woodford, R. B., Harmel, M. H., Freyhan, F. A., Appel, K. E., and Schmidt, C. F. Cerebral blood flow and metabolism in schizophrenia. The effects of barbiturate semi-narcosis, insulin coma, and electroshock. *Am. J. Psychiatry* 104:765, 1948.

35. Sloviter, H. A., and Kamimoto, T. The isolated, perfused rat brain preparation metabolizes mannose but not maltose. *J. Neurochem.* 17:1109, 1970.

36. Sloviter, H. A., and Suhara, K. A brain-infusion method for demonstrating utilization of glycerol by rabbit brain in vivo. *J. Appl. Physiol.* 23:792, 1967.

37. Raskin, N. H., and Sokoloff, L. Alcohol dehydrogenase activity in rat brain and liver. *J. Neurochem.* 17:1677, 1970.

38. Itoh, T., and Quastel, J. H. Acetoacetate metabolism in infant and adult rat brain in vitro. *Biochem. J.* 116:641, 1970.

39. Pull, I., and McIlwain, H. 3-Hydroxybutyrate dehydrogenase of rat brain on dietary change and during maturation. *J. Neurochem.* 18:1163, 1971.

40. Williamson, D. H., Bates, M. W., Page, M. A., and Krebs, H. A. Activities of enzymes involved in acetoacetate utilization in adult mammalian tissues. *Biochem. J.* 121:41, 1971.

41. Owen, O. E., Morgan, A. P., Kemp, H. G., Sullivan, J. M., Herrera, M. G., and Cahill, G. F., Jr. Brain metabolism during fasting. *J. Clin. Invest.* 46:1589, 1967.

*42. Krebs, H. A., Williamson, D. H., Bates, M. W., Page, M. A., and Hawkins, R. A. The role of ketone bodies in caloric homeostasis. *Adv. Enzyme Regul.* 9:387, 1971.

*43. Sokoloff, L. Metabolism of ketone bodies by the brain. *Annu. Rev. Med.* 24:271, 1973.

44. Klee, C. B., and Sokoloff, L. Changes in D(-)-β-hydroxybutyric dehydrogenase activity during brain maturation in the rat. *J. Biol. Chem.* 242:3880, 1967.

45. Sloviter, H. A. Personal communication, 1971.

46. Himwich, H. E., and Fazekas, J. F. Comparative studies of the metabolism of the brain of infant and adult dogs. *Am. J. Physiol.* 132:454, 1941.

47. Kennedy, C., Grave, G. D., Jehle, J. W., and Sokoloff, L. Changes in blood flow in the component structures of the dog brain during postnatal maturation. *J. Neurochem.* 19:2423, 1972.

*48. Kety, S. S. Human cerebral blood flow and oxygen consumption as related to aging. *Res. Publ. Assoc. Res. Nerv. Ment. Dis.* 35:31, 1956.

49. Sokoloff, L. Cerebral circulatory and metabolic changes associated with aging. *Res. Publ. Assoc. Res. Nerv. Ment. Dis.* 41:237, 1966.

50. Larrabee, M. G., Ramos, J. F., and Bülbring, E. Effects of anesthetics on oxygen consumption and synaptic transmission in sympathetic ganglia. *J. Cell. Comp. Physiol.* 40:461, 1952.

51. Dumke, P. R., and Schmidt, C. F. Quantative measurements of cerebral blood flow in the Macacque monkey. *Am. J. Physiol.* 138:421, 1943.

*52. Sokoloff, L. Cerebral Blood Flow and Energy Metabolism in Convulsive Disorders. In Jasper, H. H., Ward, A. A., and Pope, A. (Eds.), *Basic Mechanisms of the Epilepsies.* Boston: Little, Brown, 1969.

*53. Sokoloff, L. Neurophysiology and Neurochemistry of Coma. In Polli, E. (Ed.), *Neurochemistry of Hepatic Coma.* New York/Basel: S. Karger, 1971.

*54. Sokoloff, L. Cerebral Circulation and Behavior in Man: Strategy and Findings. In Mandell, A. J., and Mandell, M. P. (Eds.), *Psychochemical Research in Man.* New York: Academic, 1969.

55. Mangold, R., Sokoloff, L., Conner, E. L., Kleinerman, J., Therman, P. G., and Kety, S. S. The effects of sleep and lack of sleep on the cerebral circulation and metabolism of normal young men. *J. Clin. Invest.* 34:1092, 1955.

56. Reivich, M., Isaacs, G., Evarts, E. V., and Kety, S. S. The effect of slow wave sleep and REM sleep on regional cerebral blood flow in cats. *J. Neurochem.* 15:301, 1968.

57. Lassen, N. A., Munck, O., and Tottey, E. R. Mental function and cerebral oxygen consumption in organic dementia. *Arch. Neurol. Psychiatry* 77:126, 1957.

*58. Sokoloff, L. Control of Cerebral Blood Flow: The Effects of Anesthetic Agents. In Papper, E. M., and Kitz, R. J. (Eds.), *Uptake and Distribution of Anesthetic Agents.* New York: McGraw-Hill, 1963.

59. Kety, S. S., Polis, B. D., Nadler, C. S., and Schmidt, C. F. The blood flow and oxygen consumption of the human brain in diabetic acidosis and coma. *J. Clin. Invest.* 27:500, 1948.

60. Bessman, S. P., and Bessman, A. N. The cerebral and peripheral uptake of ammonia in liver disease with an hypothesis for the mechanism of hepatic coma. *J. Clin. Invest.* 34:622, 1955.

Chapter 20

Blood-Brain-CSF Barriers
Robert Katzman

TRANSPORT MECHANISMS

Although the brain obtains its required constant supply of oxygen and glucose from the bloodstream, other substances are not readily absorbed from the blood. For example, fructose cannot be substituted for glucose in the treatment of hypoglycemic coma, although brain slices will metabolize it as well as glucose. The central nervous system of mammals appears to require an ultrastable internal environment in order to function effectively, and there are special controls involved in the transport of many materials in the CNS. The sum total of these special transport mechanisms is called the *blood-brain barrier.* Similar special transport mechanisms between blood and cerebrospinal fluid (CSF) constitute the *blood-CSF barrier.* These topics have been reviewed comprehensively [1, 2].

Many substances either do not cross brain capillaries (e.g., acidic dyes) or they cross at a very slow rate (e.g., ions). The classic experiments of Paul Ehrlich in 1882 are said to be the first to demonstrate that animals given vital dyes become intensely stained by the dye in all parts of the body except the brain. In 1909 Goldman studied the vital staining by trypan blue and again noted that, after parenteral administration of a dye, only the brain remains unstained. He found that, even after injection of such large amounts of trypan blue that the animal tissues became intensely stained, the brain remained "snow-white." When he instilled a small amount of trypan blue into the subarachnoid space, however, the brain became intensely stained. The processes that hindered the movement of trypan blue from the bloodstream to the CNS but did not hinder the movement of the dye from the spinal fluid into the brain became known as the *blood-brain barrier.* It is now known that trypan blue quickly binds to albumin in the bloodstream. Hence, the impermeant molecule studied by Goldman was the albumin-dye complex. The morphological basis of this blood-brain barrier to proteins will be discussed shortly.

The concentration of many substances in brain and CSF is independent of their concentration in the blood. Homeostatic mechanisms produce an ultrastable internal environment in the brain as other mechanisms maintain the stable internal environment of the body. The stability of the internal environment may be due either to special transport processes (utilizing mediated transport, active clearance, and other mechanisms) or to the dynamic state of cerebral metabolites. It may be worthwhile to review briefly some of the mechanisms involved in transport. (See [3] and Chap. 6.)

1. *Bulk Flow.* Bulk flow means that solutes of various sizes move together with the solvent as a bulk liquid. This concept is important in discussing circulation and absorption in the CSF. Cerebrospinal fluid circulates through the ventricular subarachnoid spaces and is absorbed into the bloodstream as a bulk fluid. Because bulk flow implies that solutes of various sizes move collectively with the solvent, all as a

single body, it is convenient to measure the bulk absorption of CSF into the blood-stream by measuring the rate of clearance of an inert large molecular-weight tracer, such as blue dextran, radiolabeled serum albumin, or inulin. These tracers do not diffuse into adjacent brain tissue at an appreciable rate. It has been suggested that there may also be bulk flow of brain extracellular fluid through the brain, but this has not yet been unequivocally demonstrated.

2. *Diffusion.* Diffusion consists in the movement of a solute from a region of higher to one of lower concentration as a result of the random motion of the solute molecule. Diffusion occurs both in the CSF and in the extracellular spaces of the brain. In addition, diffusion across plasma membranes is important, especially diffusion across the membranes of the capillary endothelial cells. This is the major route of movement of water, urea, and gases into the brain.

3. *Pinocytosis.* In this process, fluid that includes large molecules and even particles is engulfed by invaginating cell membranes, forming a vesicle which then separates from the membrane. This vesicle can now transport its contents across the cell. Under ordinary conditions, pinocytosis is a slow and uncertain process in brain capillaries and probably moves few molecules.

4. *Mediated Transport.* Mediated transport requires a carrier molecule with specific sites for the substrate involved; this permits the substrate molecule to move readily across either the plasma membrane of a single cell or a cellular membrane composed of a sheet of cells in continuity. When the limited number of sites on the carrier molecule are filled, as, for example, when the concentration of substrate molecules on one side of the membrane is sufficient, the number of molecules transported can no longer increase. With further increases of the substrate molecule, the carrier is then saturated and transport is independent of concentration. The kinetics of carrier-substrate transport are identical with those of enzyme-substrate complexing. Moreover, mediated transport implies that another molecule of sufficiently close chemical and steric similarity may occupy the site intended for a given substrate molecule. In the mammalian blood-brain or blood-CSF barrier, such carriers are hypothesized to explain the stereospecificity and concentration independence of transport of hexoses and amino acids, but as yet carrier molecules per se have not been isolated.

5. *Active Transport.* Active transport implies that energy is utilized in the transport of a molecule. To prove the existence of active transport, either the energy-utilizing process must be specifically identified or it must be shown that the molecule is transported against an electrochemical gradient. Although it is possible that the transport of many molecules by carriers does require energy, this must be demonstrated for each molecular species studied. The best examples of active transport are the halides and the small organic molecules which can be actively cleared from CSF, even though the blood level of such molecules is much higher than that in CSF.

6. *Stability Due to Transport Processes.* The question arises of how the various transport processes can be combined to provide a high degree of stability for the constituents of the CSF and the brain extracellular fluid. Bradbury [4] has recently defined stability of the blood-CSF systems as follows. If a substance is present in CSF at concentration C_{CSF} and in plasma at concentration C_{pl}, stability occurs

when, as a result of a change in plasma concentration, a new steady state is reached so that

$$\Delta C_{CSF} < \Delta C_{pl} \tag{20-1}$$

At steady state the flux of this substance from plasma to CSF, J_{in}, must equal its flux out, J_{out}, so that for any change in plasma concentration, ΔC_{pl}, stability of CSF will occur when

$$\Delta J_{in}/\Delta C_{pl} < \Delta J_{out}/\Delta C_{CSF} \tag{20-2}$$

J_{in} and J_{out} represent transport processes that need not be identical. For instance, one might be passive and one active. If the carrier involved in J_{in} is saturated at usual plasma concentrations, then the ratio $\Delta J_{in}/\Delta C_{pl}$ will approach zero. Such carrier mediated transport is probably the most common mechanism controlling the flow of nonlipid soluble substances from the capillary lumen to the brain, but carrier systems have also been found to operate for outward flux. Here, the greatest stability is achieved when the carrier system operates well below saturation, so that the ratio $\Delta J_{out}/\Delta C_{CSF}$ is a positive number. Such asymmetric carrier mechanisms have been implicated in the maintenance of the control of the stability of K^+ in CSF and may also exist for other molecules.

7. *Stability Due to Dynamic State of Cerebral Metabolites.* The processes involved in maintaining constant brain levels of molecules that are metabolized within the brain system are obviously complex and depend upon the interplay of dynamic processes. For example, the level of dopamine in brain tissue is relatively constant under normal conditions. This level is the result of an equilibrium between the synthesis of dopamine via tyrosine hydroxylase and DOPA decarboxylase; its degradation by dopamine β-hydroxylase, monoamine oxidase, and catechol-O-methyl transferase; and its storage in granules, which isolates it from these degradative enzymes. Synthesis is usually rate limited by the activity of tyrosine hydroxylase. If this is bypassed by administration of large amounts of the amino acid L-DOPA, the brain dopamine will increase. If dopamine in storage granules is released by the administration of reserpine, it will be destroyed quickly by the degradative enzymes, and the total brain level will fall dramatically. Hence, the interplay of factors in remaining levels of metabolizable compounds is exceedingly complex and relatively difficult to analyze.

CHARACTERISTICS OF THE BLOOD-BRAIN BARRIER

Morphology

For many years there was confusion about the nature of the blood-brain barrier since it did not seem conceivable that the capillaries within the brain could be different from capillaries elsewhere in the body. Ultrastructural studies using markers such as peroxidase, the reaction product of which can be stained in situ with osmium, have shown, however, that the capillary endothelium is a continuous layer, with tight junctions between contiguous cells that do not permit the passage of protein markers from the capillary lumen to the basement membrane that surrounds the capillary endothelium [5]. Studies using microperoxidase have shown that molecules with molecular weights as low as 2000 are excluded by these tight junctions [6]. Although it has not been possible to establish morphologically whether the tight junctions also exclude small hydrophilic molecules of the molecular weight of sugars and amino acids, physiological data suggest that they do and that the primary movement of such substances is via mediated transport. In contrast, if the protein marker enters the extracellular space of the brain (for example, if it is placed in the subarachnoid or ventricular space), it will diffuse through this extracellular space until it reaches the capillary endothelium. Under these circumstances, the tight junctions will prevent the molecules from diffusing into the capillary lumen.

Diffusion
Gases such as CO_2, O_2, N_2O, Xe, Kr, and volatile anesthetics diffuse rapidly into brain. As a consequence, the rate at which concentration in brain comes into equilibrium with plasma concentration is limited primarily by the cerebral blood flow. Hence, the inert gases — N_2O, Xe, and Kr — have been widely used to measure cerebral blood flow by both the Kety technique for total blood flow and the Lassen-Ingvar technique for regional blood flow.

Water is a most important substance entering the brain by diffusion. When deuterium oxide (D_2O) is administered intravenously as a tracer, the half-time of exchange of brain water varies between 12 and 25 sec, depending upon the vascularity of the region studied. This rate of exchange is rapid compared with the rate of exchange of most solutes, but it is limited both by the permeability of the capillary epithelium to H_2O and by the rate of cerebral blood flow [7]. In fact, the calculated permeability constant of the cerebral capillary wall to the diffusion of D_2O is about the same as that estimated for the diffusion of water across the lipid membranes [8].

Water also moves freely into or out of the brain as the osmolality of the plasma changes. This phenomenon is clinically useful, since intracranial pressure can be reduced by the dehydration of the brain after plasma tonicity has been elevated by such substances as mannitol. It has been calculated that when plasma osmolality is raised from 310 milliosmols (mOsm) to 344 mOsm, for example, a 10 percent shrinkage of the brain will result, with half of the shrinkage taking place in 12 min. The permeability constant calculated for this osmotic flow of water is slightly larger than that for diffusional flow. However, the movement of water under osmotic load in the example given is slower than diffusional flow because the concentration gradient for the osmotic load is only 34 mOsm, but, when D_2O is used to measure diffusion, the gradient is 55 Osm, the same as for H_2O.

Lipid-soluble substances also move freely, diffusing across the plasma membranes in proportion to their lipid solubility. The permeability of such alcohols as ethanol is especially great. The permeability constant of such substances as thiopental and aminopyrine is similar to that of antipyrine; barbital is somewhat slower, and urea and salicylic acid are still slower. In a compound such as salicylic acid, the un-ionized form diffuses across the membranes, and hence the dissociation constant and the pH of the blood are of importance [9].

It should be noted that, whereas many drugs may depend upon their lipid solubility for entry into the brain, amphetamine may enter the brain by a saturable but nonstereospecific mediated transport system [10].

Mediated Transport

There is evidence that glucose, amino acids, and ions are transported by carrier systems. Due to the steric specificity of the glucose transport system, D-glucose, but not L-glucose, will enter the brain. Such hexoses as mannose and maltose will also be transported rapidly into the brain; the uptake of galactose is intermediate, whereas fructose is taken up very slowly [11]. 2-Deoxyglucose, on the other hand, is taken up quickly and will competitively inhibit the transport of glucose. The 2-deoxyglucose is then phosphorylated, but not further metabolized. If 2-deoxyglucose is used in tracer quantities, the amount of the phosphorylated tracer in the brain reflects the rate of glucose uptake and metabolism (see Chaps. 19 and 31). In addition, there is a separate stereospecific transport system for L-lactate and such other low-weight monocarboxylic acidic glucose metabolites as acetate and pyruvate. The rate of entry of these substances is significantly lower than that of glucose [12, 13].

There is great variability in the rate of movement of amino acids into the brain [11]. Phenylalanine, leucine, tyrosine, isoleucine, tryptophan, methionine, histidine, and DOPA may enter as rapidly as glucose, whereas alanine, proline, glutamic acid, aspartic acid, γ-aminobutyric acid, and glycine are virtually excluded. Lysine, arginine, threonine, and serine are intermediate in their rate of uptake. Competitive inhibition is easy to demonstrate. Although there may be several carriers of amino acids across the brain capillary, there appears to be a single carrier of large neutral amino acids with properties similar to the L (for leucine) transport system defined by Christensen [14] for various mammalian cellular and organic systems [15, 16]. It should be noted that both the essential amino acids and the amino acids serving as precursors for catecholamine and indoleamine synthesis are transported readily, whereas amino acids synthesized readily from glucose metabolites, including those amino acids that act as neurotransmitters, are virtually excluded from the brain.

Simple charged ions will exchange with brain ions, although the exchange is noticeably slower than with other tissues. For example, intravenously administered $^{42}K^+$ exchanges with muscle in 1 hr, but K^+ exchange in brain is only half completed in 24 to 36 hr. The rate of K^+ flux shows little change as plasma K^+ is varied. Ca^{2+} and Mg^{2+} exchange as slowly as does K^+. Na^+ is somewhat faster, with half exchange occurring in 3 to 8 hr, but Na^+ exchange is much slower between blood and brain than between blood and the Na^+ of other tissues.

A good contrast is found between the effects of CO_2, a gas which diffuses freely, and H^+, which moves very slowly into brain. Consequently, the brain pH will reflect blood P_{CO_2} rather than blood pH. In a patient with a metabolic acidosis and a secondary respiratory alkalosis, the brain then will tend to become alkalotic.

It is not known which of these carrier-mediated transport processes are immediately linked to metabolism. In many tissues, the transport of such organic molecules as amino acids requires the presence of Na^+; whether this is true for brain tissue has not been established.

In addition to carrier-mediated transport into brain tissue, carrier-mediated transport out of such tissue has been established for some amino acids. Mediated transport of organic acids and certain halogen ions from CSF into the blood has been well established. Evidence shows that the organic acid and halogen transport systems also serve to transport molecules from brain into blood [17, 18].

Enzymatic Barriers in Capillary Endothelium

For at least two molecules, γ-aminobutyrate and L-DOPA, transport between blood and brain tissue is retarded by enzymatic degradation of these substances within the capillary endothelium. In both instances, the rate of degradation is such that the administration of very large amounts of γ-aminobutyrate or L-DOPA will cause an increase in the amount of these substances reaching brain tissue [19]. Such enzymatic degradation can be demonstrated histochemically. The enzyme γ-glutamyl transpeptidase, which may function in other tissues in the transfer of amino acids, is present in significant activity in cerebral capillaries. The relationship of this finding to the transport of amino acids across the blood-brain barrier has not yet been established [20].

Permeability of Capillary Endothelium

The existence of the tight endothelial junctions impedes water-soluble molecules that are much larger than urea molecules. The evidence from ultrastructural studies shows that molecules with molecular weights as low as 2000 do not enter. Physiological studies in which the osmotic effect of different solutes upon the brain has been monitored have shown that there is little diffusion of hexoses [21]. The uptake of hexoses noted previously is presumably via mediated transport.

Because of the failure of proteins to move into the brain, molecules that are bound to protein are, therefore, impermeant, even if intrinsically lipid soluble. Examples of these are dyes such as trypan blue and the important molecule bilirubin. If sulfa drugs are administered to an infant or young animal with elevated bilirubin, the bilirubin will be displaced from the albumin by the sulfa drug. This can lead to an increase in kernicterus in jaundiced infants. Some movement of impermeant molecules can occur by pinocytosis, but this is slight in the cerebral capillary endothelium as compared with that in other capillaries.

INFLUENCES ON BLOOD-BRAIN BARRIER

Development

Many of the features of the blood-brain barrier, including the presence of tight junctions in the capillary endothelium and the exclusion of protein molecules, are present in newborn animals. Transport of certain substances — for example, simple ions — is moderately faster in the newborn animal. There is a very great increase in the rate of uptake of certain actively metabolized substances in the rapidly growing brain, which may result either from transport per se or from the high rate of turnover of the metabolites.

Chemicals and Drugs

As already discussed, competitive inhibitors may interfere with carrier-mediated transport of molecules into the brain. Drugs such as dilute mercuric chloride, given intravascularly, may increase the permeability of the blood-brain barrier. There is some possibility that greatly elevated levels of PCO_2 may also increase blood-brain barrier permeability [22].

Hyperosmolarity

Intracarotid injection of extremely hyperosmotic solutions disrupts the blood-brain barrier by osmotic shrinkage of capillary endothelial cells and may produce cerebral edema and focal necrosis. It has been demonstrated, however, that this effect can be controlled and that, by the injection of small amounts of such hyperosmotic solutions as 2 M urea or 5% NaCl, a reversible opening of the blood-brain barrier can be produced [23]. Apparently, there is both opening of tight junctions secondary to the osmotic shrinkage of cells and an increase in pinocytotic activity [24, 25]. Such intracarotid injections permit an increased uptake of tracers that ordinarily are excluded, e.g., horseradish peroxidase. Surprisingly, an increase in facilitated transport of glucose has been reported [26]. Whether this ability to control the blood-brain barrier reversibly will be important in the use of chemotherapeutic agents normally excluded by the barrier has not yet been determined. Contrast media used routinely for cerebral angiography are perhaps hypertonic enough to produce similar changes during angiography.

Pathology

A characteristic of almost any focal injury to the brain — whether it is produced by excessive convulsions, knife wounds, freezing, heating, electrical currents, tumors, inflammatory agents, toxins, or other causes — is the breakdown of the blood-brain barrier with subsequent diffusion of protein molecules into brain tissue. This phenomenon has been used widely to trace such areas of injury, since administration of such dyes as trypan blue bound to protein, or fluorescein bound to protein, enables histological identification of those areas in which the blood-brain barrier has broken down. The mechanism of traumatic injury to the capillary endothelium is still under investigation. Apparently, the tight junctions between cells are sometimes ruptured; in other instances, the capillary endothelial cells are simply torn, so that plasma proteins may pour into the area of injury. But in some instances, at least in certain brain tumors, the capillary endothelium thins out, forming fenestrations across which proteins may move.

REGIONAL DIFFERENCES IN THE BLOOD-BRAIN BARRIER

There are, of course, significant differences in the distribution of metabolites throughout the brain. Hence, the blood-brain barrier, in a broad sense of the term, must be different in different regions. Movement of substances into given regions will depend upon such factors as capillary density. Most important, however, are various regions that are "excluded" from the blood-brain barrier. These are regions in which the capillary endothelium contains fenestrations across which proteins and small organic molecules may move from the blood into the adjacent tissue. Examples of such areas are the area postrema, the median eminence of the hypothalamus, the line of attachment of the choroid plexus, and the pineal gland. These areas may have special functional significance in that they may be where the brain samples the contents of the blood. The area postrema is close to what has been called the "vomiting" center of the brain, and the hypothalamus is involved in the regulation of the body's metabolic activity. Therefore, these areas that lack the blood-brain barrier may provide sites at which neuronal receptors may sample plasma directly.

Physiologists have been aware that such substances as radio-labeled serum albumin, which are excluded from entering the brain at the capillary endothelium, nevertheless accumulate throughout the cerebrum in quantities just above those which can be accounted for by the vascular volume of the brain. Recently, it has been demonstrated that some small arterioles (15 to 30μ in diameter), located primarily near the sulci at the cerebral cortex and scattered through the diencephalon and neighboring regions, will transport such high molecular-weight compounds as horseradish peroxidase through their endothelial cells by the process of pinocytosis. These same arterioles possess tight junctions as does the capillary endothelium [27]. On a quantitative basis, such transport is not sufficient to circumvent the blood-brain barrier, but it may account for the small amounts of serum protein that enter the CSF.

CHARACTERISTICS OF BLOOD-CSF BARRIER

Composition of CSF
Cerebrospinal fluid has been characterized as an "ideal" physiological solution. It differs from plasma in that it is almost free of protein, and it differs from an ultrafiltrate of plasma by maintaining the concentrations of various ions at different levels (Table 20-1).

CSF Formation
CSF is constantly being formed and removed. The major site of CSF formation is in the choroid plexuses of the ventricles. In addition, a significant extrachoroidal formation of CSF has been demonstrated. In normal subjects, the rate of CSF secretion has been estimated to be 0.3 to 0.4 milliliter per minute, about one-third the rate at which urine is formed. The fluid elaborated in the ventricles contains only 5 to 10 milligrams of protein per 100 ml.

There is considerable evidence that the choroid plexus acts to secrete the CSF. Histochemical and electron-microscopic investigations indicate that the cells have morphological features similar to those of other secretory cells. Formation of CSF within the ventricles has been deduced from the fact that, if the foramen is obstructed, fluid accumulates rapidly and the ventricle enlarges. From time to time, neurosurgeons have reported seeing drops of fluid form on the surfaces of the choroid plexus.

Table 20-1. Typical Plasma and CSF Levels of Various Substances[a]

Substance	Plasma	CSF
Na^+	145.0	150.0
K^+	4.8	2.9
Ca^{2+}	5.2	2.3
Mg^{2+}	1.7	2.3
Cl^-	108.0	130.0
HCO_3^-	27.4	21.0
Lactate	7.9	2.6
PO_4^{3-}	1.8	0.5
Protein	7000.0	20.0
Glucose	95.0	60.0

[a]Protein and glucose concentrations in mg/100 ml; all others in mEq/liter.

The formation of CSF by the choroid plexus has been studied more directly. Welch [28] has been able to cannulate the artery and vein of a single choroid plexus in the rabbit. By measuring the average differences of radioisotopes and of the hematocrit, he demonstrated that the formation of CSF and the movement of $^{24}Na^+$ are stopped when a carbonic anhydrase inhibitor (diamox) is applied to the plexus. Thus, CSF is not simply a serum ultrafiltrate but is controlled by enzymatic processes. Ames et al. [29] were able to collect and analyze the fluid formed at the choroid plexus after filling the ventricle with oil. In the cat, they found that the concentration of electrolytes, particularly K^+, Ca^{2+}, and Mg^{2+}, differs from that of a plasma ultrafiltrate. However, the choroid plexus fluid was again found to change very slightly in its electrolytic concentration by the time it reached the cisterna magna.

The rate of formation of CSF has been measured by various means. One simple measurement, carried out more than 30 years ago by Masserman, was to determine the time for CSF replacement after a known amount had been drained. The rate in man was 0.35 milliliter per minute. This measurement was criticized on the basis that the drainage altered the intracranial pressure relationships and, therefore, probably altered (increased) the rate of CSF formation. Recently, a sophisticated method of measuring CSF formation was introduced by Pappenheimer and co-workers [30, 31]. In this method, simulated spinal fluid is perfused between the ventricle and the cisterna magna, and inulin is added to the perfusion fluid. Because inulin diffuses very slowly into tissue during such a perfusion, the dilution of inulin is taken as a measure of the rate of formation of new spinal fluid. Therefore, if the perfusion at a rate of V_i ml per minute is carried out until a steady state is reached and if the initial concentration of inulin is C_i and the outflow concentration of inulin is C_o, the rate of formation of spinal fluid, V_f, is given by the equation:

$$V_f = V_i(C_i - C_o)/C_o \tag{20-3}$$

Such perfusions have been carried out in a wide variety of species. Typical rates of spinal fluid formation are: rabbit, 0.001 ml per min; cat, 0.02 ml per min; rhesus monkey, 0.08 ml per min; and goat, 0.19 ml per min. Recently, both Cutler et al. [32] and Rubin et al. [33] carried out measurements in patients undergoing ventriculocisternal perfusion with antitumor drugs and found an average value of 0.37 ml per minute. This corresponds rather closely to the value previously determined from drainage experiments.

By use of the ventriculocisternal perfusion method, it has been found that CSF formation decreases slightly as intracranial pressure increases. Moreover, CSF is formed at a normal rate and osmolality even when the fluid perfused through the ventricle is moderately hypertonic or hypotonic. However, when the perfused fluid is very hypotonic, CSF formation may cease [34]. CSF formation is also reduced if serum osmolality is raised.

The total volume of CSF is not precisely measurable. It is estimated to be 100 to 150 ml in normal adults.

The circulation of CSF begins with the elaboration of fluid in the ventricles. It is aided by arterial pulsations of the choroid plexuses. These pulsations are transmitted throughout the CSF and can be seen on the manometer at the time of lumbar spinal tap. The usual manometer, however, is filled with spinal fluid, and this displacement of fluid tends to dampen the pulsations. More prominent pulsations are recorded if a strain gauge is used.

The fluid circulation is from the lateral ventricles into the third ventricle and then into the fourth ventricle. If obstructions are placed at the foramen between these ventricles, the ventricle upstream from the obstruction will enlarge significantly, producing obstructive hydrocephalus. Thus, if a foramen of Monro is obstructed, the lateral ventricle will enlarge. If the aqueduct is obstructed, both lateral ventricles and the third ventricle will enlarge. The fluid passes from the fourth ventricle to the cisterna magna and then circulates into the cerebral and spinal subarachnoid spaces.

The CSF is removed at multiple sites. Among the important sites are cranial and spinal nerve root sheaths and the villi and granulations over the large venous sinuses in the skull. There is evidence that the villi act as valves, permitting the one-way flow of CSF from the subarachnoid spaces into sinuses. It has been suggested that some absorption of CSF may occur via pial vessels or by unidirectional pinocytotic activity in capillaries adjacent to the Virchow-Robin spaces at the cortical surface, but this has not been confirmed. Occasionally disease processes will affect these removal sites. For example, obliteration of the subarachnoid space following inflammatory processes or thrombosis of the sinuses will prevent the clearance of fluid. When this occurs, CSF pressure will increase, and hydrocephalus may develop without any obstruction in the ventricular foramina. This is called communicating hydrocephalus. Under such circumstances, some CSF absorption may take place across the ventricular ependyma; in addition, some CSF may circulate to the lumbar subarachnoid space via a dilated central canal [35].

Physiologically, absorption appears to be via a valvular mechanism. CSF absorption does not occur until CSF pressure exceeds the pressure within the sinuses.

Once this threshold pressure has been reached, the rate of absorption is roughly proportional to the difference between CSF and sinus pressures. A normal human can absorb CSF at a rate up to four to six times the normal rate of CSF formation with only a moderate increase in intracranial pressure. This phenomenon has been used in the development of a constant infusion manometric test for estimation of CSF absorptive capacity [36] . The CSF spaces also show elasticity or distensibility: that is, the volume of the space changes in a nonlinear fashion with changes in CSF pressure.

Throughout the circulation of CSF, exchange of substances occurs between CSF and blood. This exchange appears to involve metabolic processes that serve to maintain the concentration of substances within the CSF at relatively fixed values.

Similarity to the Blood-Brain Barrier

The movement of substances from the blood into CSF is, in many ways, analogous to that from blood into the brain, even though there are differences in the development of the choroid plexus and of the capillary system in the brain. There is free movement of such molecules as water, gases, and lipid-soluble substances from the blood into the CSF. Substances important for metabolism and maintenance of CSF electrolytes, such as glucose, amino acids, and cations (including K^+, Ca^{2+}, and Mg^{2+}), are transported by saturable carrier-mediated processes. Although it has been supposed that Cl^- content of CSF was determined by passive movement of this anion, it is now evident that a Cl^- "pump" may exist. The transport of Na^+ from blood into CSF is significantly reduced when carbonic anhydrase is inhibited. With Na^+, however, it has not been possible to demonstrate the phenomenon of saturability, because blood Na^+ cannot be elevated safely much above its usual concentrations. Finally, such macromolecules as proteins and most hexoses other than glucose are impermeable and do not enter the CSF.

Active Transport from CSF and Brain

Active transport from CSF has been especially well studied because it is possible to manipulate the concentration gradient of substances between the CSF and the bloodstream. It has been shown that iodide and thiocyanate, transported from the CSF by saturable carrier mechanisms, can be inhibited competitively (for example, perchlorate will inhibit iodide transfer) and must be active because the transport can be carried out against unfavorable electrochemical gradients. A special relationship for choroid plexus in iodide clearance is proposed, based upon the very active accumulation of anions that occurs in choroid plexus in vitro. The combination of the bulk absorption of CSF and the active transport of molecules from it has been termed the "sink" function of the CSF. This implies that molecules reaching the extracellular fluid of the brain may diffuse into CSF and then be removed either by bulk absorption, by active transport, or by both mechanisms. It has been postulated that this will explain such phenomena as the low concentration of iodide in both brain and CSF, as compared with plasma concentrations. The possibility has now been established for iodide, however, that brain capillaries may transport the same substances out of the brain as the choroid plexus does out of the CSF.

Another important transport system is that involved in clearing weak organic acids out of the CSF. Among the molecules cleared by this mechanism are penicillin, diodrast, and such metabolites as HVA and 5-HIAA. This clearance system also shows the phenomenon of saturability and transport against an unfavorable CSF-blood gradient. Probenecid is an effective inhibitor of this system. Again, the choroid plexus can accumulate these molecules in vitro, and there is some suggestive evidence that it plays a role in vivo. In addition, however, rapid clearance of 5-HIAA by probenecid-sensitive transport mechanisms is shown in the cerebral subarachnoid space [18]. There is also evidence that amino acids are cleared from CSF to blood by analogous transport processes [37].

CSF-BRAIN INTERFACE

The absence of tight junctions between some ependymal and pial cells permits diffusion of proteins and other hydrophilic molecules from the CSF into the brain, and vice versa. Frequently, the diffusional gradients near the interface are very steep, and many substances penetrate only superficially. Some recent evidence shows that the diffusion across the ependyma is slightly faster than that across the pial-glial surface; however, molecules do move across both surfaces by diffusion.

Quantitative studies of the movement of substances between CSF and brain have led some investigators to postulate that the concentration of ions in extracellular fluid in brain is similar to that of CSF. Because the concentrations of K^+, Ca^{2+}, and Mg^{2+} in CSF are quite different from those in plasma, the verification of this postulate was considered to be of physiological significance (see first edition of this text). It has now been shown, using specific ion-exchanger electrodes that can measure K^+ activity in brain extracellular space, that in fact this postulate was correct, that brain extracellular K is at a concentration of about 3 mM, similar to CSF and different from plasma concentration. Moreover, brain extracellular K is extremely stable, except under such severe experimental conditions as spreading depression or death (when it may rise to values as high as 80 mM) or sustained seizure activity (when it may rise to levels of about 10 to 12 mM). Only small shifts, usually less than 1 mM, are found as a result of local increases in cerebral activity, such as are produced by evoked cortical activity [38]. Thus one of the functions of CSF, in addition to mechanical protection of the brain and its "sink" function, may be to provide a reservoir of extracellular fluid.

The CSF-brain interface may play a special role in the regulation of respiration. There is suggestive evidence that some respiratory neurons or their processes are located close enough to the ependymal surface to be responsive both to the HCO_3^- level in the CSF and to capillary P_{CO_2}.

ANTIBIOTICS

The movement of penicillin, perhaps the most important of the antibiotics, into the brain is controlled largely by the probenecid-sensitive, organic-acid transport system in both the choroid plexus and brain capillaries. As a consequence, in the normal state, penicillin concentrations in the CSF are much lower than those in plasma. This is, in fact, of considerable benefit because penicillin is toxic to cerebral tissue and may produce seizures at relatively low concentrations. During active bacterial infections of the meninges of the brain, however, there are alterations in the blood-brain barrier, and the penicillin concentrations will increase at the site of inflammation. Similar mechanisms may control such compounds as ampicillin and penicillin G. The low concentration of cephalothin in the CSF might be due to a similar mechanism, but this has not been studied explicitly. However, it is known that the concentration of cephalothin does not increase reliably during infection and, therefore, it is not the optimum drug to use in patients with meningitis. The concentration of gentamicin in CSF is so low that, if this drug is required, it is often given intrathecally. In contrast, certain broad-spectrum antibiotics, such as chloramphenicol and tetracycline, are transported readily into the brain and, in fact, the concentration in brain tissue may be much higher than that in plasma.

REFERENCES

  *1. Katzman, R., and Pappius, H. M. *Brain Electrolytes and Fluid Metabolism.* Baltimore: Williams & Wilkins, 1973.
  *2. Lajtha, A., and Ford, D. H. (Eds.). *Brain Barrier Systems.* Amsterdam: Elsevier, 1968.
  *3. Davson, H. *Physiology of the Cerebrospinal Fluid.* Boston: Little, Brown, 1967.
   4. Bradbury, M. W. B., and Stulcova, B. Efflux mechanism contributing to the stability of the potassium concentration in cerebrospinal fluid. *J. Physiol.* (Lond.) 208:415, 1970.
   5. Reese, T. S., and Karnovsky, M. J. Fine structural localization of a blood-brain barrier to exogenous peroxidase. *J. Cell Biol.* 34:207, 1967.
   6. Reese, T. S., Feder, N., and Brightman, M. W. Electron microscopic study of the blood-brain and blood−cerebrospinal fluid barriers with microperoxidase. *J. Neuropathol. Exp. Neurol.* 30:137, 1971.
   7. Raichle, M. E., Eichling, J. O., and Grubb, R. L. Brain permeability of water. *Arch. Neurol.* 30:319, 1974.
   8. Katzman, R., Schimmel, H., and Wilson, C. E. Diffusion of inulin as a measure of extracellular fluid space in brain. *Proc. Rudolph Virchow Med. Soc. N.Y.* 26:254, 1968.
   9. Crone, C. The permeability of brain capillaries to non-electrolytes. *Acta Physiol. Scand.* 64:407, 1965.
  10. Pardridge, W. M., and Connor, J. D. Saturable transport of amphetamine across the blood-brain barrier. *Experientia* 29:302, 1972.
  11. Oldendorf, W. H. Brain uptake of radiolabeled amino acids, amines, and hexoses after arterial injection. *Am. J. Physiol.* 221:1629, 1971.
  12. Nemoto, E. M., and Sveringhaus, J. W. Stereospecificity permeability of rat blood-brain barrier to lactic acid. *Stroke* 5:81, 1974.·
  13. Oldendorf, W. H. Carrier mediated blood-brain transport of short-chain mono-carboxylic organic acids. *Am. J. Physiol.* 224:1450, 1973.
  14. Christensen, H. N. Nature and Roles of Receptor Sites for Amino Acid Transport. In Ebadi, M. S., and Costa, E. (Eds.), *Advances in Biochemical Psychopharmacology.* Vol. 4. New York: Raven, 1972.

*Asterisks denote key references.

15. Wade, L. A., and Katzman, R. Rat brain regional uptake and decarboxylation of L-DOPA following carotid injection. *Am. J. Physiol.* 228:352, 1975.
16. Yudilevich, D. L., De Rose, N., and Sepúlveda, F. V. Facilitated transport of amino acids through the blood-brain barrier of the dog studied in a single capillary circulation. *Brain Res.* 44:569, 1972.
17. Davson, H., Domer, F. R., and Hollingsworth, J. R. The mechanism of drainage of the cerebrospinal fluid. *Brain* 96:329, 1973.
18. Wolfson, L. I., Katzman, R., and Escriva, A. Clearance of amine metabolites from the cerebrospinal fluid: The brain as a "sink." *Neurology* 24:772, 1974.
19. Owman, C., and Rosengren, E. Dopamine formation in brain capillaries – an enzymatic blood-brain barrier mechanism. *J. Neurochem.* 14:547, 1967.
20. Orlowski, M., Sessa, G., and Green, J. P. γ-Glutamyl transpeptidase in brain capillaries: Possible site of a blood-brain barrier for amino acids. *Science* 184:66, 1974.
21. Fenstermacher, J. D., and Johnson, J. A. Filtration and reflection coefficients of the rabbit blood-brain barrier. *Am. J. Physiol.* 211:341, 1966.
22. Mayer, S., Maickel, R. P., and Brodie, B. B. Kinetics of penetration of drugs and other foreign compounds into cerebrospinal fluid and brain. *J. Pharmacol.* 127:205, 1959.
23. Rapoport, S. I., Hori, M., and Klatzo, I. Testing of a hypothesis for osmotic opening of the blood-brain barrier. *Am. J. Physiol.* 223:323, 1972.
24. Brightman, M. W., Hori, M., Rapoport, S. I., Reese, T. S., and Westergaard, E. Osmotic opening of tight junctions in cerebral endothelium. *J. Comp. Neurol.* 152:317, 1973.
25. Sterrett, P. R., Thompson, A. M., Chapman, A. L., and Matzke, H. A. The effects of hyperosmolarity on the blood-brain barrier. A morphological and physiological correlation. *Brain Res.* 77:281, 1974.
26. Spatz, M., Rap, Z. M., Rapoport, S. I., and Klatzo, I. The Effects of Hypertonic Urea on the Blood-Brain Barrier and on the Glucose Transport in the Brain. In Reulen, H. J., and Schürmann, K. (Eds.), *Steroids and Brain Edema.* Proceedings of an International Workshop. New York: Springer-Verlag, 1972.
27. Westergaard, E., and Brightman, M. W. Transport of proteins across normal cerebral arterioles. *J. Comp. Neurol.* 152:17, 1973.
28. Welch, K. Secretion of cerebrospinal fluid by choroid plexus of the rabbit. *Am. J. Physiol.* 205:617, 1963.
29. Ames, A., Sakanoue, M., and Endo, S. Na, K, Ca, Mg, and Cl concentrations in choroid plexus of fluid and cisternal fluid compared with plasma ultrafiltrate. *J. Neurophysiol.* 27:672, 1964.
30. Pappenheimer, J. R., Heisey, S. R., and Jordan, E. F. Active transport of diodrast and phenolsulfonphthalein from cerebrospinal fluid to blood. *Am. J. Physiol.* 200:1, 1961.
31. Heisey, S. R., Held, D., and Pappenheimer, J. R. Bulk flow and diffusion in the cerebrospinal fluid system of the goat. *Am. J. Physiol.* 203:775, 1962.
32. Cutler, R. W. P., Page, L., Galicich, J., and Watters, G. V. Formation and absorption of cerebrospinal fluid in man. *Brain* 91:707, 1968.
33. Rubin, R. C., Henderson, E. S., Ommaya, A. K., Walker, M. D., and Rall, D. P. The production of cerebrospinal fluid in man and its modification by acetazolamide. *J. Neurosurg.* 25:430, 1966.
34. Hochwald, G. M., Wald, A., DiMattio, J., and Malhan, C. The effects of serum osmolarity on cerebrospinal fluid volume flow. *Life Sci.* 15:1309, 1975.
35. Eisenberg, H. M., McLennan, J. E., and Welch, K. Ventricular perfusion in cats with kaolin-induced hydrocephalus. *J. Neurosurg.* 41:20, 1974.

36. Katzman, R., and Hussey, F. A simple constant-infusion manometric test for measurement of CSF absorption. 1. Rationale and method. *Neurology* 20:534, 1970.

37. Lorenzo, A. V. Amino acid transport mechanisms of the cerebrospinal fluid. *Fed. Proc.* 33:2079, 1974.

38. Katzman, R., and Grossman, R. Neuronal Activity and Potassium Movement. In Ingvar, D. H., and Lassen, N. A. (Eds.), *Brain Work.* Alfred Benzon Symposium VIII. Copenhagen: Munksgaard, 1975.

Chapter 21
Axoplasmic Transport
Sidney Ochs

The long extension of the axon requires a continual downflow of proteins, including enzymes, organelles, and other components supplied from the cell body. Some of the materials transported in the fibers are components needed to replace constituents of the membrane and organelles of the fiber; others participate in the local metabolism of the fiber; yet others are transmitter or transmitter-related components supplied to the terminals or carried out into the postsynaptic cells to act as trophic substances that control or modify their functional state. A study of axoplasmic transport requires a consideration of both the spatial and the temporal distribution of various components within the neuron and an understanding of this process is a necessary prelude to a full neurochemical analysis of nerve tissue.

Until recently, it was possible to consider the metabolism of the peripheral nerve solely in terms of those mechanisms that subserve the excitation and propagation of the action potential. A sodium pump in the membrane is required to maintain the ionic asymmetry on which the resting potential and the excitation of the nerve impulse depend, the pump supplied by energy in the form of ATP. Another energy demand in nerve must now be considered: that required to maintain the mechanism underlying axoplasmic transport.

METHODS OF STUDY AND MATERIALS TRANSPORTED
The early development of our concept of transport of material in nerves and its characteristics has been described in general reviews [1–11]. In this section, we outline the techniques in use to study transport and its general characteristics. First, we consider transport in the usual outward or anterograde direction away from the cell body. Then we discuss transport back toward the cell body, in the retrograde direction.

Anterograde Transport
If a nerve is cut, constricted, or ligated, an increased bulging of the nerve and its fibers is seen, over a period of days or weeks, just above the level of interruption. This phenomenon has been termed "damming" by Weiss and Hiscoe [1], and they relate it to a continual outward movement of all the axonal contents [1a]. In the dammed region, which usually extends only a few millimeters (mm) above a ligation, various organelles and materials are found to accumulate within the fibers. These include mitochondria and lysosomes, with their various enzymes and components [12]. In addition, an increase in soluble proteins and water is found both within and between the fibers in the dammed region. These changes must be dealt with in estimating the increases of a component in terms of weight or protein.

The pattern of accumulation above the ligations within double-ligated nerve segments and depletion in the rest of the segment of nerve has shown a fast transport of various specific components. Norepinephrine, the sympathetic transmitter seen present in granule form as a fluorescent material in formaldehyde-treated preparations, was shown to have a fast rate of transport in double-ligated nerves [5]. Lubińska and her colleagues found that acetylcholinesterase (AChE) in double-ligated nerves also moves at a fast rate [1a], a rate found to be close to that of labeled proteins [13]. This is seen in frog nerve [12a] as well as in mammalian nerve.

A more direct measure of transport rates and the properties of transport within the fibers is by the use of isotopically labeled precursors. These include ^{32}P as orthophosphate, ^3H- or ^{14}C-labeled amino acids, and, more recently, various other precursors. When injected systemically, they may be taken up from the circulation by the nerve cell bodies and, after their incorporation, labeled phospholipids, phosphoproteins, various proteins, and polypeptides are found to move down the fibers. At various times after injection, the nerves are removed and cut into small sections lengthwise for counting or for preparing autoradiographs. The amount of activity is displayed as a function of distance to show the outflow pattern and to derive the rate of transport from it.

When ^3H-leucine and autoradiography were used, an intraaxonal location of the labeled proteins transported in the fibers was seen. In early studies, only a slow rate of 1−2 mm per day was found [14]. By injecting the labeled precursor directly into the nervous system near the cell bodies, the blood-brain barrier can be bypassed and a much higher level of incorporation attained. As a result, a higher level of activity is transported into the fibers [6a]. Another advantage of this technique is that it greatly reduces the adventitious incorporation of the precursor by the Schwann cells. After direct injection into the cord or after direct uptake of the precursors [^{32}P- or low levels of ^3H-amino acids) by brain stem cell bodies, a declining exponential curve of outflow is seen in the nerve. Its slope becomes shallower with time, which indicates a slow outflow with a rate variously estimated from several millimeters per day to 80 mm per day [15].

Using a higher concentration of ^3H-leucine and sufficiently long lengths of nerve, a much faster outflow was found [6a]. This is seen as a characteristic crest of activity after injecting the motor-neuron region of the lumbar seventh (L7) ventral horn of the spinal cord or the L7 dorsal root ganglion supplying the cat sciatic nerve (Fig. 21-1). The precursor, ^3H-leucine or ^3H-lysine, is taken up relatively quickly and incorporated by the L7 motor neuron or dorsal root ganglion cells within a matter of some 15 minutes (Fig. 21-1). Labeled proteins and polypeptides then move down inside the fibers of the sciatic nerve to give rise to the characteristic crest indicative of fast transport. The crest position was found to progress outward in the sciatic nerve linearly with time at a rate of 410 ± 50 mm per day [6a]. This same rate is seen in the sciatic nerves of mammalian species ranging in size from the rat to the goat, and it does not depend on the functionality of the nerve, sensory or motor. It was shown by autoradiography that the same rate obtains for myelinated fibers ranging in diameter from 3 μ to 22 μ, and the same fast rate also was found in the smaller nonmyelinated fibers. The same fast rate of close to 410 mm per day was

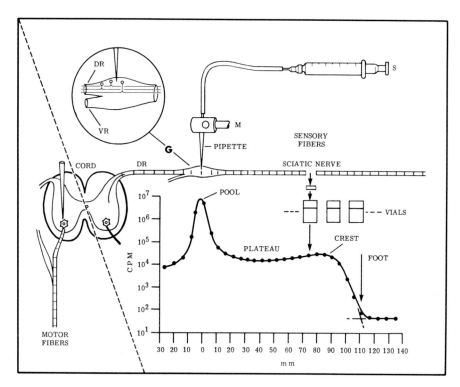

Figure 21-1
Technique of isotope labeling and outflow pattern of fast transport. The L7 dorsal root ganglion is shown with a pipette held by a micromanipulator (M) inserted for injection of ³H-leucine using a polyethylene tube and syringe (S). (Insert shows ganglion with T-shaped nerve cells.) The dorsal root (DR), ganglion (G), and sciatic nerve are cut into 5-mm segments. Each segment is placed in a vial and solubilized, and scintillation fluid added and counted. The display of activity in counts per minute (CPM) is shown, using a logarithmic scale on the ordinate and the distance in millimeters on the abscissa, taking zero as the ganglion. A typical pattern for a 6.5-hour downflow is shown. A high level of activity remains in the ganglion and falls distally to a plateau before rising to a crest. A sharp drop is seen at the front of the crest down to the baseline level at its foot. At the left of the slanting dashed line (left), an injection of the L7 ventral horn is shown for uptake of precursor by the motor neurons. The ventral root (VR) and sciatic nerve are cut and measured for outflow, as in the case of ganglion injection.

computed in the fibers of nonmammalian species — the garfish olfactory [16, 16a] and frog sciatic nerve [17] — when a correction was made to extrapolate to the rate at 38° C. The rate is changed greatly by temperature with a Q_{10} of 2 to 3.5 in all the various transport systems studied [6]. This suggests, as will be shown later under Mechanism of Transport, an enzymatic utilization of energy by the axoplasmic-transport mechanism.

Transport is also studied in the visual system with ³H-leucine and other labeled amino acids injected into the eye for uptake by the ganglion cells of the retina. After its synthesis into proteins in these cells, transport in the optic nerve is shown by the time and the pattern of labeled activity that accumulates in the nerve endings of the ganglion cells in the tectum (fish and bird) [2] or lateral geniculate (mammal) [18] of the opposite side. (In the fish, bird, and rodent, the optic fibers from each eye cross almost entirely to terminate in the opposite tectum or lateral geniculate.) The precursor leaks from the eye into the circulation to a considerable degree and so gains entry to the terminals to become locally synthesized in the tectum or lateral geniculate. Therefore, transport is determined by subtracting the activity in the curve of locally incorporated activity on the ipsilateral side, which does not receive axonally transported components, from that of the contralateral eye, which does. The difference represents the transported components. ^{3}H-Proline is a better precursor to use in the visual system, as it does not readily pass the blood-brain barrier to become locally incorporated [19]. In studies of the visual system, at least two and, in some cases, several more waves of accumulation of labeled materials have been seen, which suggests one or more slow and fast transport systems.

When amino acids are used as precursors, a wide range of proteins and polypeptides is transported within the fibers. This was shown by analyses of sciatic-nerve homogenates or of the tectum or lateral geniculate taken at various times after injections of the labeled precursor. Differential centrifugation and gel filtration of the high-speed supernatant is used to isolate various protein and polypeptide components. A wide range of nerve components is found to be transported at both fast and slow rates [20]. Less than about nine hours after the injection of precursor, and corresponding to fast transport, the particulate fraction is seen to be labeled to a somewhat greater extent than other components. These "other" components include soluble proteins of high molecular weight, from 450,000 to 50,000, polypeptides of molecular weight below 10,000, and some free leucine. During slow transport, in nerves taken one to seven days after injection, a greater degree of labeling in soluble proteins of higher molecular weight is found [20].

When ^{3}H-glucosamine or ^{3}H-fucose is injected as a precursor, a fast downflow of labeled glycoproteins and glycolipids is found [21, 21a]. A rapid transport of sulfated mucopolysaccharide proteins also was found after [35]SO_4 was injected [19]. These results point to a possible transport of these components to the terminals, where they are relocated in the synaptic cleft and play a role in synaptic function. A fast downflow of phospholipids was also reported when ^{3}H-leucine or ^{3}H-choline was used as a precursor [22]. This finding, and the indication of transport using ^{3}H-cholesterol as a precursor [23], suggest a downflow of components that have a turnover in the membrane or membrane-bound organelles. The association of particulates with fast transport was seen when dark-field or Nomarski optics microscopy was used [24]. Relatively large elongated organelles that could only be mitochondria were seen to move within the fibers in both the forward and retrograde direction at rates approaching that of fast transport of labeled proteins. On the other hand, several independent studies using other techniques have shown that the mitochondria have a slow net downflow in nerve fibers [10c]. This can be explained by the mechanism proposed below as underlying axoplasmic transport.

A transport of components into the synaptic terminals was seen after brains were injected with ^{3}H-leucine and the synaptosomes were isolated by differential centrifugation [25]. During a period of days, a gradual accumulation of labeled proteins was found in the synaptosomal membrane fraction. The specific activity of labeled soluble proteins was seen to rise gradually with time, as expected of their slow transport into the terminals. A local synthesis of protein in the terminals is less likely to account for this pattern (see Chap. 16). Recently, studies using electron-microscopic autoradiography have indicated a fast transport of some labeled material into the synaptosomes.

Another technique of interest is the intracellular injection of a labeled amino acid, such as ^{3}H-glycine, directly into the cell body. After its incorporation into protein, grains are seen in autoradiograph preparations indicating the spread both down the axon and into the dendrites [26]. By this means, the full extent of the dendritic arborization may be traced. A similar result may arise after Procion yellow is injected into the cell body and transported into the dendrites.

Retrograde Transport

Retrograde transport was first seen by Lubińska as an accumulation of AChE distal to a nerve ligation [1a]. Enzymatic markers for mitochondria showed that this organelle and other substances are transported in the retrograde direction [8]. In mammalian nerves, accumulations distal to ligations cannot be followed profitably much beyond two days when Wallerian degeneration ensues, but the retrograde transport of AChE is sufficiently fast so that its rate can be assessed readily by using double ligations. By this means, its retrograde transport has been estimated to have a rate half that of forward transport, or close to 200 mm per day [13].

The fiber terminals do not incorporate labeled amino acids with a subsequent retrograde transport of labeled proteins. Exogenous markers have, however, clearly shown retrograde transport. Horseradish peroxidase (HRP) is taken up by nerve fiber terminals and is found located within the cell bodies soon afterward [8]. The rate of its retrograde movement was shown recently by LaVail and LaVail [26a] to be at least 84 mm per day; it may possibly have the still faster rate determined for AChE. For example, HRP was shown to be taken up by damaged nerve fibers from their cut ends, and this extends its usefulness as a neuroanatomical tracer (see section Fiber Tracing Using Transport Autoradiography, page 441).

Retrograde transport may have a special role in the regulation of nerve function. The chromatolysis that is seen several days after transection of the nerve fibers and that lasts some months is associated with increased levels of nucleic acids and proteins in the nerve cell bodies. This suggests that a "signal" substance which normally is transported up the fibers acts as a negative feedback substance controlling protein synthesis in the cell, whereas its absence leads to an increased level of protein synthesis.

Conflicting reports of an increase, decrease, or a lack of change in the rate of axoplasmic transport during chromatolysis have been published. A recent study [26b] of the rate of crest movement in nerves supplied by L7 dorsal root ganglion cells that are made to undergo chromatolysis shows little change in the rate or even in the amount of labeled activity transported in those nerves as compared to controls. The levels of synthesis in the cell body seem to be relatively independent of the rate of transport in the axon. This concept is called the "independence principle" for transport.

MECHANISM OF TRANSPORT

A Model for Transport

The materials transported rapidly are, as noted in the preceding section, heterogeneous both with respect to number and to the molecular weights of components. This requires a common carrier, such as has been proposed in the "transport filament" model [10c]. The various materials transported are considered to be bound to transport filaments that are then moved along the microtubules (Chap. 1) by means of cross-bridges (Fig. 21-2). This model was suggested by the sliding-filament theory for muscle contraction. The microtubules may be the stationary elements along which the transport filaments are moved by making and breaking cross bridges. The energy is most likely supplied by ATP, as will be described in the next section.

Rapid transport is a characteristic of materials that have high affinities for the transport filaments. As a further extension of this hypothesis, we could consider that the more slowly moving components have a greater tendency to dissociate from the transport filaments [10c].

Metabolism and Fast Axoplasmic Transport

Oxidative Phosphorylation

Fast axoplasmic transport is maintained with its usual form and rate in vitro as in vivo. For in vitro studies, the precursor ^{3}H-leucine is injected into the L7 dorsal root ganglia and two or three hours of downflow are allowed in vivo before the sciatic nerves are removed and placed in chambers equilibrated with moist, 95 percent O_2 + 5 percent CO_2 at a temperature of 38° C [6]. The continuation of transport in vitro is shown by the movement of the crest down the nerve at its usual rate [27]. (See also Fig. 21-3.) No special requirement for any of the Krebs-Ringer ions was found for

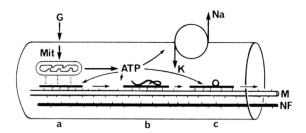

Figure 21-2
Transport filament hypothesis. Glucose (G) enters the fiber and after oxidative phosphorylation in the mitochondrion (Mit), ATP is produced. A pool of ATP supplies energy to the sodium pump, which controls both the level of Na^+ and K^+ in the fiber and the movement of "transport filaments," shown as black bars. The various components bound to the filaments and carried down the fiber include: the mitochondria (a) attached, as indicated by dashed lines, to the transport filament and giving rise to a fast to-and-fro movement (though with a slow net forward movement); soluble protein (b) shown as a folded or globular configuration; and other polypeptides and small particulates (c). Thus a wide range of component types and sizes is carried down the fiber at the same fast rate. Cross-bridges between the transport filament and the microtubules effect the movement when supplied by ATP.

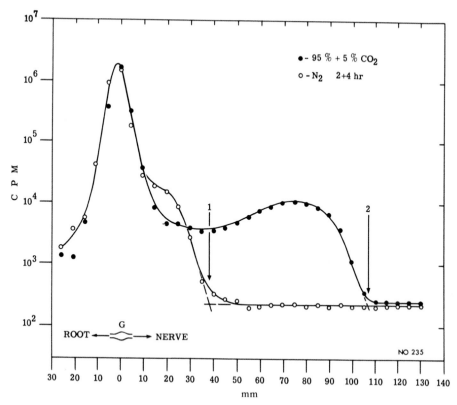

Figure 21-3

Block of in vitro downflow by N_2. In vitro transport shown in nerves removed from an animal 2 hours after injecting the L7 dorsal root ganglia with ^3H-leucine. One nerve (•) was placed in a chamber containing moist, 95 percent O_2 + 5 percent CO_2 for an additional 4 hours and kept at 38° C. The rate of transport in vitro was that expected of fast transport in the animal (arrow 2). The other nerve (o) was similarly treated except that it was exposed to N_2 in a chamber for 4 hours. The crest advanced no further than that of the 2 hours of downflow (arrow 1) that had taken place in the animal.

transport to continue. An isotonic sucrose (300 mM) solution can maintain fast axoplasmic transport in vitro for periods up to eight hours or more, indicating the presence of an endogenous store of metabolite in the fibers. A similar lack of need for exogenous metabolites was noted for particulate movement in vitro in studies using Normarski optics microscopy [24].

The dependence of transport on oxidative metabolism was shown in in-vitro preparations by the block, within 15 minutes, of fast axoplasmic transport produced by N_2 anoxia (Fig. 21-3), cyanide (CN), azide, and dinitrophenol (DNP) [6]. At that time, the total ~P, the level of ATP and creatine phosphate (CP), fell from 1.2 to 1.4 μM per g to approximately half that value, 0.6 μM per g [28]. The ~P remaining is not available to the fast axoplasmic transport system; perhaps it is compartmented in Schwann cells or in some equally inaccessible axonal site.

The available ATP appears to be in a pool shared by the transport mechanism and the sodium pump. When ~P falls to half its control level, both axoplasmic transport and the sodium pump are blocked. This was shown by the failure of action potentials elicited from sciatic nerves in chambers where they could be stimulated and action potentials recorded while fast axoplasmic transport was also going on [10c]. Hodgkin and Keynes [28a] found that CN, azide, and DNP block oxidative metabolism in the squid giant axon, but that resting-membrane potentials and action potentials remain present for considerably longer periods of time — well over 90 minutes, as compared to the block in 15 minutes seen in mammalian nerve. The longer survival time of the giant axons is explained by their very much larger axoplasmic volumes and the correspondingly longer time required for the concentrations of Na^+ and K^+ within them to change enough to result in reduced resting-membrane potentials and a block of action potentials.

Glycolysis

When nerves transporting labeled proteins in vitro were exposed to iodoacetic acid (IAA) in concentrations of 2 to 10 mM, a gradual block of transport occurred, with a complete block after 1.5 to 2.0 hr. This protracted time to block does not represent a delayed entry of IAA into the nerve fibers. A complete inhibition of glyceraldehyde-2-phosphate dehydrogenase (GAPD) activity takes place in the nerve after only 10 min of exposure to IAA [6].

Arrest of fast axoplasmic transport, seen within 15 minutes after oxidative phosphorylation is blocked, is to be contrasted with the 1.5 to 2.0 hr when glycolysis is blocked with IAA. The longer survival may be explained by an entry of metabolites such as α-ketoglutarate and acetyl-CoA into the citric acid cycle below the level of glycolysis block. These can maintain a supply of ATP for an additional period of time until, when such metabolites are no longer available in adequate amount, ATP production fails and fast axoplasmic transport stops. The level of ~P in nerve exposed to IAA was found to drop to half control levels at the time fast axoplasmic transport was blocked at 1.5 to 2.0 hr, which shows the correlation of ATP with transport.

If pyruvate or lactate is added to nerves blocked with IAA, the effect of the glycolysis block is reversed and an almost normal appearing crest of fast-transported activity is seen to move at its usual rate and with levels of ~P close to normal. Pyruvate is approximately 20 times more effective in restoring transport than is L-lactate. The lactic dehydrogenase extracted from cat sciatic nerve was found by Khan and Sabri [28b] to have an isoenzyme pattern closer to the heart than to the skeletal-muscle pattern, which would account for the utilization of lactate by nerve.

Glycolysis block by fluoride also causes an interruption of fast axoplasmic transport, as shown by using the accumulation of norepinephrine at a ligation as the measure of fast axoplasmic transport in vitro [29]. Additionally, in fluoride-blocked nerves, pyruvate added to the medium reverses the block of glycolysis and restores axoplasmic transport.

Citric Acid Cycle

The citric acid cycle may be blocked by fluoracetate (FA) after its conversion to fluorcitrate. Nerves maintained in vitro and exposed to FA in concentration of

4 to 10 mM showed a well-defined block of fast axoplasmic transport in approximately 1.0 to 1.5 hr [6, 6a]. Like the block of glycolysis with IAA, this blocking time is significantly longer than the 15 minutes required for a block of oxidative metabolism. This suggests, in analogy to the prolonged time for the block of IAA, that some citric acid cycle intermediates can continue to supply the oxidative chain until, when eventually exhausted, the production of ATP fails. When fast axoplasmic transport is blocked by FA, the level of ~P falls to half of control level at about one hour, again showing the correlation of ATP with fast axoplasmic transport.

Local Anoxia and Energy Stores
The supply of ATP is maintained locally all along the nerve fiber. This was indicated by producing a local block of transport in vitro by covering a small region of the nerve with plastic strips just forward of the advancing crest. The strips contain petrolatum jelly, which prevents entry of oxygen into that region [6, 6a]. The block of fast axoplasmic transport produced in this fashion was shown by the damming of activity just above the anoxic region, with lack of entry of labeled components into the anoxic region. The damming also showed that the labeled materials do not diffuse or move to any significant distance in the fiber without a supply of ~P energy. The results further indicate that oxygen and ATP, as well, do not diffuse to any extent from the nearby oxygenated nerve into the covered portion. If they did, transport would occur and labeled materials would penetrate into that region.

When the local anoxic block was relieved by removing the strips covering the nerves after a period of time, fast axoplasmic transport resumed without delay. Transport recovers after anoxia lasting up to approximately 1.5 hr. As expected, ATP production and its utilization by the transport mechanism also showed a prompt return. After anoxia lasted 1.75 hr and longer, an apparent block of transport was seen. However, transport can resume if long enough times are allowed for recovery. Using cuff compression to produce anoxia in vivo, it was found that recovery could take place after anoxia had lasted up to 6 hr if several days were allowed for recovery [29a].

Energy Transduction for Axoplasmic Flow
There is evidence that ATP is the source of energy for axoplasmic transport. The existence of ATPase activity with actomyosin-like properties, an ATPase stimulated by Mg^{2+} or Ca^{2+} found in peripheral nerve [30], further suggests such a relationship to axoplasmic transport.

MICROTUBULES

Properties
Following the introduction of glutaraldehyde as a fixative for electron microscopy, the ubiquity of microtubules in cells was recognized. These organelles exist in the nerve fibers as long tubular structures about 240 Å in diameter (Chap. 1) [11, 11a]. The walls of the microtubules are composed of globular protein subunits termed *tubulin*.

The subunits are dimers of 120,000 molecular weight (mol wt), composed of A and B tubulin protomers, each about 55,000 in mol wt. The subunits have a regular packing; 13 of them ring the wall with a helical pitch. They appear to extend longitudinally in the wall as chains of protofilaments. In cross section or at intervals along the length of the microtubules, spurlike projections or crossarms are seen to extend from them.

Two classes of microtubules are recognized: the stable form found, for example, in cilia; the labile form in dividing cells and in axons and dendrites. Advantage is taken of this temperature lability to purify the tubulin extracted from brain in order to study its properties chemically. The microtubules are disassembled at low temperatures, and they reassemble on rewarming. By successive disassembly and reassembly along with centrifugation and the use of other separation techniques, tubulin may be purified [11, 11a].

In mammalian nerves, fast axoplasmic transport in vitro is blocked by bringing the nerves to temperatures below 11°C [27]. Such "cold blocks" are reversible; a return of the nerves to a temperature of 38° C, after a cold block that lasts some 12 to 18 hr, is followed by a resumption of fast axoplasmic transport. If the microtubules are disassembled during cold block, they can reassemble on rewarming. A small lag of half an hour to an hour in the resumption of transport is the time presumably required for reassembly and for the transport filaments to find their way back into the reassembled microtubules.

An action of cyclic AMP (cAMP) on the growth of neurites has been noted in tissue culture [31], and protein kinase activity dependent on cAMP has been ascribed to the tubulin [32]. Cyclic AMP may also play a role in the turnover of tubulin in the microtubules or in the transport process itself (see Chap. 12). As yet, this is not well understood.

The relation of microtubules to the neurofilaments — organelles that are 80 Å in diameter and consist of double-helical chains of globular protein — and to the smaller microfilaments found in the nerve fibers is as yet unknown. The microfilaments have been implicated in neurite elongation and are reported to be selectively blocked by cytochalasin B.

Effects of Mitosis-Blocking Agents

Colchicine and the vinca alkaloids, vinblastine and vincristine, can bind to the tubular subunits. This property is used as an identifying feature, i.e., tubulin is a colchicine- or vinblastine-binding protein. These agents also act to disrupt the microtubules within cells. This disassembly is thought to occur as a result of the dynamic equilibrium existing in the fibers between tubulin subunits and microtubules. By binding to tubulin, colchicine and the vinca alkaloids cause the microtubules to disassemble. Such a disruption occurs readily in the microtubules that comprise the spindles of dividing cells, so these agents are also referred to as mitotic blocking agents. It was on this basis that colchicine was first used and shown by Dahlström to interrupt fast axoplasmic transport. A disassembly of microtubules into their tubulin subunits, however, may not account for the block of axoplasmic transport in all nerves. Electron microscopy shows that colchicine blocks transport in crayfish ventral-cord axons without disrupting microtubular morphology [33]. In other nerves, colchicine and vinblastine disrupt microtubules [34]. Disassembly may also occur with some volatile anesthetics, such as halothane [35], and some of the local anesthetics, such as lidocaine [36]. A transformation of microtubules to larger microtubules may also be seen with some of the anesthetics [37]. Disassembly takes place at concentrations greater than that required to block transport.

Routing

A selective transfer of materials to the various branches of the same nerve cell is termed *routing*. This phenomenon was seen in the two branches of the T-shaped dorsal-root ganglion neuron which has one fiber branch entering the root, the other the sciatic nerve (cf. insert, Fig. 21-1). After the ganglion had been injected with ^{3}H-leucine, the same rate of fast axoplasmic transport was found in each of the two branches but with three to five times more labeled activity moving down the peripheral nerve than into the dorsal root branch [10c]. The asymmetry in the amount of labeled proteins transported into these two branches could not be accounted for by differences in their overall diameters or the numbers of microtubules or neurofilaments in the two fiber branches. Such findings suggest that more filaments are moved down the nerve branch, as compared to the dorsal-root branch.

A special routing of different materials would be expected. The peripheral nerve branch requires components destined for the sensory receptors; the branch entering the cord requires that transmitter materials be contributed to the presynaptic terminals to effect synaptic transmission. Thus, at least some small part of the materials transported to each branch differ in their composition in order to serve these two ends.

INTERRELATION OF MEMBRANE AND TRANSPORT MECHANISMS

Little evidence of an interaction between the mechanisms subserving membrane excitability and that of fast axoplasmic transport has been found. When tetrodotoxin or procaine is used to block electrical excitation, fast axoplasmic transport continues as usual [6]. The nerve membrane can be completely depolarized in vitro by replacing Na^+ with K^+ or by replacing all ions in the medium and using a 300 mM solution of sucrose instead of Ringer's solution; this has no effect on the usual pattern and rate of fast axoplasmic transport. Some local anesthetics, however, may have a more complex action. At low concentrations, lidocaine and halothane block excitability without affecting transport, but at higher concentrations they can enter the fiber and block the transport mechanism. Recently, batrachotoxin, an agent acting to block excitability, was found to block fast axoplasmic transport at concentrations of less than 0.2 μM [38]. The block of transport was not due to an increased entry of Na^+; it took place in Na^+-free media as well. Block of transport by batrachotoxin differs in this respect from its known action of blocking excitability by increasing Na^+ permeability. Probably it acts on the transport mechanism itself.

An interaction between membrane events and the transport mechanism has not been substantiated. Repetitive stimulation of nerves in vitro has no effect on fast axoplasmic transport, even at rates up to 350 pulses per second [38a].

Fate of Transported Materials

The fate of various materials synthesized in the cell body and transported in the fibers is schematized in Figure 21-4. A number of points regarding the synthesis of materials in the cell body (Region I) and the fate of materials supplying the fiber (Region II) have been noted throughout this chapter. Some special aspects are pointed out in this section with reference to processes occurring at the nerve terminal, at the neuromuscular junction in this schematization (Region III).

Transmitters

The relationship between the amount of transmitter carried by axoplasmic transport into the terminal (Fig. 21-4, arrow 12) and the amount of transmitter synthesized locally in the terminals (Fig. 21-4, arrow 11) remains to be determined. Synthesis of ACh from choline taken up locally at the motor nerve terminal is known (Chap. 9), and in the sympathetic nervous system norepinephrine (Chap. 10) is, to a considerable extent, taken back up into the nerve terminals to be used again for transmission [4, 4a]. It is of interest in this regard that enzymes involved in such local synthesis, e.g., dopamine β-hydroxylase and tyrosine hydroxylase, are also carried down the nerve fibers by axoplasmic transport, the former at a fast, the latter at a slower, rate [4, 4a].

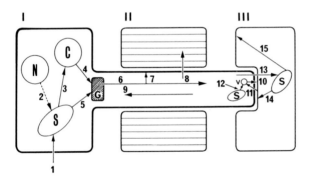

Figure 21-4
Diagrammatic representation of the neuron. (I) cell body; (II) nerve fiber; (III) terminal region with the neuron-muscular junction and a muscle fiber shown. In (I), upon entry of precursor (1), the nucleus (N) controls (2) synthesis (S) with an arrow (3) showing a compartmentalization (C) of synthesized material. Materials move from the synthesis site and the compartment into the axon (4, 5). (G) represents a gate, possibly the Golgi apparatus, controlling the egress of materials into the axon. The transport system in the fiber (II) moves materials down the nerve fiber (6). Arrows represent some components that supply the membrane (7) and the Schwann cell (8). The myelin sheath is represented by parallel horizontal lines. Retrograde transport is shown by the arrow directed toward the cell body (9). In the terminal region (III), vesicles (V) involved in transmission are shown (10) with a reconstitution of their transmitter content (11). Synthesis of components in their termini is indicated by (S) with some contribution of downflowing control materials (12). An arrow from the nerve terminal entering the muscle fiber (13) indicates the flow of trophic materials, some of which control synthesis (S) in the muscle fiber. Receptor proteins are shown by arrows (14) and (15) inserted into the membrane at the usual site (14) and, after denervation, at new (15) receptor sites.

Trophic Control Substances

Materials move out of the nerve terminal (Fig. 21-4, arrow 13) to enter the post-synaptic cell to control or modify its function. These materials are termed *trophic substances* [10]. Alterations in the excitability and metabolism of muscle have long been known to follow upon the transection of the nerves that innervate them. Direct evidence for a translocation of labeled materials from the nerve is limited at present. Autoradiographs have suggested an entry of labeled proteins into muscle. Labeled components have been found in the visual cortex of brain after retinal uptake indicating passage of compounds through the lateral geniculate synapse. How much of that activity represents a specific transfer and how much a nonspecific spread from the terminals is at present uncertain [10c].

That trophic materials control the rate of muscle contraction is shown by the conversion of fast-contracting to slow muscle and slow-contracting to fast muscle by cutting their nerves and cross-innervating the muscles [39]. A later-occurring alteration of metabolism is shown when the nerve is cut far from the muscle rather than close to it. The time of onset of fibrillation also depends on the length of nerve left attached to the muscle [10a]. A similar lag in the development of tetrodotoxin-resistant action potentials in muscle and a decrease in resting membrane potential that follows nerve transection [10b] were also seen when cutting the nerve close to and far from the muscle.

A trophic factor in nerve that controls the development of ACh receptors in the membrane has been studied extensively [40–42]. Cutting the nerve causes a spread of ACh sensitivity in the muscle membrane outside the end-plate region (Fig. 21-4, arrow 15) to which it is usually restricted in the adult muscle fiber (Fig. 21-4, arrow 14). Puromycin and cycloheximide prevent the development of new receptor when added to the medium in which denervated muscle has been cultured. Actinomycin D was also effective in blocking the appearance of new receptor [40]. These findings suggest that the trophic substances normally act to repress the synthesis of receptor protein at the DNA-RNA level in the muscle fiber.

FIBER TRACING USING TRANSPORT AUTORADIOGRAPHY

Axoplasmic transport combined with autoradiography is now a valuable neuro-anatomical technique used to trace fiber pathways within the CNS. After the injection and incorporation of a labeled amino acid, such as ^{3}H-leucine, by nerve cell bodies, the subsequent transport of labeled proteins in their axons is used to show the distribution of those fibers by means of autoradiography. The advantage of this "transport autoradiography" technique over the usual neuroanatomical methods, in which lesions are made and the degenerated fibers traced, is that isotope is not taken up by nearby fibers. When lesions are made, fibers are often adventitiously included; their degeneration, along with that of the intended group to be studied, often makes the subsequent analysis difficult.

In addition, when transport autoradiography is compared with the Nauta method, the best of the staining methods used in degeneration studies, some projections were seen with transport autoradiography that were not seen in the Nauta preparations

[43, 44]. Another interesting aspect of the autoradiographic technique is that, when a brief time is allowed, fast transport carries labeled activity into the terminal endings. These bouton termini are seen to be labeled in autoradiographs while their axons show only a low level of activity. By this means, the pattern of terminations of the neuronal boutons can be determined. With more time allowed for slow transport, activity is evident in the fibers, which can be traced down to their synaptic terminations.

The technique of transport autoradiography has been used to complement evidence of transport in the nigrostriatal system where dopaminergic fibers were shown by the accumulation of dopamine in fibers above a cut (those still connected to the nigral cell bodies) and its depletion in the fibers of the striatum disconnected from their cell bodies. Both anterograde transport shown by the autoradiography technique and retrograde transport following uptake of horseradish peroxidase by nerve terminals are now extensively used in neuroanatomical studies [45].

REFERENCES

1. Weiss, Paul, and Hiscoe, H. B. Experiments in the mechanism of nerve growth. *J. Exp. Zool.* 107:315, 1948.

1a. Lubińska, L. On axoplasmic flow. *Int. Rev. Neurobiol.* 17:241, 1975.

*2. Grafstein, B. Axonal Transport: Communication Between Soma and Synapse. In Costa, E., and Greengard, P. (Eds.), *Advances in Biochemical Psychopharmacology.* Vol. 1. New York: Raven, 1969, p. 11.

*3. Lasek, R. J. Protein transport in neurons. *Int. Rev. Neurobiol.* 13:289, 1970.

*4. Dahlström, A. Axoplasmic transport with particular respect to adrenergic neurons. *Philos. Trans. R. Soc. Lond. [Biol.]* 261:325, 1971.

*4a. Banks, P., and Mayor, D. Intra-axonal transport in noradrenergic neurons in the sympathetic nervous system. *Biochem. Soc. Symp.* 36:133, 1973.

*5. Jeffrey, P. L., and Austin, L. Axoplasmic transport. *Prog. Neurobiol.* 2:205, 1973.

*6. Ochs, S. Energy metabolism and supply of nerve by axoplasmic transport. *Fed. Proc.* 33:1049, 1974.

*6a. Ochs, S. Fast Axoplasmic Transport – Energy and Metabolism and Mechanism. In Hubbard, J. I. (Ed.), *The Vertebrate Peripheral Nervous System.* New York: Plenum, 1974, p. 47.

*7. Barondes, S. Neuronal Proteins: Synthesis, Transport and Neuronal Regulation. In Schneider, D. J., Angeletti, R. H., Bradshaw, R. A., Grasso, A., and Moore, B. W. (Eds.), *Proteins of the Nervous System.* New York: Raven, 1973.

*8. Kristensson, K., and Olsson, Y. Diffusion pathways and retrograde axonal transport of protein tracers in peripheral nerves. *Prog. Neurobiol.* 1:87, 1973.

*9. Harris, A. J. Inductive functions of the nervous system. *Annu. Rev. Physiol.* 36:251, 1974.

*10. Drachman, D. B. (Ed.). Trophic functions of the neuron. *Ann. N.Y. Acad. Sci.* Vol. 228, 1974.

10a. Thesleff, S. Physiological effects of denervation of muscle [10], p. 89.

10b. Albuquerque, E. X., Warnick, J. E., Sansone, F. M., and Onur, R. The effects of vinblastine and colchicine on neural regulation of muscle [10], p. 89.

10c. Ochs, S. Systems of material transport in nerve fibers (axoplasmic transport) related to nerve function and trophic control [10], p. 203.

*Asterisks denote key references.

*11.   Wilson, L. (Symposium Chairman). Pharmacological and biochemical properties of microtubule proteins. *Fed. Proc.* 33:151, 1974.

11a.   Soifer, D. (Ed.). The biology of cytoplasmic microtubules. *Ann. N.Y. Acad. Sci.* Vol. 253, 1975.

12.   Martinez, A. J., and Friede, R. L. Accumulation of axoplasmic organelles in swollen nerve fibers. *Brain Res.* 19:183, 1970.

12a.   Partlow, L. M., Ross, C. D., Motwani, R., and McDougal, D. B., Jr. Transport of axonal enzymes in surviving segments of frog sciatic nerves. *J. Gen. Physiol.* 60:388, 1972.

13.   Ranish, N., and Ochs, S. Fast axoplasmic transport of acetylcholinesterase in mammalian nerve fibers. *J. Neurochem.* 19:2641, 1972.

14.   Droz, B., and Leblond, C. P. Migration of proteins along the axons of the sciatic nerve. *Science* 137:1047, 1962.

15.   Miani, N. Analysis of the somato-axonal movement of phospholipids in the vagus and hypoglossal nerves. *J. Neurochem.* 10:859, 1963.

16.   Gross, G. W., and Beidler, L. M. Fast axonal transport in the C-fibers of the garfish olfactory nerve. *J. Neurobiol.* 4:413, 1973.

16a.   Gross, G. W. The effect of temperature on the rapid axoplasmic transport in C-fibers. *Brain Res.* 56:359, 1973.

17.   Edström, A., and Hansom, M. Temperature effects on fast axonal transport of proteins in vitro in frog sciatic nerves. *Brain Res.* 58:345, 1973.

18.   Hendrickson, A. E. Electron microscopic distribution of axoplasmic transport. *J. Comp. Neurol.* 144:381, 1972.

19.   Elam, J. S., and Agranoff, B. W. Transport of proteins and sulfated mucopolysaccharides in the goldfish visual system. *J. Neurobiol.* 2:379, 1971.

20.   Sabri, M. I., and Ochs, S. Characterization of fast and slow transported proteins in dorsal root and sciatic nerve of cat. *J. Neurobiol.* 4:145, 1973.

21.   Karlsson, J. O., and Sjöstrand, J. Rapid intracellular transport of fucose-containing glycoproteins in retinal ganglion cells. *J. Neurochem.* 18:2209, 1971.

21a.   Forman, D., Grafstein, B., and McEwen, B. S. Rapid axonal transport of (^{3}H) fucosyl glycoproteins in the goldfish optic system. *Brain Res.* 48:327, 1972.

22.   Abe, T., Haga, T., and Kukokawa, M. Rapid transport of phosphatidylcholine occurring simultaneously with protein transport in the frog sciatic nerve. *Biochem. J.* 136:731, 1973.

23.   Rostas, J. A. P., McGregor, A., Jeffrey, P. L., and Austin, L. Transport of cholesterol in the chick optic system. *J. Neurochem.* 24:295, 1975.

24.   Kirkpatrick, J. B., Bray, J. J., and Palmer, S. M. Visualization of axoplasmic flow in vitro by Nomarski microscopy. Comparison to rapid flow of radioactive proteins. *Brain Res.* 43:1, 1972.

25.   Barondes, S. H. Delayed appearances of labeled protein in isolated nerve endings and axoplasmic flow. *Science* 146:779, 1964.

26.   Lux, H. D., Schubert, P., Kreutzberg, G. W., and Globus, A. Excitation and axonal flow: Autoradiographic study on motoneurons intracellularly injected with a ^{3}H-amino acid. *Exp. Brain Res.* 10:197, 1970.

26a.   LaVail, J. H., and LaVail, M. M. The retrograde intraaxonal transport of horseradish peroxidase in the chick visual system: A light and electron microscope study. *J. Comp. Neurol.* 157:303, 1974.

26b.   Ochs, S. Fast axoplasmic transport in the fibers of chromatalyzed neurons. *J. Physiol.*, in press, 1976.

27.   Ochs, S., and Smith, C. Low temperature slowing and cold-block of fast axoplasmic transport in mammalian nerves in vitro. *J. Neurobiol.* 6:85, 1975.

28.   Sabri, M. I., and Ochs, S. Relation of ATP and creatine phosphate to fast axoplasmic transport in mammalian nerve. *J. Neurochem.* 19:2821, 1972.

28a. Hodgkin, A. L., and Keynes, R. D. Active transport of cations in giant axons from *Sepia* and *Loligo. J. Physiol.* 128:423, 1955.

28b. Kahn, M. A., and Sabri, M. I. Unpublished experiments, 1972.

29. Banks, P., Mayor, D., and Mraz, P. Metabolic aspects of the synthesis and intra-axonal transport of noradrenaline storage vesicles. *J. Physiol.* 229:383, 1973.

29a. Leoni, J., and Ochs, S. Reversibility of fast axoplasmic transport following different durations of anoxic block in vitro and in vivo. *Abst. Soc. Neurosci.* 3:147, 1973.

30. Khan, M. A., and Ochs, S. Magnesium or calcium activated ATPase in mammalian nerve. *Brain Res.* 80:413, 1974.

31. Roisen, F., and Murphey, R. A. Neurite development in vitro: II. The role of microfilaments and microtubules in dibutyryl adenosine 3',5'-cyclic monophosphate and nerve growth factor stimulated maturation. *J. Neurobiol.* 4:397, 1973.

32. Soifer, D. Enzymatic activity in tubulin preparations: Cyclic-AMP dependent protein kinase activity of brain microtubule protein. *J. Neurochem.* 24:21, 1975.

33. Fernandez, H. L., Huneeus, F. C., and Davison, P. F. Studies on the mechanism of axoplasmic transport in the crayfish cord. *J. Neurobiol.* 1:395, 1970.

34. Banks, P., Mayor, D., Mitchell, M., and Tomlinson, O. Studies on the translocation of noradrenaline-containing vesicles in post-ganglionic sympathetic neurones in vitro. Inhibition of movement by colchicine and vinblastine and evidence for the involvement of axonal microtubules. *J. Physiol.* 216:625, 1971.

35. Kennedy, R. D., Fink, B. R., and Byers, M. R. The effect of halothane on rapid axonal transport in the rabbit vagus. *Anesthesiology* 36:433, 1972.

36. Byers, M. R., Hendrickson, A. E., Fink, B. R., Kennedy, R. D., and Middaugh, M. E. Effects of lidocaine on axonal morphology, microtubules, and rapid transport in rabbit vagus nerve in vitro. *J. Neurobiol.* 4:125, 1973.

37. Hinkley, R. E., and Samson, F. E. Anesthetic-induced transformation of axonal microtubules. *J. Cell Biol.* 53:258, 1972.

38. Jankowska, E., Lubińska, L., and Niemierko, D. Translocation of AChE-containing particles in the axoplasm during nerve activity. *Comp. Biochem. Physiol.* 28:907, 1969.

38a. Worth, R., and Ochs, S. Batrachotoxin block of fast axoplasmic transport in mammalian nerve fibers. *Science* 187:1087, 1975.

39. Buller, A. J., Eccles, J. C., and Eccles, R. M. Differentiation of fast and slow muscles in the cat hind limb. *J. Physiol.* 150:399, 1960.

40. Fambrough, D. M. Acetylcholine sensitivity of muscle fiber membranes: Mechanism of regulation by motoneurons. *Science* 168:372, 1970.

41. Drachman, D. B. Neurotrophic regulation of muscle cholinesterase: Effects of botulinum toxin and denervation. *J. Physiol.* 226:619, 1972.

42. Pilar, G., and Landmesser, L. Axotomy mimicked by localized colchicine application. *Science* 177:1116, 1972.

42a. Perisic, M., and Cuénod, M. Synaptic transmission depressed by colchicine blockade of axoplasmic flow. *Science* 175:1140, 1972.

43. Lasek, R., Joseph, B. S., and Whitlock, D. G. Evaluation of a radioautographic neuroanatomical tracing method. *Brain Res.* 8:319, 1968.

44. Cowan, W., Gottlieb, D., Hendrickson, A., Price, J., and Woolsey, T. The autoradiographic demonstration of axonal connections in the CNS. *Brain Res.* 37:21, 1972.

45. International Symposium. The use of axonal transport for studies of neuronal connectivity, Gwatt-Thun, Switzerland, July 2–4, 1974. *Brain Res.* 85:201, 1975.

Chapter 22

Hypothalamic-Pituitary Regulation

George J. Siegel
Joseph S. Eisenman

The neuroendocrine system is broadly defined as those structures in the central nervous system that are concerned with the regulation of endocrine function. With the known exceptions of the direct innervations of the adrenal medulla and the pineal gland, CNS influences on endocrine functions are mediated by hypothalamo-hypophysial connections. The main paths in mammals are: (1) the neural tracts from the supraoptic and paraventricular nuclei to the posterior pituitary lobe and (2) the portal circulatory link between the median eminence and the anterior pituitary lobe or adenohypophysis. At these interfaces between the CNS and the endocrine system, coding of information is transformed from neural action potentials to the secretion of specific chemicals into the bloodstream. In the first case, oxytocin and vasopressin are liberated from nerve terminals in the neurohypophysis (posterior pituitary lobe) into the systemic circulation. In the second instance, hypothalamic nerve terminals in the median eminence secrete several different hypophysiotropic hormones into the pituitary portal system. These hormones then act on the adenohypophysis. Numerous volumes and reviews on this subject have been published, and the interested reader is referred to their tables of contents and bibliographies [1–13]. This chapter emphasizes the more recent citations.

NEUROHYPOPHYSIS

The hypothalamo-neurohypophysial complex consists of the supraoptic and paraventricular nuclei, their axons coursing in the supraoptico-neurohypophysial tract through the infundibular stalk to the neurohypophysis (posterior pituitary lobe), and the posterior lobe itself. Within the posterior lobe, the axons end adjacent to perivascular spaces. Glial cells, or pituicytes, are found intermingled among the axons and nerve terminals.

Neurohypophysial extracts from different vertebrate species have yielded several nonapeptide hormones that are classified into two groups according to their main biological activity: the oxytocic–milk-ejecting and vasopressor–antidiuretic groups. Mammalian pituitaries contain oxytocin, which represents the first group, and arginine-vasopressin or lysine-vasopressin, which represent the second group [7]. The chemistry and synthesis of these peptides and their analogs originated from the pioneering work by du Vigneaud and his school. Detailed reviews of the chemistry, biology, and pharmacology of these peptides have been compiled [3].

Neurosecretion

Early workers who extracted vasopressin and oxytocin from the posterior pituitary assumed that the pituicytes were the source of these factors and that these cells were controlled by hypothalamic neurons. Largely through the work of Scharrer, Scharrer, and Bargmann, it was recognized that this assumption is incorrect. The pituicytes are glial, not glandular, cells, and the terminals of the supraoptico-hypophysial tract are not in apposition to the pituicytes but rather to vascular channels [8]. These authors developed the concept that the hypothalamic neurons themselves are the sites of hormone synthesis and that their terminals represent hormone storage sites.

Strong support for this concept was obtained by using Gomori histological stains, which identify aggregates of proteinaceous material; such aggregates, or granules, are typical of products destined for secretion within glandular cells. In these studies, Bargmann traced secretory material from the perikarya within the hypothalamic nuclei along axons to their terminals in the posterior lobe. Subsequent experiments showed that the quantity of secretory material that was demonstrated as Gomoripositive granules correlated with the amount of antidiuretic activity in the hypothalamo-neurohypophysial complex. Furthermore, when the pituitary stalk was severed, secretory material accumulated rostral to the severance. Bargmann and Scharrer postulated that "the pars nervosa of the vertebrate hypophysis stores, but does not produce, the stainable material which it contains. This material originates in the neurosecretory cells of the nuclei supraopticus and paraventricularis in the higher vertebrates . . . ; it passes to the pars nervosa by way of the hypothalamo-hypophysial tracts. There is evidence that this stainable material carries the antidiuretic, oxytocic, and vasopressor principles from the site of origin in the hypothalamic nuclei to the place of storage in the pars nervosa" [14].

The production and release of hormones by neurons that have the cytological features of secretory cells is called *neurosecretion*; it has been well characterized in the neurohypophysial system. Gomori-positive secretory granules have been isolated from hypothalamic-median eminence tissue as well as from the posterior pituitary. These membrane-bound granules, 1000 to 3000 Å in diameter, are rich in hormone activity. Within the granules, the peptide hormones are noncovalently bonded to a carrier protein, neurophysin, which actually constitutes the bulk of the Gomoripositive material.

Biosynthesis

Biochemical studies show that the hypothalamic-median eminence tissue is the primary site for vasopressin and neurophysin biosynthesis. For example, the specific radioactivity of vasopressin and neurophysin isolated from dog hypothalamic tissue after infusion of ^{35}S-cysteine into the third ventricle is several times greater than in the posterior lobe. In addition, guinea pig hypothalamic-median eminence tissue, but not neurohypophysial tissue, has been found capable of vasopressin biosynthesis in vitro. However, it is considered likely that some modifications of the hormone or its complex with neurophysin may take place during transport within the axon or in the axon terminal. Information on the initial steps in vasopressin biosynthesis suggests there is a precursor molecule that is biologically inactive [7, 15].

Neurophysin

The carrier protein neurophysin appears to be synthesized simultaneously with the peptide hormones and is also released with them. Neurophysin has the property of binding vasopressin and oxytocin, and it presumably functions in the transport and storage of these hormones [7]. Neurophysin isolated from bovine neurosecretory granules consists of two major proteins of similar amino acid composition that are referred to as neurophysin-I and neurophysin-II. Both neurophysins are capable of binding vasopressin and oxytocin. However, there is evidence that, in vivo, neurophysin-I is associated in hypothalamic cells with oxytocin and neurophysin-II with vasopressin [16].

Corroborating the concepts of the Scharrers and Bargmann, immunohistochemical methods have shown the distribution of neurophysin-I in neuronal perikarya within the supraoptic and paraventricular nuclei, in their axons coursing through the hypothalamus, and in the nerve terminals within the posterior pituitary. No reaction has been observed in the pituitary glial cells or in other portions of the brain [17].

The complete amino acid sequences of bovine, porcine, and ovine neurophysins have been determined [18, 18a, 19] and the complete covalent structure of bovine neurophysin-II has been elucidated [18]. The structures of bovine and human neurophysins-I are partially completed [20]. These proteins are in the range of 10,000 molecular weight and consist of about 92 to 97 amino acid residues. All these polypeptides exhibit close homology in their structures, particularly in the central portions, which are believed to contain the hormone-binding sites.

Neural Regulation of Hormone Release

The hormones of the posterior lobe are released in response to a variety of stimuli. Osmotic stimuli, such as infusion of hyperosmotic solutions into the carotid artery, or changes in vascular volume by graded hemorrhages, lead mainly to release of vasopressin. Pain and certain emotional states may also affect vasopressin release. The antidiuretic and pressor activities of vasopressin help maintain constant blood volume and osmotic content, as well as blood pressure. Oxytocin, on the other hand, is released mainly in response to suckling, coitus, and parturition, with the attendant effects of milk ejection and uterine contractions. Under some conditions, however, oxytocic and vasopressor activities in the plasma are increased simultaneously. Hence, the pathways for regulation of these two principles, although independent, show some convergence.

It was deduced from the classic experiments of Verney that osmoreceptors exist in the anterior hypothalamus, which is the location of the supraoptic and paraventricular nuclei. In addition, there is evidence that several groups of receptors within the cardiovascular system participate in regulating vasopressin release in response to changes in volume, pressure, and oxygenation of the blood. These receptors apparently are important in integrating the regulation of blood pressure and blood volume. The physiology and pharmacology of such receptors have been reviewed in detail [2, 3, 9].

Little is known about the precise neuronal pathways that mediate the various physiological responses, but the final path is obviously through the hypothalamus.

It has been amply demonstrated that acute secretory stimuli and electrical stimulation of the supraoptic and paraventricular nuclei and pituitary stalk result in rapid release of vasopressin and oxytocin. The supraoptic nucleus is more concerned with vasopressin and the paraventricular nucleus more with oxytocin, although this demarcation is not complete [16]. Stimuli which provoke hormone release also increase the firing frequency of neurons in these hypothalamic nuclei. The stimuli excite hypothalamic units in which potentials also can be evoked antidromically by electrical stimulation of the pituitary stalk. This latter procedure identifies hypothalamic neurons afferent to the pituitary. Iontophoretic application of acetylcholine increases and norepinephrine decreases single-unit firing rates of antidromically identified neurosecretory cells in the cat supraoptic nucleus [21].

Excitation-Secretion Coupling

The last link in the regulation of hormone release is mediated by the propagated action potentials in the supraoptico-neurohypophysial tract. The mechanism for the coupling of neuronal conduction to secretion appears to depend upon depolarization that leads to Ca^{2+} uptake into the neuron. It has been found that, in the presence of Ca^{2+}, the isolated pars nervosa fails to respond to osmotic stimuli but does release hormone upon electrical stimulation or increased K^+ concentration. The hormone release is strictly dependent on Ca^{2+} and is associated with Ca^{2+} uptake. The precise role of Ca^{2+} in the release process has not been defined. Although Ca^{2+} inhibits the binding of hormone to neurophysin, the physiological role of this action is unknown, and there is no evidence that this effect constitutes the Ca^{2+} function in the physiological release mechanism. On the contrary, there is evidence that neurophysin is released together with hormone, and one hypothesis holds that the entire granule content is discharged in a process of exocytosis [7, 10]. The role of acetylcholine in hormone release appears to be mainly through its neurotransmitter action on the hypothalamic neurons because the isolated neurohypophysis fails to respond to acetylcholine. The physiological role of the acetylcholine found in the neural lobe is not known [22].

   The neurosecretory function of the hypothalamic-neurohypophysial tract provides an important model for the study of neuronal control of biochemical events; the output of vasopressin, for example, is specific for this precisely localized nerve bundle and is sensitive to fairly well-defined stimuli, such as dehydration. Hypothalamic-median eminence slices from guinea pig deprived of water for 4 days incorporate two to five times more radioactivity into vasopressin than do similar slices from nondehydrated animals. Neurons of rat supraoptic nuclei show increases in RNA content and in numbers of ribosomes when the animals are subjected to chronic osmotic stimuli or dehydration. Thus, prolonged neural receptor activity is capable of inducing major synthesis, not just of an end product, but of the entire system of participating biosynthetic machinery [7]. Attempting to uncover the intervening biochemical events directly induced by membrane potential changes is one of the most exciting problems in modern neurobiology.

ADENOHYPOPHYSIS

Hypothalamic-Adenohypophysial Communication

The anterior pituitary secretes the tropic hormones that control the function of the gonads, adrenal, thyroid, and mammary glands, as well as secreting growth hormone (somatotropin), which has widespread actions in the body (Table 22-1). The reader is referred to standard texts on endocrinology for full accounts of their actions on target organs.

Although many of the pathways have not been well defined anatomically, there is abundant physiological evidence for neural input to the hypothalamus from other CNS areas. Stimulation of peripheral nerves, midbrain reticular formation, or limbic structures evokes changes in the activity of hypothalamic neurons. Examples of neural inputs influencing endocrine function include cold-induced thyroxine secretion, stress-induced adrenocorticoid secretion, and ovulation induced by mechanical stimulation of the vagina. Many neural paths converge upon the hypothalamus, which is then the final link in controlling the adenohypophysis.

Hypothalamic neurons send their axons to the median eminence or infundibulum, where they end in close proximity to the capillary loops of the portal system. The median eminence is not hypothalamic tissue, but rather is a part of the neurohypophysis. It contains mainly the longitudinally coursing axons of hypothalamic neurons, glial cells, and the capillary loops of the portal system. It does not appear to have neuronal cell bodies. The blood-brain barrier is absent here, as it is in the neural lobe of the pituitary gland. This feature is most important because it allows substances released from the axon terminals in these areas to diffuse into the circulation: into the pituitary portal system from the median eminence and into the general circulation from the neurohypophysis. The absence of a blood-brain barrier also means that the

Table 22-1. Hypothalamic and Pituitary Hormones

Hypothalamic Hormones	Pituitary Hormones	Target Organs
	Adenohypophysis	
LH-RF and FSH-RF	LH and FSH	Gonads
PIF and PRF	Prolactin	Mammary glands
CRF	ACTH	Adrenal cortex
TRF	Thyrotropin	Thyroid gland
GH-RF and GH-IF	Growth hormone	Many tissues
	Pars Intermedia	
MSH-IF and MSH-RF	MSH	Melanocytes
	Neurohypophysis	
Oxytocin	(Site for storage and secretion)	Mammary glands, uterus
Vasopressin		Kidneys, vascular smooth muscle

axon terminals will be exposed to blood-borne compounds and that the pituitary can
be exposed to CSF. Whether this has any functional significance is not known [23].

Isolation of the pituitary gland from the hypothalamus by stalk section or by
transplantation to another site in the body depresses the secretion of the adenohypophysial tropic hormones except prolactin secretion, which is increased. Electrical stimulation in the median eminence induces release of the tropic hormones
except prolactin, the secretion of which is inhibited. Thus it is clear that the neural
terminals in the median eminence contain substances — either releasing factors (RF)
or release-inhibiting factors (IF) — which either stimulate or inhibit secretion of the
anterior pituitary tropic hormones. The net effect of the hypothalamic influence is
to inhibit prolactin, but to stimulate the other adenohypophysial hormones [11].

Because the median eminence forms the final common pathway for control of
the anterior pituitary lobe, most attempts to isolate and purify the releasing factors
begin with extracts of the stalk-median eminence (SME) tissue. The disadvantage in
using this material is that all of the releasing factors are found in it and the purification procedures must separate the individual factors from each other as well as from
inactive components. An advantage in using SME extracts is that the specific activity
of the releasing factors may be much higher than in the diffusely organized hypothalamic nuclei.

Injection of SME extracts in animals previously prepared by section of the pituitary stalk causes inhibition of prolactin secretion and release of all the other tropic
hormones. Addition of these extracts to pituitary glands in vitro produces similar
effects and supports the conclusions based on hypothalamic stalk section and electrical stimulation. Collectively, the various factors may be called *hypothalamic
hypophysiotropic hormones* (HHH).

General Features of the Hypothalamic Hormones

The hypothalamic areas — which, when stimulated or destroyed, most clearly influence adenohypophysial function — have been collectively termed by Halász the
hypophysiotropic area. This area includes the ventromedial tuberal hypothalamus
(median eminence, arcuate nucleus, and parts of the ventromedial nucleus) and the
ventromedial anterior hypothalamic and suprachiasmatic portions. The somata of
neurons that synthesize the releasing factors appear to be situated in various parts
of the hypophysiotropic area. For some releasing factors, the cells seem to be
localized in a fairly discrete area; for others, they are more widely dispersed.

One method for determining the locations of these neurons is to place a lesion
in a part of the hypothalamus and, after allowing a few days for degeneration to
progress, to assay the concentration of releasing factor in the median eminence.
By using this technique, it was found that follicle-stimulating hormone releasing
factor (FSH-RF) disappeared from the median eminence after lesions were made
in the area of the paraventricular nuclei. Luteinizing hormone releasing factor
(LH-RF) can be decreased by lesions in two distinct regions: the suprachiasmatic-
preoptic portion or the arcuate nucleus ventromedial portion. On the other hand,
some reduction in thyrotropin releasing factor (TRF) in the median eminence took
place after lesions were made anywhere in the ventral tuberal area. The ablation

method has the disadvantage of possibly interrupting neural pathways that impinge upon, or lead from, the hormone-secreting cells. Direct localization of sites of synthesis and storage of LH-RF is under intensive investigation, employing techniques of immunohistochemistry and radioimmunoassay [24, 25]. Within the hypothalamus, TRF appears to be concentrated discretely in the ventromedial, dorsomedial, and preoptic areas and the median eminence [26, 26a]. Within the median eminence, LH-RF has been reported variably to be either in nerve endings or in specialized ependymal cells (tanycytes) that have elongated processes that intermingle with nerve endings [27, 28].

The secretory response of the anterior lobe to the hypophysiotropic hormones has a short latency (5 to 10 min). The need for Ca^{2+} for release of TSH, LH, and FSH from pituitary glands in vitro has been demonstrated. A role for $3',5'$-cyclic AMP has been proposed, based on the observations that this substance or its analogs induce release of several hormones from pituitary glands in vitro and that theophylline (a drug that prevents inactivation of $3',5'$-cyclic AMP by inhibiting phosphodiesterase) mimics or potentiates the action of hypothalamic factors [6]. Prolactin secretion, on the other hand, is reduced by injection of dibutyryl cAMP or dopaminergic drugs into the third ventricles of rats. Dopaminergic blocking agents or prostaglandin PGE_1 (see Chap. 13) inhibit the response to dopamine, but not to dibutyryl cAMP [29]. Thus the adenylcyclase system may also inhibit hormone secretion in appropriate cells (see Chap. 12).

Most evidence supports the view that the hypophysiotropic hormones primarily cause release of stored tropins, but they may also act to increase the rate of synthesis of the adenohypophysial hormones. This effect, however, may be secondary to the discharge of the hormone store.

Synaptic Transmitters in Neuroendocrine Control

In common with other CNS neurons, hypothalamic cells are acted upon by presumed synaptic transmitter compounds. Histofluorescence techniques have demonstrated the presence of norepinephrine, dopamine, and serotonin (5-HT) in the axon terminals that end in most parts of the hypothalamus. The cell bodies of these neurons are situated in various structures of the limbic and reticular activating systems. Amine-containing somata have also been observed in some hypothalamic areas, particularly in the arcuate nucleus, from which arises a dopaminergic pathway — the tubero-infundibular tract — that traverses the median eminence and infundibulum to terminate at the portal plexus [6]. Acetylcholinesterase has been demonstrated histochemically in several hypothalamic areas, indicating the presence of cholinergic pathways, as well. The hypothalamus also contains histamine in high concentrations. Thus, many of the substances commonly implicated in synaptic transmission are found in the hypothalamus; some or all of these undoubtedly act as synaptic transmitters in the hypothalamic neural pathways [12].

One special feature that deserves consideration here is the role of the dopaminergic terminations of the tubero-infundibular tract within the median eminence. Much of the dopamine in the median eminence is contained in terminals that end on other synaptic boutons. This is of interest because studies on excised hypothalamic fragments have shown that dopamine causes the secretion of LH-RF and FSH-RF. Activity in this pathway might affect the secretion of releasing factors via axo-axonal junctions; dopamine and other substances, acting on the terminals of the neurosecretory cells, could modify the amount of releasing factor secreted when the neurosecretory cell discharges [30]. Dopaminergic terminals also end on capillary loops of the portal system, and dopamine might have some action on the anterior pituitary gland.

Feedback Control of Neuroendocrine Function

A general principle that applies to a greater or lesser degree to all endocrine glands is that the control of secretory activity involves a feedback loop. That is, some change in the periphery related to secretory activity is detected by the control mechanism, and this information is used to adjust the output of the control mechanism in an appropriate way. Negative or inhibitory feedback is the usual means for maintaining a particular level of output in the face of uncontrolled or unpredictable disturbances. For example, an increase in the blood level of adrenocorticosteroids will cause a decrease in the release of adrenocorticotropic hormone (ACTH) from the anterior pituitary lobe. This, in turn, will decrease the secretion level of adreno-corticosteroids toward its original value. Negative feedback by target-organ hor-mones is also exhibited by the gonadal steroids and thyroxine.

Three pituitary hormones do not act on endocrine glands as their target organs and so may not be controlled by a negative-feedback loop of the type described: growth hormone, prolactin, and melanocyte-stimulating hormone. Hypothalamic control of the secretion of these hormones appears to include a release-inhibiting factor, as well as a releasing factor. The evolution of this dual control may be related to the absence of negative feedback by a target-organ hormone.

There is also evidence for positive or stimulatory feedback, especially in the con-trol of gonadotropic hormones. Thus implantation of estrogen in the rat pituitary during a critical period of the estrus cycle causes advancement of ovulation, pre-sumably due to an increase in the plasma level of LH, which, by its action on the developing ovarian follicle, leads to a further increase in the estrogen level [1]. Unlike the negative-feedback system, positive-feedback controls are inherently unstable in that, once started, they will shift rapidly to a high level of activity. The major advantage of positive feedback is that it forces the system to attain this level of activity very rapidly. For example, estrogen stimulation of LH secretion may be responsible for the peak of LH output that triggers ovulation at the end of follicular maturation (Chap. 35). An inhibitory mechanism is required for turning off a posi-tive-feedback control loop. LH can inhibit its own secretion, so this effect may represent such a mechanism.

The common means of monitoring endocrine output uses receptors sensitive to the levels of the various hormones in the general circulation. For many of the target-organ secretions, sensitivity to hormonal levels has been found in the anterior pitu-itary lobe itself. In addition, the hypothalamus and other areas in the CNS contain cells sensitive to these hormones (Chap. 35). The hypothalamus has also been shown to be sensitive to the tropic hormones secreted by the anterior pituitary lobe. Feedback loops involving these structures are described below. A schematic diagram will be found in Figure 35-1.

Long Feedback Loops

Basal secretion of tropic hormones does not depend on intact pathways from the hypothalamus to the hypophysis. After transection of the pituitary stalk, the levels of target-gland hormones drop significantly, but low levels of secretion are maintained. Only after complete hypophysectomy does total atrophy of the peripheral organs take place. Thus, even in the absence of stimulation by hypothalamic releasing factors, the pituitary gland will secrete tropic hormones. Furthermore, a measure of feedback control exists after stalk section. For example, repeated injection of exogenous thyroxine into an animal with an isolated pituitary will cause atrophy of its thyroid gland, just as in the intact animal. This is taken to indicate that the isolated pituitary gland, in addition to maintaining a low rate of TSH secretion, is sensitive to the level of thyroxine in the general circulation. Increasing the thyroxine level inhibits the TSH release and produces atrophy of the target organ. After stalk section, however, the system is unresponsive to many of the stimuli that would have increased thyroid activity in the intact animal, such as environmental stress. The input to the pituitary-thyroid axis from these stimuli is neural and is mediated by the hypophysial portal system. Similar observations have been made for the ACTH and the gonadotropic hormones. Interestingly, the thyroid response to physical trauma (laparotomy) is maintained in animals with transected pituitary stalks. In this instance, the tissue damage may produce a change in some humoral or hormonal agent that can act directly on the isolated pituitary gland.

It appears that one level of feedback control of the anterior pituitary lobe involves a reciprocal interaction between the pituitary gland and its target organs, this interaction taking place in the anterior lobe itself. However, the level of activity and the dynamic range of control are greatly enhanced by the addition of hypothalamic input. All neurally mediated stimuli to pituitary endocrine activity are channeled through the hypothalamus and the hypothalamo-hypophysial interface.

Another level of long feedback input is to the hypothalamus itself. Hypothalamic neurons "sense" circulating hormone levels and modify the synthesis or release of the hypophysiotropic hormones. It is not known if the detector cells are identical with those producing the releasing factors or if specialized detector cells are present. Although it is experimentally difficult to demonstrate hormone specificity at the cellular level, it may be presumed that independent control of the tropic hormones requires independent detector mechanisms for each control loop.

The existence of detector cells has been determined by implanting target-organ hormones into various parts of the hypothalamus and noting the effect on the release of tropic hormones. Areas of the anterior hypothalamus are sensitive to ovarian steroids, whereas portions of the medial hypothalamus appear sensitive to adrenocortical hormones. There is still some doubt as to the existence of somatotropin and thyroxine sensitivity in the hypothalamus. Precise localization of hormonal sensitivity is difficult to achieve by this method because of diffusion of the implanted materials.

Short Feedback Loops

Increasing evidence shows that the hypothalamus contains elements sensitive to the adenohypophysial tropic hormones themselves. This feedback loop has been termed *internal*, or *short*, as opposed to the external, longer feedback loop from target organ to the hypothalamus or the anterior lobe. There is evidence that each of the tropic hormones except prolactin may act on the hypothalamus in such an inhibitory short loop. Prolactin, on the other hand, stimulates PIF. There is some evidence that FSH-RF can inhibit its own secretion within the hypothalamus. This would constitute an ultrashort feedback loop. Another form of ultrashort feedback loop that has been proposed is an inhibitory action of the adenohypophysial tropins themselves on the adenohypophysis.

Neural "Override" or "Reset" of the Control System

The negative feedback controls discussed above provide a mechanism for maintaining endocrine activity at some fixed level in the face of various uncontrollable disturbances that might change that level. This is the essence of a feedback control mechanism. Obviously, situations arise in which the body economy requires a change in endocrine gland output. In such situations, it becomes necessary to "override" the basic control loop and to drive the hormonal concentration to a new level. This may be accomplished by a simple override: a parallel input to the hypothalamo-hypophysial pathways that would activate tropic hormone release independently of (or in spite of) feedback inhibition. Another mechanism would involve changing the operating level, or set-point, of the controller so that activity is controlled at a new level. This reset mechanism is comparable to resetting the thermostat of a home heating system to give a different room temperature; the override technique is comparable to building a fire and ignoring the thermostat.

The adjustments of endocrine activity in response to neurally mediated inputs probably represent control of both these types. Examples include the thyroid response to cold, the adrenal response to stress, and the gonadotropin response to coitus in animals that ovulate nonspontaneously.

In most instances, studies have not been performed to distinguish override from reset mechanisms. Simple override can probably account for most transient responses to environmental stimuli. There are two cases in which a reset mechanism has been postulated: the adrenocortical response to chronic stress and the increase in gonadotropin secretion at puberty. In the immature animal, the gonadotropin feedback control seems to operate to keep ovarian steroid levels low; the hypothalamic steroid-sensitive cells respond to low concentrations of hormone by inhibiting the secretion of gonadotropin releasing factors. At puberty, some mechanism, presumably neural, decreases the sensitivity of the detector cells, thus allowing circulating sex hormone levels to rise. Various CNS areas, other than the hypothalamus, which play important roles in the endocrine system are taken up in Chapter 35.

Gonadotropin Releasing Factors (FSH-RF and LH-RF)

Numerous investigators have demonstrated FSH-RF and LH-RF in both hypothalamic and SME extracts from a number of species, including rats, sheep, pigs, cattle, and humans. Extracts obtained from some species have been shown to have physiological activity in man, thus indicating a lack of species specificity [11].

Schally and his co-workers achieved purification of LH-RF from porcine hypothalami and found that this hormone is a decapeptide of the following structure: (pyro)Glu–His–Trp–Ser–Tyr–Gly–Leu–Arg–Pro–Gly–NH$_2$. The purified product also has potent FSH releasing activity. The identical decapeptide synthesized in the laboratory also has both LH and FSH releasing activities [31]. Analogs of the decapeptide that inhibit LH responses also inhibit FSH responses [32], and antisera to LH-RF conjugated with albumin prevent preovulatory surges of both LH and FSH in rats [33]. Because of the dual activity, this factor is often called gonadotropin releasing factor (Gn-RF).

It is not yet certain, however, that the identical molecule represents both the physiological FSH-RF and LH-RF. It is known, for example, that there are differences in LH and FSH responses to estrogen [34, 35]. Such differences, on the other hand, might also be explained by presuming differential effects of circulating hormones on the sensitivity of pituitary LH and FSH secreting cells.

It is well known that ovulation, even in cyclically ovulatory animals, depends on hypothalamic control of the release of gonadotropin from the pituitary. Ovulation may be induced by local vaginal stimulation with a glass rod or by electrical stimulation of the preoptic area in the brain. In both cases, ovulation is accompanied by increased multiunit neuronal activity in the arcuate nucleus-median eminence area, which is recorded by stereotactically implaced electrodes [36]. In addition, the plasma levels of both LH and FSH, as measured by radioimmunoassay, are elevated after the hypothalamus has been stimulated electrically in a zone extending from the septal region through the preoptic nucleus, anterior hypothalamic area, and median eminence-arcuate complex [37]. When synthetic antagonists of LH-RF are administered to rats, ovulation that otherwise is induced by exogenous LH-RF is blocked [38]. Earlier studies have shown that anterior deafferentation of the medial basal hypothalamus blocks ovulation, despite the maintenance of basal gonadotropin secretion. According to the current view, the neural signals involved in producing ovulation (which requires a surge of gonadotropins at a critical period of estrus) activate neuronal somata in the preoptic area to increase the synthesis or release of the releasing factors. Based on physiological studies, gonadal steroids, acting within the hypothalamus, appear to play an important role in the control of Gn-Rf [39].

Gonadotropin secretion is subject to both stimulatory and inhibitory feedback regulation by the ovarian steroids, estrogen and progesterone. Feedback control in general is discussed in the preceding section, and some specific examples are cited here.

Elevated plasma levels of the ovarian steroids, probably acting together, are important in suppressing FSH and LH secretion in the postovulatory phase of estrus and during pregnancy. This represents a long, or external, feedback loop. In ovariectomized rats, subcutaneous administration of ovulation-inhibiting steroids lowers the plasma concentrations of FSH and LH. In such animals, these substances do not block the stimulatory effects of FSH-RF and LH-RF on the pituitary, indicating that the steroids have an action on the hypothalamus or, perhaps, on some proximal brain center. When the ovaries have not previously been removed, progesterone suppresses the action of administered releasing factors, showing that, under certain conditions, the pituitary itself may be sensitive to target-organ feedback [40]. In addition, recent evidence indicates that, in female rats, estrogen effects on pituitary sensitivity to LRF are biphasic, with the net effect depending on the time that has elapsed after estradiol injection [35].

An example of the short-loop feedback mechanism is provided by the experiments in which LH, when implanted into the median eminence of castrated female rats, produces decreases in pituitary and plasma levels of LH. This action of LH might be involved in the rapid return of LH secretion to basal levels after ovulation. The ultra-short feedback loop was postulated by Motta and colleagues [2, 6], who demonstrated that when rat SME extract is injected subcutaneously into male rats that have been castrated and hypophysectomized, the hypothalamic stores of FSH-RF are reduced.

The effects of gonadal steroids on development of sexual maturity and behavior are discussed in Chapter 35. It would appear that, in addition to effects on other neuronal events, hormone secretion in both the hypothalamus and the pituitary is responsive to ovarian hormones and may be either stimulated or inhibited. These various responses depend upon the background of neuronal and endocrine events in a manner that has not yet been elucidated.

Prolactin Inhibitory Factor (PIF)

Maintenance of lactation in the postpartum state depends on pituitary secretion of prolactin. This hormone acts mainly on the mammary glands (in promoting pubertal development and lactation) and on the corpus luteum (in sustaining the formed luteum and its secretion of the ovarian steroids). There is ample evidence that the suckling stimulus required to maintain prolactin secretion and lactation in the postpartum period is mediated through neural pathways that ascend in the spinal cord. Electrical stimulation of the lateral mesencephalic tegmentum, the rostral-central gray matter, and the posterior hypothalamus has induced lactogenesis in pseudo-pregnant rabbits [41].

In mammals, it is well substantiated that the predominant or net effect of the hypothalamus on prolactin secretion is one of inhibition. When the pituitary is removed from hypothalamic control by section of the hypophysial stalk, transplantation, or incubation of the pituitary in vitro, prolactin secretion is greatly increased. This is in contrast to the secretions of the other pituitary hormones, which are decreased under these circumstances. In avian species, however, the hypothalamus appears to exert a net stimulatory control over prolactin secretion.

Factors that either stimulate (PRF) or inhibit (PIF) the release of prolactin have been reported in SME extracts from mammalian species. The inhibitory factor is of low molecular weight, but there is controversy as to whether it is represented by catecholamines present in such extracts from porcine tissue [42] or by peptide components, as found in bovine extracts [43]. A releasing factor, also of low molecular weight, has been reported in bovine extracts [43]. TRF is known to have prolactin releasing activity, but the physiological significance of this is not known [44]. Difficulties in bioassay of these factors may arise because the pituitary sensitivity to the factors depends on various conditions such as the hormonal status of the bio-assay animal [45] and the composition of infusion media [42].

Under physiological circumstances, suckling maintains a high serum prolactin level. When the litter of a lactating rat is removed for a 12-hr interval, her serum prolactin falls. After 30 min of suckling, the serum prolactin again rises while, at the same time, the pituitary is depleted of prolactin. However, after prolonged suckling (more than 3 hr) prolactin contents in both the serum and the pituitary remain high [46]. It would appear from these data that an acute suckling stimulus primarily increases the release of prolactin, and prolonged stimuli increase synthesis as well.

If PIF is injected intraperitoneally 1 to 2 min before the 30-min suckling period, the prolactin depletion in the pituitary and the prolactin increase in serum are com-pletely prevented. This effect is used as a bioassay of PIF activity. Another bio-assay technique takes advantage of the fact that when the cervix in rats, for example, is stimulated with a vibrating glass rod for 2 min, pituitary depletion of prolactin takes place within 30 min. This effect is also blocked by prior administration of PIF. SME extracts can also block the pituitary depletion of prolactin that occurs subsequent to bleeding or severe stress. However, as noted above, pituitary pro-lactin content is not always inversely related to serum prolactin levels. A lack of species specificity among mammalian sources is indicated by the fact that ovine, bovine, and porcine hypothalamic extracts have exhibited PIF activity in rats.

Radioimmunological measurements of levels of prolactin in the serum of rats during their various reproductive phases have shown that the levels are highest during estrus, tend to be slightly lower during proestrus and metestrus, and are about 60 percent lower during diestrus [46]. Thus, the serum prolactin tends to be highest when the estrogen levels are high and low when the estrogen levels are low. Estrogen administration results in elevated serum and pituitary prolactin. In vivo, estrogen or progesterone implants in the anterior hypothalamus or estrogen implants in the hypophysis deplete PIF from the rat hypothalamus and stimulate prolactin secretion. Estrogen, but not progesterone, also stimulates release of prolactin from anterior pituitary tissue incubated in vitro. These observations show that estrogen acts both on the hypothalamus to reduce PIF and directly on the pituitary to increase prolactin release. Progesterone, on the other hand, appears to act only on the hypothalamus. High concentrations of progesterone inhibit pro-lactin secretion. These data are useful in explaining the high levels of prolactin in serum during estrus (high estrogen) and the low levels during most of pregnancy (low estrogen coupled with very high progesterone) in the rat.

Prolactin secretion may also be regulated in a short feedback loop through stimulation of PIF. Prolactin implants in the median eminence lead to depressed prolactin levels in serum, decreased prolactin content in the pituitary, elevated FSH and LH in serum, and the termination of pseudopregnancy [47].

Dopamine or norepinephrine can cause reductions in basal secretion of prolactin and in TRF-stimulated prolactin release when applied to pituitary tissue in vitro or infused into the third ventricle or hypophysial portal vein in vivo [29, 44, 48]. Catecholamines do not alter the effects of TRF on thyroid hormone, however.

Agents such as apomorphine that stimulate dopamine receptors have been found to depress prolactin secretion from pituitary fragments in vitro, and dopamine blocking agents such as haloperidol and Pimozide prevent the dopamine effect on pituitary tissue. Thus catecholamines, particularly dopamine, can act directly on prolactin secreting cells in the pituitary but may also be involved elsewhere in the neural control. Whether dopamine is the physiological PIF remains to be established.

Thyrotropin Releasing Factor (TRF)

Several hypothalamic structures have been implicated in thyrotropin regulation, including the suprachiasmatic, paraventricular, and arcuate-ventromedial hypothalamic areas. Destruction of these regions is associated with decreased thyroid function and decreased hypothalamic TRF. Electrical stimulation of the medial-basal hypothalamus in rats produces an increase in serum TSH, as measured directly by radioimmunoassay. As mentioned before, however, radioimmunoassay of TRF in dissected nuclei of the rat hypothalamus shows this factor to be somewhat more restricted in distribution. The role of other hypothalamic nuclei in regulating thyroid function is not yet understood. The neural pathways, however, appear to involve serotonin rather than catecholamines [44].

Thyroxine inhibits the release of thyrotropin from pituitary tissue incubated in vitro and blocks the stimulating action of TRF on the pituitary both in vitro and in vivo. The thyroxine inhibition of TRF action appears to depend on synthesis of a specific protein because its inhibition can be blocked by actinomycin D and puromycin (see McKenzie, Adiga, and Solomon in [6]). In the condition of cold stress in rats, exogenous thyroxine abolishes the elevation in serum TSH but not in that of TRF. Experimentally induced hypothyroidism or hyperthyroidism in rats produces no changes in serum levels of TRF [49]. These examples illustrate the inhibitory effect of thyroxine directly on the thyrotropin-secreting cell in the pituitary and, as yet, there is no distinct evidence of thyroxine effects on the hypothalamus.

An interesting point is that somatostatin, which inhibits growth hormone release (see below), also inhibits TSH responses to TRF [50]. TRF also causes the release of growth hormone, which effect is suppressed in rats by triiodothyronine [51]. The effect of TRF on prolactin release has already been mentioned. These facts suggest complex interlocking regulations of the endocrine system at the hypothalamic-pituitary interface.

TRF is the first of the releasing factors to be identified chemically and synthesized in vitro. Schally and Burgus and their colleagues discovered that porcine and ovine TRF consist of the identical tripeptide: L-(pyro)Glu–L-His–L-Pro–NH$_2$. Biological activity depends on cyclization of the glutamate [11]. The synthetic hormone is active orally and parenterally in mice, rats, and humans. It stimulates TSH release in vivo and from anterior pituitary tissue incubated in vitro; both actions are blocked by triiodothyronine. When administered to euthyroid humans, it elicits increases of serum thyrotropin within 5 minutes.

The demonstration of TRF biosynthesis in vitro by rat hypothalamic fragments has also been accomplished [52]. The sole precursors of a TRF that corresponds chromatographically to the structure described above were found to be glutamate, histidine, and proline, indicating that the glutamate was cyclized and the proline amidated biologically. The TRF-synthesizing tissue was found to be widely distributed in the hypothalamus, including the ventral and dorsal portions, as well as in the median eminence.

Growth Hormone Releasing Factor (GH-RF)

The normal maintenance of growth hormone secretion by the pituitary depends upon the integrity of the hypothalamus and the hypothalamic-pituitary portal circulation. Destruction of the pituitary stalk or median eminence blocks growth hormone secretion in response to (1) insulin-induced hypoglycemia, (2) blocking of glucose metabolism with 2-deoxyglucose, and (3) arginine infusion. Bilateral lesions limited to the ventromedial nuclei of the hypothalamus are sufficient to cause growth impairment and decreased plasma and pituitary growth hormone, as measured by radioimmunoassay in weanling rats. Electrical stimulation in the ventromedial nuclei, but not in the lateral hypothalamus, of rats evokes increases in growth hormone in plasma within 5 min [53]. Although some specific cell group or fiber bundle subserving growth hormone regulation may be situated in these nuclei, possible roles of other hypothalamic structures are not yet known.

Injections of SME extracts cause depletion of growth hormone in pituitary and elevation of the levels in plasma, as measured by bioassay techniques. Taken together, these observations indicate that the net hypothalamic influence on the pituitary is one of stimulation of growth hormone release. Although there is wide agreement that SME extracts contain a growth hormone releasing factor (GH-RF), discrepancies between bioassay and radioimmunoassay for growth hormone and the possible multiplicity of such factors have complicated the task of purifying the GH-RF [54, 55].

Somatostatin, as has been mentioned, inhibits growth hormone release (GH-IF, SRIF). This factor has been purified from ovine SME extracts and synthesized in the laboratory [56, 57]. It is a tetradecapeptide of the following structure: H–Ala–Gly–Cys–Lys–Asn–Phe–Phe–Trp–Lys–Thr–Phe–Thr–Ser–Cys–OH. The native peptide is obtained in the oxidized form, cyclized between the cysteine residues; the active synthetic product is reduced and linear.

GH-IF suppresses basal levels of growth hormone release and inhibits release stimulated by dibutyryl cAMP or theophylline in vitro and in vivo. The factor is effective in various species, including the human [58].

GH-IF is potent also in reducing the release of TSH stimulated by TRF, high K^+, or theophylline in vitro and by TRF in vivo. The effect on TSH is rapid and reversible in contrast to the inhibition by thyroxine, which appears to involve protein synthesis, as mentioned earlier. Prolactin release is reduced by GH-IF in vitro but not in vivo. No effect has been found upon LH release [58]. GH-IF acts also on other glands to inhibit release of gastrin, insulin, and glucagon [59].

Within the hypothalamus, GH-IF is found mainly in the arcuate nucleus and median eminence, but, as discussed later, it is encountered also throughout the brain [60].

There is evidence for a feedback loop for growth hormone regulation. In monkeys, growth hormone release provoked by the administration of insulin or vasopressin is impaired by prior infusions of human growth hormone. In addition, implantation of growth hormone in various regions of the brain causes decreases in pituitary weight and growth hormone content. Hormone secretion in response to electrical stimulation of the rat hypothalamus is greater in those animals with initially lower growth hormone levels in serum [53]. This inhibitory feedback effect may be on either the pituitary or the hypothalamus.

It should be noted that other factors enter into the regulation of growth hormone secretion. In men, for example, estrogen potentiates the growth hormone release stimulated by arginine infusion; thyroid hormone and glucocorticoids appear to be important in the response of the pituitary to GH-RF [58].

Little is known of the neural circuits that participate in the activation of GH-RF, although there is evidence that catecholamines are involved, possibly as neurotransmitters. It has been found that epinephrine, norepinephrine, and dopamine, but not serotonin, cause depletion of pituitary growth hormone when injected in rat lateral ventricles. The catecholamines also lead to the disappearance of GH-RF activity from the hypothalamus and, in hypophysectomized rats, its appearance in the plasma. It is known that such agents as reserpine and the phenothiazines, which deplete monoamines, suppress the release of hormone that normally follows insulin-induced hypoglycemia. In addition, α-adrenergic blocking agents decrease, while β-adrenergic blocking agents increase, growth hormone releasing responses to provocative tests. L-DOPA administration in humans or dogs produces increased growth hormone levels in the serum, and this can be blocked by GH-IF [61, 62]. The effect of L-DOPA may be mediated either as dopamine or norepinephrine in the hypothalamus or pituitary or both.

Corticotropin Releasing Factor (CRF)

Corticosteroid secretion is increased in response to a variety of stress stimuli including specific physical stress, such as limb fracture, pain, or bleeding, and emotional stress, such as fear or anxiety. The adrenal hormones are intimately related to emotional reactions, as discussed in Chapter 35.

A number of nervous system structures have been implicated in ACTH regulation. Steroid responses to a specific stress, such as limb fracture, can be abolished by contralateral spinal cord hemisection. Stimulation in the brain-stem reticular formation can evoke steroidal output.

Although the net influence of the hypothalamic efferent path is stimulatory to the adenohypophysis, the hypothalamus itself appears subject to a net tonic inhibitory input. Complete isolation of a hypothalamic island, with preserved median eminence and stalk, leads to increased ACTH secretion. Inhibitory zones may be situated in portions of the hippocampus and brain stem.

There is evidence that the neuronal path that sets the diurnal rhythm of steroid secretion is different from the one that mediates stress-induced responses. For example, destruction of periventricular and arcuate nuclei of the anterior hypothalamus of animals blocks the afternoon rise in corticosteroids but not the rise subsequent to acute stress. Destruction of the posterior tuber cinereum, on the other hand, blocks the stress response but not the circadian variation. The afternoon rise in corticosteroids is also abolished by Halász knife cuts through the anterior hypothalamus, presumably because of transection of the anterior input to the hypothalamus [2].

There is still considerable controversy over the identity and the production site of corticotropin releasing factor (CRF) [63]. Posterior-lobe extracts contain at least four factors with CRF activity: vasopressin, β-CRF (related to vasopressin), and α_1-CRF and α_2-CRF (which are related to alpha melanocyte-stimulating hormone, α-MSH). In addition, CRF activity can be obtained from hypothalamic extracts free of vasopressin. The hypothalamic CRF can be separated from other releasing factors and the neurohypophysial hormones by gel filtration; in contrast to vasopressin, it is resistant to inactivation by thioglycollate, which reduces disulfide bonds. It appears to be a peptide since it is inactivated by proteolytic digestion. It is believed that the hypothalamic CRF, rather than those in posterior lobe extracts, is more important physiologically (see Ganong in [6]). The site of CRF production within the hypothalamus has not been localized. Upon subcellular fractionation, it is associated with synaptosome and granule fractions, which suggests that it is stored in granules within nerve endings [30].

Little is known of the physiological functions of the other factors that show CRF activity. Of these, vasopressin is the most studied [64]. Vasopressin activates adrenal responses when given to animals with a sectioned median eminence, and it causes release of ACTH in vitro and from pituitary tumor after injection into the anterior lobe. Thus it has a direct action on the adenohypophysis. There is evidence, however, that at least a portion of the vasopressin effects may result from potentiation of hypothalamic CRF.

It is well known that corticosteroids exert an inhibitory feedback control of ACTH secretion, which is an example of an external, or long, feedback loop. Whether the physiological sites of this cortisone action are within the brain, the pituitary, or both, is still conjectural. Cortisone receptor sites within the brain are discussed in Chapter 35. It is clear that corticosteroids are capable of acting directly on the pituitary to block the action of administered CRF. This has been shown in vivo in rats and dogs via intravenous and intrapituitary injections of dexamethasone, a synthetic corticosteroid, and in vitro with rat pituitary tissue. The dexamethasone suppression of the CRF effect on rat tissue may be abolished by prior treatment with actinomycin D, either in vivo or in vitro, thus implicating DNA-dependent RNA synthesis in the dexamethasone block.

In view of the multiplicity of neural pathways that regulate ACTH secretion through both inhibitory and stimulatory influences, it is not surprising that several classes of putative neurotransmitters are capable of modifying pituitary-adrenal function [12, 65]. Stimulation of cortisone secretion has been produced by implants of carbachol, norepinephrine, serotonin, and γ-aminobutyrate in various loci within the cat hypothalamus and limbic system. Although the various zones show some selectivity in their responses to the different agents, the median eminence is responsive to all four. Intraventricular norepinephrine, dopamine, and carbachol also produce ACTH secretion.

On the other hand, some evidence suggests that norepinephrine may inhibit CRF secretion. For instance, administration of reserpine causes, first, a strong activation of ACTH release; second, the blockade of response to stress and a reduction in hypothalamic CRF content. These data have been interpreted to mean that amine depletion induced by reserpine releases the CRF secretion from some inhibitory influence and that the CRF stores are finally exhausted. This concept is supported by additional observations that norepinephrine inhibits secretion of CRF stimulated by iontophoresis of acetylcholine in the rat hypothalamus. The norepinephrine effect is reduced by phentolamine, an α-blocking agent. The acetylcholine stimulation is suppressed by either hexamethonium or atropine [66].

Hypothalamic Hormones in the Brain

Most interesting has been the demonstration of substantial amounts of TRF [67] and GH-IF [60] throughout the brain of rats and of TRF and LH-RF in the pineal gland of several species [68]. It is not known whether these factors are synthesized or are taken up from the extracellular fluid by these cells. TRF has been found also in the CSF [23]. Although the physiological significance of these distributions is not known, there is information that some of the hypothalamic peptides can alter behavior [69]. In addition, GH-IF has CNS-depressing effects whereas TRF has CNS-stimulating effects [60]. An important implication is the possible use of TRF to counteract phenobarbital-induced coma and the use of GH-IF as an anticonvulsant.

PARS INTERMEDIA

Control of Melanocyte-Stimulating Hormone (MSH) Release

In amphibians, reptiles, and fish, MSH is important in adaptive color changes of skin. In these animals, MSH produces dispersion of melanin within melanophores, which results in darkening of the skin. The biological role of MSH in mammals is not known, although behavioral effects of this hormone can be demonstrated (Chap. 35). MSH is present in mammalian species, nevertheless, and increasing attention is being given to studying the relationship of melanocyte-stimulating hormone to hypothalamic function.

MSH is produced in the pars intermedia of the pituitary. Studies primarily of amphibian tissue have shown that the anatomical communication of the hypothalamus with this lobe differs from that with the anterior pituitary in two notable features: (1) generally, the pars intermedia is poorly vascularized, and no significant blood flow from the median eminence has been found to enter the intermedia; and (2) the intermedia, in contrast to the pars anterior, is invaded by many nonmyelinated nerve fibers derived from the hypothalamus, most of which contain one of two kinds of granules – a large, peptidergic type (1000 to 3000 Å) or a smaller type (700 Å) – and appear to form synapses with intermedia cells and, presumably, control their hormone release (see Etkin in Vol. II of [5]).

The release of MSH from mammalian and amphibian pituitary glands is subject to a net inhibitory control by the hypothalamus which resembles that of prolactin release. Thus, in rats and frogs, destruction of the hypothalamus or transplantations of the pituitary to sites remote from the hypothalamus result in decreased pituitary MSH content and, in the frog, in darkening of the skin. Both releasing factors

(MSH-RF) and release-inhibiting factors (MSH-IF) are found, however, in hypo-thalamic extracts. It would appear that the regulation of MSH release in amphibian and mammalian species depends on the balance between these inhibitory and stimu-latory factors [4].

Some insight into a possible enzymatic control of this balance and the possible structure of MSH-IF has been gained from studies by Celis, Taleisnik, and Walter [11, 70]. These workers have found that incubation of oxytocin with a rat SME microsome preparation blocks the usual MSH-RF activity exhibited by these prep-arations. They employed a number of acyclic oxytocin intermediates and other neurohypophysial hormones and analogs and concluded that the incubation product responsible for the inhibition of MSH release should be L-prolyl—L-leucyl—glycin-amide, the C-terminal portion of oxytocin. Subsequently, it was found that rabbit and rat hypothalami contain membrane-bound exopeptidase activity that produces stepwise cleavage of oxytocin, yielding Pro—Leu—Gly—NH_2 as one of the products. On the other hand, when supernatant fractions are incubated with oxytocin, Leu—Gly—NH_2 and Gly—NH_2 are yielded more rapidly [70].

The C-terminal tripeptide derived from oxytocin shows biphasic effects on MSH release from rat pituitary in vitro, inhibition in nanogram quantities, and stimulation in much higher doses (see Vivas and Celis in [71]).

In addition, a mitochondrial-bound enzyme was found in SME extracts from female rats. Upon incubation with microsomes and oxytocin, the enzyme prevents the appearance of MSH-IF activity. This factor varies with the estrous phase, being lower during proestrus and diestrus and highest during estrus. It is possible that estrogen, which causes depletion of pituitary MSH, inhibits the net activity or syn-thesis of MSH-IF through activation of such a factor. Celis et al. proposed that oxytocin functions as a "prohormone," which yields MSH-IF to an extent depend-ing on the balance between the activities of two different enzymes. Several candi-dates for MSH-RF have been found and include the N-terminal penta- and tri-pep-tides of oxytocin (see Celis, Nakagawa, and Walter in [71]).

REFERENCES

 *1. Davidson, J. M., Weick, R. F., Smith, E. R., and Dominguez, R. Feedback mechanisms in relation to ovulation. *Fed. Proc.* 29:1900, 1970.
 *2. Ganong, W. F., and Martini, L. (Eds.). *Frontiers in Neuroendocrinology, 1969 and 1973.* New York: Oxford University Press, 1969 and 1973.
 *3. Heller, H., and Pickering, B. T. (Eds.). *Pharmacology of the Endocrine Sys-tem and Related Drugs: The Neurohypophysis.* Vol. 1. Oxford: Pergamon, 1970.
 *4. Kastin, A. J., and Schally, A. V. MSH Release in Mammals. In Riley, V. (Ed.), *Pigmentation: Its Genesis and Biologic Control.* New York: Appleton-Century-Crofts, 1971.
 *5. Martini, L., and Ganong, W. F. (Eds.). *Neuroendocrinology.* New York: Academic, Vol. I, 1966, and Vol. II, 1967.
 *6. Martini, L., Motta, M., and Fraschini, F. (Eds.). *The Hypothalamus.* New York: Academic, 1970.

*Asterisks denote key references.

*7.   Sachs, H. Neurosecretion. In Lajtha, A. (Ed.), *Handbook of Neurochemistry*. Vol. 4. New York: Plenum, 1970, p. 373.

*8.   Scharrer, E., and Scharrer, B. Hormones produced by neurosecretory cells. *Recent Prog. Horm. Res.* 10:183, 1954.

*9.   Share, L. Vasopressin, its bioassay and the physiological control of its release. *Am. J. Med.* 42:701, 1967.

*10. Douglas, W. W. How do neurones secrete peptides? Exocytosis and its consequences, including "synaptic vesicle" formation, in the hypothalamo-neuro-hypophyseal system. *Prog. Brain Res.* 39:21, 1973.

*11. Schally, A. V., Arimura, A., and Kastin, A. J. Hypothalamic regulatory hormones. *Science* 179:341, 1973.

*12. Ganong, W. F. The role of catecholamines and acetylcholine in the regulation of endocrine function. *Life Sci.* 15:1401, 1974.

*13. Lederis, K., and Cooper, K. E. (Eds.). *Recent Studies of Hypothalamic Function.* Basel: S. Karger, 1974.

14.   Bargmann, W., and Scharrer, E. The site of origin of the hormones of the posterior pituitary. *Am. Sci.* 39:255, 1951.

15.   Pearson, D., Shainberg, A., Malamed, S., and Sachs, H. The hypothalamo-neurohypophysial complex in organ culture: Effects of metabolic inhibitors, biologic and pharmacologic agents. *Endocrinology* 96:994, 1975.

16.   Zimmerman, E., Robinson, A., Husain, M., Acosta, M., Frantz, A., and Sawyer, W. Neurohypophysial peptides in the bovine hypothalamus: The relationship of neurophysin I to oxytocin and neurophysin II to vasopressin in supraoptic and paraventricular regions. *Endocrinology* 95:931, 1974.

17.   Zimmerman, E., Hsu, K., Robinson, A., Carmel, P., Frantz, A., and Tannenbaum, M. Studies of neurophysin secreting neurons with immunoperoxidase techniques employing antibody to bovine neurophysin-I. Light microscopic findings in monkey and bovine tissues. *Endocrinology* 92:931, 1973.

18.   Schlesinger, D., Frangione, B., and Walter, R. Covalent structure of bovine neurophysin-II: Localization of the disulfide bonds. *Proc. Natl. Acad. Sci. U.S.A.* 69:3350, 1972.

18a. Walter, R., Schlesinger, D. H., Schwartz, I. L., and Capra, J. D. Complete amino acid sequence of bovine neurophysin II. *Biochem. Biophys. Res. Commun.* 44:293, 1971.

19.   Wuu, T. C., Crumm, S., and Saffran, M. Amino acid sequence of porcine neurophysin-I. *J. Biol. Chem.* 246:6043, 1971.

20.   North, W., Walter, R., Schlesinger, D., Breslow, E., and Capra, J. Structural studies of bovine neurophysin-I. *Ann. N. Y. Acad. Sci.* 248:408, 1975.

21.   Barker, J. L., Crayton, J. W., and Nicoll, R. A. Supraoptic neurosecretory cells: Adrenergic and cholinergic sensitivity. *Science* 171:208, 1971.

22.   Lederis, K., and Livingston, A. Neuronal and subcellular localization of acetylcholine in the posterior pituitary of the rabbit. *J. Physiol.* (Lond.) 210:187, 1970.

*23. Kendall, J. W., Jacobs, J. J., and Kramer, R. M. Studies on the Transport of Hormones from the Cerebrospinal Fluid to Hypothalamus and Pituitary. In Knigge, K. M., Scott, D. E., and Weindl, A. (Eds.), *Brain-Endocrine Interaction.* Basel: S. Karger, 1972, p. 342.

24.   Wheaton, J. E., Krulich, L., and McCann, S. M. Localization of luteinizing hormone-releasing hormone in the preoptic area and hypothalamus of the rat using radioimmunoassay. *Endocrinology* 97:30, 1975.

25.   Sétáló, G., Vigh, S., Schally, A. V., Arimura, A., and Flerkó, B. LH-RH-containing neural elements in the rat hypothalamus. *Endocrinology* 96:135, 1975.

*26. Krulich, L., Quijada, M., Hefco, E., and Sundberg, D. Localization of thyro-tropin-releasing factor (TRF) in the hypothalamus of the rat. *Endocrinology* 95:9, 1974.

26a. Brownstein, M. J., Palkovits, M., Saavedra, J. M., Bassini, R. M., and Utiger, R. D. Thyrotropin-releasing hormone in specific nuclei of rat brain. *Science* 185:267, 1974.

27. Pelletier, G., Labrie, F., Puviani, R., Arimura, A., and Schally, A. Immuno-histochemical localization of luteinizing hormone-releasing hormone in the rat median eminence. *Endocrinology* 95:314, 1974.

28. Zimmerman, E., Hsu, K., Ferin, M., and Kozlowski, G. Localization of gona-dotropin-releasing hormone (Gn-RH) in the hypothalamus of the mouse by immunoperoxidase technique. *Endocrinology* 95:1, 1974.

29. Ojeda, S., Harms, P., and McCann, S. Possible role of cyclic AMP and prosta-glandin E in the dopaminergic control of prolactin release. *Endocrinology* 95:1694, 1974.

30. Edwardson, J. A., and Bennett, G. W. Modulation of corticotrophin-releasing factor release from hypothalamic synaptosomes. *Nature* 251:425, 1974.

31. Schally, A. W., Arimura, A., Kastin, A. J., Matsuo, H., Baba, Y., Redding, T., Nair, R. M. G., Debeljuk, L., and White, W. Gonadotropin-releasing hormone: One polypeptide regulates secretion of luteinizing and follicle-stimulating hormones. *Science* 173:1036, 1971.

32. Arimura, A., Debeljuk, L., and Schally, A. Blockade of the preovulatory surge of LH and FSH and of ovulation by anti-LH-RH serum in rats. *Endocrinology* 95:323, 1974.

33. Vilchez-Martinez, J. A., Schally, A. V., Coy, D. H., Coy, E. J., Miller, C. M., and Arimura, A. An in vivo assay for anti-LH-RH and anti-FSH-RH activity of inhibitory analogues of LH-RH. *Endocrinology* 96:1130, 1975.

34. Dierschke, D., Weiss, G., and Knobil, E. Sexual maturation in the female rhesus monkey and the development of estrogen-induced gonadotropic hor-mone release. *Endocrinology* 94:198, 1974.

35. Libertun, C., Orias, R., and McCann, S. Biphasic effect of estrogen on the sensitivity of the pituitary to luteinizing hormone-releasing factor (LRF). *Endocrinology* 94:1094, 1974.

36. Sawyer, C. H. Electrophysiological correlates of release of pituitary ovulating hormones. *Fed. Proc.* 29:1895, 1970.

37. Clemens, J. A., Shaar, C. J., Kleber, J. W., and Tandy, W. A. Areas of the brain stimulatory to LH and FSH secretion. *Endocrinology* 88:180, 1971.

38. Vilchez-Martinez, J., Schally, A., Coy, D., Coy, E., Debeljuk, L., and Arimura, A. In vivo inhibition of LH release by a synthetic antagonist of LH-releasing hormone (LH-RH). *Endocrinology* 95:213, 1974.

39. Araki, S., Ferin, M., Zimmerman, E. A., and Vande Wiele, R. L. Ovarian modulation of immunoreactive gonadotropins-releasing hormone (Gn-RH) in the rat brain: Evidence for a differential effect on the anterior and mid-hypothalamus. *Endocrinology* 96:644, 1975.

40. Arimura, A., and Schally, A. V. Progesterone suppression of LH-releasing hormone-induced stimulation of LH release in rats. *Endocrinology* 87:653, 1970.

41. Tindal, J. S., and Knaggs, G. S. An ascending pathway for release of prolactin in the brain of the rabbit. *J. Endocrinol.* 45:111, 1969.

42. Takahara, J., Arimura, A., and Schally, A. Suppression of prolactin release by a purified porcine PIF preparation and catecholamines infused into a rat hypo-physial portal vessel. *Endocrinology* 95:462, 1974.

43. Dular, R., LaBella, F., Vivian, S., and Eddie, L. Purification of prolactin-releasing and inhibiting factors from beef. *Endocrinology* 94:563, 1974.

44. Chen, H. J., and Meites, J. Effects of biogenic amines and TRH on release of prolactin and TSH in the rat. *Endocrinology* 96:10, 1975.

45. Milmore, J. E., and Reece, R. P. Effects of porcine hypothalamic extract on prolactin release in the rat. *Endocrinology* 96:732, 1975.

46. Amenomori, Y., Chen, C. L., and Meites, J. Serum prolactin levels in rats during different reproductive states. *Endocrinology* 86:506, 1970.

47. Voogt, J. L., and Meites, J. Effects of an implant of prolactin in median eminence of pseudopregnant rats on serum and pituitary LH, FSH and prolactin. *Endocrinology* 88:286, 1971.

48. Takahara, J., Arimura, A., and Schally, A. Effect of catecholamines on the TRH-stimulated release of prolactin and growth hormone from sheep pituitaries in vitro. *Endocrinology* 95:1490, 1974.

49. Montoya, E., Seibel, M. J., and Wilber, J. F. Thyrotropin-releasing hormone secretory physiology: Studies by radioimmunoassay and affinity chromatography. *Endocrinology* 96:1413, 1975.

50. Jackson, I., and Reichlin, S. Thyrotropin-releasing hormone (TRH): Distribution in the brain, blood and urine of the rat. *Life Sci.* 14:2259, 1974.

51. Kato, Y., Chihara, K., Maeda, K., Ohgo, S., Okanishi, Y., and Imura, H. Plasma growth hormone responses to thyrotropin-releasing hormone in the urethane-anesthetized rat. *Endocrinology* 96:1114, 1975.

52. Mitnick, M., and Reichlin, S. Thyrotropin-releasing hormone: Biosynthesis by rat hypothalamic fragments in vitro. *Science* 172:1241, 1971.

53. Frohman, L. A., Bernardis, L. L., and Kant, K. J. Hypothalamic stimulation of growth hormone secretion. *Science* 162:580, 1968.

54. Takahara, J., Arimura, A., and Schally, A. Assessment of GH releasing hormone activity in sephadex-separated fractions of porcine hypothalamic extracts by hypophysial portal vessel infusion in the rat. *Acta Endocrinol.* 78:428, 1975.

55. Johansson, K. N., Currie, B. L., Folkers, K., and Bowers, C. Y. Identification and purification of factor B-GHRH from hypothalami which releases growth hormone. *Biochem. Biophys. Res. Commun.* 60:610, 1974.

56. Brazeau, P., Vale, W., Burgus, R., Ling, N., Butcher, M., Rivier, J., and Guillemin, R. Hypothalamic polypeptide that inhibits the secretion of immunoreactive pituitary growth hormone. *Science* 179:77, 1973.

57. Brazeau, P., Rivier, J., Vale, W., and Guillemin, R. Inhibition of growth hormone secretion in the rat by synthetic somatostatin. *Endocrinology* 94:184, 1974.

58. Vale, W., Rivier, C., Brazeau, P., and Guillemin, R. Effects of somatostatin on the secretion of thyrotropin and prolactin. *Endocrinology* 95:968, 1974.

59. Hayes, J. R., Johnson, D. G., Koerker, D., and Williams, R. H. Inhibition of gastrin release by somatostatin in vitro. *Endocrinology* 96:1374, 1975.

60. Brownstein, M., Arimura, A., Sato, H., Schally, A. V., and Kizer, J. S. The regional distribution of somatostatin in the rat brain. *Endocrinology* 96:1456, 1975.

61. Lovinger, R., Connors, M., Kaplan, S., Ganong, W., and Grumbach, M. Effect of L-dihydroxyphenylalanine (L-dopa), anesthesia and surgical stress on the secretion of growth hormone in the dog. *Endocrinology* 95:1317, 1974.

62. Lovinger, R., Boryczka, A., Shackelford, R., Kaplan, S., Ganong, W., and Grumbach, M. Effect of synthetic somatotropin release inhibiting factor on the increase in plasma growth hormone elicited by L-dopa in the dog. *Endocrinology* 95:943, 1974.

63. Portanova, R., and Sayers, G. Isolated pituitary cells: CRF-like activity of neurohypophysial and related polypeptides. *Proc. Soc. Exp. Biol. Med.* 143:661, 1973.
64. Rivier, C., Vale, W., and Guillemin, R. An in vivo corticotropin-releasing factor (CRF) assay based on plasma levels of radioimmunoassayable ACTH. *Proc. Soc. Exp. Biol. Med.* 142:842, 1973.
65. Krieger, H. P., and Krieger, D. T. Chemical stimulation of the brain: Effect on adrenal corticoid release. *Am. J. Physiol.* 218:1632, 1970.
66. Hillhouse, E. W., Burden, J., and Jones, M. T. The effect of various putative neurotransmitters on the release of corticotropin releasing hormone from the hypothalamus of the rat in vitro. *Neuroendocrinology* 17:1, 1975.
67. Oliver, C., Eskay, R., Ben-Jonathan, N., and Porter, J. Distribution and concentration of TRH in the rat brain. *Endocrinology* 95:540, 1974.
68. White, W., Hedlund, M., Weber, G., Rippel, R., Johnson, E., and Wilber, J. The pineal gland: A supplemental source of hypothalamic-releasing hormones. *Endocrinology* 94:1422, 1974.
69. Brown, M., and Vale, W. Central nervous system effects of hypothalamic peptides. *Endocrinology* 96:1333, 1975.
70. Walter, R., Griffiths, E., and Hooper, K. Production of MSH-release-inhibiting hormone by a particulate preparation of hypothalami: Mechanisms of oxytocin inactivation. *Brain Res.* 60:449, 1973.
*71. Vivas, A., and Celis, M. Effect of Pro—Leu—Gly—NH$_2$ and Tocinoic Acid on the Secretion of Melanocyte-Stimulating Hormone from Rat Pituitaries Incubated in Vitro. In Walter, R., and Meienhofe, J. (Eds.), *Peptides: Chemistry, Structure and Biology.* Ann Arbor: Ann Arbor Sciences Publishers, 1976.

Part Two
Medical Neurochemistry

Chapter 23

Biochemistry of Muscle and of Muscle Disorders

Frederick Samaha

John Gergely

Consideration of the biochemical aspects of muscle is virtually impossible without taking a close look at those structures that distinguish muscle cells from other cells, viz., those structures that constitute the contractile machinery. For recent reviews, see [1–8]. Light microscopists have long known that the physiological unit of muscle, the cell or fiber, contains typical repeating structures along its length. These repeating units are known as sarcomeres and are separated from each other by Z discs. Within each sarcomere can be distinguished the A and I bands; the A band, lying between two I bands, occupies the center of each sarcomere and is highly birefringent. Within the A band, a central lighter zone, the H zone, can be seen; and in the center of the H zone is the darker M band. The Z discs are at the centers of the I bands. The contractile material is subdivided into smaller units — myofibrils separated by mitochondria and sarcoplasmic reticulum. The muscle cell is surrounded by a plasma membrane which, together with the various connective tissue elements and collagen filaments, forms the sarcolemma. The interior of the resting cell is maintained by the plasma membrane at an electrical potential of about 100 millivolts more negative than the exterior. When the muscle is stimulated by its nerve, activation of the contractile machinery results in contraction and tension development. One aspect of the coupling of excitation and contraction involves the so-called transverse, or T, tubules, elements of the muscle cell that are continuous with the plasma membrane [7]. These tubules are seen as openings on the surface of the muscle cell, disposed either at the level of the Z bands or at the junction of the A and I bands, depending on the species; and the depolarization of the membrane spreads activity along the tubules to the interior of the fiber.

At the same time that the ultrastructure of muscle was clarified, a profound change took place in our understanding of the contractile machinery itself [8]. Owing to the work of H. E. Huxley and J. Hanson [9] and A. F. Huxley and Niedergerke [10] the typical striation pattern of voluntary muscle can now be attributed to a regular arrangement of two sets of filaments, and contraction can be attributed to the relative sliding motion of these filaments (Fig. 23-1). The thin filaments (diameter of about 80 Å) appear to be attached to the Z bands and are found in the I band. The second set of so-called thick filaments (diameter about 150 Å) occupies the A band. The thick filaments seem to be connected crosswise by some material in the M zone. In cross section the thick filaments constitute a hexagonal lattice. In vertebrate muscle the thin filaments occupy the centers of the triangles formed by the thick filaments. With the abandonment of the view that muscle contains continuous filaments running from one end of the cell to the other and with the acceptance of the existence of two sets of discrete discontinuous filaments came the recognition that (1) the two kinds of filaments become cross-linked only upon excitation, and (2) contraction of muscle does not depend on shortening

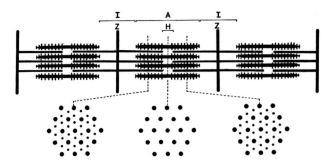

Figure 23-1
Schematic representation of structure of striated muscle. Actin-containing thin filaments originate at Z lines. Note thick myosin-containing filaments that bear cross-bridges. The M disc lies
in the center of the H band. (See text.) (Reproduced with permission from Huxley, H. E., The
mechanism of muscle contraction. *Science* 164:1356, 1969. Copyright 1969. The American
Association for the Advancement of Science.)

on an ultramicroscopic scale in the length of the filaments but rather to the relative
motion of the two sets of filaments (sliding filament mechanism).

The length of the muscle depends on the length of the sarcomeres and, in turn,
the variation in sarcomere length is based upon the variation in overlap between the
thin and thick filaments [8]. High-resolution micrographs have shown that cross-
bridges emanate from the thick filaments, and it is thought that, in active muscle,
these structures are responsible for the links with thin filaments. Table 23-1 shows
the protein constituents of the various structures of the contractile apparatus. There
is good evidence that myosin is the chief constituent of thick filaments and actin the
chief constituent of thin filaments. A recently discovered component of the thick
filaments is the so-called C protein [12], the role of which is still unknown.

It is now also accepted that tropomyosin and a complex of three subunits [13]
subsumed under the name troponin [6a] are present in the thin filaments and play

Table 23-1. Myofibrillar Proteins

Protein	Localization	Function
Myosin	A band	Contraction; ATPase
C protein	Thick filaments	Structural?
Actin	I band (partial overlap with thick filaments)	Contraction
Tropomyosin	Thin filaments	Regulation of actin-myosin interaction; confer Ca^{2+} requirement
Troponin	Thin filaments	
M protein	Center of thick filaments	Structural?
α-actinin	Z band	Structural?

an important role in the regulation of muscle contraction. The proteins constituting the M and the Z bands have not been fully characterized. There is evidence, however, that α-actinin is present in the Z bands [14] and in the abnormal rodlike bodies connected to the Z bands that are found in nemaline myopathy, a muscle disorder [14a].

ACTIN

Electron micrographs show that thin filaments of muscle contain globular subunits about 55 Å in diameter, arranged in a double helical structure [15]. These are identified with the protein actin that can be extracted from muscle that has been treated with acetone. The complete amino acid sequence of actin, which contains 3-methyl histidine, an unusual amino acid, has recently been determined [16]. The molecular weight of actin is about 42,000. On addition of salts to a solution of actin, there results a drastic change in viscosity, and negatively stained electron micrographs reveal the presence of double helical filaments that are essentially identical with the thin filaments. The two physical states of actin characterized by low and high viscosity are referred to respectively as G- and F-actin (G = globular; F = fibrous). The G → F transition is referred to as *polymerization*. The nucleotide in G-actin is ATP [17]; that in F-actin is ADP. The transformation of ATP to ADP takes place during polymerization. According to our current knowledge, the transformation of ATP to ADP that accompanies the polymerization of actin is not involved in muscle contraction. This reaction presumably takes place when actin filaments are laid down in the course of development, growth, or regeneration [18].

MYOSIN

Myosin, the chief constituent of the thick filaments, contrasts with actin in almost every respect. Myosin is a highly asymmetrical molecule having an overall length of about 1500 Å and a molecular weight of about 500,000 (Fig. 23-2). Its lateral dimension varies between about 20 Å and 100 Å. In contrast to actin, which is made up of a single polypeptide chain, myosin consists of several chains. There are two so-called heavy chains, each with a molecular weight of about 200,000, that run from one end of the molecule to the other. Along most of the length of the molecule, the two chains are intertwined to form a double α-helix; at one end of the molecule they separate, each forming an essentially globular portion. The two globular portions (known as HMM S-1) contain the sites responsible for the biological activity of myosin, that is, the ability to hydrolyze ATP and to combine with actin. In addition to the two main heavy chains, each myosin molecule contains four light chains. Two of the light chains are associated with each globular region, and have apparent molecular weights of the order of 17,000 to 27,000 [19]. An interesting chemical feature of the heavy chain of myosin is the presence of methylated amino acids — lysine and histidine. The latter is present in myosin of adult fast-twitch muscle but is absent from myosin in embryonic muscle, adult slow-twitch, and cardiac muscle [20–22].

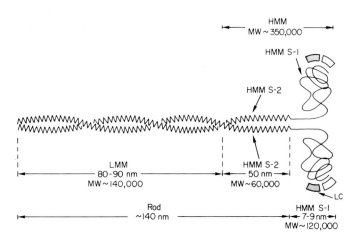

Figure 23-2
Schematic representation of the structure of the myosin molecule, based chiefly on Lowey, S., Slayter, H. S., Weeds, A. G., and Baker, H. Substructure of the myosin molecule. I. Subfragments of myosin by enzymic degradation. *J. Mol. Biol.* 42:1, 1969. The rod portion of the molecule has a coiled α-helical structure. Hinge regions postulated in the mechanism of contraction are at the junction of HMM S-1 and HMM S-2 and of HMM S-2 and LMM. It should be noted that HMM S-1 has one chief polypeptide chain while other fragments have two. Note the light chains (LC) in the head region. The scheme suggests the presence of two different subunits in each HMM S-1. (Reproduced with permission. Walton, J. N. (Ed.), *Disorders of Voluntary Muscle*, 3d ed. Edinburgh: Churchill Livingstone, 1974.)

Myosin in fast-twitch muscle contains three types of light chains, designated as LC_1, LC_2, and LC_3 in order of increasing speed of migration in SDS polyacrylamide gel electrophoresis. Myosin in cardiac and slow-twitch muscle contains only two types of light chains, whose mobilities are similar to those of LC_1 and LC_2, respectively, of fast-muscle myosin [19, 23, 24].

The stoichiometric relation of the various light chains seems to indicate that each molecule contains two LC_2's. In fast-muscle myosin, the sum of $LC_1 + LC_3$ is two per molecule. At present, one can only speculate on the role the light chains play in the functional properties of myosin [25], and it is not clear if in a given myosin molecule LC_1 and LC_2 can occur together or if LC_1 is always present with LC_1 and LC_3 with LC_3. In slow and cardiac-muscle myosin, it appears that there are a pair of LC_1's and a pair of LC_2's per molecule. The light chains are useful markers of the type of myosin present and have been used to study the transformation of myosin from one type to another under experimentally changed conditions of innervation or activity (see below) [26, 27].

The peptide chains of myosin can be separated from each other by reagents that are known to destroy hydrophobic interactions among proteins, such as guanidine, urea, and sodium dodecyl sulfate. The latter, in conjunction with gel electrophoresis, has been useful in studying the light chains.

By using various proteolytic enzymes, it has been possible to isolate the α-helical rod structure of light meromyosin (LMM) as well as fragments containing the globular active portions of the molecule [heavy meromyosin (HMM), and HMM subfragment-1 (S-1)] [27a].

Aggregation of Myosin

Myosin molecules have a tendency to form end-to-end aggregates involving the LMM rods, which then grow into larger structures (the thick filament). The polarity of the myosin molecules is reversed on either side of the central portion of the filament. The globular ends of the molecules form projections on the aggregates similar to those seen on the thick filaments [9]. The central 0.2 μm portion of the thick filament is devoid of cross-bridges (Fig. 23-2). The use of fluorescent antibodies in combination with electron-microscopic studies has greatly elucidated the structure of the myosin filaments, and the data suggest that the headpieces are attached to the filaments by means of flexible hinges [11, 28] (Fig. 23-2). This has important implications for the possible molecular mechanisms of contraction discussed below. According to x-ray data, the cross-bridges on the thick filaments are arranged in a helical fashion. The cross-bridges emerge at levels separated by 143 nm; the number of bridges at each level has not been definitively determined. Estimates range from 2 to 4, the most likely number being 3 [29–31].

MYOSIN-ACTIN INTERACTION AND ATPase ACTIVITY

The discovery by Engelhardt and Ljubimova [32] of the ATPase activity of myosin led to the recognition of the important interrelations between the structural and functional aspects of this protein and its role in muscle contraction. The protein originally termed "myosin" was, in the light of our current knowledge, a complex of actin and myosin [33]. The ATPase activity of myosin itself is stimulated by Ca^{2+} and is low in Mg^{2+}-containing media. If purified actin is added to myosin at low ionic strength in the presence of Mg^{2+}, considerable activation of ATPase activity takes place. This activation is also accompanied by a remarkable change in the physical state of the system. Turbidity increases and, depending on the concentration, there results the so-called superprecipitation. The latter refers to the appearance of a flocculent precipitate, which often shrinks into a contracted plug. Glycerol-extracted muscle fibers have also been found useful for studying the interaction of myosin and actin without destroying the spatial relation existing in intact muscle. These fibers lack the energy-supply system and the excitation-contraction coupling mechanism of intact muscle. Addition of ATP, however, elicits contraction accompanied by the hydrolysis of the ATP.

The combination of actin and myosin can also be observed in solutions of high ionic strength, as indicated by an increase in viscosity. Addition of ATP to this system results in a lowering of viscosity and a decrease of light scattering by the actomyosin solution, both of which are attributable to the dissociation of actomyosin into actin and myosin.

The dissociating effect of ATP, which can also be observed under some conditions at low ionic strength, and the stimulation of the myosin ATPase activity by actin are important links in our understanding of the mechanism of muscle contraction. Kinetic measurements [5] have suggested that, whether actin is present or not, the actual splitting of the phosphate bond in ATP is carried out by myosin without combining with actin. If myosin alone carries out the hydrolysis in the presence of Mg^{2+}, the rate of release of ADP, and perhaps of phosphate, is slow. Combination with actin accelerates the release of ADP. The complex is dissociated by ATP and the cycle starts again. It should be noted that, in this view, the direct effect of ATP is always dissociation of actomyosin; the combination of the two proteins results from an inherent affinity between the two. How these processes lead to contraction is discussed later in this chapter.

TROPOMYOSIN AND TROPONIN

Within recent years it has become apparent that regulatory proteins in the thin filaments exert a great influence on the interaction of actin and myosin. Of these proteins, tropomyosin had been known for a long time, but its role in muscle contraction or its localization had not been elucidated [34]. It is now reasonably clear that tropomyosin, an α-helical protein consisting of two polypeptide chains, is associated with the thin filaments. Interestingly, a protein very similar to vertebrate tropomyosin, the so-called tropomyosin A, or paramyosin, is present in molluscan muscles in close association with myosin rather than with actin. The function of tropomyosin A in mollusks may be related to the so-called catch — the ability of molluscan muscles to maintain tension over long periods of time without expenditures of energy.

The other component is troponin, a complex of three proteins [6, 35] (Fig. 23-3). If the tropomyosin-troponin complex is present, actin cannot stimulate the ATPase activity of myosin unless the concentration of free Ca^{2+} exceeds about 10^{-6} M. The system consisting solely of purified actin and myosin does not show the dependence on Ca^{2+}. Thus the actin-myosin interaction becomes controlled by Ca^{2+} in the presence of the regulatory troponin-tropomyosin complex. This can be demonstrated readily in vitro; the Ca^{2+} concentration can be altered by varying the ratio of total Ca^{2+} added and of chelators such as EGTA. In vivo, the interaction of actin and myosin is regulated by the intracellular concentration of Ca^{2+}. It has been proposed that, of the three proteins in troponin, one anchors it to tropomyosin (TnT), one is

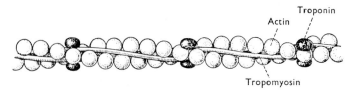

Troponin

Actin

Tropomyosin

Figure 23-3
Model of arrangement of actin, tropomyosin, and troponin in the thin filament. Note that troponin itself is a complex of three proteins. Tropomyosin is close to the groove of the actin filaments in relaxed muscle. (Reproduced with permission from Ebashi, S., Endo, M., and Otsuki, I. Control of muscle contraction. *Q. Rev. Biophys.* 2:351, 1969. Cambridge University Press.)

responsible for combination with Ca^{2+} (TnC), and the third (TnI) binds to a site made up of actin and tropomyosin when Ca^{2+} is absent [36, 37]. If Ca^{2+} binds to TnC, TnI is released, and, as x-ray evidence indicates, tropomyosin changes its position within the thin filament to permit the combination of myosin with actin. As discussed below, the structure involved in the regulation of the intracellular Ca^{2+} level is the sarcoplasmic reticulum, a closed compartment within the cell, and it plays an important role in the mechanism of excitation-contraction coupling.

EXCITATION-CONTRACTION COUPLING

Motor nerve impulses cause the release of acetylcholine at the neuromuscular junction. Acetylcholine initiates depolarization of the muscle cell membrane. Depolarization of the membrane penetrates into the interior of the cell via the transverse tubules that are continuous with the outer membrane (Fig. 23-4). The sarcoplasmic

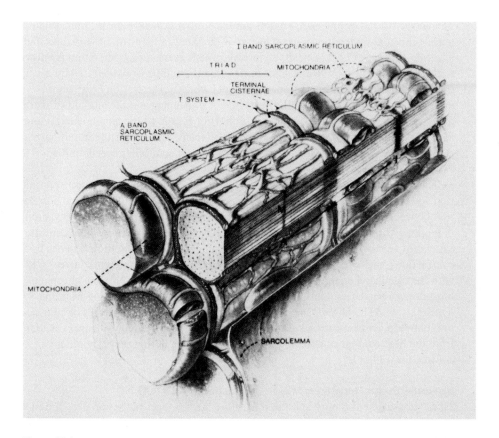

Figure 23-4
A schematic drawing of part of a mammalian skeletal muscle fiber showing the relationship of the sarcoplasmic reticulum, terminal cisternae, T system, and mitochondria to a few myofibrils. (Reproduced with permission from Eisenberg, B. R., Kuda, A. H., and Peter, J. B. Stereological analysis of mammalian skeletal muscle. I. Soleus muscle of the adult guinea pig. *J. Cell Biol.* 60:732, 1974.)

reticulum is in close contact with, but distinct from, the transverse tubules [3, 7]. Together, they are the elements that, in appropriately oriented sections, form the so-called triads noted in electron micrographs. It is currently supposed that the sarcoplasmic reticulum stores Ca^{2+} in relaxed muscle and releases it into the sarcoplasm upon depolarization of the cell membrane and the transverse tubular system. The sarcoplasmic reticulum maintains the low intracellular concentration of resting muscle by means of an ATP-dependent Ca^{2+} pump (Chap. 6). The transport protein appears to be identical with the ATPase isolated from sarcoplasmic reticulum membrane. Another protein component of the membrane (the calcium-binding protein, calsequestrin) may be important for the Ca^{2+} storage and release process [3, 38].

MUSCLE CONTRACTION

Energetics
The classic studies on the energetics of muscle contraction have shown that, when muscle shortens under a load, extra energy is liberated in the form of work and that a certain amount of heat is inevitably evolved. This extra energy liberation is known as the Fenn effect. A. V. Hill sought to describe the total energy liberated by a shortening muscle as a sum of three terms: (1) work; (2) activation heat, whose magnitude is independent of both the degree of shortening and the amount of work done; and (3) shortening heat, which is proportional only to the length changes and independent of the load and, hence, of work. Recent studies by Hill himself have shown that this analysis of the energy balance may be somewhat oversimplified [40, 41].

There is now general agreement that ATP hydrolysis accompanies muscle contraction and is the immediate source of its energy [42]. Although there is good correlation between the amount of ATP plus phosphocreatine broken down and the amount of work performed at various lengths and speeds of contraction, no valid chemical equivalent for the shortening heat in a single twitch has been found. In a series of contractions, the total energy liberated by a muscle, i.e., heat and work, agrees well with the calculated energy release (based on in vitro data) from creatine phosphate breakdown. The latter continually rephosphorylates the ADP resulting from the hydrolysis of ATP. Discrepancies, however, still exist in the early stages of contraction between actual measured chemical energy changes and energy changes calculated from the heat content of compounds known to change during the early phase of contraction [39, 39a].

Molecular Events Underlying Muscle Contraction
The sliding-filament theory and the role of the cross-bridges in tension production are supported by the agreement between the experimentally determined tension of single muscle fibers as a function of length and the tension that would follow from the sliding theory on the assumption that tension is proportional to the number of links formed between the thick and thin filaments [11].

X-ray work has shown that the cross-bridges of the thick filaments undergo a transient movement when muscle contracts [29]. Differences in the orientation of the cross-bridges can be seen clearly when electron micrographs of relaxed and rigor insect muscle are compared [43].

That the interaction of the cross-bridges with the actin filaments results in a unidirectional movement, contraction, seems to be based on two things: First, the myosin filaments change their polarity in the middle of the sarcomere owing to the end-to-end aggregation of the constituent molecules; second, there is a built-in polarity in the actin filaments on each side of the Z band, as shown in electron micrographs by the "arrowheads" formed when heavy meromyosin (HMM) or subfragment-1 complexes with actin [8].

X-ray diffraction studies indicate that the distances among actin and myosin filaments increase as the sarcomere shortens. The flexible attachments of the myosin heads to the rod portions, discussed earlier, make it possible for the cross-bridges to interact with actin across various distances. The driving force of the contraction is most likely the interaction between actin and the S-1 portion of the myosin molecule [44]. ATP apparently plays a role in the dissociation of the links between actin and myosin. ATP, bound to myosin, is then hydrolyzed, and ADP remains bound [5]. A new interaction at a different actin site can then take place, resulting in the displacement of ADP and a small relative movement between the two filaments caused by a changed angle between the attached myosin head and actin. The pitch of the actin helix is slightly different from that of the cross-bridge helix, so a slight movement at an attached bridge would create a favorable situation for attachment of another bridge several actin units away on the same filament (Fig. 23-5).

MUSCLE METABOLISM

Muscle utilizes energy made available in the form of ATP, which is hydrolyzed to ADP and inorganic phosphate. The task of muscle metabolism, apart from the necessity of producing the specific constituents required for the construction of structural components, is to produce ATP. The metabolic pathways involved in the production of ATP in muscle are not essentially different from those present in other tissues, including nerve. There are, however, some features of muscle metabolism that are closely related to the mechanism by which contraction is initiated and others that are subject to control by metabolites arising in the course of contraction.

It has long been known that active muscle increases its metabolism by a factor greater than 100. This is particularly true in muscles that contain "white" fibers, which are involved in rapid bursts of activity [45]. In these muscles, the chief source of energy is the anaerobic breakdown of glycogen, and the enzymatic control of glycogen metabolism furnishes a good illustration of the complex way in which regulation takes place. Apart from specific mechanisms discussed below, the increase in ADP and inorganic phosphate itself serves to increase the rate of ATP synthesis by providing more substrate.

Muscle Phosphorylase

The breakdown of glycogen is catalyzed by the enzyme phosphorylase, the properties and structure of which have been clarified recently in considerable detail (see

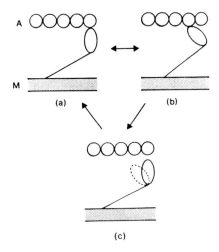

Figure 23-5
Model of interaction of the myosin head with actin. (a) Attachment of myosin to actin, (b) tilt-
ing of myosin head, (c) detachment produced by ATP followed by its hydrolysis. On the basis of
in vitro kinetic studies, the myosin species present in (a) + (b) carries the ADP·P product complex.
Whether the dissociated myosin head can oscillate between the perpendicular and tilted position or
whether it is locked in the perpendicular position is not finally settled. Similarly, the precise point
at which the product leaves the actin-myosin complex requires clarification. The attached head
may oscillate between positions (a) and (b). Based on Huxley, H. E., The mechanism of muscle
contraction. *Science* 164:1356, 1969, and Huxley, A. F., and Simmons, R. M., Proposed mechanism of force generation in striated muscle. *Nature* 233:533, 1971. (Reproduced with permission
from Watson, J. N. (Ed.), *Disorders of Voluntary Muscle,* 3d ed. Edinburgh: Churchill Livingstone,
1974.)

Chap. 14). Phosphorylase *b*, the inactive form, is transformed into phosphorylase *a*,
the active form, by means of phosphorylation. ATP is the phosphate donor, and the
reaction is catalyzed by another enzyme, phosphorylase kinase (for a recent review
see [46]). The latter enzyme exists also in an inactive and an active form, and the
activation again is produced by phosphorylation catalyzed by a protein kinase. This
enzyme, in turn, is activated by cyclic AMP, which is formed from ATP under the
influence of adenyl cyclase (Chap. 12). The cyclase responds to hormones and neurohumoral agents such as epinephrine. It has been demonstrated in vivo that stimulation
of muscle produces an increase in the active form of phosphorylase *a*. Although
epinephrine stimulates muscles, no actual increase in the phosphorylated form of
phosphorylase kinase has been demonstrated. Phosphorylated forms of troponin
subunits and of one of the myosin light chains have been found, but the functional
significance of this process remains to be established [46].

    As has been discussed, the onset of activity in muscle is mediated by the release
of Ca^{2+} from the sarcoplasmic reticulum. It now appears that Ca^{2+}, in roughly the
same concentration as that required for the activation of the actomyosin system,
also increases the activity of the activated phosphorylase kinase by increasing the
affinity to phosphorylase *b*. Thus, as the actomyosin system becomes active, phosphorylase kinase produced phosphorylase at a higher rate. As muscle activity

proceeds, more phosphorylated glucose, which in turn is broken down through the glycolytic pathway, is made available from glycogen. It has also been shown that phosphorylase phosphatase, the enzyme that reverses the activation of phosphorylase, is inactivated by Ca^{2+}. Thus the economy of the operation of the system is ensured by shutting off the activation mechanism of phosphorylase while there is no need for higher activity. Recent studies have shown that these control features are particularly well established in the complex formed by glycogen and the enzymes involved in glycogen metabolism. It is likely that these enzymes are present in vivo in the form of such a complex.

Glycogen Synthetase
A different enzyme system is involved in the biosynthesis of glycogen (see Chap. 14). This enzyme, glycogen synthetase, exists in two forms, one of which is active under physiological conditions, whereas the other is inactive. The same protein kinase that activates phosphorylase kinase also phosphorylates glycogen synthetase I. In this case, however, the I form is the active form, and phosphorylation by the kinase converts it into an inactive D form. This ensures that, when greater muscle activity requires increased release of phosphorylated glucose units from glycogen, the reverse process, which is catalyzed by the synthetase, is shut off [47].

Phosphofructokinase
The activity of another enzyme of the glycolytic pathway, phosphofructokinase, which introduces a second phosphate into the hexose molecule prior to its breakdown into trioses, is under the control of the concentration of various metabolites, e.g., inorganic phosphate and AMP. Similar allosteric regulation applies to the normally inactive phosphorylase b that can be activated by high levels of AMP. The extent to which these regulations play a role under physiological conditions is not clear. Several muscle diseases are attributable to abnormalities in their metabolic processes, including the absence of certain specific enzymes involved in glycogen metabolism (see below).

Myokinase and Creatine Kinase
Myokinase catalyzes the reversible reaction

$$2 \text{ ADP} \rightleftharpoons \text{ATP} + \text{AMP}$$

This reaction assures better utilization of the energy stored in ATP by permitting the hydrolysis of both high-energy phosphate bonds. Creatine kinase catalyzes the reaction

$$\text{creatine phosphate} + \text{ADP} \rightleftharpoons \text{ATP} + \text{creatine}$$

Because creatine phosphate is the chief store of high-energy phosphates in muscle, this reaction permits the rephosphorylation of ADP to ATP, the immediate source of energy in contraction. During rest, the metabolic processes, in turn, replenish the creatine phosphate stores. Interestingly, this reaction created considerable difficulty in an attempt to demonstrate the breakdown of ATP during single twitch because ATP broken down to ADP was immediately rephosphorylated to ATP, and the small change in creatine phosphate could not be detected. Fortunately, the creatine kinase can be blocked selectively by small amounts of dinitrofluorobenzene, and under that condition a decrease in ATP and formation of ADP could be measured [42].

Biosynthesis of Specialized Muscle Proteins

The biosynthesis of myofibrillar proteins is a problem of considerable interest for understanding a number of normal and pathological — developmental, adaptive, and regenerative — processes. The basic mechanism of protein synthesis in muscle is the same as in other tissues, including nervous tissue (see Chaps. 16 and 17). The role of the various polysome fractions reflects the size of the polypeptide chain encoded in the messenger RNA. On this basis, Heywood and Rich [48] isolated separate polysomal fractions capable of in vitro synthesis of myosin, actin, and tropomyosin chains. Further work has shown that distinct polysomal fractions and, therefore, distinct messenger RNAs are involved in the biosynthesis of the several peptide chains (heavy and light) of myosin [49, 50]. This implies the existence of separate genetic coding for different subunits of myosin and, in view of the different types of muscle, the existence of multiple genes for analogous subunits.

The chemical signal for the increase in protein synthesis in work-induced hypertrophy is not known. Creatine has recently been suggested as a possible mediation of this effect [51].

Oxidative Metabolism

During sustained activity, particularly in those muscles that are rich in mitochondria (see below) and in heart muscle, oxidative processes dominate. These involve the oxidation, via the Krebs cycle (see Chap. 14), of pyruvate formed in the anaerobic breakdown of glucose or glycogen and the oxidation of fatty acids which, as has now been generally recognized, serve as the chief source of energy in oxidative metabolism. Citrate, an intermediate in the Krebs cycle, is an allosteric inhibitor of phosphofructokinase.

COMPARISON OF MUSCLE TYPES

It has been known for about a hundred years that striated muscles differ in their velocity of contraction and that correlations exist among the color of the muscle, its velocity of contraction, and the stimulation frequency required to produce a tetanus. In general, red muscles are slow and require higher frequencies to produce tetanus than do white muscles. Even if this discussion is restricted to mammalian muscles, no simple classification based on color, histochemical enzyme reactions, or speed of contraction shows the complexity of the structure and the full multiplicity of the differentiating criteria [52]. It should be emphasized that, as a rule, no muscle as a whole can be considered red or white and that the properties of each individual muscle are determined by the distribution of various fiber types to which the criteria listed in Table 23-2 apply. There are wide differences when one compares various species. For instance, the soleus muscle in rabbit consists almost entirely of typically slow fibers, whereas in rat and man the soleus contains a mixture of fiber types. Differences in speed of contraction, unless one refers to the intrinsic shortening speed of the sarcomere [53], are partly due to differences in the contractile apparatus, partly to differences in the speed of Ca^{2+} release and uptake by the sarcoplasmic reticulum [54].

It appears that the differences between types of muscle depend not only on the presence of certain enzymes in large or small amounts: Different types of muscle may contain the same enzyme in different forms. The different forms of an enzyme catalyzing the same reaction have been called isoenzymes, or isozymes. The first molecule thus identified was lactic dehydrogenase [55], which consists of four subunits. Each subunit can be either the so-called H type, which is predominant in heart muscle, or the M type, predominant in white skeletal muscle. A higher content of the H type has been found in embryonic muscle, although differences exist with respect to species and the muscles examined. Various muscles of the adult contain the two kinds of subunits in various proportions, again with species variations, and the complexes, of which there can be five, differ in their electrophoretic mobility.

During recent years, myosin has been discovered to exist in different forms [56, 57]. The ATPase activity of myosin from a typical slow, or red, muscle is lower than that from fast, or white, muscle, and a striking difference exists with respect to its stability at alkaline pH. White-muscle myosin can be incubated for 10 minutes at pH 9 without any loss of activity, whereas myosin from red muscle loses most of its activity when so treated. More subtle differences exist with respect to stability at low pH. Both kinds of myosin lose activity at acid pH, but the red-muscle type loses less than the white. These in vitro differences in myosin stability at different pHs have been used recently for the histochemical classification of various fiber types. The degree of correlation between staining for oxidative enzymes and the pH effect on myosin ATPase varies, depending on the species.

Differences in the light chains of different types of myosin and in the methylated histidine content of the heavy chain have been discussed above. A striking difference between myosin from white and red muscles can be demonstrated also by looking at the electron micrographs of negatively stained aggregates of the rodlike portion of the myosin molecule, LMM [58]. Whereas LMM aggregates from white-muscle myosin show a striation pattern consisting of strongly stained wider segments alternating with lightly stained narrower segments with a main period of 430 Å, the situation is reversed in LMM aggregates from red-muscle myosin. The lighter segments are wider than the electron-dense segments, and within the lightly stained zones are four distinct darkly stained lines. The precise molecular basis of this difference is not known, but it seems that it can serve as a useful "molecular" fingerprint of the two types of myosin. Differences in fiber types have recently been found in other myofibrillar proteins, including tropomyosin and subunits of troponin [59].

A final word of clarification may be in order. Very often the literature refers to slow muscles as "tonic" and fast muscles as "phasic." This identification is not

Table 23-2. Terminologies Employed for Description of Fibers of Mammalian Muscle[a]

Authors	Method	Fiber Spectrum		
Dubowitz and Pearse,[b] Engel[c]	Oxidative enzymes, phosphorylase	I	II	II
Stein and Padykula[d]	Mitochondrial distribution, ATPase	B	C	A
Romanul[e]	Histochemical profile	III	II	I
Padykula and Gauthier[f]	Mitochondrial distribution	Intermediate	Red	White
Gauthier[g]	Z line width	Red	Intermediate	White
Brooke and Kaiser[h]		I	IIA	IIB
Yellin and Guth[i]	Oxidative enzymes, ATPase	β	αβ	α
Ashmore and Doerr[j]		β-red	α-red	α-white
Burke et al.[k]	Motor unit physiology and histochemistry	S	FR	FF
Peter et al.[l]	Homogeneous muscle: physiology, histochemistry, biochemistry	Slow-twitch oxidative	Fast-twitch oxidative-glycolytic	Fast-twitch glycolytic

[a]Reproduced with permission from Eisenberg, B. A. *Excerpta Med. Int. Congr. Ser.* 333:258, 1974.
[b]Dubowitz, V., and Pearse, A. G. E. *Histochemie* 2:105, 1960.
[c]Engel, W. K. *Neurology* 12:778, 1962.
[d]Stein, J. M., and Padykula, H. A. *Am. J. Anat.* 110:103, 1962.
[e]Romanul, F. C. A. *Arch. Neurol.* 11:355, 1964.
[f]Padykula, H. A., and Gauthier, G. F. *Excerpta Med. Int. Congr. Ser.* 147:117, 1967.
[g]Gauthier, G. F. *Z. Zellforsch.* 95:462, 1969.
[h]Brooke, M. H., and Kaiser, K. K. *Arch. Neurol.* 23:369, 1970.
[i]Yellin, H., and Guth, L. *Exp. Neurol.* 26:424, 1970.
[j]Ashmore, C. R., and Doerr, L. *Exp. Neurol.* 31:408, 1971.
[k]Burke, R. E., Levine, D. N., Zajac, F. E., Tsairis, P., and Engel, W. K. *Physiol. Rev.* 52:129, 1972.
[l]Peter, J. P., Barnard, R. J., Edgerton, V. R., Gillespie, C. A., and Stempel, K. E. *Biochemistry* 11:2627, 1972.

correct for most vertebrate muscles. Most slow and fast muscles of vertebrates are so-called phasic muscles that exhibit a spread of action potential and exhibit innervation by a single endplate, the so-called en plaque innervation. On the other hand, true tonic muscles, such as the latissimus dorsi anterior of the chicken and some mammalian eye muscles, have multiple innervation, called en grappe, and there is no spreading action potential.

EFFECT OF CHANGES IN ACTIVITY AND INNERVATION

Starting with the work of Eccles and his colleagues [60, 61], several authors have reported that, when a fast muscle is cross-reinnervated by a nerve that originally supplied a slow muscle, it acquires properties characteristic of a slow muscle; reciprocal changes take place in a slow muscle that has been cross-reinnervated by a fast muscle nerve. Changes in contractile speed are accompanied by corresponding changes in both the myosin ATPase activity [62, 63] and the protein subunit pattern [27, 64]. Changes have also been observed in the pattern of metabolic enzymes [64–67] and in the activity of the sarcoplasmic reticulum [68].

The work of Salmons and Vrbova [69] shows that, even with undisturbed nerve-muscle connections, changes in physiological parameters can be brought about if the pattern of neural activity reaching the muscle is changed. When the motor nerve is stimulated continuously over a period of weeks, imposing on the fast muscle a pattern of activity similar to that normally reaching a slow muscle, a marked slowing of the time course of isometric contraction and relaxation ensues. Such stimulation also produces changes in the subunit pattern of myosin, the ATPase activity of myosin, the staining pattern of LMM paracrystals, and the Ca^{2+} uptake of the sarcoplasmic reticulum. The changes correspond to an essentially constant fast → slow transformation. The biochemical changes in myosin are paralleled by changes in the histochemical ATPase reaction as well as by changes in the glycolytic oxidative enzyme pattern [70, 71].

Clearly, the fact that changes in the neural activity pattern, with an undisturbed nerve-muscle connection, can alter the physiological and biochemical properties of a muscle raises many interesting questions concerning the so-called trophic effects of the motor nerve (see, e.g., [72]). More work will be required to differentiate genuine neural effects related to the type of nerve from those effects originating in the neural activity pattern.

BIOCHEMICAL ASPECTS OF MUSCLE DISEASE

The biochemical information presented in the first part of this chapter has formed the basis for many investigators to attempt to understand the defects in diseased muscle. Although the basic abnormalities are known in only a very few instances, a significant amount of work has been done on diseased muscle. In this section, biochemical observations on some of the most common syndromes are presented. Other more recently opened subjects of biochemical investigations in muscle disease involve abnormal mitochondrial function, lipid metabolism, and carnitine metabolism

[72a, 72b]. This section is not all-inclusive, and the interested reader is advised to consult the references for more details.

The Muscular Dystrophies

The muscular dystrophies are usually considered to be a group of genetic disorders in which there is progressive degeneration of skeletal muscle with no evidence of alteration of central or peripheral nervous system function to account for the muscle deterioration. The dystrophies are at present classified according to clinical, genetic, and histopathological criteria. The primary molecular abnormality or abnormalities have not been determined in any of these disorders. In this classification, the "pure" dystrophies are separated from the myotonic disorders, even though the term "dystrophia myotonia" has been retained [2].

Myogenic Defects

The myogenic view of the disease considers the possibilities that the primary defect is either in the contractile apparatus itself or in the various membrane components of the muscle.

It has been shown that the speed of contraction is slow and the contraction time is prolonged in Duchenne dystrophic muscle. Since in normal muscle there is a correlation between speed of contraction, actomyosin ATPase activity, and superprecipitation [56, 73], such studies on Duchenne dystrophic muscle have been performed with isolated actomyosin from patients with Duchenne dystrophy. The studies have shown that the initial rate of superprecipitation is slow and that the ATPase activity is low [74, 75]. Similar changes noted in myotonic dystrophy are correlated with degenerative changes [76]. Although the actomyosin ATPase activity may be the rate-limiting factor in the speed of contraction, there is no evidence that this abnormality of actomyosin is a primary event in Duchenne dystrophic muscle. Other studies show that the pH stability of Duchenne dystrophic actomyosin ATPase is normal [75] and that the immunological properties of myosin from Duchenne, facioscapulohumeral, and limb-girdle dystrophies and from myotonic dystrophy are normal [77].

Duchenne dystrophic myofibril and actomyosin preparations contain the regulatory proteins tropomyosin and troponin in one-third to one-half of the normal amounts, but no abnormalities were found when these proteins were studied by sodium dodecyl sulfate polyacrylamide gel electrophoresis [78]. Thus in the dystrophic myofibril, tropomyosin and troponin components were identical in number and in molecular weight with those of normal myofibrils.

Abnormalities in the sarcoplasmic reticulum of Duchenne dystrophic muscle, reflected in the depressed rate and total uptake of calcium by the sarcoplasmic reticulum, suggest participation of this system in the altered pattern of contraction [78–81]. Peter and Worsfold have shown that Duchenne dystrophic sarcotubular vesicles have diminished ability to remove calcium from very dilute solutions [81]. Similar studies on sarcoplasmic reticulum of patients with myotonic dystrophy yielded no changes from normal [80]. Whatever these changes in sarcoplasmic reticulum represent, the underlying molecular abnormalities of Duchenne dystrophy are unknown.

Tropic Influence of Nerve on Muscle
as a Cause of Muscular Dystrophy

The profound influence exerted by the nerve on properties of the muscle (see above), whether one postulates the existence of trophic substance(s) or ascribes the effects to the activity pattern imposed by the nerve on the muscle, has raised speculation that some muscular dystrophies may be caused by an abnormal trophic influence of nerve on muscle. The studies by McComas and his co-workers, indicating a loss of motor units in muscle of patients with Duchenne, myotonic, limb-girdle, and facioscapulo-humeral dystrophies, would suggest that the dystrophies may be caused by some disordered function of motor neurons [82]. However, several investigators have not been able to confirm these observations. In addition, one must take into account that no structural abnormalities have been found in motor axon terminals, and a normal number of motor neurons have been found at autopsy in cases of Duchenne dystrophy.

Muscle Microcirculation Abnormalities

Although vasomotor disturbances and prolonged arm-to-arm circulation times have been found in patients with and carriers of Duchenne dystrophy, the ischemia hypothesis for muscular dystrophy is based mostly on experimental animal models. Histologically, a group pattern of degeneration may be seen as the earliest morphological sign of disease in Duchenne dystrophy [83]. Similar, as well as more advanced, lesions can be produced in animals by arterial embolization with dextran beads, aortic ligation, and 5-hydroxytryptamine, and a combination of 5-hydroxytryptamine and imipramine [83–86]. However, a recent study of the fine structure of 71 capillaries from nine cases of Duchenne dystrophy provided no anatomical evidence that Duchenne dystrophy may be caused by a primary abnormality of the muscle microcirculation [87].

Elevation of Muscle Enzymes in Serum

A well-known expression of the dystrophic process is the release of some soluble enzymes from within the muscle cell into the serum. The observation was first made by Sibley and Lehninger [93] in 1949 with respect to aldolase, but the list of such enzymes has grown to include lactic dehydrogenase, glutamic oxaloacetic transaminase, alpha-glucan phosphorylase, phosphoglucomutase, glycerophosphate isomerase, and creatine phosphokinase [2]. Although elevated levels of enzymes in the serum may be found in several types of muscular dystrophy, they are most striking in the Duchenne form. The highest levels are encountered in the early stages of the Duchenne dystrophy, and the values decline with the progression of the disease.

Since the enzyme creatine phosphokinase is confined mostly to muscle and central nervous system tissues, it has proved to be the most sensitive enzyme assay used in the diagnosis of Duchenne dystrophy. Creatine phosphokinase usually exists in muscle as a dimer (MM) of two identical peptide chains while creatine phosphokinase found in the central nervous system is composed of a dimer of a different type (BB). The enzyme found elevated in the serum of patients with Duchenne dystrophy is usually of the MM type. In some patients with muscle disease, a significant amount of the MB form has been detected in the serum. Since the level of serum creatine phosphokinase may be elevated in about three-fourths of female carriers of Duchenne muscular dystrophy, this assay has been found useful in genetic counseling [2].

Most recently the enzyme pyruvate kinase has been found to be elevated in the serum of patients with muscular dystrophy and has also been shown to be at least as useful as the creatine phosphokinase in discovering the carriers of Duchenne muscular dystrophy [88].

Although many investigators feel that most of the leakage of muscle enzymes into the serum in Duchenne muscular dystrophy is secondary to degenerative changes of the muscle cell and its membrane, the possibility of a basic sarcolemmal membrane defect in the muscle of patients with Duchenne dystrophy cannot be ruled out at this time.

Myotonic Dystrophy

Myotonic dystrophy, first described by Steinert in 1909, presents with a unique constellation of abnormalities apart from other dystrophies. It is an autosomal dominantly inherited systemic disease in which myotonia and muscular atrophy are prominent features [2]. The dystrophic features primarily involve the distal musculature of the upper and lower extremities and the facial and neck muscles. As progression occurs, other muscles become involved.

Clinical myotonia appears as a failure of relaxation after firm contraction of muscle. The phenomenon responsible for the myotonia resides in the sarcolemma, which repetitively depolarizes after sustained contraction, keeping the contractile mechanism in the contracted state. This muscle membrane abnormality is easily demonstrated by electromyographic examination and is a fundamental sign in establishing the diagnosis.

Cardiac involvement is common, and about 65 percent of the patients have electrocardiographic abnormalities, usually in the form of conduction defects. A host of other abnormalities in tissues other than muscles have been described, such as posterior subcapsular cataracts in about 90 percent of cases, impairment of maximum breathing capacity, testicular atrophy, and bone abnormalities. In addition, accelerated breakdown of immunoglobulin G and abnormally elevated fasting plasma insulin levels and exaggerated plasma insulin responses to both orally and intravenously administered glucose have been found [2]. In general, these patients show normal sensitivity to exogenous insulin and a normal disappearance curve of it. Often there is selective failure of androgenic function and occasionally thyroid activity is impaired. There is a significant incidence of brain abnormalities in these patients, and progressive dementia has been described.

For some years it has been known that the inhibitor of desmosterol reductase, azocholesterol, can cause myotonia in animals and in humans [89]. In addition, many rats so treated develop severe cataracts and the $(Na^+ + K^+)$-ATPase activity is increased in the erythrocyte membrane and sarcolemma [90]. Electromyographic studies showing repetitive depolarization of the sarcolemma and the abnormalities induced by azocholesterol suggest that the basic abnormality in myotonic dystrophy might be a general defect in cell membranes. In 1973 Roses and Appel [91] demonstrated an abnormality in the membrane-bound protein kinase of erythrocyte ghosts from patients with myotonic dystrophy. They showed that erythrocyte ghosts frozen at $-20°$ for one week phosphorylated membrane proteins at one-half the rate of normal. On the other hand, freshly prepared red cell ghosts from normal and myotonic dystrophic patients showed no difference with regard to membrane protein phosphorylation. Neither were differences noted in red cell membrane protein phosphorylation when the erythrocytes were frozen and immediately thawed and studied. Another study [92] has demonstrated that electron-spin resonance spectra of spin-labeled myotonic erythrocyte membranes are distinguishable from those of normal erythrocytes. These studies, although intriguing, must be carried further before a correlation can be understood between the abnormality described in the erythrocyte membrane and the pathology that is present in many organs of the myotonic dystrophic patient.

The Periodic Paralyses
Members of certain families suffer from recurrent attacks of flaccid paralysis which, because of the inheritance pattern, has become known as familial periodic paralysis. Within this major syndrome, two clear subcategories have been identified with regard to the alterations in serum potassium levels during the onset of an attack [2]. In one category, attacks are associated with a decrease in serum potassium levels. In the second clinical category, the episodes of flaccid paralysis are associated with a rise in serum potassium. In view of the many variables that affect the serum potassium level, a single measurement is not always helpful in identifying the hypokalemic and hyperkalemic forms of periodic paralysis. For example, the serum potassium may be normal or low during some phase of the weakness in the hyperkalemic forms. A more reliable approach has been the provocation of an attack by lowering the serum potassium levels with glucose and insulin or by elevating serum potassium levels with the administration of potassium per os. In almost all (one exception has been reported) patients suffering from flaccid paralysis, provocative tests permitted categorization as hypo- or hyperkalemic periodic paralysis.

Whether the alteration in the serum potassium level is or is not the primary event precipitating paralysis in either syndrome is not known. From the clinical point of view, however, the division of the familial periodic paralyses into these two categories allows other features to be identified as being associated with either the hyper- or hypokalemic form (Table 23-3). The onset of hypokalemic periodic paralysis usually is in the second decade. The inheritance is usually autosomal dominant, but sporadic instances have been recorded in approximately 10 to 20 percent of cases. On rare occasions myotonia may be identified clinically or by electromyographic studies in these patients. In some areas of the world, such as China and Japan, there is a relatively high incidence of sporadic cases of hypokalemic periodic paralysis occurring in association with thyrotoxicosis [94]. Whether the abnormality of thyroid function is the primary factor in these cases of episodic paralysis or whether the oriental people have a predisposition to developing periodic paralysis that can be triggered by thyrotoxicosis is not known.

In the case of hyperkalemic periodic paralysis, the onset is usually in the first decade and almost all cases are inherited as an autosomal dominant (Table 23-3).

Table 23-3. Clinical Description of Familial Periodic Paralyses

Features	Hypokalemic Periodic Paralysis	Hyperkalemic Periodic Paralysis
Onset	Second decade	First decade
Inheritance	Autosomal dominant 10–20% sporadic	Autosomal dominant
Provoking factors	Exercise followed by rest Stress Alcohol Agents lowering serum K^+	Exercise followed by rest Stress Alcohol Agents raising serum K^+
Serum K^+ during attacks	Usually low	Usually rises May be high or low
Myotonia	Rare	Present nearly 100%
High Na^+ intake	Detrimental	Beneficial
Low Na^+ intake	Beneficial	Detrimental
Resting membrane potential	Low	Low
Associated with thyroid disease	Yes	No
Vacuolar myopathy	Yes	Yes
Chronic myopathy	Yes	Yes

Nearly all patients with hyperkalemic periodic paralysis will manifest myotonia on clinical observation or through electromyographic examination. There has been no consistent incidence of cases of hyperkalemic periodic paralysis in association with thyrotoxicosis.

In both syndromes, provoking factors are exercise followed by rest, stress, and alcohol. In addition, in both instances a vacuolar myopathy in association with a chronic progressive myopathy has been described.

Several abnormalities associated with and preceding the episodes of flaccid paralysis are known. In both instances the onset of an attack is accompanied by an unresponsiveness of the muscle membrane to nerve stimulation [94]. In patients with hyperkalemic periodic paralysis, at the time of attacks and between attacks, the muscle resting membrane potential is lower than normal [95, 96]. The change of the resting membrane potential from −72 to −45 millivolts during an attack suggests that the muscle weakness and loss of excitability are a consequence of a depolarizing block of the muscle fiber membrane. In the case of hyperkalemic periodic paralysis a detailed study of the electrical properties of the muscle membrane in a biopsy sample indicated that the muscle cell was unable to maintain a normal resting membrane potential [97].

From these studies, it is apparent that there is a defect in the muscle membrane in both syndromes. In both forms of periodic paralysis, muscle samples obtained between attacks demonstrated normal ($Na^+ + K^+$)-ATPase activity [98]. In addition, in vivo studies on patients with hypokalemic periodic paralysis have shown that between attacks the muscle resting membrane potential may be normal. Thus the abnormality, which leads to changes in membrane potential and to paralysis, must be intermittent. The work of Engel and Lambert [99] showed that, in hypokalemic periodic paralysis, a paralyzed muscle fiber with the sarcolemma removed can be made to contract by applying calcium directly to the myofibrils. Those studies also indicate that the defect is confined to the muscle membrane and that the molecular mechanism of contraction is intact at the time that paralysis is present. Cardiac arrhythmias have been reported in both syndromes, indicating that the effect extends to cardiac muscle.

Although both syndromes seem to involve a pathological change in the muscle membrane, the processes that lead to paralysis in each case appear to be quite different. Potassium ingestion provokes an attack in one patient, and lowering serum potassium provokes an attack in the other. The severity and frequency of attacks in the hyperkalemic form may be reduced with high sodium intake or increased by a low sodium diet. Attacks in the hypokalemic patient are improved by a low sodium diet and aggravated by high sodium intake. It is, however, important to note that both forms of periodic paralysis respond with decreased attacks when a diuretic such as acetazolamide is administered.

Glycogen Storage Diseases That Affect Muscle Function

In 1952 G. T. Cori and C. F. Cori demonstrated a specific deficiency of glucose 6-phosphatase in a patient with the hepatic form of glycogen storage disease [100]. Although at least eight groups of inherited abnormalities of glycogen metabolism can be identified clinically or biochemically in man (Table 23-2) [94], only three, predominantly involving muscle, will be discussed here. Additional discussions of glycogen metabolism are found in Chapters 14 and 25.

Type 2 Glycogen Storage Disease (Pompe's Disease)

Patients with Pompe's disease [101] usually present in early infancy with profound hypotonia. The inheritance pattern is autosomal recessive. Beside the obvious clinical involvement of the skeletal musculature, striking enlargement of the heart, electrocardiographic abnormalities, and cardiac failure without cyanosis clearly indicate that the heart also is involved. The average life span of these patients is about 5½ months with death caused by respiratory failure. Postmortem examination has shown increased glycogen concentrations in muscle, liver, heart, glial cells, nuclei of the brain stem, and anterior horn cells of the spinal cord. In 1963, Hers [102] demonstrated that tissues from normal persons contain an α-1,4-glucosidase and that this enzyme is absent in patients with type 2 glycogen storage disease. Electron microscopic study showed the presence of two fractions of glycogen, one freely dispersed within the cell and the other segregated in vacuoles surrounded by a single membrane [103]. The vacuoles appear to be lysosomes engorged with glycogen that cannot be broken down because of the lack in the lysosomes of α-1,4-glucosidase required for catabolism. There is no interference with the breakdown of free glycogen in the cytoplasm because the phosphorylase present is capable of hydrolyzing the 1-4 linkages in glycogen. Although this enzyme is absent in a number of tissues in these patients, the deficiency is not always manifested in white blood cells, so the presence of enzyme in these cells does not rule out the disease. Milder variants may be found in older children or in adults with mild proximal limb weakness.

Type 5 Glycogen Storage Disease (McArdle's Disease)

The clinical presentation of patients with McArdle's disease depends upon their age and the severity of the case [2]. From childhood to adolescence the patients may have few complaints except for some evidence of increased fatigability. Between 20 and 40 years of age, severe muscle cramps and myoglobulinuria develop when the patients perform strenuous exercise. Cramped muscles in these patients do not exhibit action potentials on the electromyogram and so are believed to be contractures. If patients with McArdle's syndrome continue the exercise after the appearance of fatigue and cramps, the discomfort may disappear as the patient seems to get his "second wind." Pernow et al. [104] found that this phenomenon was closely related to levels of free fatty acids and increased blood flow. They postulated that these changes improved the energy supply for the muscle. After strenuous exercise, however, the serum levels of such enzymes as the creatine phosphokinase, lactate dehydrogenase, and aldolase may rise dramatically, indicating either some membrane incompetence or necrosis of muscle cells. Mild to moderate exercise may be tolerated with no symptoms. Over the age of 40, the patient may have progressive wasting and weakness of muscles.

This disorder was first delineated clinically by McArdle in 1951, when he studied a patient who, during ischemic arm exercise, had no increase in venous lactate [105]. He suggested that a defect was present in the degradation of glycogen to lactate. Subsequently, Mommaerts et al. [106] and Schmid et al. [107] demonstrated the lack of muscle phosphorylase in the form of glycogen storage disease.

A relatively simple diagnostic test, the forearm ischemic exercise test, involves exercising the forearm while the arterial inflow is occluded by means of an inflated arm cuff. Samples of the venous blood from that limb are taken for measurements of pyruvate and lactate. This is a useful clinical test for identifying various disorders. In a normal person, ischemic exercise causes a rise in venous lactate approximately 2½ to 5 times preexercise levels. In a patient with McArdle's disease, there is usually little or no rise in venous lactate with exercise [105]. The most definitive diagnostic test is the biochemical demonstration of a deficiency of phosphorylase in a muscle biopsy sample. Absence of phosphorylase and a moderate accumulation of glycogen can also be shown histochemically.

Type 7 Glycogen Storage Disease

In this form of glycogen storage disease, the clinical story is similar to that of McArdle's syndrome [108]. Again, the ischemic arm exercise test results in no elevation of venous lactate, indicating a block in the production of lactate, but phosphorylase is present. In the six reported cases of this syndrome the molecular alterations that have been described include a near absence of muscle phosphofructokinase, an increase in glucose 6-phosphate and fructose 6-phosphate, and low muscle fructose 1-6-diphosphate [108–111]. In addition, the patients showed a deficiency in the muscle form of phosphofructokinase in red blood cells. A similar reduction in the muscle form of this enzyme in the erythrocytes of three parents, the occurrence of the disease in siblings of both sexes, and consanguinity in one set of parents suggest an autosomal recessive inheritance.

Myasthenia Gravis

Earlier studies on myasthenia gravis indicated a possible presynaptic defect at the neuromuscular junction. Recent studies have produced important information that disputes this view [112—114], although a presynaptic defect is not necessarily excluded.

In 1973 Patrick and Lindstrom [112] injected rabbits with acetylcholine receptor highly purified from the electric organ of *Electrophorus electricus* and emulsified in Freund's adjuvant. This resulted in the production of precipitating antibody to the acetylcholine receptor and to a flaccid paralysis in the rabbits. Abnormal electromyograms characteristic of neuromuscular blockade were observed, and treatment of the rabbits with an anticholinesterase (edrophonium) improved their motor power and reversed the neuromuscular blockade. The involvement of the postsynaptic area is further supported by a study in which rats injected with α-bungarotoxin from the Formosan cobra [113] developed a postsynaptic blockade with physiological and pharmacological properties typical of myasthenia gravis.

In another study, based on the binding of α-bungarotoxin, myasthenic muscles were shown to contain fewer junctional acetylcholine receptors than does normal muscle [114].

The suggestion of the involvement of the postsynaptic area in myasthenia gravis, based on animal studies which show that antibodies to acetylcholine receptor sites cause abnormalities similar to those seen in myasthenia gravis, are supported with direct human studies. Almon and Appel [115] have recently shown that 17 of 25 patients with myasthenia gravis have an IgG antibody in their sera that binds to the acetylcholine receptor macromolecular complex in human muscle; it may interfere with the accessibility of the receptor site to acetylcholine. Apparently, serum from some myasthenia gravis patients can block α-bungarotoxin binding to human muscle fibers [116]. Repeated injections into mice of globulin fractions of serum from myasthenia gravis patients can produce some of the characteristic features of the disease [117]. The precise role of these humoral factors in the pathogenesis and course of the disease remains to be elucidated.

REFERENCES

*1.   Drabikowski, W., Strzelecka-Golaszewska, H., and Carafoli, E. (Eds.). *Calcium Binding Proteins.* Amsterdam: Elsevier, 1974.

*2.   Walton, J. N. (Ed.). *Disorders of Voluntary Muscle.* 3d ed. Edinburgh: Churchill, 1974.

*3.   Martonosi, A. Biochemical and clinical aspects of sarcoplasmic reticulum function. *Curr. Top. Membr. Transp.* 3:83, 1972.

*4.   The mechanism of muscle contraction. *Cold Spring Harbor Symp. Quant. Biol.* Vol. 37, 1972.

*5.   Taylor, E. W. Mechanisms of actomyosin ATPase and the problem of muscle contraction. *Curr. Top. Bioenergetics* 5:201, 1973.

*6.   Weber, A., and Murray, J. M. Molecular control mechanisms in muscle contraction. *Physiol. Rev.* 53:612, 1973.

*Asterisks denote key references.

6a.   Ebashi, S., Endo, M., and Ohtsuki, I. Control of muscle contraction. *Q. Rev. Biophys.* 2:351, 1969.

*7.   Franzini-Armstrong, C. Membranous Systems in Muscle Fibers. In Bourne, G. H. (Ed.), *The Structure and Function of Muscle.* Vol. II, Pt. 2. New York: Academic, 1972, p. 531.

*8.   Huxley, H. E. Molecular Basis of Contraction in Cross-Striated Muscles. In Bourne, G. H. (Ed.), *The Structure and Function of Muscle.* Vol. II, Pt. 1. New York: Academic, 1972.

9.   Huxley, H. E., and Hanson, J. Changes in the cross-striations of muscle during contraction and stretch and their structural interpretation. *Nature* 173:973, 1954.

10.   Huxley, A. F., and Niedergerke, R. Structural changes in muscle during contraction. *Nature* 173:971, 1954.

11.   Gordon, A. M., Huxley, A. F., and Julian, F. J. The variation in isometric tension with sarcomere lengths in vertebrate muscle fibers. *J. Physiol.* (Lond.) 184:170, 1966.

12.   Offer, G., Moos, C., and Starr, R. A new protein of the thick filaments of vertebrate skeletal myofibrils. *J. Mol. Biol.* 74:653, 1973.

13.   Greaser, M., and Gergely, J. Reconstitution of troponin activity from three protein components. *J. Biol. Chem.* 246:4226, 1971.

14.   Goll, D., Mommaerts, W. F. H. M., Reedy, M. K., and Seraydarian, K. Studies on α-actinin-like protein liberated during trypsin digestion of α-actinin and of myofibrils. *Biochim. Biophys. Acta* 175:174, 1969.

14a.   Sugita, H., Masaki, T., Ebashi, S., and Pearson, C. M. Staining of the nemaline rod by fluorescent antibody against iso-actinin. *Proc. Jpn. Acad.* 50:237, 1974.

15.   Hanson, J., and Lowy, J. The structure of F-actin and of actin filaments isolated from muscle. *J. Mol. Biol.* 6:46, 1963.

16.   Elzinga, M., Collins, J. H., Keuhl, W. M., and Adelstein, R. S. Complete amino acid sequence of actin of rabbit skeletal muscle. *Proc. Natl. Acad. Sci. U.S.A.* 70:2687, 1973.

17.   Straub, F. B., and Feuer, G. Adenosine triphosphate: The functional group of actin. *Biochim. Biophys. Acta* 4:455, 1950.

18.   Martonosi, A., Gouvea, M. A., and Gergely, J. Studies on actin. III. G-F transformation of actin and muscular contraction (experiments in vivo). *J. Biol. Chem.* 235:1707, 1960.

19.   Sarkar, S., Sreter, F. A., and Gergely, J. Light chains of myosins from fast, slow and cardiac muscles. *Proc. Natl. Acad. Sci. U.S.A.* 68:946, 1971.

20.   Kuehl, W. M., and Adelstein, R. S. The absence of 3-methylhistidine in red, cardiac and fetal myosin. *Biochem. Biophys. Res. Commun.* 39:956, 1970.

21.   Trayer, I. P., Harris, C. I., and Perry, S. V. 3-Methyl histidine and adult and foetal forms of skeletal muscle myosin. *Nature* 217:452, 1968.

22.   Huszar, G., and Elzinga, M. ε-N-methyl lysine in myosin. *Nature* 223:834, 1969.

23.   Lowey, S., and Risby, D. Light chains from fast and slow muscle myosins. *Nature* 234:81, 1971.

24.   Frank, G., and Weeds, A. G. The amino acid sequence of some alkali light chains of rabbit skeletal muscle myosin. *Eur. J. Biochem.* 44:317, 1974.

25.   Kendrick-Jones, J. Role of myosin light chains in calcium regulation. *Nature* 249:631, 1974.

26.   Sreter, F. A., Romanul, F. C. A., Salmons, S., and Gergely, J. The Effect of a Changed Pattern of Activity on Some Biochemical Characteristics of Muscle. In Milhorat, A. T. (Ed.), *International Conference on Exploratory Concepts in Muscular Dystrophy,* II. New York: Elsevier, 1974, p. 338.

27.   Sreter, F. A., Gergely, J., and Luff, A. L. The effect of cross reinnervation on the synthesis of myosin light chains. *Biochem. Biophys. Res. Commun.* 56:84, 1974.

27a. Lowey, S., Slayter, H. S., Weeds, A. G., and Baker, H. Substructure of the myosin molecule. I. Subfragments of myosin by enzymic degradation. *J. Mol. Biol.* 42:1, 1969.

27b. Huxley, H. E. The mechanism of muscle contraction. *Science* 164:1356, 1969.

28.   Pepe, F. A. The myosin filament: II. Interaction between myosin and actin filaments observed using antibody staining in fluorescent and electron microscopy. *J. Mol. Biol.* 27:227, 1967.

29.   Huxley, H. E., and Brown, W. The low angle x-ray diagram of vertebrate striated muscles and its behaviour during contraction and rigor. *J. Mol. Biol.* 30:383, 1967.

30.   Tregear, R. T., and Squire, J. M. Myosin content and filament structure in smooth and striated muscle. *J. Mol. Biol.* 77:279, 1973.

31.   Morimoto, K., and Harrington, W. F. Substructure of the thick filament of vertebrate striated muscle. *J. Mol. Biol.* 83:83, 1974.

32.   Engelhardt, W. A., and Ljubimova, M. N. Myosin and adenosine triphosphatase. *Nature* 144:668, 1939.

33.   Szent-Györgyi, A. Studies on muscle. *Acta Physiol. Scand.* 9 (Suppl. 25), 1945.

34.   Bailey, K. Tropomyosin: A new asymmetrical protein component of the muscle fibril. *Biochem. J.* 43:271, 1948.

35.   Greaser, M. L., Yamaguchi, M., Brekke, C., Potter, J. D., and Gergely, J. Troponin subunits and their interactions. *Cold Spring Harbor Symp. Quant. Biol.* 37:235, 1972.

36.   Potter, J. D., and Gergely, J. Troponin, tropomyosin and actin interactions in the Ca^{2+} regulation of muscle contraction. *Biochemistry* 13:2697, 1974.

37.   Hitchcock, S. E. Regulation of muscle contraction: Binding of troponin and its components of actin and tropomyosin. *Eur. J. Biochem.* 52:255, 1975.

38.   MacLennan, D. H. Resolution of the calcium transport system of sarcoplasmic reticulum. *Can. J. Biochem.* 53:251, 1975.

39.   Gilbert, C., Kretzschmar, K. M., and Wilkie, D. R. Heat, work and phosphocreatine splitting during muscular contraction. *Cold Spring Harbor Symp. Quant. Biol.* 37:613, 1972.

39a. Homsher, E., Rall, J. A., Wallner, A., and Ricchiuti, N. V. Energy liberation and chemical change in frog skeletal muscle during single isometric tetanic contractions. *J. Gen. Physiol.* 65:1, 1975.

40.   Hill, A. V. The effect of load on the heat of shortening of muscle. *Proc. R. Soc. Lond. [Biol.]* Ser. B. 159:297, 1964.

41.   Hill, A. V. The variation of total heat production in a twitch with velocity of shortening. *Proc. R. Soc. Lond. [Biol.]* Ser. B. 159:596, 1964.

42.   Cain, D. F., and Davies, R. E. Breakdown of adenosine triphosphate during a single contraction of working muscle. *Biochem. Biophys. Res. Commun.* 8:361, 1962.

43.   Reedy, M. K., Holmes, K. C., and Tregear, R. T. Induced changes in orientation of the cross bridges of glycerinated insect flight muscle. *Nature* 207:1276, 1965.

44.   Huxley, A. F., and Simmons, R. M. Mechanical transients and the origin of muscular force. *Cold Spring Harbor Symp. Quant. Biol.* 37:669, 1972.

*45. Huxley, A. F., and Simmons, R. M. Proposed mechanism of force generation in striated muscle. *Nature* 233:533, 1971.

*46. Gross, S. R., and Mayer, S. E. Phosphorylase kinase mediating the effects of cyclic AMP in muscle. *Metabolism* 24:369, 1975.

47. Soderling, T. R., Hickenbottom, J. P., Reimann, E. M., Hunkeler, F. L., Walsh, D., and Krebs, E. G. Inactivation of glycogen synthetase and activation of phosphorylase kinase by muscle adenosine 3-,5-monophosphate-dependent protein kinase. *J. Biol. Chem.* 245:6317, 1970.

48. Heywood, S. M., and Rich, A. In vitro synthesis of native myosin, actin and tropomyosin from embryonic chick polysomes. *Proc. Natl. Acad. Sci. U.S.A.* 59:590, 1968.

49. Sarkar, S., and Cooke, P. In vitro synthesis of light and heavy polypeptide chains of myosin. *Biochem. Biophys. Res. Commun.* 41:918, 1970.

50. Low, R. B., Vournakis, J. N., and Rich, A. Identification of separate polysomes active in the synthesis of the light and heavy chains of myosin. *Biochemistry* 10:1813, 1971.

51. Morales, M. F., Ingwall, J. S., Stockdale, F. E., McKay, R., Brivio-Haugland, R., and Kenyon, G. L. Creatine and Muscle Protein Synthesis. In Milhorat, A. T. (Ed.), *International Conference on Exploratory Concepts in Muscular Dystrophy,* II. New York: Elsevier, 1974, p. 212.

52. Gauthier, G. F. Fiber Types in Mammalian Skeletal Muscle and Their Relationship to Functional Activity. In Pearson, A. M. (Ed.), *Advances in Muscle Biology,* Vol. II. East Lansing, Mich.: Michigan State University Press, 1975.

53. Close, R. I. Dynamic properties of mammalian skeletal muscle. *Physiol. Rev.* 52:129, 1972.

54. Close, R. I. Specialization Among Fast-Twitch Muscles. In Milhorat, A. T. (Ed.), *International Conference on Exploratory Concepts in Muscular Dystrophy,* II. New York: Elsevier, 1974, p. 309.

55. Kaplan, N. O., and Goodfriend, J. L. Role of the types of lactic dehydrogenase. *Enzyme Regulation* 2:203, 1964.

56. Barany, M. Activation of myosin correlated with speed of muscle shortening. *J. Gen. Physiol.* 50:197, 1967.

57. Sreter, F. A., Seidel, J., and Gergely, J. Studies on myosin from red and white skeletal muscle of the rabbit: I. Adenosine triphosphate activity. *J. Biol. Chem.* 241:5772, 1966.

58. Nakamura, A., Sreter, F., and Gergely, J. Comparative studies of light meromyosin paracrystals derived from red, white and cardiac muscle myosins. *J. Cell. Biol.* 49:883, 1971.

59. Perry, S. V. Variation in the Contractile and Regulatory Proteins of the Myofibril with Muscle Type. In Milhorat, A. T. (Ed.), *International Conference on Exploratory Concepts in Muscular Dystrophy,* II. New York: Elsevier, 1974, p. 319.

60. Eccles, J. C., Eccles, R. M., and Lundberg, A. The action potentials of the α motor neurones supplying fast and slow muscle. *J. Physiol.* (Lond.) 142:275, 1958.

61. Buller, A. J., Eccles, J. C., and Eccles, R. M. Differentiation of fast and slow muscles in the cat hind leg. *J. Physiol.* (Lond.) 150:399, 1960.

62. Barany, M., and Close, R. I. The transformation of myosin in cross-innervated rat muscle. *J. Physiol.* (Lond.) 213:455, 1971.

63. Buller, A. J., Mommaerts, W. F. H. M., and Seraydarian, K. Enzymatic properties of myosin in fast and slow twitch muscles of the cat following cross-innervation. *J. Physiol.* (Lond.) 205:581, 1969.

64. Weeds, A. G., Trentham, D. R., Kean, C. J. C., and Buller, A. J. Myosin from cross-reinnervated cat muscles. *Nature* 247:135, 1974.

65. Dubowitz, V. Cross-innervated mammalian skeletal muscle: Histochemical, physiological and biochemical observations. *J. Physiol.* (Lond.) 193:481, 1967.

66. Romanul, F. C. A., and Van Der Meulen, J. P. Slow and fast muscles after cross innervation: Enzymatic and physiological changes. *Arch. Neurol.* 17:387, 1967.

67. Guth, L., Watson, P. K., and Brown, W. C. Effects of cross reinnervation on some chemical properties of red and white muscles of rat and cat. *Exp. Neurol.* 20:52, 1968.

68. Mommaerts, W. F. H. M., Buller, A. J., and Seraydarian, K. The modification of some biochemical properties of·muscle by cross-innervation. *Proc. Natl. Acad. Sci. U.S.A.* 64:128, 1969.

69. Salmons, S., and Vrbova, G. The influence of activity on some contractile characteristics of mammalian fast and slow muscle. *J. Physiol.* (Lond.) 201:535, 1969.

70. Romanul, F. C. A., Sreter, F. A., Salmons, S., and Gergely, J. The Effect of a Changed Pattern of Activity on Histochemical Characteristics of Muscle Fibers. In Milhorat, A. T. (Ed.), *International Conference on Exploratory Concepts in Muscular Dystrophy,* II. New York: Elsevier, 1974, p. 344.

71. Pette, D., Smith, M. E., Staudte, H. W., and Vrbova, G. Effects of long-term electrical stimulation on some contractile and metabolic characteristics of fast rabbit muscles. *Pfluegers Arch.* 338:252, 1973.

*72. Guth, L. Trophic influences of nerve on muscle. *Physiol. Rev.* 48:645, 1968.

72a. Dimauro, S., Schotland, D. L., Bonilla, E., Lee, C. P., Dimauro, P. M. M., and Scarpa, A. Mitochondrial Myopathies: Which and How Many? In Milhorat, A. T. (Ed.), *International Conference on Exploratory Concepts in Muscular Dystrophy,* II. New York: Elsevier, 1974, p. 506.

72b. Engel, A. G., Angelini, C., and Nelson, R. A. Identification of Carnitine Deficiency as a Cause of Human Lipid Storage Myopathy. In Milhorat, A. T. (Ed.), *see* [72a], p. 601.

73. Samaha, F. J., and Thies, W. H. Superprecipitation and ATPase activity in normal, newborn and denervated muscle. *Exp. Neurol.* 38:398, 1973.

74. Furukawa, T., and Peter, J. B. Superprecipitation and adenosine triphosphatase activity of myosin B in Duchenne muscular dystrophy. *Neurology* 21:920, 1971.

75. Samaha, F. J. Actomyosin alterations in Duchenne muscular dystrophy. *Arch. Neurol.* 28:405, 1973.

76. Samaha, F. J., Schroeder, J. M., Rebeiz, J., and Adams, R. D. Studies on myotonia. *Arch. Neurol.* 17:22, 1967.

77. Penn, A. S., Cloak, R. A., and Rowland, L. P. Myosin from normal and dystrophic human muscle. *Arch. Neurol.* 27:159, 1972.

78. Samaha, F. J. Tropomyosin and troponin in normal and dystrophic human muscle. *Arch. Neurol.* 26:547, 1972.

79. Sugita, H., Okomoto, K., and Ebashi, S. Some observations on the microsome fraction of biopsied muscle from patients with progressive muscular dystrophy. *Proc. Jpn. Acad.* 42:295, 1966.

80. Samaha, F. J., and Gergely, J. Biochemical abnormalities of the sarcoplasmic reticulum in muscular dystrophy. *N. Engl. J. Med.* 280:184, 1969.

81. Peter, J. B., and Worsfold, M. Muscular dystrophy and other myopathies: Sarcotubular vesicles in early disease. *Biochem. Med.* 2:364, 1969.

82. McComas, A. J., Sica, R. E. P., and Campbell, M. J. "Sick" motoneurones. A unifying concept of muscle disease. *Lancet* 1:321, 1971.

83. Engel, W. K. Duchenne Muscular Dystrophy: A Histologically Based Ischemia Hypothesis and Comparison with Experimental Ischemia Myopathy. In Pearson, C. M., and Mostofi, F. K. (Eds.), *The Striated Muscle.* Baltimore: Williams & Wilkins, 1973, p. 453.

84. Mendell, J. R., Engel, W. K., and Derrer, E. C. Duchenne muscular dystrophy: Functional ischemia reproduces its characteristic lesions. *Science* 172:1143, 1971.

85. Mendell, J. R., Engel, W. K., and Derrer, E. C. Increased plasma enzyme concentrations in rats with functional ischemia of muscle provide a possible model of Duchenne muscular dystrophy. *Nature* 239:522, 1972.

86. Parker, J. M., and Mendell, J. R. Proximal myopathy induced by 5-HT-imipramine simulates Duchenne dystrophy. *Nature* 247:103, 1974.

87. Jerusalem, F., Engel, A. G., and Gomez, M. R. Duchenne dystrophy: I. Morphometric study of the muscule microvasculature. *Brain* 97:115, 1974.

88. Alberts, M. D., and Samaha, F. J. Serum pyruvate kinase in muscle disease and carrier states. *Neurology* 24:462, 1974.

89. Somers, J. E., and Winer, N. Reversible myopathy and myotonia following administration of a hypocholesterolemic agent. *Neurology* 16:761, 1966.

90. Peter, J. B., Andiman, R. M., Bowman, R. L., and Nagatomo, T. Myotonia induced by diazacholesterol: Increased ($Na^+ + K^+$)-ATPase activity of erythrocyte ghosts and development of cataracts. *Exp. Neurol.* 41:738, 1973.

91. Roses, A. D., and Appel, S. H. Protein kinase activity in erythrocyte ghosts of patients with myotonic muscular dystrophy. *Proc. Natl. Acad. Sci. U.S.A.* 70:1855, 1973.

92. Butterfield, D. A., Chesnut, D. B., Roses, A. D., and Appel, S. H. Electron spin resonance studies of erythrocytes from patients with myotonic muscular dystrophy. *Proc. Natl. Acad. Sci. U.S.A.* 71:909, 1974.

93. Sibley, J. A., and Lehninger, A. L. Aldolase in the serum and tissues of tumor-bearing animals. *J. Natl. Cancer Inst.* 9:203, 1949.

*94. Stanbury, J. B., Wyngaarden, J. B., and Frederickson, D. S. (Eds.). *The Metabolic Basis of Inherited Disease*. New York: McGraw-Hill, 1972.

95. Creutzfeldt, O. D., Abbott, B. C., Fowler, W. M., and Pearson, C. M. Muscle membrane potentials in episodic adynamia. *Electroencephalogr. Clin. Neurophysiol.* 15:508, 1963.

96. Brooks, J. E. Hyperkalemic periodic paralysis. Intracellular Electromyographic studies. *Arch. Neurol.* 20:13, 1969.

97. Hofmann, W. W., and Smith, R. A. Hypokalaemic periodic paralysis studied in vitro. *Brain* 93:445, 1970.

98. Samaha, F. J. Sodium-potassium adenosine triphosphate in diseased muscle. Studies on periodic paralysis, myasthenia gravis and Eaton-Lambert syndrome. *Neurology* 19:551, 1969.

99. Engel, A. G., and Lambert, E. H. Calcium activation of electrically inexcitable muscle fibers in primary hypokalemic periodic paralysis. *Neurology* 19:851, 1969.

100. Cori, G. T., and Cori, C. F. Glucose-6-phosphatase of the liver in glycogen storage disease. *J. Biol. Chem.* 199:661, 1952.

101. Pompe, J. C. Over idiopatische hypertrophie van het hart. *Ned. Tijdschr. Geneeskd.* 76:304, 1932.

102. Hers, H. G. Alpha-glucosidase deficiency in generalized glycogen storage disease (Pompe). *Biochem. J.* 86:11, 1963.

103. Baudhuin, P., Hers, H. G., and Loeb, H. An electronmicroscopic and biochemical study of type II glycogenosis. *Lab. Invest.* 13:1139, 1964.

104. Pernow, B. B., Havel, R. J., and Jennings, D. B. The second wind phenomenon in McArdle's syndrome. *Acta Med. Scand.* [*Suppl.*] 472:294, 1967.

105. McArdle, B. Myopathy due to a defect in muscle glycogen breakdown. *Clin. Sci.* 10:13, 1951.

106. Mommaerts. W. F. H. M., Illingworth, B., Pearson, G. M., Guillory, P. J., and Seraydarian, K. A. Functional disorder of muscle associated with the absence of phosphorylase. *Proc. Natl. Acad. Sci. U.S.A.* 45:791, 1959.

107.   Schmid, R., and Mahler, R. Chronic progressive myopathy with myoglo-binuria: Demonstration of a glycogenolytic defect in muscle. *J. Clin. Invest.* 38:1044, 1959.
108.   Tarui, S., Okuno, G., Ikura, Y., Tanaka, T., Masami, S., and Nishikawa, M. Phosphofructokinase deficiency in skeletal muscle: A new type of glyco-genosis. *Biochem. Biophys. Res. Commun.* 19:517, 1965.
109.   Layzer, R. B., Rowland, L. P., and Ranney, H. M. Muscle phosphofructo-kinase deficiency. *Arch. Neurol.* 17:512, 1967.
110.   Serratrice, G. Forme myopathique du déficit en phosphofructokinase. *Rev. Neurol.* 120:271, 1969.
111.   Tobin, W. E., Huijing, F., Porro, R. S., and Salzman, R. T. Muscle phospho-fructokinase deficiency. *Arch. Neurol.* 28:128, 1973.
112.   Patrick, J., and Lindstrom, J. Autoimmune response to acetylcholine receptor. *Science* 180:871, 1973.
113.   Satyamurti, S., Drachman, D. B., and Slone, F. Blockade of acetylcholine receptors: A model of myasthenia gravis. *Science* 187:955, 1975.
114.   Fambrough, D. M., Drachman, D. B., and Satyamurti, S. Neuromuscular junction in myasthenia gravis: Decreased acetylcholine receptors. *Science* 182:293, 1973.
115.   Almon, R. R., and Appel, S. H. Serum acetylcholine receptor antibodies in myasthenia gravis. *Ann. N. Y. Acad. Sci.* 1976 (in press).
116.   Bender, A. N., Engel, W. K., Ringel, S. P., Daniels, M. P., and Vogel, Z. Myasthenia gravis: A serum factor blocking acetylcholine receptors of the human neuromuscular junction. *Lancet* 1:607, 1975.
117.   Toyka, K. V., Drachman, D. B., Pestronk, A., and Kao, I. Myasthenia gravis: Passive transfer from man to mouse. *Science* 190:397, 1975.

Chapter 24
Disorders of Amino Acid Metabolism
Y. Edward Hsia

Inherited biochemical disorders share a common mechanism. A structurally or quantitatively defective protein – an enzyme, for example, or a membrane-transport protein – is formed from a mutant gene via its primary gene product, messenger RNA. The consequent interruption of biochemical pathways can cause widespread repercussions in systemic metabolism, related to accumulation of substances proximal to the abnormality and depletion of substances distal to the abnormality. Some of the enzyme blocks are closely related to compounds with potent neurochemical properties (Chaps. 10 and 11), which could account for specific neurotoxic effects. Accumulation of substances proximal to the abnormality can result in toxicity; depletion of substances distal to the abnormality can result in deficiency.

The inherited disorders of amino acid metabolism [1] and of their organic acid derivatives [2] are illustrated in Figures 24-1 to 24-5 [3]. These illustrations contain almost all the aminoacidopathies of neurological significance, save for those of proline metabolism. Together, these illustrations provide an overall perspective of individual biochemical disorders in the context of the major biochemical pathways of intermediary metabolism. (For those of carbohydrate metabolism see Chapter 25.)

DISTURBANCES OF BRAIN NUTRITION
Whenever the concentration of any single plasma amino acid is significantly altered, the active transport of other individual amino acids across the blood-brain barrier may be seriously imbalanced (Chap. 14). Experimentally, the uptake of labeled amino acids across the blood-brain barrier varies with the amino acid and is strongly influenced by the concentration of the other amino acids. The essential amino acids, plus tyrosine, and the precursors of the biogenic amines are most actively taken up by the brain, whereas the nonessential amino acids, notably those with neurotransmitter properties, such as aspartate, glutamate, and glycine, are taken up least actively [4–6].

ABNORMALITIES OF AMINO ACID TRANSPORT

Hartnup Disease
This is a particularly apt example of systemic amino acid imbalance with indirect nutritional consequences on neurological function. The biochemical lesion in this condition is a defect of membrane transport of neutral amino acids affecting only the intestinal mucosa and renal tubule. Tryptophan malabsorption and tryptophanuria are prominent features of this condition, but the transport defect involves all the cyclic and neutral amino acids in greater or lesser degree. This lesion can be benign, and individuals with the disorder often remain asymptomatic. A well-recognized complication of Hartnup disease, however, is a pellagralike syndrome. Pellagra, characterized by a photosensitive erythematous skin rash, gastrointestinal

500

upset, ataxia, and dementia, is caused classically by nutritional lack of niacin, a precursor of nicotinamide (Chap. 29). In man, an estimated half of the usual requirement for nicotinamide is synthesized endogenously from dietary tryptophan, so patients with Hartnup disease are likely to have nicotinamide deficiency. Therapeutic doses of niacin will reverse the cutaneous and neurological lesions, confirming that in Hartnup disease the pellagralike syndrome is an indirect consequence of the transport defect. Other transport abnormalities are shown in Table 24-1.

METABOLIC ACIDOSIS

Severe metabolic acidosis, often together with ketosis or lactic acidosis, occurs in many disturbances of carbohydrate metabolism (Chap. 25) and of amino acid metabolism (Table 24-2). These conditions produce vomiting, altered respiratory excursions, and clouding of consciousness, leading in extreme acidosis to irreversible brain damage or to coma and death; chronic acidosis may also result in general debility and malnutritional brain damage. It is conceivable that moderately severe, but brief, periods of acidosis do not affect brain function seriously. For instance, one child who succumbed to severe acidosis, without being retarded in any way, appeared to have a deficiency of succinyl Co-A acetoacetyl-CoA transferase [7] (Fig. 24-1, reaction 5). The intense recurrent acidosis suffered by this patient and by patients with pyroglutamicaciduria (see under Glutathione later in this chapter) argue against irreversible short-term neurotoxicity of acidosis per se.

BRANCHED-CHAIN AMINO ACIDS

In disorders of branched-chain amino acid catabolism (Fig. 24-2), the nonspecific observations of acute irritability, hypertonicity, and drowsiness are attributable to severe acidosis or to hyperammonemia (see p. 509), but a common mechanism for the brain damage is suggested by the failure of postnatal myelination in several of the disorders. Cerebellar ataxia has been reported in some patients, but not consistently in any single disorder.

Maple Syrup Urine Disease

This was the first of this group of disorders to be recognized. The basic enzyme defect is of branched-chain ketoacid decarboxylase, a multienzyme complex (Fig. 24-2, reaction 3) analogous to pyruvate dehydrogenase. The intermediate steps are similar and depend on the same cofactors: thiamine pyrophosphate, lipoic acid, nicotinamide adenine dinucleotide, flavine adenine dinucleotide, and coenzyme A. A single enzyme complex appears to serve all three branched-chain amino acids. The classical variant of this disease is usually lethal in early infancy unless treated with restricted intake of leucine, isoleucine, and valine. In untreated patients who survive longer, there is defective myelination, which has been reproduced experimentally in rat cerebellum cultures by exposure to α-ketoisocaproic acid in concentrations comparable to those found in the blood of patients (and in the brain of one patient) with maple syrup urine disease [7a]. This same derivative of leucine has been shown experimentally to inhibit oxidation of pyruvate and of α-ketoglutarate [8,9], suggesting a possible mechanism for lactic acidosis

Table 24-1. Inherited Abnormalities of Amino Acid Transport

Condition	Location of Lesion	Amino Acids Involved	Neurological Features	Comments
Hartnup disease	Intestine and kidney	Tryptophan and neutral amino acids	Ataxia, dementia, psychosis	Causes nicotinamide deficiency
Blue diaper syndrome[a]	Intestine	Tryptophan	Irritability	Hypercalcemia, indolyluria
Oasthouse urine disease[a]	Intestine and kidney	Methionine	Seizures, retardation, hyperpnea	White hair, odd smell, α-hydroxybutyric aciduria
Cystinuria	Intestine and kidney	Cysteine and dibasic amino acids	Possible liability to mental illness	Renal stones
Isolated cystinuria	Kidney	Cysteine	Probably benign	–
Pancreatitis and cystine-lysinuria	Kidney	Lysine and cysteine	None	Hereditary pancreatitis
Lysinuric protein intolerance	Intestine and kidney	Dibasic amino acids	Abdominal cramps, hyperammonemia	Occasional growth retardation
Iminoglycinuria	Kidney	Proline, hydroxyproline, and glycine	Benign	–
Glycinuria	Kidney	Glycine	Benign	Kidney oxalate stones
β-Aminoisobutyric aciduria	Kidney	β-Aminoisobutyrate	Benign	A common normal variant
Hyperornithinemia and homocitrullinemia[a]	Mitochondria	Ornithine	Hyperammonemia	–
Folate malabsorption	Intestine	Pteroylglutamates	Retardation, athetosis, seizures	Megaloblastic anemia
Vitamin B_{12} malabsorption	Intestine	Vitamin B_{12}	Growth retardation, risk of brain damage	Megaloblastic anemia
Intrinsic factor deficiency	Intestine	Vitamin B_{12}	Growth retardation, risk of brain damage	Megaloblastic anemia
Transcobalamin II deficiency	Extracellular	Vitamin B_{12}	Growth retardation, risk of brain damage	Megaloblastic anemia

[a]Disorders described in only one or two families.

Table 24-2. Disorders of Amino Acids and Organic Acids that Cause Metabolic Acidosis and Are Associated with Neurological Lesions[a]

Condition	Location of Defective Enzyme	Neurological Lesions
Maple syrup urine disease	Fig. 24-2, reaction 3	Hypotonia, rigidity, retardation, failure of myelination
Isovalerica acidemia	Fig. 24-2, reaction 4	Drowsiness, cerebellar ataxia, hyperreflexia, psychomotor retardation
Biotin responsive β-methylcrotonyl-glycinemia[b]	Fig. 24-2, reaction 5	Motoneuron degeneration in one; irritability, lethargy, and retardation in another; benign in the third
β-Ketothiolase deficiency[b]	Fig. 24-2, reaction 6	Retardation, hyperammonemia in one; benign in the second
Propionica acidemia	Fig. 24-2, reaction 7	Lethargy, prostration, seizures, retardation; hyperammonemia in some
Biotin-responsive variant	Fig. 24-2, reaction 7	Lethargy, hypotonia, athetosis, retardation
Methylmalonic acidemia	Fig. 24-2, reactions 8, 9, 10	As for propionicacidemia
Methylmalonic acidemia with homocystinuria	Fig. 24-2, reaction 10	May have cerebellar lesions, severe brain damage, convulsions
Pyroglutamic acidemia	Fig. 24-3, reaction 1	Retardation, athetosis, and tetraplegia in one of three known patients

[a]For disorders of carbohydrate metabolism and the hyperalaninemic lactic acidoses, see Chapter 25.
[b]Disorders described in only one or two families.

and ketoacidosis in this disease. The odor of maple syrup characteristic of this disease must be from a fragrant ketoacid or ester, the smell being not unlike that of α-ketobutyrate. There are milder, intermittent, and thiamine-responsive variants.

Hypervalinemia and Hyperleucine-Isoleucinemia

Hypervalinemia (reaction 2) and hyperleucine-isoleucinemia [1] (reaction 1) are shown in Figure 24-2 and Table 24-4. Perhaps the imbalanced accumulation of valine or of leucine and isoleucine prevents normal metabolism or normal transport of other amino acids into the brain [6]. Rats force-fed a diet low only in valine became very weak but were not weakened by a diet low in valine and leucine [10], confirming that imbalance may be more harmful than combined deficiencies.

Isovalerica Acidemia

This disorder [2] (Fig. 24-2, reaction 4) is characterized by severe ketoacidosis associated with an offensive odor, like that of cheese or sweaty feet, that emanates from free isovaleric acid. Although acute symptoms were attributable to the recurrent

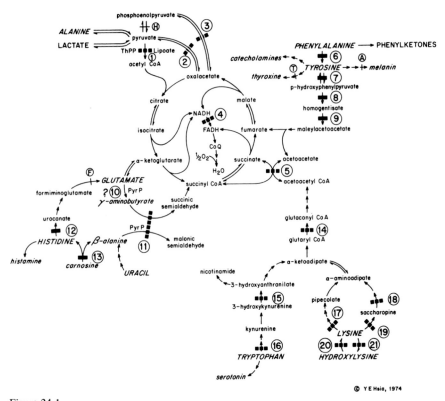

Figure 24-1
Metabolic map of citric acid cycle and metabolism of some amino acids. Solid bars indicate the sites of enzyme defects, interrupted bars designate postulated blocks. Amino acids are in capitalized italics; other important metabolites are also capitalized; neuroactive compounds and biogenic amines are in lowercase italics. The sites of metabolic blocks causing albinism are labeled A, and defects of thyroid hormone synthesis are labeled T. The symbol F locates the site of action of forminofolate transferase on histidine catabolism. Coenzymes relevant to the cause of treatment of a metabolic disorder are included. (See also Tables 24-3 and 24-4.)

Enzymes: (1) pyruvate dehydrogenase complex; (2) pyruvate carboxylase; (3) phosphoenolpyruvate carboxykinase; (4) undetermined; (5) succinyl CoA-3-ketoacid CoA transferase; (6) phenylalanine hydroxylase; (7) cytosol tyrosine aminotransferase; (8) *p*-hydroxyphenylpyruvate oxidase; (9) homogentisic acid oxidase; (10) glutamate decarboxylase; (11) ? β-alanine aminotransferase, ? γ-aminobutyric aminotransferase; (12) histidase; (13) carnosinase; (14) glutaryl CoA dehydrogenase; (15) kynureninase; (16) undetermined; (17) lysine dehydrogenase; (18) undetermined; (19) lysine α-ketoglutarate reductase; (20) undetermined; (21) lysine hydroxylase. (From Hsia, Y. E. Inherited Metabolic Disorders. In Avery, G. (Ed.), *Neonatology.* Philadelphia: Lippincott, 1975. Reproduced by permission of the author.)

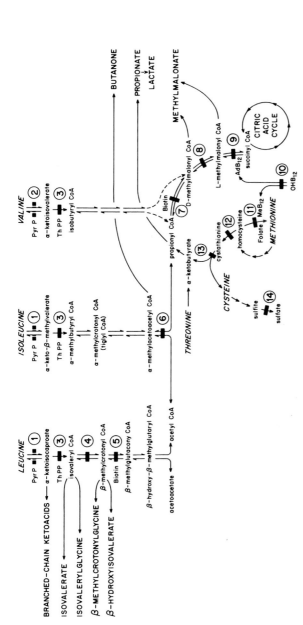

Figure 24-2

Metabolic map of branched-chain amino acids and sulfur amino acids. The catabolism of leucine, isoleucine, and valine descend to common intermediary metabolites from the top. Organic acids and other compounds excreted in these metabolic disorders are indicated on the left or right. The catabolism of threonine and methionine, with the pathways of vitamin B12 coenzyme synthesis, are below. Metabolic blocks at 6, 7, 8, 9 have all produced the syndrome of ketotic hyperglycinemia. (See also Tables 24-2 and 24-4.)

Enzymes: (1) ? branched-chain amino acid transaminase; (2) ? valine transaminase; (3) branched-chain ketoacid decarboxylase complex; (4) isovaleryl CoA dehydrogenase; (5) β-methylcrotonyl CoA carboxylase; (6) β-ketothiolase; (7) propionyl CoA carboxylase; (8) methylmalonyl CoA racemase; (9) methylmalonyl CoA carbonylmutase; (10) steps in vitamin B12 coenzyme synthesis (Fig. 24-5); (12) cystathionine synthase; (13) cystathionase; (14) sulfite oxidase. (From Hsia, Y. E. Inherited Metabolic Disorders. In Avery, G. (Ed.), *Neonatology*. Philadelphia: Lippincott, 1975. Reproduced by permission of the author.)

attacks of acidosis, yet the first reported patients had only mild mental retardation and no other neurological symptoms. Half of the known patients have died in infancy, but a few have survived with normal intelligence. One patient with isovaleric acidemia was intact neurologically until smitten by a series of critical acidotic episodes late in the stormy course of her illness [11]. Isovaleric acid itself has been shown to be neurotoxic when given to experimental animals, producing lethargy and coma, and free isovaleric acid levels in the serum of patients have correlated well with their symptoms. In fact, Krieger and Tanaka [12] have shown that administration of glycine will enhance the formation of the conjugate isovalerylglycine, with improved tolerance to dietary protein. Thus, glycine can be protective in this disorder.

THE KETOTIC HYPERGLYCINEMIC SYNDROMES

A series of organic acidemias in the pathway of isoleucine and valine catabolism (Fig. 24-2, reactions 6, 10) produce a common syndrome [2, 13] of protein-induced ketoacidosis with intermittent hyperglycinemia, low white blood cells and blood platelets, and occasionally hyperammonemia [14]. These patients may succumb to acute metabolic imbalance in infancy; they may survive with episodic attacks of ketoacidosis or hyperammonemia; or they may have moderate retardation and seizures without serious metabolic symptoms.

Alpha-Methylacetoacetyl CoA Thiolase Deficiency

This disorder [2] (Fig. 24-2, reaction 6) has been associated with two clinical syndromes. In one syndrome, ketoacidosis alone occurred in three families; in another syndrome, a single patient had hyperammonemia, hyperglycinemia, and developmental retardation.

Propionic Acidemia

In this condition [2] (Fig. 24-2, reaction 7) ketoacidosis, hyperglycinemia, and occasionally severe hyperammonemia are produced [14]. The disorder has been associated clinically with serious neurological damage in many of those who survived. In some patients, this neurotoxicity can be attributed to hyperammonemia, but in patients without hyperammonemia, the brain damage may be caused by propionate, by a precursor, or conceivably by the intermittent hyperglycinemia. The first patient reported with propionic acidemia had abnormal odd-chain and branched-chain fatty acids in the liver. The incorporation of these fatty acids into myelin and other neurolipids theoretically could impair the function of the neurolipids. Neurotoxicity of propionate or of a precursor seems possible. Many precursors and by-products of propionate have been identified in these patients, ranging from tiglic acid to β-hydroxypropionate and methylcitrate [13]. Some patients with propionic acidemia and intermittent hyperglycinemia have moderate mental retardation and seizures without overt episodes of ketoacidosis. This raises the possibility that hyperglycinemia is as neurotoxic in these patients as in patients with primary defects of glycine catabolism [15] (see p. 259). Against a neurotoxic role for either propionate or glycine is the finding of two untreated patients with propionic acidemia who were neurologically intact [16]. The sibling of one severely retarded athetotic child has been treated by restricting toxic precursors from birth, permitting her to survive with superior intelligence [17].

Methylmalonic Acidemias

These disorders are of special neurochemical interest because the enzyme methyl-malonyl carbonylmutase, deficient in primary methylmalonicacidemia (Fig. 24-2, reaction 9), is the only mammalian enzyme known to require the $5'$-deoxyadenosyl-cobalamin derivative of vitamin B_{12} as a cofactor [13]. In one inherited abnormality of vitamin B_{12} metabolism (Fig. 24-2, reaction 10), there is also a deficiency in production of methylcobalamin, a cofactor for homocysteine-methionine methyltransferase (Fig. 24-2, reaction 11, and Fig. 24-4, reaction 6). The effects of these inherited metabolic defects can be compared and contrasted with acquired vitamin B_{12} deficiency, pernicious anemia, in which there is degeneration of the long neuronal tracts of the spinal column, peripheral neuropathy, psychological deterioration, and occasionally amblyopia (Chap. 29). In pernicious anemia there is a metabolic block in both the methylmalonate oxidation and homocysteine-methionine methyl-transfer pathways, and patients have elevated urine methylmalonate excretion.

In methylmalonyl racemase and methylmalonylcarbonylmutase deficiencies (Fig. 24-2, reactions 8, 9) and in vitamin-B_{12}-responsive methylmalonic acidemia (Fig. 24-2, reaction 10), the patients have the syndrome of ketotic hyperglycinemia, occasionally with hyperammonemia. Urine methylmalonate excretion is generally elevated many times higher than in pernicious anemia. In one patient with vitamin-B_{12}-responsive methylmalonic acidemia, there was impressive clinical improvement, including improved developmental quotient, following treatment with large doses of vitamin B_{12}, despite the fact that urine methylmalonate output, although lowered, was still far higher than in patients with pernicious anemia. Observations on this patient indicate that methylmalonate was not neurotoxic, although sural nerve biopsies of patients with pernicious anemia have shown that ^{14}C-propionate is incorporated into odd-chain fatty acids and branched-chain fatty acids (see Chap. 29). When this patient was ill or exposed to a large oral load of iso-leucine, he excreted abnormal branched-chain ketones, including butanone (Fig. 24-2) and higher analog. In rats, inhalation of methyl-butyl ketone (hexanone) and possibly also methyl-ethyl ketone (butanone) caused acute muscular weakness and axonal hypertrophy, with beading and degenerative changes in the sural nerve [18]. Therefore, the branched-chain ketones may have specific neurotoxic effects.

Methylmalonic Acidemia with Homocystinuria

Caused by defective formation of both vitamin B_{12} cofactors [13] (Fig. 24-2, reaction 10), this disorder has been associated with inanition, convulsions, and death at age two months in one boy; with very mild mental retardation in one of two affected brothers; and with severe mental retardation progressing to dementia in one girl who died at age seven. Postmortem examination of the girl's brain revealed cerebral atrophy and histological changes like those observed in pernicious anemia [19]. Chemical analysis of the neurolipids in the infant boy revealed abnormal branched-chain and odd-chain fatty acids in the phosphatides of his brain, spinal cord, and sciatic nerve [20]. It is curious that, apart from the girl, none of these patients had the severe disturbances of hemopoiesis with megaloblastic anemia which are the hallmarks of pernicious anemia.

GLUTATHIONE
This tripeptide protects red cells from oxidative hemolysis and, in its reduced state, is present in high concentration in the brain, although it is rapidly oxidized after death. The enzyme γ-glutamyltranspeptidase (Fig. 24-3, reaction 3) has high activity in the choroid plexus of the ventricles, the ciliary bodies of the eye, and has been found histochemically on the surface of certain neurons [21]. The functional role of glutathione in the nervous system is unclear, with no proved relationship to amino acid transport. Figure 24-3 illustrates Meister's postulated γ-glutamyl cycle by which amino acids conjugated with glutamate could be transported across membranes by a membrane-bound enzyme, γ-glutamyltranspeptidase. Release of the amino acid after traversing the membrane would depend on the action of γ-glutamylcyclotransferase. The pyroglutamate formed by the cyclotransferase is recycled back to glutathione in three steps, each step dependent on the hydrolysis of ATP to ADP and phosphate.

   Glutathione is a potent reducing agent, being reversibly oxidized to a disulfide. In the red blood cells, enzymatic deficiencies of carbohydrate metabolism (that decrease the supply of ATP), of glutathione oxidation, and of glutathione synthesis, all lead to hemolytic anemias. Neurological abnormalities are associated with two abnormalities of glutathione synthesis and one of glutathione degradation.

Pyroglutamic Acidemia
This disorder has been recognized in one severely retarded man with athetosis and tetraplegia and in two neurologically intact infant sisters. They all excreted many grams of pyroglutamate (5-oxoproline) in the urine daily and had a severe metabolic acidosis attributable to massive accumulation of pyroglutamate in the blood (2 to 5 mM).

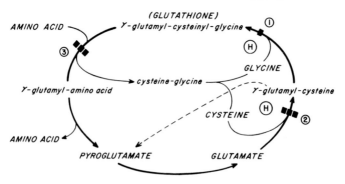

Figure 24-3
Glutathione metabolism. Enzymes: (1) glutathione synthetase; (2) γ-glutamylcysteine synthetase; (3) γ-glutamyltranspeptidase. γ-Glutamylcyclotransferase (from γ-glutamyl—amino acid and also from γ-glutamyl-cysteine. The symbol H represents enzyme defects causing hemolytic anemia) catalyzes the formation of pyroglutamate. (See also Table 24-4.)

The younger patients were found to have low red-cell glutathione and hemolytic anemia, and enzyme studies showed a deficiency not of 5-oxoprolinase, but of glutathione synthetase [22] (Fig. 24-3, reaction 1). In pyroglutamic aciduria, γ-glutamylcysteine seems to be an alternate substrate for the transpeptidase and the cyclotransferase, so that pyroglutamate accumulates beyond the capacity of 5-oxoprolinase to catabolize it.

Gamma-Glutamylcysteine Synthetase Deficiency (Fig. 24-3, reaction 2)
This was associated in one adult woman with mild hemolytic anemia, psychotic behavior, and signs of spinocerebellar degeneration [23]. Her brother showed similar symptoms, except for the psychotic behavior, but he had abnormal electromyographic changes. Both patients showed evidence of mental deterioration and exhibited generalized aminoaciduria. Glutathione was reduced in the blood cells and muscles of both patients, and γ-glutamylcysteine synthetase was less than 7 percent of control activity in the red cells of both patients. The aminoaciduria may be secondary to defective glutathione-mediated amino acid transport in the kidney, but the relation between the metabolic disorder and the neurological abnormalities is not clear.

Glutathionemia (Fig. 24-3, reaction 3)
Caused by deficiency of glutathione transpeptidase, this disease has been reported in one moderately retarded man [24]. Again, there have been no investigations to elucidate why he was retarded. He had normal renal tubular reabsorption of amino acids [25] despite the absence of the enzyme, which is supposed to be a mediator of the transport of amino acids across cell membranes.

AMMONIA, THE UREA CYCLE, AND DIBASIC AMINO ACIDS
Ammonia is normally present in relatively constant concentrations in the brain, and its turnover is closely regulated by the active enzymes that synthesize and transfer amino groups from glutamine, glutamate, and aspartate [14, 26] (Table 24-3 and Fig. 24-4). These are discussed in Chapters 11 and 14.

Clinically, severe hyperammonemia in infants produces vomiting, lethargy, alternating hypertonia and hypotonia, sometimes a coarse tremor reminiscent of the asterixis of hepatic coma, and, finally, coma and a decerebrate state with or without seizures. In older children, moderate hyperammonemia has milder manifestations of similar symptoms, and there is often a characteristic aversion to protein-rich foods. The symptoms are episodic and are aggravated by high protein intake, constipation, or intercurrent infection, each of which causes hyperammonemia in susceptible patients.

Ammonia accumulates in liver disease (Chap. 14); in the ketotic hyperglycinemia syndromes; in disorders of the urea-cycle enzymes; and in some disorders of the dibasic amino acids. How hyperammonemia arises in the ketotic hyperglycinemic syndrome is not clear; presumably some by-product of propionate metabolism inhibits one of the metabolic pathways of ammonia detoxification. Among the dibasic amino acids, ornithine and arginine participate in the urea cycle, and lysine loading results in hyperammonemia in patients with hyperlysinemia (Table 24-3) (see p. 514).

Table 24-3. Hyperammonemia Caused by Disorders of Amino Acid and Intermediary Metabolism

Condition	Location of Defective Enzyme	Comments
Disorders of Arginine-Urea Cycle		
Carbamyl phosphate synthetase deficiency[a]	Fig. 24-4, reaction 1	Several variants reported, ranging from lethal neonatal syndrome to milder syndromes; each variant was unique to one or two cases
Ornithine transcarbamylase deficiency	Fig. 24-4, reaction 2	X-linked dominant, variable severity in females, lethal in neonatal period in males. Variants are known
Citrullinemia	Fig. 24-4, reaction 3	Variable severity, ranging from lethal neonatal syndrome to milder syndromes
Argininosuccinic aciduria	Fig. 24-4, reaction 4	Hyperammonemia is inconstant and moderate. Neurological toxicity may be due to argininosuccinate, with ataxia. Psychoses may occur
Argininemia[a]	Fig. 24-4, reaction 5	Variants reported. Hyperammonemia is mild or absent; patients have all been retarded
Disorders of Dibasic Amino Acids		
Hyperornithinemia[a]	? Ornithine decarboxylase	One boy with mild hyperammonemia, myoclonic spasms, and moderate retardation. No hyperammonemia in other forms of hyperornithinemia
Hyperlysinemia[a]	Fig. 24-1, reaction 17	Moderate hyperammonemia, spasticity, seizures
Hyperlysinuria	? Transport defect	Postprandial hyperammonemia, severe mental retardation
Lysinuric dibasic aminoaciduria	? Transport defect	Variable hyperammonemia; episodic abdominal upset, with pain and distension; inconstant hyperammonemia
Hyperornithinemia and homocitrullinemia	? Mitochondrial transport	Variable retardation, seizures, and ataxia in one family; severe retardation and stupor in another

Also disorders of branched-chain amino acids, see Table 24-2 and Fig. 24-2.

[a]Disorders described in only one or two families.

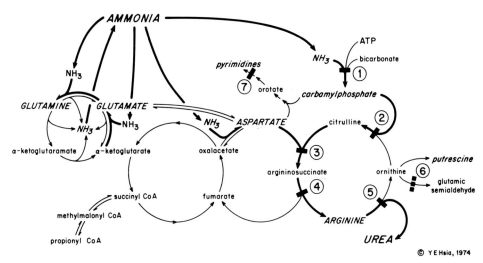

Figure 24-4

Ammonia metabolism. (1) Carbamylphosphate synthetase; (2) ornithine transcarbamylase; (3) argininosuccinate synthetase; (4) argininosuccinase; (5) arginase; (6) ornithine transaminase; (7) orotidine-5'-phosphate pyrophosphorylase and orotidine-5'-phosphate decarboxylase. (See also Table 24-3.) (From Hsia, Y. E. Inherited Metabolic Disorders. In Avery, G. (Ed.), *Neonatology*. Philadelphia: Lippincott, 1975. Reproduced by permission of the author.)

Synergism between ammonia and other metabolites could also account for neurotoxicity in some inborn errors of metabolism. Subtoxic levels of ammonia in mice interact with mercaptans or with short-chain fatty acids to precipitate lethargy and coma [14], so a combination of metabolites could produce serious neurotoxic effects without markedly elevated concentrations of any single agent. Typical histological changes in astrocytes have been reported in these hyperammonemic patients.

THE UREA CYCLE

As illustrated in Figure 24-4, detoxification of ammonia is by immediate reversible conversion to glutamate, glutamine, and aspartate and by eventual fixation as urea via the Krebs-Henseleit urea cycle. Both aspartate and carbamyl phosphate feed nitrogen into the cycle, and the cycle synthesizes arginine, urea, and ornithine (which is a precursor of some neuroactive diamines such as putrescine). Surprisingly, in defects of the urea-cycle enzymes, urea biosynthesis is never totally deficient, perhaps because total blockage of urea biosynthesis is not compatible with fetal survival.

Orotic aciduria can arise in defects of the distal urea cycle enzymes because intramitochondrial accumulation of carbamyl phosphate will result in its leakage into the cytoplasm [14], where it is converted to carbamyl aspartate, the rate-limiting step in pyrimidine biosynthesis, thus accelerating orotate production (Chap. 27). This concept is confirmed by the fact that rats on an arginine-free diet have decreased capacity to synthesize urea, and develop orotic aciduria [27].

Carbamyl Phosphate Synthetase Deficiency [14, 28]
(Fig. 24-4, reaction 1)
This deficiency has been reported in several patients with hyperammonemia. Clinical severity and the degrees of enzyme deficiency varied considerably. The exclusively hepatic mitochondrial enzyme, carbamyl phosphate synthetase I, preferentially accepts ammonia as a substrate and is the primer for urea biosynthesis. In addition, a ubiquitous cytoplasmic enzyme, carbamyl phosphate synthetase II, preferentially accepts glutamine as a substrate and is the primer for pyrimidine biosynthesis. Unless assays have differentiated carefully between these two enzymes, the degree of deficiency of the hepatic mitochondrial enzyme may be miscalculated. Further confusion has arisen because secondary deficiency of the mitochondrial enzyme may result from low protein diets or from toxic inhibition, as has been found in one patient with methylmalonicaciduria and in another who was thought to have primary carbamyl phosphate synthetase deficiency [29] but who probably has a mitochondrial ornithine transport defect (see p. 513).

Ornithine Transcarbamylase Deficiency (Fig. 24-4, reaction 2)
Here is an excellent example of a systemic metabolic disorder that produces major toxicity in the central nervous system. Some patients have prominent secondary orotic aciduria and, in severely affected patients, amino acid imbalances appear, particularly after liver-cell damage and hepatic necrosis. In this disorder, hyperammonemia appears to be primarily responsible for both the neurotoxicity and the hepatotoxicity.

Its X-linked dominant pattern of inheritance has been reported in more than 50 patients. It is lethal in affected males and expresses various symptoms in affected females, who risk transmitting the condition to their offspring. Variants of the disorder have been described that may be related to different mutations of the enzyme [30]. Therapy with low-protein diet and lactulose in affected females has been satisfactory, but no therapy has yet succeeded in salvaging an affected male [31] except in mild variants.

Citrullinemia [32, 33]
Caused by deficiency of argininosuccinate synthetase (Fig. 24-4, reaction 3), citrullinemia can be fatal in early infancy as a result of acute liver failure, or it can cause severe neurological dysfunction in late infancy. At least six infants with citrullinemia presented with lethargy, followed by hypertonia, convulsions, and coma in the first few days of life. They had marked elevation of ammonia and citrulline in body fluids, and several also developed acute hepatic necrosis. A similar number of infants showed vomiting, irritability, and seizures beginning a few months after birth, followed by severe mental retardation and sometimes ataxia; their electroencephalograms often were abnormal. Pathologically, some of these patients were found to have cerebral atrophy and histological liver damage. Neurotoxicity in this condition may be caused mostly by hyperammonemia, but argininosuccinate synthetase, an enzyme normally found in brain tissue, was defective in these patients, and the subsequent massive elevation of citrulline in brain and cerebrospinal fluid may also have been toxic. The

clinical variability of citrullinemia is illustrated by the finding of a retarded adult and a neurologically intact boy, neither of whom had hyperammonemia. Biochemical variants of argininosuccinate synthetase deficiency have been demonstrated [33].

Argininosuccinic Aciduria
This is the result of deficiency of argininosuccinase (Fig. 24-4, reaction 4) and is inconstantly associated with hyperammonemia [32]. It can be lethal in early infancy, but the majority of the more than 50 patients reported have survived longer, although almost invariably with retardation and seizures and usually with intermittent ataxia. These patients have elevated argininosuccinate in cerebrospinal fluid with concentrations that are triple its concentrations in the blood. There is suggestive evidence for arginine deficiency in these patients, so the enzyme defect in the condition may be toxic both because of a toxic precursor and a deficient product. The best therapy, therefore, may be restricted protein intake, supplemented with extra arginine.

Hyperornithinemia
This has been claimed to be due to ornithine decarboxylase deficiency in one retarded boy with myoclonic seizures and postprandial hyperammonemia [33a]. Two other retarded patients with milder hyperornithinemia and similar symptoms had reduced hepatic ornithine ketoacid transaminase (Fig. 24-4, reaction 6). Another remarkable form of hyperornithinemia has been observed in 22 Scandinavian patients with tapeto-retinal degeneration. Patients had gyrate atrophy of the retina and choroid, extinguished electroretinogram, and progressive loss of vision, but no other consistent neurological abnormalities. A defect in the ornithine ketoacid transaminase is suspected, but not proved [33b].

Hyperornithinemia and homocitrullinemia, with hyperammonemia, was reported in two families. It was thought to be due to ornithine decarboxylase deficiency, but is probably a mitochondrial transport defect. Studies of six members of one Canadian family [29], all of whom were afflicted with variable retardation, seizures, and ataxia, showed decreased carbamyl phosphate synthetase I activity in leukocytes. Tissue from one liver biopsy also contained decreased activity of this enzyme and, in addition, exhibited abnormal mitochondria on electron microscopy. A severely retarded British woman with episodes of stupor and limited motor ability also had ornithinemia, homocitrullinemia, and hyperammonemia [34]. Both studies [29, 34] reported protein intolerance and increased homocitrullinuria after protein or lysine ingestion. The British patient exhibited improved tolerance to protein and lower ammonia levels when her diet was supplemented with either ornithine or arginine, even though the former increased her hyperornithinemia.

The intramitochondrial enzyme ornithine transcarbamylase can also transcarbamylate lysine into homocitrulline. Hence, impaired transport of ornithine into the mitochondrion could result in impeded ammonia detoxification; accumulation of ornithine and homocitrulline; and improved protein tolerance on a diet supplemented with ornithine or arginine [29, 34].

LYSINE AND HYDROXYLYSINE

Lysine is an essential amino acid that does not participate in the transamination reactions of the general amino acid pool. It can be carbamylated to form homo-citrulline, which, in turn, can form homoarginine in a manner analogous to arginine formation from citrulline (Fig. 24-4). Lysine is catabolized by at least two pathways (Fig. 24-1): one via ϵ-N-acetyllysine and pipecolate and the other via saccharopine. Both pathways form α-aminoadipate, which then joins the degradative pathway for tryptophan at α-ketoadipate. Hydroxylysine is formed by hydroxylation of lysyl residues in the polypeptides of collagen. No specific neurochemical action is known for lysine, hydroxylysine, or their derivatives except that pipecolate and its deriva-tives are structurally similar to some hypnotic or cerebrotoxic agents such as piperidine.

Various forms of hyperlysinemia are listed in Table 24-4. These conditions affecting lysine catabolism, together with the hyperornithinemia due to a mito-chondrial-membrane transport block, suggest a definite but inconstant relationship between lysine and ammonia metabolism. The neurotoxicity of hyperlysinemia is still obscure.

Lysinuric Familial Protein Intolerance

This is associated with intermittent hyperammonemia [35]. Almost two dozen patients, mostly interrelated, have been reported from Finland and elsewhere. Typically, the patients have an aversion to protein and develop abdominal cramps, diarrhea, and vomiting after protein ingestion, sometimes accompanied by hyper-ammonemia and obtundation of consciousness. They are thin, small in stature; they may have mild mental retardation or, often, enlarged livers. Blood urea, lysine, and arginine are low; urine lysine, arginine, and sometimes cystine are elevated. Present evidence is that the defect is in the transport of dibasic amino acids by renal and intestinal mucosa. Oral loading with lysine has aggravated abdominal symptoms, and intravenous infusion of lysine or arginine, or both, showed a defect in their common renal reabsorptive mechanism. Three of five patients improved clinically on long-term supplementation with arginine, so in this condition, too, there may be a secondary arginine deficiency, which could explain the hyperammonemia.

Pipecolatemia

One patient with this disease has been described as exhibiting irritability, inten-tion tremor, nystagmus, and hypotonia, progressing to paralysis and death in the third year [36]; four patients have been reported with the cerebrohepatorenal syn-drome [37]. The first patient had elevated brain pipecolate but low brain homo-carnosine (γ-aminobutyrylhistidine). His brain showed scattered demyelination, with gliosis and accumulation of lipid; his muscles were atrophied, and both muscles and liver were fibrotic.

The cerebrohepatorenal syndrome, although associated with pipecolatemia [37], has also been related to a block in mitochondrial oxidation between the succinic dehydrogenase flavoprotein and coenzyme Q [38] (Fig. 24-1).

TRYPTOPHAN, XANTHURENATE, AND GLUTARATE

Tryptophan is an essential amino acid [4] that is hydroxylated and decarboxylated to yield the biogenic amine serotonin [39] (Chap. 10). Its catabolism is via the kynurenines to α-ketoadipate, glutaryl CoA, and eventually acetoacetyl CoA (Fig. 24-1).

Kynureninase (Fig. 24-1, reaction 15) is a pyridoxal phosphate-requiring enzyme. In vitamin B_6 deficiency there is increased urinary excretion of hydroxykynurenine and of its by-product, xanthurenic acid. Clinically, vitamin B_6 deficiency causes convulsions and other neurological disturbances in infants (Chap. 29). Table 24-4 lists several rare disorders in this group, including tryptophanemia [40], xanthurenic aciduria, hydroxykynurenic acidemia, α-ketoadipic acidemia [41], and glutaric acidemia [42, 43].

GLUTAMATE, GABA, AND BETA-ALANINE

The metabolism and functional significance of glutamate and its derivatives are discussed in Chapters 11 and 14. Metabolic disturbances within this group have been reported in a few cases (Table 24-4).

Glutamic Acidemia

The disease has been reported as an isolated finding in Menkes' syndrome, in which the relation of its elevation to copper deficiency and profound mental retardation and seizures is obscure. Excessive ingestion of glutamate as a food additive causes acute symptoms, similar to those induced by acetylcholine administration. Experimental long-term administration of glutamate to young animals produces damage to retinal ganglion cells and results in neuronal necrosis in the hypothalamus, arcuate nucleus, and elsewhere in the developing brain, paradoxically without measurable increase in brain glutamate levels [44].

Pyridoxine-Responsive Seizures

This disorder, presumed to be due to a mutation that affects the interaction of glutamic acid decarboxylase with its cofactor pyridoxal phosphate (Fig. 24-1, reaction 10), is discussed in Chapter 29.

Beta-Alaninemia

This has been reported in one infant who had somnolence from birth and developed seizures that were unresponsive to anticonvulsants. β-Alanine and GABA were detected in high concentration in the brain, cerebrospinal fluid, blood, and urine of this patient. This is one basis for the postulate that β-alanine and GABA share the same transaminase (Fig. 24-1, reaction 11). Carnosine and β-alanylhistidine also were increased in the brain.

Carnosinemia

This has been described in a few patients [45]. It is caused by the absence of carnosinase (Fig. 24-1, reaction 13). Carnosine is synthesized in brain and muscle from

Table 24-4. Other Amino Acid Disorders with Neurological Lesions

Condition	Location of Defective Enzyme	Neurological Lesions	Comments
Branched-Chain Amino Acids			
Hypervalinemia[a]	Fig. 24-2, reaction 2	Lethargy, retardation	
Hyperleucine-isoleucinemia[a]	Fig. 24-2, reaction 1	Retardation, seizures, retinal degeneration, deafness	Also had prolinemia type II
Glutathione			
Pyroglutamic acidemia	Fig. 24-3, reaction 1	Retardation, athetosis, tetraplegia in one adult	Severe acidosis
γ-Glutamyl-cysteine synthetase deficiency[a]	Fig. 24-3, reaction 2	Spinocerebellar degeneration, psychotic behavior in one patient; myopathy in the other	
Glutathionemia[a]	Fig. 24-3 reaction 3	Moderately retarded adult	
Dibasic Amino Acids			
Ornithinemia[a]	Fig. 24-4, reaction 6	Lethargy, ataxia, myoclonic seizures, retardation	
Gyrate atrophy with ornithinemia	? Fig. 24-4, reaction 6	Choroido-retinal degeneration	
Persistent hyperlysinemia[a]	Fig. 24-1, reaction 19	Two of three sisters were normal	No hyperammonemia
Hyperlysinemia[a]	?	Severe retardation, hypotonia	No hyperammonemia, lax ligaments
Saccharopinemia and lysinemia[a]	Fig. 24-1, reaction 18	Severe retardation	No hyperammonemia, cerebrospinal fluid lysine increased
Pipecolatemia[a]	?	Irritability, tremor, hypotonia, nystagmus, paralysis	Low brain homocarnosine
Cerebrohepato-renal syndrome	?	Hypotonia, retardation	Biochemical lesion might be in electron transfer system
Hydroxylysinuria	Fig. 24-1, reaction 2	Retardation, myoclonic seizures	
α-Aminoadipic acidemia[a]	?	Borderline intelligence in one of two affected brothers	

[a]Disorders described in only one or two families. *(cont.)*

Table 24-4 (*cont.*)

Condition	Location of Defective Enzyme	Neurological Lesions	Comments
Tryptophan and Metabolites			
Tryptophanemia[a]	Fig. 24-1, reaction 16	Retardation, cerebellar ataxia	Photosensitive rash
Xanthurenic aciduria	Fig. 24-1, reaction 15	Some patients had subnormal intelligence	One variant is responsive to vitamin B_6
Hydroxykynurenic acidemia[a]	Fig. 24-1, reaction 15	Mildly retarded girl	Unresponsive to vitamin B_6
α-Ketoadipic acidemia[a]	?	One of two affected brothers was retarded, self-abusive, and without speech	
Glutaric acidemia[a]	? Fig. 24-1, reaction 14	Dystonia, athetosis, retardation	
Glutamate and Metabolites			
Glutametemia	Nongenetic	? Neonatal susceptibility to neuronal necrosis and retinal degeneration	Also reported in Menkes' syndrome
Pyridoxine-dependent seizures	? Fig. 24-1, reaction 10	Neonatal seizures, retardation	Dramatically responsive to vitamin B_6
β-Alaninemia[a]	Fig. 24-1, reaction 11	Somnolence, seizures	
Carnosinemia	Fig. 24-1, reaction 13	Retardation, seizures, lethargy, spasticity	
Homocarnosinosis[a]	?	Spastic tetraparesis, retardation	
Phenylalanine and Tyrosine			
Phenylketonuria	Fig. 24-1, reation 6	Retardation, microcephaly, hyperactivity, occasional seizures, psychoses	Treatable by restricted dietary phenylalanine
Maternal phenylketonuria		Embryopathy and intrauterine brain damage	
Hyperphenylalaninemia variants	Fig. 24-1, reaction 6	Probably benign	
Dihydrobiopterin reductase deficiency[a]	(See text)	Early retardation, myoclonus, seizures, hypotonia, chorea	

(*cont.*)

Table 24-4 (*cont.*)

Condition	Location of Defective Enzyme	Neurological Lesions	Comments
Albinisms[a]: Cross syndrome		Microcephaly, retardation, athetosis	Microphthalmia. Other variants have no consistent neurological lesions
Tyrosine aminotransferase deficiency	Fig. 24-1, reaction 7	Agitation, tics, moderate retardation	Corneal ulcers, keratosis palmoplantaris
Neonatal tyrosinemia	Fig. 24-1, reaction 8	Risk of some retardation	Immature enzyme responsive to ascorbate
Hereditary tyrosinemia	Fig. 24-1, reaction 8	Irritability, mild retardation	Liver and kidney damage
Alcaptonuria	Fig. 24-1, reaction 9	None	Urine turns black, arthropathy
Sulfur Amino Acids and Metabolites			
Methioninemia[a]	Fig. 24-6, reaction 5	One healthy infant	
Homocystinemia	Fig. 24-2, reaction 12	Borderline retardation and mental instability in some patients	Marfanoid habitus, ectopia lentis, thromboembolic complications Vitamin B_6 responsive variant
Cystathioninemia	Fig. 24-2, reaction 13	Benign	Vitamin B_6 responsive variant
Cystinosis	? Lysosomal transport	Photophobia, retinopathy	Progressive nephropathy, aminoaciduria.
Taurine deficiency[a]	?	Depression, insomnia, dysphagia, visual abnormality, parkinsonism, respiratory failure	Three brothers, their mother, and her brother were similarly affected
β-Mercaptolactate cysteine disulfiduria	?	One retarded man with seizures, one retarded man without seizures, two sisters who were normal	Second man had ectopia lentis
Sulfite oxidase deficiency[a]	Fig. 24-2, reaction 14	Severe retardation, multiple abnormalities, decerebrate rigidity	Ectopia lentis

[a]Disorders described in only one or two families.

(*cont.*)

Table 24-4 (*cont.*)

Condition	Location of Defective Enzyme	Neurological Lesions	Comments
Glycine and Sarcosine			
Nonketotic hyperglycinemia	Fig. 24-6, reaction 7	Lethargy, hypotonia, myoclonus, seizures, severe retardation in survivors, hypertonia and decerebrate rigidity	
Other glycinemic syndromes[a]	?	Extremely variable	
Sarcosinemia	Sarcosine dehydrogenase	Inconsistent association with retardation	May be benign
Proline and Hydroxyproline			
Type I hyperprolinemia	Proline oxidase	Possible retardation or seizures	Renal disease may be coincidental
Type II hyperprolinemia	Pyrolline carboxylate dehydrogenase	Retardation, seizures	
Hydroxyprolinemia[a]	? Hydroxyproline oxidase	Retardation in two of three known cases	
Iminopeptiduria[a]	Prolidase	Borderline intelligence	Abnormal collagen
Folate Metabolites (and Histidine)			
Dihydrofolate reductase deficiency[a]	Fig. 24-6, reaction 1	None	Megaloblastic anemia
Cyclohydrolase deficiency[a]	Fig. 24-6, reaction 3	Retardation	Probably nonexistent entity
Formiminotransferase deficiency	Fig. 24-6, reaction 4	Retardation, cerebral atrophy; another two were very clumsy; one was normal; three had speech problems	Histidine catabolism is blocked
Histidinemia	Fig. 24-1, reaction 12	Perhaps speech problems and mild retardation	May be benign
Homocysteine-methionine methyltransferase deficiency	Fig. 24-6, reaction 6	Retardation, cerebral atrophy	
$N^{5,10}$ methylene-tetrahydrofolate reductase deficiency[a]	Fig. 24-6, reaction 8	Two sisters were retarded; the third patient was hypotonic with seizures; one of the sisters had psychotic episodes	Patients had homocystinemia. The psychotic episodes appeared to respond to folate therapy

[a]Disorders described in only one or two families.

β-alanine and histidine; the same enzyme will synthesize homocarnosine from
γ-aminobutyrate and histidine. The compound anserine, β-alanylmethylhistidine,
is found in the muscle of many mammals, but is not synthesized in human tissues.
The role of these dipeptides in brain and muscle is not known, although anserine
and carnosine are both potent activators of muscle ATPase [1]. All but one of the
affected patients were retarded and had seizure disorders; some also were lethargic
and spastic. Autopsy findings in one affected patient showed demyelination of
white matter, loss of Purkinje fibers, cortical atrophy, unidentified "spheroids" in
the cerebral gray matter, severe axonal degeneration, and atrophic fibrosis of muscle
fibers [46].

Homocarnosinosis

This has been reported briefly in one adult woman with spastic tetraparesis and
mental retardation [47]. Homocarnosine in her cerebrospinal fluid was increased
10 to 20 times. This patient had normal carnosine levels, so the homocarnosine
elevation could not have been the result of carnosinase deficiency. This provides
circumstantial evidence for a separate homocarnosinase enzyme. It is intriguing
that spasticity occurred in this condition and in carnosinemia. Homocarnosine, a
derivative of GABA, has no known neurochemical function. It is normally found in
brain and cerebrospinal fluid, and is reported to raise the electroconvulsive threshold
of dog brain when injected into the cerebrospinal fluid. Homocarnosine was reduced
in the brain of the original patient with pipecolatemia (see p. 514).

PHENYLALANINE AND TYROSINE

These aromatic amino acids are of major significance neurochemically because they
are the precursors of the catecholamines [4, 39] (Chap. 10) and of thyroid hormone.
Phenylalanine has only one major catabolic reaction, hydroxylation to tyrosine.
Tyrosine can be decarboxylated to tyramine, hydroxylated to dihydroxyphenylala-
nine (DOPA), iodinated to the iodotyrosines and thyroid hormone, or catabolized via
p-hydroxyphenylpyruvic acid to homogentisic acid and maleylacetoacetate (Fig. 24-1).

The hydroxylases for these aromatic amino acids (and tryptophan) have common
properties [48]: cross-specificity for aromatic amino acid derivatives; complex substrate and product-inhibition behavior; requirement for a reduced pteridine cofactor
and for molecular oxygen. Phenylalanine hydroxylase is a microsomal enzyme present
only in liver, pancreas, and kidney, but not in nervous tissue. Tyrosine hydroxylase
is present in noradrenergic and adrenergic neurons of the central and sympathetic
nervous systems. It is the rate-limiting enzyme for catecholamine biosynthesis.
Distribution of tryptophan hydroxylase in the brain closely corresponds to that of
serotonin and presumably is localized in serotoninergic neurons. It is the rate-limiting
enzyme for serotonin biosynthesis.

The natural pteridine cofactor for phenylalanine hydroxylase is tetrahydrobiopterin
(Fig. 24-5), which is formed from 7,8-dihydrobiopterin catalyzed by a dihydrofolate
reductase (see p. 530; also Fig. 24-6, reaction 1). During the phenylalanine hydroxylase
reaction, tetrahydrobiopterin is oxidized to the quinonoid form of dihydrobiopterin.

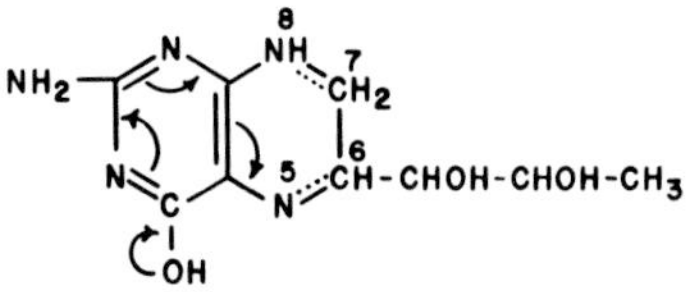

Tetrahydrobiopterin

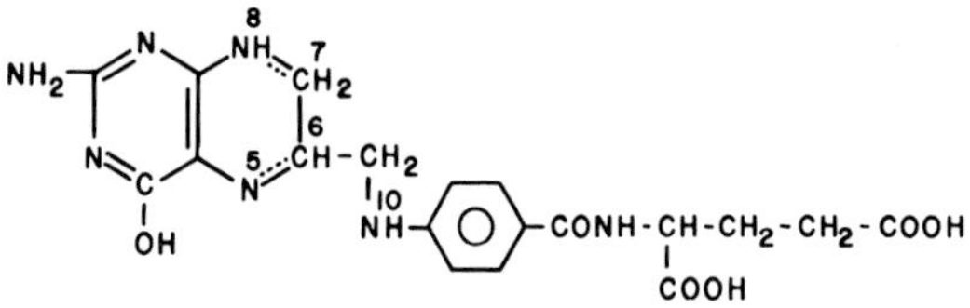

Tetrahydrofolate (monoglutamate form)

Figure 24-5
Molecular structure of tetrahydrobiopterin and tetrahydrofolate. Biopterin, like folate, has the double bonds within the heterocyclic nucleus indicated by the dotted lines. Dihydrobiopterin and dihydrofolate have reduction of the bond between C^7 and N^8. The quinonoid form of dihydrobiopterin has the rearrangements indicated by the arrows plus a double bond probably between C^7 and N^8. The substitutions on tetrahydrofolate in Figure 24-6 are at positions N^5 and N^{10}.

Tetrahydrobiopterin is regenerated by a specific dihydropteridine reductase. The aromatic amino acid hydroxylases will accept many artificial alternate cofactors, but their physical properties are considerably modified unless the natural cofactor is used. Although it has not been proved, tetrahydrobiopterin is assumed to be the natural cofactor for the brain tyrosine and tryptophan hydroxylases, too.

Phenylketonuria (PKU)

This is the commonest of the inborn errors of metabolism that threaten neurological integrity. In the United States alone, it affects about 1 in 20,000 of all infants born [5]. The genetic defect in phenylketonuria is of hepatic phenylalanine hydroxylase [49, 50] (Fig. 24-1, reaction 6). Patients with PKU have no detectable immunoreactive phenylalanine hydroxylase, but careful assays in one patient's liver showed only 0.27 percent of normal enzyme activity, with altered substrate affinities. This confirms that the genetic defect is a structural mutation which seriously compromises the function of this enzyme, including loss of its immunoreactive properties. In this condition, there is an insidious onset of mental retardation, leading to microcephaly and a hyperactive, aggressive, hypertonic state. Patients often have seizures, eczema, pigment dilution, and are permeated with a pungent, mousy odor. Histochemically, there is generalized failure of myelination (Chap. 28), but the composition of the myelin, despite conflicting reports, is probably normal.

PKU is the first of the inborn errors of metabolism shown to respond successfully to treatment by diet restriction [3, 51] and for which large scale presymptomatic screening programs have been effective [52]. Soon after birth, infants with PKU have abnormal elevations of serum phenylalanine that can be detected in a drop of dried blood by use of one of several efficient techniques. If PKU is confirmed, these patients should be given a restricted diet that limits their serum phenylalanine to between two and five times normal. On this regimen, PKU children are protected from severe brain damage, but only if the diet is started in early infancy [53]. The hyper-active, aggressive behavior of older, retarded, PKU patients may be slightly ameliorated through diet therapy. Inappropriate therapy by overzealous restriction or imbalance of the diet can lead to malnutrition, which can itself cause mental retardation. For instance, in children with PKU, tyrosine becomes an essential amino acid and must be supplied in sufficient quantity to fulfill all nutritional needs.

Exceptionally, a few adults have been identified who have had all the metabolic abnormalities of PKU but have escaped with intact intelligence. Some of these indi-viduals have had severely retarded siblings and parents who were chemically hetero-zygous for PKU [54]. Unrecognized adult PKU has three important implications for the neurochemical understanding of this condition. First, unidentified innate bio-chemical differences or exceptional environmental experiences have somehow pro-tected these individuals during their vulnerable period of brain growth. Second, some of these adults have had major psychotic illnesses, for which a causal relation-ship to their biochemical abnormality could be responsible. Third, adult PKU women, whether brain-damaged or unscathed, offer a hostile intrauterine environment for their offspring.

Maternal PKU provides compelling substantiation of the concept that the neuro-toxicity of PKU is secondary to changes in the *milieu intérieur.* PKU women have a high risk of having spontaneous abortions or children born with congenital mal-formations. In addition, almost all children born to these mothers have had growth failure, microcephaly, and severe mental retardation. Diet restriction during pregnancy has reportedly protected the human fetus from intrauterine brain damage on at least two occasions.

The neurological consequences of PKU must be produced indirectly, via changes in composition of the extracellular fluid, because phenylalanine hydroxylase is not found in neural tissue; dietary corrective measures can avert the brain damage in this condition; and maternal PKU is toxic to the fetus.

Possible neurochemical mechanisms for the brain damage, however, are still con-fusing. Gaull et al. have recently presented a detailed review of many possible neuro-toxic mechanisms responsible for the brain damage in PKU, among which are the following [54]:

1. Extreme hyperphenylalanemia opens up several alternate minor pathways for phenylalanine metabolism, and abnormal concentrations of phenylpyruvic acid, phenyllactic acid, phenylacetic acid, o-hydroxyphenylacetic acid, and phenylethyl-amine, among others, all are increased in the body fluids of untreated PKU patients. Some of these compounds may have physiological roles in the nervous system, and some of them do have neuropharmacological properties.

2. Relative dietary deficiency of tyrosine in PKU is aggravated by the inhibitory effect of high phenylalanine levels on tyrosine transport [6]. This may deprive the

brain of an adequate supply of the neuroactive derivatives of tyrosine, including tyramine and its hydroxylated product octopamine as well as the catecholamines. This deficiency state might prevent brain growth and impede normal neurotransmission.

3. Increased concentrations of phenylalanine also inhibit transport of the other neutral amino acids [6], producing an imbalance of amino acids in the brain that could upset regulatory control of neuroactive derivatives and interfere with protein synthesis.

4. Phenylalanine or its by-products may inhibit key metabolic pathways:

    a. Tyrosine hydroxylase will accept phenylalanine as an alternate substrate [4, 39] but does not accelerate brain tyrosine formation in PKU because elevated phenylalanine inhibits that enzyme [48]. This inhibition could block catecholamine synthesis. In support of this thesis, catecholamine concentrations were much lower in the caudate nucleus, brain stem, and cortex of four PKU patients than in controls who were mentally retarded from other causes [55].

    b. Tryptophan metabolism is altered in untreated patients with PKU and in experimental hyperphenylalaninemia. Tryptophan hydroxylase is inhibited by excess of phenylalanine. (Although phenylpyruvate also inhibits aromatic amino acid decarboxylase, the concentration of phenylpyruvate required exceeds that found in patients with phenylketonuria.) Although much of the disturbance of tryptophan metabolism is probably systemic, brain serotonin levels can be reduced experimentally [56] and have been found reduced in the brains of four PKU patients with reduced catecholamines [55]. The deaminated breakdown products of serotonin reportedly were low in the cerebrospinal fluid of untreated phenylketonuric patients but rose with dietary restriction of phenylalanine.

    c. Energy metabolism is suppressed in PKU, perhaps by inhibition of pyruvate kinase. The synthesis of ATP for general brain metabolism may be compromised by this inhibition, and its degradation may be accelerated, as phenylalanine stimulates Na^+, K^+-dependent ATPase.

5. Protein synthesis in the brain may be impaired seriously by the marked imbalance of amino acid concentrations and by disaggregation of polyribosomes. This disaggregation was shown experimentally to occur in the brains of immature rats with hyperphenylalaninemia with suppressed incorporation of labeled amino acids into protein [57]. The defect in myelination found in the brains of PKU patients may reflect generalized inhibition of protein synthesis or specific inhibition of proteolipid synthesis.

Experimental models for PKU have been generally unsatisfactory. Simply feeding phenylalanine produces combined elevations of tyrosine as well as phenylalanine; specific inhibitors of phenylalanine hydroxylase such as p-chlorophenylalanine may have other pharmacological actions [5]; in vitro tests of the neurotoxicity of phenylalanine on single enzymes or single pathways cannot duplicate the complex metabolic disturbances known to occur in PKU; no animal is known that has a mutation affecting this enzyme alone; measurement of the biochemical effects of experimental PKU could never be fully correlated with the intellectual impairment and behavioral changes that occur in the human species [58]. Perhaps the most convincing animal model is that for maternal PKU. If a pregnant rat is fed large amounts of phenyl-alanine with or without a phenylalanine hydroxylase inhibitor, the effects of this maternal hyperphenylalaninemia can be analyzed in her fetuses or after birth in her offspring [57].

Atypical Phenylketonuria

Atypical phenylketonuria or benign hyperphenylalaninemia variants have been discovered as a result of screening programs that select young infants with high blood phenylalanine. Neonatal hyperphenylalaninemia can be secondary to any cause of tyrosinemia, including prematurity; other hyperphenylalaninemia variants have been transient or more persistent but with lower blood phenylalanine than in classical phenylketonuria. Immunoreactive hepatic phenylalanine hydroxylase was reduced but not absent in one patient with atypical hyperphenylalaninemia [48]; the enzyme protein may be altered structurally in hyperphenylalaninemia. Whether these hyperphenylalaninemia variants are indeed benign has not yet been resolved satisfactorily. Whatever the neurotoxic mechanism might be in PKU, if it were directly proportional to the degree of hyperphenylalaninemia, the potential toxicity of the milder variants could be predicted according to the phenylalanine level found in blood during the vulnerable period of brain growth.

Maternal hyperphenylalaninemia has been innocuous in some pregnancies of women with moderate hyperphenylalaninemia.

Dihydrobiopterin Reductase Deficiency

This recently recognized variant of PKU is unresponsive to dietary phenylalanine control and so has important neurochemical implications [59, 60]. These patients were identified as having chemical PKU in the newborn period, but they had gener-alized and myoclonic seizures, hypotonia, choreiform movements, and early onset of progressive mental retardation, despite adequate dietary regulation of their blood phenylalanine levels. Liver biopsies in two patients showed normal activity of phenyl-alanine hydroxylase, and enzyme assays on one patient confirmed that dihydropteri-dine reductase activity was less than 1 percent of normal in liver, brain, and other tissues. This absence prevents the regeneration of tetrahydrobiopterin (see p. 520).

Deficiency of dihydrobiopterin reductase also prevents the hydroxylation of tyrosine and of tryptophan and the biosynthesis of catecholamines and serotonin. In one patient, dopamine, serotonin, and their metabolites were low in various tissues, including cerebrospinal fluid and brain, but, oddly, norepinephrine metabolites were normal. These findings confirm that neurotransmitter biosynthesis is severely impaired in this disorder, with serious neurological complications. It has been suggested that replacement therapy with L-DOPA and 5-hydroxytryptophan should be attempted.

TYROSINE

Tyrosine Aminotransferase Deficiency

Tyrosine aminotransferase deficiency (Fig. 24-1, reaction 7) causes tyrosinemia, with painful dendritic corneal ulcers, keratosis palmoplantaris, agitation, tics, and moderate mental retardation [61, 62]. The biochemical defect is in the cytosol enzyme, which accepts the cosubstrate α-ketoglutarate to form p-hydroxyphenyl-pyruvate and glutamate, requiring pyridoxal phosphate as a cofactor. Because mitochondrial tyrosine aminotransferase (15 percent of the total enzyme in the liver) remains intact, these patients can still produce p-hydroxyphenylpyruvate and p-hydroxyphenyllactate, which appear in the urine in this syndrome. In the brain, tyrosine aminotransferase is mainly mitochondrial, and its role in tyrosine metabolism appears to be overshadowed by that of tyrosine hydroxylase, so degradation of tyrosine is superseded in the brain by catecholamine biosynthesis. It is not clear how neurotoxicity is produced in this condition, but dietary restriction to control tyrosine levels in blood has improved at least the ocular and cutaneous manifestations of this condition, although behavioral improvement could have resulted from symptomatic relief of the distressing eye and skin lesions [62, 63].

Neonatal Tyrosinemia

Not at all uncommon in premature infants, this disorder occurs in many full-term infants as well. There is increased urinary p-hydroxyphenylpyruvic, p-hydroxyphenyllactic, and p-hydroxyphenylacetic acids, probably due to deficiency of p-hydroxyphenylpyruvate oxidase (Fig. 24-1, reaction 8). This enzyme has low activity in the fetus and is subject to substrate inhibition, so developmental immaturity of the enzyme is worsened by hypertyrosinemia. Administration of ascorbic acid, a reducing substance which stimulates the enzyme, will improve or curtail the syndrome. Whether transient neonatal tyrosinemia is benign or neurotoxic has not been fully established, but one careful study has shown that almost 10 percent of affected infants may end with mental retardation [64].

Albinism

This is due to defective melanosome synthesis of melanin, in some variants caused by tyrosinase deficiency (Fig. 24-1, reaction A). Curiously, melanin formation in the central nervous system, e.g., in the substantia nigra, is never affected. Mammals [64a] with generalized albinism have been found to have failure of decussation of the optic pathways to the lateral geniculate bodies. In patients with generalized albinism there is physiological evidence of faulty lateralization of visual stimuli.

THE SULFUR AMINO ACIDS: METHIONINE, CYSTEINE, TAURINE

Methionine has great biological significance, because S-adenosylmethionine (Fig. 24-6) is the major methyl donor in many biochemical reactions, including the synthesis of such biogenic amines as choline and epinephrine; the inactivation of such neurotransmitters as serotonin and the catecholamines; and the methylation of creatine, of

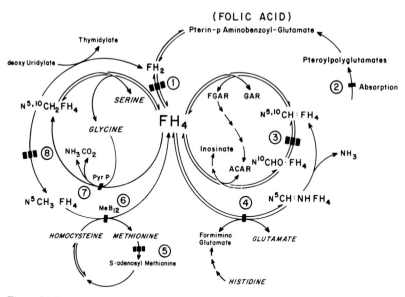

Figure 24-6

Folate and glycine metabolism. The relationships of derivatives of FH_4 to glycine metabolism are indicated in the circle on the left and to purine biosynthesis in the circle on the right. Acronyms: ACAR, 5-aminoimidazole-4-carboxyamide ribinucleotide; FGAR, N-formylglycinamide ribonucleotide; GAR, glycinamide ribonucleotide.

Enzymes: (1) Dihydrofolate reductase; (2) folate absorption; (3) cyclohydrolase; (4) form-iminotransferase; (5) S-adenosylmethionine synthetase; (6) homocysteine-methionine methyltransferase; (7) glycine cleavage enzyme; (8) $N^{5,10}$ methylene-FH_4 reductase.

proteins, and of nucleotides. It may also be a precursor of polyamines [65]. S-adenosylhomocysteine, the product of these reactions, can be recycled to methionine via homocysteine by methyltetrahydrofolate methyltransferase (Fig. 24-2, reaction 11, and Fig. 24-5, reaction 6) or by hepatic betaine-homocysteine methyltransferase (see also Chap. 29). By this means, methionine, an essential amino acid, is regenerated with one-carbon fragments from serine and glycine (Fig. 24-5) for its methylation reactions. Methionine deficiency interferes with growth; in excess, it exacerbates the psychotic behavior of chronic schizophrenics. Hypermethioninemia occurs clinically in acute liver failure, and results also from blocked methionine or homocysteine metabolism.

Homocysteine reacts with serine to form cystathionine (Fig. 24-2, reaction 12), a compound with no other known biological function. Cystathionine is found in high concentration in the brain, predominantly in white matter. It is found also in liver, kidney, and muscle. It is hydrolyzed by cystathionase to cysteine and homoserine (Fig. 24-2, reaction 13). Homoserine, like threonine, is a precursor of α-ketobutyrate and is catabolized via the propionate pathway (Fig. 24-2).

Cysteine is an important component of proteins, glutathione, coenzyme A, and many other biologically active compounds. The further catabolism of cysteine is either by removal and eventual oxidation of the sulfur radical by inorganic sulfite oxidase (Fig. 24-2, reaction 14) to sulfate, which accounts for 80 percent of urinary sulfur, or by oxidation to cysteinsulfinate, decarboxylation to hypotaurine, and oxidation to taurine.

Taurine is found in very high concentration in the brain, especially in the cerebellar cortex. It may act as an inhibitory modulator or transmitter of synaptic transmission [66]. It is also present in high concentration in the retina, in muscle, and in the liver, where it forms bile salts that participate in fat absorption and are excreted in the stool.

Methioninemia

Caused by hepatic deficiency of methionine adenosyltransferase activity, methioninemia (Fig. 24-6, reaction 5) has been found in one clinically healthy normal infant [67]. S-adenosylmethionine has a vital role in transmethylation reactions, so it is surprising that this condition is compatible with survival, let alone with no neurological abnormality. The biochemical abnormalities were a persistent methioninemia, thirty times normal, which was nevertheless nontoxic; no change in the other sulfur amino acids; and a high serum folate. The only enzyme found to be low was methionine adenosyltransferase, which was about 10 percent of normal in two liver biopsies. Perhaps the partial enzyme deficiency was overcome by the huge excess of methionine.

Homocystinemia

Caused by cystathionase synthase deficiency, homocystinemia (Fig. 24-2, reaction 12) results in tissue accumulation of homocysteine and methionine, and depletion of cystathionine, with urinary excretion of homocysteine and even traces of S-adenosylhomocysteine. Patients with homocystinuria share with Marfan's syndrome a characteristic lanky body habitus, long extremities and digits, and abnormal lens capsular ligaments that predispose to dislocation of the ocular lens. Homocystinuric patients often suffer from mental retardation, are liable to psychiatric disturbances, and have a propensity for thromboembolic occlusion of arteries and veins. In homocystinemia it has been speculated that an abnormality of connective tissue might be the cause for lens dislocation and joint laxity and perhaps for endothelial vulnerability to fibrin deposition, too. The brain damage might then be due to multiple repeated small cerebral thromboses. This hypothetical sequence has been apparently verified by finding in patients an abnormal consumption of platelets that could be duplicated experimentally by continuous infusion of homocysteine into baboons [68].

Vitamin responsiveness to large doses of pyridoxal phosphate has been demonstrated biochemically in some patients, with or without enzymatic evidence of enhanced cystathionine synthase activity, indicating there are variants of this disorder. Some patients had abnormal liver mitochondria [69].

Neurochemically, toxicity from cystathionine deficiency or from other disturbances of homocysteine metabolism has neither been proved nor disproved. Dietary treatment with low methionine intake has not produced any clinical improvement; dietary treatment with serine and cysteine to stimulate reverse synthesis of homocysteine via cystathionase has been recommended [65].

Homocystinuria due to abnormalities of vitamin B_{12} metabolism and folate metabolism are discussed on pages 507 and 531.

Cystathioninemia

Deficiency of cystathionase (Fig. 24-2, reaction 13) causes this disorder, which is associated with abnormally elevated concentrations of cystathionine in the brain. Patients are often asymptomatic, but some have had severe mental retardation, and a few have had seizures. The association of this condition with neurological abnormalities has been considered to be coincidental. There are vitamin B_6-responsive and unresponsive variants.

Cystinosis

Cystinosis is not deficiency of an enzyme of intermediary metabolism, but almost certainly an abnormality of lysosomal transport of cysteine or its disulfide cystine. The severe variant is nephrotoxic, producing generalized aminoaciduria and leading to renal failure by the second decade of life. Apart from patchy depigmentation of the retina, and photophobia possibly related to cystine crystal deposition in the cornea, this condition is benign neurologically. Unlike many of the other lysosomal storage diseases (Chaps. 26 and 27), in this disease there is no brain involvement.

Taurine Deficiency

This has been reported in three brothers; their mother, maternal uncle, and maternal grandfather had similar clinical symptoms [66]. Presentation was not until midadulthood, with mental depression, insomnia, anorexia or dysphagia, dyspnea, and loss of visual depth perception; they then developed signs of parkinsonism, mental confusion, and, finally, all the brothers died of respiratory failure. Only one brother had detailed biochemical investigations, and these showed less than half of normal taurine levels in plasma and very low cerebrospinal fluid taurine. At autopsy, his brain taurine was reduced in all regions tested, particularly the cerebellum, compared with controls or patients with Huntington's and Parkinson's diseases. There were no abnormalities of other metabolites of sulfur amino acids or of γ-aminobutyrate. Histologically there was distinct depigmentation of the substantia nigra, with extensive loss of neurons and some gliosis in his brain and that of one brother. A younger sister has become depressed and is developing low plasma taurine concentrations, but has shown no measurable metabolic abnormality in response to an oral taurine loading test.

Taurine has inhibitory neurotransmitter properties in the central nervous system and retina, so the findings in this patient are tantalizingly suggestive of an inborn error of cysteine conversion to taurine with very specific neurochemical consequences. In kittens, dietary taurine deficiency produces a specific retinal degeneration [70], which demonstrates that taurine does have an essential functional role in the nervous system. Because oral taurine in man has had no untoward effects and can enter the central nervous system, taurine therapy should be tried if other patients are ever found with this disorder.

Related Disorders

Other rare disorders in this group (Table 24-4) are β-mercaptolactate, cysteine disulfiduria, and sulfite oxidase deficiency [71].

GLYCINE AND SARCOSINE

Glycine is abundantly present in most tissue fluids and cells. Its general metabolism has been extensively studied, and it participates in innumerable metabolic reactions. In the nervous system, it can be formed from glyoxylate and serine, but it is formed only slowly from glucose (Chap. 14). Glycine degradation in the brain is primarily by glycine cleavage enzyme (Fig. 24-6). It is a putative inhibitory neurotransmitter, present in very high concentration in spinal gray matter associated with inhibitory interneurons (Chap. 11).

Hyperglycinemia

Hyperglycinemia (Fig. 24-6, reaction 7) is caused by deficiency of glycine-cleavage enzyme; it must be differentiated from a host of genetic and nongenetic causes of hyperglycinemia, including the disorders of branched-chain amino acid catabolism (see p. 507 ff.) [15, 72]. The patients with defective glycine cleavage have early onset of lethargy, hypotonia, myoclonic jerks, and generalized seizures that are unresponsive to anticonvulsant therapy, but they have no attacks of ketoacidosis or hyperammonemia. Mental retardation appears in those who survive beyond infancy. In the later stages of the condition, hypertonia and decerebrate rigidity may supervene. Tracer studies have shown delayed oxidation of glycine and a defect in liver- and brain-cleavage enzyme. The activity of this enzyme was almost equally decreased in one patient with the ketotic hyperglycinemia syndrome due to methylmalonicacidemia. Suppressed enzyme activity in the ketotic hyperglycinemia syndromes would explain the presence of hyperglycinemia in metabolic disorders of branched-chain amino acid and propionate catabolism, but an inhibitory mechanism is not known.

Glycine concentrations in the urine and blood are markedly elevated in all the conditions causing hyperglycinemia. Cerebrospinal fluid and glycine concentrations in brain, however, were found to be very elevated only in patients with nonketotic hyperglycinemia. In two such patients, examined postmortem, glycine-cleavage enzyme activity was undetectable in brain, although the activity in liver was decreased to just a third of normal. The distribution of excess glycine in these patients was in every region of the brain that was analyzed [72].

Hyperglycinuria

This is a prominent feature of any cause of generalized aminoaciduria, ranging from immaturity to drug toxicity and includes the aminoacidurias secondary to galactosemia (Chap. 25), Wilson's disease (Chap. 32), and cystinosis. The inherited transport defect of glycine alone is rare; iminoglycinuria, where it is associated with prolinuria, is benign. The shared renal transport mechanisms for proline, hydroxyproline, and glycine explain the presence of glycinuria in the hyperprolinemias.

Glycinuria and glycinemia occasionally have been reported in individual patients with neurological disorders, such as a family with spinal cord lesions [73] and one patient with an oculocerebral syndrome [74].

THE IMINOACIDS PROLINE AND HYDROXYPROLINE

Proline is a nonessential amino acid synthesized from glutamate via pyrolline-carboxylate. It is degraded by independent catabolic enzymes, mitochondrial proline oxidase, and cytosol pyrolline-carboxylate dehydrogenase, through the same intermediate back to glutamate. This degradation can occur in brain tissue. Hydroxyproline is found mostly in collagen, where it is formed by hydroxylation of proline residues during collagen synthesis. Degradation is via oxidation to hydroxypyrolline-carboxylate, then via hydroxyglutamate and α-hydroxy-γ-ketoglutarate to glyoxylate and pyruvate. Most iminoacid is excreted as oligopeptides; at least one transport system for the free iminoacids is shared with glycine.

Type I hyperprolinemia, a proline-oxidase deficiency, has had coincidental association with renal disease and possibly mental retardation or seizures. Type II hyperprolinemia, a pyrolline-carboxylate dehydrogenase deficiency, has been more frequently associated with mental retardation and seizures and causes greater hyperprolinemia. In both conditions, there is overflow renal prolinuria with competitive saturation of the common reabsorptive mechanism it shares with hydroxyproline and glycine.

Hydroxyprolinemia and iminopeptiduria [75] have also been reported (Table 24-4). Iminopeptiduria occurs also in other conditions as a nonspecific expression of disturbed collagen metabolism.

FOLATE

Folic acid (Fig. 24-6), pteroylglutamic acid, is present in food mainly as polyglutamates. Absorption of these polyglutamates is preceded by hydrolysis to the monoglutamate form by the enzyme conjugase. Extracellular folic acid is then reduced intracellularly to tetrahydrofolate (FH_4), which serves as a cofactor that transfers one-carbon units in many key aspects of intermediary metabolism, including glycine catabolism (in the left circle in Fig. 24-6) and purine biosynthesis (in the right circle in Fig. 24-6). Folic acid deficiency, like vitamin B_{12} deficiency, produces megaloblastic anemia, but usually without serious neurological consequences (Chap. 29). Cerebrospinal fluid folates are normally three times the concentration of blood folates. Every cell probably has the capacity to synthesize and recycle the folate intermediates necessary for its own metabolism, although nerve cells are supposed to lack dihydrofolate reductase activity and so would be unable to utilize oxidized folic acid.

Congenital Malabsorption of Folate

Congenital malabsorption of folate (Fig. 24-6, reaction 2), unlike folic acid deficiency in older patients, presents in infancy with ataxia or athetosis, seizures, and mental retardation, as well as megaloblastic anemia [76, 77]. One older patient was found to have punctate calcification of the basal ganglia. The condition has been reported only four times, in three families. One patient had no neurological abnormality apart from poor school performance but had been treated with large doses of folate parenterally from the age of three months. The severe neurological damage in this disorder may be due to susceptibility of the developing brain to folate deficiency, but in one patient the seizures were aggravated and in another they were reduced by folate treatment.

Formiminotransferase Deficiency

Formiminotransferase deficiency (Fig. 24-6, reaction 4) blocks the catabolism of histidine and causes a large increase in urinary formiminoglutamate excretion [78, 79]. Four Japanese patients had elevated serum folate, formiminoglutamicaciduria exacerbated by histidine loading, mental retardation, abnormal electroencephalograms, and cerebral cortical atrophy. One of two Swiss sisters had much worse formiminoglutamicaciduria, normal serum folate, retarded speech, and borderline I.Q.; her sister had similar chemical findings, but had perfect speech and intelligence. A Canadian brother and sister had persistent formiminoglutamicaciduria, normal serum folate, delayed speech, hypotonia, and clumsiness. Aminoimidazole carboximide, the riboside of which is converted to inosinic acid by an enzyme requiring N^{10} formyl-FH_4 as a cofactor (Fig. 24-6), was present in excess in the urine of three of the Japanese patients.

Analysis of the structure of myelin has revealed abnormally low hydroxy fatty acid content of cerebroside in one patient who presented in infancy with a megaloblastic anemia that improved with folate or pyridoxine therapy [79a].

$N^{5,10}$ Methylene-FH_4 Reductase Deficiency

This blocks the synthesis of methyl-FH_4 [80, 81] (Fig. 24-6, reaction 8). It has aroused considerable interest because one patient had psychiatric disturbances that fluctuated with periods on and off folic acid therapy. This condition has been found in two affected sisters and one unrelated boy. The patients had elevated homocysteine in the urine and plasma, marginally low plasma methionine and serum folate, but no measurements were made of amino acids in cerebrospinal fluid. The older of the sisters was retarded mentally, had an abnormal electroencephalogram, and showed features of catatonic schizophrenia at the age of 14. Upon therapy with pyridoxine and then folate, her mental status improved, but peripheral neuropathy became manifest. Off therapy, she relapsed, and improved on reinstitution of pyridoxine and folate, or of folate alone, which suppressed her homocysteine production at the same time that her mental derangement was improving. Platelet monoamine oxidase was decreased in the older woman during a relapse, but was normal in her sister. This sister, also retarded, had no psychotic symptoms, and the boy presented with proximal muscle weakness, abnormal electroencephalogram and seizures, but no reported mental instability.

The enzyme activity, demonstrated in liver and cultured skin fibroblasts, was less than 20 percent of normal. FAD coenzyme addition doubled enzyme activity in cells from the sisters and from controls, but stimulated it fivefold in cells from the affected male. Because methyl-FH_4 is a cofactor for the vitamin B_{12}-dependent regeneration of methionine from homocysteine (Fig. 24-6, reaction 6), its decrease in this condition would interrupt the resupply of S-adenosylmethionine for other methyltransfer reactions. Metabolically, degradation of administered methionine to sulfate in these patients was normal, and their cultured cells did not grow in a medium when homocysteine was substituted for methionine. This confirmed that remethylation of homocysteine to methionine was impaired in this condition.

Related Disorders

Other disorders in this group (Fig. 24-6) include single reported cases of dihydrofolate reductase (reaction 1), cyclohydrolase (reaction 3), and homocysteine methionine methyltransferase (reaction 6) deficiencies.

PORPHYRINS

Heme is a tetrapyrrole ring with a central iron atom that can convert molecular oxygen to its first activated state. It is a prosthetic group for many enzyme reactions involving electron transfer, oxygen transport, activation or breakdown of peroxide, and protein synthesis [82]. The porphyrin tetrapyrroles (Fig. 24-7) are synthesized by the union of glycine with succinyl-CoA to form δ-aminolevulinate (δ-ALA), the asymmetrical condensation of two molecules of δ-ALA to form the pyrrole porphobilinogen, polymerization of four porphobilinogen rings to form urobilinogens, partial breakdown of the pyrrole side chains to convert the urobilinogens to coproporphyrins and protoporphyrins, and finally incorporation of iron. Only the first enzyme, δ-ALA synthetase, and the last few enzymes are mitochondrial (Fig. 24-7). δ-ALA synthetase, the rate-limiting enzyme, has a half-life of just one or two hours; it is repressed by heme and is induced by steroid hormones, drugs, alcohol, and many other agents. Heme is an essential component of all mitochondrial electron-transfer systems in every cell, but the major portion of heme production is in the erythrocyte series for hemoglobin synthesis.

Degradation of heme is only partial, iron is released for reutilization, and the porphyrin ring is broken to form biliverdin and bilirubin, which then is excreted as a conjugate with glucuronic acid, mainly in the bile. Hyperbilirubinemia is neurotoxic to the brain of the premature or newborn infant.

Porphyria

Porphyria is caused by exogenous toxins such as lead (Chap. 33) and alcohol, metabolic diseases such as tyrosinemia, and several inherited abnormalities of porphyrin synthesis (Table 24-5). Unlike almost all other clinically significant inherited enzyme abnormalities, these are anabolic (cf. deficient ganglioside biosynthesis in Chap. 26), and most are autosomal dominant disorders. A few of these have neurological and psychological manifestations, often precipitated by medications or other exogenous factors.

Acute Intermittent Porphyria (Swedish Type)

This is associated with up to 50-fold overproduction of porphyrins, 4-fold increase of δ-ALA synthetase activity, half of normal uroporphobilinogen I synthetase activity (Fig. 24-7, reaction 1) and decreased steroid 5 α-reductase activity [82–84]. Affected individuals may remain asymptomatic, but postpubertally are prone to acute attacks that are often precipitated by (a) any drug that is capable of inducing δ-ALA synthetase, including the barbiturates and alcohol; (b) febrile illnesses; (c) exogenous or endogenous steroid hormones, particularly estrogens; and (d) starvation.

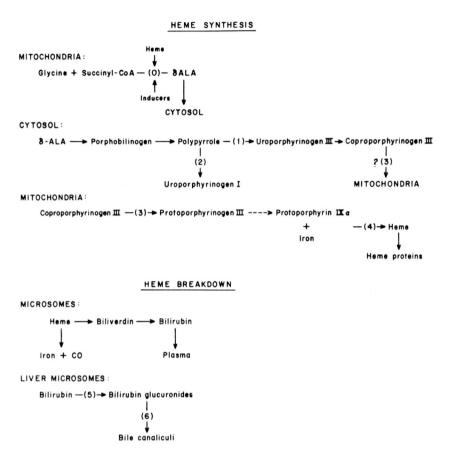

Figure 24-7
Heme (and porphyrin) metabolism. δ-Aminolevulinic acid synthetase, the rate-limiting enzyme
for heme biosynthesis, is designated by (o). It is a repressible (↓) and inducible (↑) enzyme.
    Enzymes: (1) Uroporphobilinigen I synthetase; (2) uroporphyrinogen III cosynthetase;
(3) coproporphyrinogen III oxidase (transport ?); (4) ferrochelatase; (5) bilirubin glucuronyl-
transferase; (6) transport into bile canaliculi. (See also Table 24-5.)

A unifying concept of the protean symptoms in this disorder is that they are all
neuropathic. (*a*) Abdominal pain and gastrointestinal upset may be caused by autonomic nervous system disturbances, which would also account for the sweating, labile
blood pressure, urinary retention, and skin blanching that occur in the acute attacks
or crises. (*b*) When peripheral neuropathy appears, it can have any distribution; it may
be sensory, with pain and parasthesia, or motor, with weakness and paralysis. If respiratory muscles are weakened, life becomes endangered. (*c*) Hypothalamic damage and
inappropriate secretion of antidiuretic hormone probably account for the episodes of
water retention and hyponatremia associated with attacks; disturbances of consciousness may also arise from damage to the brain stem or hypothalamus. (*d*) Mental illness
occurs in up to one-third of all affected patients. Some of the acute symptoms of

Table 24-5. The Porphyrias

| Type | Postulated Location of Metabolic Block (Fig. 24-7) | Clinical Features | | | | | Mode of Inheritance |
		Skin Photosensitivity	Hemolytic Anemia	Liver Toxicity	Neurological Involvement	
Acute intermittent porphyria (Swedish type)	1	None	None	None	Severe acute attacks	Autosomal dominant
Congenital erythropoietic porphyria (Günther's disease)	2	Bullous scarring	Usual	None	None	Autosomal recessive
Porphyria cutanea tarda	?	Bullous scarring	None	Frequent	None	Unclear
Hereditary coproporphyria	3	Mild	None	None	Milder acute attacks	Autosomal dominant
Variegate porphyria (South African type)	4	Mild scarring	None	None	Infrequent attacks	Autosomal dominant
Erythropoietic protoporphyria	4	Solar urticaria	Minimal	Present	None	Autosomal dominant

irritability and irrationality may be from discomfort and from the disturbed fluid and electrolyte balance, but other forms of organic brain syndrome, ranging from depression through anxiety states to frank schizophrenia, have been observed.

Unlike the other porphyrias, there are no skin rashes, hemolytic anemia, or liver abnormalities in acute intermittent porphyria.

Metabolically, there is increased urinary excretion of δ-ALA and porphobilinogen, especially during attacks, with increase of these compounds in the cerebrospinal fluid too. The neurochemical basis for the toxicity is a complete mystery, but may be due to toxic effects of precursor accumulation or to deficiency of heme production. δ-ALA has been shown to inhibit human brain ATPase and porphobilinogen to inhibit rat presynaptic transmission. Although there is no absolute correlation between the degree of metabolic abnormality and the neurological symptoms in a given patient, urine excretion of porphyrin precursors tends to fluctuate in correspondence with attacks.

Biochemically, δ-ALA synthetase is actively repressed by heme and may be derepressed if heme production is diminished in this condition. Activity of urobilinogen I synthetase activity is only half of normal, and might become rate limiting for heme biosynthesis. Based on observations in one patient and his two parents, steroid 5 α-reductase has been proposed to be an independent genetic mutation, and symptomatic disease would appear only when both abnormalities coincided in a patient [83]. Reduced steroid 5 α-reductase activity may explain the vulnerability of affected individuals to estrogen steroids, since degradation to 5 β-H derivatives that would induce δ-ALA synthetase could be enhanced if 5 α-H derivatives were less readily formed.

Treatment is primarily by avoidance of precipitating factors such as known toxic drugs or starvation; generous carbohydrate intake is protective. Infusions of heme or its ferrihydroxy form, hematin, have suppressed red-cell plasma and urine porphyrins in this condition [85]. This approach should suppress excess production of precursors and supply the putatively missing product, heme. Even if there is systemic biochemical improvement, however, there is no evidence yet that infused heme can penetrate into nerve cells or improve neuronal function.

Hereditary Coproporphyria

Hereditary coproporphyria appears to be the result of a defect either of membrane transport of coproporphyrin III into mitochondria or of coproporphyrin oxidation (Fig. 24-7, reaction 3), because affected patients excrete an excess of urinary δ-ALA, porphobilinogen, and coproporphyrins but no maturer products of heme biosynthesis [82]. Affected individuals (Table 24-5) can have acute neuropathic attacks as in acute intermittent porphyria, but, unlike that condition, there can be photosensitive skin rashes.

Metabolically there is about 50-fold increased coproporphyrin excretion, mainly in the stool, with inconstant urinary excretion of δ-ALA, porphobilinogen, and coproporphyrins that tends to be increased during acute attacks.

Variegate Porphyria (South African Type)

This might be due to a defect of iron incorporation into heme by ferrochelatase or of one of the last enzymes of protoporphyrin synthesis [82]. Affected individuals

(Table 24-5) have highly variable clinical manifestations including acute neuropathic attacks and a scarring photosensitivity of the skin. Intriguing circumstantial evidence has been presented that the British Royal House of Hanover, and notably George III, may have suffered from either variegate or acute intermittent porphyria. The evidence includes the diagnosis of variegate porphyria in two of his direct descendents. British colonial America could have had a very different history but for this genetic disease.

Metabolically, there is intermittent urinary excretion of δ-ALA, porphobilinogen, and coproporphyrins, with increased fecal coproporphyrins and protoporphyrins. During attacks, urine and fecal uroporphyrins are much increased. These findings are compatible with blocked biosynthesis of heme from protoporphyrin and derepressed δ-ALA synthetase, but no enzyme deficiency has been identified as yet for this disorder.

BILIRUBIN

Basal ganglia degeneration can be caused by many diseases, including toxic accumulation of bilirubin and of copper (Chap. 32). Neonatal hyperbilirubinemia, usually resulting from prematurity or one of the neonatal hemolytic anemias, is characterized by high concentrations of unconjugated bilirubin, which is less efficiently excluded by an immature blood-brain barrier, resulting in bilirubin staining of the brain, particularly of the basal ganglia. Bilirubin appears to be toxic to the central nervous system, producing staining and cystic degeneration of the basal ganglia (kernicterus) with disorders of movement, mental retardation, and deafness.

Crigler-Najjar Syndrome

The Crigler-Najjar syndrome is a genetic deficiency of bilirubin conjugation that generally causes kernicterus in infancy (Fig. 24-7, reaction 5). Two patients, however, after being neurologically intact for over 15 years, developed disorders of movement, myotonia, intellectual deterioration, and seizures. One was improved by phototherapy [86] ; the other eventually lapsed into coma and died. The autopsy findings showed neuronal loss and gliosis of the thalamus, and considerable neuronal loss in the basal ganglia and cerebellum [87]. Although there was no yellow staining of the brain, these patients can be considered to have had late onset of bilirubin neurotoxicity equivalent to neonatal kernicterus.

TREATMENT

Treatment must be a prime consideration [51]. Specific therapeutic approaches have been discussed in relation to the various disorders, but curative replacement or restoration of a defective protein is seldom possible (see Chaps. 26 and 27). Better understanding of the neurochemical disturbances in these inborn errors of metabolism may eventually produce rational therapies that will benefit both the common and the rare disorders.

Acknowledgment. This work has been supported by the PHS Center Grant GM-20124-03, Training Grant HD-00198-08, Children's Clinical Research Grant RR-125, Amino Acid Grant AM-09527, and a National Foundation–March of Dimes Medical Service Grant C-143.

REFERENCES

*1. Scriver, C. R., and Rosenberg, L. E. *Amino Acid Metabolism and Its Disorders.* Philadelphia: Saunders, 1973.

*2. Tanaka, K. Disorders of Organic Acid Metabolism. In Gaull, G. E. (Ed.), *The Biology of Brain Dysfunction.* Vol. III. New York: Plenum, 1975, p. 145.

3. Hsia, Y. E. Inherited Metabolic Disorders. In Avery, G. (Ed.), *Neonatology,* Philadelphia: Lippincott, 1975, p. 437.

*4. Ciba Foundation Symposium 22 (new series). *Aromatic Amino Acids in the Brain.* Amsterdam: Associated Scientific Publishers, 1974.

*5. Gaull, G. E., Tallan, H. H., Lajtha, A., and Rassin, D. K. Pathogenesis of Brain Dysfunction in Inborn Errors of Amino Acid Metabolism. In Gaull, G. E. (Ed.), *The Biology of Brain Dysfunction.* Vol. III. New York: Plenum, 1975, p. 47.

6. Oldendorf, W. H. Saturation of blood brain barrier transport of amino acids in phenylketonuria. *Arch. Neurol.* 28:45, 1973.

7. Tyson, T. J., and Cornblath, M. Succinyl-CoA: 3-Ketoacid CoA-transferase deficiency. *J. Clin. Invest.* 51:493, 1973.

7a. Silberberg, D. H. Maple syrup urine disease metabolites studied in cerebellum cultures. *J. Neurochem.* 16:1141, 1969.

8. Bowden, J. A., Brestel, E. P., Cope, W. T., McArthur, C. L., Westfall, D. N., and Fried, M. α-Ketoisocaproic acid inhibition of pyruvate and α-ketoglutarate oxidative decarboxylation in rat liver slices. *Biochem. Med.* 4:69, 1970.

9. Bowden, J. A., McArthur, C. L., and Fried, M. Metabolic diseases and mental retardation. I. The chick embryo as a model system in branched-chain ketoaciduria; the effect of α-ketoisocaproic acid. *Int. J. Biochem.* 5:391, 1974.

10. Kimura, T., and Tahara, M. Effect of force-feeding diets lacking leucine, valine, isoleucine, threonine or methionine on amino acid catabolism in rats. *J. Nutr.* 101:1647, 1971.

11. Guibaud, P., Divry, P., Dubois, Y., Collombel, C., and Larbre, F. Une observation d'acidemie isovalerique. *Arch. Fr. Pediatr.* 30:633, 1973.

12. Krieger, I., and Tanaka, K. Therapeutic effects of glycine in isovaleric acidemia. *Pediatr. Res.* 10:25, 1976.

13. Ando, T., and Nyhan, W. L. Propionic Acidemia and the Ketotic Hyperglycinemia Syndrome. In Nyhan, W. L. (Ed.), *Heritable Disorders of Amino Acid Metabolism.* New York: Wiley, 1974, p. 37.

*14. Hsia, Y. E. Inherited hyperammonemic syndromes. *Gastroenterology* 67:347, 1974.

15. Perry, T. L., Urquhart, N., McLean, J., Evans, M. E., Hansen, S., Davidson, A. G. F., Applegarth, D. A., MacLeod, P. J., and Lock, J. E. Nonketotic hyperglycinemia: Glycine accumulation due to absence of glycine cleavage in brain. *N. Engl. J. Med.* 292:1269, 1975.

16. Paulsen, E. P., and Hsia, Y. E. Asymptomatic propionicacidemia: Variability of clinical expression in a Mennonite kindred. *Am. J. Hum. Genet.* 26:66a, 1974 (abstr.).

17. Brandt, I. K., Hsia, Y. E., Clement, D. H., and Provence, S. A. Propionicacidemia (ketotic hyperglycinemia): Dietary treatment resulting in normal growth and development. *Pediatrics* 53:391, 1974.

18. Duckett, S., Williams, N., and Francis, S. Peripheral neuropathy associated with inhalation of methyl-*n*-butyl ketone. *Experientia* 30:1283, 1974.

*Asterisks denote key references.

19.   Dillon, M. J., England, J. M., Gompertz, D., Goodey, P. A., Grant, D. B., Hussein, H. A.-A., Linnell, J. C., Matthews, D. M., Mudd, S. H., Newns, G. H., Seakins, J. W. T., Uhlendorf, B. W., and Wise, I. J. Mental retardation, megaloblastic anemia, methylmalonic aciduria and abnormal homocysteine metabolism due to an error in vitamin B_{12} metabolism. *Clin. Sci.* 47:43, 1974.

20.   Kishimoto, Y., Williams, M., Moser, H. W., Hignite, C., and Bieman, K. Branched-chain and odd-numbered fatty acids and aldehydes in the nervous system of a patient with deranged vitamin B_{12} metabolism. *J. Lipid Res.* 14:69, 1973.

21.   Meister, A. An Enzymatic Basis for a Blood-Brain Barrier? The γ-Glutamyl Cycle — Background and Considerations Relating to Amino Acid Transport in the Brain. In Plum, F. (Ed.), *Brain Dysfunction in Metabolic Disorders.* New York: Raven, 1974, p. 273.

22.   Wellner, V. P., Sekura, R., Meister, A., and Larsson, A. Glutathione synthetase deficiency, an inborn error of metabolism involving the γ-glutamyl cycle in patients with 5-oxoprolinuria (pyroglutamic aciduria). *Proc. Natl. Acad. Sci. U.S.A.* 71:2505, 1974.

23.   Richards, F., Cooper, M. R., Pearce, L. A., Cowan, R. J., and Spurr, C. L. Familial spinocerebellar degeneration, hemolytic anemia, and glutathione deficiency. *Arch. Intern. Med.* 134:534, 1974.

24.   Goodman, S. I., Mace, J. W., and Pollak, S. Serum gamma-glutamyl transpeptidase deficiency. *Lancet* 1:234, 1971.

25.   Schulman, J. D., Patrick, A. D., Goodman, S. I., Tietze, F., and Butler, J. Gamma-glutamyl transpeptidase (GGTPase): Investigations in normals and patients with inborn errors of sulfur metabolism. *Pediatr. Res.* 9:355, 1975 (abstr.).

26.   Krebs, H. A., Hems, R., and Lund, P. Some regulatory mechanisms in the synthesis of urea in the mammalian liver. *Adv. Enzyme Regul.* 11:361, 1973.

27.   Milner, J. A., and Visek, W. J. Orotate, citrate, and urea excretion in rats fed various levels of arginine. *Proc. Soc. Exp. Biol. Med.* 147:757, 1974.

28.   Batshaw, M., Brusilow, S., and Walser, M. Keto acid therapy of carbamyl phosphate synthetase deficiency. *N. Engl. J. Med.* 292:1085, 1975.

29.   Gatfield, P. D., Taller, E., Wolfe, D. M., and Haust, M. D. Hyperornithinemia, hyperammonemia and homocitrullinemia associated with decreased carbamyl phosphate synthetase I activity. *Pediatr. Res.* 9:488, 1975.

30.   Cathelineau, L., Saudubray, J.-M., and Polonovski, C. Heterogenous mutations of the structural gene of human ornithine carbamyl transferase as observed in five personal cases. *Enzyme* 18:103, 1974.

31.   Gelehrter, T. D., and Rosenberg, L. E. Ornithine transcarbamylase deficiency: Unsuccessful therapy of neonatal hyperammonemia with N-carbamyl-l-glutamate and l-arginine. *N. Engl. J. Med.* 292:351, 1975.

32.   Backman, C. Urea Cycle. In Nyhan, W. L. (Ed.), *Heritable Disorders of Amino Acid Metabolism.* New York: Wiley, 1974, p. 361.

33.   Kennaway, N. G., Harwood, P. J., Rambert, D. A., Koler, R. D., and Buist, N. R. M. Citrullinemia: Evidence for genetic heterogeneity. *Pediatr. Res.* 9:554, 1975.

33a. Shih, V. E., and Mandell, R. Metabolic defect in hyperornithinemia. *Lancet* 2:1522, 1974.

33b. Takki, K. Gyrate Atrophy of the Choroid and Retina Associated with Hyperornithinaemia. Ph.D. thesis. University of Helsinki, 1975.

34.   Fell, V., Pollitt, R. J., Sampson, G. A., and Wright, T. Ornithinemia, hyperammonemia, and homocitrullinuria: A disease associated with mental retardation and possibly caused by defective mitochondrial transport. *Am. J. Dis. Child.* 127:752, 1974.

35. Simell, O., and Perheentupa, J. Renal handling of diamino acids in lysinuric protein intolerance. *J. Clin. Invest.* 54:9, 1974.
36. Gatfield, P. D., Taller, E., Hinton, G. G., Wallace, A. C., Abdelnour, G. M., and Haust, M. D. Hyperpipecolatemia: A new metabolic disorder associated with neuropathy and hepatomegaly. *Can. Med. Assoc. J.* 99:1215, 1968.
37. Danks, D. M., Tippett, P., Adams, C., and Campbell, P. Cerebro-hepato-renal syndrome of Zellweger: A report of eight cases with comments upon the incidence, the liver lesion, and a fault in pipecolic acid metabolism. *J. Pediatr.* 86:382, 1975.
38. Goldfischer, S., Moore, C. L., Johnson, A. B., Spiro, A. J., Valsamis, M. P., Wisniewski, H. K., Ritch, R. H., Norton, W. T., Rapin, I., and Gartner, L. M. Peroxisomal and mitochondrial defects in the cerebro-hepato-renal syndrome. *Science* 182:62, 1973.
*39. Cooper, J. R., Bloom, F. E., and Roth, R. H. *The Biochemical Basis of Neuropharmacology.* New York: Oxford, 1974.
40. Wong, P. W. K., Forman, P., and Justice, P. A new inborn error of tryptophan metabolism. *Pediatr. Res.* 9:358, 1975 (abstr.).
41. Wilson, R. W., Wilson, C. M., Gates, S. C., and Higgins, J. V. α-Ketoadipic aciduria: A description of a new metabolic error in lysine-tryptophan degradation. *Pediatr. Res.* 9:522, 1975.
42. Goodman, S. I., Markey, S. P., Moe, P. G., Miles, B. S., and Teng, C. C. Glutaric aciduria: A "new" disorder of amino acid metabolism. *Biochem. Med.* 12:12, 1975.
43. Goodman, S. I., Stokke, O., and Moe, P. G. Glutaric aciduria due to glutaryl-CoA dehydrogenase deficiency: Inhibition of brain glutamate decarboxylase by glutaric, β-hydroxyglutaric, and glutaconic acids. *Pediatr. Res.* 9:382, 1975 (abstr.).
44. Olney, J. W. Toxic Effects of Glutamate and Related Amino Acids on the Developing Central Nervous System. In Nyhan, W. L. (Ed.), *Heritable Disorders of Amino Acid Metabolism.* New York: Wiley, 1974, p. 501.
45. Murphey, W. H., Lindmark, D. G., Patchen, L. I., Housler, M. E., Harrod, E. K., and Mosovich, L. Serum carnosinase deficiency concomitant with mental retardation. *Pediatr. Res.* 7:601, 1973.
46. Terplan, K. L., and Cares, H. L. Histopathology of the nervous system in carnosinase enzyme deficiency with mental retardation. *Neurology (Minneap.)* 22:644, 1972.
47. Gjessing, L. R., and Sjaastad, O. Homocarnosinosis: A new metabolic disorder associated with spasticity and mental retardation. *Lancet* 2:1028, 1974.
48. Kaufman, S. Properties of the Pterin-Dependent Aromatic Amino Acid Hydroxylases. In *Aromatic Amino Acids in the Brain.* Ciba Symposium 22. Amsterdam: Assoc. Scientific Publishers, 1974, p. 86.
49. Friedman, P. A., Kaufman, S., and Kang, E. S. Nature of the molecular defect in phenylketonuria and hyperphenylalaninemia. *Nature* 240:157, 1972.
50. Friedman, P. A., Fisher, D. B., Kang, E. S., and Kaufman, S. Detection of hepatic phenylalanine 4-hydroxylase in classical phenylketonuria. *Proc. Natl. Acad. Sci. U.S.A.* 70:552, 1973.
*51. Hsia, Y. E. Treatment in Genetic Diseases. In Milunsky, A. (Ed.), *The Prevention of Genetic Disease and Mental Retardation.* Philadelphia: Saunders, 1975, p. 277.
52. Bush, J. W., Chen, M. M., and Patrick, D. L. Health Status Index in Cost Effectiveness: Analysis of PKU Program. In Berg, R. L. (Ed.), *Health Status Indexes.* Chicago: Hospital Research and Educational Trust, 1973, p. 174.

53. Smith, I., and Wolff, O. H. Natural history of phenylketonuria and influence of early treatment. *Lancet* 2:540, 1974.

54. Perry, T. L., Hansen, S., Tischler, B., Richards, F. M., and Sokol, M. Unrecognized adult phenylketonuria: Implications for obstetrics and psychiatry. *N. Engl. J. Med.* 289:395, 1973.

55. McKean, C. M. The effects of high phenylalanine concentrations on serotonin and catecholamine metabolism in human brain. *Brain Res.* 47:469, 1972.

56. Loo, Y. H. Serotonin deficiency in experimental hyperphenylalaninemia. *J. Neurochem.* 23:139, 1974.

57. Copenhaver, J. H., Vacanti, J. P., and Carver, M. J. Experimental maternal hyperphenylalaninemia: Disaggregation of fetal brain ribosomes. *J. Neurochem.* 21:273, 1973.

58. Andersen, A. E., Rowe, V., and Guroff, G. The enduring behavioral changes in rats with experimental phenylketonuria. *Proc. Natl. Acad. Sci. U.S.A.* 71:21, 1974.

59. Butler, I. J., Krumholz, A., Holtzman, N. A., Koslow, S. H., Kaufman, S., and Milstein, S. Dihydrobiopterine reductase deficiency variant of phenylketonuria: A disorder of neurotransmitters. *Arch. Neurol.* 32:350, 1975 (abstr.).

60. Smith, I., Clayton, B. E., and Wolff, O. H. New variant of phenylketonuria with progressive neurological illness unresponsive to phenylalanine restriction. *Lancet* 1:1108, 1975.

61. Goldsmith, L. A., Kang, E., Bienfang, D. C., Jimbow, K., Gerald, P., and Baden, H. P. Tyrosinemia with plantar and palmar keratosis and keratitis. *J. Pediatr.* 83:798, 1973.

62. Zaleski, W. A., and Hill, A. Tyrosinosis: A new variant. *Can. Med. Assoc. J.* 108:477, 1973.

63. Morrow, G., and Barness, L. A. Malignant tyrosinemia with apparent dietary cure. *Pediatr. Res.* 8:436, 1974 (abstr.).

64. Menkes, J. H., Welcher, D. W., Levi, H. S., Dallas, J., and Gretsky, N. E. Relationship of elevated blood tyrosine to the ultimate intellectual performance of premature infants. *Pediatrics* 49:218, 1972.

64a. Guillery, R. W. Visual pathways in albinos. *Sci. Am.* 230(5):44, 1974.

65. Finkelstein, J. D. Methionine metabolism in mammals: The biochemical basis for homocystinuria. *Metabolism* 23:387, 1974.

66. Perry, T. L., Bratly, P. J. A., Hansen, S., Kennedy, J., Urquhart, N., and Dolman, C. L. Hereditary mental depression and parkinsonism with taurine deficiency. *Arch. Neurol.* 32:108, 1975.

67. Gaull, G. E., and Tallan, H. H. Methionine adenosyltransferase deficiency: New enzymatic defect associated with hypermethioninemia. *Science* 186:59, 1974.

68. Harker, L. A., Slichter, S. J., Scott, C. R., and Ross, R. Homocystinuria: Vascular injury and arterial thrombosis. *N. Engl. J. Med.* 291:537, 1974.

69. Gaull, G., Sturman, J. A., and Schaffuer, F. Homocystinuria due to cystathionine synthase deficiency: Enzymatic and ultrastructural studies. *J. Pediatr.* 84:381, 1974.

70. Hayes, K. C., Carey, R. E., and Schmidt, S. Y. Retinal degeneration associated with taurine deficiency in the cat. *Science* 188:949, 1975.

71. Crawhall, J. C. The Unknown Disorders of Sulfur Amino Acid Metabolism: β-Mercaptolactate-Cysteine Disulfiduria, Sulfite Oxidase Deficiency, The Methionine Malabsorption Syndrome, and Hypermethioninemia. In Nyhan, W. L. (Ed.), *Heritable Disorders of Amino Acid Metabolism.* New York: Wiley, 1974, p. 467.

72. Tada, K., Corbeel, L. M., Eeckels, R., and Eggermont, E. A block in glycine cleavage reaction as a common mechanism in ketotic and nonketotic hyperglycinemia. *Pediatr. Res.* 8:721, 1974.
73. Bank, W. J., and Morrow, G. A familial spinal cord disorder with hyperglycinemia. *Arch. Neurol.* 27:136, 1972.
74. Balci, S., Say, B., and Firat, T. Corneal opacity, microphthalmia, mental retardation, microcephaly and generalized muscular spasticity associated with hyperglycinemia. *Clin. Genet.* 5:36, 1974.
75. Powell, G. F., Rasco, M. A., and Maniscalco, R. M. A prolidase deficiency in man with iminopeptiduria. *Metabolism* 23:505, 1974.
76. Lanzkowsky, P. Congenital malabsorption of folate. *Am. J. Med.* 48:580, 1970.
77. Santiago-Borrero, P. J., Santini, R., Perez-Santiago, E., and Maldonado, N. Congenital isolated defect of folic acid absorption. *J. Pediatr.* 82:450, 1973.
78. Niederwieser, A., Giliberti, P., Tatasouvic, A., Pluznik, S., Steinmann, B., and Baerlocher, K. Folic acid non-dependent formiminoglutamic aciduria in two siblings. *Clin. Chim. Acta* 54:293, 1974.
79. Perry, T. L., Applegarth, D. A., Evans, M. E., Hansen, S., and Jellum, E. Metabolic studies of a family with massive formiminoglutamic aciduria. *Pediatr. Res.* 9:117, 1975.
79a. Chida, N., and Arakawa, T. Decrease in long-chain hydroxy fatty acids of myelin cerebroside in formiminotransferase deficiency. *Tokoku J. Med.* 108:279, 1972.
80. Mudd, S. H., Uhlendorf, B., Freeman, J. M., Finkelstein, J. D., and Shih, V. E. Homocystinuria associated with decreased methylenetetrahydrofolate reductase activity. *Biochem. Biophys. Res. Comm.* 46:905, 1972.
81. Freeman, J. M., Finkelstein, J. D., and Mudd, S. H. Folate-responsive homocystinuria and "schizophrenia," a defect in methylation due to deficient 5,10-methylenetetrahydrofolate reductase activity. *N. Engl. J. Med.* 292:491, 1975.
82. Tschudy, D. P. Porphyrin Metabolism and the Porphyrias. In Bondy, P. K., and Rosenberg, L. E. (Eds.), *Duncan's Diseases of Metabolism.* 7th ed. Philadelphia: Saunders, 1974, p. 775.
83. Sassa, S., Granick, S., Bickers, D. R., Bradlow, H. L., and Kappas, A. A microassay for uroporphyrinogen I synthase, one of the abnormal enzyme activities in acute intermittent porphyria, and its application to the study of the genetics of this disease. *Proc. Natl. Acad. Sci. U.S.A.* 71:732, 1974.
84. Watson, C. J., Dhar, G. J., Bossenmaier, I., Cardinal, R., and Petryka, Z. J. Effect of hematin in acute porphyric relapse. *Ann. Intern. Med.* 79:80, 1973.
85. Watson, C. J., Bossenmaier, I., Cardinal, R., and Petryka, Z. J. Repression by hematin of porphyrin biosynthesis in erythrocyte precursors in congenital erythropoietic porphyria. *Proc. Natl. Acad. Sci. U.S.A.* 71:278, 1974.
86. Blashke, T., Berk, P. D., Scharschmidt, B. F., Guyther, J. R., Vergalla, J. M., and Waggoner, J. G. Crigler-Najjar syndrome: An unusual course with development of neurologic damage at age eighteen. *Pediatr. Res.* 8:573, 1974.
87. Gardner, W. A., and Konigsmark, B. W. Familial nonhemolytic jaundice: Bilirubinosis and encephalopathy. *Pediatrics* 43:365, 1969.

Chapter 25

Defects in Carbohydrate Metabolism

David B. McDougal, Jr.

It is a surprising fact that, although glucose has long been established as the primary source of energy for brain (Chap. 19), the mechanisms by which disorders of carbohydrate metabolism produce their effects on neural function are still essentially unknown. Because these disorders may afford opportunities that cannot be duplicated in the laboratory, their study in human beings can be expected to make unique contributions to the understanding of carbohydrate metabolism and its role in the function of the nervous system.

The commonest disorders of carbohydrate metabolism affect the use or storage of glucose. Diabetes mellitus is discussed in detail below. Other disorders, such as galactosemia and several disorders of hepatic glycogen metabolism, appear to act on the nervous system by depriving it of glucose.

A handful of rare diseases are characterized by the accumulation of pyruvic or lactic acids, or both, in blood or urine. In some of these the nervous system is permanently affected; in others the nervous system appears to be normal once the metabolic acidosis is controlled. The reader should consult Chapter 14 for a general review of carbohydrate metabolism.

DIABETES MELLITUS

Certainly the commonest disease of carbohydrate metabolism is diabetes mellitus, which is characterized by elevated blood glucose levels and inappropriately low plasma insulin. This disorder affects both the peripheral and the central nervous system. The disease is heritable, but the mode of inheritance has not been completely worked out and may not be the same for different individuals.

Encephalopathy

An untreated or inadequately treated diabetic may go, more or less suddenly, into a state of metabolic acidosis and then into coma. The clinical chemistry includes marked elevation of the blood and urine glucose levels, ketone bodies in the urine, and depressed plasma P_{CO_2} and pH. Occasionally coma is seen with very high blood glucose levels in the absence of acidosis. This has been attributed to hyperosmolarity.

Some patients who have had diabetes for many years suffer central nervous system effects which are considered by some to constitute a true diabetic encephalopathy [1]. These patients often lose tendon reflexes and many have mental disturbances, sometimes severe. Neurological symptoms and signs include attacks of vertigo, transitory hemiparesis, intellectual impairment, dyspraxia, incoordination, and orthostatic hypotension.

A relatively constant pattern of pathological changes is observed in diabetic encephalopathy: diffuse degenerative abnormalities, often with severe pseudocalcinosis of
the cerebellum, demyelination of cranial nerves, leptomeningeal fibrosis, and angi-
opathy.

Experimentally, the effect of diabetic acidosis on cerebral metabolites in mice
has been studied in animals in a state of relative immobility before coma or loss of
righting reflexes supervened [2]. Cerebral phosphocreatine levels were elevated 70
percent, glucose-6-P was elevated, and fructose diphosphate was depressed, suggesting depressed cerebral and glycolytic activity. Glycogen levels were decreased in
most animals despite a tenfold increase in brain glucose. Lactate levels were decreased,
but pyruvate levels were unchanged.

In an effort to ascertain to what extent these changes could be attributed to the
dehydration that accompanies diabetic acidosis, mannitol loading combined with
water deprivation was used to dehydrate animals. The changes were similar to those
found in diabetes produced by alloxan except that the decrease in lactate was accompanied by a decrease in pyruvate. The brain glucose level was not increased.

In another type of experiment, acute diabetes was produced by the injection of
antiinsulin serum, which promptly resulted in hyperglycemia. During the first day,
cerebral glycogen rose 50 percent, and pyruvate rose 40 percent.

It may be concluded that, as far as these studies go, changes observed in experimental diabetes are those expected of depressed neuronal activity, dehydration, and
hyperglucosemia. The only apparent exception is the elevation of brain pyruvate
soon after the administration of antiinsulin serum and its maintenance at control
levels after alloxan treatment. This suggests an increase in the NAD^+/NADH ratio,
but the change has not been explained. Despite considerable interference with glucose (and other) metabolism, both in the brain and elsewhere, the brain clearly has
no difficulty in maintaining its energy supply (Table 25-1).

Table 25-1. The Effect of Various Experimentally Produced Alterations of Glucose Metabolism
upon Cerebral Levels of ATP and P-Creatine.[a]

	ATP	P-Creatine (mmol per kg)	Glucose
Diabetes[b]	2.98 ± 0.05	4.86 ± 0.015	38 ± 4
Control[b]	2.80 ± 0.02	2.86 ± 0.05	9.2 ± 0.3
Hypoglycemia[c]			
Stupor	3.02 ± 0.03	4.50 ± 0.08	0.07 ± 0.01
Convulsions	2.98 ± 0.12	4.08 ± 0.30	0.07 ± 0.01
Control [c]	2.90 ± 0.04	4.08 ± 0.07	1.66 ± 0.08
Galactosemia[d]	1.67 ± 0.06	1.70 ± 0.07	0.07 ± 0.05
Control [d]	1.95 ± 0.12	1.92 ± 0.17	1.2 ± 0.4

[a]In none of these conditions does there appear to be serious interference with energy supply.
The rise in P-creatine in diabetes suggests a reduction in energy use.
[b]Whole mouse brain [2].
[c]Frontal cerebral cortex of mouse brain [27].
[d]Whole chick brain [16].

Neuropathy

Symptoms and signs of peripheral nerve involvement may occur early in the development of diabetes in some individuals, or may be a late manifestation [3]. Any nerve may be involved, and symptoms and signs may be sensory, motor, or autonomic. Tendon jerks may disappear, and neurotropic arthropathy and perforating ulcers of the skin are not uncommon. There is some correlation between these effects and control of the diabetic state, so that the symptoms and signs tend to improve with better diabetic control, and even to disappear. This is not always the case, unfortunately, and signs may persist despite apparently good control.

Physiological studies of nerve and muscle in patients with diabetic neuropathy have shown reduced conduction velocity in motor and sensory nerves [4]. It is said that stimulation of nerve trunks in diabetic patients of long standing requires higher voltages [3]. The data supporting this statement have not been published. It is also said, without support, that the elevated threshold is indicative of "poor axoplasmic flow" [3].

Experimentally, a 25 to 30 percent reduction of conduction velocity in nerve has been observed in rats made diabetic by subtotal pancreatectomy or by injection of alloxan [5]. In this experiment, insulin treatment of diabetic animals was unsuccessful in restoring the conduction velocity to normal. The nerves were found to have reduced levels of myoinositol in animals made diabetic by injections of alloxan or streotozotocin. When myoinositol was added to the diet of diabetic animals, conduction velocity in nerve returned to normal, although the severity of the diabetes was not otherwise reduced [6]. In an earlier study, myoinositol levels in the sciatic nerves of rats treated with alloxan were reduced by 40 percent, whether the animals became diabetic or not [7]. The nerves of the diabetic animals had greatly elevated fructose and sorbitol levels, 10 and 20 times higher, respectively, than the levels in the alloxan-treated, nondiabetic controls. Brain-level elevations of these substances were less dramatic (about fourfold).

A crucial question for understanding the pathogenesis of neural and cerebral dysfunction in diabetes is whether insulin affects neural metabolism directly. Several studies have been performed in an attempt to answer this question. The results of a careful clinical study suggest that insulin enhances glucose transport into brain [8]. In mice, too, increased blood-to-brain glucose transport is suggested because insulin produces an increase in the ratio of brain glucose to blood glucose [9]. On the basis of studies of water and electrolytes in brain, it has been suggested that insulin may enhance uptake of both K^+ and Na^+ into cerebral cells [10, 11]. It is only fair to add that effects of insulin on cerebral glucose and cation metabolism have also been questioned, based on experiments with the isolated perfused rat brain [12]. Whether any portion of the disordered function of brain and nerve in diabetes is the direct result of insulin lack cannot be answered at present.

GALACTOSEMIA

Galactosemia is an inherited defect of galactose metabolism in which galactose-1-P uridylyl transferase activity is deficient (Table 25-2, Fig. 25-1) [13]. Whether the enzyme protein is entirely lacking has not been determined.

Table 25-2. Enzyme Deficits in Carbohydrate Metabolism, Most with Neurological Consequences.

Disorder	References	Enzyme Affected and Number in Fig. 25-1	Principal Organ Affected	Neurological Involvement
Aglycogenosis	[20, 23]	1. Glycogen synthetase	Liver, muscle	Hypoglycemia
Glycogenoses				
I. Von Gierke's	[20, 21]	8. Glucose 6-phosphatase	Liver, kidney	Severe hypoglycemia, convulsions
II. Pompe's	[20]	Lysosomal acid maltase	General	Hypotonia, loss of motor skills
III. Cori's	[20, 21]	3, 4. Debranching enzymes	General, liver (leukocytes)	Mild hypoglycemia
IV. Andersen's	[20, 21]	2. Branching enzyme	General	
V. McArdle's	[20, 21]	5. Phosphorylase	Muscle	
VI. Hers'	[20, 23]	5. Phosphorylase	Liver, leukocytes	Mild hypoglycemia
VII.	[20, 21]	11. Phosphofructokinase	Muscle, erythrocytes	
VIII.	[20, 23]	9. Phosphohexose isomerase	Muscle	
VIII or IX.	[20, 23]	6. Phosphorylase b kinase	Liver, leukocytes	Mild hypoglycemia
Ketotic hypoglycemia	[24]	10. Fructose 1,6-diphosphatase	Liver, intestine	Severe hypoglycemia
Galactosemia	[13]	7. Galactose-1-P uridylyl-transferase	Liver, lens (erythrocytes)	Severe hypoglycemia, retardation, seizures
Hereditary fructose intolerance	[22]	12. Fructose-1-P aldolase	Liver	Hypoglycemia, vomiting
Familial lactic acidosis	[30, 41, 42]	14. Pyruvate carboxylase?	Liver	Acidosis, hypotonia, tachypnea, convulsions
Subacute necrotizing encephalomyelopathy (Leigh's)	[31, 34]	Inhibitor of thiamine triphosphate formation? Pyruvate carboxylase?	Brain	Retardation, weakness, ataxia, cranial nerve palsies
Lactic acidosis	[36—40]	13. Pyruvate decarboxylase	Liver, brain	Ataxia, confusion, choreoathetosis

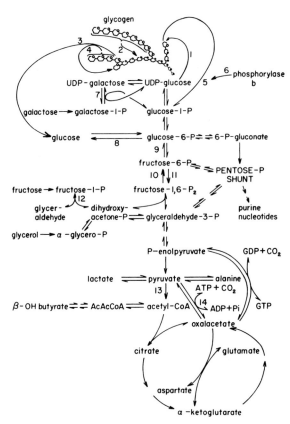

Figure 25-1
Overall scheme of glucose metabolism. Some aspects of galactose and fructose metabolism are also shown. Numbers refer to reactions catalyzed by enzymes, absence of which leads to one of the glycogenoses (Table 25-2, Chap. 23, and [20, 21]), and reactions catalyzed by other enzymes mentioned in the text or in Table 25-2. The reader is reminded that the activities of some of the enzymes shown are significant only in extracerebral tissues and that reactions of the same name may, in different tissues, be catalyzed by very different proteins. Two arrows in sequence indicate multiple steps.

Vomiting or diarrhea occurs within a few days after the patient first ingests milk. Jaundice or hepatomegaly occurs early. There may be ascites. Cataracts have been seen soon after birth. After a few months, if the disease is not recognized, mental retardation occurs.

Chemically, liver-function tests give evidence of derangement, blood galactose is elevated, and galactosuria, albuminuria, and amino aciduria are present. Rarely, blood glucose levels may be depressed. Red-cell galactose-1-P content is elevated in these patients, and the demonstration that the red cells lack galactose-1-P uridylyl transferase activity is diagnostic.

The mechanism by which the derangement of neural function is brought about has been the subject of considerable experimental work. There are no known specific inhibitors of galactose-1-P uridylyl transferase. It is possible, however, to elevate cerebral galactose levels in experimental animals enormously by giving them large amounts of galactose in the diet. In one experiment with rats [7], for example, feeding galactose for 5 weeks produced a serum galactose level of 10 mM and a brain level of 5 mmol per kg. The galactitol level in brain was 9 mmol per kg (in control animals, the level was less than 0.04 mmol per kg). In nerve, galactitol levels reached 17 mmol per kg. Despite these changes, the animals appeared to be in good health, except that they developed cataracts.

In young chickens, however, galactose feeding soon produces signs of cerebral dysfunction, according to two studies. They have high cerebral levels of galactose and galactitol (as much as 15 to 20 mmol per kg) and relatively high levels of galactose-1-P (0.2 to 1 mmol per kg) [14]. After 50 hours on the diet and after allowing for the blood content of the brain (1.75 percent), the brain glucose level was 0.07 mmol per kg, only 6 percent of control, despite normal blood glucose levels. As might have been expected, concentrations of glycolytic metabolites were also reduced, and brain glycogen was 1/3 of control [15]. Studies with radioactive glucose showed that penetration of glucose from blood to brain was reduced by at least 90 percent. Relatively rapid phosphorylation of galactose occurs under these conditions in vivo, and yet the galactose-1-P concentration, after an initial rise in the first few hours of feeding, falls gradually during the next two days. The existence of a "futile cycle" that uses ATP to phosphorylate galactose was postulated, and has been further supported by the discovery of a brain enzyme that hydrolyzes galactose-1-P [14]. No evidence was found for the participation of the cycle when the cerebral metabolic rate was measured, late in the course of the intoxication (after two days). However, early in the disease there is a small drop in cerebral ATP and a small rise in AMP at the time when cerebral galactose-1-P levels are highest and before the levels of glycolytic intermediates or glucose have started to fall. This tends to substantiate the notion of an energy-wasting process. Whether the galactose-1-P phosphatase in brain is sufficiently active to account for the observed turnover of galactose-1-P has not been reported. In any event, the disturbance of energy metabolism does not appear to be severe, even when the cerebral glucose levels are exceedingly low (Table 25-1).

Because there is good evidence that glucose penetration into the brain is severely inhibited by the high plasma galactose levels, the effects of a large dose of intraperitoneal glucose were examined in chicks [16]. The animals showed at least a partial reversal of symptoms, concomitant with an increase toward control levels of intracerebral glucose, lactate, and fructose 1,6-P_2, and an increase in intracerebral ATP and P-creatine. Cerebral galactose, galactitol, and galactose-1-P levels were unchanged.

Plasma hyperosmolality, equivalent to that seen in galactose-fed chicks, induced by NaCl or xylose in the drinking water produced no apparent change in cerebral energy status [16]. Although xylose feeding (but not NaCl) produced a decrease in cerebral glucose to 1/5 of control levels and an increase in xylose and xylitol, lactate and fructose-1,6-P_2 levels were unchanged and cerebral glycogen was nearly 80 percent of control. The chicks showed no sign of neurotoxicity.

Whether the galactose-fed chick is a good model for the galactosemic child has been debated [17, 18]. One criticism has been that the chick recovers promptly when galactose feeding is discontinued, whereas the galactosemic child does not. However, psychological and neuropathological studies of chickens treated in infancy with galactose and reared to adulthood have not been reported.

HYPOGLYCEMIA

The principal symptoms of hypoglycemia are referable to the nervous system (Chap. 19). The causes may originate almost anywhere in the body [19]. For example, a variety of tumors produce insulin and, consequently, hypoglycemia. In addition, large malignant tumors are sometimes associated with hypoglycemia. It is thought that such a tumor may use large amounts of glucose and, at the same time, may be the source of such substances as tryptophan and its metabolites which inhibit hepatic gluconeogenesis at the level of phosphoenolpyruvate formation. Some of the tumors have been found to contain increased levels of insulinlike activity, although circulating insulinlike activity is almost never elevated.

Diffuse liver disease, such as acute viral or toxic necrosis or the severe chronic passive congestion associated with heart failure, may give rise to hypoglycemia. In such cases, hepatic output of glucose is reduced because of impaired gluconeogenesis, glycogenesis, and glycogenolysis.

Hypopituitarism is occasionally associated with hypoglycemia as a consequence of increased sensitivity to insulin. Adrenocortical failure is commonly marked by hypoglycemia because the glucocorticoids stimulate some of the key enzymes for gluconeogenesis in the liver. Furthermore, the glucocorticoids play a permissive role in the action of glucagon and epinephrine in producing gluconeogenesis and glycogenolysis.

Hypoglycemia is also a feature of a variety of specific hepatic enzyme deficiencies. In three glycogen-storage diseases (type I, type III, and type VI, Table 25-2) the capacity to convert glycogen to glucose, and hence to elevate blood glucose levels, is impaired [20, 21]. Although the hypoglycemia produced is severe in glucose 6-phosphatase deficiency, often accompanied by convulsions, permanent cerebral damage appears to be rare in this disorder. Hypoglycemia is less severe in the type III glycogenosis and least severe in phosphorylase deficiency (see also Chap. 23).

Hepatic fructose-1-P aldolase deficiency [22] (Table 25-2), galactose-1-P uridylyl transferase deficiency (see above), glycogen synthetase deficiency [23] (Table 25-2), fructose 1,6-diphosphatase deficiency [24] (Table 25-2), and maple syrup urine disease (deficiency of branched chain α-keto acid decarboxylases, see Chap. 24) are also characterized by hypoglycemia.

Finally, there is a miscellaneous group of conditions of which hypoglycemia is a more or less constant accompaniment: alcohol ingestion, hyperinsulinism after gastrectomy, reactive hypoglycemia in early diabetes mellitus, and leucine hypersensitivity in children. The list may be extended almost indefinitely by reference to any medical textbook.

Pathophysiology

A rapidly falling blood glucose level will evoke epinephrine secretion. The chain of events probably involves activation of the adrenal medulla by way of the sympathetic outflow from the central nervous system, perhaps by means of glucose-sensitive neurons in the hypothalamus. The elevated epinephrine level in blood produces symptoms of sweating, weakness, tachycardia, hunger, and "inward trembling." If hypoglycemia persists, symptoms of brain involvement will be added: headache, blurred vision, diplopia, confusion, coma, and convulsions. When hypoglycemia develops slowly, epinephrine secretion and its attendant symptoms may not be evoked but the signs of brain involvement develop in any event. Chronic or repeated hypoglycemia may lead to permanent brain damage [24a].

Experimental

It seems evident that, if an organ derives most of its energy from glucose, that organ will show signs of malfunction when glucose is withheld. Clearly this is true in brain. It would appear to follow that, if such an organ manifests signs of malfunction during glucose deprivation, the malfunction is the result of energy lack. In brain, at least, this does not appear to be so.

In rats, electroencephalographic evidence of cerebral malfunction occurs at about the time that the calculated intracellular glucose concentration falls to zero [25, 26]. Citric acid cycle intermediates are reduced about 20 percent and pyruvate, glucose-6-P, and glycogen are somewhat lower. But there is no depression of ATP or P-creatine levels, and AMP, a more sensitive indicator of difficulty with the maintenance of ATP levels, does not rise. Even when the animals start to convulse, adenine nucleotide and P-creatine levels remain unchanged, although there is marked depletion of metabolites all along the glycolytic path and around the citric acid cycle, and glycogen has fallen to 30 percent. When electrical activity finally disappeared from the EEG, the energy levels in brain fell precipitously. It is unfortunate that these chemical studies were done on anesthetized animals so that the correlation had to be made between EEG and chemistry rather than between behavior and chemistry. In the anesthetized animal, furthermore, the cerebral metabolic rate may be considerably reduced (Chaps. 19 and 31). A regional study of the brain in hypoglycemic unanesthetized mice suggests that symptoms probably develop before the intracellular glucose has completely disappeared [27]. Otherwise, the results of this regional study were similar to those obtained with rats, with no changes in cerebral energy levels at times when cerebral function is severely impaired (Table 25-1).

During hypoglycemia in man and, presumably, in mouse, cerebral glucose and oxygen consumption are reduced (Chap. 19). But there is a parallel reduction in the rate of energy utilization [27] with a consequent preservation of steady-state cerebral ATP and P-creatine levels (in mouse and, presumably, in man). Therefore hypoglycemia appears to reduce the rate of cerebral function (hence the confusion and, finally, coma), but, unless there is a very local energy deficit, as, for example, in synaptic terminals, some metabolic defect other than that of energy supply is the basis of the functional derangement.

To date, the evidence favoring a pivotal role for any of the possible metabolic defects studied is inconclusive. For example, it has been known for many years that in hypoglycemia cerebral aspartate levels rise, and glutamate levels fall. But the changes lag behind the expression of EEG signs of cerebral involvement [25, 26]. Other amino acids also change: Alanine rises, glutamine falls, but the timing of the chemical changes, with respect to that of the EEG signs, does not suggest a precursor-product relationship. When all electrical activity disappears from the EEG, cerebral ammonia levels rise tenfold. Intracellular pH, as indicated by the creatine kinase equilibrium and by the bicarbonate-to-carbonic acid ratio, remains constant throughout the course of the hypoglycemia. The lactate-to-pyruvate ratio falls relatively early in the course of the experiment. It is conceivable that the depression in free cytoplasmic NADH which this suggests could have functional repercussions. What they might be is unknown.

Obviously, if insulin has a direct effect on nervous tissue [8–11], the hypoglycemia produced by insulin may be different from that occurring as the result of other causes.

The production of hypoglycemia in hereditary fructose intolerance has been approached experimentally using fructose loading in normal animals. The apparent explanation is based upon the multiplicity of events that occur in the liver after injection of fructose, including increased fructose-1-P and IMP levels, decreased P_i, inhibition of fructose-1-P aldolase by IMP, inactivation of liver phosphorylase, and inhibition of fructokinase, and fructose diphosphate aldolase and phosphorylase by fructose-1-P (for a brief up-to-date review, see [28, 29])

DISORDERS OF PYRUVATE AND LACTATE METABOLISM

Lactic acidosis has been seen in a variety of clinical conditions. Some instances are obvious, as when oxygenation of the patient is impaired because of pulmonary or cardiac disease. Lactic acidosis is also seen in some disorders of amino acid and glycogen metabolism. In a small group of patients, however, the causes of the acidosis are far from clear. Only now is some order being brought to a group of patients who are characterized by metabolic acidosis and sometimes physical retardation and signs of central nervous system disorder. Schärer [30] believes that congenital lactic acidosis can be divided into four types: (1) transitory, seen in full-term newborn infants who recover spontaneously in a few months; (2) early fatal, exemplified by one infant who died a week after birth; (3) slowly progressive, with death in one to four years; and (4) late onset, beginning in the second year of life and usually having a fatal outcome. In all of these types, the most frequent clinical manifestations are periodic attacks of dyspnea, tachypnea and hyperpnea, lethargy, hypotonia, and often obesity. Twitching, convulsions, and psychomotor retardation are not uncommon. Biochemically, there is a metabolic acidosis with elevated blood lactate from 5 to 10 times normal. The rise in pyruvate is often less striking. Lactate levels are elevated in the urine and, less frequently, urinary α-ketoglutarate and amino acids are elevated.

Leigh's Disease: Subacute Necrotizing Encephalomyelopathy
This disease of children is characterized pathologically by symmetrical lesions of the mesencephalic tegmentum, pons, medulla, and sometimes the spinal cord (see also Chap. 29). The distribution of lesions is reminiscent of that seen in Wernicke's encephalopathy (see Chap. 29). Clinically, the patients often show failure of nervous and mental development, with difficulty in standing, sitting, and swallowing. Motor weakness is frequent. Ataxia and abnormalities of the pupillary reflexes are reported less frequently. Cranial nerve palsies are common [31].

Chemically, lactic acidosis is seen frequently, but it tends to be relatively mild and episodic [30]. Subnormal levels of thiamine triphosphate are reported from post-mortem specimens of brain, but not of liver [32], and an inhibitor of cerebral thiamine pyrophosphate:ATP phosphotransferase has been found in the blood and urine of children thought to be suffering from this disease and in that of their relatives [33], but rarely in patients with other diseases.

At present, the status of this disorder is not clear. In some, the phosphotransferase inhibitor has been found; in others, an apparent defect in pyruvate carboxylase (Fig. 25-1; also see below) has been reported [34]. In one of the latter patients, the pyruvate carboxylase activity of a liver biopsy sample was within normal limits early in the disease, but nearly absent from a sample obtained at autopsy [35]. Therefore, the loss of the enzyme, at least in this case, cannot have been the cause of the disease.

Pyruvate Decarboxylase Deficiency
Two children have been described whose clinical histories were marked by repeated episodes of cerebellar ataxia combined with choreoathetosis or confusion which usually followed nonspecific febrile illnesses or other stresses [36–39]. One of these patients is mildly retarded, but the other appears to be above average in intelligence, although even between attacks he is somewhat clumsy.

Biochemically, the blood pyruvate and urine alanine levels were elevated. Blood alanine and lactate were also somewhat increased. Blood glucose levels were within the normal range, and intravenous glucose tolerance tests were also normal. No change was seen in blood glucose levels after an oral glucose load, however, suggesting a defect in intestinal glucose absorption. The rate of decarboxylation of pyruvate by whole white cells and fibroblasts and by cell-free preparations of fibroblasts was 25 percent or less than that of the lowest of several preparations from normal subjects (Table 25-2, Fig. 25-1). When glutamate, palmitate, or acetate decarboxylation was studied, the patients' rates were not different from those of controls. Each patient had one parent whose cells showed pyruvate decarboxylation rates intermediate between those of the patient and those of the control. The other parent's cells gave rates equal to the lowest control rates.

Thiamine deficiency did not appear to be a factor, since addition of thiamine pyrophosphate stimulated pyruvate oxidation in cell-free preparations from 20 to 40 percent in both controls and in one patient, but not in the other.

Quite a different sort of patient was described recently by Farrell et al. [40]. An infant with congenital lactic acidosis failed to thrive and died at six months of age. Biochemical examination of brain and liver showed that pyruvate dehydrogenase activity was completely missing, due to a lack of pyruvate decarboxylase. The activities of both dihydrolipoyl transacetylase and dihydrolipoyl dehydrogenase were present in normal amounts. Perhaps the most remarkable thing about this infant was that he survived so long with a defect of this magnitude.

Pyruvate Carboxylase Deficiency

Several patients have been described as being deficient in hepatic pyruvate carboxylase activity (Fig. 25-1, Table 25-2). As an example, one had hypoglycemia and convulsions early in life with psychomotor retardation [41]. Acidemia was severe, and pyruvate, lactate, and alanine in blood were all three to ten times normal levels. Blood thiamine levels were adjudged to be normal. (The data as reported do not inspire confidence because the control values for free and total thiamine were identical.) Pyruvate decarboxylation by white blood cells and fibroblasts was normal. The patient's liver was not tested for its capacity to decarboxylate pyruvate. In normal livers, pyruvate carboxylase activity was thought to be the result of at least two enzymes, one with a K_m for pyruvate of 0.2 to 0.4 mM, the other with a K_m 10 times as high. The activity of the high K_m enzyme appeared to be normal in the patient's liver, but there was no evidence for the presence of the enzyme with the low K_m.

Despite the lack of evidence of thiamine deficiency, the patient was put on massive doses of thiamine. The acidemia improved. (There was actually a phase of alkalosis following initiation of thiamine treatment.) Mental retardation persisted, however.

Another patient, seen at age 9, had normal blood glucose, but also showed mental retardation and motor dysfunction [42]. Blood pyruvate levels were two to three times normal, and lactate and alanine levels were at the upper limits of normal. This patient had deficient hepatic pyruvate carboxylase activity (20 percent of controls). A kinetic study was not performed, but the rate of incorporation of pyruvate-2-^{14}C into glycogen by liver slices was only 5 percent of that of a normal control preparation. The patient's pyruvate decarboxylase activity (in liver) was not different from that of two normal controls.

A recent careful study of partially purified pyruvate carboxylase from human liver [43] has illuminated its properties and called sharply into question previous measurements of its activity in human liver samples obtained by biopsy or at autopsy. These workers found that the enzyme could be made to act at 1/5 of its maximal activity in the absence of acetyl coenzyme A (acetyl-CoA) by increasing substrate levels greatly. In the presence of acetyl coenzyme A, however, the affinity of the enzyme for its substrates ($MgATP^{2-}$, HCO_3^-, and pyruvate) is enhanced about 30 times. The affinity of the human enzyme for acetyl coenzyme A was only 1/6 that of the enzyme from chicken liver (which shows no activity without acetyl-CoA). The enzyme appears to have two binding sites for acetyl-CoA and two or more for pyruvate. Scrutton and White [43] point out that, when the enzyme activity is measured in crude tissue preparations by means of the two-step assay used in the studies of patients already cited, precautions should be taken to assure a constant level of acetyl-CoA during the first step. In an assay of crude tissue, acetyl-CoA would be produced by the action of pyruvate dehydrogenase and destroyed by hydrolysis and by reaction via citrate synthetase with oxalacetate, a product of the pyruvate carboxylase reaction. Therefore the apparent pyruvate carboxylase activity found in the tissue would be determined in part by the balance of influences acting on the acetyl-CoA level during the assay as well as by the amount of pyruvate carboxylase present.

The questions raised by this study have not yet been examined in a patient in whom the conventional pyruvate carboxylase assay suggests a deficit. The patient reported by Yoshida et al. [42] was said to have hepatic pyruvate decarboxylase activity within normal limits. Willems et al. [44] report the case of a baby with mental and motor retardation, microcephaly, and hypotonia, who had metabolic acidosis, blood pyruvate levels three to four times higher than normal, and somewhat elevated blood lactate. There was some improvement after massive doses of thiamine were administered. At autopsy, the pyruvate decarboxylase activity was found to be greatly reduced in the liver (it was normal in white blood cells and fibroblasts), and also pyruvate carboxylase activity was reduced to 20 percent of normal. Although the data might suggest a double enzyme deficiency, other factors seem likely to account for the low pyruvate carboxylase activity seen in this patient [43].

REFERENCES

*1.   Reske-Nielsen, E., and Lundbaek, K. Diabetic Encephalopathy. In Pfeiffer, E. F. (Ed.), *Handbook of Diabetes Mellitus*. Vol. II. Munich: Lehmann, 1971, p. 719.

2.   Thurston, J. H., Hauhart, R. E., Jones, E. M., and Ater, J. L. Effects of alloxan diabetes, anti-insulin serum diabetes, and non-diabetic dehydration on brain carbohydrate and energy metabolism in young mice. *J. Biol. Chem.* 250:1751, 1975.

*3.   Schneider, T. Diabetic Neuropathy. In Pfeiffer, E. F. (Ed.), *Handbook of Diabetes Mellitus*. Vol. II. Munich: Lehmann, 1971, p. 607.

4.   Mulder, D. W., Lambert, E., Bastron, J. A., and Sprague, K. G. The neuropathies associated with diabetes mellitus. *Neurology* 11:275, 1961.

5.   Eliasson, S. G. Nerve conduction changes in experimental diabetes. *J. Clin. Invest.* 43:2353, 1964.

6.   Greene, D. A., De Jesus, P. V., Jr., and Winegrad, A. I. Effects of insulin and dietary myoinositol on impaired peripheral motor nerve conduction velocity in acute streptozotocin diabetes. *J. Clin. Invest.* 55:1326, 1975.

7.   Stewart, M. A., Sherman, W. R., Kurien, M. M., Moonsammy, G. I., and Wisgerhof, M. Polyol accumulations in nervous tissue of rats with experimental diabetes and galactosemia. *J. Neurochem.* 14:1057, 1967.

8.   Goffstein, U., Held, K., Sebening, H., and Walpurger, G. Glucose verbrauch des Gehirns nach intravenösen Infusionen von Glucose, Glucagon und Glucose-Insulin. *Klin. Wochenschr.* 43:965, 1965.

9.   Nelson, S. R., Schulz, D. W., Passonneau, J. V., and Lowry, O. H. Control of glycogen levels in brain. *J. Neurochem.* 15:1274, 1968.

10.   Arieff, A. I., Doerner, T., Zelig, H., and Massry, S. G. Evidence for a direct effect of insulin on electrolyte transport in brain. *J. Clin. Invest.* 54:654, 1974.

11.   Arieff, A. J., and Kleeman, C. R. Studies on mechanisms of cerebral edema in diabetic comas: Effects of hyperglycemia and rapid lowering of plasma glucose in normal rabbits. *J. Clin. Invest.* 52:571, 1973.

*Asterisks denote key references.

12. Sloviter, H. A., and Yamada, H. Absence of a direct effect of insulin on metabolism of the isolated perfused rat brain. *J. Neurochem.* 18:1269, 1971.

*13. Segal, S. Disorders of Galactose Metabolism. In Stanbury, J. B., Wyngaarden, J. B., and Fredrickson, D. S. (Eds.), *The Metabolic Basis of Inherited Disease,* 3d ed. New York: McGraw-Hill, 1972.

14. Kozak, L. P., and Wells, W. W. Studies on the metabolic determinants of D-galactose-induced neurotoxicity in the chick. *J. Neurochem.* 18:2217, 1971.

15. Granett, S. E., Kozak, L. P., McIntyre, J. P., and Wells, W. W. Studies on cerebral energy metabolism during the course of galactose neurotoxicity in chicks. *J. Neurochem.* 19:1659, 1972.

16. Knull, H. R., and Wells, W. W. Recovery from galactose-induced neurotoxicity in the chick by the administration of glucose. *J. Neurochem.* 20:415, 1973.

17. Knull, H. R., Wells, W. W., and Kozak, L. P. Galactose toxicity in the chick: Hyperosmolality or depressed brain energy reserves. *Science* 176:815, 1972.

18. Malone, J. I., Wells, H. J., and Segal, S. Galactose toxicity in the chick: Hyperosmolality or depressed brain energy reserves. *Science* 176:816, 1972.

*19. Conn, J. W., and Pek, S. Current Concepts on Spontaneous Hypoglycemia. *Scope Monograph.* Kalamazoo: Upjohn, 1970.

*20. Howell, R. R. The Glycogen Storage Diseases. In Stanbury, J. B., Wyngaarden, J. B., and Fredrickson, D. S. (Eds.), *The Metabolic Basis of Inherited Disease,* 3d ed. New York: McGraw-Hill, 1972.

*21. Brown, B. I., and Brown, D. H. Glycogen-Storage Diseases: Types I, III, IV, V, VII and Unclassified Glycogenoses. In Dickens, F., Randle, P. J., and Whelan, W. J. (Eds.), *Carbohydrate Metabolism and Its Disorders.* London: Academic, 1968.

22. Froesch, E. R. Essential Fructosuria and Hereditary Fructose Intolerance. In Stanbury, J. B., Wyngaarden, J. B., and Fredrickson, D. S. (Eds.), *The Metabolic Basis of Inherited Disease,* 3d ed. New York: McGraw-Hill, 1972.

23. Sidbury, J. B., Jr. The Glycogenoses. In Gardiner, L. I. (Ed.), *Endocrine and Genetic Diseases of Childhood.* Philadelphia: Saunders, 1969.

24. Baker, L., and Winegrad, A. I. Fasting hypoglycemia and metabolic acidosis associated with deficiency of hepatic fructose-1,6-diphosphatase activity. *Lancet* 2:13, 1970.

24a. Blau, A., Reider, N., and Bender, M. B. B. Extrapyramidal syndrome and encephalographic pictures of progressive internal hydrocephalus in chronic hypoglycemia. *Ann. Intern. Med.* 10:910, 1936.

25. Lewis, L. W., Ljunggren, B., Ratcheson, R. A., and Siesjö, B. K. Cerebral energy state in insulin induced hypoglycemia, related to blood glucose and to EEG. *J. Neurochem.* 23:673, 1974.

26. Lewis, L. D., Ljunggren, B., Norberg, K., and Siesjö, P. K. Changes in carbohydrate substrates, amino acids, and ammonia in the brain during hypoglycemia. *J. Neurochem.* 23:659, 1974.

27. Ferrendelli, J. A., and Chang, M.-M. Brain metabolism during hypoglycemia. *Arch. Neurol.* 28:173, 1973.

28. Thurston, J. H., Jones, E. M., and Hauhart, R. E. Decrease and inhibition of liver glycogen phosphorylase after fructose: An experimental model for the study of hereditary fructose intolerance. *Diabetes* 23:597, 1974.

29. Thurston, J. H., Jones, E. M., and Hauhart, R. E. Failure of adrenaline to induce hyperglycemia after fructose injection in young mice. *Biochem. J.* 148:149, 1975.

*30.   Schärer, K. Congenital Lactic Acidosis. In Stern, J., and Toothill, C. (Eds.), *Organic Acidurias.* Edinburgh: Churchill-Livingstone, 1972.

*31.   Cooper, J. R., and Pincus, J. H. Thiamine Triphosphate Deficiency in Leigh's Disease (Subacute Necrotizing Encephalomyelopathy). In Hommes, F. A., and van den Berg, C. J. (Eds.), *Inborn Errors of Metabolism.* London: Academic, 1973.

32.   Cooper, J. R., Itokawa, Y., and Pincus, J. H. Thiamine triphosphate deficiency in subacute necrotizing encephalomyelopathy. *Science* 164:74, 1969.

33.   Pincus, J. H., Cooper, J. R., Piros, K., and Turner, V. Specificity of the urine inhibitor test for Leigh's disease. *Neurology* 24:885, 1974.

34.   de Groot, C. J., Jouxis, J. H. P., and Hommes, F. A. Further Studies on Leigh's Encephalomyelopathy. In Stern, J., and Toothill, C. (Eds.), *Organic Acidurias.* Edinburgh: Churchill-Livingstone, 1972.

35.   Grover, W. D., Auerbach, V. H., and Patel, M. J. Biochemical studies and therapy in subacute necrotizing encephalomyelopathy (Leigh's disease). *J. Pediatr.* 81:39, 1972.

36.   Blass, J. P., Avigan, J., and Uhlendorf, B. W. A defect in pyruvate decarboxylase in a child with an intermittent movement disorder. *J. Clin. Invest.* 49:423, 1970.

37.   Blass, J. P., Kark, A. P., and Engel, W. P. Clinical studies of a patient with pyruvate decarboxylase deficiency. *Arch. Neurol.* 25:449, 1971.

38.   Lonsdale, D., Faulkner, W. R., Price, J. W., and Smeby, R. R. Intermittent cerebellar ataxia associated with hyperpyruvic acidemia, hyperalaninemia and hyperalaninuria. *Pediatrics* 43:1025, 1969.

39.   Blass, J. P., Lonsdale, D., Uhlendorf, B. W., and Horn, H. Intermittent ataxia with pyruvate-decarboxylase deficiency. *Lancet* 1:1302, 1971.

40.   Farrell, D. F., Clark, A. F., Scott, C. R., and Wennberg, R. P. Absence of pyruvate decarboxylase activity in man: A cause of congenital lactic acidosis. *Science* 187:1082, 1975.

41.   Brunette, M. G., Delvin, E., Hazel, B., and Scriver, C. R. Thiamine responsive lactic acidosis in a patient with deficient low K_m pyruvate carboxylase activity in liver. *Pediatrics* 50:702, 1972.

42.   Yoshida, T., Tada, K., Konno, T., and Arakawa, T. Hyperalaninemia with pyruvicemia due to pyruvate carboxylase deficiency of the liver. *Tohoku J. Exp. Med.* 99:121, 1969.

43.   Scrutton, M. C., and White, M. D. Purification and properties of human liver pyruvate carboxylase. *Biochem. Med.* 9:271, 1974.

44.   Willems, J. L., Monnens, L. A. H., Trijbels, J. M. F., Sengers, R. C. A., and Veerkamp, J. H. Pyruvate decarboxylase deficiency in liver. *N. Engl. J. Med.* 290:406, 1974.

Chapter 26

Sphingolipidoses and Other Lipid Metabolic Disorders
Roscoe O. Brady

A major group of inherited metabolic disorders that are characterized by profound deleterious effects on the central nervous system is distinguished by the accumulation of excessive quantities of lipids in various organs and tissues throughout the body of the afflicted individuals. There are a dozen conditions of this type whose abnormal biochemistry and mode of inheritance are now known. Each of these diseases is caused by an inherited deficiency of a particular enzyme required for the catabolism of a specific lipid. Most of these disorders are inherited as autosomal recessive traits; therefore both parents, who are perfectly healthy, must be heterozygotes in order to produce an affected offspring. When two carriers mate, there is one chance in four that the fetus will be affected; two will be heterozygotes like the parents, and one will not be involved at all. This distribution of the harmful genes occurs with each pregnancy. On the other hand, one of the lipid-storage diseases known as Fabry's disease is transmitted as an X-linked recessive condition. Here only the female need be a heterozygote to produce an affected (hemizygote) male child. Half of her sons will have the disorder; half of her daughters will be carriers; the others will not be involved.

PATHWAYS OF SPHINGOLIPID CATABOLISM
Most of the disorders considered in this chapter are caused by defects in catabolism of one of a specific group of lipids in which one portion of the molecule is called ceramide (Cer). Ceramide is a long-chain fatty acyl derivative of the amino alcohol sphingosine $(CH_3 - (CH_2)_{12} - CH = CH - CH(OH) - CH(NH - CO - R) - CH_2OH)$. Various oligosaccharides or a molecule of phosphorylcholine may be joined to carbon atom one of ceramide. The nomenclature for this group of substances is described in Chapter 15.

The major lipid of myelin on a weight basis is galactocerebroside (Cer-Gal), which consists of one molecule each of sphingosine, fatty acid, and galactose. The ceramide tetrasaccharide called globoside is the major neutral glycolipid of erythrocyte membranes (Fig. 26-1). Acidic glycolipids called gangliosides have from one to four molecules of N-acetylneuraminic acid in addition to ceramide and an oligosaccharide portion. These substances that contain a tetrahexosyl oligosaccharide moiety are highly concentrated in brain and appear to play a major role in the development of the central nervous system and, in particular, in the formation of the myelin sheath. Gangliosides are sequentially degraded by a group of catabolic enzymes, indicated in schematic form in Figure 26-2. The individual disorders of lipid catabolism are described below.

GAUCHER'S DISEASE
Gaucher's disease, like many other lipid-storage diseases, bears the name of the clinician who first described the clinical manifestations. Gaucher's disease is an autosomal recessive disorder, and patients exhibiting it have been divided into three clinical

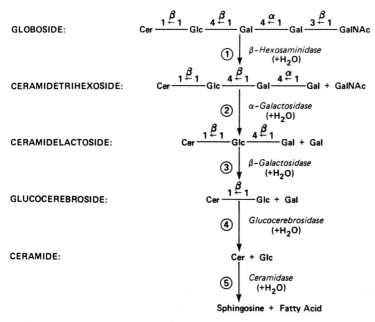

Figure 26-1
Pathway of catabolism of globoside, the major neutral glycolipid of erythrocyte stroma. Cer = ceramide; Glc = glucose; Gal = galactose; GalNAc = N-acetylgalactosamine.

categories. The first, called Type I, is the "adult" form of Gaucher's disease, manifested by hepatosplenomegaly, a hemorrhagic diathesis, and extreme pain and pathological fractures of the long bones, vertebrae, and pelvis, caused by the osteoporosis that attends the infiltration of glycolipid-storing cells in the marrow. Patients with Type II, the infantile form, have an early onset of the systemic manifestations that occur in patients with Type I. Those with Type II also have severe mental retardation because of involvement of the central nervous system. Patients with Type III, the juvenile form, have the systemic signs and symptoms of Type I; they develop seizures and show gradual deterioration of the central nervous system, usually beginning in their teens.

An excessive quantity of the glucocerebroside (Cer-Glc) is found in the organs and tissues of patients with this disease. The accumulation is caused by a deficiency of the β-glucosidase, which catalyzes Step 4, as shown in Figure 26-1 [1]. Patients with the infantile form of the disease have virtually no detectable glucocerebrosidase activity in their tissues [2]. Patients with the adult form always have some residual glucocerebrosidase, and in this type it is usually greater than 25 percent and may be as high as 44 percent of that in normal individuals. Patients with Type III Gaucher's disease can have up to 20 percent of normal glucocerebrosidase activity, although it is generally somewhat less than this but more than in Type II patients.

Most of the glucocerebroside that accumulates in the reticuloendothelial cells of the liver, spleen, and bone marrow appears to be derived primarily from senescent leukocytes and erythrocytes. The principal neutral glycolipid in leukocytes is

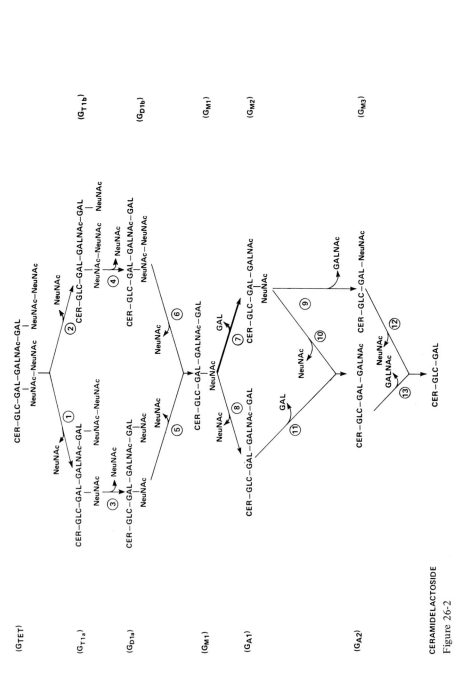

Figure 26-2

Pathways of catabolism of gangliosides. The major gangliosides of brain are G_{TET}, G_{T1a}, G_{T1b}, G_{D1a}, G_{D1b}, and G_{M1}. Abbreviations as in Figure 26-1. NeuNAc = N-acetylneuraminic acid.

ceramidelactoside, and its catabolism (via Steps, 3, 4, and 5 in Fig. 26-1) is impaired by the deficiency of glucocerebrosidase (Step 4). In like fashion, the catabolism of erythrocyte globoside proceeds through reactions 1 through 3, but cannot be further metabolized at a normal rate by patients with Gaucher's disease. In addition, erythrocytes contain hematoside (G_{M3}, Fig. 26-2), which also contributes to the quantity of glycolipid that must be removed. Actually, erythrocytes have some ceramidetrihexoside (cf. Fig. 26-1) and glucocerebroside itself in association with their membranes. Leukocytes also have a small quantity of glucocerebroside.

The cause of the central nervous system damage in patients with Type II and Type III Gaucher's disease is believed to be an impairment of the catabolism of glucocerebroside derived from the metabolism of gangliosides (Fig. 26-2 and Fig. 26-1, Steps 3–5). Ganglioside turnover is very rapid in the neonatal period and thereafter slows considerably. It is assumed that the residual glucocerebrosidase activity in the brain of patients with the adult form of Gaucher's disease is sufficient to catabolize the glucocerebroside derived from gangliosides, and the nerve cells in Type I patients do not have a pathological accumulation of this lipid.

The diagnosis of homozygotes and the detection of heterozygotes can be carried out easily through determinations of glucocerebrosidase activity by using washed leukocytes or by enzyme assays in extracts of cultured skin fibroblasts [3]. Glucocerebroside labeled with ^{14}C in the D-glucose portion of the molecule is the preferred substrate; however, 4-methylumbelliferyl-β-D-glycopyranoside has also been employed for diagnosis [4, 5].

The therapy of Gaucher's disease has recently been investigated. When glucocerebrosidase isolated from human placental tissue was injected into two patients with Gaucher's disease, there was a 26 percent reduction in the amount of stored glucocerebroside in their liver [6]. The elevated level of erythrocyte glucocerebroside in these patients returned to normal after glucocerebrosidase had been administered. This effect persisted over a period of many months [7]. These findings provide much hope for the amelioration of the adult form of Gaucher's disease by enzyme-replacement therapy. Glucocerebrosidase assays can be performed on extracts of cultured amniotic cells [8], and, if indicated, genetic counseling is available for Type II and Type III patients.

NIEMANN-PICK DISEASE

Niemann-Pick disease also is transmitted as an autosomal recessive trait. Patients with this disorder have been subdivided into four categories on the basis of the clinical findings. Type A is the severe infantile form, with extensive neurological involvement, emaciation, hepatosplenomegaly, and foam cells in the bone marrow. Patients with Type B have organomegaly but are generally without central nervous system involvement. Patients with Type C have both organomegaly and neurological abnormalities, the latter appearing in late childhood or in the teens. Patients with Type D have organomegaly and central nervous system damage resembling that in Type C, but the ancestry of these individuals is confined to the Nova Scotia area.

The enzymatic defect in Niemann-Pick disease is a deficiency of sphingomyelinase, which causes the accumulation of sphingomyelin throughout the body [9a]:

$$\text{Cer-phosphorylcholine} + H_2O \xrightarrow{\text{sphingomyelinase}} \text{Cer} + \text{phosphorylcholine}$$

Sphingomyelin is a ubiquitous component of the membranes of cells. Therefore its accumulation is assumed to be a consequence of normal cellular turnover.

Homozygotes and heterozygotes may be diagnosed by using radioactive sphingomyelin labeled in the choline portion of the molecule with sonicated leukocyte preparations or extracts of cultured skin fibroblasts [3]. A chromogenic substrate proposed several years ago [3] has recently been synthesized. Gal et al. have found this substance reliable also for the diagnosis of Niemann-Pick patients and for the detection of heterozygotes [9b].

Sphingomyelinase has recently been isolated from human placental tissue (Pentchev, P. G., Brady, R. O., Hibbert, S. R., and Gal, A. E., in preparation). So far, however, no enzyme replacement trials have been carried out. It seems unlikely that it will benefit patients with Types A and C (and Type D), although it may be helpful for those with Type B Niemann-Pick disease. Because of the wide distribution of sphingomyelin, extensive investigations in experimental animals will have to be carried out before sphingomyelinase can be administered to humans.

The prenatal diagnosis of Niemann-Pick disease is now available by using the radioactive substrate [10]. It is reasonable to assume that the newly developed assay with the chromogenic substrate will be equally reliable for monitoring pregnancies at risk for Niemann-Pick disease.

GLOBOID LEUKODYSTROPHY (KRABBE'S DISEASE)

This autosomal recessive disorder is characterized by hyperirritability, hyperesthesia, and episodic fever, which begins about the fourth or fifth month of age. These symptoms are followed by convulsions, mental retardation, hyperactivity, blindness, and deafness. There is no organomegaly or net accumulation of a specific lipid in the brain. The metabolic defect is a deficiency of a β-galactosidase, which catalyzes the hydrolysis of galactocerebroside [11]:

$$\text{Cer-Gal} + H_2O \xrightarrow{\text{galactocerebroside-}\beta\text{-galactosidase}} \text{Cer} + \text{Gal}$$

A diagnostic test is available that employs galactocerebroside labeled with ^3H in the hexose moiety [12]. Washed leukocytes, cultured skin fibroblasts, and even serum samples may be used for this test [13]. Prenatal detection of Krabbe's disease is feasible through enzyme assays on cultured amniotic cells [14]. There is no specific therapy at this time, although allosteric activation of the mutated galactocerebrosidase has been proposed [15].

METACHROMATIC LEUKODYSTROPHY

Patients with the most common form of this autosomal recessive disorder show progressive flaccidity and weakness of the arms and legs between 12 and 18 months of age. These signs are followed by loss of ability to stand plus mental retardation, which is manifested initially by a loss of speech. The clinical aspects of the disorder progress to blindness and complete mental retardation. In other patients, many of the same signs and symptoms appear in the early teens and a slower clinical course is followed.

The underlying pathological chemistry in this disorder is the accumulation of large quantities of sulfatide, the 3-O-sulfate ester of galactocerebroside. Sections of peripheral nerves show brownish yellow metachromatic deposits when stained with cresyl violet dye. The cause of the accumulation is a deficiency of a sulfatase, which catalyzes the following reaction [16]:

$$\text{Cer-Gal-SO}_3\text{H}_2 + \text{H}_2\text{O} \xrightarrow{\text{sulfatidase}} \text{Cer-Gal} + \text{H}_2\text{SO}_4$$

Sulfatidase was partially purified from hog kidney, and evidence was obtained for the requirement of a heat-stable factor for the hydrolysis of the sulfated glycolipid [17] in buffers (ionic strength $\geqslant 0.2$) with osmolarity in the physiological range [18].

The diagnosis of homozygotes and heterozygous carriers can be carried out easily on washed leukocytes [19] or on cultured skin fibroblasts [20] by assaying the hydrolysis of nitrocatechol sulfate, as described by Austin and co-workers [21].

The therapy of patients with metachromatic leukodystrophy has been undertaken by intravenous and intrathecal injection of crude preparations of arylsulfatase A. There was no improvement in the patients' conditions, and in several instances there were severe pyrogenic reactions. Genetic counseling is available, based on the determination of arylsulfatase activity in cultured amniotic cells [20].

In a recent important contribution from Austin's laboratory, convincing evidence was presented for the presence of a novel protein in tissues from patients with metachromatic leukodystrophy. Although catalytically inactive, this protein cross-reacted with antibody against normal human arylsulfatase A [22]. This is the first evidence of a catalytically inactive protein in patients with lipid-storage diseases, and it provides strong support for the occurrence of structural mutations in the enzymes of patients with such disorders.

CERAMIDELACTOSIDE LIPIDOSIS

A few years ago Dawson and Stein described a patient [23a] in the early teens with progressive mental retardation, hepatosplenomegaly, and lipid-storage cells in the bone marrow. The principal accumulating lipid was reported to be ceramidelactoside (cf. Fig. 26-1). At first, the disorder was attributed to a deficiency of the β-galactosidase, which catalyzes Step 3 in Figure 26-1. Although this type of enzymatic defect would appear to be logical, subsequent reexamination of enzyme activities in this patient's cultured skin fibroblasts showed, instead, a partial deficiency in sphingomyelinase activity [23b].

FABRY'S DISEASE

Fabry's disease is inherited as an X-linked recessive characteristic, and the major clinical manifestations occur primarily in men. The afflicted males have reddish purple maculopapular lesions in the skin over the buttocks, inguinal region, and scrotum. Fabry patients experience excruciating attacks of peripheral neuralgia in the hands and feet. They are also unable to sweat. Ceramidetrihexoside accumulates progressively (cf. Fig. 26-1) in the walls of the blood vessels and in the glomeruli of the kidneys. Males with the disorder usually experience complete renal failure in their late 40s or early 50s. Some have myocardial infarctions or cerebrovascular thromboses caused by the extensive arteriosclerosis that attends the deposition of lipid in the blood vessels. Heterozygous females may exhibit some of the manifestations of the disease, including corneal opacification and signs of renal impairment. The symptoms are usually milder in the females, although several women have recently been reported with severe manifestations of the disease.

The metabolic defect is a deficiency of the α-galactosidase that catalyzes Step 2 in Figure 26-1 [24]. The reaction may be assayed through the use of ceramidetrihexoside labeled with ^{3}H in the terminal molecule of galactose [12]. The principal source of the accumulating lipid is believed to be globoside and ceramidetrihexoside derived from the stroma of senescent erythrocytes (Fig. 26-1). A diagnosis can be made by using artificial chromogenic or fluorogenic galactopyranosides as substrates to determine α-galactosidase activity in leukocytes or cultured skin fibroblasts [25]. Heterozygotes may also be detected by enzyme assays with these substrates, and a reliable procedure has been developed for the prenatal diagnosis of Fabry's disease [26].

The therapy for Fabry's disease has recently come under investigation in several laboratories. One of the primary considerations is the amelioration of renal failure, and kidney transplantation has been used with some degree of success. Reports of the efficacy or lack of benefit from this procedure are cited [27]. A more direct approach to the correction of the enzymatic deficiency was recently undertaken by using intravenous infusion of purified ceramidetrihexosidase isolated from human placental tissue [27]. Two males with the disorder received a highly purified preparation of the enzyme. They tolerated the procedure without untoward effects. The ceramidetrihexosidase was rapidly cleared from the circulation, and much of it appeared in the liver. The elevated level of ceramidetrihexoside in the blood was dramatically lowered after the enzyme had been administered. Several observations make it seem likely that the placental enzyme exerted its catabolic effect extracirculatorily: (1) the enzyme is catalytically active at moderately acid pHs and is virtually inactive at the pH of blood; (2) the enzyme had almost completely disappeared from the circulation before the plasma ceramidetrihexoside decreased; (3) there was no elevation of plasma ceramidelactoside, the immediate product of the reaction at a time when ceramidetrihexoside was falling. Another potentially important observation was made in the course of these investigations. There appeared to be three or four times more α-galactosidase activity in the postinfusion liver biopsy sample from a patient than actually was administered. This finding suggests that the placental enzyme had activated the patient's mutated catalytically inactive protein. If this observation is substantiated by further investigation, it will provide further encouragement for direct enzyme-replacement therapy.

TAY-SACHS DISEASE

Tay-Sachs disease, a lipid-storage disease, is transmitted as an autosomal recessive trait. There are several clinical and biochemically distinct forms of Tay-Sachs disease. The signs and symptoms of the classic infantile form of the disorder are restricted to the central nervous system. These patients appear normal for the first five or six months of life and then fail to develop motor and mental capacities properly. Convulsions, apathy, and blindness follow. Death usually occurs about the third year. A cherry-red spot is present in the macular region of the eye. Other patients may have delayed onset of these manifestations until between 12 and 18 months of age. Here the progression is slower, with death occurring at 5 or 6 years of age. Occasionally, a patient is seen with a very early onset and rapid progression of the signs and symptoms, along with some hepatomegaly. This is the so-called Sandhoff-Jatzkewitz form of Tay-Sachs disease.

The metabolic defect in all of the different clinical forms of Tay-Sachs disease is a deficiency of the hexosaminidase that catalyzes Step 9 in Figure 26-2 [28, 29]. The enzyme is normally found in brain lysosomes. Its absence in patients with Tay-Sachs disease causes lipids and protein in the form of membranous cytoplasmic bodies to accumulate within these organelles. The catabolism of ganglioside G_{M2} by purified enzymes requires the presence of a suitable detergent [30] or heat-stable factor [31]. The latter substance appears to be nonspecific since it is also effective in other glycolipid-hydrolyzing enzyme systems. It may be a natural detergent; however, its presence has not yet been demonstrated in lysosomes that are fully capable of catalyzing the hydrolysis of G_{M2} without an added detergent or activating factor.

The diagnosis of *most* homozygotes and heterozygotes for Tay-Sachs disease can be made through the use of the fluorogenic substrate 4-methylumbelliferyl-β-D-N-acetylglucosaminide [32]. Patients with the most frequently seen form of the disease lack hexosaminidase A — one of two normally occurring hexosaminidase "isozymes." Hexosaminidase B activity in these patients is higher than normal. Patients with the Sandhoff-Jatzkewitz form have no detectable hexosaminidase activity in their tissues. Patients with still another mutation show good activity with the fluorogenic substrate but cannot catabolize G_{M2}. Because of these variations, much caution must be exercised in detecting heterozygotes and homozygotes via artificial substrates. It has recently been shown, furthermore, that there are perfectly normal, but rare, individuals who lack any detectable hexosaminidase A activity in their tissues but who nevertheless can catabolize G_{M2} [33]. If one were to use an artificial chromogenic or fluorogenic substrate for prenatal detection, as commonly practiced today, such a fetus would be classified mistakenly as a Tay-Sachs homozygote.

The therapy of Tay-Sachs disease has been investigated through the intravenous administration of purified hexosaminidase A, isolated from human urine, to an infant with this disorder [34]. A normal level of hexosaminidase A was quickly reached in the bloodstream after the infusion. None of the hexosaminidase A appeared to cross the blood-brain barrier, however, since there was no detectable increase of hexosaminidase activity in brain biopsy specimens or in the cerebrospinal fluid. Thus, replacement therapy by intravenous administration of exogenous enzyme, which appears very promising for patients with Gaucher's disease [6] and Fabry's disease [27], does not seem likely to be of benefit for Tay-Sachs disease. The development of additional procedures, such as a means of opening the blood-brain barrier temporarily or modifying the enzyme so it will traverse the barrier, will be required before enzyme replacement for this and other CNS disorders becomes feasible. Before these measures can be undertaken, critical experiments with neuronal cells in culture should be carried out to determine whether the cells can pinocytize exogenous hexosaminidase and other lipid-hydrolyzing enzymes from various sources. The elucidation of the behavior of neural cells in this regard is an important aspect of neurobiology.

GENERALIZED GANGLIOSIDOSIS
This inherited disorder is also transmitted as an autosomal recessive trait. Patients with generalized (G_{M1}) gangliosidoses exhibit severe mental deterioration and frequently have a cherry-red spot in the retina. They also have hepatosplenomegaly, bony abnormalities, and foam cells in the marrow and viscera. At least two clinical forms — the classic infantile type and the juvenile form — have been described on the basis of the age of onset and rapidity of progression of the disease.

Generalized gangliosidosis is the result of a deficiency of the β-galactosidase that catalyzes Step 7 in Figure 26-2 [35]. In some patients, acid mucopolysaccharides also accumulate because of the drastic diminution of total β-galactosidase activity. The generalized reduction of β-galactosidase makes the diagnosis of homozygotes and heterozygotes readily available through the use of such chromogenic or fluorogenic substrates as p-nitrophenyl-β-D-galactopyranoside or 4-methylumbelliferyl-β-D-galactopyranoside.

An interesting feature of generalized gangliosidosis is that the activities of such other neutral lipid β-galactosidases as galactocerebrosidase and ceramidelactosidase must be accommodated within the residual 7 to 9 percent of total normal β-galactosidase activity in the tissues of patients. In fact, the activity of these enzymes, as well as that of glucocerebrosidase and ceramidetrihexosidase, may be increased as much as sixfold over that in brain-tissue samples from human controls. These observations have at least two implications. First, there is a compensatory increase in lysosomal hydrolases in the brains of patients with generalized gangliosidosis (and other lipid-storage diseases, as well), and, second, the major portion of tissue β-galactosidase activity is involved in ganglioside turnover. The latter point suggests that there are aspects of ganglioside turnover, particularly in parenchymal organs, the metabolic relevance of which is still unrecognized.

There is no known therapy for this G_{M1}-gangliosidosis. However, genetic counseling may be indicated because cultured normal amniotic cells have readily demonstrable β-galactosidase activity with artificial substrates.

FARBER'S DISEASE
The signs and symptoms of Farber's disease begin about the third or fourth month after birth with the onset of hoarseness and a brownish desquamating dermatitis.

Later, foam cells infiltrate the bones and joints, and a granulomatous reaction takes place in the lymph nodes, subcutaneous tissues, heart, and lungs. Central nervous system damage causes psychomotor retardation. The disorder is assumed to be transmitted as an autosomal recessive characteristic because it has been reported in both male and female infants.

The metabolic lesion in this disease appears to be a deficiency of ceramidase (Fig. 26-1, Step 5) [36]. The conditions for the assay of maximal ceramidase activity are stringent, and an important lesson can be derived from investigations of this disorder. Whenever a metabolic defect is suspected, great care should be taken to establish, in tissue samples from both animal and human controls, the optimal conditions for the assay of the enzyme in question. It is mandatory that the precise parameters for maximal catalytic activity in lipid storage diseases be determined, especially with regard to the type and concentration of detergents used to solubilize the natural substrates.

WOLMAN'S DISEASE
Probably Wolman's disease is also an autosomal recessive condition since both males and females may be afflicted. The symptoms, which appear a few weeks after birth, include vomiting, diarrhea, and abdominal distention. Hepatosplenomegaly and, usually, enlargement and calcification of the adrenals are also present. There are no signs of nervous system dysfunction. The course of the disease progresses through cachexia, and death usually occurs by the age of six months. Triglycerides and cholesterol esters accumulate in many tissues, particularly the liver, spleen, and adrenals. The enzymatic defect is a deficiency of an acid lipase; the hydrolysis of triglycerides and cholesteryl esters is normal at neutral or alkaline pH [37]. Long-chain fatty esters of p-nitrophenol appear to be useful for the diagnosis of homozygotes and may provide a convenient procedure for the identification of heterozygotes. So far there has been no attempt at enzyme-replacement therapy for this disorder, and treatment has been directed primarily toward amelioration of symptoms.

REFSUM'S SYNDROME
Refsum's syndrome is an autosomal recessive condition characterized by retinitis pigmentosa, peripheral polyneuropathy with both motor and sensory deficiencies, elevated cerebrospinal fluid protein, deafness, anosmia, pupillary abnormalities, ichthyosislike alterations of the skin, and epiphyseal dysplasia. Phytanic acid, a 20-carbon branched-chain fatty acid, accumulates in most tissues; especially large quantities of it are present in the liver and kidneys of the afflicted individuals. The major source of the accumulating substance is dietary phytanic acid and phytol; the latter compound is converted to phytanic acid in the body. The metabolic defect is an inability to oxidize phytanic acid to α-hydroxyphytanate [38]. The hydroxylase catalyzing this reaction is present normally in cultured skin fibroblasts. Homozygotes and heterozygotes can be detected by assaying the activity of the enzyme in extracts of these cells. Patients with the syndrome seem to benefit from restriction of phytol and phytanic acid in their diet.

DISORDER IN GANGLIOSIDE BIOSYNTHESIS

An entirely new type of sphingolipodystrophy is represented in a recently discovered patient whose primary neurological manifestation was severe mental retardation associated with extreme hypomyelination in the CNS. In this patient, the metabolic defect was a failure to synthesize ganglioside G_{M2} from G_{M3} [39]. There was a fourfold accumulation of G_{M3} and a total absence of all higher ganglioside homologs in the brain. This defect is not simply the reverse of Step 9 in Figure 26-2 but is caused by a deficiency of a different enzyme, which catalyzes the transfer of a molecule of N-acetylgalactosamine from UDP-N-acetylgalactosamine to ganglioside G_{M3} [40]. Such a disorder appears to be the prototype for a new field of inherited diseases of ganglioside anabolism.

REFERENCES

1. Brady, R. O., Kanfer, J. N., and Shapiro, D. Metabolism of glucocerebrosides. II. Evidence of an enzymatic deficiency in Gaucher's disease. *Biochem. Biophys. Res. Commun.* 18:221, 1965.
2. Brady, R. O., Kanfer, J. N., Bradley, R. M., and Shapiro, D. Demonstration of a deficiency of glucocerebroside-cleaving enzyme in Gaucher's disease. *J. Clin. Invest.* 45:1112, 1966.
3. Brady, R. O., Johnson, W. G., and Uhlendorf, B. W. Identification of heterozygous carriers of lipid storage diseases. *Am. J. Med.* 51:423, 1971.
4. Beutler, E., and Kuhl, W. Diagnosis of the adult type of Gaucher's disease and its carrier state by demonstration of a deficiency of β-glucosidase activity in peripheral blood leukocytes. *J. Lab. Clin. Med.* 76:747, 1970.
5. Ho, M. W., Seck, J., Schmidt, D., Veath, L., Johnson, W. G., Brady, R. O., and O'Brien, J. S. Adult Gaucher's disease: Kindred studies and demonstration of a deficiency of acid β-glucosidase in cultured fibroblasts. *Am. J. Hum. Genet.* 24:37, 1972.
6. Brady, R. O., Pentchev, P. G., Gal, A. E., Hibbert, S. R., and Dekaban, A. S. Replacement therapy for inherited enzyme deficiency: Use of purified glucocerebrosidase in Gaucher's disease. *N. Engl. J. Med.* 291:989, 1974.
7. Pentchev, P. G., Brady, R. O., Gal, A. E., and Hibbert, S. R. Replacement therapy for inherited enzyme deficiency: Sustained clearance of accumulated glucocerebroside in Gaucher's disease following infusion of purified glucocerebrosidase. *J. Mol. Med.* 1:73, 1975.
8. Schneider, E. L., Ellis, W. G., Brady, R. O., McCulloch, J. R., and Epstein, C. J. Infantile (Type II) Gaucher's disease: In utero diagnosis and fetal pathology. *J. Pediatr.* 81:1134, 1972.
9a. Brady, R. O., Kanfer, J. N., Mock, M. B., and Fredrickson, D. S. The metabolism of sphingomyelin. II. Evidence of an enzymatic deficiency in Niemann-Pick disease. *Proc. Natl. Acad. Sci. U.S.A.* 55:366, 1966.
9b. Gal, A. E., Brady, R. O., Hibbert, S. R., and Pentchev, P. G. A practical chromogenic procedure for the detection of homozygotes and heterozygous carriers of Niemann-Pick disease. *N. Engl. J. Med.* 293:632, 1975.
10. Epstein, C. J., Brady, R. O., Schneider, E. L., Bradley, R. M., and Shapiro, D. In utero diagnosis of Niemann-Pick disease. *Am. J. Hum. Genet.* 23:533, 1971.

11. Suzuki, K., and Suzuki, Y. Globoid cell leukodystrophy (Krabbe's disease): Deficiency of galactocerebroside beta-galactosidase. *Proc. Natl. Acad. Sci. U.S.A.* 66:302, 1970.
12. Radin, N. S., Hof, L., Bradley, R. M., and Brady, R. O. Lactosylceramidase: Comparison with other sphingolipid hydrolases in developing rat brain. *Brain Res.* 14:497, 1969.
13. Suzuki, Y., and Suzuki, K. Krabbe's globoid leukodystrophy: Deficiency of galactocerebrosidase in serum, leukocytes, and fibroblasts. *Science* 171:73, 1971.
14. Suzuki, K., Schneider, E. L., and Epstein, C. J. In utero diagnosis of globoid cell leukodystrophy (Krabbe's disease). *Biochem. Biophys. Res. Commun.* 45:1363, 1972.
15. Arora, R. C., and Radin, N. S. Stimulation in vitro of galactocerebroside galactosidase by *N*-decanoyl-2-amino-2-methylpropanol. *Lipids* 7:56, 1972.
16. Mehl, E., and Jatzkewitz, H. Evidence for a genetic block in metachromatic leukodystrophy (ML). *Biochem. Biophys. Res. Commun.* 19:407, 1965.
17. Mehl, E., and Jatzkewitz, H. Eine Cerebrosidsulfatase aus Schweineniere. *Z. Physiol. Chem.* 339:260, 1964.
18. Jatzkewitz, H., and Stinshoff, K. An activator of cerebroside sulfatase in human normal liver and in cases of congenital metachromatic leukodystrophy. *FEBS. Lett.* 32:129, 1973.
19. Percy, A. K., and Brady, R. O. Metachromatic leukodystrophy: Diagnosis with samples of venous blood. *Science* 161:594, 1968.
20. Kaback, M. M., and Howell, R. R. Infantile metachromatic leukodystrophy: Heterozygote detection in skin fibroblasts and possible applications to intra-uterine diagnosis. *N. Engl. J. Med.* 282:1336, 1970.
21. Austin, J., Balasubramanian, A. S., Patabiraman, T. N., Saraswathi, S., Basu, D., and Bachhawat, B. K. A controlled study of enzymatic activities in three human disorders of glycolipid metabolism. *J. Neurochem.* 10:805, 1963.
22. Stumpf, D., Neuwelt, E., Austin, J., and Kohler, P. Metachromatic leuko-dystrophy (MLD). X. Immunological studies of the abnormal sulfatase A. *Arch. Neurol.* 25:427, 1971.
23a. Dawson, G., and Stein, A. O. Lactosyl ceramidosis: Catabolic defect of glycosphingolipid metabolism. *Science* 170:556, 1970.
23b. Wenger, D., Sattler, M., Clark, C., Tanaka, H., Suzuki, K., and Dawson, G. Lactosyl ceramidosis: Normal activity for two lactosyl ceramide β-galacto-sidases. *Science* 188:1310, 1975.
24. Brady, R. O., Gal, A. E., Bradley, R. M., Martensson, E., and Laster, L. Enzymatic defect in Fabry's disease: Ceramidetrihexosidase deficiency. *N. Engl. J. Med.* 276:1163, 1967.
25. Kint, J. A. Fabry's disease: Alpha-galactosidase deficiency. *Science* 167:1268, 1970.
26. Brady, R. O., Uhlendorf, B. W., and Jacobson, C. B. Fabry's disease: Ante-natal detection. *Science* 172:174, 1971.
27. Brady, R. O., Tallman, J. F., Johnson, W. G., Gal, A. E., Leahy, W. R., Quirk, J. M., and Dekaban, A. S. Replacement therapy for inherited enzyme deficiency: Use of purified ceramidetrihexosidase in Fabry's disease. *N. Engl. J. Med.* 289:9, 1973.
28. Kolodny, E. H., Brady, R. O., and Volk, B. W. Demonstration of an alteration of ganglioside metabolism in Tay-Sachs disease. *Biochem. Biophys. Res. Commun.* 37:526, 1969.

29.   Tallman, J. F., Johnson, W. G., and Brady, R. O. The metabolism of Tay-Sachs ganglioside: Catabolic studies with lysosomal enzymes from normal and Tay-Sachs brain tissue. *J. Clin. Invest.* 51:2339, 1972.

30.   Tallman, J. F., Brady, R. O., Quirk, J. M., Villalba, M., and Gal, A. E. Isolation and relationship of human hexosaminidases. *J. Biol. Chem.* 249:3489, 1974.

31.   Li, Y.-T., Mazzotta, M. Y., Wan, C.-C., Orth, R., and Li, S.-C. Hydrolysis of Tay-Sachs ganglioside by β-hexosaminidase A of human liver and urine. *J. Biol. Chem.* 248:7512, 1973.

32.   Okada, S., and O'Brien, J. S. Tay-Sachs disease: Generalized absence of a beta-hexosaminidase component. *Science* 165:698, 1969.

33.   Tallman, J. F., Brady, R. O., Navon, R., and Padeh, B. Ganglioside catabolism in hexosaminidase A deficient adults. *Nature* 252:254, 1974.

34.   Johnson, W. G., Desnick, R. L., Long, D. M., Sharp, H. L., Krivit, W., Brady, B., and Brady, R. O. Intravenous Injection of Purified Hexosaminidase A into a Patient with Tay-Sachs Disease. In Bergsma, D. (Ed.), *Enzyme Therapy in Genetic Diseases,* Birth Defects Original Article Series. Vol. 9, No. 2. Baltimore: Williams & Wilkins, 1973, p. 120.

*35.   O'Brien, J. S. GM_1 Gangliosidosis. In Stanbury, J. B., Wyngaarden, J. B., and Fredrickson, D. S. (Eds.), *The Metabolic Basis of Inherited Disease,* 3d ed. New York: McGraw-Hill, 1972, p. 639.

36.   Sugita, M., Dulaney, J. T., and Moser, H. W. Ceramidase deficiency in Farber's disease (lipogranulomatosis). *Science* 178:1100, 1972.

*37.   Patrick, A. D., and Lake, B. D. Wolman's Disease. In Hers, H. G., and Van Hoof, F. (Eds.), *Lysosomes and Storage Diseases.* New York: Academic, 1973, p. 453.

*38.   Steinberg, D. Phytanic Acid Storage Disease: Refsum's Syndrome. In Stanbury, J. B., Wyngaarden, J. B., and Fredrickson, D. S. (Eds.), *The Metabolic Basis of Inherited Disease,* 3d ed. New York: McGraw-Hill, 1972, p. 833.

39.   Max, S. R., Maclaren, N. K., Brady, R. O., Bradley, R. M., Rennels, M. B., Tanaka, J., Garcia, J. H., and Cornblath, M. GM_3 (hematoside) sphingolipodystrophy. *N. Engl. J. Med.* 291:929, 1974.

40.   Fishman, P. H., Max, S. R., Tallman, J. F., Brady, R. O., Maclaren, N. K., and Cornblath, M. Deficient ganglioside biosynthesis: A novel human sphingolipidosis. *Science* 187:68, 1975.

*Asterisks denote key references.

Chapter 27

Genetic Mucopolysaccharidoses and Mucolipidoses

Larry J. Shapiro
Elizabeth F. Neufeld

The mucopolysaccharidoses and the mucolipidoses are related groups of inherited human diseases caused by abnormal lysosomal function and storage of mucopolysaccharides. They are rare, affecting collectively perhaps only one in 20,000 liveborn infants [1]. Although some of these disorders have long been recognized by virtue of their dramatic clinical expression, it is only recently that a biochemical explanation for their pathogenesis has been found. As a result of the demonstration that these conditions are caused by a deficiency of various lysosomal hydrolase activities, understanding the normal physiological processes has been enhanced. Practical application of this knowledge has resulted in more accurate diagnosis and counseling, as well as in prenatal diagnosis.

THE BIOCHEMICAL BASIS OF THE MUCOPOLYSACCHARIDOSES

Mucopolysaccharides (also called glycosaminoglycans) are polymeric molecules composed of carbohydrate chains in which amino sugars and uronic acids alternate. The chains are sulfated in varying degrees and are connected to protein backbones by specific linkage regions [2]. These polymers are widely distributed in mammalian tissues and, together with collagen, constitute most of the intercellular matrix. Several of the mucopolysaccharides are major constituents of cartilage, skin, and blood vessels, whereas one, heparan sulfate, is a small but rapidly turning-over component of all cell membranes [3]. No doubt such ubiquitous molecules are involved in the pathogenesis of a number of disease states, but only inherited disorders of mucopolysaccharide catabolism will be considered here.

The disorders now known as the mucopolysaccharidoses were clearly described clinically in the first half of this century [1] but were considered derangements of lipid metabolism ("lipochondrodystrophies"). It was only after Brante [4], Brown [5], and Meyer et al. [6] identified the stored material as mucopolysaccharide rather than lipid that understanding the biochemical basis of these conditions became possible. The discovery by Dorfman and Lorincz [7] of mucopolysacchariduria allowed relatively easy identification of affected patients. From careful comparison of clinical features, mode of inheritance, and chemistry of excreted mucopolysaccharides, McKusick [8] proposed a systematic classification that was adopted widely. This classification was subsequently revised as information about the basic defect in each disorder became available. The revised classification of McKusick [1] will be used in this chapter. Each mucopolysaccharidosis (MPS) is designated by an eponym and a Roman numeral followed, where further subdivision is warranted, by a capital letter, e.g., the Hurler syndrome, MPS IH.

Mucopolysaccharidoses as Catabolic Disorders

On the basis of pathological findings in the Hurler syndrome (the prototype of mucopolysaccharidoses), van Hoof and Hers [9] suggested that the disease might be a lysosomal-storage disorder. Danes and Bearn [10] found that fibroblasts derived from mucopolysaccharidosis patients stored mucopolysaccharide in culture. The use of radioactive sulfate to measure the fate of mucopolysaccharide in such cultured fibroblasts showed that the storage was a result of faulty degradation, rather than of an increased rate of synthesis, decreased secretion, or abnormal structure of the accumulated polymer [11, 12].

Through subsequent work, a specific deficiency of a lysosomal hydrolase has been described for each of the classical phenotypic syndromes [13]. In each case, the enzyme is one involved in the degradation of the polymers that accumulate and are excreted in the genetic mucopolysaccharidoses: dermatan sulfate, heparan sulfate, or both (Figs. 27-1 and 27-2). The best interpretation of available data is that the normal degradation of dermatan sulfate and heparan sulfate proceeds unidirectionally from the nonreducing end of the carbohydrate chain by the sequential actions of lysosomal exoglycosidases and exosulfatases. The absence of any one of these enzymes results in a block in catabolism, although a limited degradation by endoglycosidases may occur in certain tissues. Each of the inherited disorders of mucopolysaccharide metabolism corresponds to the absence of activity of one of these enzymes. The relationship is analogous to that among the disorders of ganglioside catabolism described in Chapter 26. It is clear that there are still several linkages in the known structure of these mucopolysaccharides for which as yet undiscovered hydrolases probably exist, and it is reasonable to speculate that in the future defects of these enzymes will be recognized in the patients with "unclassifiable" mucopolysaccharidoses.

It is important to note that the structures of dermatan and heparan sulfate given in Figures 27-1 and 27-2 represent all the sugar and sulfate residues known to occur in these polymers; the polymers are not, however, regular, repeating units of the sequence shown, as there is considerable variation along the chain.

The spectrum of clinical phenotypes that can be associated with each enzymatic error is very great. On the other hand, as might be anticipated with a multienzyme pathway, patients with different enzymatic lesions may appear phenotypically similar to the clinician. This complexity requires an integrated biochemical, clinical, and genetic approach to be used in order to give accurate prognostic information and counseling.

Alpha-L-Iduronidase Deficiency

Patients deficient in the enzyme α-L-iduronidase [14, 15] are variably affected. At one end of the spectrum are those with classical Hurler syndrome (MPS IH) [1, 2, 16]. The disorder has its apparent clinical onset in infancy, although pathological and biochemical manifestations may be recognized much earlier. The patients have diminished linear growth after approximately the first year, resulting in significantly shortened stature, and have severe psychomotor retardation. The characteristic facial appearance of these patients, along with the excretion of dermatan sulfate

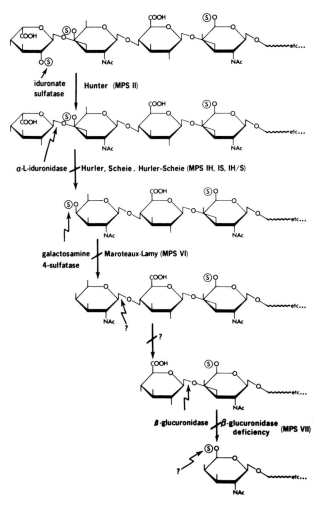

Figure 27-1
Degradation of dermatan sulfate. This is a schematic representation of the structure to indicate every known linkage. Known hydrolytic enzymes and their corresponding deficiency diseases are indicated. Anticipated enzyme activities, which should be required for degradation of the polymer but of which the existence has not been demonstrated, are indicated by question marks.

and heparan sulfate (usually in a 2:1 ratio) in the urine, are major criteria for establishing the clinical diagnosis. Hepatosplenomegaly is prominent, as is opacification of the corneas. There is often retinal degeneration as well as optic atrophy, which may be the result of increased intracranial pressure. Characteristic skeletal manifestations are noted clinically and radiographically. These include "dysostosis multiplex," kyphosis with a hump deformity, and stiff joints. Deafness and hydrocephalus occur frequently, the latter probably due to meningeal infiltration with mucopolysaccharide. Mucopolysaccharide deposition in walls of blood vessels and heart valves frequently leads

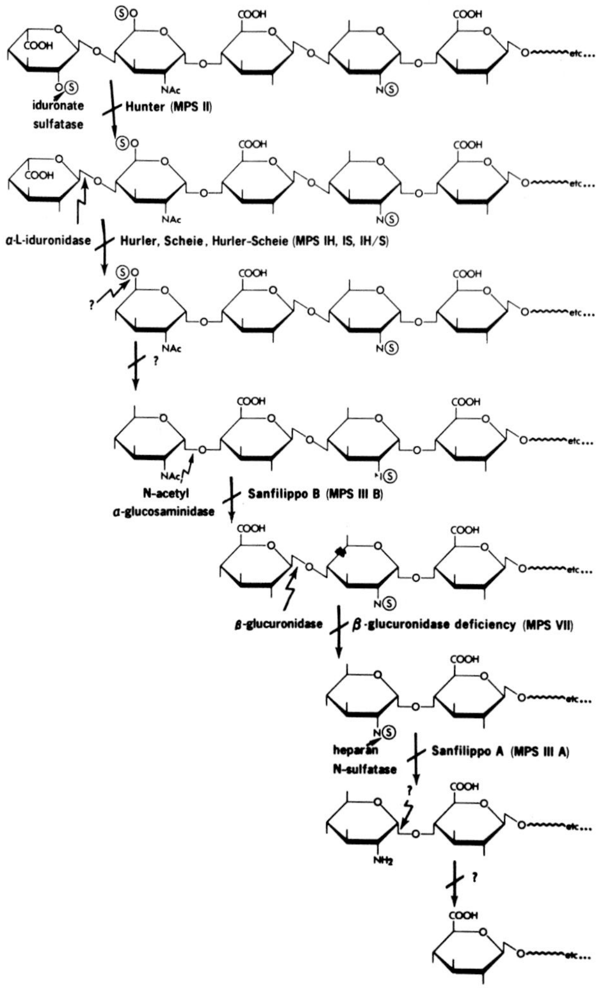

Figure 27-2
Degradation of heparan sulfate. Schematic representation is analogous to that in Figure 27-1.

to cardiac complications. Congestive heart failure and repeated pulmonary infections are common causes of death, usually before the age of 10.

Patients who had been classified clinically as having a distinct disorder, the Scheie syndrome (MPS IS), have also been found to be deficient in α-L-iduronidase activity. The Scheie and Hurler syndromes represent two disorders within the same clinical spectrum. The Scheie syndrome is compatible with a normal life span and normal intelligence. Stature is near normal. Clinical problems include clouding of the cornea and retinitis pigmentosa, aortic valve involvement, and hand deformities associated with median nerve entrapment.

Recently, attention has been drawn to a number of patients who are also devoid of α-L-iduronidase activity and whose phenotype is intermediate between the Hurler and Scheie syndromes [17]. The onset of clinical symptoms occurs later than in the Hurler syndromes and development is less affected. Somatic stigmata are, however, considerably more severe than in the Scheie syndrome, and a number of these patients have developed hydrocephalus with subsequent neurological deterioration. The Scheie patients and those of intermediate type also have mucopolysacchariduria, the quantity or distribution of which bears no direct relationship to the severity of the clinical disease.

The relationships among the mutations in these diseases, all of which are inherited in autosomal recessive fashion, is uncertain. It has been suggested that the Hurler syndrome and the Scheie syndrome are due to homozygosity of two different sets of alleles at the same genetic locus and that the intermediate phenotype, or Hurler-Scheie genetic compound (MPS IH/S), carries a Scheie allele and a Hurler allele at this locus [17]. This would be analogous to hemoglobinopathies in which one can identify individuals with two different mutant alleles at the same locus (e.g., hemoglobin S-C disease). Alternative explanations are possible, however, and there may well be more than two mutant alleles involved. Resolution of this question must await development of techniques for characterizing the molecular structure of the mutant iduronidases assumed to be present in these disorders.

Iduronate Sulfatase Deficiency

The gene corresponding to iduronate sulfatase [18], or to some function essential for its expression, resides on the X-chromosome. Deficiency of this enzyme is inherited as a sex-linked recessive trait, the clinical expression of which is the Hunter syndrome (MPS II), described below [1, 2]. Heterozygous females have been, with one known exception [19], clinically normal, although clones of deficient cells can be identified in fibroblast cultures derived from these individuals, as predicted by the Lyon hypothesis [20].

The clinical spectrum encompassed by iduronate sulfatase deficiency is again very broad. Mild and severe variants have been described; because they occur consistently within families having several affected males, they are thought to constitute distinct genetic entities [16]. Once again, this may be caused by different mutational alterations in the same gene. Severely affected individuals may be similar in appearance to patients with the Hurler syndrome. At times, however, the somatic manifestations may be milder, and central nervous system dysfunction may predominate to the extent of causing confusion with Sanfilippo patients. Hunter syndrome is distinguished from Hurler syndrome by the absence of corneal clouding, although retinal lesions do occur. Characteristic nodular infiltrations in the skin are frequently seen. The milder variant may have normal or near normal intelligence, and live well into adulthood; however, only one Hunter patient is known to have reproduced. Dermatan sulfate and heparan sulfate are excreted in the urine, usually in equal amounts though an occasional patient will excrete heparan sulfate only.

Heparan *N*-Sulfatase and
N-acetyl alpha-Glucosaminidase Deficiencies

Heparan *N*-sulfatase [21, 22] and *N*-acetyl-α-glucosaminidase [23, 24] are two of the enzymes involved in the degradation of heparan sulfate. Deficiency of these activities results in the Sanfilippo A (MPS IIIA) or the Sanfilippo B (MPS IIIB) syndromes, respectively [1]. Heparan sulfate is the only mucopolysaccharide excreted in these disorders. Heparansulfaturia may be greatly underestimated for methodological reasons, however, and even its absence should not exclude a probable diagnosis in the context of an appropriate clinical situation. The two disorders are indistinguishable on clinical grounds but are distinguishable by enzyme assays. In general, somatic manifestations are milder than in the other mucopolysaccharidoses, but, again, there is some overlap. The corneas are clear, and organomegaly and cardiac manifestations are variable. Early psychomotor development is often normal, but later in childhood progressive loss of intellectual and motor skills results in severe impairment and eventual loss of most cortical functions. Seizures are common.

Both of these conditions are inherited in an autosomal recessive fashion. Mild variants have not yet been identified.

Galactosamine 4-Sulfatase Deficiency

Galactosamine 4-sulfatase activity may be a property of a well-known lysosomal enzyme, arylsulfatase B. The physiological role of this hydrolase is the degradation of appropriate residues of dermatan sulfate [25–28], and the clinical sequela of its deficiency is the Maroteaux-Lamy syndrome (MPS VI) [1, 16]. As in MPS I and MPS II, the manifestations of the Maroteaux-Lamy syndrome are variable. The index case [1] most closely resembles the Hurler-Scheie compound; more severe variants resemble the Hurler syndrome, whereas milder variants most closely resemble in their clinical features patients with the Scheie syndrome or those with mucolipidosis III (see below). Characteristic large white-cell inclusions are said to be striking in this disease, in contrast to the less impressive bodies seen in cells of patients with other mucopolysaccharidoses. Only dermatan sulfate appears in the urine of affected individuals because the chemical linkage involved is unique to this polymer. The disease is inherited in an autosomal recessive fashion.

Beta-Glucuronidase Deficiency

β-Glucuronidase is a well-known lysosomal enzyme, and its involvement in the catabolism of mucopolysaccharides only recently came to light with the description of a patient deficient in this activity [29, 43]. The index case had unusual facies, skeletal deformities, hepatosplenomegaly, and delayed development. A few more patients have been described since then, with widely different manifestations [30].

This disorder (MPS VII) is of some historic interest because it was the first mucopolysaccharidosis to have its enzymatic basis elucidated and its autosomal recessive mode of inheritance established on the basis of enzyme levels in close relatives. The mucopolysacchariduria is puzzling: Dermatan sulfate and heparan sulfate excretion have been reported, although one would predict from chemical consideration that fragments of chondroitin sulfate and hyaluronic acid should likewise be excreted.

The Morquio Syndrome

The Morquio syndrome (MPS IV) is diagnosed as a genetic mucopolysaccharidosis on the basis of pathologic findings, corneal clouding, and mucopolysacchariduria [1, 16]. It differs from the other mucopolysaccharidoses in that keratan sulfate is the polymer excreted in the urine and that abnormal metabolism of sulfated mucopolysaccharide is not readily demonstrated in cultured fibroblasts. This may be due to the absence of keratan sulfate synthesis by skin fibroblasts. Recent evidence has shown a deficiency in a specific sulfatase in fibroblasts from Morquio patients [31].

Clinically, patients have a spondyloepiphyseal dysplasia with pronounced short stature. As cartilage and cornea are the only tissues that produce keratan sulfate, it might be expected that these would be the principal sites of pathological involvement, as a result of disturbed metabolism of this polymer. Pectus carinatum deformity and marked kyphoscoliosis are prominent. Neurosensory deafness is reported commonly. A major complication is spinal cord and medullary compression, caused by odontoid hypoplasia and subsequent atlanto-axial subluxation. Intelligence is generally normal.

The Morquio syndrome is inherited in an autosomal recessive manner. Because keratan sulfate contains a number of unique linkages, one should anticipate genetic heterogeneity in this disease.

THE BIOCHEMICAL BASIS OF THE MUCOLIPIDOSES

A number of patients have been described with diseases that clinically resemble the mucopolysaccharidoses, but without mucopolysacchariduria. A superficial similarity to disorders of sphingolipid metabolism as well has led to their classification as "mucolipidoses" [32]. Two of these conditions will be considered: mucolipidosis II (inclusion cell, or I-cell, disease) and mucolipidosis III (pseudo-Hurler polydystrophy). Both are inherited in an autosomal recessive manner.

I-cell disease derives its name from the grossly enlarged lysosomes within the fibroblasts of affected patients. By phase-contrast microscopy, these enlarged lysosomes are visible as dark inclusions. The clinical phenotype is similar to the Hurler syndrome, although even more severe and with obvious manifestations in the neonatal period [1]. Skeletal dysplasia and hypertrophy of the gums are striking. Psychomotor retardation is severe. Most patients die in early childhood.

Mucolipidosis III is milder and evolves more slowly [1]. Corneal clouding and skeletal radiographic changes are milder than in the Hurler syndrome, and intelligence is normal or minimally impaired. Stature is short. On clinical grounds, the disease may be confused with the Maroteaux-Lamy or Hurler-Scheie syndrome.

Fibroblasts cultured from skin of these two kinds of patients are deficient in a whole panel of lysosomal hydrolases, as shown in Table 27-1 [33–36]. Because of this multiple enzyme deficiency, the fibroblasts accumulate a variety of undegraded material. The interesting observation has been made that the culture fluid surrounding these fibroblasts, as well as serum, urine, and cerebrospinal fluid from the affected individuals, contains markedly elevated levels of some of the enzymes that are deficient intracellularly [35–37]. Further investigations suggest that, in mucolipidoses II and III, a single pleiotropic gene defect results in the faulty packing of a

Table 27-1. Enzyme Deficiences in Fibroblasts of Patients with Mucolipidoses II

	Normal	Partly Deficient[a]	Profoundly Deficient[a]
Enzymes degrading mucopolysaccharides			
α-L-Iduronidase			X
Iduronate sulfatase		X	
β-Glucuronidase		X	
Heparan N-sulfatase		X	
Enzymes degrading glycolipids and glycoproteins			
α-Galactosidase		X	
β-Galactosidase			X
β-Glucosidase	X		
β-Hexosaminidase		X	
α-L-Fucosidase			X
α-Mannosidase		X	
Arylsulfatase A			X
Other lysosomal enzymes			
Acid phosphatase	X		

[a]*Profoundly deficient* indicates activity between 0% and 10% of the normal level; *partly deficient* indicates 10% to 40% of the normal level. Similar findings are obtained in fibroblasts of patients with mucolipidosis III. In contrast to the enzyme deficiencies of mucopolysaccharidoses, the defect of the mucolipidoses is not expressed in all cells. Liver and leukocytes have normal enzyme activity.

number of hydrolases [38]. A special structural feature, or "recognition marker," is apparently required for the appropriate localization of certain hydrolases within lysosomes [39]. The two diseases may arise from separate defects in the posttranscriptional generation of such recognition markers, so that the hydrolases do not become fixed in lysosomes.

CLINICAL IMPLICATIONS

The discovery of the biochemical defect in each of the mucopolysaccharidoses and in mucolipidoses II and III permits the dissemination of accurate diagnostic and counseling information. Prenatal diagnosis with selective abortion can be offered to families desiring this option [40]. Total prevention of these disorders by amniocentesis and selective abortion is not possible, however, because the families who risk producing affected offspring can be ascertained only through the production of a previously affected child (except in the case of the X-linked Hunter syndrome). Heterozygote detection for the autosomally recessive mucopolysaccharidoses and mucolipidoses, analogous to that established for Tay-Sachs disease [41], does not appear likely. This is because of the large variation of normal values of the relevant lysosomal enzymes and the lack of a defined and limited population suitable for screening.

A significant problem is heterozygote detection in female relatives of Hunter patients. Although demonstration of deficient and nondeficient fibroblast clones, in accord with the Lyon hypothesis, is proof of the carrier state, the absence of abnormal clones does not guarantee normality; the abnormal cell type might have been selectively lost in vivo as well as during culture [42]. At present, potential carriers of the Hunter syndrome must be counseled on the basis of statistical risks.

Hope of replacement therapy for mucopolysaccharidosis patients has been encouraged by the ease with which their fibroblasts can be cured in vitro when the missing enzyme is added to the medium [15, 43]. In the absence of highly purified enzymes for administration to patients, whole or fractionated plasma as well as leukocytes has been used. Though early experiments were considered promising [44, 45], continued evaluation has resulted in a much less optimistic outlook [46, 47]. The therapeutic failures are attributable in part to the low levels of the relevant enzymes in plasma and other blood fractions used. An additional difficulty lies in the requirement for enzymes to have a special chemical feature ("recognition marker," see above) in order to enter cells efficiently and lodge in lysosomes [39]. Plasma enzymes may lack this marker [48]. Infusion of purified enzymes possessing the marker to ensure high uptake in the target tissue, perhaps modified for greater stability in vivo, will no doubt receive attention in the near future.

CONCLUSION

The genetic mucopolysaccharidoses are caused by inherited deficiencies of lysosomal glycosidases or sulfatases required for mucopolysaccharide catabolism. Through the study of these disease states, the normal degradative pathways have been elucidated. Further investigation of the defect in mucolipidoses should yield information about the processes by which the lysosomal enzymes arrive at their proper intracellular location. From a clinical standpoint, the diversity of phenotypes of individuals sharing a common enzymatic deficiency still requires explanation, and patients with as yet undescribed defects of mucopolysaccharide catabolism must be expected.

REFERENCES

1. McKusick, V. A. *Heritable Disorders of Connective Tissue,* 4th ed. St. Louis: Mosby, 1972, p. 521.
*2. Dorfman, A., and Matalon, R. The Metabolic Basis of Inherited Disease. In Stanbury, J. B., Wyngaarden, J. B., and Fredrickson, D. S. (Eds.), *The Mucopolysaccharidoses,* 3d ed. New York: McGraw-Hill, 1972, p. 1218.
3. Kraemer, P. M. Heparan sulfates of cultured cells. *Biochemistry* 10:1445, 1971.
4. Brante, G. Gargoylism: A mucopolysaccharidosis. *Scand. J. Clin. Lab. Invest.* 4:43, 1952.
5. Brown, D. H. Tissue storage of mucopolysaccharides in Hurler-Pfaundler's disease. *Proc. Natl. Acad. Sci. U.S.A.* 43:738, 1957.

*Asterisks denote key references.

6. Meyer, K., Hoffman, P., Linker, A., Grumbach, M., and Sampson, P. Sulfated mucopolysaccharides of urine and organs in gargoylism (Hurler's syndrome). *Proc. Soc. Exp. Biol. Med.* 102:587, 1959.

7. Dorfman, A., and Lorincz, A. E. Occurrence of urinary acid mucopolysaccharide in the Hurler syndrome. *Proc. Natl. Acad. Sci. U.S.A.* 43:443, 1957.

8. McKusick, V. A. Heritable Disorders of Connective Tissue. In Stanbury, J. B., Wyngaarden, J. B., and Fredrickson, D. S. (Eds.), *The Mucopolysaccharidoses,* 3d ed. St. Louis: Mosby, 1966, p. 325.

9. Van Hoff, F., and Hers, H. G. L'ultrastructure des cellules hépatiques dans la maladie de Hurler (Gargoylisme). *C. R. Acad. Sci. [D.]* (Paris) 259:1281, 1964.

10. Danes, B. S., and Bearn, A. G. Hurler's syndrome; demonstration of an inherited disorder of connective tissue in cell culture. *Science* 149:987, 1965.

11. Fratantoni, J. C., Hall, C. W., and Neufeld, E. F. The defect in Hurler's and Hunter's syndromes: Faulty degradation of mucopolysaccharide. *Proc. Natl. Acad. Sci. U.S.A.* 60:699, 1968.

*12. Neufeld, E. F., and Cantz, M. The Mucopolysaccharidoses Studied in Cell Culture. In Hers, H. G., and Van Hoof, F. (Eds.), *Lysosomes and Storage Diseases.* New York: Academic, 1973, p. 262.

*13. Neufeld, E. F. The biochemical basis for mucopolysaccharidoses and mucolipidoses. *Prog. Med. Genet.* 10:81, 1974.

14. Matalon, R., and Dorfman, A. Hurler's syndrome, an α-L-iduronidase deficiency. *Biochem. Biophys. Res. Commun.* 47:959, 1972.

15. Bach, G., Friedman, R., Weissmann, B., and Neufeld, E. F. The defect in the Hurler and Scheie syndromes: Deficiency of α-L-iduronidase. *Proc. Natl. Acad. Sci. U.S.A.* 69:2048, 1972.

*16. Spranger, J. W. The systemic mucopolysaccharidoses. *Ergebn. Inn. Med. Kinderheilk.* 32:165, 1972.

17. McKusick, V. A., Howell, R. R., Hussels, I. E., Neufeld, E. F., and Stevenson, R. Allelism, non-allelism and genetic compounds among the mucopolysaccharidoses. *Lancet* 1:993, 1972.

18. Bach, G., Eisenberg, F., Jr., Cantz, M., and Neufeld, E. F. The defect in the Hunter syndrome: Deficiency of sulfoiduronate sulfatase. *Proc. Natl. Acad. Sci. U.S.A.* 70:2134, 1973.

19. Milunsky, A., and Neufeld, E. F. The Hunter syndrome in a 46 XX girl. *N. Engl. J. Med.* 288:106, 1973.

20. Danes, B. S., and Bearn, A. G. Hurler's syndrome: A genetic study of clones in cell culture with particular reference to the Lyon hypothesis. *J. Exp. Med.* 126:509, 1967.

21. Kresse, H. Mucopolysaccharidosis III A (Sanfilippo A disease): Deficiency of a heparin sulfamidase in skin fibroblasts and leukocytes. *Biochem. Biophys. Res. Commun.* 54:1111, 1973.

22. Matalon, R., and Dorfman, A. Sanfilippo A syndrome: Sulfamidase deficiency in cultured skin fibroblasts and liver. *J. Clin. Invest.* 54:907, 1974.

23. O'Brien, J. S. Sanfilippo syndrome: Profound deficiency of alpha-acetylglucosaminidase activity in organs and skin fibroblasts from type B patients. *Proc. Natl. Acad. Sci. U.S.A.* 69:1720, 1972.

24. von Figura, K., and Kresse, H. The Sanfilippo B corrective factor: An N-acetyl-α-glucosaminidase. *Biochem. Biophys. Res. Commun.* 48:262, 1972.

25. Stumpf, D. A., Austin, J. H., Crocker, A. C., and La France, M. Mucopolysaccharidosis type VI (Maroteaux-Lamy syndrome). I. Sulfatase B deficiency in tissues. *Am. J. Dis. Child.* 126:747, 1973.

26. O'Brien, J. F., Cantz, M., and Spranger, J. W. Maroteaux-Lamy disease (muco-polysaccharidosis VI), subtype A: Deficiency of a *N*-acetylgalactosamine-4-sulfatase. *Biochem. Biophys. Res. Commun.* 60:1170, 1974.
27. Matalon, R., Arbogast, B., and Dorfman, A. Deficiency of chondroitin sulfate *N*-acetylgalactosamine-4-sulfatase in Maroteaux-Lamy syndrome. *Biochem. Biophys. Res. Commun.* 61:1450, 1974.
28. Stevens, R. L., Fluharty, A. G., Fung, D., Peak, S., and Kihara, H. UDP-*N*-acetyl-galactosamine-4-sulfate sulfohydrolase activity of human arylsulfatase B. *Fed. Proc.* 23:635, 1975.
29. Sly, W. S., Quinton, B. A., McAllister, W. H., and Rimoin, D. L. β-Glucuronidase deficiency: Report of clinical, radiologic, and biochemical features of a new mucopolysaccharidosis. *J. Pediatr.* 82:249, 1973.
30. Beaudet, A. C. DiFerrante, N., Ferry, G., Nichols, B. L., and Mullins, C. E. Variation in the phenotypic expression of β-glucuronidase deficiency. *J. Pediatr.* 86:388, 1975.
31. Matalon, R., Arbogast, B., Justice, P., Brandt, I., and Dorfman, A. Morquio's syndrome: Deficiency of a chondroitin sulfate *N*-acetylhexosamine sulfate sulfatase. *Biochem. Biophys. Res. Commun.* 61:759, 1974.
*32. Spranger, J. W., and Wiedemann, H. R. The genetic mucolipidoses: Diagnosis and differential diagnosis. *Humangenetik* 9:113, 1970.
33. Lightbody, J., Wiesmann, U., Hadorn, B., and Herschkowitz, N. I-cell disease: Multiple lysosomal enzyme defect. *Lancet* 1:451, 1971.
34. Leroy, J. G., Ho, M. W., MacBrinn, M. C., Zielke, K., Jacobs, J., and O'Brien, J. S. I-cell disease: Biochemical studies. *Pediatr. Res.* 6:752, 1972.
35. Glaser, J. H., McAllister, W. H., and Sly, W. S. Genetic heterogeneity in multiple lysosomal hydrolase deficiency. *J. Pediatr.* 85:192, 1974.
36. Thomas, G. J., Taylor, H. A., Reynolds, L. W., and Miller, C. S. Mucolipidosis III (pseudoHurler polydystrophy): Multiple lysosomal enzyme abnormalities in serum and cultured fibroblasts. *Pediatr. Res.* 7:751, 1973.
37. Wiesmann, U., Vasella, F., and Herschkowitz, N. I-cell disease: Leakage of lysosomal enzymes into extracellular fluids. *N. Engl. J. Med.* 285:1090, 1971.
38. Hickman, S., and Neufeld, E. F. A hypothesis for I-cell disease: Defective hydrolases that do not enter lysosomes. *Biochem. Biophys. Res. Commun.* 49:992, 1972.
39. Hickman, S., Shapiro, L. J., and Neufeld, E. F. A recognition marker required for uptake of a lysosomal enzyme by cultured fibroblasts. *Biochem. Biophys. Res. Commun.* 57:55, 1974.
40. Fratantoni, J. C., Neufeld, E. F., Uhlendorf, B. W., and Jacobson, C. B. Intra-uterine diagnosis of the Hurler and Hunter syndromes. *N. Engl. J. Med.* 280:686, 1969.
41. Kaback, M. M., Zeiger, R. S., Reynolds, L. W., and Sonneborn, M. Approaches to the control and prevention of Tay-Sachs disease. *Prog. Med. Genet.* 10:103, 1974.
42. Nyhan, W. L., Bakay, B., Connor, J. D., Marks, J. F., and Keele, D. Hemizygous expression of glucose 6-phosphate dehydrogenase in erythrocytes of hetero-zygotes for the Lesch-Nyhan syndrome. *Proc. Natl. Acad. Sci. U.S.A.* 65:214, 1970.
43. Hall, C. W., Cantz, M., and Neufeld, E. F. A β-glucuronidase deficiency muco-polysaccharidosis: Studies in cultured fibroblasts. *Arch. Biochem. Biophys.* 155:32, 1973.

44. DiFerrante, N., Nichols, B. G., Donnelly, P. V., Neri, G., Hrgovčic, R., and Berglund, R. K. Induced degradation of glycosaminoglycans in Hurler's and Hunter's syndromes by plasma infusion. *Proc. Natl. Acad. Sci. U.S.A.* 68:303, 1971.

45. Knudson, A. G., DiFerrante, N., and Curtis, J. E. Effect of leukocyte transfusion in a child with type II mucopolysaccharidosis. *Proc. Natl. Acad. Sci. U.S.A.* 68:1738, 1971.

*46. Crocker, A. C. Plasma infusion therapy for Hurler's syndrome. *Pediatrics* 50:683, 1972.

47. Moser, H. W., O'Brien, J. S., Atkins, L., Fuller, T. C., Kliman, A., Janowska, S., Russell, P. S., Bartsocas, C. S., Cosimi, B., and Dulaney, J. T. Infusion of normal HL-A identical leukocytes in Sanfilippo disease type B. *Arch. Neurol.* 31:329, 1974.

48. Brot, F. E., Glaser, J. H., Roozen, K. J., and Sly, W. S. In vitro correction of deficient human fibroblasts by β-glucuronidase from different human sources. *Biochem. Biophys. Res. Commun.* 57:1, 1974.

Chapter 28

Diseases Involving Myelin

Pierre Morell
Murray B. Bornstein
Cedric S. Raine

The title of this chapter has been selected to emphasize that myelin cannot in any way be considered an isolated entity. For its maintenance in the peripheral nervous system (PNS), myelin depends upon the normal functioning of a Schwann cell and, in the central nervous system (CNS), an oligodendrocyte. The integrity of the myelin sheaths is also dependent upon the viability of the axons that they ensheath and the neuronal cell bodies from which the axons emanate. It is well known, for example, that neuronal death inevitably leads to subsequent degeneration of axons and myelin. Although many conditions are recognized in which preferential loss of myelin occurs, damage to myelin is a common consequence of a multitude of unrelated pathological stigmata (e.g., genetically determined disorders, viral infection, toxic agents, neoplasia, trauma, and anoxia) that happen to affect myelin or myelin-supporting cells either selectively or initially.

CLASSIFICATION

Many of the diseases involving demyelination can be subdivided into primary and secondary categories on the basis of morphological observations. Primary demyelination involves the early destruction of myelin with relative sparing of axons. Subsequently, other structures may be affected. Secondary demyelination includes those disorders in which myelin is involved after damage to neurons or axons. The classification scheme detailed in this chapter is based on etiology as well as comparative neuropathology. Most (but not all) of the disorders in the first three categories are primary demyelination disorders whereas the fourth includes disorders involving secondary demyelination. All four will be discussed from both the pathological and the biochemical standpoints.

ACQUIRED INFLAMMATORY DEMYELINATING DISEASES

Probably the most complete outline of human diseases in this category has been given by Adams and Sidman [1]. In contrast to the metabolic disorders of myelin, nervous system damage in this group of demyelinating diseases is more specifically directed against myelin, and there is relatively little damage to other parenchymal elements. With one exception, lesion formation appears to be related to an immunological response. It is not always clear, however, whether the immunological activity is autoimmune in nature or whether it is related primarily to an antecedent viral infection. Nor is the extent of damage directly ascribable to putative viral agents known at the present time.

Pathology of the Acquired Inflammatory Demyelinating Diseases
The following pathological criteria apply to virtually all the acquired demyelinating diseases with one notable exception, progressive multifocal leukoencephalopathy, in which lesions are noninflammatory. The extent to which criteria (a) through (d), below, are fulfilled depends on the phase of the disease:

a. perivenular demyelination;
b. perivenular inflammation;
c. relative sparing of axons;
d. macrophage activity in lesions;
e. disseminated lesions;
f. pia-arachnoid inflammation;
g. sudanophilic deposits, presumably products of myelin degradation.

Of the many human demyelinating diseases encountered, the following serve as type examples:

1. *Multiple Sclerosis (MS)*, the paradigm of the demyelinating diseases [2] and the most common, typically showing a chronic relapsing course. Large plaques of primary demyelination, frequently in a periventricular distribution, are typical. Some variants exist by pathological criteria: for example, Devic's disease. The active lesions are inflammatory and are believed to have an immunogenical basis which may be linked to an antecedent viral infection. Genetic factors are considered possible but not proved. A distinction is often made between the chronic and acute forms of MS. The so-called acute form is rarer, presents a more fulminant picture, and is of shorter duration. Lesions are more inflammatory than in chronic MS and a viral infection is also suspected.

2. *Acute Disseminated Encephalomyelitis,* also called postinfectious or postimmunization encephalitis. This represents a group of disorders of either viral, immunological, or mixed viral-immunological etiology. The condition is most commonly related to a spontaneous viral infection (e.g., measles, smallpox, or chicken pox) [2].

3. *Acute Hemorrhagic Leukoencephalopathy,* a rare condition in which demyelination is accompanied by multiple and focal hemorrhages and inflammation. A viral infection is implicated, and the disease is considered by some [3] to represent a hyperacute form of acute disseminated encephalomyelitis.

4. *Progressive Multifocal Leukoencephalopathy (PML)*, a rare demyelinating disease that is usually associated with disorders of the reticuloendothelial system, neoplasias, and immunosuppressive therapy [4]. Lesions are noninflammatory and are believed to be etiologically related to a papovavirus infection.

5. *Landry-Guillain-Barré Syndrome,* an inflammatory disease of the peripheral nervous system that is encountered in many forms, ranging from acute-monophasic to chronic-remitting [5].

A number of spontaneously occurring animal models (see [6] for a more complete listing and description) are recognized in which inflammatory changes, primarily demyelination and, frequently, a viral infection, are the major components of the disease process.

6. *Canine Distemper Encephalomyelitis,* a viral-induced, CNS demyelinating disease in dogs in which lesions show a strong inflammatory response [6]. Some similarities to acute disseminated encephalomyelitis exist.

7. *Coon-Hound Paralysis*, claimed by some to be the animal counterpart of the Landry-Guillain-Barré syndrome [7].
8. *Marek's Disease*, a lymphomatous condition in chickens associated with demyelination in the PNS and caused by herpes virus [6].
9. *Visna*, a viral-induced demyelinating disease of sheep, claimed by some to be of relevance to the study of MS. Other workers consider visna a degenerative CNS condition, not specifically demyelinative in type [6].

Certain experimental demyelinating models in animals are of considerable interest to neurochemists. Because the disease process can be initiated when desired and the course of the disease is known, experiments to elucidate the metabolic events associated with the disease process are possible.

10. *Experimental Allergic Encephalomyelitis (EAE)*, an autoimmune demyelinating disease inducible in most animal species by the inoculation of CNS material. The disease is usually acute and has certain morphological features reminiscent of acute MS and human disseminated encephalomyelitis [8].
11. *Experimental Allergic Neuritis (EAN)*, the PNS counterpart of EAE, caused by sensitization against PNS material. This model is relevant to the study of Landry-Guillain-Barré syndrome (see p. 590).
12. *Mouse Hepatitis Virus Encephalomyelitis*, caused by a neurotropic corona virus strain. Considered to be a model for the study of acute disseminated encephalomyelitis and progressive multifocal leukoencephalopathy [9].

Biochemistry of Demyelinating Diseases — General Comments

The analytical findings of whole brain (or white matter, which usually shows more pronounced symptoms) are usually similar in demyelinating diseases, regardless of the etiology of the disorder. The water content of white matter increases markedly because myelin (which is relatively rich in solids) is lost, leaving behind relatively more hydrated material. Other changes include a decrease of myelin constituents — especially proteolipid protein, cerebroside, ethanolamine phosphatides, and cholesterol — and the presence of cholesterol esters. Table 28-1 details the results from a genetically determined metabolic disorder (Schilder's disease) [10] and demyelination secondary to a viral infection, subacute sclerosing panencephalitis (SSPE) [11]. These changes can often be explained by the breakdown and gradual loss of myelin and its replacement by tissue components (e.g., extracellular fluid, astrocytes, inflammatory cells) which are more hydrated, relatively lipid poor, and free of myelin-specific constituents. The magnitude of these changes varies considerably from specimen to specimen in the same disease, variations undoubtedly related to the severity, duration, and activity of the disease process.

Biochemistry of the Acquired Inflammatory Demyelinating Diseases

Neurochemical investigation of the acquired inflammatory demyelinating diseases must be described in the context of both the concomitant immunological response and the pathophysiological response to the disease. These primary demyelinating disorders (with the exception of progressive multifocal leukoencephalopathy) are characterized pathologically by perivascular cuffing early in the disease process.

Table 28-1. Human White Matter Composition in Two Diseases Compared with Controls

	SSPE[a]		Schilder's[b]	
	Control	Patient	Control	Patient
Water[c]	72	88.5	72.5	84.3
Proteolipid[d]			8.2	1.4
Total lipid[d]	60	23.4	56.3	34.7
Cholesterol	14	6.1	14.4	9.3
Cholesterol ester	0.2	1.8		9.9
Cerebroside	13	0.8		
Sulfatide	3.5	3	10.6	0.8
Total phospholipid	30	15	25	8.5
Ethanolamine phosphatides	10.2	4.7	7.9	1.3
Lecithin	7.5	4.8	7.3	2.7
Sphingomyelin	5.4	2.9	4.2	1.6
Serine phosphatides	5.9	1.7	5.0	1.6
Phosphatidylinositol	0.8	0.7	0.3	0.7

[a] Data recalculated from reference [11].
[b] Data recalculated from reference [10].
[c] Percentage wet weight.
[d] Proteolipid, total lipid, and individual lipids as percentage of white matter dry weight.

This fact, coupled with studies that directly demonstrate blood-brain barrier damage in EAE, implies the participation of hematogenous elements in the destruction of myelin. Locally derived macrophages are also involved, and it is sometimes difficult to quantitate the relative role of the two classes of cells. It has been definitively demonstrated that, during EAE and EAN, hematogenously derived macrophages attack myelin and, among the human diseases, a similar phenomenon has been seen only in Landry-Guillain-Barré syndrome. In these latter diseases, pieces of myelin are stripped from axons and engulfed by phagocytic cells. Frequently, however, myelin degradation is seen in EAE whereby myelin is lost from axons without the direct participation of invading cells. Once inside macrophages, the myelin presumably is then degraded by lysosomal enzymes. Cholesterol, however, is apparently one myelin constituent that cannot be degraded to smaller units and is esterified, often remaining in phagocytes for some time at the site of the lesion. Cholesterol esters are essentially absent in mature brain so that their presence in myelin disorders is considered indicative of recent demyelination. Such compounds are also responsible for the neutral fat-staining, or sudanophilia, demonstrated histochemically in many demyelinating diseases.

The presence of macrophages and cholesterol esters in a particular region of the brain may be a relatively transient phase in the disease process. In demyelinated MS plaques, the center of the plaque might not contain macrophages or cholesterol esters whereas the margins of the plaque frequently have both. Much of our knowledge of the disease process relating to inflammatory demyelinating disorders comes from work with the experimental model described in the following section.

Experimental Allergic Encephalomyelitis and Multiple Sclerosis

The first suggestion of a possible autoallergic cause for multiple sclerosis arose from observations of patients who had been treated with the Pasteur antirabies vaccine. A small minority of these patients developed an acute, severe, sometimes fatal, postinoculation encephalomyelitis. These lesions somewhat resembled those in MS brain, and they were unrelated to the pathological changes observed in an untreated case of rabies. It was later shown that the demyelination was due to a delayed hypersensitivity response to brain tissue contained in the inoculum and not to the virus.

The laboratory counterpart to the clinical observations was a series of studies demonstrating that animals produce organ-specific complement-fixing antibrain antibodies in response to inoculations of CNS tissue and that a series of such inoculations could lead to neurological symptoms of paralysis, ataxia, and sphincter disturbance. This condition was histologically accompanied by a characteristic lesion consisting of perivascular and diffuse lymphocytic infiltrations as well as demyelination with sparing of axons. If nerve tissue was emulsified with Freund's adjuvant — a combination of mineral oil, killed mycobacteria, and a binding agent — the allergenic potency of the tissue was increased, and the onset of the disease appeared within two to three weeks, rather than months, after inoculation. The assumed relation of EAE to MS is based largely on the similarities between the histological lesions in EAE cases and those seen in some acute forms of MS.

Much effort has been spent on isolating the specific CNS components responsible for induction of the disease process, an effort which gained momentum when Kies and Alvord succeeded in establishing a quantitative assay system based on clinical symptoms and histopathological lesions in guinea pigs. Eventually, the major encephalitogenic component was identified as a myelin protein, a basic protein with a molecular weight of about 18,000. As a result of work by Eyelar, Hashim, Brostoff, and co-workers, Carnegie and co-workers, and Kibler, Shapira, and co-workers, the complete sequences of bovine and human basic proteins have been elucidated [12–15]. It is of interest that the sequence of this protein is highly conserved during the evolution of the high mammalian species; bovine and human proteins differ by only eleven residues. The complete sequences of these proteins are shown in Figure 28-1. There is some disagreement about certain parts of this sequence, but the discrepancies are relatively minor. The sequence is similar in different species, but different portions of the molecule are antigenic in different species. A peptide of nine amino acids (114 to 122 in Figure 28-1) has been isolated and shown to produce EAE in guinea pigs [14]. A study using chemically synthesized peptides demonstrated that the minimal sequence requirement for this peptide to induce EAE was Trp– ... – ... – ... – ... –Gln– Lys (Arg); various amino acid substitutions for the amino acids between tryptophan and glutamine do not eliminate encephalitogenic activity [14]. Peptides containing this sequence induce EAE in rabbits and guinea pigs but are inactive in monkeys. Another peptide (44 to 89 in Figure 28-1) is active in rabbits but inactive in guinea pigs and has been studied and sequenced [14, 16]. Still a third EAE-inducing peptide, active in monkeys but not in rabbits and guinea pigs, has also been characterized (134 to 170 in Figure 28-1) [14]. The interest in the chemistry of this protein is

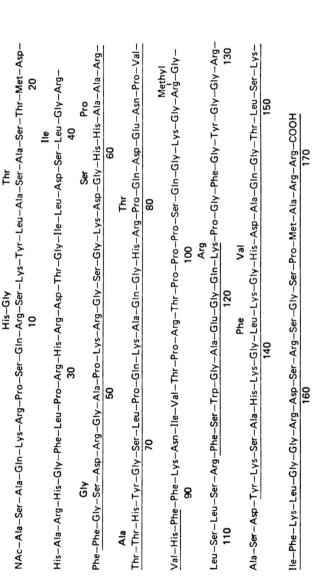

Figure 28-1

Amino acid sequence of basic protein of bovine myelin. Substitutions in human basic protein are shown above the appropriate position of the bovine sequence. Between positions 10 and 11, a His—Gly dipeptide occurs in the human protein. (After Eylar [14].)

closely related to the possibility of developing modified amino acid sequences which might suppress the immune reaction (see p. 590).

The obligate participation of the immune system in the disease process of EAE was shown by demonstrating passive transfer of the disease by the use of either viable lymph node cells or circulating cells obtained directly from the peripheral blood of various animals during the development of EAE [17]. The immune system of mammals involves both B cells, which produce humoral antibodies, and T cells, which are responsible for cell-mediated responses [18]. The disease process in EAE is primarily T cell mediated, and, in fact, the disease can be induced in normal animals by injection of circulating blood leukocytes from animals with EAE [19]. There is also evidence showing that B cells play a part in EAE, and the participation of circulating antibody must be considered. One test for circulating factors in EAE has been the use of organotypic cultures of CNS tissue. For this, fragments of rat cerebellum are cultured in vitro in such a way that the various cellular elements interact to form myelin [20]. The presence of factors that interfere with myelination can be determined directly by microscopic comparison with control cultures. Samples of serum and lymphocytes from EAE-affected animals, whether obtained before or after the onset of clinical signs, produce a specific and characteristic pattern of demyelination [20]. EAE serum also severely inhibits incorporation of ^{35}S into sulfatide (a relatively myelin-specific lipid), while not affecting several other enzyme activities [21]. It is of considerable significance that MS serum also has such demyelinating effects in cultures [20]. It should be noted that consistent demyelination in tissue cultures by EAE serum was observed only when such serum was obtained from animals immunized against whole CNS tissue [22]. Serum from animals immunized against purified basic protein, although containing antibasic protein antibodies, often does not cause demyelination. This observation is one of many emphasizing that the term *EAE* really describes a class of experimental disorders. The clinical course, the histopathology, and the details of the immune mechanisms involved may vary greatly depending on the conditions used to induce the disease process [8]. There is great species variability; the disease is usually terminal in monkeys (and lesions similar to those in human MS cases are found) whereas rats (with less severe lesions) generally recover. There is even considerable strain difference; Lewis rats show a more severe form of the disease than do Wistar rats. Although discussion of different types of EAE is beyond the scope of this book, it is of importance to note that a chronic form of EAE can be induced that occasionally follows a fluctuating clinical course, thus approaching the relapsing course of MS in humans [8].

Analytical studies on CNS tissue from animals with EAE are somewhat contradictory. One study has shown a marked depression (20 to 30 percent) in levels of myelin-specific lipids [23]. Such changes are much more marked in the spinal cord than in brain and vary during the course of the disease (acute and recovery stages in rats). No change, however, in the amount of myelin recovered from EAE rats has been observed. Indeed, by histological criteria, only a small amount of the total myelin is involved [24]. Such differences may be due to the variable course and severity of the disorder as a function of changes in protocol of the induction process. There does not appear to be any marked difference between the composition of CNS myelin from EAE animals and that from controls.

Proteolytic enzymes have been implicated in the hydrolysis of myelin basic proteins during EAE [17]. The levels of acid proteinase, neutral proteinase, and cathepsin A are increased in the vicinity of lesions in monkeys with EAE [24]. It is likely that some of these enzymes are released from mononuclear cells including lymphocytes. There is also evidence that levels of enzymes involved in catabolism of myelin lipids are elevated in brain tissue from rats with acute EAE [25].

As tested on spinal-cord slices from rats with acute EAE, changes in metabolism have been observed [26] that involve depressed incorporation of radioactive precursors into lipids and increased protein synthesis. Part of the increased rate of protein synthesis is due to the metabolism of invading mononuclear cells whereas part represents increased incorporation into myelin proteins. Further studies relating to metabolism of myelin components in the acute stages and, equally important, during the recovery stage, may cast light on the capacity of oligodendroglial cells to recover from injury and synthesize new myelin.

Extensive biochemical investigations on autopsy materials from MS brains have been carried out. White matter shows a decreased yield of myelin, an expected decrease of myelin-specific components, and a build-up of cholesterol esters.

Many laboratories have investigated the composition of myelin from normal appearing areas of MS brain in the hope of finding some abnormality that might be related to the specific destruction of myelin in this disease. These reports are often contradictory; abnormalities in phospholipid fatty acid composition and alterations in cerebroside-to-sulfatide ratio have been reported, as well as a reduction in the level of myelin basic protein. The most recent very careful studies, however, report no significant compositional differences between control and MS brain [27], and it appears that any consistent compositional abnormalities remain to be demonstrated.

Various histochemical and biochemical techniques have been used to demonstrate that the periphery of active plaques is greatly enriched in acid-proteinase activity and that these enzyme levels correlate with the loss of basic protein in these microscopic regions [13]. Further studies along these lines could clarify the relationship between EAE and MS with respect to the role of phagocytes in the breakdown of myelin.

It is now well established that both humoral- and cell-mediated factors are involved in producing the observed demyelination in EAE. Evidence is contradictory regarding a similar role for the immune system in relation to the pathogenesis observed in cases of MS. Some of the positive evidence is discussed below. MS patients have a significantly elevated level of a particular immunoglobulin IgG in the cerebrospinal fluid. In fact, this is a standard diagnostic test for the disorder [28]. There is evidence that some IgG is specifically localized in the region of MS plaques [28]. These data suggest involvement of humoral factors (as does some of the tissue-culture work discussed on p. 587).

Data also imply cellular hypersensitivity to CNS antigen by lymphocytes of MS patients. It is known that lymphocytes isolated from guinea pigs with EAE produce a soluble factor that inhibits migration of macrophages from normal animals. Production of this migration inhibition factor (MIF) is assayed by comparing the rates of migration of macrophages out of a capillary tube into a medium containing cell-free supernatant obtained from lymphocytes cultured in the presence or absence of myelin basic protein. In a significant number of cases, lymphocytes from MS patients produce migration inhibition factor while the lymphocytes from control patients do not [13, 29]. Still another immunological result (of significance, as it pertains to a possible slow-virus etiology for MS) is a specific increase in cerebrospinal fluid antibodies against measles antigen in MS patients [30]. There also appears to be a suppression of cellular immunity in MS patients; they have a specific lack of cellular recognition for measles antigen [31]. Despite the positive data presented above and that from other studies, there is still "a paucity of evidence that cell-mediated immune responses directed against neuroantigens have a pathogenic role in MS" [32].

The most successful attempts to determine the genetic factors involved in pathogenesis of MS are based on measurements of the frequency of histocompatibility antigens in populations of MS patients. The HL-A system, which determines certain glycoproteins on the surface of most cell types, is the major identifiable histocompatibility site in man (see [33] for a review of this topic). At each of the two major subloci of this system there are a number of different alleles. An individual heterozygote at both loci would express four major HL-A specificities (two maternal and two paternal). Different antigens can be distinguished on the basis of cytotoxicity tests involving interactions with certain standard antisera (prepared from patients who have undergone an immune challenge by virtue of multiple pregnancies, whole blood transfusion, organ-graft rejection, etc.).

In studies conducted in northern Europe, MS patients as a group have an increased frequency of HL-A3 and HL-A7 (from the first and second allelic series respectively) and consequently a decreased frequency of other antigens, in particular HL-A2 and HL-A12 [34]. These immunogenetic studies demonstrate "weak" relationships (i.e., although they are highly significant statistically, they are clearly not cause-and-effect relationships). For example, in a group of Danish MS patients, the frequency of HL-A3 was found to be 36.4 percent whereas a control population had a 26.97 percent frequency of this antigen [34].

Another immunological determinant closely linked genetically to the HL-A sites is the LD series (lymphocyte-determined factors). These are defined by reaction of one population of lymphocytes in response to "foreign" determinants on another population of lymphocytes. The complex methodology and genetic analysis involved in studies of this antigen are not as advanced as with the serum-defined system, but it is noteworthy that a marked increase in the frequency of the LD-7 antigen is observed in MS patients [34].

The operational hypothesis derived from the above data is that the possession of a certain genotype increases susceptibility to MS. We can only speculate about the molecular basis of this correlation of HL-A antigen distribution with increased susceptibility to MS; possible explanations are (1) certain HL-A antigens may function more efficiently than others as receptors for a virus that triggers an autoimmune reaction; or conversely (2) some HL-A antigens have a structure that induces immunological tolerance of such a virus; or (3) the HL-A antigen might not be involved at all, but happens to be genetically linked, and in linkage disequilibrium, with an immuno-responsive gene that plays a part in MS. These investigations may provide a basis for explaining various reports that indicate a weak trend toward familial aggregation of MS cases.

As has been discussed previously, the fact that myelin is destroyed in a particular disease does not imply that this is the initial lesion or the primary cause of malfunction of the nervous system. The CNS tissue-culture system offers a model system for correlating demyelination with other pathological events. Neurons in the cultured fragments form synaptic interrelationships and develop complex patterns of bio-electrical activity. The complex responses characteristic of synaptic transmission often disappear within a few minutes after the application of EAE or MS serum, but simple axon spikes can still be evoked. The complex polysynaptic functions may return within minutes after removal of the offending serum. Recently, it has been noted that normal sera may also block the electrical responses of CNS in vitro [35, 36]. The factors in both normal and EAE sera are complement-dependent, thermolabile globulins [35]. In addition, the depressant factors in normal sera are all irreversibly inactivated by heat, whereas about one-half of the tests involving heated EAE and MS sera showed restoration of activity by the addition of fresh guinea-pig serum [36]. The complex polysynaptic functions return within minutes after removal of the offending serum. Neurochemical investigations of models of this sort still await the development of sufficiently sensitive techniques.

Much effort has been devoted to developing methods for suppression of EAE. Various immunosuppressive agents are effective in such treatment, but the action of these compounds is limited in time, and they may have toxic side effects. In principle, the most effective approach is to counteract the sensitized lymphocytes in situ. This can be done by giving inoculations of the encephalitogen itself [12, 13, 37, 38], either in a native or a modified form and often with incomplete Freund's adjuvant. The mechanism by which the excess of antigen protects the nervous system is not known. An obviously oversimplified working hypothesis is that circulating lymphocytes interact with the antigen, and the receptor sites to in situ encephalitogen are blocked.

Other Acquired Demyelinating Disorders

Experimental allergic neuritis (EAN) is the experimental analog of the Landry-Guillain-Barré syndrome in man and the PNS counterpart of EAE. It is induced by injection of whole peripheral nerve tissue emulsified in an adjuvant. There is evidence that a particular protein of PNS myelin (not present in CNS myelin), the P_2 protein, is a component of the primary antigenic stimulus [39]. This involvement is not as clear-cut as that of basic protein in induction of EAE. There have been few neurochemical studies of the Landry-Guillain-Barré syndrome, the human analog of EAN, because, with proper treatment, this disorder is usually not fatal, and there is almost complete recovery. With regard to these and other PNS disorders that are discussed in following sections, it should be kept in mind that CNS and PNS myelin, although morphologically related, differ significantly in chemical composition, especially with regard to protein composition (see Chap. 4).

There have not yet been extensive neurochemical studies of the experimentally induced animal virus diseases; relevant virology and pathology have been reviewed recently [40].

GENETICALLY DETERMINED METABOLIC DISORDERS OF MYELIN

This large grouping of diseases consists of degenerative disorders of CNS white matter, most of which have their onset in humans before the age of ten years. The various members of this class of myelin disorders have been well outlined by Poser [41], and in Vinken and Bruyn [42]. These diseases are classified as inborn errors of metabolism (presumably due to a genetic insufficiency of a particular enzyme) because they are familial.

Pathological Criteria for Genetic Metabolical Diseases of Myelin

1. Loss of myelin from areas of white matter.
2. Concomitant loss of axons.
3. Preservation of axotomized neurons.
4. Diffuse and symmetrical lesions.
5. Lack of inflammatory changes (except adrenoleukodystrophy, see below). Macrophages, however, are present (except in the hypomyelinating disorders, see below).

In addition, a morphological distinction can be made between those diseases in which there is apparent accumulation of an abnormal storage product (e.g., metachromatic leukodystrophy) and genetically determined hypomyelinating disorders in which there is an extreme paucity of myelin formation (e.g., Pelizaeus-Merzbacher disease). Meaningful distinctions within the group of metabolic disorders are possible by biochemical criteria (see below).

The examples cited do not represent all the leukodystrophies, and each example may well have several variants. It must be emphasized that the pathology observed in the human disorders describes the end results of the disease process.

1. *Metachromatic Leukodystrophy (MLD)*, a disorder (see Chap. 26) characterized by loss of myelin and the presence of spherical granular masses within oligodendroglial cells, neurons, and macrophages as well as free in the white matter. These granules and inclusions are metachromatic. In contrast to most leukodystrophies, there is striking involvement of the PNS. An adult variant of this disorder exists.
2. *Krabbe's (Globoid) Leukodystrophy,* in which normal white matter is deficient due to widespread diffuse demyelination (see Chap. 26). There is astrocytic gliosis and, most striking, characteristic abnormal mononuclear or multinuclear globoid cells scattered in white matter. In some cases, these cells may constitute up to one-half of the total white matter.
3. *Adrenoleukodystrophy (Sex-Linked Schilder's Disease)*, a disease, usually of young males, differing from other leukodystrophies by the presence of inflammatory changes in the CNS lesions. Involvement of the adrenals is obligatory. The morphological picture suggests a lipid-storage disorder [42a].

4. *Refsum's Disease,* a PNS-myelin disorder characterized by swelling of the peripheral nerves, thinning of the myelin sheath, and myelin degradation. CNS lesions are also present (see Chap. 26).

5. *Pelizaeus-Merzbacher Disease (Sudanophilic Leukodystrophy),* an early appearing, slowly progressive CNS degenerative condition resulting in a lack of proper myelination (hypomyelination) [43].

6. *Alexander's Disease,* a disease usually apparent in the first year of life, with a variable course thereafter [44].

7. *Canavan's Disease (Spongy Degeneration of the White Matter),* a disease of infants resulting in widespread edema of white matter [45]. Myelin is diminished in certain brain areas, but without the accumulation of sudanophilic myelin products that is characteristic of many metabolic disorders of myelin.

8. *Nonspecific Hypomyelination,* a wide variety of chronic metabolic insults, if present at a particular early developmental stage (the first few years of life), that result in a decrease in the amount of myelin deposited. This grouping is based primarily on chemical data since quantitation of myelin deficits is difficult by morphological techniques. Histological and ultrastructural data indicate, however, a thinner myelin sheath in certain of these disorders. Genetically determined causes of hypomyelination include disorders of amino acid metabolism (phenylketonuria), familial hypothyroidism, and disorders involving copper metabolism (trichopoliodystrophy).

Genetic disorders involving myelin have been studied in animals, notably several murine mutants.

9. *Myelin Defects.* Among the genetically determined hypomyelinating disorders in mice are quaking mouse and jimpy mouse [46], which show deficiencies in myelin production. Ultrastructurally, adult quaking animals show an impediment of myelination, thin myelin sheaths, lack of myelin compaction with pockets of oligodendroglial cell cytoplasm evident, and uneven growth of lateral loops, all of which are typical of the immature (5-to-10-day-old) mouse. Jimpy mice show severe hypomyelination with sudanophilic deposits. Murine muscular dystrophy is characterized by regional deficiencies in myelin in the PNS.

10. *Globoid Leukodystrophy.* Several breeds of dogs and cats are known to carry genetic factors for a disease analogous to human Krabbe's disease [47].

Biochemical Studies of Genetic Disorders of Myelin

Compositional studies of brain, especially of isolated myelin from autopsy material, have proved a powerful tool in the study of genetic disorders relating to the myelin sheath. On the basis of such work, we feel justified in subclassifying these disorders largely on the basis of chemical criteria. The demyelination resulting from genetic disorders is often accompanied by abnormal composition of white matter, increase in water, decrease in myelin constituents, and the presence of cholesterol esters. More useful distinctions can be made by studies of myelin isolated from autopsy material. Chemical analysis indicates that pathological myelin can be divided into two categories, specific and nonspecific. The specific category is characterized by an abnormal myelin that contains an accumulation of a particular compound due to a primary genetic lesion which prevents its catabolism. Disorders in the second category are characterized by the presence of myelin in which the compositional abnormality is not directly related to the genetic lesion. A third category of genetic disorders, established on the basis of both chemical and morphological observation, is hypomyelination, an impediment in normal myelinogenesis.

Diseases in which myelin shows a specific chemical pathology can also be correctly referred to as dysmyelinating in accordance with the definition of Poser [41], which states that this category would include those inborn errors of metabolism in which "myelin initially formed is abnormally constituted, thus inherently unstable, vulnerable and liable to degeneration." The student should be warned that the term "dysmyelinating disorder" is usually rather loosely and incorrectly used in the clinical literature. According to the original definition, the term has primarily a chemical basis, and the common use of "dysmyelination" to describe morphological observations has no precise meaning.

Metachromatic leukodystrophy (MLD) is a sphingolipidosis characterized at a molecular level by an inability to catabolize sulfatide (Chap. 26). Morphological observations are consistent with the hypothesis that a certain amount of myelin (normal by morphological criteria but presumably abnormal chemically) is formed initially. With time, however, the sulfatide which would normally be catabolized (albeit at a very slow rate) continues to accumulate in myelin (Table 28-2) and oligodendroglial cells as its synthesis proceeds. One theory is that the molecular architecture of the myelin is disturbed to such a point that it becomes unstable and breaks down. Sulfatide also accumulates in the oligodendroglial cells and macrophages and presumably gives rise to the metachromatic bodies that are characteristic of this disorder.

Table 28-2. Human Myelin Composition in Two Diseases Compared with Controls[a]

Lipids[b]	Control	Spongy Degeneration	SSPE	MLD
Total lipid, % dry weight	70.0	63.8	73.7	63.2
Cholesterol	27.7	58.0	43.7	21.2
Cerebrosides	22.7	8.0	18.8	9.0
Sulfatides	3.8	2.0	2.8	28.4
Total phospholipids	43.1	33.4	36.6	36.1
Ethanolamine phosphatides	15.6	9.8	9.7	8.1
Plasmalogen[c]	12.3		9.1	5.3
Lecithin	11.2	11.3	10.4	10.7
Sphingomyelin	7.9	5.9	8.8	7.1
Serine phosphatides	4.8	5.5	4.6	3.8
Phosphatidylinositol	0.6	0.8	1.4	3.1

[a]The values listed are from the laboratory of Dr. Norton (reported in the previous edition of this book) except for the spongy degeneration data [48].
[b]Individual lipids expressed as weight percentage of total lipid.
[c]Most of the plasmalogen is phosphatidal ethanolamine and is also included in the ethanolamine phosphatides column.

In Refsum's disease, a genetic disorder clearly classified as dysmyelinating, the ability to degrade branched-chain fatty acids by oxidation is lacking (Chap. 26). Although the defective enzyme is not essential for degradation of normal fatty acids, it is necessary for degradation of phytanic acid (3,7,11,14-tetramethylhexadecanoic acid). Phytanic acid has been found to accumulate in myelin phospholipids as it does in other tissues of the body. The effects are more serious in the PNS than in the CNS, perhaps because some of this fatty acid is excluded by the blood-brain barrier. The possibility cannot be eliminated that some dysmyelinating disorders have not been classified as such because current analytical methods are not sensitive enough to detect the myelin abnormality.

Globoid cell leukodystrophy is also a sphingolipidosis, characterized at a genetic level by a β-galactosidase deficit and the consequent inability to catabolize cerebroside (Chap. 26). Curiously, excess amounts of cerebroside are not found upon whole-brain analysis. The ratio of cerebroside to sulfatide is abnormally high, however, indicating a disproportionate preservation of cerebroside, rather than lack of sulfatide. The excess cerebroside accumulates in a specific storage cell, the globoid cell. Poser [41] considered this to be a dysmyelinating disease, and one would expect that the myelin would contain abnormal accumulations of cerebroside (analogous with the findings in MLD, see above). However, this is not the case; the very small amounts of myelin that can be isolated from brains of such patients is normal. The reason for this is not clear, but it is believed that the myelin present is formed during early myelination, before the enzyme deficiency is fully manifested. It has been postulated that the accumulation of cerebroside then elicits the globoid body response, followed by death of oligodendroglia and little further production of myelin, normal or otherwise. Thus, this disease might better be classified as a hypomyelinating disease.

In the three disorders discussed above, the primary genetic lesion is known. The enzyme deficit is expressed in all cells, so it is possible to utilize fibroblasts or circulating blood elements for diagnosis, detection of heterozygotes, and related studies. Floating cells of fetal origin can be obtained early in pregnancy by amniocentesis and cultured for prenatal diagnosis.

The genetic disorders characterized by the accumulation of "nonspecific" pathological myelin include adrenoleukodystrophy [42a], Canavan's disease or spongy degeneration [48] (Table 28-2), and probably a number of disorders where the myelin has not yet been analyzed. The primary genetic lesion is not known in these diseases. It is important to note that myelin isolated from these two disorders and myelin isolated from patients with a wide variety of disorders involving secondary demyelination (see p. 598) have similar "abnormal" chemical compositions. Certain experimental disorders induced in animals by toxic agents show the same type of abnormality with respect to isolated myelin. In each case, the isolated pathological myelin has a grossly normal ultrastructural appearance. It has much more cholesterol, however, less cerebroside, and less phosphatidalethanolamine than does normal human myelin. When a considerable accumulation of cholesterol esters is found in whole brain, such compounds are not found in the myelin itself. The purification of myelin involves discontinuous sucrose gradients and a light fraction, containing cholesterol esters, is separated from the denser myelin. The abnormal myelin probably represents a partially degraded form. It is possible that the degradation involves locally derived macrophages.

A third category of genetic disorders, hypomyelination, includes the very rare Pelizaeus-Merzbacher disease in humans, characterized ultrastructurally by failure of mature myelin to accumulate [43]. It bears some structural resemblances to the disorder of the murine mutant quaking mouse. Certain disorders of amino acid metabolism result in diminished levels of myelin. Globoid cell leukodystrophy (see p. 592) can also be considered to be in the category of hypomyelinating disorders. In these disorders, there is a great diminution in amounts of myelin constituents, often without the presence of phagocytic cells.

A number of animal models of such disorders are available, most notably the murine mutants. Although oligodendroglial cells appear relatively normal in the adult "quaking" mouse, myelinated axons have only a few lamellae of myelin. The ultrastructural picture as well as the composition of isolated myelin of adult quaking mouse is characteristic of the developing 7- to 10-day-old CNS of normal mice [49, 50]. Enzymatic activity for cerebroside formation (UDP-galactose-ceramide galactosyl transferase) is more than 60 percent depressed in the mutants as compared with control littermates at various points, but the regulatory enzymes that are involved in biosynthesis of sphingolipids found in gray matter are completely normal [51]. Two other murine mutants (jimpy and MD), characterized histologically by pathologically low levels of myelin, are defective in enzymatic activity for the biosynthesis of myelin-specific lipids.

Although these observations are certainly relevant to an understanding of the pathology involved, there is no reason to believe that any of the enzymatic deficits are primary. The murine mutants are powerful tools for elucidating details of myelin assembly by contrasting these genetically perturbed animals with controls in various metabolic experiments. A variety of other interesting animal models exist for study of hypomyelinating disorders. Border disease (hypomyelinogenesis congenita) of sheep is of interest because, although the amount of myelin in the CNS is decreased, in vitro assay indicates an increased role of cerebroside synthesis [52]. This is an acquired congenital disorder, induced by inoculating pregnant ewes with a suspension of brain from affected lambs. It is mentioned in this section because, with regard to cerebroside synthesis, it contrasts with the genetic hypomyelinating disorder of mice.

Certain disorders of amino acid metabolism cause some hypomyelination. Phenylketonuria (see Chap. 24) is quantitatively the most significant condition. Upon histological examination, the amount of myelin in brains of patients suffering from phenylketonuria appears abnormally low, and the white matter may be up to 40 percent deficient (as compared to controls) in such myelin-specific components as proteolipid protein and cerebroside [53]. One possible explanation of the myelin deficit is that the high phenylalanine levels depress uptake of tyrosine into brain by competing for a common active transport system. Because of the lack of this essential amino acid, protein synthesis is severely depressed and myelin, which is usually synthesized rapidly early in life, is not formed. Another explanation may be that the alternate pathways used for phenylalanine metabolism (present normally but of quantitative significance only in this disease state) cause significant accumulation of phenylpyruvic acid and other metabolites which may be toxic. In either case, there is depression of protein synthesis during the period of maximal myelin accumulation. A further complication exists: At autopsy, untreated cases show significant demyelination. This has led to suggestions that phenylketonuria is a dysmyelinating disorder, but (as in MS) consistent evidence of significant abnormalities in myelin from such patients remains to be demonstrated.

The model system of experimental hyperphenylalaninemia (induced by injecting animals with large doses of phenylalanine) is characterized by a deficit in myelin accumulation and a decreased ability, both in vitro and in vivo, to incorporate radioactive precursors into cerebral lipids and proteins. A deficit in long-chain and unsaturated fatty acids of myelin, possibly caused by accumulation of abnormal metabolites of phenylalanine metabolism, also has been demonstrated [54]. This model is not identical with phenylketonuria because, since phenylalanine hydroxylase is present at normal levels, both phenylalanine and tyrosine levels are elevated.

A related model for phenylketonuria is obtained by injections of parachlorophenylalanine, which inhibits phenylalanine hydroxylase, causing a rise in blood phenylalanine levels. More importantly, tyrosine levels are not significantly elevated, thus approximating the human disorder more closely. Again, there is both a deficit in myelin accumulation and decreased uptake of radioactive precursors into brain lipids and proteins [55]. The incorporation into myelin proteolipid protein is depressed more severely than is incorporation into nonmyelin proteins. This inhibition may not be the result of the phenylalanine itself but of the abnormal deaminated metabolites that accumulate.

ACQUIRED TOXIC AND NUTRITIONAL METABOLIC DISORDERS OF MYELIN

Acquired Toxic Metabolic Disorders

Certain toxic agents to which humans have been accidentally exposed are associated with primary demyelination. A few such agents, notably diphtheria toxin, hexachlorophene, triethyltin sulfate, and cuprizone have been studied experimentally in animals. Ultrastructural observations vary with the toxic agent used.

In man, diphtheritic neuropathy is a possible complication of *Corynebacterium diphtheriae* infection. Paralysis, often following a terminal course, is induced by the bacterial exotoxin. There is swelling and vacuolation, followed by fragmentation of myelin sheaths in the peripheral nervous system, leading to segmental demyelination. A similar disorder can be induced in animals by injection of the toxin.

Hexachlorophene is a recent addition to the list of myelinotoxic compounds. Overexposure of infants in hospital nurseries to hexachlorophene causes a spongy change in myelinated areas in the CNS and PNS. In experimental animal studies, these neurotoxins cause a spongy degeneration of myelin and an edema similar to that seen in the metabolic disorder of Canavan's disease.

Extensive chemical studies have been done on the triethyltin model, brought about by the inclusion of this compound in the drinking water of rats. Acute intoxication involves severe myelin edema, specifically in the CNS, where myelin sheaths become highly vacuolated. The splitting in the myelin is associated with the intraperiod line. Chronic administration of triethyltin may result in a loss of up to one-half of the total myelin, relative to untreated control animals [56]. That myelin which is isolated is of the nonspecific pathological type (see p. 594). This massive myelin loss is not accompanied by inflammatory cells, and the levels of various proteinases are not elevated as they are in such disorders as EAE. There is also no buildup of sudanophilic deposits of degraded myelin. The mechanism by which myelin is damaged under these conditions is unknown, but it clearly contrasts with the situation observed in the acquired demyelinating disorders. Interestingly, there may be almost complete recovery of myelin after triethyltin is removed from drinking water. In vitro metabolic studies demonstrated a greatly increased incorporation of leucine into myelin proteins during chronic administration of triethyltin [57]. Incorporation of acetate into lipid was depressed initially; later in the course of the disease it was somewhat elevated. The goal of the increasing number of metabolic experiments of this type is to elucidate the mechanism of myelin destruction and, perhaps even more important, to study the degree of recovery possible and the mechanism involved.

In neurochemical studies of this type, "loss of myelin" is an operational term, i.e., myelin is not recovered by the standard isolation procedures. The myelin need not be degraded to molecular constituents; it may only be broken down to vesicles too small (or too different in density) to be isolated with normal myelin. Indeed, when animals are treated with hexachlorophene a "floating fraction," somewhat lighter than normal myelin but containing many myelin constituents, can be recovered [58].

Lead encephalopathy, an acquired toxic disorder leading to hypomyelination, has been studied in detail (Chap. 33). The disorder was induced in developing rats by adding lead carbonate to the diet of nursing mothers. The hypomyelination observed appeared to be related primarily to retarded growth and maturation of the neuron [59]. Axons were reduced in size, but the previously established relationship between axon size and number of myelin lamellae in the regions studied was the same in experimental and control animals. Chemical studies indicated that the number of glial cells was not affected, nor was the myelin composition. These results are in contrast to the starvation model, in which glial cell proliferation is decreased and myelinogenesis retarded (see below).

Certain drugs, although not normally a risk factor for human populations, have been used to perturb myelin synthesis with a view toward gaining understanding of normal myelin metabolism. Many inhibitors of cholesterol synthesis have been studied pharmacologically as being of potential utility in treatment of arteriosclerosis. Cholesterol is a major component of myelin, and application of inhibitory drugs during the time of myelin deposition may have relatively specific effects on myelin. The drug AY9944, an inhibitor of $\Delta 7$ reductase, causes 7-dehydrocholesterol to accumulate in myelin, taking the place of normal cholesterol. Not only is the isolated myelin abnormal; its yield is also greatly reduced, implying that a deficiency of a particular lipid will limit the assembly of myelin [60]. In vitro experiments show a decreased uptake of radioactive glucose into myelin lipids and proteins.

Acquired Nutritional Disorders

As was discussed in Chapters 4 and 18, much of the CNS myelin in mammals is formed during a relatively restricted time period, corresponding to the first few years of life in humans and between 15 and 30 days in rats. Just before this rapid deposition of myelin, there is a burst of oligodendroglial cell proliferation. During these restricted periods of time, a major part of the metabolic activity of brain and, even more important, possibly most of the brain's protein and lipid synthetic capacity are involved in myelinogenesis. This developmental phenomenon has practical input for the understanding of hypomyelination. Any metabolic insult during this "vulnerable period" may lead to a preferential reduction in myelin formation (Chap. 18). A model system for such studies is obtained by restricting access of rat pups to the mother and so inducing undernourishment. Starvation of rats from birth onward leads to a deficit in myelin lipids and a reduced amount of isolatable myelin compared with normally fed littermate controls. The size of whole brain is also somewhat reduced, but it is clear that the depression of myelin-specific lipids and proteins is greater than that of other brain components. The implication is that, during starvation, there is preferential depression in synthesis of myelin-specific components. This has been directly demonstrated by the use of in vivo isotope experiments [61]. Possibly not only is there depression in the amount of myelin deposited (in part due to a lessened number of oligodendroglial cells) but also the developmental program with regard to myelinogenesis is somewhat delayed.

An experimental dietary copper-deficiency model has been used with developing animals. A highly significant decrease in myelinogenesis is observed [62], possibly secondary to the generalized metabolic insult caused by decreased activity of the copper-requiring cytochrome oxidase. This system may be an experimental analog to the sex-linked disorder called trichopoliodystrophy (Menkes' kinky hair disease), in which there appears to be a disorder of copper metabolism.

DISORDERS WITH SECONDARY INVOLVEMENT OF MYELIN

A large number of disorders of the human nervous system that preferentially cause lesions in gray matter, in particular neurons, and that are associated with "accidental" brain damage (infarcts, trauma, tumors, etc.), eventually result in regions of demyelination. Such diseases can be of viral (e.g., subacute sclerosing panencephalitis), genetic (Tay-Sachs disease), or of unknown etiology (amyotrophic lateral sclerosis).

The first human disease to be studied with respect to myelin composition during secondary demyelination was subacute sclerosing panencephalitis (SSPE), a CNS disease caused by a defective measles-virus infection [63]. It is probable that this disease involves destruction of both neurons and oligodendroglia. The brain white matter shows the typical changes for severe demyelination. The isolated myelin has a grossly normal ultrastructural appearance and a normal lipid-protein ratio. However, it was found to have much more cholesterol, less cerebroside, and less ethanolamine phosphatides than normal human myelin (Table 28-1). No cholesterol esters were found in the myelin, although they were abundant in the white matter. Such abnormal myelin has also been isolated from such sphingolipidoses as Tay-Sachs disease, generalized gangliosidosis, and Niemann-Pick disease (see Chap. 26).

The archetypical model for secondary demyelination is Wallerian degeneration. When a nerve (in the PNS) or a myelinated tract (in the CNS) is sectioned surgically, the proximal segment not infrequently survives and shows regeneration. In the distal segment, Wallerian degeneration occurs, and both axons and myelin degenerate rapidly. Debris is phagocytosed by the myelinating cell and by macrophages. Eventually, the denervated region becomes replaced by scar tissue. From such experiments, it is clear that the integrity of the myelin sheath depends on continued contact with a viable axon. The nature of this interrelationship is not known. Any disease that results in a general impairment of neuronal function also will result in axon degeneration and cause onset of myelin breakdown secondary to the neuronal damage. Most biochemical studies on Wallerian degeneration have been done on whole sciatic nerve because isolation of myelin from PNS tissue is difficult. There is a rapid loss of myelin-specific lipids within a period of a week or two. Loss of myelin-specific proteins proceeds even more rapidly, and the disappearance of the major PNS myelin protein may precede slightly the loss of the slower-moving basic protein [64, 65]. There is also a concomitant increase in many lysosomal enzymes [66].

Paradoxically, there is also an increase in certain enzymes implicated in myelinogenesis; e.g., NADP$^+$-dependent isocitrate dehydrogenase increases greatly during Wallerian degeneration. This may be related to the Schwann cell proliferation that takes place as these cells phagocytose the myelin they originally synthesized. Cholesterol esters accumulate and cholesterol esterase activity decreases during the degenerative phase following lesion [67]. It is not clear whether the decrease in cholesterol esterase activity reflects decreased synthesis of myelin (developmentally, this enzyme is implicated in myelinogenesis in the CNS), or whether the decrease reflects only the loss of myelin since a considerable portion of this enzyme (as studied in the CNS) may be localized in myelin [68].

As indicated earlier, analytical studies of the chemical composition of tissue during a disease process may be difficult to interpret because the analysis is indicative of the entire past history of the tissue with regard to total accumulation of any compound, no matter at what time point it is studied. Metabolic studies, in vitro, with incorporation of labeled precursors into the myelin isolated from degenerating rat sciatic nerve, make possible investigations in which various stages of tissue injury and consequent block in the synthesis of new myelin can be separated from a recovery stage, when new myelin sheath is being formed.

Recently, Wallerian degeneration has been studied in the central nervous system by enucleating eyes in rats and examining the optic nerve at different times [69]. The degeneration of CNS myelin is a much slower process than is PNS myelin degeneration and takes place within macrophages (not within the myelin synthesizing cells as in the PNS). Using this system, it has been demonstrated that the myelin isolated after enucleation, although present in decreasing amounts as degeneration progresses, does not differ significantly in composition from control myelin. It is not of the nonspecific pathological type encountered in other disease processes that lead to secondary demyelination. This may indicate a real difference in the degenerative processes during surgically induced Wallerian degeneration and the secondary demyelination that follows naturally occurring disease processes (also generally referred to as "Wallerian" degeneration by neuropathological criteria). Another explanation might be that, because of the morphological uniformity of the optic nerve, the existence of the degraded myelin (nonspecific pathological myelin) might be a transitory process and so not have been detected.

PRESENT PROBLEMS AND FUTURE SOLUTIONS
Neurochemical studies are often hampered by the anatomical complexity of the CNS. Research often must be carried out with whole brain or with large areas of brain that contain a great number of functional units and many different cell types. Future analysis both of autopsy material and animal systems will utilize better application of currently available microanalytical methods to the analysis of myelin components of specific brain regions. This may require the development of new methods to isolate small amounts of myelin. In situ localization of certain components by immunological methods may also prove useful.

In order to study the dynamic aspects of the biochemistry of diseases that involve myelin, animal model systems are required. More sophisticated application of currently available isotope methodology may prove very useful to study such systems. In particular, double-label isotope methodology can eliminate the need for determinations of specific activity that would require considerable amounts of material and so make possible work with restricted brain regions. Closer correlation of ultrastructural observations, including those from autoradiography, with biochemical results should be possible by using such techniques.

Finally, it appears likely that tissue-culture methodology will play an increasingly important role in future investigations of myelin metabolism and pathology. In myelin disorders involving primary genetic defects in lipid metabolism, studies of the enzyme defect can be conducted in fibroblast culture or in non-CNS tissue from affected patients. Myelin development can be directly investigated in animal systems by using fetal brain explant cultures. Application of microtechniques and isotope methodology to such investigations has been of great value. The full power of any such approach will not be attained, however, until a defined oligodendroglial cell line can be cultured and stimulated so that the cells produce myelin synchronously.

ACKNOWLEDGMENTS
During the course of writing this review, P.M. was partially supported by USPHS grants HD-03110 and NS-11615; C.S.R. by USPHS grant NS-08952 and Research

Career Development Award NS-70265; and M.B.B. by USPHS grant NS-06735 and National Multiple Sclerosis Society grant 433. Dr. Bornstein is a Kennedy Scholar of the Rose F. Kennedy Center for Research in Mental Retardation and Human Development.

REFERENCES

*1.   Adams, R. D., and Sidman, R. L. *Introduction to Neuropathology.* New York: McGraw-Hill, 1968.

2.   Adams, R. D., and Kubik, C. S. The morbid anatomy of the demyelinative diseases. *Am. J. Med.* 12:510, 1952.

*3.   Johnson, R. T., and Weiner, L. P. The Role of Viral Infections in Demyelinating Diseases. In Wolfgram, F., Ellison, G. W., Stevens, J. G., and Andrews, J. M. (Eds.), *Multiple Sclerosis: Immunology, Virology and Ultrastructure.* New York: Academic, 1972, p. 245.

4.   Richardson, E. P., Jr. Progressive Multifocal Leukoencephalopathy. In Vinken, R. J., and Bruyn, G. W. (Eds.), *Handbook of Clinical Neurology: Multiple Sclerosis and Demyelinating Diseases.* Vol. 9. Amsterdam: North-Holland, 1970, p. 485.

5.   Asbury, A. K., Arnason, B. W. G., and Adams, R. D. The inflammatory lesion in idiopathic polyneuritis. *Medicine* 48:173, 1969.

6.   *Animal Models of Human Disease.* Washington, D.C.: Armed Forces Institute of Pathology Registry of Comparative Pathology, 1972.

7.   Cummings, J. F., and Haas, D. C. Coonhound paralysis: An acute idiopathic polyradiculoneuritis in dogs resembling the Landry-Guillain-Barré syndrome. *J. Neurol. Sci.* 4:51, 1967.

*8.   Raine, C. S. Experimental Allergic Encephalomyelitis and Related Conditions. In Zimmerman, H. M. (Ed.), *Progress in Neuropathology.* Vol. 3. New York: Grune & Stratton, in press.

9.   Lampert, P. W., Sims, J. K., and Kniazeff, A. J. Mechanism of demyelination in JHM virus encephalomyelitis: Electron microscopic studies. *Acta Neuropathol.* 24:76, 1973.

*10.   Suzuki, Y., Tucker, S. H., Rorke, L. B., and Suzuki, K. Ultrastructural and biochemical studies of Schilder's Disease. II. Biochemistry. *J. Neuropathol. Exp. Neurol.* 29:405, 1970.

*11.   Svennerholm, L., Haltia, M., and Sourander, P. Chronic sclerosing panencephalitis. II. A neurochemical study. *Acta Neuropathol.* 14:293, 1970.

*12.   Kies, M. W. Experimental Allergic Encephalomyelitis. In Gaull, G. E. (Ed.), *Biology of Brain Dysfunction.* Vol. 2. New York: Plenum, 1973, p. 185.

13.   Rauch, H. C., and Einstein, E. R. Specific Brain Proteins: A Biochemical and Immunological Review. In Ehrenpresis, S., and Kopin, I. (Eds.), *Reviews of Neuroscience.* Vol. 1. New York: Raven, 1974, p. 283.

*14.   Eylar, E. H. Myelin Specific Proteins. In Scheider, D. J. (Ed.), *Proteins of the Nervous System.* New York: Raven, 1973, p. 27.

15.   Carnegie, P. R., and Dunkley, P. R. Basic Proteins of Central and Peripheral Nervous System Myelin. In Agranoff, B. W., and Aprison, M. H. (Eds.), *Advances in Neurochemistry.* Vol. 1. New York: Plenum, 1975, p. 95.

16.   Kibler, F. R., Shapira, R., McKneally, S., Jenkins, J., Selden, P., and Chou, F. Encephalitogenic protein: Structure. *Science* 164:577, 1969.

17.   Paterson, P. Y. Experimental Autoimmune (Allergic) Encephalomyelitis. In Miescher, P. A., and Muller-Eberhard, H. U. (Eds.), *Textbook of Immunopathology.* New York: Grune & Stratton, 1968, p. 132.

*Asterisks denote key references.

18. Roitt, I. M. *Essential Immunology,* 2d ed. Oxford: Blackwell Scientific, 1974.

19. Wenk, E. J., Levine, S., and Warren, B. Passive transfer of allergic encephalomyelitis with blood leucocytes. *Nature* 214:803, 1967.

*20. Bornstein, M. B. Immunopathology of Demyelinating Disorders Examined in Organotypic Cultures of Mammalian Central Nervous Tissue. In Zimmerman, H. M. (Ed.), *Progress in Neuropathology.* New York: Grune & Stratton, 1972, p. 69.

21. Fry, J. M., Lehrer, G. M., and Bornstein, M. B. Experimental inhibition of myelination in spinal cord tissue cultures: Enzyme assays. *J. Neurobiol.* 4:453, 1973.

22. Seil, R. J., Falk, G. A., and Kies, M. W. The in vitro demyelinating activity of sera from guinea pigs sensitized with whole CNS and with purified encephalitogen. *Exp. Neurol.* 22:545, 1968.

23. Maggio, B., Cumar, F. A., and Maccioni, H. J. Lipid content in brain and spinal cord during experimental allergic encephalomyelitis in rats. *J. Neurochem.* 19:1031, 1972.

*24. Smith, M. E., Sedgwick, L. M., and Tagg, J. S. Proteolytic enzymes and experimental demyelination in the rat and monkey. *J. Neurochem.* 23:965, 1974.

25. Maggio, B., Maccioni, H. J., and Cumar, F. A. Arylsulphatase A (EC 3.1.6.1) activity in rat central nervous system during experimental allergic encephalomyelitis. *J. Neurochem.* 20:503, 1973.

26. Smith, M. E., and Rauch, H. C. Metabolic activity of CNS proteins in rats and monkeys with experimental allergic encephalomyelitis (EAE). *J. Neurochem.* 23:775, 1974.

*27. Suzuki, K., Kamoshita, S., Eto, Y., Tourtellotte, W. W., and Gonatas, J. O. Myelin in multiple sclerosis: Composition of myelin from normal-appearing white matter. *Arch. Neurol.* 28:293, 1973.

28. Tourtellotte, W. W. Interaction of Local Central Nervous System Immunity and Systemic Immunity in the Spread of Multiple Sclerosis Demyelination. In Wolfgram, F., Ellison, G. W., Stevens, J. G., and Andrews, J. M. (Eds.), *Multiple Sclerosis.* New York: Academic, 1972, p. 285.

29. Rocklin, R. R. What Is the Role of the Lymphocyte in Multiple Sclerosis? In Wolfgram, F., Ellison, G. W., Stevens, J. G., and Andrews, J. M. (Eds.), *Multiple Sclerosis.* New York: Academic, 1972, p. 365.

30. Brown, P., Cathala, F., Gajdusek, D. C., and Gibbs, C. J. Measles antibodies in the cerebrospinal fluid of patients with multiple sclerosis. *Proc. Soc. Exp. Biol. Med.* 137:956, 1971.

31. Utermohlen, V., and Zabriskie, J. B. A suppression of cellular immunity in patients with multiple sclerosis. *J. Exp. Med.* 138:1591, 1973.

*32. Paterson, P. Y. Multiple sclerosis: An immunologic reassessment. *J. Chronic Dis.* 26:119, 1973.

33. Ward, F. E., and Amos, D. B. Immunogenetics of the HL-A system. *Physiol. Rev.* 55:206, 1975.

34. Jersild, C., Dupont, B., Fog, T., Platz, P. J., and Svejgaard, A. Histocompatibility determinants in multiple sclerosis. *Transplant. Rev.* 22:148, 1974.

35. Seil, F. J., Smith, M. E., Leiman, A. L., and Kelly, J. M., III. Myelination inhibiting and neuroelectric blocking factors in experimental allergic encephalomyelitis. *Science* 187:951, 1975.

36. Crain, S. M., Bornstein, M. B., and Lennon, V. Depression of complex bioelectric discharges in cerebral tissue cultures by thermolabile complement-dependent serum factors. *Exp. Neurol.* 49:330, 1975.

37. Eylar, E. H., Jackson, J., Rothenberg, R., and Brostoff, S. W. Suppression of the immune response: Reversal of the disease state with antigen in allergic encephalomyelitis. *Nature* 236:74, 1972.

38. Driscoll, B. R., Kies, M. W., and Alvord, E. C., Jr. Successful treatment of experimental allergic encephalomyelitis (EAE) in guinea pigs with homologous myelin basic protein. *J. Immunol.* 112:392, 1974.

39. Brostoff, S. W., Sacks, H., Dal Canto, M., Johnson, A. B., Raine, C. S., and Wisniewski, H. The P_2 protein of bovine root myelin: Isolation and some chemical and immunological properties. *J. Neurochem.* 23:1037, 1974.

*40. Weiner, L. P., Johnson, R. T., and Herndon, R. M. Viral infections and demyelinating diseases. *N. Engl. J. Med.* 288:1103, 1973.

41. Poser, C. M. Diseases of the Myelin Sheath. In Minckler, J. (Ed.), *Pathology of the Nervous System.* Vol. 1. New York: McGraw-Hill, 1968, p. 676.

42. Vinken, P. J., and Bruyn, G. W. (Eds.). *Handbook of Clinical Neurology: Leucodystrophies and Poliodystrophies.* Vol. 10. Amsterdam: North-Holland, 1970.

42a. Schaumberg, H. H., Powers, J. M., Raine, C. S., Suzuki, K., and Richardson, E. P., Jr. Adrenoleukodystrophy. *Arch. Neurol.* 32:577, 1975.

43. Watanabe, I., Patel, V., Goebel, H. H., Siakotos, A. N., Zeman, W., DeMyer, W., and Dyer, J. S. Early lesion of Pelizaeus-Merzbacher disease: Electron microscopic and biochemical study. *J. Neuropathol. Exp. Neurol.* 32:313, 1972.

44. Herndon, R. M., Rubenstein, L. J., Freedman, J. M., and Mathieson, G. Light and electron microscopic observations on Rosenthal fibers in Alexander's disease and in multiple sclerosis. *J. Neuropathol. Exp. Neurol.* 29:524, 1970.

45. van Bogaert, L. Spongy degeneration of the brain. In Vinken, P. J., and Bruyn, G. W. (Eds.), *Handbook of Clinical Neurology: Leucodystrophies and Poliodystrophies.* Vol. 10. Amsterdam: North-Holland, 1970, p. 203.

46. Sidman, R. L., Dickie, M. M., and Appel, S. H. Mutant mice (quaking and jimpy) with deficient myelination in the central nervous system. *Science* 144:309, 1964.

47. Fletcher, R. F., Lee, D. G., and Hammer, R. F. Ultrastructure of globoid leukodystrophy in the dog. *Am. J. Vet. Res.* 32:177, 1971.

48. Kamoshita, S., Rapin, I., Suzuki, K., and Suzuki, K. Spongy degeneration of the brain. A chemical study of two cases including isolation and characterization of myelin. *Neurology* 18:975, 1968.

49. Greenfield, S., Norton, W. T., and Morell, P. Quaking mouse: Isolation and characterization of myelin protein. *J. Neurochem.* 18:2119, 1971.

50. Singh, H., Spritz, N., and Geyer, B. Studies of brain myelin in the "quaking mouse." *J. Lipid Res.* 12:473, 1971.

51. Costantino-Ceccarini, E., and Morell, P. Quaking mouse: In vitro studies of brain sphingolipid biosynthesis. *Brain Res.* 29:75, 1971.

52. Patterson, D. S. P., Terlecki, S., Foulkes, J. A., Sweasey, D., and Glancy, E. M. Spinal cord lipids and myelin composition in border disease (hypomyelino-genesis congenita) of lambs. *J. Neurochem.* 24:513, 1975.

53. Prensky, A. L., Carr, S., and Moser, H. W. Development of myelin in inherited disorders of amino acid metabolism. *Arch. Neurol.* 19:552, 1968.

54. Johnson, R. C., and Shah, S. M. Effect of hyperphenylalaninemia on fatty acid composition of lipids of rat brain myelin. *J. Neurochem.* 21:1225, 1973.

55. Grundt, I. K., and Hole, K. *p*-Chlorophenylalanine treatment in developing rats: Protein and lipids in whole brain and myelin. *Brain Res.* 74:269, 1974.

56. Eto, Y., Suzuki, K., and Suzuki, K. Lipid composition of rat brain myelin triethyl tin-induced edema. *J. Lipid Res.* 12:570, 1971.

57. Smith, M. E. Studies on the mechanism of demyelination: Triethyl tin-induced demyelination. *J. Neurochem.* 21:357, 1973.

58. Matthieu, J. M., Zimmerman, A. W., Webster, H. deF., Ulsamer, A. G., Brady, R. O., and Quarles, R. H. Hexachlorophene intoxication: Characterization of myelin and myelin related fractions in the rat during early postnatal development. *Exp. Neurol.* 45:558, 1974.

59. Krigman, M. R., Druse, M. J., Traylor, T. D., Wilson, M. H., Newell, L. R., and Hogan, E. L. Lead encephalopathy in the developing rat: Effect upon myelination. *J. Neuropathol. Exp. Neurol.* 33:58, 1974.

60. Smith, M. E., and Hasinoff, C. M. Inhibitors of cholesterol synthesis and myelin formation. *Lipids* 5:665, 1970.

61. Wiggins, R. C., Benjamins, J. A., Krigman, M. R., and Morell, P. Synthesis of myelin proteins during starvation. *Brain Res.* 80:345, 1974.

62. Prohaska, J. R., and Wells, W. W. Copper deficiency in the developing rat brain: A possible model for Menkes' steely-hair disease. *J. Neurochem.* 23:91, 1974.

63. Norton, W. T., Poduslo, S. E., and Suzuki, K. Subacute sclerosing leucoencephalitis. II. Chemical studies including abnormal myelin and an abnormal ganglioside pattern. *J. Neuropathol. Exp. Neurol.* 25:582, 1966.

64. Adams, C. W. M., Csejtey, J., Hallpike, J. F., and Bayliss, O. B. Histochemistry of myelin: XV. Changes in the myelin proteins of the peripheral nerve undergoing Wallerian degeneration – electrophoretic and microdensitometric observations. *J. Neurochem.* 19:2043, 1972.

65. Wood, J. G., and Dawson, R. M. C. Lipid and protein changes in sciatic nerve during Wallerian degeneration. *J. Neurochem.* 22:631, 1974.

66. McCaman, R. E., and Robins, E. Quantitative biochemical studies of Wallerian degeneration in the peripheral and central nervous systems: II. Twelve enzymes. *J. Neurochem.* 5:32, 1959.

67. Mezei, C. Cholesterol esters and hydrolytic cholesterol esterase during Wallerian degeneration. *J. Neurochem.* 17:1163, 1970.

68. Eto, Y., and Suzuki, K. Cholesterol ester metabolism in rat brain: A cholesterol ester hydrolase specifically localized in the myelin sheath. *J. Biol. Chem.* 248:1986, 1973.

69. Bignami, A., and Eng, K. F. Biochemical studies of myelin in Wallerian degeneration of rat optic nerve. *J. Neurochem.* 20:165, 1973.

Chapter 29
Vitamin and Nutritional Deficiencies

Pierre M. Dreyfus
Stanley E. Geel

Vitamins are indispensable to the normal metabolic activity of the nervous system. A severe deficiency of these fundamental nutrients results in the improper function of a number of enzyme systems that are essential to the synthesis of basic constituents or to the removal of potentially toxic metabolites, thus interfering with normal development, growth, and function of the nervous system.

The most obvious cause for depletion of vitamins and other nutrients is an inadequate or improper diet. Vitamin deficiency can occur, however, even in the presence of an adequate diet when the absorption of nutrients is limited. Vitamins, in order to be metabolically useful, must first be absorbed in the gastrointestinal tract and transported by the plasma. Then they must gain entry into cells, be converted into their active cofactor form, and interact with specific apoenzyme proteins. A number of genetically determined diseases — referred to as vitamin-responsive or vitamin-dependent states — have been identified. In these, at least one of the steps involved in vitamin utilization or the conversion of vitamins to active forms is faulty or incomplete [1].

Among the various vitamins, some appear to be more essential to the well-being of the nervous system than do others. Despite increasing knowledge regarding the role of vitamins in general metabolism, the specific mechanisms by which a deficiency state affects the development and function of the nervous system remain essentially unknown.

Vitamin deprivation leads to several readily identifiable biochemical lesions, which invariably antedate clinical manifestations and histopathological alterations. In most instances, the correlation between the duration or the severity of these biochemical lesions and the irreversibility of symptoms or tissue changes has not been established, nor has it been possible to elucidate which of the various biochemical lesions are responsible for the neurological manifestations.

Knowledge concerning the neurochemical changes that attend nutritional disorders of the nervous system has been gleaned mainly from the experimentally induced deficiency of single vitamins, without other associated variables, or by the use of specific antivitamins. The results of these studies, for the most part, bear little or no resemblance to the naturally occurring disorders, which are usually the result of combined multiple deficiencies. Any discussion of the neurochemistry of vitamin and other nutritional deficiencies relies, therefore, mainly on information derived from studies on experimental animals; speculations regarding human disease states are based on fragmentary clinical and biochemical observations.

An attempt will be made to summarize, rather than to review in detail, current knowledge of the effects of vitamin and nutritional deprivation on the nervous system.

605

THIAMINE

Thiamine (vitamin B_1), in its phosphorylated form (thiamine diphosphate, also known as thiamine pyrophosphate or cocarboxylase), acts as cofactor in two major enzyme systems that are present in all mammalian tissues, including the nervous system (see also Chap. 14). The first relates to glycolysis and involves the oxidative decarboxylation of pyruvic and α-ketoglutaric acid. The second concerns two transketolation steps of the phosphogluconate pathway (hexose monophosphate shunt), an alternate route of carbohydrate metabolism of importance in biosynthetic mechanisms (see Chap. 14):

ribose 5-phosphate + xylulose 5-phosphate
$\rightleftharpoons$ sedoheptulose 7-phosphate + glyceraldehyde 3-phosphate
xylulose 5-phosphate + erythrose 4-phosphate
$\rightleftharpoons$ fructose 6-phosphate + glyceraldehyde 3-phosphate

In addition to its function as a coenzyme, thiamine has been shown to play a unique role in the function of excitable membranes [2]. Experimentally induced thiamine deficiency in animals bears certain clinical and pathological similarities to naturally occurring thiamine deficiency in man, as exemplified by the Wernicke-Korsakoff syndrome, an affliction of the central nervous system, and beriberi, a disease of the peripheral nervous system. The histopathological changes observed in animals and man assume a fairly constant topography. Although the afflicted regions or parts of the nervous system differ from species to species, the lesions appear to have a definite predilection for bilaterally symmetrical parts of the central nervous system and distal parts of the peripheral nerves. Neurochemical data gathered to date have only partially elucidated some of the pathophysiological mechanisms that underlie this apparently selective vulnerability [3].

Much of the current knowledge concerning thiamine, its role in intermediary carbohydrate metabolism and some of the metabolic consequences brought about by its deficiency in diets, was first elucidated by the classical studies of Peters and his colleagues [4]. The concept of the "biochemical lesion," first articulated in relation to thiamine deficiency and the nervous system by this group, has formed the basis for more recent biochemical investigations.

Experimental Thiamine Depletion

Brain thiamine in normal rat and human brain tends to be fairly evenly distributed, the content being higher in cerebral cortex and cerebellum than in white matter [5]. In human brain, the mammillary bodies, structures which are invariably afflicted in Wernicke's disease, appear to have particularly high concentrations of the vitamin. The monophosphate and diphosphate esters make up about 80 percent of the total thiamine in brain; the triphosphate ester and free thiamine each constitute approximately 10 percent of the total thiamine in that organ. Observations made on rat brain during progressive depletion have shown that brain thiamine begins to fall during the second week of deprivation, experiencing its greatest drop during that week. Total brain thiamine has to be reduced to less than 20 percent of normal before severe neurological signs of deficiency (ataxia, loss of equilibrium, opisthotonus, loss of righting reflexes) and irreversible tissue changes (pannecrosis in the lateral pontine tegmentum) become manifest. Complete reversal of neurological manifestations is noted when brain thiamine returns to 30 percent of normal. These observations suggest that normal cerebral tissue has a substantial reserve of this vitamin. It has been demonstrated that, during progressive depletion, it is predominantly brain thiamine diphosphate that is decreased, whereas free thiamine and thiamine mono-phosphate remain fairly constant and thiamine triphosphate tends to increase. It is of interest that the loss of thiamine diphosphate is greater in the pons, the site of major pathological changes, than it is in any other part of rat brain [6].

Thiamine-Dependent Enzyme Systems
Studies of thiamine-dependent enzyme systems in normal and deficient rat nervous systems suggest that selective vulnerability and the character of histopathological changes may be attributed to the enzymatic topography of the normal nervous system and to the effect of vitamin deprivation on these enzyme systems. In normal adult rat brain, the activity of pyruvic dehydrogenase (the enzyme responsible for the decarboxylation of pyruvate) is highest in areas richly endowed with neurons, such as the cerebellum and the cerebral cortex, and is considerably lower in the spinal cord (white matter). During progressive vitamin depletion, no appreciable decreases in activity can be measured in the brain when the animals are vitamin deficient yet asymptomatic. Even at the most advanced stage of deficiency, when the animals exhibit severe neurological impairment, only the brain stem shows a 25 percent reduction in enzymatic activity. By contrast, even at the asymptomatic stage, the ability to decarboxylate pyruvate is sharply decreased in other organs, such as the heart, liver, and kidney. Lactate and pyruvate levels, measured in various parts of the rat nervous system during progressive depletion, generally tend to mirror the enzymatic defect [7]. Normal brain transketolase activity is highest in the spinal cord (white matter) and brain stem (pons), whereas cellular aggregates, such as the cerebral cortex and the caudate nucleus, have the lowest enzymatic activity. During progressive vitamin depletion, the lateral pontine tegmentum, where the histological changes are most pronounced, exhibits a significant decrease in transketolase activity when compared with other parts of the brain. Severe signs and symptoms of deficiency and irreversible pathological changes become manifest when brain transketolase activity has been reduced by 58 percent. It appears that the defect in transketolation tends to reverse more slowly than does the abnormality of pyruvate metabolism.

Thiamine Deficiency in Developing Brain
Studies performed on immature thiamine-deficient rats during a period when the brain develops rapidly and is most susceptible to nutritional insults have failed to demonstrate changes in myelin lipids or other lipid components that could be separated from the effects of simple undernutrition in either the whole brain or the areas most vulnerable to the lack of thiamine. Thiamine deficiency induced during pregnancy has resulted in offspring with reduced body and brain growth.

Brain DNA concentration tended to be higher and protein-to-DNA ratio lower, suggesting smaller cell size. RNA levels expressed per unit DNA were significantly reduced. Ganglioside concentrations tended to be elevated, cholesterol levels were normal, and cerebroside and total sphingolipid (minus gangliosides) concentrations were markedly reduced. Total phospholipid concentration and the distribution of individual phospholipids were normal. Similar results were noted in pair-fed control animals which were undernourished yet adequately supplied with thiamine [8]. These studies support the notion that thiamine is not related to myelin or its formation in the nervous system. Barchi and Braun [9] could not find any correlation between the content of thiamine and the state of myelination of the nervous system of developing animals, nor did they detect reduced levels of the vitamin in demyelinated areas of human brain afflicted with multiple sclerosis.

Experimental evidence suggests that thiamine-dependent metabolism differs in the nervous systems of the developing and the mature rat. During fetal development, transketolase activity is half of that measured in adult brain. Within the first 10 days of life, the activity rises sharply to about twice the fetal level, reaching a plateau, or adult level, thereafter (Fig. 29-1). When the results are expressed in terms of the DNA or cellular content of the samples, a similar curve is obtained. The rise in brain transketolase activity during the first 10 days of life correlates with the rise of enzymes involved in oxidative phosphorylation and glycolysis. Transketolase activity may reflect an increase in glial cell duplication, proliferation, and migration. Animals born to thiamine-deprived mothers tend to be smaller, and their growth rate is retarded. The cerebral transketolase activity of suckling rats is considerably more depressed than is that of their deficient mothers. This may be a reflection of enzyme

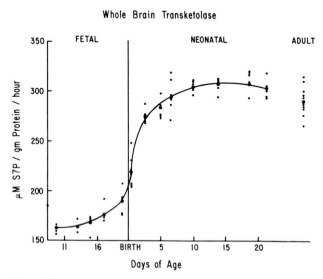

Figure 29-1
Whole-brain transketolase activity as a function of age.

immaturity or of a difference in coenzyme binding in newborn rat brain, rather than of a selective reduction of the vitamin in neonatal brain.

Biochemical Pathology

The extensive investigations carried out by Peters and his colleagues on abnormal pyruvate metabolism in thiamine deficiency and its consequences on brain metabolism as a whole [4] seemed to provide a logical explanation for the neurological symptoms and signs as well as for the observed histopathological changes. More detailed scrutiny of this biochemical lesion, however, has failed to substantiate the previously postulated theories. Theoretically, severe thiamine deficiency caused by faulty pyruvate and alphaketoglutaric acid metabolism should lead to impaired synthesis of both lipids and acetylcholine and to the generation of ATP, particularly in the most severely involved parts of the brain, e.g., the brain stem. Regional concentrations of ATP have been shown to be normal or slightly elevated in thiamine deficiency [7]. The levels of acetylcholine have been found to be normal by some investigators [10]. By contrast, acetylcholine levels have also been noted, by other investigators, to be significantly reduced [11]. These investigators have also found reduced levels of acetyl-coenzyme-A in the face of normal pyruvate dehydrogenase, choline acetyltransferase, and acetylcholine esterase activities. ATP and acetylcholine turnover rates and subcellular pools have not been estimated as yet.

It is generally accepted that the pentose phosphate pathway plays an important role in synthetic mechanisms by virtue of NADPH production for lipid synthesis and ribose production for nucleic acid synthesis. Whereas this metabolic pathway is operative in the central nervous system, some controversy remains regarding its importance. In the adult brain, it probably plays a relatively minor role, although it may assume greater importance in the developing and chronically depleted nervous system. In severe thiamine deficiency, decreased NADPH production could impair fatty acid and nucleic acid synthesis and the conversion of oxidized glutathione to its reduced form (GSH). The latter has been shown to be reduced in the brain stem of symptomatic thiamine-deficient animals [7]. No significant alterations in the profile of cerebral fatty acid have been demonstrated. RNA, and possibly protein, synthesis has been reported to be affected [12].

Neurochemical Studies in Man

Limited neurochemical observations have been made on the human nervous system. Estimation of transketolase activity in normal brain obtained six hours after sudden death has shown high transketolase activity in the mammillary body, a structure invariably affected in cases of Wernicke's disease, and in other structures of the brain stem and diencephalon that are not involved in this disease entity [3]. To date, it has not been possible to define the "biochemical lesion" in patients suffering from thiamine deficiency. Evidence of a severe impairment of tissue transketolase activity has been obtained, however, by simple biochemical measurements of blood. Determinations of pyruvate levels in blood before and after the administration of glucose, for many years the standard method of estimating thiamine deficiency, have proved to lack specificity. Other metabolic disorders can cause an elevation of pyruvate levels in blood. Determinations of transketolase levels in blood have been found to reflect, in a highly sensitive and specific manner, the state of thiamine nutrition in man [13]. Transketolase assays not only reveal the levels of available coenzyme, they also differentiate an acute from a chronic state of deficiency by reflecting the levels of available apoenzyme. Blood transketolase assays provide clinical and biochemical evidence that at least some of the signs and symptoms of Wernicke's disease, particularly ophthalmoplegia, result from a specific lack of thiamine and that their prompt reversibility must be attributed to the presence of a "biochemical lesion" that antedates irreversible histopathological alterations [14]. It has been demonstrated that a deficiency of magnesium interferes with the restoration of transketolase activity by the addition, in vitro, of thiamine diphosphate to thiamine-deficient red blood cells [15].

Role of Thiamine in Membrane Function
Since von Muralt [2] first demonstrated that thiamine, in addition to its function as a coenzyme, plays a unique role in nerve excitation, a number of important observations have been made which have immediate relevance to neurological disease in man. Experimental evidence has shown that electrical stimulation of peripheral nerves and spinal cord promotes the release of thiamine [16]. It has also been shown that physiological concentrations of such drugs as ouabain and tetrodotoxin have a similar effect [16]. It appears that the release involves thiamine triphosphate predominantly and that the phenomenon emanates from a subcellular fraction rich in axonal membranes [16]. Pyrithiamine, a thiamine analog that produces neurological symptoms of deficiency in experimental animals, has a profound and irreversible effect on the electrical activity of nerve preparations. This appears to be due to a displacement of the vitamin rather than to inhibition of thiamine-dependent enzyme systems.

Recent studies have suggested that either thiamine diphosphate or triphosphate, both of which occupy a site on the nerve membrane, is involved in the sodium-gating mechanisms of excitable membranes and that agents such as pyrithiamine and other analogs interfere with membrane sites catalyzing the voltage-dependent changes in membrane permeability to sodium [17]. There appears to be general agreement that thiamine phosphatases and phosphotransferases, which regulate the tissue levels of thiamine and its phosphoesters, are localized in axonal membranes [16]. The effects of thiamine deficiency on membrane function could readily explain the dramatic evolution of the clinical events and some of the subtle, early, ultrastructural changes as well as the lack of demonstrable histological changes noted by conventional methods in most severely affected experimental animals. It is conceivable that the thiamine-deficient membrane is unable to maintain osmotic gradients and that a failure of the energy-dependent component of glial-electrolyte and water transport ensues as the result of relatively acute thiamine deficiency. Chronic depletion of the vitamin may induce other irreversible physiological and histopathological changes.

Thiamine-Responsive Diseases
At least three thiamine-responsive diseases affecting the nervous system have been identified. In many cases of subacute necrotizing encephalomyelopathy, or Leigh's

disease [18], a genetically determined disease of infants that resembles Wernicke's disease, there appears to be a total lack of triphosphothiamine in the brain. Furthermore, patients afflicted with the disease and some normal carriers seem to elaborate a factor found in blood, urine, and cerebrospinal fluid that inhibits thiamine pyrophosphate-ATP phosphoryl transferase, the enzyme responsible for the synthesis of thiamine triphosphate in brain. In some cases of Leigh's disease, the inborn error of metabolism has been shown to consist of a deficiency of pyruvic carboxylase [19], suggesting that variants of the disease exist. An inherited defect in pyruvate oxidation associated with intermittent ataxia has been reported by Blass and his colleagues [20]. In these patients, a defect in the thiamine-dependent first enzyme of the pyruvate dehydrogenase complex — pyruvic decarboxylase — has been found. Finally, it has been demonstrated that thiamine diphosphate increases the activity of the branched-chain α-keto acid dehydrogenase complex of hepatic tissue and fibroblasts cultured from the skin of individuals afflicted with maple syrup urine disease, an inherited disorder characterized by mental retardation and neurological deterioration [21]. The oral administration of thiamine and the dietary restriction of branched-chain amino acids, begun soon after birth, have proved beneficial. It has been shown that α-ketoisocaproic acid, the derivative of leucine, one of the three branched-chain amino acids that accumulate in the plasma of affected children, inhibits the decarboxylation of pyruvate in rat brain [22].

Although thiamine deficiency has been the subject of extensive investigation, large gaps exist in our knowledge concerning the sequence of events that ultimately leads to severe neurological dysfunction and irreversible histopathological changes.

VITAMIN B$_6$ (PYRIDOXINE)

Several forms of vitamin B$_6$ and its active phosphorylated coenzymes have been identified in mammalian tissue, including brain. The vitamin B$_6$ group bears the collective name of pyridoxine, and consists of pyridoxal, pyridoxol, and pyridoxamine. The predominant coenzyme forms of the vitamin in animal tissue are the phosphates of pyridoxal and pyridoxamine. The relative content and the rate of disappearance of these coenzymes during states of deficiency vary from species to species. Pyridoxal 5-phosphotransferase is responsible for the phosphorylation of the vitamin in brain and other organs. Isonicotinic acid hydrazide, the drug used in the treatment of tuberculosis, and 4-deoxypyridoxine, a vitamin B$_6$ antagonist, interfere with the phosphorylating enzyme and cause a decrease in the tissue levels of pyridoxal phosphate.

The dietary deprivation of pyridoxine has been shown to lower the seizure threshold in a variety of mammalian species, including man; the young appear to be more susceptible to convulsions than do the older and more mature (see Chap. 11). Pyridoxine deficiency in pigs is said to result in both convulsions and ataxia, the latter being caused by pathological changes in peripheral nerves, posterior root ganglia, and the posterior funiculi of the spinal cord.

Pyridoxine-Dependent Enzymes

The coenzyme forms of vitamin B_6 act as catalysts in a number of important enzymatic reactions related to the synthesis, catabolism, and transport of amino acids and the metabolism of glycogen and unsaturated fatty acids [23]. Many pyridoxine-dependent enzyme systems have been identified in the nervous system, yet knowledge concerning the biochemical pathology of vitamin B_6 deficiency continues to be limited.

Pyridoxine-dependent enzymes in the nervous system fall into two major categories: transaminases and L-amino acid decarboxylases (see Chap. 11). Some of these enzymes are involved in the γ-aminobutyric acid shunt, an alternate oxidative pathway restricted to nervous tissue, in which α-ketoglutaric acid is metabolized to succinate by way of glutamic and γ-aminobutyric acid. Vitamin B_6 deprivation leads to significant enzymatic depression in all tissues. The affinity of the coenzyme for its apoenzyme varies from enzyme to enzyme. Decarboxylases tend to have a lower affinity for coenzyme than do other enzymes; thus, the decarboxylases tend to be more readily affected than are the transaminases. Severe vitamin deprivation also results in a decrease in enzyme protein. The addition of excess pyridoxal phosphate to a vitamin-deficient enzyme preparation in vitro fails to restore complete activity; apoenzyme production stimulated by the addition of excess vitamin B_6 to a normal tissue extract can be inhibited by puromycin. Thus it would appear that pyridoxal phosphate regulates intracellular enzyme synthesis. It is generally believed that organs or cells with a high rate of protein turnover are most sensitive to pyridoxine depletion. Finally, pyridoxine has been shown to be required for cellular proliferation and for the synthesis of specific proteins involved in immunological reactions [24].

Of the various pyridoxine-dependent enzymes, two decarboxylases appear to be of particular importance to the integrity of neuronal function. The first, glutamic decarboxylase, is responsible for the production of the neuroinhibitor γ-aminobutyric acid from glutamic acid (Chap. 11). Although generally believed to be restricted to neurons, this enzyme may also be active in other mammalian tissue. In rat CNS, enzymatic activity is highest in the hypothalamus and midbrain and lowest in the spinal cord (see Table 29-1). Although significant enzymatic activity can be demonstrated in human cerebral cortex (32 μmol of glutamic acid decarboxylated per gram of protein per hour), none can be measured in white matter.

The second enzyme, 5-hydroxytryptophan decarboxylase, has been shown to be localized in nerve terminals and is involved in the synthesis of serotonin (5-hydroxytryptamine). The same enzyme may be involved in the decarboxylation of L-DOPA to dopamine (see also Chaps. 10 and 32). Exogenous pyridoxine has been found to reduce the effectiveness of L-DOPA in treating Parkinson's disease, presumably because of increased peripheral decarboxylation of L-DOPA. In the rat, the activity of this enzyme appears to be highest in the caudate nucleus, the hypothalamus, and the midbrain, which may be rich in serotoninergic terminals. The activity is surprisingly low in cerebral and cerebellar cortex (see Table 29-2).

Experimental Pyridoxine Deficiency

When weanling rats are fed a diet deficient in vitamin B_6 for several weeks, severe depression of growth and acrodynia of paws, nose, ears, and tail ensue. The animals

Table 29-1. Distribution of Glutamic Decarboxylase Activity in the Nervous System of Normal and Pyridoxine-Deficient Rats

Area Sampled	Micromoles Glutamic Acid Decarboxylated/Gram Protein/Hour	
	Normal[a]	Deficient[a]
Caudate nucleus	73.0	32.2
Cerebral cortex	74.3	34.0
Thalamus	96.7	33.8
Hypothalamus	101.2	47.2
Cerebellar hemisphere	65.8	42.2
Midbrain	100.7	42.8
Spinal cord	31.8	15.2

[a]Means of 7 animals. All determinations carried out in triplicate.
Source: Dreyfus, P. M., Meier, F., and York, C. Unpublished data, 1967.

Table 29-2. Distribution of 5-Hydroxytryptophan Decarboxylase Activity in the Nervous System of Normal and Pyridoxine-Deficient Rats

Area Sampled	Nanomoles ^{14}C-Serotonin Formed/ Gram Protein/Hour	
	Normal[a]	Deficient[a]
Caudate nucleus	4,417.9	2,390.1
Cerebral cortex	621.6	99.5
Thalamus	1,319.3	428.0
Hypothalamus	4,332.0	1,767.2
Cerebellar hemisphere	474.7	165.2
Midbrain	4,204.1	805.3
Spinal cord	1,150.1	226.1
Liver	13,494.6	1,722.2

[a]Means of 7 animals. All determinations carried out in triplicate.
Source: Dreyfus, P. M., Meier, F., and York, C. Unpublished data, 1967.

demonstrate unusual irritability; however, they rarely suffer convulsive seizures or motor weaknesses. This is in sharp contrast to the frequent seizures observed in pyridoxine-deficient newborn rats. When the activity of serum glutamic oxalo-acetic transaminase reaches 50 percent of normal, glutamic decarboxylase and 5-hydroxytryptophan decarboxylase activities are sharply reduced. The loss of glutamic decarboxylase activity is 36 to 65 percent, while that of 5-hydroxytryptophan decarboxylase is 46 to 84 percent, depending upon the area of the nervous system (see Tables 29-1 and 29-2). Decreased enzymatic activity is even greater in pyridoxine-deficient neonatal animals that convulse, suggesting that there is a correlation between the seizure threshold and the level of activity of these two brain enzymes that are dependent on vitamin B_6 [25]. The DNA, RNA, and protein content of the brain of immature pyridoxine-deficient rats has been found to be reduced significantly, although it remains normal in deficient adult animals [26].

B_6 Deficiency in Man

Evidence of vitamin B_6 deficiency in man can be obtained from a number of biochemical determinations on blood and urine. None is thought to specifically reflect involvement of the nervous system. Pyridoxal estimations in blood and cerebrospinal fluid are of limited usefulness. Under normal circumstances, vitamin B_6 and its derivatives can be measured in the urine; none can be found during severe depletion, regardless of the patient's protein intake. Pyridoxine deficiency causes increased urinary excretion of the tryptophan metabolites, xanthurenic and kynurenic acids, after a high dose of tryptophan. Decreased activity of 5-hydroxytryptophan decarboxylase results in decreased excretion of 5-hydroxyindoleacetic acid in the urine, and faulty activity of cysteine sulfinic acid or cysteic acid decarboxylase causes decreased taurine levels in urine. Vitamin B_6 deficiency affects cystathionine cleavage, with resultant cystathioninuria. Occasionally, reduced activity of serum glutamic oxaloacetic and glutamic pyruvic acid transaminases and oxaluria can be demonstrated.

In recent years, pyridoxine dependency has been the subject of a number of investigations (see Chap. 24). This is a familial disorder of infants, characterized by seizures and a high daily requirement of vitamin B_6 (10 mg per day), in which no specific metabolic defect has yet been demonstrated. Although the seizures respond promptly to the administration of pyridoxine and possibly to γ-aminobutyric acid, biochemical evidence of vitamin deficiency is lacking. An abnormality of the glutamic acid decarboxylase system has been postulated [27]. Other pyridoxine-responsive genetic diseases affecting the nervous system have been identified [28]. Homocystinuria, cystathioninuria, and xanthurenic aciduria, associated with mental retardation and occasionally with seizures, may be amenable to effective treatment with large doses of vitamin B_6, which have been shown to enhance the residual activity of the affected enzymes [1].

VITAMIN B_{12} AND FOLIC ACID

Although a great deal is known about the role of vitamin B_{12} in biochemical reactions, virtually nothing is known about its function in the nervous system. However, biochemical studies on the nervous tissue of animals suffering from experimentally induced vitamin B_{12} deficiency may assume greater importance in view of the recent discovery that Rhesus monkeys maintained on a deficient diet for 4 years develop neurological manifestations and demyelinating lesions in the central nervous system [29].

Histopathology in B_{12} Deficiency

In man the insufficient absorption of vitamin B_{12} may result in definite histopathological changes, i.e., those seen in subacute degeneration of the spinal cord, optic nerves, cerebral white matter, and peripheral nerves. This failure of absorption may be caused by the absence of intrinsic factor, disease of the gastrointestinal tract (postgastrectomy, sprue, fish-tapeworm infestation), or vegetarianism. The earliest visible change in the affected parts consists of swelling of individual myelinated nerve fibers in small foci. This is followed by coalescence of the lesions into larger, irregular, spongy, honeycomblike zones of demyelination. Fibers with the largest diameter seem to be predominantly affected, and axis cylinders tend to be spared. Although demyelination appears to be the primary lesion, the possibility of initial involvement of axonal metabolism and axoplasmic flow has not been entirely excluded.

B_{12}-Dependent Enzymes

Only microorganisms that inhabit the mammalian gastrointestinal tract have the capacity to synthesize vitamin B_{12}. Although large amounts of the vitamin are manufactured in the rumen of some herbivorous animals, humans depend upon a dietary source, largely in the form of meat. The natural vitamin cyanocobalamin must be metabolized by the body to its active coenzyme form, 5-deoxyadenosylcobalamin. To date, only two reactions dependent on vitamin B_{12} have been demonstrated in human and other mammalian tissue (Table 29-3) [30]. The isomerization of L-methylmalonyl-CoA to succinyl-CoA requires the active coenzyme form of the vitamin, whereas the transmethylation of homocysteine to methionine probably depends upon another coenzyme form of the vitamin, methyl-B_{12}. The enzymes catalyzing these two reactions are L-methylmalonyl-CoA mutase and N^5-methyltetrahydrofolate homocysteine methyl transferase. Ribonucleotide reductase, an enzyme of great importance in the synthesis of DNA in microorganisms, has not as yet been measured in mammalian tissue, and virtually nothing is known about B_{12}-dependent enzyme systems involved in sulfhydryl reduction.

B_{12} Deficiency and Methylmalonyl-CoA Metabolism

Vitamin B_{12} deficiency in humans and experimental animals results in decreased conversion of methylmalonyl-CoA to succinyl-CoA in the tissues and the urinary excretion of methylmalonic acid ensues. It is not known to what extent reduced levels of vitamin B_{12} coenzyme and consequent reduction in methylmalonyl-CoA mutase activity are responsible for the neurological complications of vitamin B_{12} deficiency. The dietary restriction of vitamin B_{12} results in decreased coenzyme levels in the liver, kidney, and brain of rats and miniature pigs [31], the percentage decrease being approximately the same in each organ. When rats or miniature pigs are fed a diet free of vitamin B_{12}, methylmalonuria is detectable after 4 to 6 weeks. The presence of this organic acid in the urine of animals and humans has been found to be

Table 29-3. Vitamin B_{12}-Dependent Reactions in Mammalian Tissue

Reaction	Active Cofactor
Methylmalonyl-CoA mutase	B_{12} coenzyme
Ribonucleotide reductase (?)	B_{12} coenzyme
Methionine synthetase	Vitamin B_{12}
Sulfhydryl reduction	B_{12} coenzyme (?)

both a specific and a sensitive index of vitamin B_{12} deficiency; it is, in fact, a more reliable indicator than are levels of B_{12} in serum. Levels of methylmalonic acid in cerebrospinal fluid have been found to exceed those in plasma, suggesting that this acid may be elaborated by cerebral tissue as a consequence of faulty propionate metabolism in the brain [32].

Although rats kept on a deficient diet for 8 to 12 months appear clinically to be intact, miniature pigs show a significant reduction in food consumption and body weight, generalized weakness, and intermittent lethargy after 3 months of deprivation. Brain, spinal cord, and peripheral nerves examined by standard-light microscopic methods reveal no histopathological alterations.

It has been shown that, in deficient liver, kidney, and brain, methylmalonyl-CoA mutase activity is reduced; methylmalonyl-CoA is hydrolyzed to methylmalonic acid rather than to succinyl-CoA. The rate of disappearance of methylmalonyl-CoA, however, is essentially the same in normal and deficient tissue [31]. These observations suggest that, in the deficient state, adaptive mechanisms exist that rid the organism of high levels of methylmalonyl-CoA.

Recent investigations into fatty acids and their metabolism in sural nerve samples obtained from patients afflicted with pernicious anemia have revealed the presence of branched-chain (C_{15}) and odd-chain (C_{17}) fatty acids. In addition, the mean content of total lipids and fatty acids, as well as the in vitro rate of synthesis of fatty acids, was significantly reduced in nerve samples from pernicious anemia patients [33]. It has been shown that methylmalonyl-CoA inhibits both acetyl-CoA carboxylase and fatty acid synthetase activity [34]. These findings may explain, in part, the structural changes observed in central and peripheral myelin.

B_{12} Deficiency and Methionine Metabolism

The enzyme N^5-methyltetrahydrofolate homocysteine methyl transferase is involved in the synthesis of methionine from homocysteine. The remethylation of homocysteine to methionine is accomplished together with the regeneration of tetrahydrofolate from N^5-methyltetrahydrofolate. The reaction requires trace amounts of S-adenosylmethionine and a vitamin form of B_{12}, most likely methyl-B_{12}. The activity of the enzyme has been estimated in liver and brain of both humans and rats [35]. In deficient rat liver and brain, enzymatic activity is drastically reduced (Table 29-4). The addition in vitro of excess vitamin B_{12} (cyanocobalamin) partially restores the activity of the enzyme in brain, but has no significant effect on liver-enzyme activity.

The stimulation in vitro of enzymatic activity in normal and deficient brains suggests that this tissue contains excessive amounts of apoprotein. This may represent a protective mechanism against vitamin B_{12} deficiency. A substantial decrease in enzymatic activity may result in a critical reduction of methionine synthesis in neural tissue, which, in humans, could lead to neurological manifestations and histopathological changes in the nervous system.

Even though methionine is considered an essential dietary amino acid, metabolic pathways obviously exist for its synthesis at the cellular level. Methionine is an important constituent of neural protein and may also be a constituent of proteolipids. The

Table 29-4. Activity of N^5-Methyltetrahydrofolate Homocysteine Methyl Transferase in Normal and Vitamin B_{12}-Deficient Rat Liver and Brain[a]

	Controls (6)		Deficient (6)	
	Plain	+B_{12}[b]	Plain	+B_{12}[b]
Liver	246.8 ± 8	248.2 ± 10	36.8 ± 5	48.8 ± 7
Brain	212.0 ± 10	341　± 35	47.0 ± 3	97.3 ± 10

[a]Animals on deficient diet for 8 months. Results expressed as CPM/mg protein/hr ± SEM. Numbers in parentheses represent number of animals. All determinations were carried out in triplicate.
[b]Cyanocobalamin (0.3 μmol) was added to the incubation mixture.
Source: Dreyfus, P. M., and Gross, C. Unpublished data, 1968.

rate of its incorporation into protein is rapid; it is greatest in those regions of the nervous system that contain the highest density of cells, and it is more active in microsomes than in either nuclei or mitochondria. Profound changes in concentrations of amino acid in brain affect the flux of amino acid into protein and influence protein synthesis and degradation. It can be postulated that, in the nervous system, prolonged and severe vitamin B_{12} deprivation results in a critical reduction of methionine levels in tissue, which, in turn, may alter protein synthesis, turnover, and, conceivably, structure and function. In support of this contention, it has been demonstrated that when 1-aminocyclopentane carboxylic acid (a powerful inhibitor of the transmethylation reaction that converts homocysteine to methionine) is administered to adult mice, ataxia and paralysis, as well as a spongiform demyelination of the spinal cord, result [36].

Other Disorders Related to B_{12}

Vitamin B_{12} deficiency has been shown to cause hydrocephalus in newborn rats. It is believed that the anatomical defect is caused by stenosis of the aqueduct of Sylvius, associated with aplasia of the subcommissural organ, a special group of columnar ependymal cells in the roof of the aqueduct and the posterior part of the third ventricle [37]. It has been postulated that the congenital abnormality may be the result of faulty ribonucleic acid metabolism in the newborn brain. Levels of RNA per cell are decreased, whereas the amount of DNA remains normal.

Two distinct types of genetically determined vitamin B_{12}-dependent methylmalonicacidurias associated with developmental arrest, ataxia, and coma have been identified [38] (see also Chap. 24). Examination of the brain, spinal cord, and sciatic nerve obtained from an infant who died of methylmalonicaciduria in which vitamin B_{12} was not converted to deoxyadenosyl B_{12} revealed an accumulation of branched-chain and odd-numbered fatty acids. The former were identified as a mixture of methylhexadecanoic acids. The phosphatides separated from the spinal cord were shown to contain methyl-substituted palmitic acid, with the highest concentration occurring on the beta position of the phosphatidylcholine [39].

Folic Acid Deficiency

Folic acid depletion occurs as a consequence of dietary deficiency, which is sometimes caused by unmet increased needs during pregnancy or by gastrointestinal
defects in absorption. The neurological manifestations engendered in some patients
by folic acid depletion alone are thought to be similar to those caused by the lack
of vitamin B_{12} (see Chap. 24). In general, however, most folate-deficient patients
are neurologically intact.

It is now well established that patients receiving anticonvulsant medications,
particularly diphenylhydantoin, may develop folate deficiency. Conversely, the
administration of folic acid significantly reduces the levels of these drugs in blood
[40]. Presumably the vitamin and the drug compete for intestinal absorption.

Folic acid and its derivatives are involved in the formation of methionine from
homocysteine and in the synthesis of purines and pyrimidines and, hence, of RNA
and DNA. Folate is therefore fundamental to normal cell division and growth. The
active coenzyme form of folate contains four additional hydrogen atoms (tetrahydrofolate). Folic acid reductase permits the reduction of folate to tetrahydrofolate.
Various derivatives of tetrahydrofolate have been identified, including methyl-,
methenyl-, methylene-, hydroxymethyl-, formyl-, and formiminotetrahydrofolate.
In general, the various forms of folate are instrumental in the transfer of single
carbon units. As yet, the specific role of folic acid in cerebral metabolism has not
been elucidated although it is interesting to note that folate levels in cerebrospinal
fluid are two to three times those in serum, even in states of folate deficiency [32].
Clinical studies have shown that anticonvulsant therapy does not alter the relationship between serum and cerebrospinal fluid folate levels.

The neurological complications of pernicious anemia may be precipitated when
folic acid is administered in order to treat the anemia. This phenomenon remains
essentially unexplained.

NIACIN, PANTOTHENIC ACID, AND VITAMIN E

Niacin

In the nervous system, as in other tissues, nicotinic acid, or niacin, is a constituent
of the two coenzymes that transfer hydrogen or electrons. These coenzymes are
nicotinamide adenine dinucleotide (NAD) and nicotinamide adenine dinucleotide
phosphate (NADP). Both of these nucleotides are essential to a number of important enzymatic reactions involved in carbohydrate, fatty acid, and glutathione
metabolism. The significance of these biochemical reactions has been covered in
other chapters (see Chaps. 14 and 15).

Most mammalian cells synthesize nicotinic acid from tryptophan. It is not known
whether this synthesis can also take place in the cells of the nervous system. In man,
a deficiency of niacin is responsible for the neurological and psychological manifestations of pellagra: encephalopathy associated with signs of spinal cord and peripheral
nerve involvement. The pathological changes observed in this disorder are known as
central chromatolysis, or central neuritis, and consist of a characteristic degeneration
of the large pyramidal cells (Betz cells) of the motor cortex.

A deficiency of niacin brings about visible alterations of neuronal Nissl substance, but the mechanism involved is not yet known. It has been demonstrated that niacin deficiency causes a reduction of cerebral NAD and NADP levels and decreased activity of the enzymes that depend upon these nucleotides. No specific neurochemical lesion, however, has as yet been identified. It is well recognized that, in patients afflicted with pellagra, NAD levels in red blood cells are reduced and the urinary excretion product of niacin, N-methylnicotinamide, is diminished. The administration of niacin to pellagrins and to niacin-deficient animals causes the nucleotide content of the blood and the tissues to rise temporarily above normal levels. An experimental myelopathy has been produced in rats by the administration of 6-aminonicotinamide, an antimetabolite of nicotinamide. This agent, which is said to produce structural changes in glial cells, interferes with the activity of the pentose phosphate pathway [41].

Pantothenic Acid

Pantothenic acid, a constituent of coenzyme-A (CoA), is widely distributed in mammalian tissue. Pantothenic acid differs from other vitamins, however, in that it is not the active unit of CoA. In addition to pantothenic acid, CoA contains ribose, adenine, phosphoric acid, and β-mercaptoethylamine. The sulfhydryl group of the latter is the site that links acid and acetyl groups.

As part of CoA, pantothenic acid participates in a variety of biochemical reactions involved in fatty acid synthesis and oxidation as well as in the metabolism of steroids and acetylcholine. Generally speaking, the reactions mediated by CoA fall into two categories: acetokinases and transacetylases [42]. The vitamin is also essential to the formation of certain amide and peptide linkages.

Relatively high concentrations of pantothenic acid exist in the brain, which does not readily yield its stores. Experimentally induced pantothenic acid deficiency in animals results in lesions of the peripheral nerves. In man, pantothenic acid deficiency causes numbness and tingling of hands and feet and occasionally a "burning-foot" syndrome. Biochemical changes that have been noted in such patients are characterized by the impaired ability to acetylate p-aminobenzoic acid and by a decline in blood levels of cholesterol and its esters. In addition, evidence of adrenal-cortical hypofunction has been described [43]. It has been noted that a deficiency of the vitamin in experimental animals results in deranged synthesis of acetylcholine, cholesterol, glucosamine, and galactosamine, in faulty fatty acid oxidation, and in reduced energy production. More specific information concerning the neurochemistry of pantothenic acid deficiency is lacking.

Vitamin E

Alpha-tocopherol, the major component of vitamin E, is generously distributed in nature. Investigations aimed at delineating the metabolic role of tocopherols strongly suggest that the vitamin acts as a nonspecific antioxidant and, as such, prevents the peroxidation of polyunsaturated fatty acids, a nonenzymatic reaction that occurs normally in the course of intracellular metabolism. The resultant products of peroxidation are highly damaging to the cell. It is of interest that human infant brain, which is relatively abundant in highly unsaturated fatty acids, contains less tocopherol than does any other tissue, including adipose tissue. It is therefore conceivable that the requirements for vitamin E are higher in infants than in adults. There is evidence that the antioxidant effect of tocopherol preserves reduced glutathione (GSH) stability through the maintenance of sulfhydryl bonds and that the vitamin inhibits membrane adenosine triphosphatase. Thus it may also play a role in the stabilization of biological membranes.

Experimentally induced vitamin E deficiency in animals has resulted in a number of interesting observations. Myocardial degeneration, a dystrophic process of voluntary muscles, encephalomalacia, axonal dystrophy (particularly pronounced in the dorsal funiculi of the spinal cord), and ceroid (a form of lipochrome pigment) deposits in muscles and neurons have all been described. The persistent inability to absorb fats is the most likely mechanism by which human subjects become deficient in this fat-soluble vitamin. In children, the most common underlying causes of deficiency are celiac disease, cystic fibrosis, and biliary atresia. In the adult, sprue and chronic pancreatitis have been the most frequent offenders. On occasion, a progressive myelopathy, ceroid deposition in smooth muscles, and skeletal muscle lesions that resemble muscular dystrophy have been reported.

Recently, Zeeman [44] has suggested that the administration of a mixture of vitamin E, ascorbic acid, methionine, and butylated hydroxytoluene has a beneficial effect on the course of Batten's disease, a juvenile form of amaurotic familial idiocy. It has been postulated that, in the course of this disease, the massive peroxidation of essential fatty acids leads to the neuronal accumulation of cross-linked polymers in the form of lipopigments, lipofuscin, and ceroid [44]. Normal levels of blood and tissue alpha-tocopherol have been found in patients afflicted with this disorder.

PROTEIN-CALORIE MALNUTRITION

According to current estimates, approximately one-fourth of the world population may be suffering from varying forms of undernutrition. Although the problem in general is cause for concern, particular significance is attached to the proportion of this population represented by children. The possibility that undernutrition during a period of functional development of the brain may seriously impair subsequent intellectual function [45, 46] has provided a considerable impetus to the study of a causal relationship at the molecular level. Two extreme forms of infant malnutrition are classified clinically as kwashiorkor, produced by a deficiency of protein, and the more severe marasmus, resulting from a lack of both protein and carbohydrate. It is clear that the intermediate forms existing between these two extreme conditions affect the greatest percentage of the undernourished population.

A unifying concept that has emerged is that the brain is markedly susceptible to environmental insults during a period of intense functional differentiation [47, 48] (see also Chap. 18). In the rat, this corresponds to a period of ontogenesis between birth and about 21 days of age, when the proliferation of neuronal processes and myelinogenesis is most rapid. An analogous period in the human is from approximately 30 weeks of gestation to 18 to 24 months of age. Most of our knowledge about neurochemical changes accompanying undernutrition during early life has been derived from animal experimentation. Often, however, too little regard has been given to the timing and duration of experimental undernutrition to make it comparable to that occurring in humans. In addition, the variety of procedures employed to produce nutrient deficiency in animals — restricting periods of suckling, increasing litter size, and decreasing protein alone or protein and calories — makes comparison and interpretation of results difficult. A further complicating factor often lost sight of is the heterogeneous nature of the brain, which is reflected in differences in cellularity and regional rates of development. Animal models have, however, provided valuable information, much of which may be extrapolated to the human condition.

Animal Studies

Protein-calorie malnutrition in the rat during early life induces structural defects in the CNS, the most prominent of which are increased cell density, impaired axonal and dendritic aborization, and diminished synaptogenesis [49–52]. The total brain and regional DNA content is decreased [53, 54]; at the same time, there is a marked suppression of the rate of DNA synthesis [55]. A combined histological and microchemical study of the cerebral cortex suggests that the alteration of DNA levels in undernourished suckling rats results from delayed glial cell proliferation and migration [56].

Several lines of evidence have demonstrated the adverse effects of undernutrition on myelinogenesis. The diminished recovery [57] and histological appearance of myelin [58] in undernourished infant rats are reflected in the reduced whole brain and regional concentration [8, 59] and synthesis [60] of myelin-associated lipids.

The modification of certain key constituents of the brain by undernutrition may conceivably mediate changes in behavior. Neonatal undernutrition in rats severely retards the conversion of glucose to amino acids [61]; protein malnutrition in immature pigs [62] and rats [63] alters the levels and metabolism of amino acids. The incorporation of precursors into cerebral RNA of perinatally malnourished mice [64] and postnatally undernourished rats [65] is diminished. General undernutrition, as well as severe protein malnutrition, in developing rats specifically alters the activity of enzymes associated with nerve endings [54, 66, 67] and is responsible for alterations in the levels of central neurotransmitters [54, 68].

Whether learning capacity is impaired by nutritional inadequacy in early life is a question whose answer has thus far remained elusive. The performance of undernourished rats is inferior in behavioral tests [69, 70], but it has been suggested that the results of tests designed to measure intellectual performance may be confounded by emotional and motivational defects [71].

A further important consideration is the extent of reversibility of changes induced by nutritional deficiencies. In this respect, a distinction must be made between those biochemical alterations that represent a temporary adaptive response and those due to pathological change, which are not reversible by nutritional rehabilitation. Most evidence favors the view that undernutrition during a critical phase of brain development induces biochemical changes that are not reversible by subsequent reinstitution of an adequate diet, even for a prolonged period.

The hormonal status of nutritionally deprived animals is frequently overlooked [72], and a diminished thyroid function [73] can be expected to contribute to the modification of brain development and mental capacity.

Human Studies

Neurochemical data on malnourished children are obviously scarce and, when available, are derived from a limited population whose clinical history is frequently unknown. In addition, the data are often compared to an inadequate control population.

Behavioral studies in undernourished children have demonstrated an altered intellectual function [45, 46]. The persistence of the defect may be directly related to the age of onset of nutritional deprivation. In most cases, undernutrition in children is accompanied by a multitude of variables, including disease, cultural differences, and psychological factors. Thus the apparent effect of malnutrition on learning performance in children has been considered to arise from coexisting variables other than differences in intellectual capacity.

CONCLUSION

Many important findings have emerged from animal and human studies. There appears to be a marked correlation between the extent and persistence of cerebral dysfunction and the age of onset and duration of malnutrition. It seems well established that certain molecular indices of central nervous system structure and function are irreversibly altered by undernutrition. In the future, controlled studies will be necessary to establish whether specific biochemical alterations attributable to nutritional deprivation are reflected in defective learning ability and diminished intellectual development.

REFERENCES

*1. Rosenberg, L. E. Vitamin-Responsive Inherited Diseases Affecting the Nervous System. In Plum, F. (Ed.), *Brain Dysfunction in Metabolic Disorders.* New York: Raven, 1974.

*2. von Muralt, A. The role of thiamine (vitamin B_1) in nervous excitation. *Exp. Cell Res.* Suppl. 5, 72, 1958.

*3. Dreyfus, P. M. Transketolase Activity in the Nervous System. In Wohlstenholme, G. E. W. (Ed.), *Thiamine Deficiency: Biochemical Lesions and Their Clinical Significance.* Boston: Little, Brown, 1967.

*4. Peters, R. A. *Biochemical Lesions and Lethal Synthesis.* London: Pergamon, 1963.

5. Dreyfus, P. M. The quantitative histochemical distribution of thiamine in normal rat brain. *J. Neurochem.* 4:183, 1959.

6. Pincus, J. H., and Grove, I. Distribution of thiamine phosphate esters in normal and thiamine deficient brain. *Exp. Neurol.* 28:477, 1970.

*Asterisks denote key references.

7. McCandless, D. W., and Schenker, S. Encephalopathy of thiamine deficiency: Studies of intracerebral mechanisms. *J. Clin. Invest.* 47:2268, 1968.

8. Geel, S. E., and Dreyfus, P. M. Brain lipid composition of immature thiamine-deficient and undernourished rats. *J. Neurochem.* 24:353, 1975.

9. Barchi, R. L., and Braun, P. E. Thiamine in neural membranes. A developmental approach. *Brain Res.* 35:622, 1971.

10. Speeg, K. V., Chen, D., McCandless, D. W., and Schenker, S. *Proc. Soc. Exp. Biol. Med.* 134:1005, 1970.

11. Heinrich, C. P., Stadler, H., and Weiser, H. The effect of thiamine deficiency on the acetylcoenzyme A and acetylcholine levels in the rat brain. *J. Neurochem.* 21:1273, 1973.

12. Reinauer, H., and Hollmann, S. Zur coenzymbedingten Induktion der pyruvat-dehydrogenase bei thiaminmangel. *Hoppe-Seylers Z. Physiol. Chem.* 350:40, 1969.

13. Brin, M. Erythrocyte transketolase in early thiamine deficiency. In Wuest, H. M. (Ed.), Unsolved Problems of Thiamine. *Ann. N.Y. Acad. Sci.* 98:528, 1962.

14. Dreyfus, P. M. Clinical application of blood transketolase determinations. *N. Engl. J. Med.* 267:596, 1962.

15. Zieve, L. Influence of magnesium deficiency on the utilization of thiamine. *Ann. N.Y. Acad. Sci.* 162:732, 1969.

16. Itokawa, Y., Schulz, R. A., and Cooper, J. R. Thiamine in nerve membranes. *Biochim. Biophys. Acta* 266:293, 1972.

17. Barchi, R. Unpublished data. 1974.

*18. Pincus, J. H. Subacute necrotizing encephalomyelopathy (Leigh's disease): A consideration of clinical features and etiology. *Dev. Med. Child Neurol.* 14:87, 1972.

19. Tang, T. T., Good, T. A., Dyken, P. R., Johnsen, S. D., McCreadie, S. R., Sy, S. T., Lardy, H. A., and Rudolph, F. B. Pathogenesis of Leigh's encephalomyelopathy. *J. Pediatr.* 81:189, 1972.

20. Blass, J. P., Avigan, J., and Uhlendorf, B. W. A defect in pyruvate decarboxylase in a child with an intermittent movement disorder. *J. Clin. Invest.* 49:423, 1970.

21. Elsas, L. J., and Danner, D. J. Unpublished data. 1974.

22. Bowden, J. A., McArthur, C. L., and Fried, M. The inhibition of pyruvate decarboxylation in rat brain by α-ketoisocaproic acid. *Biochem. Med.* 5:101, 1971.

*23. Williams, M. A. Vitamin B_6 and amino acids: Recent research in animals. *Vitam. Horm.* 22:561, 1964.

*24. Axelrod, A. E., and Trakatellis, A. C. Relationship of pyridoxine to immunological phenomena. *Vitam. Horm.* 22:591, 1964.

25. Wiss, O., and Weber, F. Biochemical pathology of vitamin B_6 deficiency. *Vitam. Horm.* 22:495, 1964.

26. Bhagavan, H. N., and Coursin, D. B. Effects of pyridoxine on nucleic acid and protein contents of brain and liver in rats. *Int. J. Vitam. Nutr. Res.* 41:419, 1971.

27. Scriver, C. R., and Whelan, D. T. Glutamic acid decarboxylase in mammalian tissue outside the central nervous system and its possible relevance to hereditary vitamin B_6 dependency with seizures. *Ann. N. Y. Acad. Sci.* 166:83, 1969.

*28. Mudd, S. H. Pyridoxine-responsive genetic disease. *Fed. Proc.* 30:970, 1971.

29. Agamanolis, D. P., Victor, M., Chester, E. M., Banker, B. Q., Kark, J. A., Hines, J. D., and Harris, J. W. Neuropathological Changes in Vitamin B_1-Deficient Monkeys. Proceedings of the American Association of Pathologists, May–June 1975, p. 70, New York.

*30. Stadtman, T. C. Vitamin B_{12}. *Science* 171:859, 1971.

31. Cardinale, G. J., Dreyfus, P. M., Auld, P., and Abeles, R. H. Experimental vitamin B_{12} deficiency. *Arch. Biochem. Biophys.* 131:92, 1969.

32. Girdwood, R. H. Abnormalities of vitamin B_{12} and folic acid metabolism: Their influence on the nervous system. *Proc. Nutr. Soc.* 27:101, 1968.

33. Frenkel, E. P. Abnormal fatty acid metabolism in peripheral nerves of patients with pernicious anemia. *J. Clin. Invest.* 52:1237, 1973.

34. Frenkel, E. P., Kitchens, R. L., and Johnston, J. M. The effect of vitamin B_{12} deprivation on the enzymes of fatty acid synthesis. *J. Biol. Chem.* 248:7540. 1973.

35. Levy, H. L., Mudd, S. H., Schulman, J. D., Dreyfus, P. M., and Abeles, R. H. A derangement in B_{12} metabolism associated with homocystinemia, cystathioninemia, hypomethioninemia and methylmalonic aciduria. *Am. J. Med.* 48:390, 1970.

36. Gandy, G., Jacobson, W., and Sidman, R. Inhibition of transmethylation reaction in the central nervous system: An experimental model for subacute combined degeneration of the cord. *J. Physiol.* (Lond.) 233:1P, 1973.

37. Woodard, J. C., and Newberne, P. M. The pathogenesis of hydrocephalus in newborn rats deficient in vitamin B_{12}. *J. Embryol. Exp. Morphol.* 17:177, 1967.

*38. Mahoney, M. J., and Rosenberg, L. E. Inherited defects of B_{12} metabolism. *Am. J. Med.* 48:584, 1970.

39. Kishimoto, Y., Williams, M., Moser, H. W., Hignite, C., and Biemann, K. Branched-chain and odd-numbered fatty acids and aldehydes in the nervous system of a patient with deranged vitamin B_{12} metabolism. *J. Lipid Res.* 14:69, 1973.

40. Bayliss, E. M., Crowley, J. M., Preece, J. M., Sylvester, P. E., and Marks, V. Influence of folic acid on blood-phenytoin levels. *Lancet* 1:62, 1971.

41. Herken, H., Lange, K., Kolbe, H., and Keller, K. Antimetabolic Action on the Pentose Phosphate Pathway in the Central Nervous System Induced by 6-Aminonicotinamide. In Genazzani, E., and Herken, H. (Eds.), *Central Nervous System — Studies on Metabolic Regulation and Function.* Berlin: Springer-Verlag, 1974.

42. Novelli, G. D. Metabolic functions of pantothenic acid. *Physiol. Rev.* 33:525, 1953.

43. Bean, W. B., and Hodges, R. E. Pantothenic acid deficiency induced in human subjects. *Proc. Soc. Exp. Biol. Med.* 86:693, 1954.

44. Zeeman, W. Studies in the neuronal ceroid-lipofuscinoses. *J. Neuropathol. Exp. Neurol.* 33:1, 1974.

*45. Tizard, J. Early malnutrition, growth and mental development in man. *Br. Med. Bull.* 30:169, 1974.

*46. Cravioto, J., Hambraeus, L., and Valquist, B. (Eds.). *Early Malnutrition and Mental Development.* Symp. Swedish Nutr. Found. Vol. XII. Uppsala: Alqvist and Wiksell, 1974.

*47. Dobbing, J. Undernutrition and the Developing Brain. In Lajtha, A. (Ed.), *Handbook of Neurochemistry.* Vol. 6. New York: Plenum, 1971, p. 255.

*48. Dobbing, J., and Smart, J. L. Vulnerability of developing brain and behavior. *Br. Med. Bull.* 30:164, 1974.

49. Cragg, B. G. The development of cortical synapses during starvation in the rat. *Brain* 95:143, 1972.

50. Gambetti, P., Autilio-Gambetti, L., Rizzuto, N., Shafer, B., and Pfaff, L. Synapses and malnutrition: Quantitative ultrastructural study of rat cerebral cortex. *Exp. Neurol.* 43:464, 1974.

51. Stewart, R. J. C., Merat, A., and Dickerson, J. W. T. Effect of a low protein diet in mother rats on the structure of the brains of the offspring. *Biol. Neonate* 25:125, 1974.

52. Salas, M., Diaz, S., and Nieto, A. Effects of neonatal food deprivation on cortical spines and dendritic development of the rat. *Brain Res.* 73:139, 1974.

*53. Winick, M. Nutrition and nerve cell growth. *Fed. Proc.* 29:1510, 1970.

54. Sobotka, T. J., Cook, M. P., and Brodie, R. E. Neonatal malnutrition: Neurochemical, hormonal and behavioral manifestations. *Brain Res.* 65:443, 1974.

55. Lewis, P. D., Balazs, R., Patel, A. J., and Johnson, A. L. The effect of undernutrition in early life on cell generation in the rat brain. *Brain Res.* 83:235, 1975.

56. Bass, N. H., Netsky, M. G., and Young, E. Effect of neonatal malnutrition on developing cerebrum. I. Microchemical and histologic study of cellular differentiation in the rat. *Arch. Neurol.* 23:289, 1970.

57. Fishman, M. A., Madyastha, P., and Prensky, A. L. The effect of undernutrition on the development of myelin in the rat CNS. *Lipids* 6:458, 1971.

58. Bass, N. H., Netsky, M. G., and Young, E. Effect of neonatal malnutrition on developing cerebrum. II. Microchemical and histologic study of myelin formation in the rat. *Arch. Neurol.* 23:303, 1970.

59. Geison, R. L., and Waisman, H. A. Effects of nutritional status on rat brain malnutrition as increased bv lipid composition. *J. Nutr.* 100:315, 1970.

60. Chase, H. P., Dorsey, J., and McKhann, G. M. The effect of malnutrition on the synthesis of a myelin lipid. *Pediatrics* 40:551, 1967.

*61. Balazs, R., and Patel, A. J. Factors Affecting the Biochemical Maturation of the Brain: Effect of Undernutrition During Early Life. In Ford, D. H. (Ed.), *Neurobiological Aspects of Maturation and Ageing.* Vol. 40. Amsterdam: Elsevier, 1973, p. 115.

62. Badger, T. M., and Tumbleson, M. E. Protein-caloric malnutrition in young miniature swine: Brain free amino acids. *J. Nutr.* 104:1329, 1974.

63. Roach, M. K., Corbin, J., and Pennington, W. Effect of undernutrition on amino acid compartmentation in the developing rat brain. *J. Neurochem.* 22:521, 1974.

64. Lee, C. J. Biosynthesis and characteristics of brain protein and ribonucleic acid in mice subjected to neonatal infection on undernutrition. *J. Biol. Chem.* 245:1998, 1970.

65. De Guglielmone, A. E. R., Soto, A. M., and Duvilanski, B. H. Neonatal undernutrition and RNA synthesis in developing rat brain. *J. Neurochem.* 22:529, 1974.

66. Adlard, P. P., and Dobbing, J. Vulnerability of developing brain. V. Effects of total and postnatal undernutrition on regional brain enzyme activities in three-week-old rats. *Pediatr. Res.* 6:38, 1972.

67. Rajalakshmi, R., Parameswaran, M., Telang, S. D., and Ramakrishnan, C. V. Effects of undernutrition and protein deficiency on glutamate dehydrogenase and decarboxylase in rat brain. *J. Neurochem.* 23:129, 1974.

68. Shoemaker, W. J., and Wurtman, R. J. Effect of perinatal undernutrition on the metabolism of catecholamines in the rat brain. *J. Nutr.* 103:1537, 1973.

69. Baird, A., Widdowson, E. M., and Cowley, J. J. Effects of caloric and protein deficiencies early in life on the subsequent learning ability of rats. *Br. J. Nutr.* 25:391, 1971.

70. Wells, A. M., Geist, C. R., and Zimmermann, R. R. Influence of environmental and nutritional factors on problem solving in the rat. *Percept. Mot. Skills* 35:235, 1972.

*71. Levitsky, D. A. Early malnutrition and behavior. *N. Y. State J. Med.* 71:350, 1971.
*72. Gardner, L. I., and Amacher, P. Endocrine aspects of malnutrition. *Kroc Found. Symp. No. 1.* Santa Ynez, Calif.: Kroc Foundation, 1973, p. 1.
 73. Shambaugh, G. E., and Wilbur, J. F. The effect of caloric deprivation upon thyroid function in the neonatal rat. *Endocrinology* 94:1145, 1974.

Chapter 30

Cerebral Edema

Donald B. Tower

Edema is a frequent and important complication of trauma, vascular accidents, tumors, and various inflammatory or toxic states of the central nervous system, and it is often a major sequela of surgery on the brain and spinal cord. Clinical recognition of edema probably dates from the time of Hippocrates, but our understanding of the nature and therapy of the various types of edema is still very incomplete. The clinical picture of cerebral edema comprises general signs and symptoms of increased intracranial pressure (especially impaired consciousness), various and inconstant focal cerebral signs, and secondary complications, such as signs of cerebral herniation. These signs and symptoms may also characterize the primary, underlying pathological condition, so the diagnosis of edema as a complication and a critical evaluation of responses to therapy are often problematical. Moreover, the more objective means of ascertaining the presence and extent of the edema are seldom applicable to the acute clinical emergency. Most of the following discussion is based on studies in various animal species of experimentally induced forms of cerebral edema that resemble, but are not necessarily identical to, their respective clinical counterparts.

In the older literature, especially from Germany, a distinction was made between brain swelling (*Hirnschwellung*), or an increase in bulk of the brain with essentially no interstitial or intracellular fluid component, and brain edema (*Hirnödem*). Most investigators no longer consider this to be a useful or even a valid distinction. The two phenomena are usually viewed as different phases of the same process, and the two terms are often used interchangeably. In the biochemical literature, swelling is the term more commonly employed to denote imbibition of extra fluid by tissue samples. A more recent classification has been proposed [1], subdividing cerebral edema into the vasogenic type (primarily extracellular in location) and the cytotoxic type (primarily intracellular in location).

The earlier concepts of cerebral edema, based on gross changes of brain volume and fluid content [2, 3], have been supplanted by the recognition that edema usually involves white matter rather selectively, although slices of cerebral cortex swell readily [4, 5]. Selective edema of subcortical white matter was demonstrated in the classic study on human brain tumors by Stewart-Wallace [6] and by more recent experimental studies by Aleu et al. [7] and Pappius and colleagues [8], among others. Much of the more recent work on the clinical, pathological, and experimental aspects of edema of the central nervous system was reviewed at a symposium held in Vienna in 1965. For details beyond the scope of this chapter, the reader is referred to the published proceedings of that symposium [9] or to the latest monograph on brain fluids and electrolytes by Katzman and Pappius [10].

QUANTIFICATION

In Vivo

The usual procedure for the definition of edema in vivo involves sampling the edematous area of brain (or spinal cord) and a control area (preferably the homologous contralateral area) and determining the respective dry-weight percentages. This is most conveniently done by weighing each fresh tissue sample (wet weight) in a tared container, drying the sample at approximately 100°C to constant weight (usually within 12 to 24 hr), and reweighing to obtain the weight of the residue (dry weight). The percentage is simply calculated as:

$$(\text{dry weight/wet weight}) \times 100 = \%\text{ dry weight (P)} \tag{30-1}$$

The percentage of swelling or edema (or of shrinkage) can then be calculated by the formula of Elliott and Jasper [3]:

$$(P_{control} - P_{exptl})/P_{exptl} \times 100 = \%\text{ swelling (or shrinkage)} \tag{30-2}$$

If there is an appreciable content of solids (e.g., protein) in the edema fluid, a more accurate calculation is given by:

$$(P_{control} - P_{exptl})/(P_{exptl} - P_F) \times 100 = \%\text{ swelling (or shrinkage)} \tag{30-2a}$$

where P_F = dry-weight percentage of the edema fluid. Elliott and Jasper [3] point out that it is misleading to report results in terms of percentage of water in the tissue, because a change in water content from 80 to 81 percent represents a swelling and hence an increase in weight of 5.3 percent. The foregoing method of calculation is usually necessary for studies in vivo because it is frequently difficult or impossible to obtain meaningful differences in tissue weights directly.

On the other hand, the foregoing procedures assume a uniform distribution of fluid within the respective normal and edematous samples. Such may not always be the case, especially when effects of therapeutic measures are being evaluated. In these cases, actual weights of control and experimental hemispheres should be obtained to supplement the dry-weight data. An example of this approach is the recent study by Pappius and McCann [11] on the effects of steroids on experimentally induced cerebral edema in cats.

Additional insight into the nature of the edema fluid can be obtained by analyses of the tissue samples for proteins and electrolytes (Na^+, K^+, and Cl^-). Initially, such data will be obtained per unit of fresh (or wet) weight of tissue sample. In this form, the data may not be as meaningful as when expressed per unit of dry weight or as ratios (e.g., Na^+/K^+). In many cases, the edema fluid may be enriched in only one monovalent cation, with a consequent change in the Na^+/K^+ ratio. In such cases, the calculation in terms of dry weight will verify which ion remained unaffected by the influx of edema fluid.

In Vitro

For slices of cerebral tissues studied in vitro, a different approach is required. Under most circumstances, once a slice has been cut, its dry weight (solids) will remain relatively constant. Hence, wet weights (W) must be taken initially and after the experimental period to assess the changes of fluid content within the tissue slice:

$$(W_{final} - W_{initial})/W_{initial} \times 100 = \% \text{ swelling} \qquad (30\text{-}3)$$

Of course, dry weights should also be obtained, whenever feasible, to check any significant change in the content of solids in such samples. Some investigators prefer to report their data for fluid content of slices as weight or volume of fluid per unit of dry weight. Differences between control and experimental samples can be appreciated readily, but, unless the absolute values for wet and dry weights are reported, one may be unable to convert such data to percentage swelling.

TYPES OF CEREBRAL EDEMA

There is a distinct contrast between the edema of cerebral tissues encountered clinically or experimentally in vivo and that seen commonly in preparations in vitro. Swelling of cerebral tissues in vitro is characteristically of gray matter, whereas most types of edema in vivo are localized principally to white matter, usually in association with disruption of the blood-brain barrier. However, there may be some relationship between swelling in high K^+ media in vitro and cerebral edema in vivo.

Edema Associated with Tumors

Historically, the recognition of white matter as a site of predilection for many types of edema of neural tissues came from the studies by Stewart-Wallace [6] on the brains of patients with intracranial tumors. In these brains, the swelling of cerebral cortex was less than 5 percent, but in subcortical white matter homolateral or adjacent to the tumor, the swelling averaged 77 percent (Table 30-1). Several subsequent studies utilizing experimental models have confirmed this observation either morphologically or biochemically. The experimental models include implanted tumors, the

Table 30-1. Types of Edema in Cerebral Tissues in Vivo

Species	Type of Lesion	Cerebral Cortex Dry Weight (%)			Subcortical White Dry Weight (%)		
		Control	Exptl.	Swelling (%)	Control	Exptl.	Swelling (%)
Man [6][a]	Tumor	17.4	16.9	< 5	30.7	17.4	77
Cat [8]	Freezing lesion	19.3	18.3	5.5	31.9	21.4	49
Rabbit [7, 12]	Triethyl tin	20.2	20.0	0	30.8	22.9	34
Cat [13]	Circulatory arrest	16.0	13.5	18.5[b]	29.6	29.0	0

[a]References to sources of data are given in brackets.
[b]For the monkey, the comparable value is 22.2% swelling [14].

use of intracranial balloons, and the implantation of such hydrophilic materials as psillium seeds, polysaccharides, or proteins. Despite the focal and often variable nature of the edema in these preparations, there is good agreement among investigators that the edema fluid resembles a plasma exudate with a relatively high content of Na^+ and Cl^- and a demonstrable content of protein, notably albumin [12]. The presence of this protein exudate is indicative of the disruption of the blood-brain barrier (see Chap. 20).

Freezing Lesions
A similar type of edema is seen in the cerebral white matter of animals after the brain has been subjected to a localized freezing lesion. The technique involves application of a metal plate, cooled to $-50°C$ or below, to the intact skull or directly to the surface of the brain. This produces a local necrotizing lesion that breaches the blood-brain barrier. In 24 hr after placement of the lesion, edema fluid spreads from the white matter of the traumatized gyrus to involve adjacent white matter extensively, the rate of spread being a function primarily of the systolic blood pressure. The process seems to be most consistent with an extravascular migration of plasma-like fluid via extracellular routes to the white matter beneath and surrounding the lesion [15]. This form of edema has been termed vasogenic edema [1]. Morphologically and biochemically, the freezing lesion closely resembles traumatic lesions of the brain, such as a stab wound, as well as the edema around tumors and foreign substances.

From the behavior of edema fluid around tumors or the freezing lesion, it seems that the interstitial spaces of cerebral gray matter are much less distensible than are those of cerebral white matter, so the exudates from the plasma that penetrate the breached blood-brain barrier tend to collect in the latter location.

Direct analyses of cerebral tissues from the control hemisphere and from the hemisphere in which the freezing lesion was placed demonstrate minimal swelling (about 5 percent) of cerebral cortex, but very pronounced swelling (about 50 percent) of subcortical white matter on the side of the lesion (see Table 30-1). The edema fluid consists principally of Na^+ and Cl^- plus an appreciable content of protein (Table 30-2). The composition of the edema fluid probably reflects some dilution and contamination with intracellular (nonedema) fluid during the isolation procedure [16], but it still primarily resembles a plasma exudate. An additional indication of plasma as the source for the protein in the edema fluid is provided by the selective uptake of radioactively labeled albumin into the edema fluid [12] in quantities that closely parallel the increases in volume and weight of the edematous hemisphere [11].

Triethyl Tin Edema
Both dialkyl and trialkyl tin compounds are fatally toxic for a variety of mammalian species, including man. Only the trialkyl form is neurotoxic, however, producing edema of subcortical white matter that is distinct in several ways from the examples already described. Some swelling of astrocytes is seen in cerebral cortex, but the striking abnormality is an accumulation of fluid within the myelin sheaths of nerve

Table 30-2. Approximate Composition of Edema Fluids

Preparations	Swelling (%)	Na^+ (mM)	Cl^- (mM)	K^+ (mM)	Albumin (g/liter)
Rabbit: subcortical white matter; triethyl tin [7][a]	34	133	115	–[b]	0
Dog: subcortical white matter; freezing lesion [16]	43	124	87	59	19
Cat: cerebral cortex; slices incubated in 54 mM K^+ [17]	33	39	135	108	–

[a] References to sources of data are given in brackets.
[b] Per unit dry weight, the K^+ was identical to control values; hence the K^+ concentration in the edema fluid was presumably close to that of plasma (about 3 mM).

fibers in the subcortical white matter [7]. The edema fluid is clearly intramyelinic, splitting the myelin lamellae at the interperiod line to form large blebs, or vacuoles. No swelling of gray matter can be demonstrated quantitatively, but the subcortical white matter exhibits swelling of some 34 percent (see Table 30-1).

The edema fluid is composed essentially of Na^+ and Cl^- (see Table 30-2) with no protein or other constituents that would be expected if the fluid were a plasma exudate. Evidence for the intactness of the blood-brain barrier is provided by studies with radioactively labeled albumin and vital dyes. In contrast with most forms of cerebral edema, in which the uptake of $^{24}Na^+$ by brain is normal or increased, the uptake of $^{24}Na^+$ in triethyl tin intoxication of the brain appears to be reduced, but further analyses have revealed that the apparent reduction is a reflection of the fact that a large fraction of the brain Na^+ (presumably that within the intramyelinic blebs) is no longer exchangeable. No evidence has been obtained for impairment of the activity of the Mg^{2+}-dependent, (Na^+, K^+)-activated ATPase under these conditions [12], and the precise mechanism of action of triethyl tin remains obscure. It is known that the earliest manifestation of toxicity is inhibition of oxidative phosphorylation in cerebral mitochondria. This inhibition apparently represents two effects: an inhibition of coupled phosphorylation (analogous to the effect of oligomycin) and an alteration of anion exchange across the mitochondrial membrane with intramitochondrial accumulation of Cl^- [18]. How these latter effects of triethyl tin may relate to the intramyelinic edema is not at all clear.

Ouabain-Induced Edema

When ouabain (G-strophanthin) is introduced into the central nervous system by local application or by local perfusion, violent and usually fatal convulsions follow [19]. There is an associated leakage of K^+ from brain tissue into the perfusate and, by light microscopy, the cerebral cortex has the appearance of the classic status spongiosus, the underlying white matter being spared. Detailed examination of central nervous tissues by electron microscopy has not been reported, but Birks [20] has described perfusing the sympathetic ganglia of the cat with plasma containing a related glycoside, digoxin (5 μM), and the effects on the electron microscopic appearance of the ganglia after perfusion fixation. In ganglia perfused with plasma plus digoxin, the neurons exhibited significant swelling of the cytoplasm, as did the lumina of the rough endoplasmic reticulum and the nerve endings. The mitochondria appeared dense and shrunken. The Schwann cells appeared perfectly normal. If the ganglia were perfused with digoxin in a Cl^--free or low-Na^+ plasma, most or all of the abnormalities in the neurons were prevented.

These observations are consistent with the known effects of ouabain and related glycosides as specific inhibitors of Na^+ and K^+ transport, presumably mediated by binding to and inhibition of Mg^{2+}-dependent, (Na^+, K^+)-activated ATPase in neural membranes (Chap. 6 and [21]), and as mobilizers of cellular Ca^{2+} with subsequent accumulation in mitochondria [22]. The cellular locus of the swelling is also consistent with data from studies in vitro (see Table 30-7) that demonstrate swelling of slices of cerebral cortex incubated with ouabain, but no parallel increase in the inulin (extracellular) space of the slices, in association with the striking increase of intracellular Na^+ and Cl^- and depletion of intracellular K^+ [5, 22]. In contrast, sections of corpus callosum incubated with ouabain exhibit some extracellular swelling but no depletion of K^+, an indication that glial cells are relatively unaffected [22]. However, ouabain appears to produce edema selectively in astrocytes in cerebral cortex [23, 24].

With this background, the changes in the distribution of cerebral fluids and electrolytes caused by ouabain (0.35 μM) when applied to the surface of the brain are more understandable (Table 30-3). The involved cerebral hemisphere exhibits about 20 percent swelling and a considerable alteration of the Na^+/K^+ ratio, reflecting the depletion of tissue K^+ and its replacement by Na^+. The effect is much more pronounced if the ouabain is applied to a preexisting freezing lesion (Table 30-3), and striking focal seizures ensue.

Table 30-3. Ouabain-Induced Edema in Rat Cerebral Hemispheres in Vivo

Preparation	Dry Weight (%)[a]		Swelling (%)	Na^+/K^+ Ratio[a]	
	Control H	Exptl. H		Control H	Exptl. H
Freezing lesion[b]	23.0	16.2	42	0.49	1.4
Ouabain (3.5 X 10^{-7} M)	22.6	18.6	21.5	0.46	1.1
Freezing lesion + ouabain	22.5	14.8	52	0.51	2.3

[a] Data are given for control hemispheres (H) and for hemispheres containing the lesion and/or to which ouabain was applied (exptl. H) [25, 26].
[b] The freezing lesions were produced by local application of vinyl ether sprayed onto the cortical surface. The results are comparable to those obtained by the application of cooled metal plates.

Circulatory Arrest

When the circulation to the brain is totally arrested for 15 to 30 min, edema of cerebral cortex develops to the extent of 18 to 22 percent, without any swelling of subcortical white matter (see Table 30-1). There is an associated leakage of K^+ from the cortical tissue and replacement by Na^+ without change in tissue Cl^- [13]. The edema cannot be attributed to hypoxia since respiratory arrest does not produce significant swelling of brain. The edema is clearly intracellular since the inulin and sucrose spaces decrease to one-half or one-third of normal after circulatory arrest (Table 30-4). It has been customary to dismiss such observations as postmortem artifacts [27], but these data may prove important in the problem of cerebrovascular occlusions (stroke).

Ames and colleagues [29, 30] have described the "no reflow" phenomenon in animals subjected to cerebral ischemia. When circulation was resumed, sizable areas of brain failed to reperfuse, partly because of increased viscosity of blood and partly because of compression of capillary lumina caused by the swelling of perivascular glia and endothelial cells, with formation of blebs (endothelial ?) intruding into the capillary lumina. It has been speculated that the swelling of these cells may be caused by K^+ released by the anoxic tissue. Thus, this swelling would be similar to that seen with high-K^+ solutions in vitro. The investigators suggest that, if these changes can be minimized or reversed (e.g., by increased blood perfusion pressure), the neurons may be able to resume normal function after periods of ischemia of 15 min or longer.

More recent studies have emphasized the factor of systemic hypotension as an important concomitant of the no-reflow phenomenon [31, 32]. An interesting

Table 30-4. Respiratory vs. Circulatory Arrest in Cerebral Gray Matter

Observations	Controls (%)	Arrest (%)	References
Respiratory arrest (5−7 min)[a]			
Cat cerebral cortex (circulation intact)			
Swelling, initial	−	3.8	[27]
Swelling after survival 1 hr	−	0	
Circulatory arrest (15−30 min)			
Cat cerebral cortex (not perfused)			
Swelling	−	18.5	[13]
Monkey cerebral cortex (perfused)[b]			
Swelling	0.7	22.2	[14]
After Cl^--free perfusate	−	5.8	
Dog caudate nucleus (perfused)[c]			
Sucrose space	18.1	6.6	[28]
Inulin space	12.7	5.4	

[a]Respiration was arrested until the blood pressure failed (5−7 min) and then was resumed with 100% oxygen.
[b]Perfusion over the surface of the cortex.
[c]Ventriculocisternal perfusion.

suggestion not yet fully confirmed is that, after CNS trauma, disturbance of the autoregulation of the microcirculation in spinal cord (or brain) may be a key mechanism in the development of edema and subsequent hemorrhagic necrosis of the tissue [33]. Thus in experimental trauma to the spinal cord, ischemia develops as a result of the trauma-induced release of one or more biogenic amines, with consequent leakage of K^+ ions, edema, and further ischemia in an increasingly damaging cycle. Although still a subject of debate, the use of antimetabolites of the relevant amines has completely prevented spinal cord necrosis in some experimental situations [33].

A practical corollary to the foregoing applies to samples of brain taken after any significant interval of circulatory arrest: The associated edema and shifts of electrolytes can seriously obscure the original state of the tissue prior to stopping the circulation [13].

Water Intoxication

A specialized case of cerebral edema occurs as a result of water intoxication, with or without concomitant hyponatremia. This subject has been reviewed recently by Katzman and Pappius [10], who distinguish four similar situations: simple water intoxication, sodium depletion, the dialysis disequilibrium syndrome, and the syndrome of inappropriate secretion of antidiuretic hormone.

The immediate disturbance is an increase in brain water. After several hours, cerebral potassium is depleted in association with cerebral swelling, changes not paralleled in skeletal muscle, where the edema is not accompanied by significant loss of tissue potassium (Table 30-5). Most authors agree that, in acute situations, the cerebral swelling arising from the hyponatremia and hypoosmolarity accounts for the altered state of consciousness. More chronic preparations exhibit loss of cerebral K^+. Katzman and Pappius [10] suggest that, in the chronic preparations, the depletion of K^+ accounts for the neurological dysfunction. This latter explanation has been questioned by Rhymer and Fishman [34], who found that animals with severely depleted levels of cerebral potassium were alert when cerebral swelling was minimal but comatose when the swelling was pronounced (Table 30-6). Further studies are needed to elucidate precise explanations for these phenomena.

Table 30-5. Effects of 24-Hour Water Intoxication on Brain and Skeletal Muscle

Tissue	Percentage Changes in		
	Swelling	Tissue K^+	Tissue Na^+
Brain (cerebrum)	+10.3	−21.5	−22.6
Muscle (quadriceps)	+20.9	− 6.9	−27.5

Averages for 10 animals.
Rats were prepared for one day with subcutaneous injections twice daily of vasopressin (1 unit) plus 2.5 percent (w/v) dextrose in water equal to 6 percent of the body weight. The animals were decapitated the next morning (24 hr) for tissue sampling and analyses [34].

Table 30-6. Water Intoxication and State of Consciousness

Duration of Intoxication	State of Consciousness	Percentage Changes in		
		Cerebral Swelling	Cerebral Tissue K^+	Cerebral Tissue Na^+
4 hr	Alert	+ 7.7	0	−14.9
24 hr	Comatose	+10.3	−18.7	−22.2
28−36 hr	Alert	+ 4.6	−23.1	−17.8

Averages for 5−10 animals.
Experimental details were like those in Table 30-5 [34].

Comparative and Ontogenetic Aspects

Knowledge of edema in the central nervous system of nonmammalian species is relatively meager. The shark exemplifies how great the comparative differences may be; it is virtually impossible to produce edema in shark brain by some of the experimental procedures described above [15]. Presumably, the very different structural organization of the shark brain accounts for the failure to produce edema after placement of destructive surface lesions. Among the various mammalian species, there seems to be little qualitative or quantitative difference in the response of brain or spinal cord to various edema-producing procedures. The immature brain, however, exemplified by the neonatal kitten brain, is relatively resistant to production of edema either in vivo [15] or in vitro ([22, 35]; Table 30-7). Of the various body tissues that have been examined in vitro, slices of cerebral cortex exhibit much greater swelling under usual incubation conditions than do slices of subcortical white matter, liver, renal cortex, or diaphragm [35].

Table 30-7. Edema of Cerebral Tissues in Vitro

Preparations[a]	Swelling (%)	Na^+/K^+ Ratios
Adult cerebral cortex		
Controls	+32.7	0.75
Cl^--free media[b]	+14.4	0.58
Ouabain (10^{-5} M)	+45.1	2.75
Neonatal cerebral cortex		
Controls	− 1.2	1.05
Ouabain (10^{-5} M)	+19.6	3.1
Adult corpus callosum		
Controls	+ 6.7	1.2
Ouabain (10^{-5} M)	+15.3	1.85

[a]Slices of cat cerebral tissues were incubated for 1 hr at 37°C in 27 mM K^+ media.
[b]Isethionate (125 mM) was substituted for Cl^- [22].

SLICES OF BRAIN TISSUE INCUBATED IN VITRO

Artifacts of In Vitro Conditions

Several types of swelling characteristic of brain slices have been identified (Fig. 30-1) [4, 5, 35]. Inevitably, there is some adherent fluid from the incubation medium, amounting to 5 to 10 percent of the initial fresh weight of the tissue, that is probably associated primarily with the major cut surfaces of the slices. Swelling to this extent occurs when slices of cerebral cortex are dipped into isoosmotic media for as few as 10 sec, and it is clearly extracellular because it is accessible to both inulin and chloride.

A second type of swelling in cerebral cortical slices takes place during the immersion of slices at room temperature while the incubation flasks are being gassed. The extra fluid in the slices after immersion for 5 min totals about 15 percent of the initial fresh weight of the tissue. This "preparatory" swelling can progress under suboptimal conditions, so that after 30 min it may reach 50 percent of the initial fresh weight of the tissue. It takes place in a compartment inaccessible to inulin but accessible to chloride. This type of swelling accounts for the rather high values for slice swelling encountered in the earlier literature and in studies in which the preparation of slices is prolonged unduly. The foregoing types of swelling are primarily artifacts of in vitro conditions and probably have little direct relevance to in vivo forms of edema.

K^+- and Cl^--Dependent Swelling

A third form of swelling that occurs in vitro may be related to intracellular swelling in vivo; this form is produced by elevated concentrations of K^+ and Cl^- in the incubation medium. This swelling is presumably intracellular because it takes place in tissue compartments inaccessible to inulin and partially inaccessible to Cl^- (Fig. 30-1). Superimposition of anoxia or inhibitory concentrations of ouabain that impair energy sources or operation of the Na^+-K^+ ion pumps depolarizes the cells, enhances the

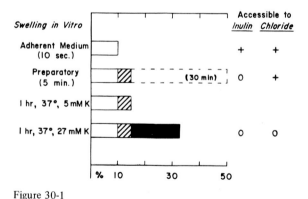

Figure 30-1
Types of swelling encountered in slices of cerebral cortex prepared and incubated in vitro. The values on the abscissa are percentages of initial fresh (wet) weight of slices. The slices were incubated or immersed in bicarbonate-buffered saline-glucose media containing 5 mM K^+ (normal) or 27 mM K^+ (high K^+). Incubations were for 1 hr at 37°C. The interpretations are based on studies by Varon and McIlwain [4], and the data are taken from the studies on cat cerebral cortex by Bourke and Tower [5]; see [5] for further details.

degree of edema, and results in gross shifts in distribution of electrolytes (see Table 30-7). Exposure of slices to inhibitory concentrations of ouabain is one of the few situations that cause slices of cerebral cortex of neonatal kitten to swell. The concomitant shift in the Na^+/K^+ ratio indicates that, despite the immaturity of these slices, the cation pumps must already be operative (see Table 30-7). For external concentrations of K^+ over about 20 mM, the swelling of slices is a linear function of increasing concentrations of K^+ in the incubation medium. This swelling occurs in a tissue compartment inaccessible to inulin, but is paralleled by an increase in fluid spaces accessible to Cl^-. The role of Cl^- is dramatically illustrated by the effect of substituting a less permeable anion, isethionate (2-hydroxyethanesulfonate), for Cl^-. Even at an external K^+ concentration of 125 mM, essentially no slice swelling occurs in the absence of Cl^- (Fig. 30-2).

Kinetic studies of these phenomena indicate that in incubated cerebral cortical slices there is a K^+-dependent, mediated transport of Cl^- into a cellular (astroglial ?) compartment and that the edema fluid in this compartment is composed primarily of K^+ and Cl^- ([17]; see Table 30-2). Presumably, these findings relate to the high-K^+ glial cells described by Kuffler and colleagues and for which they have proposed a K^+ "spatial buffering" function [36]. In their proposal, the K^+ that is released, upon stimulation of neurons, into the immediate extraneuronal environment

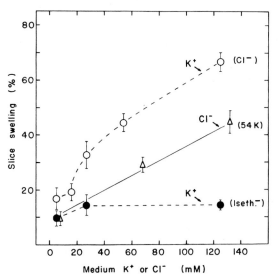

Figure 30-2
Swelling of incubated slices of cat cerebral cortex as a function of the concentration of K^+ (open circles) or of Cl^- (Δ) in the incubation medium. The effects of varying Cl^- in the medium were examined at a constant concentration of K^+ (54 mM); similarly, the effects of varying K^+ in the medium were examined at a constant concentration of Cl^- (125 mM). The effects of incubation in Cl^--free media (closed circles) were obtained by substituting isethionate (125 mM) for Cl^-. Slice swelling (ordinate) is the percentage of initial fresh (wet) weight of slices. All slices were incubated for 1 hr at 37°C in bicarbonate-buffered saline-glucose media with variations in composition as indicated. (From [5] and [17].)

is removed by being taken up into adjacent astrocytes. Hence, the effects of accumulated external K^+ on neuronal membrane potentials and excitability are minimized.

Recent studies by Bourke et al. [14] have demonstrated that many of the foregoing phenomena can be reproduced in vivo. In monkeys fitted with a plastic perfusion dome in place of skull, the perfusion over the surface of the cerebral cortex of artificial cerebrospinal fluid containing various concentrations of K^+ or Cl^-, or both, reproduced qualitatively the same K^+- and Cl^--dependent edema as that depicted in Figure 30-2. In these experiments, cerebral circulation was intact and only one cortical surface could be perfused, so the quantitative changes were somewhat less, but edema of about 20 percent could be regularly elicited during perfusion with perfusates containing Cl^- and 100 mM K^+. Most of the edema could be prevented by perfusion with Cl^--free (isethionate) perfusates and could be reversed by switching to perfusates with normal (5 mM) K^+ concentrations.

A possible enzymatic basis for the K^+-dependent transport of Cl^- into the presumed astroglial compartment is suggested from the observations by Bourke and Nelson [37]. They found that when acetazolamide (Diamox), an inhibitor of carbonic anhydrase, is administered either locally or parenterally, it inhibits the K^+-dependent swelling of primate cerebral cortex in vivo and inhibits the K^+-dependent increase of tissue Cl^- and the K^+-dependent rate of influx of ^{36}Cl into edematous cerebral cortex. Direct support for these observations has been provided by studies in tissue culture where the K^+-dependent transport of ^{36}Cl into cultured cells is demonstrable for hamster astrocytes (K_m = 38.5 mM) but not for mouse neuroblastoma cells, and the uptake of ^{36}Cl into astrocytes is competitively inhibited by acetazolamide (K_i = 27.1 mM) [38]. As already discussed, the relevance of these observations to edema associated with cerebrovascular ischemia or occlusions (stroke) should not be overlooked.

THERAPY

Osmotically Active Agents

Three principal approaches to the therapy of cerebral edema have been employed. Surgical intervention to decompress the swollen brain and relieve the increased intracranial pressure is an emergency, often a lifesaving, measure. It is obviously not a useful routine treatment, and in many cases it is only a temporary expedient. The use of osmotically active agents is an outgrowth of the classic observation of Weed and McKibben [2] that intravenous injections of hypertonic (30 percent, w/v) NaCl would shrink normal brain, whereas intravenous distilled water would cause marked cerebral edema.

Several osmotically active agents are currently employed, such as urea [39], mannitol [40], and glycerol [41]. The first two are administered intravenously (urea: 0.5 to 1.5 g per kilogram body weight in a 30 percent w/v solution; mannitol: 1.5 to 2 g per kilogram body weight in a 15 to 20 percent w/v solution) and will lower cerebrospinal fluid pressure significantly for 3 to 8 hr. Glycerol is administered orally or by gastric tube (0.5 to 0.7 g per kilogram body weight every 3 to 4 hr).

Urea is eventually distributed throughout the total body water and, in high doses, can be toxic. It is also more difficult to prepare in sterile solutions and is not stable in solution. Mannitol is distributed only in extracellular fluid and crosses the blood-brain barrier with difficulty. It is relatively nontoxic and is excreted rapidly. Glycerol is useful because of its low toxicity and because it can be employed more easily in prolonged therapy. All these agents suffer from the disadvantage that they apparently do not affect the edema fluid per se but act to relieve signs and symptoms by dehydrating normal brain [8].

Steroids

The third type of therapy involves the use of adrenocortical steroids of the glucocorticoid type. The most widely used are dexamethasone (9α-fluoro-16α-methylprednisolone) and cortisone (17-hydroxy-11-dehydrocorticosterone). Dexamethasone, a synthetic fluorinated glucocorticoid, was chosen for its antiinflammatory potency and low salt-retaining activity [42]. In clinical practice, it is usually administered initially in an intravenous 10 mg dose followed every 6 hr by an intramuscular 4 mg dose until maximum response is obtained (usually within 18 to 24 hr) or for 2 to 4 days postoperatively. Oral maintenance doses of 1 to 3 mg three times daily may be continued for longer periods [42].

Despite widespread clinical evidence for the efficacy of dexamethasone and cortisone in relieving the signs and symptoms of cerebral edema in human patients, experimental verification has been difficult to obtain [11]. However, a recent study by Pappius and McCann [11] indicates that dexamethasone significantly diminishes the edema of cerebral hemispheres traumatized by a freezing lesion (Table 30-8). In these studies, cats were pretreated with dexamethasone for 48 hr before the freezing lesion

Table 30-8. Effects of Pretreatment with Dexamethasone on Edema of Freezing Lesion in Cat Brain

Observations	Differences in Weight between Control and Edematous Hemisphere (g)[a]		Differences in Content of ^{131}I-Serum Albumin (ml serum equivalents)[a]
	24 hr after lesion	48 hr after lesion	48 hr after lesion
Untreated controls	0.62 ± 0.14 (9)	0.85 ± 0.24 (15)	0.51 ± 0.15 (9)
Dexamethasone 2.5 mg/kg[b]	0.53^{c} (8)	0.52 ± 0.19 (12)	0.36 ± 0.11 (12)
Significance (P) control vs. treated	n.s.	$P < 0.01$	$P < 0.05$

[a]The values represent means ± S.D. for the numbers of animals given in parentheses.
[b]Dexamethasone was given intraperitoneally in a dose of 1.25 mg per kilogram body weight twice daily for 48 hr prior to making the freezing lesion and continued at the same dose schedule thereafter.
[c]The S.D. for this mean was not reported [11].

was made in one hemisphere, and the animals were also injected with ^{131}I-serum albumin 48 hr before the lesion was made. Differences in hemispheral weights and in content of ^{131}I-serum albumin between the control and traumatized hemispheres were measured at various times after making the freezing lesion. At 24 hr after the lesion, the differences between the two hemispheres were similar for control (untreated) and treated animals, but after 48 hr the treated animals exhibited significantly smaller differences in weight and in ^{131}I-serum albumin between control and traumatized hemispheres (Table 30-8). By 72 hr, when the edema had begun to resolve, there were no longer any significant differences between treated and control animals. Determinations of dry weights of subcortical white matter from the two groups of animals at 48 hr demonstrated no differences; the values were 24.2 ± 2.1 percent for untreated animals ($n = 11$) and 22.8 ± 0.4 percent for dexamethasone-treated animals ($n = 4$), in comparison with normal (untraumatized) white matter in which the dry weight averaged 32.3 ± 1.7 percent ($n = 8$).

Thus, there is an apparent discrepancy between the data for dry weights and the data for interhemispheric differences in total weight and radioactive albumin content. There was no evidence for shrinkage of normal brain, so the discrepancy could most readily be explained by assuming that the treatment with dexamethasone prevented spread of the edema in the period between 24 and 48 hr after the lesion was made. This effect can be calculated to correspond to limiting the edema to approximately one-third of the hemispheral white matter, in contrast to involvement of more than one-half of the hemispheral white matter in untreated animals.

Pappius and McCann [11] also reported that, in contrast to man, treatment of cats with cortisone was ineffective under the conditions described, apparently because of poor absorption after intraperitoneal administration. This is an important observation because it emphasizes the factor of species differences that may have complicated earlier attempts to examine therapeutic efficacy experimentally. From this brief review of the current status of therapy for cerebral edema, it is obvious that additional studies are needed to establish the efficacy and precise mechanisms of action of available therapeutic agents, as well as to develop more effective therapies.

CONCLUSION

Insofar as possible, discussion in this chapter has focused on edema or swelling of neural tissues. Only passing reference has been made to the blood-brain barrier, to the various fluid compartments within cerebral tissues, or to the electrolytes and other solutes in these fluids. Little cognizance has been taken of the fact that most of the edema-producing situations in vivo are commonly characterized by the occurrence of seizures.

Such aspects are discussed in Chapters 5, 6, and 20, and those chapters should be read in conjunction with this one for a proper perspective of the normal and pathological factors responsible for the distribution of fluids and electrolytes in the central nervous system. The subject is still fraught with controversies, but it is one of the most important aspects of neurochemistry because the fluid environment surrounding neural membranes determines the viability and excitability of the neuronal and glial elements of the brain and spinal cord. The sensitivity of neural cells to even small changes in their environment accounts for the many mechanisms that operate to preserve the *milieu interne*. Such mechanisms produce or utilize the energy necessary for transporting a variety of solutes into or out of various compartments of the central nervous system, with concomitant appropriate shifts of water. The derangements of these processes that are associated with the various forms of edema promptly produce impaired consciousness, convulsions, coma, and often death — consequences that attest to the gravity of cerebral edema in clinical practice.

REFERENCES

1. Klatzo, I. Neuropathological aspects of brain edema. *J. Neuropathol. Exp. Neurol.* 26:1, 1967.
2. Weed, L. H., and McKibben, P. S. Experimental alterations of brain bulk. *Am. J. Physiol.* (Lond.) 48:531, 1919. (Report of the first experimental demonstration of alterations in brain volume by intravenous injections of hyper- or hypo-tonic solutions. An important, classic study.)
3. Elliott, K. A. C., and Jasper, H. Measurement of experimentally induced brain swelling and shrinkage. *Am. J. Physiol.* 157:122, 1949. (The classic paper on the measurement and calculation of changes in the fluid content of cerebral tissues in vivo.)
4. Varon, S., and McIlwain, H. Fluid content and compartments in isolated cerebral tissues. *J. Neurochem.* 8:262, 1961. (A definitive study identifying principal factors affecting fluid distribution in incubated slices of cerebral cortex. Clarifies earlier studies from Elliott's laboratory [*Proc. Soc. Exp. Biol. Med.* 63:234, 1946; *Can. J. Biochem. Physiol.* 34:1007, 1956; 36:217, 1958] and extended subsequently by studies reported in [5] and [35].)
5. Bourke, R. S., and Tower, D. B. Fluid compartmentation and electrolytes of cat cerebral cortex *in vitro*. I. Swelling and solute distribution in mature cerebral cortex. *J. Neurochem.* 13:1071, 1966. (Confirmation and extension of earlier studies, reported in [4], on the distribution of fluids and electrolytes in cerebral tissues. Published with two accompanying papers, one on electrolytes [same journal volume, p. 1099] and one on ontogenetic and comparative aspects; see [35].)
6. Stewart-Wallace, A. M. Biochemical study of cerebral tissue, and changes in cerebral edema. *Brain* 62:426, 1939. (The classical observations localizing edema associated with brain tumors to cerebral white matter. A study from 37 years ago that has withstood the test of time.)
7. Aleu, F. P., Katzman, R., and Terry, R. D. Structure and electrolyte analyses of cerebral edema induced by alkyl tin intoxication. *J. Neuropathol. Exp. Neurol.* 22:403, 1963. (The first definitive electron-microscopic study demonstrating the characteristic selective intramyelinic edema and evidence that the blood-brain barrier is not impaired. Confirms and extends the original light-microscopic and biochemical studies of Magee et al., *J. Pathol.* 73:107, 1957.)
8. Pappius, H. M. Biochemical studies on experimental brain edema. In [9, p. 445]. (Review of many previous studies, particularly on the edema associated with freezing lesions, and the effects of intravenous urea thereon.)
9. Klatzo, I., and Seitelberger, F. (Eds.). *Brain Edema.* New York: Springer-Verlag, 1967. (Publication of proceedings of a symposium on brain edema held in Vienna, Sept., 1965. This volume provides what is probably the best recent source of information on edema of neural tissues from clinical and

experimental points of view, with numerous contributions on morphological and biochemical aspects. Therapy is not covered. In addition to chapters cited individually in this bibliography [8, 12, 13, 15, 16, and 28], the chapters in *Brain Edema* by Hoff and Jellinger [p. 3] on clinical aspects of cerebral edema, by Feigin [p. 128] on the human pathology, and by Long et al. [p. 419] on a review of various types of experimental cerebral edema should prove useful.)

10. Katzman, R., and Pappius, H. M. *Brain Electrolytes and Fluid Metabolism.* Baltimore: Williams & Wilkins, 1973. (This monograph provides an up-to-date detailed coverage of all aspects of fluid and electrolytes in the nervous system as compiled by two experts in the field. Chapter 14 on hyponatremia and water intoxication, pp. 291–303, and Chapter 18 on cerebral edema and intracranial pressure, pp. 366–408, are most germane.)

11. Pappius, H. M., and McCann, W. P. Effects of steroids on cerebral edema in cats. *Arch. Neurol.* 20:207, 1969. (One of a series of excellent studies from Pappius' laboratory that demonstrates significant effects of dexamethasone on the development of cerebral edema after placement of a freezing lesion. The paper contains a good review of the status of steroid therapy in experimentally induced edema.)

12. Katzman, R. Biochemical correlates of cerebral edema. In [9, p. 461]. (Review of earlier work on cerebral edema produced by triethyl tin, implanted glial tumors, and implanted protein derivatives, with data on ultrastructure, nature of the edema fluid, and the involvement of the blood-brain barrier. The studies come from one of the leading groups in the field, especially for triethyl tin edema.)

13. Tower, D. B. Distribution of cerebral fluids and electrolytes *in vivo* and *in vitro*. In [9, p. 303]. (A survey of factors affecting the distribution of fluids and electrolytes in the CNS based on in vitro studies reported in [5 and 35] plus in vivo studies by Bourke et al., *Am. J. Physiol.* 208:682, 1965. In addition data on artifacts of fixation and on circulatory arrest are reported.)

14. Bourke, R. S., Nelson, K. M., Naumann, R. A., and Young, O. M. Studies on the production and subsequent reduction of swelling in primate cerebral cortex under isosmotic conditions *in vivo*. *Exp. Brain Res.* 10:427, 1970. (This paper reports the demonstration in vivo of the K- and Cl-dependent edema of cerebral cortex reported earlier [17] for in vitro preparations. Monkeys were fitted with plastic domes over the cerebral hemispheres and the surface of the cortex was perfused with artificial CSF of various compositions. The observations are particularly relevant to problems of circulatory arrest and strokes.)

15. Klatzo, I., Wisniewski, H., Steinwall, O., and Streicher, E. Dynamics of cold injury edema. In [9, p. 554]. (Review of earlier work on edema resulting from freezing lesions produced directly on the surface of the brain, with special reference to the blood-brain barrier. Studies come from the originators of this procedure and encompass structural, biochemical, comparative, and ontogenetic aspects.)

16. Clasen, R. A., Sky-Peck, H. H., Pandolfi, S., Laing, I., and Hass, G. M. The chemistry of isolated edema fluid in experimental cerebral injury. In [9, p. 536]. (Review of experiences with edema resulting from freezing lesions produced through the intact skull and details of biochemical characteristics, especially the protein content, of the edema fluid.)

17. Bourke, R. S. Studies of the development and subsequent reduction of swelling of mammalian cerebral cortex under isosmotic conditions *in vitro*. *Exp. Brain Res.* 8:232, 1969. (Together with the accompanying paper in the same journal

volume [p. 219], these studies demonstrate details of the K- and Cl-dependent edema in incubated slices of cerebral cortex, the kinetics of the mediated transport of Cl into astrocytes, and the composition of the edema fluid therein. Reference [14] reports comparable studies in vivo.)

18. Stockdale, M., Dawson, A. P., and Selwyn, M. J. Effects of trialkyltin and triphenyltin compounds on mitochondrial respiration. *Eur. J. Biochem.* 15:342, 1970. (The latest in a series of studies initiated by Aldridge and Cremer [*Biochem. J.* 61:406, 1955] on the earliest toxic effect of trialkyltin compounds — the interference with mitochondrial oxidative phosphorylation.)

19. Maccagnani, F., Bignami, A., and Palladini, G. Étude électroencéphalographique des effets de l'introduction intracrânienne d'ouabaïne chez le rat et le cobaye; mise en évidence d'un tracé périodique. *Rev. Neurol.* (Paris) 115:211, 1966. (One of the earliest contemporary reports on the clinical, EEG, and morphological effects of intracerebral ouabain. A good, detailed study extending observations originally made by Rizzolo, *Arch. Farmac. sper.* 50:16, 1929.)

20. Birks, R. I. The effects of a cardiac glycoside on subcellular structures within nerve cells and their processes in sympathetic ganglia and skeletal muscle. *Can. J. Biochem.* 40:303, 1962. (An excellent electron-microscopic study of the effects of perfusion of ganglia with plasma containing digoxin and of the prevention of the swelling of subcellular elements in the neurons by perfusion with Cl-free or low Na plasma. This is an important and unique study.)

21. Siegel, G. J., and Albers, R. W. Nucleoside Triphosphate Phosphohydrolases. In Lajtha, A. (Ed.), *Handbook of Neurochemistry.* Vol. 4. New York: Plenum, 1970, p. 13.

22. Tower, D. B. Ouabain and the distribution of calcium and magnesium in cerebral tissues *in vitro. Exp. Brain Res.* 6:273, 1968. (Evidence for mobilization by ouabain of tissue Ca and its sequestration in mitochondria; also data on comparative and ontogenetic effects of ouabain on neural tissues in vitro.)

23. Stefanelli, A., Palladini, G., and Ieradi, L. Effetto della ouabaina sul tessuto nervoso in coltura in vitro. *Experientia* 21:717, 1965. (See also [24].)

24. Renkawek, K., Palladini, G., and Ieradi, L. Morphology of glia cultured in vitro in presence of ouabain. *Brain Res.* 18:363, 1970. (These two papers [23, 24] describe studies with explant cultures from chick spinal ganglia and rat cerebellum in which only astrocytes exhibit swelling on exposure to ouabain; neurons and oligodendroglia specifically did not swell.)

25. Lewin, E. Focal "epileptogenic" lesions produced with ouabain in the rat. *Neurology* 19:310, 1969. (Abstract of studies on effects of topical ouabain, alone or after placement of a freezing lesion in cerebral cortex, on development of epileptiform activity and on the fluids and electrolytes of the experimental and control hemispheres. This paper provided data for Table 30-3 as supplemented by unpublished data made available by the author.)

26. Lewis, E., Charles, G., and McCrimmon, A. Discharging cortical lesions produced by freezing: The effect of anticonvulsants on sodium-potassium-activated ATPase, sodium, and potassium in cortex. *Neurology* 19:565, 1969. (Data on the freezing lesion to supplement those in [25].)

27. Norris, J. W., and Pappius, H. M. Cerebral water and electrolytes. Effects of asphyxia, hypoxia and hypercapnia. *Arch. Neurol.* 23:248, 1970. (One of a series of excellent studies from Pappius' laboratory that deals specifically with the effects of respiratory arrest in contradistinction to circulatory arrest. The paper contains a good review of the problem.)

28. Oppelt, W. W., and Rall, D. P. Brain extracellular space as measured by diffusion of various molecules into brain. In [9, p. 333]. (Review of the studies on extracellular space of brain as measured by ventriculo-cisternal perfusion with inulin, dextran, and sucrose. The technique was developed in Rall's laboratory and represents one of the pioneering investigations that has contributed to current concepts of fluid distribution in the brain.)

29. Ames, A., III, Wright, R. L., Kowada, M., Thurston, J. M., and Majno, G. Cerebral ischemia. II. The no-reflow phenomenon. *Am. J. Pathol.* 52:437, 1968. (See comments for the companion paper, [30].)

30. Chiang, J., Kowada, M., Ames, A., III, Wright, J. L., and Majno, G. Cerebral ischemia III. Vascular changes. *Am. J. Pathol.* 52:455, 1968. (Together with [29], describes the "no-reflow" phenomenon after experimentally induced cerebral ischemia and the associated pathological changes examined electron microscopically. These studies have much potential importance for the problems associated with cerebrovascular hypotension, transient ischemia and cerebrovascular occlusions. The therapeutic implications deserve attention.)

31. Shibata, S., Hodge, C. P., and Pappius, H. M. Cerebral Water and Electrolytes: Effects of Ischemia.(Cited in [10], pp. 384–385.)

32. Klatzo, I., Ito, U., Go, G., and Spatz, M. Observations on Experimental Cerebral Ischemia in Mongolian Gerbils. In Cervos-Navarro, J. (Ed.), *Pathology of Cerebral Microcirculation.* Berlin: de Gruyter, 1974, p. 338.

33. Osterholm, J. L. The pathophysiological response to spinal cord injury. The current status of related research. *J. Neurosurg.* 40:5, 1974. (A comprehensive review of the status of the roles of biogenic amines and autoregulation of the CNS microcirculation in the production of edema and necrosis following trauma.)

34. Rymer, M. M., and Fishman, R. A. Protective adaptation of brain to water intoxication. *Arch. Neurol.* 28:49, 1973. (Evidence in the rat for K^+ depletion as the protective mechanism in brain and for the level of cerebral swelling as the determinant of levels of consciousness.)

35. Tower, D. B., and Bourke, R. S. Fluid compartmentation and electrolytes of cat cerebral cortex *in vitro.* III. Ontogenetic and comparative aspects. *J. Neurochem.* 13:1119, 1966. (Companion paper to [5], dealing primarily with in vitro studies on fluids and electrolytes in neonatal and developing brain and in body tissues other than cerebral cortex. Contains data not available elsewhere.)

36. Orkand, R. K., Nicholls, J. G., and Kuffler, S. W. Effect of nerve impulses on the membrane potential of glial cells in the central nervous system of amphibia. *J. Neurophysiol.* 29:788, 1966. (An important contribution from Kuffler's laboratory that proposes for the high K glia a "spatial buffering" function to remove excess K from the extracellular environs of neurons. This concept has been extended to mammalian CNS by Trachtenberg and Pollen, *Science* 167:1248, 1970.)

37. Bourke, R. S., and Nelson, K. M. Further studies on the K^+-dependent swelling of primate cerebral cortex *in vivo*: The enzymatic basis of the K^+-dependent transport of chloride. *J. Neurochem.* 19:663, 1972. (Experimental demonstration in vivo of the inhibitory effect of local or parenteral acetazolamide on cortical swelling and Cl^- transport.)

38. Gill, T. H., Young, O. M., and Tower, D. B. The uptake of ^{36}Cl into astrocytes in tissue culture by a potassium-dependent, saturable process. *J. Neurochem.* 23:1011, 1974. (Direct evidence to support the previous studies on cat and primate cerebral cortex by Bourke colleagues; see [14, 17, and 37].)

39. Javid, M. Urea — new use of an old agent. Reduction of intracranial and intra-
    ocular pressure. *Surg. Clin. North Am.* 38:907, 1958. (Review of experience
    with hypertonic solutions of urea as therapy for cerebral edema, by the origi-
    nator of the use of urea for this purpose.)
40. Wise, B. L., and Chater, N. The value of hypertonic mannitol solution in
    decreasing brain mass and lowering cerebrospinal-fluid pressure. *J. Neurosurg.*
    19:1038, 1962. (The first definitive report on the efficacy of hypertonic
    solutions of mannitol for treatment of cerebral edema.)
41. Cantore, G., Guidetti, B., and Virno, M. Oral glycerol for the reduction of
    intracranial pressure. *J. Neurosurg.* 21:278, 1964. (First definitive report on
    efficacy of glycerol for treatment of cerebral edema.)
42. Galicich, J. H., and French, L. Use of dexamethasone in the treatment of
    cerebral edema resulting from brain tumors and brain surgery. *Am. Pract.*
    12:169, 1961. (First definitive report on efficacy of dexamethasone for treat-
    ment of cerebral edema, with review of previous trials of steroid therapy.)

Chapter 31

Seizures and Comatose States

Thomas E. Duffy
Fred Plum

Generalized seizures and coma represent extremes in the functional activity of the brain and its requirement for energy. Energy consumption in brain is tightly coupled to the work done. This is principally the electrochemical work associated with neuronal excitation and conduction, the uphill translocation of cations necessary to maintain cell membrane polarization, and the synthesis, release, and reuptake of synaptic transmitters. During seizures, cerebral energy consumption, as reflected either by the uptake of glucose (the chief metabolic fuel of brain) and oxygen or by the breakdown of endogenous high energy phosphate compounds, may increase as much as three- to fourfold whereas, during surgical anesthesia or coma, cerebral energy demands may fall by more than 50 percent (Chap. 19). Despite such a wide range in metabolic requirements, the cellular homeostatic mechanisms that regulate energy production tend to maintain remarkably constant the concentration of brain ATP, the energy transducer molecule. If this constancy fails, however, irreversible brain damage almost always results.

The control of energy balance is exerted both on the supply of energy precursors and on the work done by the brain, that is, on both the input and the output. Intrinsic mechanisms appropriately increase or decrease the rate of brain metabolism during periods of increased or decreased functional activity. During hypoxia or hypoglycemia, cerebral metabolic activity is depressed prior to any measurable depletion of ATP stores in tissue [1, 2], suggesting that adaptive responses can lower energy expenditure when the supply of energy precursors is threatened (see also Hypoglycemia in Chap. 25). As part of this process, the brain exerts homeostatic control over its blood supply. When cerebral functional activity increases, as during seizures, the cerebral blood flow increases commensurately, thereby increasing the delivery of substrates and oxygen to the actively metabolizing tissue. The interdependence of function, metabolism, and blood flow must somehow be coupled at the cellular-molecular level; available evidence suggests that ATP expenditure or, more specifically, the $[ATP]/[ADP][P_i]$ ratio (the "phosphate potential"), is a chief intracellular regulator of cerebral carbohydrate and energy metabolism. A change in the phosphate potential, furthermore, may generate the signal that modifies the blood supply.

SEIZURES

A seizure is the sudden onset of an intense, rapidly repetitive focal or generalized electrical discharge in the brain. Generalized cerebral seizures usually are accompanied by motor convulsions. Although epilepsy is the prototypical disease causing such paroxysmal brain dysfunction, seizures are also a common symptom of many other neurological and systemic abnormalities, including brain tumors, degenerative diseases of the CNS, high fever (in the child), or CNS infections. Experimentally seizures can be induced by several mechanisms. These include (1) repetitive electrical depolarization of cerebral neurons (electroshock); (2) interfering with the supply or processing of substrates for brain energy metabolism (hypoxia, hypoglycemia, fluoroacetate poisoning); (3) pharmacologic agents that directly stimulate the nervous system (pentylenetetrazol, flurothyl, ammonium salts); (4) agents that inhibit the uptake or metabolism of certain amino acids (methionine sulfoximine, pyridoxal phosphate antagonists); or (5) agents that antagonize the normal physiological inhibition of specific neurotransmitter molecules (picrotoxin, strychnine, bicuculline). Just as there is no single factor that serves as a common denominator for eliciting all types of seizures, so too it is unlikely that a single chemical agent triggers the abnormal electrical discharge in all cases, and none has been proposed. Consequently, the neurochemistry of epilepsy relates primarily to those events that accompany or follow secondarily from the convulsive attack but does not include its cause.

Oxidative Metabolism and Blood Flow

Generalized seizures place on brain its greatest known metabolic load. Measurements in both animals and man indicate that, during a single seizure, the overall rate of cerebral metabolism increases from 1.5 to 5 times normal, depending upon the intensity and duration of the cerebral stimulus and upon which particular substrate is taken as an index of metabolic activity (Table 31-1). Schmidt et al. [9] gave monkeys pentylenetetrazol and, by multiplying arteriovenous differences across the brain by the blood flow, calculated that, during the convulsions, the cerebral metabolic rate for oxygen $(CMRO_2)$ doubled. In paralyzed, nonanesthetized dogs and monkeys that were ventilated artificially, Plum and colleagues [8] found a twofold to fourfold increase in $CMRO_2$ during pentylenetetrazol-induced seizures. They noted that, at least during the early stages of the seizure, CO_2 production by the brain exceeded oxygen consumption so that the apparent respiratory quotient (RQ) rose to 1.45. Direct measurements of energy consumption in whole brain, cerebral cortex, or major anatomical subdivisions of the CNS in mice and rats indicate that during electroshock-induced seizures, the turnover of high energy phosphates ($\sim$P use) increases to at least two to four times the resting rate [4, 6, 10], a figure that generally corroborates the observed increase in oxygen consumption. The question remains, however, whether the increased metabolism keeps pace with cellular demands. Indeed, one

Table 31-1. Indices of Cerebral Metabolism During Generalized Seizures

Metabolic Index	Relative Increase	References
Glycolytic flux	2- to 8-fold	[3–7]
Oxygen consumption ($CMRO_2$)	2- to 4-fold	[3, 8, 9]
High-energy phosphate ($\sim$P) utilization	2- to 4-fold	[4, 6, 7, 10]
Cerebral blood flow	1.5- to 7-fold	[3, 8, 9]
Respiratory quotient (RQ)	Rises above 1.0	[3, 8]

of the tacit questions concerning the metabolism of epilepsy has been whether the seizures stop because the brain becomes energetically exhausted (see also Chap. 19).

Despite the enormous increase in cerebral oxidative metabolism caused by seizures, blood flow to the brain usually increases proportionately with or exceeds the rate of metabolism so that oxygen tension in the cerebral venous blood can actually rise [5, 8, 11]. This homeostatic response of the blood flow can be attributed to two mechanisms: One is a substantial increase in the systemic blood pressure, mediated over descending sympathetic autonomic pathways from the brain, and the other is a marked reduction in resistance of the cerebral vascular bed [8]. Ordinarily, the rate of blood flow through the normally metabolizing brain remains nearly constant in the face of wide fluctuations in blood pressure, presumably owing to the remarkable ability of cerebral arterioles to constrict or dilate appropriately whenever the pressure rises or falls (vascular autoregulation). During seizures, these vessels dilate and the cerebral vasculature becomes pressure passive, giving rise to the increased blood flow. The mechanism of this loss of autoregulation is still uncertain, but it clearly is linked to the greatly increased metabolism of the tissue (Chap. 19). Recent evidence suggests that increased tissue acidosis, secondary to alterations in intracellular glycolytic metabolism, is at least partially responsible [3, 5, 11].

Cerebral Energy Balance in Experimental Seizures

Most convulsants that cause major tonic-clonic seizures produce similar disturbances in the energy balance of the brain. Data obtained in convulsing mice, whose brains can be frozen rapidly enough to minimize postmortem artifacts, indicate that within 10 to 20 seconds after the initiation of a seizure by electroshock or the inhalation of flurothyl the concentrations of phosphocreatine and ATP in brain decrease by as much as 50 percent whereas the ADP, AMP, and inorganic phosphate levels increase correspondingly [4, 6, 7]. Simultaneously, there is an increased glycolytic flux as indicated by decreased tissue concentrations of glucose and glycogen and an increase in lactic acid [4, 6, 7]. King and colleagues [6] noted that, during electroshock-induced convulsions, lactic acid in brain increased much more than could be accounted for by decreases in the endogenous tissue glucose and glycogen stores and concluded that the rate of glucose transport into brain from blood must have been greatly facilitated, exceeding the normal rate of uptake by perhaps tenfold. Sacktor et al. [7] first demonstrated in mice that flurothyl-induced convulsions were associated with a large increase in the concentration of fructose 1,6-diphosphate in the brain and a decline in the hexose 6-phosphate level (glucose 6-phosphate and fructose 6-phosphate). They concluded that activation of phosphofructokinase was mainly responsible for the stimulation of glycolysis. King and colleagues [12, 13] observed similar changes in mouse brain during electroshock-induced seizures and noted further that the concentrations of all measured glycolytic intermediates "downstream" from fructose 1,6-diphosphate were also significantly elevated. The rise in lactate, however, was disproportionately greater than the change in pyruvate and resulted in an increased lactate/pyruvate ratio, providing evidence for an elevated $NADH/NAD^+$ ratio in the tissue cytoplasm [13].

Although total cerebral energy reserves usually decline during convulsions, decreased ATP levels during seizures have not been a universal observation. Hypoglycemic and hypoxic seizures might be expected to result from a decline in the brain's energy balance, owing to inadequate supplies of the glucose and oxygen necessary to maintain a normal rate of oxidative phosphorylation. Yet insulin-induced convulsions in rodents occur prior to any detectable change in brain ATP [2, 6, 14]. Sanders et al. [15] suggested that a decrease in the concentration of ATP in brain precedes the onset of overt convulsions in rats rendered hypoxic or treated with pentylenetetrazol, the implication being that the fall in energy reserves is the precipitating cause of the convulsions. Nelson and Mantz [16], however, observed normal ATP levels in brains of mice undergoing anoxic convulsions and suggested that the occurrence of substantial anaerobic metabolism during the time required to freeze the larger rat brain may have contributed to the changes in brain ATP reported earlier [15]. Furthermore, seizures produced by the administration of L-methionine sulfoximine, an inhibitor of glutamine synthetase, are associated not only with normal brain ATP levels but with *increased* carbohydrate stores [17]. Thus there appears to be no direct correlation between the level of cerebral energy substrates and the generation of convulsive activity. Nor is there any evidence that the spontaneous termination of seizures correlates with any change in the concentration of brain ATP [4].

When assessing the neurochemistry of seizures, a distinction must be made between the primary chemical changes in brain that follow directly from the abnormal electrical discharge and those that are partly, or perhaps entirely, secondary to the motor convulsions. Tonic-clonic convulsions prevent adequate pulmonary ventilation and give rise to intense muscular activity, both of which may deplete the oxygen content of blood. Meyer et al. [18] observed in patients undergoing electroshock therapy that, during the convulsions, oxygen tensions and pH in arterial blood both decline sharply, and similar changes in blood gases and pH have been reproduced in animals that were allowed to convulse freely [19, 20].

Contamination from chemical changes produced by convulsing muscles during experimental seizures can be largely overcome by paralysis and artificial ventilation [20]. As indicated in Table 31-2, however, seizures induced in ventilated, well-oxygenated animals given paralytic drugs are still accompanied by clear-cut changes in the glycolytic and energy metabolism of the brain. During the peak of the EEG-recorded seizure in paralyzed rats or cats, a point which corresponds temporally with the tonic phase of the convulsions in unparalyzed animals, cortical concentrations of ATP and phosphocreatine fell by approximately 15 to 25 percent, respectively. In both species, seizures caused a lactacidosis in the cerebral cortex that always was associated with an elevated lactate/pyruvate ratio and a lower tissue pH [5, 11]. These changes were calculated to have resulted in an abrupt shift of the cytoplasmic redox state (i.e., the $NAD^+/NADH$ ratio) toward increased reduction. Lack of oxygen in tissue cannot explain these metabolic alterations because cerebral venous oxygen tensions actually rose *above* control during the seizures, implying that oxygen supply to the brain was in excess of what was necessary to support oxidative metabolism. There is additional evidence pointing to this same conclusion. When Caspers

Table 31-2. Effect of Seizures on Concentrations of Metabolites in the Cerebral Cortex of
Ventilated Cats and Rats[a]

Animal and Convulsant	P-creatine	ATP	Lactate	Pyruvate	Lac/Pyr
	(mmol/kg wet weight)				
Rat (electroshock)					
Control	4.90	2.76	0.80	0.068	11.8
Seizure	4.10[b]	2.63[b]	3.62[b]	0.121[b]	29.9[b]
Cat (flurothyl)					
Control	4.40	2.28	1.26	0.071	17.7
Seizure	3.79[b]	2.00[b]	4.98[b]	0.121[b]	37.6[b]

[a] The cortices were frozen in situ by irrigation with liquid nitrogen 10 sec after the initiation of
a generalized seizure by electroshock (rats) or flurothyl (cats). Data are from Howse et al. [11]
and Duffy et al. [5].
[b] Different from the control value with $P < 0.05$.

and Speckmann [21] measured the oxygen tension of the cortical surface in rats
given pentylenetetrazol, they found that, as long as the cortical blood flow remained
high during the seizures, tissue PO_2 increased above control. Furthermore, Jöbsis
et al. [22], using the technique of reflectance fluorometry, showed that the fluorescence of the exposed cat cortex decreases during pentylenetetrazol- or strychnine-
induced seizures, suggesting that the mitochondrial pool of pyridine nucleotides had
become more oxidized. Similar redox changes have been observed in the region of
focal hippocampal seizures [23]. Thus a mild to moderate decline in the cerebral
energy balance combined with the development of a "nonhypoxic" lactacidosis in
the tissue seem to be biochemical responses that are intrinsic to the seizures themselves. The question is: How does this nonhypoxic lactacidosis come about?

Phosphate Potential and Metabolic Regulation

In brain, as in other organs, glycolysis (discussed in Chap. 14) is primarily a cytoplasmic function whose end products are pyruvate and lactate (Fig. 14-1). The
activity of lactic dehydrogenase in brain is high and localized to the cytosol, so it is
generally accepted that the lactate/pyruvate ratio at any pH is determined by the
redox state of the pyridine nucleotide couple NAD+-NADH, a reaction maintained
close to equilibrium in vivo, given in equation (31-1):

$$[NAD^+]/[NADH] = ([H^+]/K_1) \cdot ([pyruvate]/[lactate]) \qquad (31\text{-}1)$$

Krebs and Veech [24] showed that under a variety of nutritional influences the
redox state of the cytoplasmic NAD+-NADH couple in liver was determined by the
phosphorylation state of the adenine nucleotide system, i.e., the value of the ratio
$[ATP]/[ADP][HPO_4^{2-}]$, termed the phosphate potential. The link between the
phosphate potential and the pyridine nucleotide couple was provided by the demonstration that two of the intermediate reactions of glycolysis were maintained in
equilibrium. These reactions involve the conversion of glyceraldehyde 3-phosphate

to 3-phosphoglycerate, catalyzed by the enzymes glyceraldehyde 3-phosphate dehydrogenase and 3-phosphoglycerate kinase, given in equation (31-2).

$$K_2 = ([\text{3-phosphoglycerate}]/[\text{glyceraldehyde 3-phosphate}]) \\ \cdot ([\text{ATP}]/[\text{ADP}][\text{HPO}_4{}^{2-}]) \cdot ([\text{NADH}][\text{H}^+]/[\text{NAD}^+]) \quad (31\text{-}2)$$

Because the lactic dehydrogenase and glyceraldehyde 3-phosphate dehydrogenase
systems are in equilibrium with the same pool of NAD [24], combining equations
(31-1) and (31-2) yields an expression that shows the relationship between the lactate/
pyruvate ratio and the phosphate potential, given in equation (31-3).

$$K_3 = ([\text{3-phosphoglycerate}]/[\text{glyceraldehyde 3-phosphate}]) \\ \cdot ([\text{ATP}]/[\text{ADP}][\text{HPO}_4{}^{2-}]) \cdot ([\text{lactate}]/[\text{pyruvate}]) \quad (31\text{-}3)$$

Measurements of concentrations of the components of equation (31-3) in the rat cortex indicate that K_3 is near its predicted equilibrium value in vivo both in "resting"
cortex and during an electroshock-induced seizure [25]. Furthermore, during seizures, the ratio of 3-phosphoglycerate to glyceraldehyde 3-phosphate remained constant although the concentrations of both intermediates increased [25]. It follows,
therefore, that, if equilibrium is to be maintained, a fall in the phosphate potential
will be accompanied by an elevated lactate/pyruvate ratio. Since equation (31-3)
represents an equilibrium situation, the validity of the converse argument, i.e., that
increased cytoplasmic reduction raises the lactate/pyruvate ratio, causing a fall in
the phosphate potential, cannot be totally ruled out. It seems much more likely,
however, that, at least in the special case of seizures, a change in the phosphorylation
state, resulting from the increased energy demand imposed by the seizure discharge,
is the primary metabolic event.

The authors' interpretation of the homeostatic consequences of a decreased phosphate potential with cerebral lactacidosis during seizures is summarized in Figure 31-1.
The intense neural activity of the seizure discharge leads to Na^+ and K^+ shifts, which
stimulate the membrane (Na^+,K^+)-activated ATPase and bring about the breakdown
of ATP to ADP and inorganic phosphate. This sequence has several important effects.
There will be activation and deinhibition of phosphofructokinase, resulting in
increased glycolytic flux [27]. Simultaneously, the rise in tissue ADP and P_i will
stimulate mitochondrial oxidative phosphorylation, thereby increasing the rate of
oxygen consumption and CO_2 production. In addition, the decreased phosphate
potential will shift the cytoplasmic redox state toward reduction, promoting the
conversion of pyruvate into lactate and causing a nonhypoxic lactacidosis in the
tissue. The increased acidosis, in turn, will titrate tissue bicarbonate to yield CO_2
production in excess of oxygen consumption, which fits the experimental observations of an apparently increased RQ during seizures [3, 8, 11]. The released CO_2
or, more likely, the increased periarteriolar $[H^+]$ acts on vessel walls to decrease
cerebral arteriolar constriction and reduce vascular resistance. Howse et al. [11]
showed that in curarized, well-oxygenated cats maximal cerebral vasodilatation during

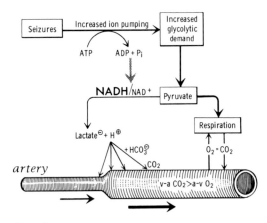

Figure 31-1
The blood flow—metabolism couple in brain, a hypothesis. From Plum et al. [26], used by permission of Raven Press.

flurothyl-induced seizures coincides precisely with the greatest reduction in the cortical-tissue pH. It seems likely that a similar kind of metabolic control of blood flow also operates during increased functional activity of the brain [28–30], although this has not yet been directly verified.

Status Epilepticus

Status epilepticus is potentially a much more serious clinical problem than the single seizure. Repeated or sustained seizures, particularly in children, are associated with a high incidence of permanent residual brain damage [31]. This limiting effect on growth and development of the brain can be reproduced experimentally in animals [32, 33]. Wasterlain and Plum [32] applied either single daily electroshock seizures or a single two-hour episode of status epilepticus to immature rats at different stages of neonatal and postnatal development and examined the effects on subsequent brain growth and chemical content. Animals undergoing convulsive seizures during the immediate 10-day postnatal period, when brain-cell mitosis in the rat is high, had a permanent reduction in brain weight and brain-cell number, as estimated from contents of DNA, RNA, protein, and cholesterol. Such animals were delayed in their achievement of behavioral developmental milestones. Animals undergoing seizures during the second 10 days of life also had permanently smaller brains but with a normal complement of DNA, suggesting a reduction of cell size without change in cell number. After 20 days of life, a time when cerebral mitotic activity has nearly ceased, neither daily seizures nor status epilepticus caused any permanent alteration in brain weight, cell number, or cell size.

In adult animals, at least, many of the acute changes of status epilepticus can be attenuated or reversed by supporting the respiration and circulation during the epileptic attack [20]. This implies that systemic effects of the convulsions (i.e., acidosis, hypotension, hypoxia) are at least partly responsible for the neurological consequences of status epilepticus, although the metabolic effects of the intense, abnormal electrical discharges must also contribute to the eventual chemical and morphological deterioration of the brain.

Brain-tissue analyses made after the experimental induction of repeated seizures in small animals indicate that the organ possesses a considerable capacity to maintain its energy balance during the early stages of status epilepticus [5, 7]. The experiment illustrated in Figure 31-2 is typical. Within 30 seconds after the first burst of EEG-recorded seizure activity in paralyzed, well-oxygenated mice treated with pentylenetetrazol, phosphocreatine levels in brain decreased and lactate began to rise. By 90 seconds after injection of the convulsant, there was a second major seizure burst, during which phosphocreatine decreased by 40 percent and ATP by 16 percent, while lactate increased by 500 percent. During the remaining 8.5 minutes of continuous epileptiform activity, only minor further changes were noted in these metabolites, suggesting that a new steady state had been reached. Animals subjected to repeated electroshocks exhibit a similar stability of the high-energy phosphate compounds in the cerebrum up to the point at which the attacks become self-generating and the transport of carbohydrate into brain begins to lag behind substrate requirements [5]. This point occurs in rodents after about 25 equally spaced successive seizures.

A progressive depletion of brain glycogen seems to be characteristic of nearly all types of repeated seizures [5, 7, 34], possibly reflecting the activation of glycogen phosphorylase by the seizure-induced rise in cyclic 3′,5′-adenosine monophosphate in the brain [35]. The significance of the loss of glycogen during status epilepticus is difficult to assess, but it suggests a metabolic change in glial cells. Most cerebral

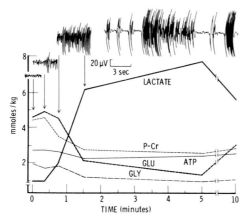

Figure 31-2
Changes in metabolite concentrations of the mouse forebrain during status epilepticus. Animals were paralyzed and artificially ventilated with oxygen. Seizures were induced with pentylenetetrazol (150 mg/kg, I.P.), and the mice were frozen after periods of continuous epileptiform activity in the EEG lasting up to 10 minutes. EEG tracings from control and treated animals at the time of freezing are shown at the top of the figure. Abbreviations: P-Cr, phosphocreatine; GLU, glucose; GLY, glycogen. Data of Duffy et al. [5]; reproduced in [26]; used by permission of Raven Press.

glycogen is confined to the cytoplasm and end feet of astrocytes, implying an intracellular concentration much greater than that of whole brain. Degradation of glycogen could lead to hyperosmolarity within the astrocytes, causing them to take up water from the extracellular space [36]. Swelling of astrocytic end feet around capillaries and neurons has been demonstrated in rat cortex after five minutes of pentylenetetrazol-induced convulsions [37], and has been implicated as the probable cause of abnormalities in the hippocampus after bicuculline-induced status epilepticus in paralyzed, well-oxygenated baboons [38]. Glial cells normally operate as the potassium, and probably the ammonium ion, sponges of the brain [39, 40], prompting some investigators to suggest that failure of the glia to translocate potassium after a seizure may predispose to the development of status epilepticus [41]. Other workers, however, have been unable to confirm the importance of extracellular potassium in the generation or termination of seizures [42], and the question remains unresolved.

COMA

Coma represents the great reduction or absence of the psychological functions of the brain. Both the neurochemistry and neurophysiology of coma can involve abnormalities that either diffusely affect the cerebral hemispheres, selectively affect the reticular formation, or involve both areas together [43]. This point is well illustrated in a patient studied by Ingvar and colleagues [44, 45]: The subject was in coma due to a destructive lesion confined to the midbrain reticular formation. Despite near-normal morphology of the cerebral hemispheres, cerebral blood flow and oxygen consumption were reduced to 25 percent of normal. Purely focal metabolic abnormalities may produce focal neurological dysfunction, but do not cause coma.

Metabolic studies of the brain of wakeful man during various psychological and physiological activities indicate that, although focal changes in metabolism accompany local increases or decreases in cerebral function [29], the metabolically more demanding integrative action of the whole brain hides these variations so that net oxygen consumption normally remains remarkably constant (Chap. 19). Earlier in this chapter, we presented the concept that homeostasis normally maintains energy levels in the brain within very narrow limits and that any deviation from the normal level represents a major threat to the brain. Available evidence in comatose conditions suggests that this is so: Comatose states that are fully reversible are associated with a normal or even supranormal level of ATP and phosphocreatine in brain, whereas states accompanied by a fall in these high-energy phosphates appear to evolve consistently into severe brain damage. As will be seen, with some forms of brain injury or intoxication, "high-energy" coma, if untreated, tends to deteriorate into "low-energy" coma, which may be the point at which permanent brain damage occurs. The sequence suggests that coma in these circumstances may reflect an initial "switching down" of synaptic activity, possibly as a protective adaptive response. Whether such a concept could extend to the effects of anesthetic agents or other nonstructural causes of coma, such as the postictal or postconcussive states, remains to be seen.

Anesthetic Coma

Anesthetic drugs produce profound, but fully reversible, depression of nervous system function. The agents vary widely in their chemical structures, so that the molecular basis of their action ordinarily cannot be predicted accurately from their structural configuration alone. Most anesthetics are lipid soluble. Indeed, lipid solubility has been proposed as a mechanism of their actions since this almost inevitably implies an effect on cell membranes [46]. Because equally lipid-soluble agents possess no anesthetic effect, however, the explanation is insufficient, although lipid solubility is probably required for these drugs to pass cerebral cell membranes to exert their effects.

Drugs that produce general anesthesia affect the cerebral circulation variably, presumably because they differ in the degree to which they affect vascular smooth muscle. Nearly all of them, with the exception of cyclohexylamine [47], reduce cerebral oxygen consumption, and most evidence indicates that this reduction occurs as much by a direct and diffuse effect on brain tissue as by any selective depression of the brainstem reticular formation. Physiologically, the anesthetic drugs vary in whether they depress reticular or somatocortical projections more [48, 49]. Phenobarbital has been demonstrated in experimental preparations to depress selectively excitatory postsynaptic potentials [50], but whether this is its major effect has not been studied, nor is it known to what degree other anesthetic agents selectively affect synaptic transmission. It seems unlikely that all anesthetics have a similar chemical effect on nervous tissue since they are so different structurally and have such different effects on metabolism.

At present, no satisfactory chemical explanation has been found for the action of any central anesthetic drug. Several physicochemical theories have been tried to relate anesthetic potency to molecular rearrangement within the cell. Perhaps the most imaginative was Pauling's [51] speculation that anesthetic agents can form hydrated microcrystals, or clathrates, within the cell that would change the impedance of synaptic regions. Unfortunately, neither this nor the other physicochemical theories have enjoyed experimental proof.

Several early workers proposed from studies in vitro that the anesthetics inhibited mitochondrial oxidation. Depression of mitochondrial respiration in vitro, however, occurs with these agents only at several times their anesthetic level in vivo [52]. The anesthetics do not directly impair oxidative phosphorylation, and several studies demonstrate that, during either barbiturate or inhalation anesthesia, whole brain levels of glycogen, glucose, and phosphocreatine are either normal or higher than normal [53, 54]. ATP, ADP, and AMP concentrations in the tissue remain normal, and lactate, if anything, is slightly reduced below waking concentrations. These several observations indicate that the reduced oxygen consumption imposed by most anesthetics is the result and not the cause of the lowered cerebral metabolic activity. During anesthesia, the brain remains in a state of energized readiness.

Hypoxic Coma

The effects of reduced oxygen supply on the brain must be separated carefully from those of ischemia (see p. 658) because failure of the circulation not only results in absolute oxygen and substrate lack but also fails to rinse the tissue of its metabolic products. Anoxia, also, and perhaps initially, affects the heart and rapidly produces systemic hypotension and cardiac arrhythmias. Accordingly, pure cerebral hypoxia causing coma is relatively uncommon in the clinical setting. Exceptions are provided by the delirium produced at high altitude, and the coma or stupor that may follow carbon-monoxide poisoning, in which high concentrations of carboxyhemoglobin block much of the oxygen transport to brain and other organs.

Chemical changes in the brain of experimental animals during controlled, selective hypoxemia [55, 56] consist initially of an accelerated glycolysis during moderate hypoxia, followed by relatively small changes in high energy reserves during progressively more profound hypoxia, and then by impairment of mitochondrial metabolism, which leads finally to energy failure (Tables 31-3 and 31-4). The lactate concentration in brain tissue begins to rise as arterial Po_2 (Pa_{O_2}) values fall below 50 mm Hg. Phosphocreatine declines at Pa_{O_2} less than 35 mm Hg, but this initial change in phosphate reserves may represent partly a pH effect on the creatine-phosphokinase equilibrium secondary to lactic acid accumulation [54].

Most human beings suffer impaired consciousness if the oxygen pressure in arterial blood is reduced acutely below about 30 mm Hg, and at this level tissue metabolites in experimental animals indicate no unequivocal change in the energy state. With more severe systemic hypoxia (Pa_{O_2} of 21 mm Hg) and as long as cerebral blood pressure is maintained, there is a markedly increased glycolysis initially, reflected by elevations of lactate and pyruvate in tissue. A substantial decline in phosphocreatine and a small change in the phosphate potential also occur, the latter due to an increase in P_i and ADP. Any further reduction of the oxygen supply — as, for example, by interference with the physiological compensatory hyperemia that normally occurs in the brain during hypoxia — induces a fall in citric acid cycle intermediates, accompanied by a substantial drop in cerebral energy reserves. The changes indicate that, despite high rates of glycolysis during hypoxia, oxygen availability finally falls sufficiently to inhibit the oxidation of pyruvate and normal mitochondrial oxidative metabolism. In keeping with the chemical observations, morphological studies

Table 31-3. Changes in Concentrations of High-Energy Phosphate Compounds of Rat Brain During Severe Hypoxia and Hypoxia-Oligemia[a]

Treatment	Pa_{O_2} (mm Hg)	ATP	ADP	AMP	P-creatine	Total $\sim$P
		(mmol/kg wet weight)				
Control	>100	3.06	0.272	0.031	4.98	11.37
Hypoxia (moderate)	28	3.07	0.342	0.039	4.31	10.79
Hypoxia (severe)	21	3.00	0.384	0.046	3.11	9.49
Hypoxia-oligemia	21	2.38	0.576	0.314	1.76	7.10

[a]Data were obtained in animals that were paralyzed and artificially ventilated with appropriate oxygen-nitrogen mixtures. Experimental animals had a clamp on one carotid artery and different levels of oxygenation for 30 min, after which the brains were frozen in situ. In this and Table 31-4, oligemia denotes the cerebral hemisphere where ipsilateral carotid artery occlusion prevented compensatory hyperemia during hypoxic exposure. Data of Salford et al. [56].

Table 31-4. Changes in Concentrations of Intermediates of Carbohydrate Metabolism in Rat Brain During Severe Hypoxia and Hypoxia-Oligemia[a]

Treatment	Pa_{O_2} (mm Hg)	Glucose	Glucose-6-P	Pyruvate	Lactate	Lactate/Pyruvate	Citrate	α-Ketoglutarate	Malate
					(mmol/kg wet weight)				
Control	>100	4.46	0.098	0.107	1.69	15.8	0.346	0.123	0.369
Hypoxia (moderate)	28	6.37	0.101	0.250	5.86	24.5	0.419	0.158	0.734
Hypoxia (severe)	21	5.56	0.147	0.297	14.31	48.3	0.341	0.115	0.865
Hypoxia-oligemia	21	4.97	0.169	0.231	20.79	94.6	0.254	0.051	0.770

[a] Animals and experimental conditions were as described in Table 31-3. Data of Salford et al. [56].

indicate that the mitochondria undergo the first permanent abnormality during profound cerebral hypoxia [57]. With prolonged less severe hypoxia, such as occurs in carbon monoxide poisoning, increased glycolosis from muscle and other systemic tissues results in a pronounced systemic lactacidosis [58], a factor incriminated as contributing to the cerebral demyelination that often occurs with severe examples of this disorder.

The exact neurochemical basis for coma in hypoxia is unclear, but it may represent an adaptive response to reduced oxygen supply. Cerebral blood flow increases markedly in hypoxia, conferring a strong homeostatic protection on the brain. Body metabolism and, perhaps, cerebral metabolism also drop. Thus, Duffy et al. [1] found evidence in mice that cerebral energy consumption declined during hypoxia before any effect on tissue concentrations of adenine nucleotides could be observed. Furthermore, Kjellmer et al. [59] found that $CMRO_2$ in lamb fetuses declined at low oxygen concentrations in blood. These findings imply that moderately severe hypoxia may initially exert an anesthetic-like action on the cerebrum, which partially protects the organ against energy failure. Johannsson and Siesjö [60], however, were unable to demonstrate a decrease in $CMRO_2$ in lightly anesthetized rats when PO_2 was reduced to 22 to 25 mm Hg, and the point is still debatable.

Oxygen lack may also affect chemical reactions other than those involved in oxidative phosphorylation. The rate-limiting steps in the synthesis of indole and catechol neurotransmitters depend on molecular oxygen (Chap. 10); Davis and Carlsson [61] observed impaired monoamine metabolism in brain at Pa_{O_2} values below about 50 mm Hg. It is unlikely that this produces an altered consciousness, however, because pharmacological measures that produce a similar impairment of monoamine synthesis have little effect on behavior [62].

Ischemic Coma

Coma results from cerebral ischemia only with either total cerebral ischemia or vascular occlusions involving the blood supply to the reticular formation. From the neurochemical standpoint, the changes in total ischemia are of greater interest (Fig. 31-3).

Consciousness is lost within 6 to 9 sec after onset of total cerebral ischemia, and the EEG disappears within 15 to 20 sec. The phosphocreatine of brain drops to unmeasurable levels within a minute, and ATP falls by 50 percent within the first minute and to near 0 by 2 min. Lactate increases to near maximal values of about 10 mmol/kg by 2 min, paralleling the rapid exhaustion of glucose and glycogen stores during the same period [63]. At one time it was believed that permanent brain damage began simultaneously with, or a very short time after, ischemic depletion of cerebral energy reserves. Biochemical studies by several laboratories [64, 65], however, indicate that the brain can regain, for at least brief periods, a considerable proportion of its energy-generating capacity and of other chemical functions after ischemic periods lasting from 15 minutes to as much as an hour. It is too early to judge the implication of these findings or whether they indicate potentially permanent recovery, but they indicate why the neurochemistry of ischemic coma presently is receiving extensive investigation.

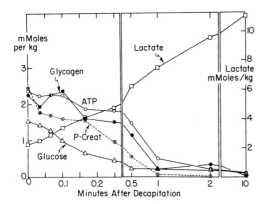

Figure 31-3
Changes in high-energy compounds and glycolytic intermediates in mouse brain during total ischemia produced by decapitation. Abbreviation: P-Creat., phosphocreatine. From Lowry et al. [63].

Hypoglycemia

Glucose is the principal substrate for cerebral oxidative metabolism (Chaps. 14 and 19). Normal blood glucose in man is about 4 mmol/l. Levels of approximately 1.5 to 2 mmol/l are associated with confusion or delirium, and levels of approximately 1 mmol/l result in deep stupor, coma, or convulsions. Transient hypoglycemic coma that lasts up to an hour usually is fully reversible in man and has been used therapeutically to treat psychosis. More prolonged coma, particularly if accompanied by convulsions, often is followed by irreversible brain damage, with the cerebral cortex suffering pathological changes similar to those incurred during severe sustained anoxia (Chaps. 19 and 25).

Early biochemical studies of hypoglycemia in man demonstrated a reduction in cerebral glucose consumption accompanied by a disproportionately smaller decrease in cerebral oxygen consumption [66]. The implication that the brain utilizes endogenous substrates for oxidative metabolism during hypoglycemia was borne out by animal experiments that indicated hypoglycemia-induced decreases in the tissue concentrations of glutamate, glutamine, alanine, and γ-aminobutyric acid (GABA) [67, 69].

Until recently, most animal experiments suggested that coma and slowing of brain metabolism during hypoglycemia were the straightforward results of substrate lack, leading to a failure of oxidative phosphorylation to generate sufficient energy to maintain brain function. More recent studies, however, indicate that whole-brain concentrations of ATP, ADP, AMP, and phosphocreatine remain at normal levels during the early stages of insulin coma, including the stage at which intermittent polyspikes or intermittent isoelectric activity develop in the EEG [2]. With glucose concentrations in blood below 3 mmol/l, intracellular glucose disappears from the brains of experimental animals and tissue concentrations of glycogen, glucose 6-phosphate, pyruvate, lactate, and citric acid cycle intermediates decline. At blood glucose levels below 1 mmol/l, concentrations of glycogen, glucose 6-phosphate, pyruvate, and lactate fall to very low levels, and calculated values for free NADH in the tissue decline sharply [68, 69]. Roughly concurrent with the development of convulsions, the concentration of the inhibitory agent GABA falls and the stimulants aspartate and ammonia increase.

Although the onset of EEG changes and stupor in animals and, presumably, insulin coma in man occur without changes in the energy state of the cerebral cortex, high-energy phosphate compounds drop rapidly, once glycogen and pyruvate fall to very low levels in the tissue [67]. This stage correlates in animals with the appearance of sustained convulsions and prolonged isoelectricity in the EEG. It is not known to what degree this stage of energy depletion can be reversed or whether it always presages permanent tissue damage. Hypothermia nearly always accompanies clinical hypoglycemia and undoubtedly partly protects the brain by reducing its rate of metabolism.

Although the decline in glycolytic and citric acid cycle intermediates, along with the changes in amino acids in brain, is a well-established sequence in prolonged and severe hypoglycemic coma, the cause of the initial functional depression is unknown (Chap. 25). Furthermore, clinical recovery after hypoglycemia occurs in animals before full restoration of all glycolytic and Krebs cycle intermediate compounds takes place [70]. The problem in deducing the functional effects of hypoglycemia from its neurochemistry may be partly methodological: Overall chemical analyses of brain tissue may not disclose local changes in concentration of such critical agents as transmitters. In this respect, it has been suggested that hypoglycemia leads to a decrease in availability of substrate necessary to synthesize acetylcholine and, perhaps, other transmitter agents [67, 69, 71].

Hepatic Coma

Two major chemical theories, not necessarily mutually exclusive, have been advanced to explain hepatic encephalopathy. One focuses on amino acid and neurotransmitter changes; the other centers around derangements in ammonia and glutamine metabolism that accompany liver disorders.

Fischer has most actively advanced the hypothesis that a "false transmitter" may explain the neurological changes of hepatic coma [72]. So far, at least, the evidence is inferential. False transmitters are conceived of as chemical molecules that may resemble normal biogenic amines and can inappropriately occupy transmitter receptor sites, thereby blocking the effects of natural transmitter action. Amino acids are likely candidates for such a role and a number of amino acids, including phenylalanine and methionine, are elevated in the blood of patients with hepatic coma. In addition, diets rich in the amino acids tyrosine, tryptophan, and phenylalanine are preferentially toxic to animals with portacaval shunts. Monoamine oxidase inhibitors are reported to accentuate hepatic stupor in man, and L-DOPA, the precursor of cerebral dopamine, lightens the coma of some patients with severe liver disease. In the brain of animals with induced hepatic coma, norepinephrine falls and octopamine and serotonin increase substantially. These several observations have led to the speculation that the decrease in norepinephrine, or an interference with its normal cerebral action, may account for some or all of the neurological symptoms of acute hepatic failure.

Abnormal ammonia metabolism has been featured in explanations of hepatic coma ever since the nineteenth century, when investigators were able to induce cerebral symptoms in animals with portacaval shunts by feeding them meat. Soon after, it was learned that orally ingested ammonium salts would do the same thing.

Ammonia is elevated in blood and brain in severe liver disease. Ammonia is generated from the degradation of amino acids in the body and from the action on the intestinal contents of urea-splitting organisms. Normally ammonia is promptly converted into urea in the liver by the Krebs-Henseleit cycle (see Fig. 24-4). Liver disease interferes with this cycle by impairing venous drainage from the intestine so that ammonia-carrying portal blood bypasses the liver to enter the systemic circulation (portal-systemic shunting) and by reducing the tissue capacity for urea synthesis.

Ammonia is constantly formed in brain by deamination, and ammonia concentrations in that organ are always 60 to 100 percent higher than in blood [73–75]. Ammonia leaves brain by diffusion as well as by conversion into glutamine. The latter is a two-stage energy-requiring process:

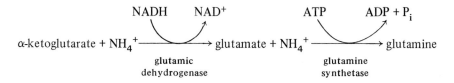

How glutamine leaves brain is unknown. Normally glutamine levels are about 5 mM in brain and 0.6 mM in CSF. With elevated blood ammonia levels, brain ammonia levels rise proportionately, and glutamine promptly increases so that, in severe liver disease, glutamine levels in CSF may reach 4 to 5 mM [76]. Moreover, an additional metabolite, α-ketoglutaramate (α-KGM), appears in CSF in significant quantities in chronic hepatic encephalopathy [76]. The mechanism of the rise of α-KGM is unknown but it is believed to be formed from glutamine by transamination with an appropriate α-keto acid, principally phenylpyruvate, glyoxylate, or pyruvate [77]. (See Fig. 31-4.) The compound can be broken down subsequently in the presence of the enzyme ω-amidase to yield α-ketoglutarate and ammonia. The entire cycling of one molecule of α-ketoglutarate through glutamine, α-ketoglutaramate, and back to α-ketoglutarate removes one molecule of ammonia from the system. Cooper and Meister [77] have suggested that sparing of α-keto acids may be the main normal function of this pathway.

Considerable physiological evidence indicates that ammonia is toxic to brain. Hyperammonemia accompanies the comas produced by inherited defects in urea synthesis and Reye's syndrome, as well as the coma of liver disease (Chap. 24). Acute hyperammonemia produces both convulsions and coma in experimental animals. Subacute or chronic hyperammonemia induces dilatation and proliferation of protoplasmic astrocytes in brain, and these cells appear to contain large amounts of glutamic dehydrogenase [78]. Glutamine metabolism is known to be compartmented in brain (Chap. 14), and it has been postulated that the compartment that involves the detoxification of ammonia lies in the protoplasmic astrocytes, which provide a biochemical defense against rising concentrations of ammonia in the adjacent extracellular fluid and neurons [40, 78, 79].

The exact mechanism by which elevated ammonia concentrations in brain cause coma is unknown. Bessman and Bessman's [80] earlier theory that α-ketoglutarate is depleted by combining with ammonia to form glutamate was not borne out by experiment: α-ketoglutarate rises, rather than declines, in brain during hyperammonemia [74, 76]. Present theories center (1) on ammonia's effect on synaptic transmission (it reduces IPSPs and interferes with the chloride pump [81]); (2) on its possible substitution for K^+ in activating the membrane $(Na^+ + K^+)$-ATPase [82]; (3) on the potential neurotoxicity of α-ketoglutaramate; and (4) on impaired cerebral energy supply.

Marked hyperammonemia produces a significant change in energy metabolism in brain. Hepatic stupor and coma in man and animals are associated with a lowered cerebral oxygen consumption [79]. Also, high concentrations of ammonia inhibit the oxidation of pyruvate in tissue slices [83]. Schenker et al. [74] first noted that coma-producing acute ammonia intoxication in rats lowered phosphocreatine and ATP contents in the brain stem. Hindfelt and Siesjö [73] obtained somewhat similar results, observing significant declines in phosphocreatine and ATP in rat brain 30 minutes after a coma-producing injection of ammonium acetate. In the latter experiments, the fall in high-energy phosphates was maximal in brain stem and cerebellum.

Ammonia-induced stupor and coma in rats with chronic portacaval shunts are associated also with a decline in energy reserves in the brain [75]. When such animals are given moderate amounts of ammonium salts, lactate and pyruvate accumulate in brain tissue, cerebral oxygen consumption declines, and cerebral concentrations of phosphocreatine and ATP fall. Concentrations of glutamate and aspartate in brain also fall, and glutamine and alanine levels rise. The changes suggest a defect or block

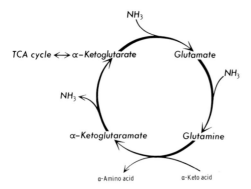

Figure 31-4
The pathway of ammonia metabolism in brain. From [76], used by permission of Raven Press.

in the oxidation of pyruvate as it enters the citric acid cycle and are consistent with an interference by ammonia with the malate-aspartate shuttle that normally transfers reducing equivalents from cytoplasm to mitochondria for oxidation. Thus, both in vitro and in vivo studies support the view that severe hyperammonemia interferes with the regulation of energy metabolism in the brain, an effect that would explain many clinical aspects of the disorder.

REFERENCES

1. Duffy, T. E., Nelson, S. R., and Lowry, O. H. Cerebral carbohydrate metabolism during acute hypoxia and recovery. *J. Neurochem.* 19:959, 1972.
2. Ferrendelli, J. A., and Chang, M.-M. Brain metabolism during hypoglycemia. *Arch. Neurol.* 28:173, 1973.
3. Brodersen, P., Paulson, O. B., Bolwig, T. G., Rogon, Z. E., Rafaelsen, O. J., and Lassen, N. A. Cerebral hyperemia in electrically induced epileptic seizures. *Arch. Neurol.* 28:334, 1973.
4. Collins, R. C., Posner, J. B., and Plum, F. Cerebral energy metabolism during electroshock seizures in mice. *Am. J. Physiol.* 218:943, 1970.
5. Duffy, T. E., Howse, D. C., and Plum, F. Cerebral energy metabolism during experimental status epilepticus. *J. Neurochem.* 24:925, 1975.
6. King, L. J., Lowry, O. H., Passonneau, J. V., and Venson, V. Effects of convulsants on energy reserves in the cerebral cortex. *J. Neurochem.* 14:599, 1967.
7. Sacktor, B., Wilson, J. E., and Tiekert, C. G. Regulation of glycolysis in brain, *in situ*, during convulsions. *J. Biol. Chem.* 241:5071, 1966.
8. Plum, F., Posner, J. B., and Troy, B. Cerebral metabolic and circulatory responses to induced convulsions in animals. *Arch. Neurol.* 18:1, 1968.
9. Schmidt, C. F., Kety, S. S., and Pennes, H. H. Gaseous metabolism of the brain of the monkey. *Am. J. Physiol.* 143:33, 1945.
10. Ferrendelli, J. A., and McDougal, D. B., Jr. The effect of electroshock on regional CNS energy reserves in mice. *J. Neurochem.* 18:1197, 1971.
11. Howse, D. C., Caronna, J. J., Duffy, T. E., and Plum, F. Cerebral energy metabolism, pH, and blood flow during seizures in the cat. *Am. J. Physiol.* 227:1444, 1974.
12. Carl, J. L., and King, L. J. Hexose and pentose phosphates in brain during convulsions. *J. Neurochem.* 17:293, 1970.
13. King, L. J., Carl, J. L., and Lao, L. Carbohydrate metabolism in brain during convulsions and its modification by phenobarbitone. *J. Neurochem.* 20:477, 1973.
14. Tarr, M., Brada, D., and Samson, F. E., Jr. Cerebral high-energy phosphates during insulin hypoglycemia. *Am. J. Physiol.* 203:690, 1962.
15. Sanders, A. P., Kramer, R. S., Woodhall, B., and Currie, W. D. Brain adenosine triphosphate: Decreased concentration precedes convulsions. *Science* 169:206, 1970.
16. Nelson, S. R., and Mantz, M.-L. Brain energy reserve levels at the onset of convulsions in hypoxic mice. *Life Sci.* 10:901, 1971.
17. Folbergrová, J., Passonneau, J. V., Lowry, O. H., and Schulz, D. W. Glycogen, ammonia and related metabolites in the brain during seizures evoked by methionine sulphoximine. *J. Neurochem.* 16:191, 1969.
18. Meyer, J. S., Gotoh, F., and Favale, E. Cerebral metabolism during epileptic seizures in man. *Electroencephalogr. Clin. Neurophysiol.* 21:10, 1966.

19. Meldrum, B. S., and Horton, R. W. Physiology of status epilepticus in primates. *Arch. Neurol.* 28:1, 1973.
20. Wasterlain, C. G. Mortality and morbidity from serial seizures. *Epilepsia* 15:155, 1974.
21. Caspers, H., and Speckmann, E. J. Cerebral PO_2, PCO_2 and pH: Changes during convulsive activity and their significance for spontaneous arrest of seizures. *Epilepsia* 13:699, 1972.
22. Jöbsis, F. F., O'Connor, M., Vitale, A., and Vreman, H. Intracellular redox changes in functioning cerebral cortex. Metabolic effects of epileptiform activity. *J. Neurophysiol.* 34:735, 1971.
23. Lewis, D. V., O'Connor, M. J., and Schuette, W. K. Oxidative metabolism during recurrent seizures in the penicillin treated hippocampus. *Electroencephalogr. Clin. Neurophysiol.* 36:347, 1974.
*24. Krebs, H. A., and Veech, R. L. Regulation of the Redox State of the Pyridine Nucleotides in Rat Liver. In Sund, H. (Ed.), *Pyridine Nucleotide-Dependent Dehydrogenases.* New York: Springer-Verlag, 1970, p. 413.
25. Howse, D. C., and Duffy, T. E. Control of the redox state of the pyridine nucleotides in the rat cerebral cortex. Effect of electroshock-induced seizures. *J. Neurochem.* 24:935, 1975.
*26. Plum, F., Howse, D. C., and Duffy, T. E. Metabolic Effects of Seizures. In Plum, F. (Ed.), *Brain Dysfunction in Metabolic Disorders. Res. Publ. Assoc. Nerv. Ment. Dis.* 53:141, 1974.
27. Lowry, O. H., and Passonneau, J. V. Kinetic evidence for multiple binding sites on phosphofructokinase. *J. Biol. Chem.* 241:2268, 1966.
28. Kennedy, C., Des Rosiers, M. H., Jehle, J. W., Reivich, M., Sharpe, F., and Sokoloff, L. Mapping of functional neural pathways by autoradiographic survey of local metabolic rate with [14C]deoxyglucose. *Science* 187:850, 1975.
29. Raichle, M. E., Grubb, R. L., Jr., Gado, M. H., Eichling, J. O., and Ter-Pogossian, M. M. *In vivo* correlation between regional cerebral blood flow and oxidative metabolism in man. *Arch. Neurol.,* in press, 1976.
30. Salford, L. G., Duffy, T. E., and Plum, F. Association of Blood Flow and Acid-Base Change in Brain During Afferent Stimulation. In Langfitt, T. W., McHenry, L. C., Jr., Reivich, M., and Wollman, H. (Eds.), *Cerebral Circulation & Metabolism.* New York, Springer-Verlag, 1975, p. 380.
31. Aicardi, J., and Chevrie, J. J. Convulsive status epilepticus in infants and children: A study of 239 cases. *Epilepsia* 11:187, 1971.
32. Wasterlain, C. G., and Plum, F. Vulnerability of developing rat brain to electroconvulsive seizures. *Arch. Neurol.* 29:38, 1973.
33. Meldrum, B. S., and Brierley, J. B. Prolonged epileptic seizures in primates. *Arch. Neurol.* 28:10, 1973.
34. Minard, F. N., Kang, C. H., and Mushahwar, I. K. The effect of periodic convulsions induced by 1,1-dimethylhydrazine on the glycogen of rat brain. *J. Neurochem.* 12:279, 1965.
35. Sattin, A. Increase in the content of adenosine-3′,5′-monophosphate in mouse forebrain during seizures and prevention of the increase by methylxanthines. *J. Neurochem.* 18:1087, 1971.
36. Friede, R. L., and van Houten, W. H. Relations between post-mortem alterations and glycolytic metabolism in the brain. *Exp. Neurol.* 4:197, 1961.
37. DeRobertis, E., Alberici, M., and Rodriguez de Lores Arnaiz, G. Astroglial swelling and phosphohydrolases in cerebral cortex of metrazol convulsant rats. *Brain Res.* 12:461, 1969.

*Asterisks denote key references.

38. Meldrum, B. S., Vigouroux, R. A., and Brierley, J. B. Systemic factors and epileptic brain damage. *Arch. Neurol.* 29:82, 1973.
39. Kuffler, S. W., Nicholls, J. G., and Orkand, R. K. Physiological properties of glial cells in the central nervous system of amphibia. *J. Neurophysiol.* 29:768, 1966.
*40. Cavanagh, J. B. Liver Bypass and the Glia. In Plum, F. (Ed.), *Brain Dysfunction in Metabolic Disorders. Res. Publ. Assoc. Nerv. Ment. Dis.* 53:13, 1974.
41. Sypert, G. W., and Ward, A. A., Jr. Changes in extracellular potassium activity during neocortical propagated seizures. *Exp. Neurol.* 45:19. 1974.
42. Moody, W. J., Jr., Futamachi, K. J., and Prince, D. A. Extracellular potassium activity during epileptogenesis. *Exp. Neurol.* 42:248, 1974.
*43. Plum, F., and Posner, J. B. *Diagnosis of Stupor and Coma,* 2d ed. Philadelphia: Davis, 1972.
44. Ingvar, D. H., Haggendal, E., Nilsson, N. J., Sourander, P., Wickbom, I., and Lassen, N. A. Cerebral circulation and metabolism in a comatose patient. *Arch. Neurol.* 11:13, 1964.
45. Ingvar, D. H., and Sourander, P. Destruction of the reticular core of the brain stem. A patho-anatomical follow-up of a case of coma of three years' duration. *Arch. Neurol.* 23:1, 1970.
*46. Meyer, H., and Overton, E. Cited by Anderson, N. B., and Amaranath, L., Anesthetic effects on transport across cell membranes. *Anesthesiology* 39:123, 1973.
47. Dawson, B., Michenfelder, J. D., and Theye, R. A. Effects of ketamine on canine cerebral blood flow and metabolism. *Anesth. Analg.* (Cleve.) 50:443, 1971.
48. Clark, D. L., Hosick, E. C., and Rosner, B. S. Neurophysiological effects of different anesthetics in unconscious man. *J. Appl. Physiol.* 31:884, 1971.
49. French, J. D., Verzeano, M., and Magoun, H. W. A neural basis of the anesthetic state. *Arch. Neurol. Psychiatr.* 69:519, 1953.
50. Hosick, E. C., Clark, D. L., Adam, N., and Rosner, B. S. Neurophysiological effects of different anesthetics in conscious man. *J. Appl. Physiol.* 31:892, 1971.
51. Pauling, L. A molecular theory of general anesthesia. *Science* 134:15, 1961.
*52. Cohen, P. J. Effect of anesthetics on mitochondrial function. *Anesthesiology* 39:152, 1973.
*53. Fink, B. R., and Haschke, R. H. Anesthetic effects on cerebral metabolism. *Anesthesiology* 39:199, 1973.
*54. Siesjö, B. K., and Plum, F. Pathophysiology of Anoxic Brain Damage. In Gaull, G. (Ed.), *Biology of Brain Dysfunction.* Vol. 1. New York: Plenum, 1973, p. 319.
*55. Siesjö, B. K., Johannsson, H., Ljunggren, B., and Norberg, K. Brain Dysfunction in Cerebral Hypoxia and Ischemia. In Plum, F. (Ed.), *Brain Dysfunction in Metabolic Disorders. Res. Publ. Assoc. Nerv. Ment. Dis.* 53:75, 1974.
56. Salford, L. G., Plum, F., and Siesjö, B. K. Graded hypoxia-oligemia in rat brain. Biochemical alterations and their implications. *Arch. Neurol.* 29:227, 1973.
57. Brown, A. W., and Brierley, J. B. The earliest alterations in rat neurons and astrocytes after anoxia-ischemia. *Acta Neuropathol.* 23:9, 1973.
58. Ginsberg, M. D., Myers, R. E., and McDonagh, B. F. Experimental carbon monoxide encephalopathy in the primate. Chemical aspects, neuropathology and physiologic correlation. *Arch. Neurol.* 30:209, 1974.

59. Kjellmer, I., Karlsson, K., Olsson, T., and Rosen, K. G. Cerebral reactions during intra-uterine asphyxia in the sheep. Circulation and oxygen consumption in the fetal brain. *Pediatr. Res.* 8:50, 1974.

60. Johannsson, H., and Siesjö, B. K. Blood flow and oxygen consumption of rat brain in profound hypoxia. *Acta Physiol. Scand.* 90:281, 1974.

61. Davis, J. N., and Carlsson, A. Effect of hypoxia on tyrosine and tryptophan hydroxylation in unanesthetized rat brain. *J. Neurochem.* 20:913, 1973.

62. Breese, G. R., and Traylor, T. D. Effect of 6-hydroxydopamine on brain nor-epinephrine and dopamine: Evidence for selective degeneration of catechol-amine neurons. *J. Pharmacol. Exp. Ther.* 174:413, 1970.

63. Lowry, O. H., Passonneau, J. V., Hasselberger, F. X., and Schulz, D. W. Effect of ischemia on known substrates and cofactors of the glycolytic pathway in brain. *J. Biol. Chem.* 239:18, 1964.

64. Hossmann, K. A., and Kleihues, P. Reversibility of ischemic brain damage. *Arch. Neurol.* 29:375, 1973.

65. Hinzen, D. H., Müller, U., Sobotka, P., Gebert, E., Lang, R., and Hirsch, H. Metabolism and function of dog's brain recovering from longtime ischemia. *Am. J. Physiol.* 223:1158, 1972.

66. Kety, S. S., Woodford, R. B., Harmel, M. H., Freyhan, M., Appel, F. A., and Schmidt, C. F. Cerebral blood flow and metabolism in schizophrenia: The effects of barbiturate semi-narcosis, insulin coma and electroshock. *Am. J. Psychiatry* 104:765, 1948.

67. Lewis, L. D., Ljunggren, B., Norberg, K., and Siesjö, B. K. Changes in carbo-hydrate substrates, amino acids and ammonia in the brain during insulin-induced hypoglycemia. *J. Neurochem.* 23:659, 1974.

68. Lewis, L. D., Ljunggren, B., Ratcheson, R. A., and Siesjö, B. K. Cerebral energy state in insulin-induced hypoglycemia, related to blood glucose and to EEG. *J. Neurochem.* 23:673, 1974.

69. Tews, J. K., Carter, S. H., and Stone, W. E. Chemical changes in the brain during insulin hypoglycemia and recovery. *J. Neurochem.* 12:679, 1965.

70. Gorell, J. M., Law, M. M., Lowry, O. H., and Ferrendelli, J. A. Cerebral cor-tical metabolism during hypoglycemia and clinical recovery. *Neurology* (Minneap.) 25:360, 1975.

71. Crossland, J., Elliott, K. A. C., and Pappius, J. M. Acetylcholine content of brain during insulin hypoglycemia. *Am. J. Physiol.* 183:32, 1955.

*72. Fischer, J. E. False Neurotransmitters and Hepatic Coma. In Plum, F. (Ed.), *Brain Dysfunction in Metabolic Disorders. Res. Publ. Assoc. Nerv. Ment. Dis.* 53:53, 1974.

73. Hindfelt, B., and Siesjö, B. K. Cerebral effects of acute ammonia intoxication. The effect upon energy metabolism. *Scand. J. Clin. Lab. Invest.* 28:365, 1971.

74. Schenker, S., McCandless, D. W., Brophy, E., and Lewis, M. Studies on the intracerebral toxicity of ammonia. *J. Clin. Invest.* 46:838, 1967.

75. Duffy, T. E., Hindfelt, B., and Plum, F. Effect of acute ammonia intoxication on cerebral metabolism in rats with portacaval shunts. *Clin. Res.* 23:393A, 1975.

*76. Duffy, T. E., Vergara, F., and Plum, F. α-Ketoglutaramate in Hepatic Encepha-lopathy. In Plum, F. (Ed.), *Brain Dysfunction in Metabolic Disorders. Res. Publ. Assoc. Nerv. Ment. Dis.* 53:39, 1974.

77. Cooper, A. J. L., and Meister, A. Isolation and properties of highly purified glutamine transaminase. *Biochemistry* 11:661, 1972.

78. Norenberg, M. D. Histochemical studies in experimental portal-systemic encephalopathy. Glutamic dehydrogenase. *Arch. Neurol.,* in press, 1976.

*79. Plum, F., and Hindfelt, B. The Neurological Complications of Liver Disease. In Vinken, P. J., and Bruyn, G. W. (Eds.), *Handbook of Clinical Neurology: Metabolic and Deficiency Diseases of the Nervous System.* Vol. 31. Amsterdam: North-Holland, 1976.

80. Bessman, S. P., and Bessman, A. N. Cerebral and peripheral uptake of ammonia in liver disease with an hypothesis for the mechanism of hepatic coma. *J. Clin. Invest.* 34:622, 1955.

81. Meyer, H., and Lux, H. D. Action of ammonium on a chloride pump: Removal of hyperpolarizing inhibition in an isolated neuron. *Pflüegers Arch.* 350:185, 1974.

82. Skou, J. C. Further investigations on a $Mg^{++}+Na^{+}$-activated adenosintriphosphatase, possibly related to the active, linked transport of Na^{+} and K^{+} across the nerve membrane. *Biochim. Biophys. Acta* 42:6, 1960.

83. McKhann, G. M., and Tower, D. B. Ammonia toxicity and cerebral metabolism. *Am. J. Physiol.* 200:420, 1961.

Chapter 32

Parkinson's Disease and Other Disorders of the Basal Ganglia

Theodore L. Sourkes

Some large, anatomically distinct masses of gray matter lie at the base of the brain and certain of these, namely the caudate nucleus, the putamen, and the globus pallidus, are collectively termed *basal ganglia*. The first two of these constitute the *striatum*, or *neostriatum*; the internal and external parts of the globus pallidus, or pallidum, are known as the *paleostriatum*. This striopallidal system is an integrative unit whose constituent parts have many connections between one another and to and from other regions of the brain. In its entirety it is equivalent to the extrapyramidal system, i.e., the structures of the brain, apart from the cerebral cortex, which send efferent fibers to the spinal cord. Among the bodily functions regulated by the extrapyramidal system are the tone and posture of the limbs.

Regulatory functions of this type have been recognized since 1911, when S. A. K. Wilson demonstrated in a series of patients the relationship between lesions of the lenticular nucleus (putamen and globus pallidus, considered as a unit), on the one hand, and rigidity, tremor, and postural defects, on the other. Wilson's disease was thus regarded as the prototype of disorders of the basal ganglia.

The most prevalent disorder of the extrapyramidal system is the "shaking palsy," first described in 1817 by the London physician James Parkinson. This disorder is now recognized to stem from degenerative changes in the substantia nigra, a portion of the brain that has extensive connections with the striatum. Besides the shaking, or tremor, of the limbs, patients with Parkinson's disease may also exhibit muscular rigidity, which leads to difficulties in walking, writing, speaking, and facial movements, as well as lethargy, or loss of volitional movement, which may progress to akinesia.

Huntington's chorea is a third important disorder of basal ganglia, with apparently specific pathology of the caudate nucleus.

BIOCHEMICAL CHARACTERISTICS OF BASAL GANGLIA AND ASSOCIATED STRUCTURES

The basal ganglia present some interesting chemical characteristics. For example, they are especially sensitive to carbon monoxide and to manganese. This sensitivity probably plays a role in the neurological complications that accompany intoxication with these chemicals. In infants, the predilection of bilirubin for the cells of the basal ganglia may result in kernicterus, or "jaundice of the nuclei of the brain." Historically, kernicterus was the first basal ganglial disease described, and Wilson made use of the parallel in his classic study.

The globus pallidus contains much iron, a property it shares with other nuclei of the extrapyramidal system: the subthalamic nucleus of Luys, the substantia nigra, the red nucleus (n. ruber), the dentate nucleus, and the inferior olive. In fact, these parts can be delineated by immersing slices of the brain in a solution of ammonium sulfide to stain for iron. It has been suggested that the metal is present in a ferritinlike molecular form.

The basal ganglia contain considerable amounts of acetylcholinesterase; this is particularly evident in the caudate nucleus. They are said also to be rich in gangliosides. The substantia nigra, the red nucleus, and the locus ceruleus each contains a distinctive pigment. The precursor of the pigmented material present in the substantia nigra and locus ceruleus is thought to be DOPA (or a catecholamine), but the specific pathways of formation of the pigments are not completely known.

DOPAMINE AND BASAL GANGLIA

The biochemical characteristics that have thrown most light on the functions of the basal ganglia and certain associated structures in man and animals concern the distribution of monoamines, particularly dopamine. The biosynthetic relationships of this compound are illustrated in Figure 32-1 (see also Chap. 10). The starting point for its biosynthesis is tyrosine, which is present in the whole brain at a level of about 1.2 mg per 100 g fresh weight of tissue, or 70 μmol per kg. The first step is catalyzed by tyrosine hydroxylase; it requires a pteridine cofactor. The product of this oxidation is DOPA (3,4-dihydroxyphenylalanine), an amino acid that is normally present only in the most minute concentrations in the body because it is so readily decarboxylated to dopamine (3,4-dihydroxyphenylethylamine). This is achieved through the action of the widely distributed enzyme DOPA decarboxylase, which also catalyzes amine formation from 5-HTP (5-hydroxytryptophan) and some other substrates. Hence, it is sometimes called either 5-HTP decarboxylase or aromatic amino acid decarboxylase. Like other amino acid decarboxylases, it requires pyridoxal phosphate as its coenzyme. It is inhibited by many compounds, some of which are effective in vivo and have even been used therapeutically in certain diseases. A few decarboxylase inhibitors are illustrated in Figure 32-2. The first of these, methyldopa, is used as an antihypertensive drug in man, but it is the only decarboxylase inhibitor with such properties. Its action in essential hypertension is mediated by

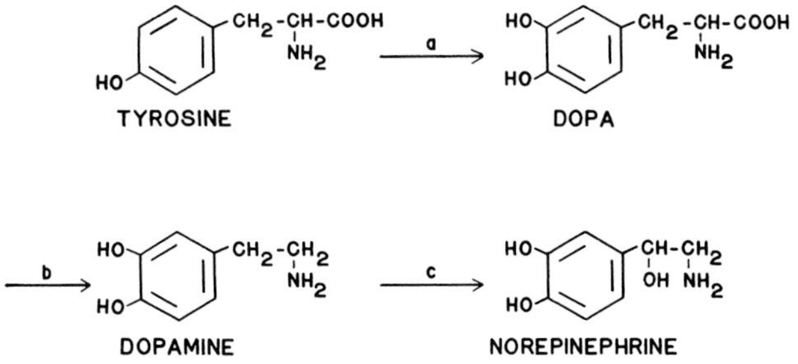

Figure 32-1
Pathway of biosynthesis of catecholamines. The enzymes catalyzing each step are: (a) tyrosine hydroxylase; (b) DOPA decarboxylase; and (c) dopamine β-hydroxylase.

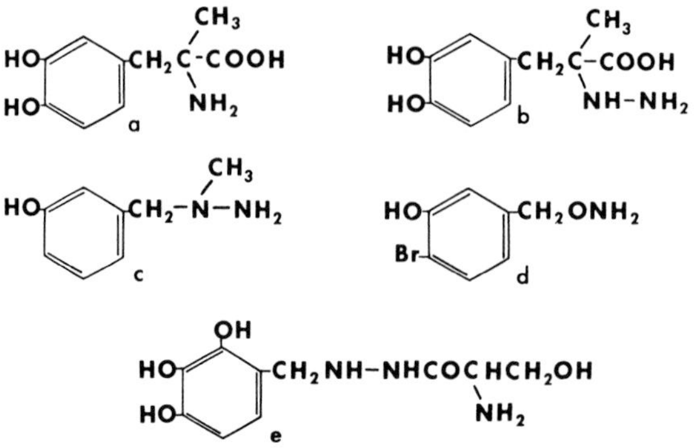

Figure 32-2
Some frequently used inhibitors of dopa decarboxylase: (*a*) methyldopa, L-alpha-methyldopa; (*b*) carbidopa, the hydrazino analog of methyldopa (MK-486); (*c*) *N*-*m*-hydroxybenzyl-*N*-methyl-hydrazine (NSD-1034); (*d*) brocresine (NSD-1055); (*e*) *N*-(DL-seryl)-*N'*-(2,3,4-trihydroxybenzyl)-hydrazine (Ro4-4602, benseracide).

mechanisms other than interference with catecholamine synthesis at the decarboxylase stage. Methyldopa has been widely used in biochemical and pharmacological studies because of important effects on monoamine metabolism. The other compounds in Figure 32-2 have also been employed extensively in research when blockade of decarboxylation is required. Carbidopa and Ro4-4602 have undergone clinical trials in Parkinson's disease.

In certain neurons of the brain, the formation of dopamine is the terminal step in biosynthesis. One method by which this has been determined makes use of the specific fluorescence of derivatives of dopamine and other monoamines developed in situ by histochemical means. By pharmacological pretreatment of animals in a manner that specifically depletes the brain of one amine or another, areas rich in a particular monoamine can be differentiated by ultraviolet histofluorescence. The fiber connections passing from the substantia nigra to the caudate nucleus belong to this group of neurons. They store dopamine in vesicles within the varicosities of the nerve terminations; the intravesicular concentration of the amine may rise to as high as 1 to 10 mg per gram fresh weight of nervous tissue, i.e., in the range of amine concentration in chromaffin tissue. Other neurons of the brain contain not only tyrosine hydroxylase and DOPA decarboxylase but also dopamine β-hydroxylase, which is found in the membrane of the storage vesicles. These are the norepinephrine-containing fibers. Under many circumstances, the biosynthesis of these monoamines is limited by the slow flux through the stage of tyrosine hydroxylase because of the low turnover rate of that enzyme.

The regional distribution of these compounds in the central nervous system is shown in Table 32-1. Dopamine and norepinephrine are irregularly distributed in the brain and spinal cord. Thus dopamine is found in certain regions along with norepinephrine, but, unlike norepinephrine, it is present in unusually high concentrations in some of the basal ganglia. On the other hand, norepinephrine, but not dopamine, is found in the spinal cord.

Table 32-1. Concentration of Catecholamines in Human Brain

Brain Region	Dopamine	Norepinephrine
Cerebral gray matter	0.02–0.17[a]	0–0.06[a]
Cerebral white matter	0.05	0
Striatum	3–8	0.06
Internal capsule	0.38	0.04
Red nucleus	1.17	0.23
Thalamus	0.30–0.46	0.05
Hypothalamus	1.12	1.11
Dentate nucleus	0.02	0–0.02
Medulla oblongata	0.17	0.14

[a]Micrograms per gram fresh weight of tissue.

Neurochemical mapping of the brain, in conjunction with histofluorescence studies, has led directly to the concept of dopamine as a neurohumor acting at certain synapses within the neostriatum and thus assisting in their neural function. This concept has drawn effectively upon our present knowledge of the function of norepinephrine as a transmitter substance at peripheral synapses served by post-ganglionic sympathetic fibers, and there is mounting evidence from many different branches of the neurosciences that dopamine plays such a role. Even though conclusive proof is still required, many investigators in the field have tacitly accepted dopamine into the family of neurohumors and speak of "dopaminergic" neurons [1–3].

The major metabolites of dopamine in the body are 3-O-methyldopamine (3-methoxytyramine) and HVA (homovanillic acid, 4-hydroxy-3-methoxyphenylacetic acid). These are depicted in Figure 32-3, which also shows the various alternative pathways of metabolism (in addition to the formation of norepinephrine). The amine and its O-methyl derivative are both subject to the action of monoamine oxidase (MAO), a flavoprotein that is present in the outer membrane of the mitochondria. Products of this oxidation include the aldehyde that corresponds to the amine substrate, hydrogen peroxide, and ammonia. Monoamines represent only a minor source of brain ammonia. Most of the aldehyde undergoes further dehydrogenation to form DOPAC (3,4-dihydroxyphenylacetic acid) or HVA. DOPAC, like dopamine, is a substrate for catechol-O-methyl transferase (COMT), an enzyme which catalyzes the transfer of the methyl group from S-adenosylmethionine to an appropriate acceptor. This process appears to be very efficient in the brain since substantial amounts of HVA are normally present in the striatum, whereas only very small quantities of DOPAC can be found there. A portion of the aldehyde undergoes reduction, catalyzed by alcohol dehydrogenase or a similar enzyme and a reduced nicotinamide coenzyme. The alcoholic products that are formed are shown in Figure 32-3; they are far less abundant as products of dopamine than are the acidic metabolites.

Biological activities have been sought for some of these metabolites of dopamine, so far with little success. Currently there is a growing interest in the possibility that the aldehyde derivatives of monoamines subserve a significant function.

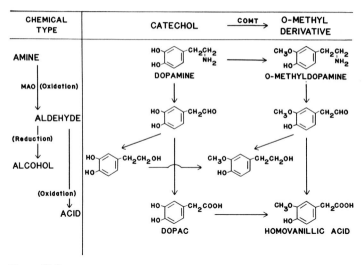

Figure 32-3
Alternative pathways of metabolism of dopamine. COMT, catechol-O-methyl transferase; MAO, monoamine oxidase; DOPAC, 3,4-dihydroxyphenylacetic acid.

Until the striking discrepancies in the regional distribution of the catecholamines were discovered (Table 32-1), dopamine was regarded as merely the precursor of norepinephrine. But the presence of dopamine in large amounts in localized areas of the brain, unaccompanied by comparable concentrations of norepinephrine, suggested a specific neural role for dopamine. As the dopamine-rich regions are preeminently the basal ganglia and some associated structures, apart from the thalamus and hypothalamus, it was natural to investigate diseases of the basal ganglia for evidence of deranged metabolism of catecholamines, in general, and dopamine, in particular. The search for such a faulty metabolism has been especially successful in Parkinson's disease, which now stands as the paradigm of cerebral diseases that involve dysfunction of monoamine-containing fibers. In that respect it may serve as a model for the investigation of monoamine function in nervous and mental diseases.

CHEMICAL PATHOLOGY OF PARKINSON'S DISEASE

The evidence for disturbed metabolism of monoamines comes from the direct measurement of these substances in the brains of patients dying of Parkinson's disease [1, 4–7]. The regional concentrations of catecholamines are shown in Table 32-2. The three monoamines that have been measured are all present in lower concentrations than those observed in normal conditions, and the decrease of dopamine is greater than that of norepinephrine and serotonin. One of the first attempts to restore normal levels of brain monoamines was to administer MAO inhibitors to patients with Parkinson's disease. MAO inhibitors prevent the oxidative deamination of the amine (see Fig. 32-3); hence, if its biosynthesis is proceeding normally, an excess of the compound could be expected to accumulate. Patients who were treated in the terminal stages of their illness with these inhibitors (some of which

Table 32-2. Catecholamines and Homovanillic Acid (HVA) in Brain in Parkinson's Disease

Brain Region	Dopamine		HVA		Norepinephrine	
	Normal[a]	P.D.	Normal	P.D.	Normal	P.D.
Caudate nucleus	3.50[b]	0.32	1.87	0.34	0.07	0.03
Putamen	3.57	0.23	2.92	0.69	0.11	0.03
Globus pallidus	0.30	0.14	1.73	0.56	0.09	0.11
Substantia nigra	0.46	0.07	1.79	0.52	0.04	0.02
Thalamus	–	–	0.35	0.21	–	–
Internal capsule (anterior limb)	–	–	1.67	0.42	–	–

[a]Normal = nonneurological cases; P.D. = Parkinson's disease.
[b]Micrograms per gram wet weight of tissue.

are used therapeutically in the treatment of mental depression) showed a significant increase in the brain concentrations of norepinephrine and serotonin but not of striatal dopamine. These results point clearly to some type of dysmetabolism of monoamines, particularly dopamine, in an important disease of the brain.

Neuropathological and necropsy findings by themselves, however, cannot establish the precise biochemical defect that causes the loss of brain dopamine. Fortunately, one can prepare models in experimental animals for comparison with the clinical state in man. This has been especially successful with monkeys, in which a small lesion is placed stereotaxically in the ventromedial tegmental area of the upper brain stem. In this way, it has been possible to induce in the animals various dyskinesias, including, significantly, tremor and hypokinesia of the limbs contralateral to the side of the lesion [4, 5, 8–11]. Histological study of the brains of these monkeys indicates that fibers from the substantia nigra to the striatum are interrupted and undergo degenerative changes. Retrograde degeneration entails loss of staining characteristics of the nigra (i.e., loss of cell bodies) accompanied by significant decreases in the concentration of catecholamines in the striatum, without involving its serotonin content. The loss of striatal monoamines is attributed, then, to anterograde degeneration of the nigrostriatal fibers, leading to the disappearance not only of the neuronal structure but also of enzymes and metabolites.

Cats, rabbits, and rats also have been used effectively in the identification of neurochemical events that accompany specific morphological changes. The primates, however, show the motor disturbances which permit neurological and neuropharmacological investigations that have a direct significance for the understanding of human diseases. There are many types of drug-induced dyskinesia and their chemical basis is being clarified [12].

Examples of the loss of enzymes and metabolites from the striatum, i.e., from the nigrostriatal terminations there, are given in [13] and Tables 32-3 and 32-4. Table 32-4 illustrates the progressive neurochemical changes accompanying the nigral degeneration as a result of experimental lesions: The greater the loss of nigral cell bodies and fibers, the greater the loss of DOPA decarboxylase and dopamine.

Table 32-3. Monoamine Functions of Striatum of Animals with Unilateral Lesions of the Nigrostriatal Tract

Function	Species	No.	Intact Side	Operated Side
Concentration of:				
Dopamine[a]	Monkey	5	5.30	0.29
Homovanillic acid[a]	Monkey	4	12.50	6.10
Norepinephrine[a]	Monkey	5	0.30	0.15
Dopamine following injection of DL-DOPA (70 mg/kg)[a]	Cat	10	17.4	6.5
Homovanillic acid in cisternal fluid[b]	Cat	4	16.7[c]	6.8[d]
Dopamine following intraventricular injection of radioactive tyrosine[e]	Monkey	3	6500	750
Activity[f] of:				
Succinic dehydrogenase	Cat	6	100	87–110
Tyrosine hydroxylase	Cat	6	100	1–34
DOPA decarboxylase	Monkey	4	100	44
Monoamine oxidase	Monkey	4	100	89

[a]Micrograms per gram fresh weight.
[b]Nanograms per milliliter.
[c]Intact animals.
[d]Bilaterally operated cats.
[e]Counts per minute (as dopamine) per gram of caudate nucleus tissue.
[f]Relative activity.

Table 32-4. Relation of Neurochemical Changes in the Striatum to Histological Changes in the Substantia Nigra After Lesioning of the Nigrostriatal Tract

Cellularity of Substantia Nigra on the Side of the Lesion	DOPA Decarboxylase Activity[a]			Dopamine Concentration[b]		
	No.	Intact Side	Lesion Side	No.	Intact Side	Lesion Side
No cell loss (or nearly none)	6	75	63	16	4.79	4.12
Partial loss of cells	4	113	31	5	2.47	1.07
Complete or nearly complete loss of cells	6	99	0	13	3.35	0.27

[a]Nanomoles of dopamine formed per hour per 100 mg tissue.
[b]Micrograms per gram fresh weight of tissue.

The decreased ability of the diseased nigra to produce dopamine in Parkinson's disease is also reflected in lower concentrations of HVA in the caudate nucleus and putamen (Table 32-2). This can be reproduced experimentally by introducing lesions of the nigrostriatal fibers in monkeys. The concentrations of the acidic metabolites of dopamine — HVA and DOPAC — then decline. Some HVA diffuses into the cerebrospinal fluid, particularly from the caudate nucleus, and samples of this fluid have easily detectable amounts of the acid. In patients undergoing brain surgery for the treatment of Parkinson's disease, the concentration of HVA is reduced far below the level found in patients who are being operated upon for reasons other than a disorder of the basal ganglia. This is clear from the data in Table 32-5. Because of the difficulty of obtaining "normal" cerebrospinal fluid (CSF), the concentration of HVA in ventricular CSF obtained from Parkinson's disease patients must be compared with

Table 32-5. Concentrations of HVA (Homovanillic Acid) and 5-HIAA (5-Hydroxyindoleacetic Acid) in CSF (ng/ml)

Diagnostic Category	HVA		5-HIAA		References
	No. of cases	Mean ± SE	No. of cases	Mean ± SE	
Lateral ventricles:					
Nonextrapyramidal disorders					
Various	18	466 ± 38	20	105 ± 11	[14]
Various	15	447 ± 40	17	111 ± 12	[15]
Temporal lobe epilepsy	6	108 ± 37	3	39 ± 15	[16]
Obsessive-compulsive neurosis	3	300 ± 16	2	37, 78	[16]
Pain syndrome	3	391 ± 41			[16]
Brain atrophy	3	62 ± 33			[16]
Extrapyramidal disorders					
Parkinson's disease	37	186 ± 17	32	59 ± 10	[14]
Parkinson's disease	53	157 ± 17	17	46 ± 7	[16]
Akinesia present	6	52 ± 24			[16]
Akinesia absent	41	169 ± 19			[16]
Attitudinal tremor	21	230 ± 19	5	77 ± 9	[16]
Attitudinal dystonia	6	225 ± 46			[16]
Double athetosis	4	391 ± 49			[16]
Lumbar space:					
Various neurological disorders	25	53 ± 6	18	31 ± 3	[16]
Various neurological disorders	10	43 ± 4	16	32 ± 2	[17]
Various neurological disorders	11	40 ± 4	11	26 ± 2	[18]
Parkinson's disease	5	15 ± 7	4	24 ± 9	[16]
Progressive supranuclear palsy	6	26 ± 9	6	20 ± 7	[16]
Temporal lobe epilepsy	12	43 ± 5	12	25 ± 2	[17]
Brain stem encephalopathy	4	32 ± 8	4	35 ± 2	[17]
Kufs' disease	2	N.D.[a]	2	32, 35	[17]
Huntington's chorea	15	23 ± 4	15	28 ± 3	[18]

[a]N.D., not detectable.

that from other groups of patients without evidence of extrapyramidal disease or degenerative changes, either focal or general. Within the category of Parkinson's syndrome, the values are much lower in patients who presented evidence of akinesia before death than in those without akinesia. This is also true for the dopamine in the caudate nucleus. The relationship is significant for akinesia, but the HVA concentration of the CSF does not seem to be related to the presence of tremor or rigidity in Parkinson's disease.

It has already been mentioned that serotonin concentrations are abnormally low in the caudate nucleus of patients who have died with Parkinson's disease. The serotonin metabolite 5-HIAA is also present in ventricular CSF in subnormal concentrations. Moreover, glutamate decarboxylase activity and the concentrations of gamma-amino-butyric acid (GABA) in the substantia nigra are lower in Parkinson's disease than in control cases.

CSF taken at the suboccipital and lumbar levels has a much lower concentration of HVA than has the ventricular fluid. Thus the ratio of mean concentrations of this substance in the CSF at ventricular, cisternal, and lumbar levels is approximately 9:4:1 [13–17]. The steep gradient is based upon two important facts: First, all the HVA originates in the brain, much of it in that portion of the caudate nucleus immediately adjacent to the lateral ventricles, i.e., the parts lining the floor of the body and anterior horn, and the roof of the inferior horn [19–21]; the remainder probably comes from dopamine-rich regions contributing HVA to the third and fourth ventricles. Although the cerebral cortex also contains some dopamine, HVA formed there and passing into the fluid bathing that region of the brain would probably not reach lower compartments of the CSF. Second, once HVA enters the CSF, an active-transport mechanism works to remove such metabolic acids as HVA, as well as 5-HIAA and 3-methoxy-4-hydroxyphenylglycol sulfate (from norepinephrine).

The compartmental ratio for 5-HIAA corresponding to that given above for HVA is 4:3:1. The lower gradient in this case arises from the fact that 5-HIAA enters the CSF at many levels, including the spinal; serotonin-containing fibers of the spinal cord make a substantial contribution to the 5-HIAA detected in the lumbar CSF, although some of this compound appears to originate from higher compartments of the CSF [21].

The concentration gradient of HVA in the CSF applies in Parkinson's disease as well (Table 32-5). The generally lowered HVA concentration, however, is not specific to Parkinson's disease. It is observed also in patients with inflammatory and other disorders of the central nervous system. Differentiation may be made by using probenecid (p-dipropylsulfamylbenzoic acid), a drug that inhibits the transport mechanism operating to remove HVA and similar acids from CSF to blood. While probenecid is present, the concentrations of HVA and 5-HIAA normally increase in the spinal fluid because their efflux is blocked. Parkinsonians tested with this drug, however, show little or no increase in HVA. Presumably nonparkinsonian patients continue to form dopamine, and thus HVA, so that, when transport of HVA from the CSF is interfered with, the acid can accumulate to some extent. This does not readily occur in Parkinson's disease because of the neuronal deficit and the consequent inability to produce dopamine, the precursor of HVA. There are several types of "probenecid tests," but, whichever is used, it is important to remember that under certain conditions the accumulation of the acidic metabolites may simply be determined by the level of probenecid attained in the CSF, regardless of other factors.

If an enzymic defect in the brain were accompanied by a comparable one in the liver, the kidneys, or both, the abnormality might be more readily detectable and so could be studied conveniently during the life of the patient. This has motivated clinical-chemical studies of Parkinson's disease, including measurement of the urinary excretion of monoamines and some of their metabolites. Such studies have shown that the excretion of epinephrine and norepinephrine is normal, but the excretion of dopamine after a loading dose is subnormal and may be especially low in those patients with the greatest degree of akinesia [7, 22]. The low values cannot be attributed, in our present state of knowledge, to variations in urinary volume, dietary factors, or a specific enzyme defect, nor can they be accounted for by a reduced turnover rate of cerebral dopamine.

Investigations with labeled dopamine indicate that these patients may convert relatively less of the amine to the norepinephrine fraction that is excreted unconjugated in the urine, with proportionately more going to other metabolites of dopamine. On the other hand, when labeled norepinephrine is infused into patients with Parkinson's disease, their pattern of excretion of urinary metabolites is similar to that of control subjects [7, 22]. It has been suggested that the abnormality with respect to free urinary dopamine results from chronic treatment with antiparkinsonian drugs; one way to study this would be to examine newly detected cases of Parkinson's disease.

L-DOPA TREATMENT OF PARKINSON'S DISEASE

An important outgrowth of catecholamine studies has been the L-DOPA treatment of Parkinson's disease [4, 5, 23, 24]. In 1961, neuropharmacological studies in patients with the disease revealed that this amino acid exerts antirigidity and antiakinesic actions; that is, it influences the two most incapacitating symptoms in Parkinson's disease. Today it is said that L-DOPA provides about 60 percent improvement in at least 50 percent of the patients with Parkinson's disease; the treatment is considered the best one available. Other antiparkinsonian drugs have had a favorable action on the tremor, but not on the other symptoms.

It is generally assumed that the therapeutic role of L-DOPA lies in the ease with with it crosses the blood-brain barrier (in contrast to the exclusion from the brain of intravenously infused dopamine) and its ability to replenish the supply of dopamine at appropriate sites. The passage of L-DOPA into the parenchymal tissue of the brain entails its transfer from blood through the endothelial cells lining the capillaries. These cells, like many other peripheral cells, contain considerable amounts of DOPA decarboxylase, so that only a portion of a given amount of L-DOPA will eventually pass into the brain. Hence, large doses (up to 8 g per day) have been used. By giving the patient inhibitors of DOPA decarboxylase, such as carbidopa and Ro4-4602, in doses that affect only the peripherally located enzyme, including that of the brain capillaries, it is possible to reduce the daily dose of L-DOPA to 2 to 3 g per day, with the same therapeutic response [25].

When L-DOPA has crossed the blood-brain barrier, it must be converted to dopamine, according to the commonly accepted hypothesis of its mode of action. This provides an apparent paradox for, if pathological processes have eliminated dopaminergic neurons, including their content of DOPA decarboxylase, how can the L-DOPA be effective? An important consideration is that studies with postmortem material have not yet revealed any cases with a total deficiency of DOPA decarboxylase in the striatum: There has always been at least a small residue of enzymic activity [26]. Thus, dopamine could be formed in the neighborhood of striatal receptors. Other possibilities have been suggested. One of these proposes that L-DOPA itself or a nondecarboxylated product acts directly on dopamine-sensitive receptors in the striatum. Other hypotheses envisage formation of a Schiff's base between L-DOPA and its transaminated product 3,4-dihydroxyphenylpyruvic acid or, analogously, between L-DOPA or dopamine and pyridoxal phosphate. In the latter cases, cyclization occurs with the formation of substituted tetrahydroisoquinolines which, it is postulated, may be the biologically effective substances. Another hypothesis is derived from the fact that cells of the striatum receive connections from many fibers, including some that contain serotonin. Hence, L-DOPA could be acted upon by the decarboxylase within those neurons, i.e., by the enzyme that normally has 5-HTP as substrate. The dopamine formed would then be available in the presynaptic space, although its path of diffusion to sensitive neurons might then be longer than usual.

There are many other speculative hypotheses concerning the action of L-DOPA in Parkinson's disease, but the choice among them must await further research. Whatever the outcome, the use of L-DOPA in this disorder represents a successful biochemical treatment of an important cerebral disease.

The actions of levodopa (the generic pharmaceutical name for L-DOPA) in Parkinson's disease or of dopamine at receptor sites in the CNS are mimicked by certain compounds, some of them with useful therapeutic actions. Amantadine is used in the treatment of Parkinson's disease as an adjunctive therapy; its action favors the release of dopamine from residual intact neurons in much the same way as amphetamine acts there. Apomorphine has a brief levodopalike activity. Its best-recognized action is at the dopamine receptor sites making up the trigger zone of the emetic center; this lies in the area postrema (region of the IV ventricle), at least in the cat. Apomorphine also is effective in certain neuroendocrine systems; for example, in man subemetic doses provoke a great increase in the concentration of growth hormone in the plasma, presumably by an action on cells producing the appropriate releasing factor. Other dopamine agonists include the ergot alkaloids, ergocornine and 2-bromo-alpha-ergocryptine (CB-154), and 6,7-dihydroxytetrahydrotetralin. Antagonists acting at dopamine receptors include the major tranquilizers (Chap. 34). Reserpine behaves differently in causing depletion of stored monoamines, including dopamine; this is achieved through blockade of the vesicular uptake mechanism for storage of monoamines.

Among the most commonly used drugs in the treatment of parkinsonism are anticholinergic agents. Their use has been rationalized in this way: In the striatum, dopamine has an inhibitory action at postsynaptic sites; if this influence is removed, as in Parkinson's disease (or by treatment with excessive amounts of neuroleptics), the cholinergic neurons in the striatum become overactive, a function which can then be regulated by anticholinergic drugs at those postsynaptic sites.

HUNTINGTON'S CHOREA

Huntington's chorea has been of special interest to the neurochemist because it is genetically transmitted and involves pathology of the striatal tissue, with loss of many neurons there [27]. Thus far, postmortem studies have revealed a small but significant decrease in the dopamine and HVA of the caudate nucleus [6]. This may result in the provision of an excess of dopamine from nigrostriatal terminations per residual striatal neuron. Other findings in necropsy material are a reduction of the glutamate decarboxylase activity in substantia nigra and pallidum and of the concentration of GABA in the same regions, as well as striatum [28–30]. If GABA normally exerts an inhibitory action over nigrostriatal function, then this deficiency in Huntington's chorea could contribute to an overproduction of dopamine at striatal synapses. Both these findings are consistent with the exacerbation of abnormal movements in Huntington's chorea caused by the administration of levodopa. Useful drugs in the treatment of this disease are those that deplete dopamine, among other monoamines, prevent its storage in nerve terminals, or block its synaptic actions. These agents are the neuroleptics, including reserpine, methyldopa, and others [31]. Thus Huntington's chorea may present certain mirror-image features of Parkinson's disease.

Extensive metabolic studies of Huntington's chorea have been carried out in a number of laboratories. Various blood lipids, serum enzymes, and urinary constituents, including trace metals and phenolic and indolic compounds, have been examined without any characteristic defect being detected. Patients with Huntington's chorea are said to have elevated levels of Mg^{2+} and Ca^{2+} in their erythrocytes, but normal levels of the two ions in serum. The plasma of fasting patients contains subnormal levels of proline, alanine, valine, isoleucine, leucine, and tyrosine, with similar, although less significant, changes in the CSF.

WILSON'S DISEASE, OR HEPATOLENTICULAR DEGENERATION

Much of our knowledge of the metabolism of copper comes from studies of experimental and farm animals [31, 32]. In recent years, the interest in Wilson's disease, a disorder in which copper accumulates in a number of organs, has helped to stimulate research into the metabolism of this metal in man. Wilson's disease is a combined brain-liver disorder characterized by progressive rigidity, intention tremor, hepatic cirrhosis of the coarse type, and recurrent hepatitis. Renamed hepatolenticular degeneration in 1921, it is recognized as a familial disorder inherited in an autosomal recessive fashion. The frequency rate of the gene is estimated to be 1 in 500. Biochemically, hepatolenticular degeneration is characterized by low concentrations of copper and ceruloplasmin in the serum, elevated excretion of the metal in the urine, and deposition of excess copper in the brain (especially in the basal ganglia) and in the liver and kidneys. In some cases, copper is deposited in the cornea, where it is reduced, forming the Kayser-Fleischer ring; this is then pathognomonic of the disease. In addition, there is a constant aminoaciduria, including excretion of some dipeptides, and sometimes abnormal excretion of monoamines and their metabolites. The amino acid levels in the plasma are normal, so the urinary findings may signify simply a renal defect caused by histotoxicity of copper [7, 12, 33, 34].

NORMAL METABOLISM OF COPPER

Copper deficiency has been demonstrated in a number of mammalian species under laboratory or range conditions, but not unequivocally in the human, even under starvation conditions, as in kwashiorkor or marasmus. It appears from many sources of evidence, however, that the metal is an actual nutritional requirement for man. About 2 mg of copper suffices daily, but most diets supply more than this. The normal infant is born with a large liver store of copper, which provides temporary protection against the deficiency of the metal in milk (just as with iron). Copper is needed for the utilization of iron in hemoglobin synthesis as well as for other functions. Conversely, iron seems to be required for the utilization of copper, so that deficiency of the one metal results in the accumulation of the other in liver or another storage organ. In some laboratory species, copper deficiency causes a microcytic hypochromic anemia. Farm animals maintained in regions where the soil (and therefore the pasturage) is deficient in copper may suffer from specific clinical disorders. In the sheep and goat, the primary symptom is anemia, but in others there is a neurological disturbance, such as "swayback" in cattle and "posterior paralysis" in swine. Myelination of nerve is delayed in lambs born of copper-deficient ewes.

The regulation of copper metabolism is attuned to the very small requirement of the metal in adult animals, so most of the copper ingested in the diet is excreted in the feces. A minute fraction of the dietary copper appears in the urine. Balance studies, however, disguise the fact that much copper is absorbed from the intestinal tract and reaches the liver in the portal circulation; a portion of it is utilized there, but the major portion is excreted in the bile. Thus, there is an enterohepatic cycle for copper. The mechanism can be overwhelmed in experimental animals by repeated parenteral administration of copper, e.g., by daily intraperitoneal injection of solutions of copper sulfate. The metal then accumulates in the liver, kidneys, and other organs, and excess copper is found in all subcellular fractions and in many proteins. The association with proteins is by no means random. Initially, the copper increases in the low-molecular-weight fraction of cytoplasmic proteins of rat liver. Only later, as the concentration of the metal rises greatly, does it appear in the largest proteins; those of intermediate size are the last to take up the copper. Under conditions of copper-loading in the rat, decreases of activity of certain hepatic enzymes can be detected. The toxic action of copper is selective, even among sulfhydryl enzymes; the present evidence runs contrary to the concept of a generalized toxic action on metabolic or respiratory processes [35].

Regulation of the entry of copper into the brain is very strict, even under the aggravated conditions of chronic copper-loading, so that the net increase of copper in that organ is significant, but small, and not of the proportions seen in hepato-lenticular degeneration in man [35].

CUPROPROTEINS
Copper is an essential constituent of several important proteins, including cytochrome-c oxidase, dopamine β-hydroxylase, uricase (in mammals below the level of the primates), ceruloplasmin, and superoxide dismutase, or cytocuprein (also named cerebrocuprein, hepatocuprein, and erythrocuprein, depending on the tissue from which it is prepared). Copper is the prosthetic group of DOPA oxidase in the melanocytes of the skin and in certain pigmented regions of the brain. Indeed, there may be as much as 0.4 mg of copper per gram of dry weight in the subthalamic nucleus, with somewhat smaller amounts in the substantia nigra and dentate nucleus. All these cuproproteins do not account for the total amount of copper in the body (100 to 150 mg in the human adult), however, and many other cuproproteins will probably be discovered.

Ceruloplasmin is a blue serum glycoprotein. Its molecular weight is about 155,000, and it contains eight atoms of copper, not all of which are equivalent in function. The divalent atoms of copper endow the protein with its blue color. The molecule displays some oxidase activity toward polyamines and polyphenols; phenylenediamine is often used as a substrate. Ceruloplasmin has a half-life in the serum of 54 hours. Its sialic acid residues seem to protect it from metabolic degradation because asialoceruloplasmin disappears from the circulation after its injection far more rapidly then does ceruloplasmin.

Normally, the serum contains about 100 μg of copper per 100 ml. Most of this (95 percent) is present as ceruloplasmin. The concentration of this protein amounts to 20 to 40 mg per 100 ml. In hepatolenticular degeneration, serum copper may fall to one-half this value or less. Ceruloplasmin does not seem to be involved in transport of the metal in the serum; this function is served by the small nonceruloplasmin fraction of serum, most of this copper being loosely bound to albumin. It has been demonstrated recently that, when ceruloplasmin is perfused through the isolated liver, there is a specific and rapid shift of transferrin from liver to the perfusate. This result is pertinent to the problem of the copper-iron interaction in metabolism.

A vexing question is that of the relationship between ceruloplasmin and hepatolenticular degeneration. The many investigations in this field have not yet provided a decisive answer, so we do not know what role, if any, the depressed level of ceruloplasmin plays in the pathogenesis of the disorder.

From time to time, a relationship between copper metabolism and schizophrenia has been claimed. For example, soon after the oxidase property of the protein became widely recognized, numerous people made studies attempting to relate the activity of serum ceruloplasmin to the psychosis, but there has been no valid evidence for such a relationship.

Some metal-chelating agents have been tested as therapeutic agents in Wilson's disease. BAL (British antilewisite) is effective in bringing about copper diuresis, but it must be given by intramuscular injection and has unpleasant side effects, which limit its use. EDTA has been tried, but it is not very effective in removing copper from the body. The most effective chelator that has been used is D-penicillamine, or 3,3-dimethylcysteine. It is given in doses up to about 3 grams per day. Its effectiveness indicates that the cytotoxicity caused by copper can be reversed, even to the extent of bringing about remission of neurological and other symptoms of hepatolenticular degeneration.

CHRONIC MANGANESE POISONING

A small portion of miners exposed to manganese dust develop "manganism." The disease is ushered in by self-limited psychiatric symptoms, followed by permanent neurological changes. The manifestations are those of extrapyramidal disease. Because they respond to treatment with L-DOPA, dopaminergic neurons of the brain are probably affected.

ACKNOWLEDGMENT

Research on the title subjects in the author's laboratory is supported by grants from the Medical Research Council of Canada.

REFERENCES

*1. Hornykiewicz, O. Dopamine (3-hydroxytyramine) and brain function. *Pharmacol. Rev.* 18:925, 1966.
*2. Sourkes, T. L. Actions of levodopa and dopamine in the central nervous system. *J.A.M.A.* 218:1909, 1971.
*3. Sourkes, T. L. Central actions of dopa and dopamine. *Rev. Can. Biol.* 31:153, 1972.
*4. De Ajuriaguerra, J., and Gauthier, G. (Eds.). *Monoamines, Noyaux Gris Centraux, et Syndrome de Parkinson.* Geneva: Georg, and Paris: Mason, 1971.
*5. Kopin, I. J. (Ed.). *Neurotransmitters.* Vol. 50. Research Publications of the Association for Research in Nervous and Mental Diseases. New York: A.R.N.M.D., 1972.
 6. Bernheimer, H., Birkmayer, W., Hornykiewicz, O., Jellinger, K., and Seitelberger, F. Brain dopamine and the syndromes of Parkinson and Huntington. *J. Neurol. Sci.* 20:415, 1973.
*7. Sourkes, T. L., Poirier, L. J., and Lal, S. Diseases of the Basal Ganglia. In Good, R. A., Day, S. B., and Yunis, J. J. (Eds.), *Molecular Pathology.* Springfield, Ill.: Thomas, 1975.
 8. Sourkes, T. L., and Poirier, L. J. Neurochemical bases of tremor and other disorders of movement. *Can. Med. Assoc. J.* 94:53, 1966.
 9. Poirier, L. J., Sourkes, T. L., Bouvier, G., Boucher, R., and Carabin, S. Striatal amines, experimental tremor and the effect of harmaline in the monkey. *Brain* 89:37, 1966.
10. Goldstein, M., Anagnoste, B., Battista, A. F., Owen, W. S., and Nakatani, S. Studies of amines in the striatum in monkeys with nigral lesions. *J. Neurochem.* 16:645, 1969.

*Asterisks denote key references.

*11. Gillingham, F. J., and Donaldson, I. M. L. (Eds.). *Third Symposium on Parkinson's Disease.* Edinburgh: Livingstone, 1969. (The basic scientific aspects cover a substantial portion of these proceedings.)

*12. Curzon, G. The biochemistry of dyskinesias. *Int. Rev. Neurobiol.* 10:323, 1967.

*13. Sourkes, T. L. Enzymology and Sites of Action of Monoamines in the Central Nervous System. In Yahr, M. D. (Ed.), *The Treatment of Parkinsonism — The Role of DOPA Decarboxylase Inhibitors.* Vol. 2. Advances in Neurology. New York: Raven, 1973, p. 13.

14. Guldberg, H. C., Turner, J. W., Hanieh, A., Ashcroft, G. W., Crawford, T. B. B., Perry, W. L. M., and Gillingham, F. J. On the occurrence of homovanillic acid and 5-hydroxyindol-3-acetic acid in the ventricular C.S.F. of patients suffering from Parkinsonism. *Confin. Neurol.* 29:73, 1967.

15. Moir, A. T. B., Ashcroft, G. W., Crawford, T. B. B., Eccleston, D., and Guldberg, H. C. Cerebral metabolites in cerebrospinal fluid as a biochemical approach to the brain. *Brain* 93:357, 1970.

16. Papeschi, R., Molina-Negro, P., Sourkes, T. L., and Erba, G. The concentration of homovanillic and 5-hydroxyindoleacetic acids in ventricular and lumbar CSF. *Neurology* (Minneap.) 22:1151, 1972.

17. Garelis, E., and Sourkes, T. L. Use of cerebrospinal fluid drawn at pneumoencephalography in the study of monoamine metabolism in man. *J. Neurol. Neurosurg. Psychiatry* 37:704, 1974.

18. Curzon, G., Gumpert, J., and Sharpe, D. Amine metabolites in the cerebrospinal fluid in Huntington's chorea. *J. Neurol. Neurosurg. Psychiatry* 35:514, 1972.

19. Sourkes, T. L. On the origin of homovanillic acid (HVA) in the cerebrospinal fluid. *J. Neural Transm.* 34:153, 1973.

20. Garelis, E., and Sourkes, T. L. Sites of origin in the central nervous system of monoamine metabolites measured in human cerebrospinal fluid. *J. Neurol. Neurosurg. Psychiatry* 36:625, 1973.

21. Garelis, E., Young, S. N., Lal, S., and Sourkes, T. L. Origin of monoamine metabolites in CSF. *Brain Res.* 79:1, 1974.

22. Sourkes, T. L. Metabolism of Monoamines in Extrapyramidal Disorders. In De Ajuriaguerra, J., and Gauthier, G. (Eds.), *Monoamines, Noyaux Gris Centraux, et Syndrome de Parkinson.* Geneva: Georg, and Paris: Masson, 1971, p. 129.

23. Calne, D. B., and Sandler, M. L-DOPA and Parkinsonism. *Nature* 226:21, 1970.

*24. Pinder, R. M. The pharmacotherapy of Parkinsonism. *Prog. Med. Chem.* 9:191, 1973.

*25. Yahr, M. D. (Ed.). *The Treatment of Parkinson's Disease — the Role of Dopa Decarboxylase Inhibitors.* Vol. 2. Advances in Neurology. New York: Raven, 1973.

26. Lloyd, K., and Hornykiewicz, O. Parkinson's disease: Activity of L-dopa decarboxylase in discrete brain regions. *Science* 170:1212, 1970.

*27. Barbeau, A., Chase, T. N., and Paulson, G. W. (Eds.). *Huntington's Chorea, 1872–1972.* Vol. 1. Advances in Neurology. New York: Raven, 1973.

28. Perry, T. L., Hansen, S., and Kloster, M. Huntington's chorea. Deficiency of γ-aminobutyric acid in brain. *N. Engl. J. Med.* 288:337, 1973.

29. McGeer, P. L., McGeer, E. G., and Fibiger, H. C. Choline acetylase and glutamic acid decarboxylase in Huntington's chorea. *Neurology* (Minneap.) 23:912, 1973.

30. Bird, E. D., Mackay, A. V. P., Rayner, C. N., and Iversen, L. L. Reduced glutamic-acid-decarboxylase activity of post-mortem brain in Huntington's chorea. *Lancet* 1:1090, 1973.

31. Sourkes, T. L., Pivnicki, D., Brown, W. T., Wiseman-Distler, M. H., Murphy, G. F., Sankoff, I., and Saint Cyr, S. A. A clinical and metabolic study of dopa (3,4-dihydroxyphenylalanine) and methyldopa in Huntington's chorea. *Psychiatr. Neurol.* (Basel) 149:7, 1965.

*32. Scheinberg, I. H., and Sternlieb, I. Copper metabolism. *Pharmacol. Rev.* 12:355, 1960.

33. Walshe, J. M. The physiology of copper in man and its relation to Wilson's disease. *Brain* 90:149, 1967.

34. Cumings, J. N. Trace metals in the brain and Wilson's disease. *J. Clin. Pathol.* 21:1, 1968.

35. Lal, S., Papeschi, R., Duncan, R. J. S., and Sourkes, T. L. Effect of copper loading on various tissue enzymes and brain monoamines in the rat. *Toxicol. Appl. Pharmacol.* 28:395, 1974.

Chapter 33
Neurotoxic Agents

Maynard M. Cohen
Francis C. G. Hoskin

Accidental exposure to naturally occurring neurotoxic agents has represented a hazard to man since prehistoric times. With development of industry and agriculture, materials damaging to the nervous system have been employed in sufficient quantity to affect the occupationally exposed, and also, potentially, to involve the general population through environmental pollution. Certain of these agents continue to be employed for homicidal or suicidal purposes. Most toxins affecting the nervous system do so by interfering with biochemical systems, some through relatively specific mechanisms. Many toxic responses are consequences of enzyme inhibition by competitive, noncompetitive, or allosteric mechanisms. A generalized toxic effect may be caused by such processes as protein denaturation or alterations in acid-base balance or ionic equilibrium. Under some circumstances, there may be interference with biochemical systems, but added factors may be necessary to produce toxicity.

In addition to direct neural action, toxins may contribute in other ways to morbidity and mortality. These include effects on cerebral blood vessels, physiological effects on cerebral centers, including depression of respiratory or circulatory centers or stimulation of neural activity, and effects on other organs and tissues.

The toxin may resemble the physiological substrate chemically, as in the case of cholinesterase inhibitors; as a consequence, it attaches to the active site and interferes with appropriate enzymic activity by preventing enzyme-substrate interaction. Substances that inhibit enzyme activity because of structural similarities to substrate are termed *antimetabolites* or *metabolic antagonists*. In competitive inhibition, the effect depends on the relative concentrations of enzyme and inhibitor and is reversible by the introduction of additional substrate. Under noncompetitive conditions, the reversal cannot be accomplished by increasing substrate concentrations. The inhibitor is bound more permanently to the enzyme, and the essential reaction proceeds when sufficient enzyme is reconstituted to carry out the process. Inhibition may result from occupation of a secondary, or "allosteric," site as well as the active catalytic site.

Nonenzyme interactions also may result in toxic effects. Many biochemical reactions essential to neural activity require nonprotein cofactors for their completion. Combination of toxin with the cofactor forms complexes that effectively prevent the cofactor from carrying out its metabolic role. Introduction of additional cofactor allows the reaction to proceed at the normal rate. The materials discussed in this chapter are those known to produce toxic neural manifestations in humans. Many other compounds are potentially toxic, but they are not considered here, either because their effect on man is, as yet, of no serious consequence or because their toxic action is not primarily on neural biochemical systems.

HEAVY METALS AND METALLOIDS

Heavy metals and metalloids are protoplasmic poisons that affect many neural enzymatic reactions. Specific effects result from inactivation of enzymes that contain sulfhydryl radicals. As a result, both glycolysis and oxidative metabolism may be altered. Not all metals and metalloids known to produce neural intoxication have been tested for toxic depression of oxidative metabolism. Many of these materials, however, have been demonstrated to interfere with the orderly progression of glycolysis into the oxidative phase. Despite common effects on sulfhydryl-containing enzymes, symptoms of intoxication are not uniform because various toxins exhibit differing affinities for the enzymes as well as differing capacities to affect other biochemical systems and other organs.

Lead

Lead is one of the toxins known earliest to man. Evidence of lead in cosmetics was found in ancient Egyptian graves, and it has been suggested that lead contamination of wines and syrup contributed to the decline of Rome. Modern industrial culture causes the urban dweller to ingest from 100 to 2,000 μg of lead daily, and inhale an additional average of 90 μg. Blood concentrations in urban dwellers usually reach 0.11 to 0.21 parts per million. Lead may be absorbed through the gastrointestinal tract, the skin, and, in vapor form, through the lungs. Large quantities are excreted in the feces, urine, and sweat.

Symptoms of intoxication appear quickly, particularly in industrial exposure, because the concentrations of lead in blood frequently border on the toxic. Individuals who use leaded paint and workers in gasoline industries, where tetraethyl lead is employed as an additive, are among the most frequently intoxicated. Pica, a craving often manifested by chewing paint that peels from cribs, walls, and other objects, is one of the leading causes of lead intoxication in children. In New York City, mass screening indicated that lead concentrations in blood reached potentially toxic levels in more than 5 percent of infants tested; in Chicago more than 1,000 of 56,000 children screened exhibited evidence of intoxication. Although lead poisoning in children has been recognized as a common problem in densely populated urban areas, evidence now points to the fact that the problem is even more widespread. In a recent study of Illinois cities of intermediate size (10,000 to 15,000 population), more than 18 percent of the children exhibited evidence of undue absorption of lead. In nearly 1 percent of the subjects examined, unequivocal evidence of lead poisoning was demonstrated by blood-lead levels exceeding 80 μg per 100 ml.

When lead enters the bloodstream, it is distributed throughout the body and is carried chiefly to storage depots in cancellous bones. Lesser concentrations are present in the kidney, liver, spleen, bone marrow, and blood. The lead stored in bone is apparently innocuous, but lead in other organs represents the toxic portion. Lead stored in bone may be mobilized by debilitating disease or acute infectious conditions and liberated into the bloodstream, producing toxicity. Lead poisoning is a hazard to the fetus and to the nursing infant since it is readily transmitted across the placenta and in particular reaches greater concentrations in the milk during late lactation [1].

In addition to the involvement of enzymes concerned with oxidative metabolism, lead also interferes with porphyrin metabolism. Significantly increased concentrations of δ-aminolevulinic acid (ALA) in the urine appear to be caused partly by the induction of enzyme ALA synthetase. This enzyme, together with pyridoxal phosphate, leads to the formation of ALA from succinyl-CoA and glycine (see Fig. 24-7). Lead inhibits ALA dehydrase, thus interfering with formation of porphobilinogen from ALA, and it is responsible for the major elevation of ALA concentration in urine. The conversion of coproporphyrinogen III to protoporphyrin-9 is mediated by the enzyme coproporphyrinogenase. This enzyme may be inhibited by lead, resulting in an increase in the urinary excretion of coproporphyrin III. The increase is relatively small in comparison with that of ALA, which is 25 to 50 times as great. Heme synthetase (ferrochelatase) also contains sulfhydryl groups. It has been suggested that inhibition of this enzyme results from lead intoxication, which diminishes the formation of heme from protoporphyrin-9.

The symptoms of lead poisoning and those of acute intermittent porphyria are often strikingly similar (see Chap. 24). Many of the laboratory findings are also similar because of deranged porphyrin metabolism resulting from lead intoxication [2]. The two conditions can be distinguished biochemically: In acute porphyria, there is a greater increase of uroporphyrin and porphobilinogen in the urine than of coproporphyrin or aminolevulinic acid. In lead poisoning, the opposite is true. Urinary coproporphyrin III is often increased up to 500 μg per 100 ml. The erythrocyte protoporphyrins, which normally amount to approximately 30 μg per 100 ml, may be increased up to 150 μg per 100 ml.

The sulfhydryl-containing enzymes are inhibited less readily by lead than by other metals such as mercury or cadmium. Lead concentrations of 10^{-3} to 10^{-4} are required to inhibit these enzymes in vitro. This appears to exceed concentrations in blood and tissue fluids of intoxicated individuals. Intracellular concentration, however, may exceed that in the blood.

Although the precise metabolic and clinical consequences of interactions of lead with other compounds have not been described in detail, some relevant data are available. Lead forms mercaptides with the sulfhydryl group of cysteine and stable complexes with side chains of such amino acids as serine and threonine. There is also complexing with a number of compounds that contain phosphate groups. In vitro, the addition of lead to adenosine monophosphate results in a precipitate of lead adenylate. When suckling rats were fed lead acetate, labeling of glutamate, aspartate, γ-aminobutyrate, and glutamine from glucose was lowered significantly. Labeling of amino acids from acetate was also diminished. The changes were suggested to be consistent with retarded development of glucose metabolism and metabolic compartmentation [3].

Because chelating agents, such as calcium disodium EDTA,* bind the toxic ions in an inactive form within their own structures, they are employed in the treatment of lead poisoning. Chelation results in the mobilization of lead, and, as a consequence, symptoms may be exacerbated during treatment. The chelated form becomes

*Calcium disodium EDTA is also effective as a diagnostic aid. When the administration of therapeutic doses markedly augments urinary lead excretion, the diagnosis of poisoning is confirmed [4].

dissociated, and potentially toxic lead is released within the body. A combination of EDTA and BAL (British anti-lewisite, 2,3-dimercaptopropanol) appears to decrease this toxic potential and is thus the treatment of choice. Large amounts of lead are in storage sites and in cellular locations not easily reached by the chelating agent, which is essentially unable to penetrate cellular membranes. As a result, a single therapeutic course may only incompletely mobilize the body burden of lead. Additional chelation may still result in mobilization for as long six months. This can be accomplished orally with penicillamine as the chelating agent.

Intoxication due to lead in organic form has been documented only as a result of exposure to tetraethyl lead, a compound that differs in its solubility, absorption characteristics, and clinical manifestations from inorganic lead. It is absorbed rapidly from the skin and is soluble in CNS structures. Tetraethyl lead is converted by liver microsomes to triethyl lead, which is presumably responsible for the toxic response.

Concentrations of triethyl lead are not as high in the brain as in other organs, but the CNS is unusually sensitive to its toxic effects. It is highly stable and remains in the brain for extended periods. Oxidative phosphorylation in vitro is inhibited by triethyl lead, resulting in lowered concentrations of cerebral phosphocreatine [5]. Oxygen consumption is depressed, and the output of carbon dioxide from glucose is decreased. Treatment with EDTA or BAL is generally ineffective because neither tetraethyl nor triethyl lead is chelated.

Mercury

Inorganic mercury poisoning occurs industrially during paper manufacture, in the preparation of chlorine, and after exposure to certain other chemical processes. In addition, acute poisoning may follow use of mercuric chloride as a local antiseptic and calomel in excess as a diuretic. The chief atmospheric source of mercury pollution is burning coal and other fossil fuels. Metallic mercury volatilizes readily at room temperature; thus it may contaminate the air, condense on skin and respiratory membranes, and be swallowed with saliva. Absorption from the skin or gastrointestinal and respiratory tracts is rapid. Elemental nonionized mercury is transported in the blood bound to plasma proteins and hemoglobin. Once incorporated, mercury is retained in the body for extensive periods. It has been found in the urine as long as six years after exposure has ceased. Inorganic mercury has a remarkable affinity for the kidney and, as a result, symptoms of inorganic intoxication relate prominently to that organ.

Under appropriate conditions, inorganic mercury is incorporated into the brain rapidly. Magos [6] noted that after mice had been exposed to mercury vapor, two-thirds of all uptake was in the nervous system. Damage to the blood-brain barrier by mercuric chloride results in increased permeability. Although concentrations in the brain are markedly less than those in the kidney, they are still approximately 10 times the plasma levels after experimental administration of a single dose of mercuric nitrate. Concentrations vary in different areas in the CNS. In rabbits examined 96 hours after a single dose of mercuric chloride, the highest neural concentrations of inorganic mercury were noted in the brain stem, followed in decreasing order by the cerebellum, cerebral cortex, and hippocampus. The variations within areas were within a relatively narrow range. There may be marked differences, however, in cellular mercury concentration. Mercury is eliminated very slowly from the CNS, particularly from certain cells in the brain stem which can contain concentrations up to 16 times the level of neighboring cells. In 17 human autopsies made at random, the concentration of mercury was greatest in the cerebellum (mean 0.28 μg per g). Concentrations of 0.17 to 0.2 μg per g were present in visual cortex, pons, and geniculate. Of the 17 brains that had no history of exceptional exposure, 2 exhibited the unusually elevated levels of 1 to 2 μg per g in many regions [7, 8].

Blood concentrations are unreliable indicators of inorganic mercury toxicity. They vary markedly among individuals with the same exposure and are even variable on different days in the same individual. Uninvolved subjects will usually have a serum concentration below 3 μg per liter. Toxic symptoms usually appear when concentrations exceed 500 μg per liter, and blood concentrations below 100 μg per liter are considered safe. Urinary excretion is also unreliable as a measure of toxicity. Symptoms have been noted in patients excreting 200 μg per liter of urine and are absent in others when excretion is as high as 1,000 μg per liter.

Inorganic mercury produces toxic effects by altering membranes, particularly through combination with $S - S$, $S - H$, and other groups. Like lead, mercury inhibits glucose oxidation through affinity for the $S - H$ enzymes, as well as for lipoic acid, coenzyme A, and panthotheine. Unlike lead, however, mercury forms complexes with the amino groups of proteins. The affinity for the $S - H$ group of proteins results in conformational changes with possible pathophysiological consequence, as in muscle phophorylase. There is no immediate inactivation of that enzyme when exposed to mercury, but a slow change occurs, indicating alteration in the protein composition. Salts of mercury also bind indoles and in vitro cleave some disulfide linkages. Mercury also binds and interferes with the activity of reduced nicotinamide adenine diphosphate (NADH). Although there is no correlation with δ-aminolevulinic acid concentrations as observed in lead poisoning, urinary coproporphyrin levels are elevated in patients with mercury intoxication.

A second form of mercurialism with different chemical and clinical manifestations results from exposure to alkyl mercury compounds, particularly the methyl and ethyl forms. Thousands of instances of intoxication have followed ingestion of contaminated seafood, as well as exposure to alkyl mercury employed in the antifungal treatment of feed grain. Under some circumstances, fish and shellfish concentrate the mercury to form a protein complex compatible with continued life and function within these organisms. When shellfish are ingested by humans, mercury may be released and produce intoxication, as has been noted in the Minamata Bay epidemics in Japan.

The alkyl mercury compounds share many of the biochemical effects of inorganic mercury since they also complex with sulfhydryl radicals. The blood-brain barrier is easily passed, and, once the brain is reached, turnover is slower than in other organs. After chronic exposure, approximately 10 percent of the total body burden of the alkyl mercury localizes in the brain. Because of a strong affinity for the sulfhydryl group of amino acids, alkyl mercury is quickly bound to protein or polypeptides and remains in the bound organic form. Less than 3 percent of all mercury is degraded to the inorganic form.

Excretion is principally into the gastrointestinal tract, mostly through biliary secretion, although there is some evidence of secretion from the intestinal mucosa. Intestinal mercury is then quickly reabsorbed into the bloodstream.

There is a significant lapse between exposure to toxic concentrations of alkyl mercurials and the appearance of symptoms. Experimentally, the brains of rats killed during the latent period and incubated in vitro exhibited no change in O_2 consumption, lactic acid, or activity of certain sulfhydryl enzymes. When the cerebral tissue was prepared from animals in which neurological symptoms had become evident, however, oxygen consumption was reduced more than 35 percent. With anaerobic incubation, lactic acid formation remained unchanged in either circumstance. There was a slight decrease of aerobic lactic acid production, however, in those animals with neurological symptoms. During the latent period, the three sulfhydryl enzymes that were studied exhibited no change. In those animals killed when neurological symptoms were present, however, succinic dehydrogenase activity was diminished. The only parameter studied in which biochemical change was observed during both the latent and symptomatic periods was inhibition of protein synthesis. Incorporation of $[U - {}^{14}C]$ leucine into protein was decreased to approximately 57 percent of control values during the latent period [9]. When ethyl mercury was added to cerebral tissue respiring in vitro at concentrations approximately 10 times that obtained in vivo, Cremer [10] noted inhibition of oxygen uptake with glucose, pyruvate, or glutamate as substrate.

The excretion of mercury in patients intoxicated with either organic or inorganic materials is augmented by the mercury-binding compounds. These include D-penicillamine, acetyl DL-penicillamine, thiol resins, and BAL. However, BAL is no longer utilized because it has been reported to increase cerebral mercury concentrations in animals receiving the methyl form [11]. The chelating agents remove the mercury irregularly from the body. When these agents have been administered, mercury concentrations in blood increase over a one- to three-day period, presumably as a result of rapid mobilization from the tissues. Subsequently, concentrations in blood decline. Thiol resins have been employed in therapy because they bind mercury within the gastrointestinal tract and are not reabsorbed. Excretion is then augmented by preventing the reabsorption of methyl mercury so that redistribution of the toxin in the body is prevented. Selenium has been observed to decrease the toxicity of mercuric chloride in rats. It is thought to complex in equimolar fashion with mercury in the blood and tissue [12], thus decreasing the availability of both elements.

Mercaptodextran was used experimentally to increase biliary excretion, but it was found to augment urinary concentration of mercury. Consequently, a thiol of smaller molecular weight, N-acetyl homocysteine, and its thiol lactone were used experimentally to remove Hg from mice. In this form, mobilization from the brain could also be accomplished [13].

Arsenic

Arsenic is ubiquitous. Its use in insecticide sprays contaminates fruits and vegetables, and certain industrial processes also contribute heavily to its dissemination. The toxic effect of arsenic results from its binding sulfhydryl radicals of enzymes, thus inactivating them. α-Keto acid decarboxylases, such as pyruvate decarboxylase, are inactivated, and the result is an inhibition of oxidative metabolism. The general reaction of an arsenoxide or arsenite with adjacent sulfhydryl groups may be written as in Formula 33-1, where R is a substituent group and Pr is the enzyme protein.

$$R - As = O \ + \ (SH)_2Pr \longrightarrow R - As \underset{S}{\overset{S}{<}} Pr + H_2O$$

Formula 33-1
General reaction of an arsenoxide or arsenite with adjacent sulfhydryl groups.

Although arsenic intoxication usually results from oral ingestion, symptoms of toxicity may result from other routes. Arsenic is absorbed rapidly from parenteral sites of administration, from skin, and from mucous membranes. It rapidly leaves the bloodstream to be stored in liver, kidneys, intestines, spleen, lymph nodes, and bone. Involvement of the blood-brain barrier by arsenic is suggested by elevation of CSF protein to concentrations between 45 and 90 mg per 100 ml of fluid in approximately 40 percent of patients. It is deposited in the hair within two weeks after administration and stays fixed in this site for years. It remains also within bone for extended periods. It is slowly eliminated in the urine and feces; excretion begins two to eight hours after arsenic enters the body. It may take 10 days before a single dose is completely excreted. After chronic arsenic administration, 70 days may be required before urinary levels return to normal. Renal excretion exceeding 0.1 mg of arsenic in 24 hours is usually abnormal. Toxic concentrations of arsenic may be present in the body even though urinary excretion levels are not elevated. When urinary concentrations are too low to be diagnostic, mobilization of tissue arsenic by a therapeutic regimen will increase concentrations to diagnostic levels. In longstanding intoxication, growing ends of the hair may be examined. Concentrations exceeding 0.1 mg per 100 g of hair indicate excessive ingestion or contamination [14]. Arsenic is radiopaque, so diagnosis may be made at times by radiographic visualization of the toxic material in the gastrointestinal tract.

Treatment for arsenic intoxication is based on providing compounds with sulfhydryl radicals that will unite with the arsenic and be excreted in urine subsequently. This can be accomplished by administering BAL.

Manganese

Manganese may be absorbed through the lungs or the gastrointestinal tract. Intoxication usually results from a combination of inhaling manganese dust and swallowing particles in mining and industry. The highest incidence of intoxication is in Chilean manganese miners; toxic symptoms occur in approximately 15 percent of exposed miners. Neurological manifestations of chronic manganese poisoning are reported to include extrapyramidal symptoms, responsive to L-DOPA (Chap. 32). Although little information on basic abnormalities of biochemical mechanisms is available, in some studies with miners injection of radioactive manganese has been used. The total body loss of the injected radioactive material occurs more rapidly in healthy miners than in those with manganese poisoning. It was suggested that concentrations in tissue are greater in healthy miners than in those intoxicated by this heavy metal. Limited studies of the effects of manganese on cerebral enzymes in rats have demonstrated reduction of activities of acetylcholinesterase and adenosine deaminase in all areas of brain examined. Monoamine oxidase activity was increased in the cerebellum and that of guanine deaminase was diminished in the remainder of the brain after cortex was removed [15].

CYANIDE AND HYDROCYANIC ACID

Cyanide combines with many enzymes that contain ferric, cupric, or zinc ions. This combination results in enzyme inhibition. The cyanide anion also forms complexes with many proteins, such as heme, and may interfere with their physiological function. The acute action of cyanide on the nervous system is caused predominately by its rapid combination with the ferric ion of cytochrome oxidase, which inhibits use of molecular oxygen in the electron transport system. The administration to experimental animals of 5 mg NaCN per kg of body weight causes cytochrome oxidase activity to decrease to approximately half of normal values, with concomitant decrease in oxygen consumption by the tissues [16]. The compounds involved in glycolysis accumulate as a result of the failure of oxidative mechanisms. Cerebral lactate and phosphopyruvate concentrations increase, and lesser increases have been noted in hexose diphosphate and phosphoglycerate. These data, together with the concomitant decrease in cerebral glycogen concentrations, have been interpreted to indicate increased glycolysis under these conditions.

The principal function of cerebral oxidative metabolism is the production of high-energy phosphate. Thus a decrease in cerebral respiratory activity is reflected in the diminution of high-energy phosphate concentrations. Cyanide intoxication is accompanied by a significant decrease in cerebral phosphocreatine and ATP. An increase in ADP parallels the decrease in ATP. The increase of inorganic phosphate concentrations, however, is even greater than the decrease in phosphocreatine.

ISONICOTINIC ACID HYDRAZIDE

Symptoms of peripheral nerve involvement were found to result from isonicotinic acid hydrazide (isoniazid) administration shortly after initiation of its use in antitubercular therapy. The similarity of isoniazid neuropathy to those produced by vitamin B_6 deficiency or by deoxypyridoxine inhibition of pyridoxine led to the recognition that isoniazid produces its neurotoxic effect by interfering with normal pyridoxine action. Isoniazid combines with pyridoxine to form the isonicotinoylhydrazine derivative of pyridoxine, which is excreted in the urine. Pyridoxine-dependent reactions, such as decarboxylation of amino acids and transaminations, are thereby inhibited. Continued therapy with isoniazid is possible if it is combined with the administration of pyridoxine [17].

FURAN

Furan derivatives possess both antimicrobial and antineoplastic activity; they interfere with enzymatic processes necessary for bacterial or tumor growth. Three furans have been employed in therapy: nitrofurantoin, nitrofurazone, and furmethanol.

Nitrofurantoin is often used against *E. coli* in urinary tract infections. When renal function is normal, antibacterial levels are achieved within the urine, but bactericidal or toxic levels are not reached in the blood. In renal dysfunction, the concentration of nitrofurantoin in blood may be sufficient to produce symptoms of neural involvement.

Nitrofurazone has been employed in the treatment of metastatic carcinoma and furmethanol has been employed against penicillin-resistant *staphylococci.* Because of the toxicity of the latter two furans, only nitrofurantoin is still used in therapy to any significant degree. Furan compounds interfere with energy metabolism in the tricarboxylic acid cycle by inhibiting formation of acetyl-CoA [18].

LATHYRISM

Ingestion of the chickpea *Lathyrus sativus* produces neurological disease. Cases of this disease are often found in humans and animals in impoverished areas of Europe and India, where consumption of chickpeas is high. A number of potent neurotoxins have been isolated from the seeds of the chickpea, only one of which appears to be responsible for the neurological involvement. This compound is β-N-oxalyl-L-α,β-diaminopropionic acid [19]. When this material is administered to experimental animals, it produces convulsions that appear to result from chronic ammonia toxicity, which, in turn, could be caused by interference with the ammonia-generating or -fixing mechanisms in the brain. This mechanism has been postulated because administration of the toxin to experimental animals causes a striking increase in glutamine concentrations in the brain. The formation of glutamine from glutamic acid is the chief mechanism of detoxification of ammonia in the brain. A significant increase in the concentrations of free ammonia in brain occurs at the onset of convulsions, with a subsequent decline to normal values. This is interpreted as detoxification through the formation of glutamine.

Another neurotoxin isolated from the chickpea, L-2,4-diaminobutyric acid, does not appear to produce nervous-system damages by direct action in the brain. The initial damage appears to be in the liver, producing a slight chronic increase of ammonia concentration in blood.

ANTICHOLINESTERASE AGENTS

Anticholinesterase agents may be subdivided on the basis of their reversible or irreversible inhibition of the enzyme acetylcholinesterase (AChE). At the cellular or whole-animal level, administration of either kind results in an accumulation of acetylcholine (ACh) in the vicinity of nerve synaptic junctions and failure of synaptic transmission (Chap. 9). Such carbamates as eserine or neostigmine exemplify toxic agents that are reversible inhibitors. The basis for their toxic action is the carbamylation of the serine hydroxyl at the active site of AChE. The enhancement of toxicity of these substances over that of other carbamyl esters is the result of inter-action of the quaternary ammonium group (of neostigmine, for example) or of the protonated tertiary amine group (of eserine, for example) of these toxic carbamates with the anionic site of AChE. This additional interaction directs the carbamyl moiety onto the serine hydroxyl of the esteratic site. The carbamylated enzyme is hydrolyzed and reactivated relatively easily on dilution [20, 21].

The irreversible anticholinesterases are most commonly exemplified by diisopro-pylphosphorofluoridate (DFP). This and similar organophosphates phosphorylate the serine at the active site of the enzyme. This configuration is also easily reversed by hydrolysis, but a competing process called "aging" occurs. In aging, another sub-stituent around the phosphorus is split off instead (an isopropyl group, in the case of DFP), resulting in a phosphorylated enzyme that is then resistant to the nucleophilic attack of water or of many of the reactivating agents designed for therapeutic use, e.g., pyridine-2-aldoxime methiodide (2-PAM) or N,N-trimethylenebis-(pyridinium-4-aldoxime) dibromide (TMB-4) [22]. However, even such an "irreversibly" inhibited AChE may, in some instances, be subject to slow reactivation [23].

Among the most toxic of the anticholinesterases are the V agents, as they have been commonly called in the research laboratories of many countries for more than a decade (although subject to a useless security until early in 1975) [24]. These combine the directing properties of the amine nitrogen of the carbamates with the potential irreversibility of compounds of the DFP type. This can be appreciated from the structural formulas of two such compounds (not technically designated as V agents, but of the same type of formula) used by one of us some years ago and shown in Formula 33-2 [25].

$$\begin{array}{c} C_2H_5 \\ \diagdown \\ C_2H_5O \end{array} P \begin{array}{c} \diagup O \\ \diagdown \\ Se-CH_2CH_2-N \end{array} \begin{array}{c} \diagup C_2H_5 \\ \diagdown \\ C_2H_5 \end{array}$$

Selenophos

$$\begin{array}{c} C_3H_7O \\ \diagdown \\ C_3H_7O \end{array} P \begin{array}{c} \diagup O \\ \diagdown \\ S-CH_2CH_2-N \end{array} \begin{array}{c} \diagup C_3H_7 \\ \diagdown \\ C_3H_7 \end{array}$$

Tetriso

Formula 33-2
Structural formulas of selenophos and tetriso.

At physiological pH, these will be protonated but will be able to penetrate skin, organs, and cell membranes by being in equilibrium with the unprotonated form.

What is written here may imply to the uninitiated that the toxicities of carbamates and organophosphates are fairly similar. Nothing could be further from the actual case as a search of the literature will reveal [26]. Many of these compounds, but not all, are detoxified by enzymes, about which more later. Their rates of penetration across the plasma membrane differ widely, and (despite the usefulness of the concept) are not a simple function of oil/water distribution coefficients. These and many more properties should be summable by quantum-mechanical treatment and probably will be one day, but as yet are not. Also, because the biochemical basis of inhibition by the carbamates and organophosphates is different in some respects, it is often difficult to find comparable data on the toxicities of these two major classes. A recent short note can be consulted for this purpose [27], but neither this nor any other data (with the exception of a few grisly reports out of the debris of World War II) are readily applicable to man. As a guideline, it may be noted that the most toxic of the organo-phosphate V agents and of the DFP-like agents (e.g., Sarin or Soman), intravenously administered to mice, have LD_{50}s of approximately 0.05 μmol per kg and 1 μmol per kg, respectively.

ANTICHOLINEACETYLASE AGENTS

As research tools, no compounds in the category of anticholineacetylase agents compare with the organophosphates or carbamates. In recent years, there are available naphthyl vinyl pyridyl (NVP) compounds [28], which are reported to be moderately potent inhibitors of choline acetylase (ChAc). It has been noted [29] that the inhibition did not appear to be easily analyzable kinetically. Results have now been obtained in this laboratory [30] that show that the NVPs do not inhibit ChAc from the tissues of the electric organ of *Electrophorus electricus,* even at concentrations approaching the limits of solubility. As reported, they do inhibit ChAc from mammalian brain. We now speculate that the seemingly anomalous inhibition kinetics referred to previously, and confirmed by us, may be caused by the normal inhibition of one or more of the isoenzymes of ChAc and the noninhibition of one or more of the other isoenzymes. It might therefore be presumed that the presence of such isoenzymes, and probably of a reserve of preformed ACh, would make the NVPs less toxic than their enzyme inhibitory properties would suggest. At present, virtually nothing is known about the toxicity of these compounds. To provide some information, results obtained in this laboratory are presented for the first time in Table 33-1. Even so, it cannot be concluded with certainty that the toxicity of the quaternary ammonium compound is directly and exclusively related to its enzyme-inhibitory properties.

ANTIRECEPTOR AGENTS

For the purposes of this chapter, we will limit ourselves to the inhibitors of the cholinergic receptor. These are indeed toxic, but our most fundamental understanding of the entire class has come from their use as research tools in the study of nerve function.

Table 33-1. Intraperitoneal Toxicities to Mice of Two Choline Acetylase Inhibitors

Compound	Dosage mg/kg	Dead/ injected
⬡⬡ – CH = CH – ⬡N·HCl	350	0/2[a]
	125	0/2
	36	0/2
⬡⬡ – CH = CH – ⬡N⁺– CH₂CH₂OH Br⁻	350	2/2[b]
	125	0/2
	36	0/2

[a]One animal showed signs of distress, but survived the 24-hr observation period.
[b]Both animals died approximately 10 min after injection.

When applied to cholinergic receptors, such terms as musarinic and nicotinic may reflect fundamental differences in molecular architecture, or may not, but they have had their origins in medical practice rather than in membrane biochemistry (see Chap. 8). It is probably misleading to treat as neurotoxic agents such compounds as atropine (a muscarinic receptor ligand) and muscarone (apparently a nicotinic receptor ligand). D-Tubocurarine, the active factor in the Indian arrow poison of legendary fame, is probably the best-known receptor ligand. Pharmacologically, it is termed a receptor inhibitor in that it blocks electrical activity without depolarizing. Clinically, it is used as a muscle relaxant. One significant recent development should be noted. D-Tubocurarine is normally depicted as a diquaternary ammonium compound. This does not now appear to be so; rather, it is monoquaternary and monotertiary [31]. This property probably explains certain observations of effects of d-tubocurarine on squid axons — effects of a solubilizing and detergentlike nature, which were not seen with dimethyl-d-tubocurarine [32]. While being synthesized by the o-methylation of d-tubocurarine, this latter compound is simultaneously N-methylated and converted into the diquaternary form.

The general subject of venoms has been covered thoroughly and need not be reviewed here [33]. The complexity of the subject may be appreciated from a consideration of the many active components of venoms, among them, lipases, esterases, lytic factors, and potential allergens. One aspect especially pertinent to this chapter is too new to have been included in any depth in the earlier reviews [29, 33]. This is the high specificity of a component of the venoms of the elapid Bungarus multicinctus for the cholinergic receptor. This specificity is evidenced, first, in the neuromuscular signs of the snakebite poisoning, more exactly in the postsynaptic responses of cells (especially the single electroplaque of the electric eel Electrophorus electricus) electrically stimulated and monitored, and finally in the specific binding of a purified component of the venom to a purified component of cholinergic cells (see Chap. 23). In the time since the neurochemical aspects of α-bungarotin binding to receptor were mentioned (and little more) in the previous edition of this book [29], the use of this substance has made advances possible in virtually all aspects of basic and clinical neurochemistry [34–36].

ANTIIONOPHORES

As toxic agents, antiionophores may be divided into "passive" agents — such as tetro-dotoxin and agents related in structure or function that interfere with ion movements down their electrochemical gradients during nerve function — and "active" agents such as ouabain and related agents, which interfere with ion pumps.

Tetrodotoxin, commonly called pufferfish poison, and saxitoxin, shellfish poison, have been treated in numerous reviews, the most recent and complete of which is that by Toshio Narahashi [37], himself a prolific contributor to the original literature. Although there are significant differences in the details of the mode of action of these two compounds as well as in their structure, basically they both act by blocking nerve action potentials. They do this by selectively inhibiting the transient sodium conductance in membrane, best visualized in voltage-clamp experiments and normally predominant in the early phase of a nerve action potential. This is accomplished at extremely low concentration, which also agrees well with the toxicity of tetro-dotoxin, in the range of 10 μg per kg.

Finally, many compounds inhibit the active ion-moving processes by inhibiting $(Na^+ + K^+)$-ATPase. Only a few of these appear to be highly specific. The best known is ouabain, a steroid glycoside of plant origin that also is called strophanthidin, from the name of the plant from which it is obtained. Ouabain and compounds of this class have marked effects, particularly on the heart, hence the name cardiac glycosides. They are not strikingly toxic on a dosage-size basis, but their mode of action is specific enough to nerve and muscle to be considered in detail in Chapter 6.

NEUROTOXIC ESTERASE

It is well known that many organophosphates cause a degeneration of some sensory and motor axons and produce permanent ataxia. The condition is seen in mammals, including man, but is especially pronounced in birds. Although many of the compounds that cause this effect are cholinesterase inhibitors, their potency as degenerative agents is not primarily a function of this property. Indeed, protection against the toxic effects is sometimes obtained with other esterase inhibitors. This seeming contradiction has led to a recent concept that appears to explain the degenerative effects [38].

Nerves appear to contain esterases that are characterized experimentally by their ability to hydrolyze phenylphenylacetate. About 10 percent of this total activity is not inhibited by paraoxon (diethyl p-nitrophenyl phosphate, a compound normally to be dealt with under the earlier heading "Anticholinesterase Agents"). About 30 percent of this, that is, 3 percent of the total, is inhibited by Mipafox (N,N'-diiso-propylphosphordiamidofluoridate). This esterase activity, which is insensitive to para-oxon and sensitive to Mipafox, is membrane bound. It does not appear to be essential to normal cell function but plays a role in the organophosphorus-induced nerve degeneration. It is termed *neurotoxic esterase*. Although the total number of compounds tested is still small, the results support the following hypothesis.

As a rule esterase inhibitors are of the general structure shown in Formula 33-3, where A and B may be the same or different and where X is a suitable leaving group.

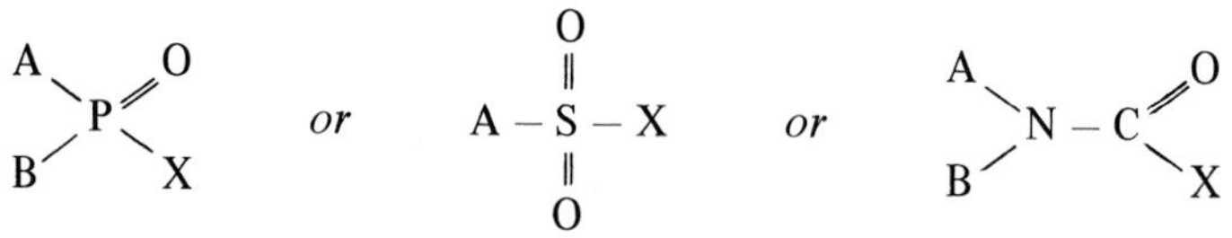

Formula 33-3
General structure of esterase inhibitors.

They also inhibit the neurotoxic esterase. If, subsequent to reaction with the enzyme, one of the A or B groups is lost by hydrolysis (equivalent to the aging process that renders inhibited acetylcholinesterase refractory to reactivation), the remaining structure can be depicted as in Formula 33-4. Compounds capable of resulting in this

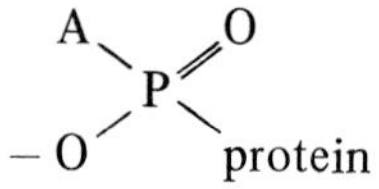

Formula 33-4
Structure of esterase inhibitor after an A or B group has been lost by hydrolysis.

product (group A compounds) are phosphates, phosphonates, and phosphoramidates and are neurotoxic in the present sense; that is, they cause nerve degeneration. It is believed that the ionized group covalently anchored in the membranes of the nerve is responsible for this degeneration. Reaction of the enzyme with group B compounds (phosphinates, sulfonates, carbamates) results in the structures shown in Formula 33-5.

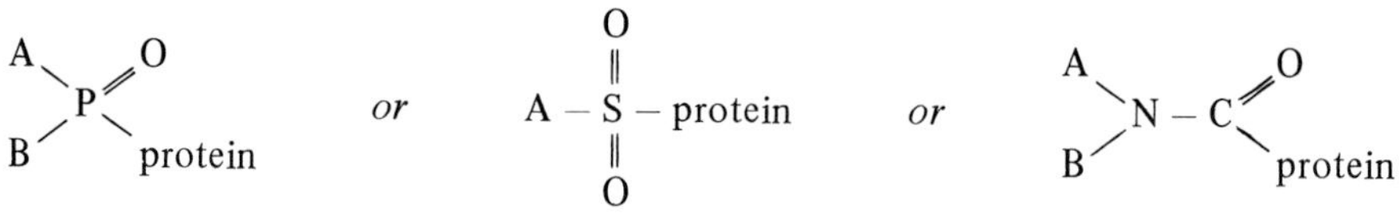

Formula 33-5
Reactions of enzyme with group B compounds.

Subsequent aging cannot occur, and nerve degeneration does not ensue. Again, this is compatible with the idea that an ionized group strategically located is responsible for the degeneration. Even more convincing, it can be shown that when group B compounds inhibit the neurotoxic esterase, such inhibition protects against nerve degeneration that might otherwise be expected to result from the subsequent administration of group A compounds.

DFPase

In 1946, an enzyme capable of hydrolyzing DFP at the P − F bond was reported in mammalian tissues and blood [39]. In the years following, enzymes capable of splitting a variety of organophosphates were reported in a small but significant number of papers. In most instances, these have continued to survive in the literature as unique enzymes or as a unique class — DFPase, phosphorylphosphatase, A-esterase, Tabunase, and so on. Their natural substrate and physiological role remain unknown, and their usual sources — liver, kidney, serum, and other organs as well, of course — have usually made it appear that they are members of a broad class of hydrolytic enzymes, perhaps analogous in enigmatic quality to the so-called pseudocholinesterase. A consequence of their presence and distribution is to add to the unpredictability of toxicities.

Recently, however, what appears to be a special case of DFPase activity has been discovered and elaborated upon and seems to hold the promise of some advance with respect to these enzymes.

In the course of examining the effects of some of the powerful irreversible cholinesterase inhibitors on nerve function, the squid giant axon was found to be capable of a high rate of hydrolysis of DFP [40]. It may be added that the effects on axonal conduction were observable at concentrations of the order of 10^{-2} M. They probably were the result not of inhibition of the acetylcholinesterase in the excitable membrane but of the production of a high level of acidity caused by the hydrolysis of DFP. Furthermore, with other organophosphorus-cholinesterase inhibitors conduction is affected even less, if at all [41].

To return to the DFPase found in squid nerve: This has since proved to be a different enzyme from the DFPase(s) of mammalian visceral organs. This particular DFPase, now termed *squid-type DFPase*, has been found in the nerves of the cephalopods — squid, cuttlefish, octopus — and not in the nerves of gastropods, annelids (even such a likely source as the giant axon of *Myxicola*), crustaceans, or vertebrates [42, 43]. It should be recognized that this negative statement is true only so far as members of these phyla or classes have been tested. This squid-type DFPase does not hydrolyze paraoxon or the V agents, with the result that these latter two are about as toxic to squid as they are to mice, whereas DFP is only a fraction as toxic to squids as to mice [27]. This information is cited to illustrate the effects of an unusual distribution of a detoxifying enzyme.

Despite the seeming uniqueness of the squid-type DFPase, it must be recognized as probable that the natural substrate is not DFP. Its localization in cephalopod nerve was originally thought to be related to its physiological role and may still prove to be. Cephalopod nerve also appears to be unique in having isethionate, $HO - CH_2CH_2SO_3{}^-$, as one of its most plentiful anions. It was speculated that the so-called DFPase might actually play a role in the metabolism of isethionate. This speculation is now put in some doubt by the finding of the squid-type DFPase in squid hepatopancreas, where isethionate is not present, and the characterization of the nerve and hepatopancreas DFPases as identical [44, 45].

In these two sections, we have considered two esterases, DFPase and neurotoxic esterase, which have in common the feature of possessing no obvious physiological function. Probably a whole spectrum of esterases share this property to some degree. Nevertheless, the generally accepted concepts of molecular biology and biochemical genetics would suggest that these enzymes do play a vital role in certain, as yet unknown biological systems.

ACKNOWLEDGMENT

This section incorporates research that was supported in part by U.S. Public Health Service Grant NS-09090.

REFERENCES

1. Krista, K., and Momčllović, B. Transport of lead 203 and calcium 47 from mother to offspring. *Arch. Environ. Health* 29:28, 1974.

2. Fromke, V. L., Lee, M. Y., and Watson, C. J. Porphyrin metabolism during versenate therapy in lead poisoning: Intoxication from an unusual source. *Ann. Intern. Med.* 70:1007, 1969.

3. Patel, A. J., Michaelson, I. A., Cremer, J. E., and Balazs, R. Changes within metabolic compartments in the brains of young rats ingesting lead. *J. Neurochem.* 22:591, 1974.

4. Whitaker, J. A., Austin, W., and Nelson, J. D. Edathamil calcium disodium (versenate) diagnostic test for lead poisoning. *Pediatrics* 29:384, 1962.

5. Cremer, J. E. Biochemical studies on the toxicity of tetraethyl lead and other organo lead compounds. *Br. J. Ind. Med.* 16:191, 1959.

6. Magos, L. Mercury-blood interaction and mercury uptake by the brain after vapor exposure. *Environ. Res.* 1:323, 1967.

*7. Webb, J. L. *Enzyme and Metabolic Inhibitors.* Vol. II. New York: Academic, 1966, Chap. 7.

8. Olszewski, W. A., Pillay, K. K. S., Glomski, C. A., and Brody, H. Mercury in the human brain. *Acta Neurol. Scand.* 50:581, 1974.

9. Cavanagh, J. B., and Chen, F. C. K. Amino acid incorporation in protein during the "silent phase" before organo-mercury and *p*-bromophenylacetylurea neuropathy in the rat. *Acta Neuropathol. (Berl.)* 19:216, 1971.

10. Cremer, J. E. The action of triethyl tin, triethyl lead, ethyl mercury and other inhibitors on the metabolism of brain and kidney slices *in vitro* using substrates labelled with ^{14}C. *J. Neurochem.* 9:289, 1962.

11. Magos, L. Effect of 2-3 dimercaptopropanol (BAL) on urinary excretion and brain content of mercury. *Br. J. Ind. Med.* 25:152, 1968.

12. Kosta, L., Byrne, A. R., and Zelenko, V. Correlation between selenium and mercury in man following exposure to inorganic mercury. *Nature* 254:238, 1975.

13. Aaseth, J. The effect of *N*-acetylhomocysteine and its thiolactone on the distribution and excretion of mercury in methyl mercuric chloride injected mice. *Acta Pharmacol. Toxicol. (Kbh.)* 36:193, 1975.

*14. Webb, J. L. *Enzyme and Metabolic Inhibitors.* Vol. III. New York: Academic, 1966, Chap. 6.

15. Sitaramayya, A., Nagar, N., and Chandra, S. U. Effect of manganese on enzymes in rat brain. *Acta Pharmacol. Toxicol. (Kbh.)* 35:185, 1974.

16. Albaum, H. G., Tepperman, J., and Bodansky, O. The *in vivo* inactivation by cyanide of cytochrome oxidase and its effect on glycolysis and on the high energy compounds in brain. *J. Biol. Chem.* 164:45, 1946.

17. Aspinall, D. L. Multiple deficiency state associated with isoniazid therapy. *Br. Med. J.* 2:1177, 1964.

18. Paul, M. F., Paul, H. D., Kopko, F., Bryson, M. J., and Harrington, C. Inhibition by furacin of citrate formation in testis preparations. *J. Biol. Chem.* 206:491, 1954.

* Asterisks denote key references.

19. Cheema, P. S., Malathi, K., Padmanaban, G., and Sarma, P. S. The neurotoxicity of β-N-oxalyl-L-α,β-diaminopropionic acid, the neurotoxin from the pulse, *Lathyrus sativus. Biochem. J.* 112:29, 1969.

*20. Cohen, J. A., and Oosterbaan, R. A. The Active Site of Acetylcholinesterase and Related Esterases and its Reactivity Toward Substrates and Inhibitors. In Koelle, G. B. (Ed.), *Handbuch der experimentellen Pharmakologie; Cholinesterases and Anticholinesterase Agents.* Berlin: Springer-Verlag, 1963.

21. Wilson, I. B., Hatch, M. A., and Ginsburg, S. Carbamylation of acetylcholinesterase. *J. Biol. Chem.* 235:2312, 1960.

22. Heilbronn-Wikstrom, E. Phosphorylated cholinesterases. *Sven. Kemisk Tidskr.* 77:11, 1965.

23. Dettbarn, W.-D., Bartels, E., Hoskin, F. C . G., and Welsch, F. Spontaneous reactivation of organophosphorus inhibited electroplax cholinesterase in relation to acetylcholine induced depolarization. *Biochem. Pharmacol.* 19:2949, 1970.

24. Sidell, F. R., and Groff, W. A. Reactivatibility of cholinesterase inhibited by VX [*S*-(2-diisopropylaminoethyl) *O*-ethyl methylphosphonothiolate] and Sarin in man. *Toxicol. Appl. Pharmacol.* 27:241, 1974. (See also *Science* 187:414, 1975.)

25. Hoskin, F. C. G., Kremzner, L. T., and Rosenberg, P. Effects of some cholinesterase inhibitors on the squid giant axon. *Biochem. Pharmacol.* 18:1727, 1969.

*26. Holmstedt, B. Structure Activity Relationships of the Organophosphorus Anticholinesterase Agents. In Koelle, G. B. (Ed.), *Handbuch der experimentellen Pharmakologie; Cholinesterases and Anticholinesterase Agents.* Berlin: Springer-Verlag, 1963.

27. Dettbarn, W.-D., and Hoskin, F. C. G. Toxicity of DFP and related compounds to squids in relation to cholinesterase inhibition and detoxifying enzyme levels. *Bull. Environ. Contam. Toxicol.* 13:133, 1975.

*28. Cavallito, C. J., White, H. L., Yun, H. S., and Foldes, F. F. Inhibitors of Choline Acetyltransferase. In Heilbronn, E. (Ed.), *Drugs Affecting Cholinergic Systems in the CNS.* Stockholm: Försvarets Forskningsanstalt (Almqvist and Wiksell, agents), 1971.

*29. Hoskin, F. C. G. Acetylcholine. In Albers, R. W., Siegel, G. J., Katzman, R., and Agranoff, B. W. (Eds.), *Basic Neurochemistry.* Boston: Little, Brown, 1972, p. 105.

30. Hopkins, G. Ph. D. Thesis: Illinois Institute of Technology, 1975. To be published.

31. Everett, A. J., Lowe, L. A., and Wilkinson, S. Revision of the structures of (+)-tubocurarine chloride and (+)-chondrocurine. *J. Chem. Soc. D* (Lond.) 16:1020, 1970.

32. Hoskin, F. C. G., and Rosenberg, P. Alteration of acetylcholine penetration into, and effects on, venom-treated squid axons by physostigmine and related compounds. *J. Gen. Physiol.* 47:1117, 1964.

*33. Simpson, L. L. (Ed.). *Neuropoisons.* New York: Plenum, 1971.

*34. Karlin, A. The acetylcholine receptor: Progress report. *Life Sci.* 14:1385, 1974.

35. Weber, M., and Changeux, J.-P. Binding of *Naja nigricolli* [^{3}H] α-toxin to membrane fragments from *Electrophorus* and *Torpedo* electric organs. *Mol. Pharmacol.* 10:15, 1974.

36. Heilbronn, E., and Mattson, C. The nicotinic cholinergic receptor protein: Improved purification method, preliminary amino acid composition and observed auto-immuno response. *J. Neurochem.* 22:315, 1974.

*37. Narahashi, T. Chemicals as tools in the study of excitable membranes. *Physiol. Rev.* 54:813, 1974.

38. Johnson, M. K. The primary biochemical lesion leading to the delayed neurotoxic effects of some organophosphorus esters. *J. Neurochem.* 23:785, 1974.

39. Mazur, A. An enzyme in animal tissue capable of hydrolyzing the phosphorus-fluorine bond of alkyl fluorophosphates. *J. Biol. Chem.* 164:271, 1946.

40. Hoskin, F. C. G., Rosenberg, P., and Brzin, M. Re-examination of the effect of DFP on electrical and cholinesterase activity of squid giant axon. *Proc. Natl. Acad. Sci. U.S.A.* 55:1231, 1966.

41. Hoskin, F. C. G. Proteins in excitable membranes. *Science* 170:1228, 1970.

42. Hoskin, F. C. G., and Long, R. L. Purification of a DFP hydrolyzing enzyme from squid head ganglion. *Arch. Biochem. Biophys.* 150:548, 1972.

43. Hoskin, F. C. G., and Brande, M. An improved sulphur assay applied to a problem of isethionate metabolism in squid axon and other nerves. *J. Neurochem.* 20:1317, 1973.

44. Hoskin, F. C. G., Pollock, M. L., and Prusch, R. D. An improved method for the measurement of $^{14}CO_2$ applied to a problem of cysteine metabolism in squid nerve. *J. Neurochem.* 25:445, 1975.

45. Garden, J. M., Hause, S. K., Hoskin, F. C. G., and Raush, A. H. Comparison of DFP-hydrolysing enzyme purified from head ganglion and hepatopancreas of squid (*Loligo pealei*) by means of isoelectric focusing. *Comp. Biochem. Physiol.* 52C:95, 1975.

Part Three

Behavioral Neurochemistry

Part Three

Chapter 34

Psychopharmacology and Biochemical Theories of Mental Disorders

Theodore L. Sourkes

Psychopharmacology is concerned with drugs that are active in the central nervous system and that influence behavior, perception, thought, and affect. In its earliest usage, the Greek word *psychopharmaka* was applied to drugs or measures with a healing action upon the psyche. In the context of modern therapeutics, the term psychopharmacology refers to the selection and use of drugs that have specific actions in the treatment of mental diseases; it also embraces the experimental study of such drugs at the biochemical, neural, and behavioral levels. The information obtained in these ways is essential to an understanding of the action of drugs on human thought and action.

The study of the biochemical effects of drugs made little progress until the discovery of the antibacterial effects of the sulfonamides. This was contemporaneous with the replacement of galenic pharmaceuticals by synthetic medicinal chemicals. The structural resemblance of sulfanilamide to *p*-aminobenzoic acid, a bacterial vitamin, led to the recognition of other comparable pairs of presumed antagonists. Antecedents can be found in Emil Fischer's concept of the precise fit of enzyme and substrate; in Ehrlich's work on the relation of drug and cell receptor; and in the concept of competitive inhibition, as first described by Quastel and Wooldridge. These concepts were further developed in their application to chemotherapy.

The antimetabolite hypothesis became a motive force in medicinal chemistry, pressing for the synthesis of new compounds to be tested pharmacologically and clinically, and it extended its influence into psychiatry. In recent years, the resemblance between the structures of the hallucinogen mescaline and the catecholamines has played a significant role in biochemical psychiatry by raising the question as to whether mescaline and similar drugs can antagonize the normal functions of catecholamines in the brain. Furthermore, although the psychological effects of the so-called phantastica had been known for many years, their significance for psychiatric research was largely dormant until the accidental discovery in 1943 of the remarkable actions of LSD-25. Minute amounts of this compound are able to bring about an acute psychological disturbance, and it was recognized that, if an exogenous chemical of this kind could evoke protean changes in subjective experience, then an endogenous chemical, resulting from some metabolic disturbance, might conceivably be responsible for the development of psychoses.

In psychiatry, drugs affecting the concentration, synthesis, binding, or release of neurohumors or neurohumorlike substances in the brain are of the utmost interest because of the important role that some of these drugs play or have played in pharmacotherapy. Chlorpromazine's adrenolytic effects are thought to consist in preventing the attachment of the sympathetic neurohumor to a specific receptor site. Reserpine causes the release of amines from the special vesicles in which most of the intracellular content of these substances is stored; it also prevents their being taken up into the vesicles from the external medium. Monoamines are subject to the action of monoamine oxidase (MAO); this enzyme is inhibited by various compounds, some of which are valuable in the treatment of mental depression. Important findings of this kind have focused the attention of large numbers of neurobiologists on the role of brain amines in mental diseases.

The neurochemist who is oriented toward pharmacology seeks another type of information that is important in understanding the mode of action of drugs and their clinical uses. That is the knowledge of the metabolism of the drugs themselves — their length of residence in the body, the blood levels attained, the chemical changes they undergo, and their mode and routes of excretion. Knowledge of drug metabolism may aid in understanding and even avoiding toxic and side actions of drugs.

An area of pharmacology in which neurochemistry aims to contribute is the classification of CNS-active drugs according to their modes of action. However, the present classification is necessarily based upon analysis by pharmacological and physiological techniques. Although the anatomical, biochemical, and physiological sites of action of many drugs in this category are being clarified, much of the terminology of psychopharmacology currently in use is subjective and admittedly is a compromise that ought not to last indefinitely.

The general pharmacological and clinical effects of psychopharmacological agents are outlined in Table 34-1. Detailed pharmacological information can be found in a number of useful monographs and symposia, some of which are described in the references [1–6]. Valuable chemical and dosage data are given in reference [7]. Pertinent biochemical actions of some psychopharmacological agents are summarized in Table 34-2 and are discussed in more detail below.

NEUROLEPTICS, OR MAJOR TRANQUILIZERS

Reserpine

Reserpine (Fig. 34-1), the first of the neuroleptics, is no longer used in psychiatry because of its side effects, such as mental depression and iatrogenic parkinsonism. Its biochemical and pharmacological actions, however, now so well correlated by the results of a vast amount of experimentation, make it a worthwhile — indeed, an essential — topic for discussion in a basic text. Its primary action is on the storage and uptake of monoamines by nerve endings and synaptic vesicles. An interesting experiment has been carried out by injecting labeled reserpine into animals and then isolating the blood platelets. These organelles are rich in serotonin, among other constituents, and may thereby serve as a model for the study of the release and uptake of monoamines by vesicles of the nervous system. In this experiment, when the membrane of the platelet is separated from the soluble contents, most of the labeled material of the platelets is found to be associated with the membrane. Similar results are obtained even if the platelets are incubated in vitro with the labeled drug. This and other experiments suggest that reserpine acts at the level of the membrane of the storage organelle by regulating movement of amines through it. Reserpine inhibits the ability of adrenal chromaffin granules to take up catecholamines by a process that is dependent on Mg^{2+} and ATP. When an animal is given a large dose of reserpine, the immediate effect on neurons is the loss of their mono-amine content. Much of the amine is released intraneuronally under the influence of reserpine; in this case, it becomes subject to the action of mitochondrial mono-amine oxidase, and ultimately it is released from the nerve ending as an acidic compound.

The action of reserpine and certain other *Rauwolfia* alkaloids on brain mono-amines is indiscriminate. Hence, this effect cannot by itself specify the amine whose loss from the brain is to be correlated with the sedative effect of the drug. A great many experiments have been carried out to establish which cerebral amine is affected by reserpine to bring about its tranquilizing action. Although much information useful to biochemical pharmacology has been acquired as a result of this effort, the issue has not yet been settled decisively. If the amines could pass the blood-brain barrier readily, they could be injected singly to determine which one is effective; but peripherally administered amines do not get into the parenchyma of the brain in appreciable amounts. Their amino acid precursors cross the blood-brain barrier much more easily, although an amino acid decarboxylase in the endo-thelial wall of the brain capillaries may limit the amount of amino acids that get into the parenchymal tissue (Chap. 20). Within the brain, an amino acid such as DOPA or 5-hydroxytryptophan is readily decarboxylated (Chap. 10), and the corre-sponding amine will be found both in regions of the brain where it is an endogenous constituent, and in others — for instance, in regions that normally do not contain that amine but possess the decarboxylase.

Experiments of this kind have shown that DOPA will temporarily reverse the sedation caused by reserpine, although 5-hydroxytryptophan will not. It has there-fore been deduced that a catecholamine, but not serotonin, is the immediately effective arousal agent. Furthermore, under experimental conditions, the β-oxida-tion of dopamine to norepinephrine is rate-limiting, so the injection of DOPA brings about large increases in the concentration of dopamine in many parts of the brain with little increase of the concentration of norepinephrine. This points to dopamine as the amine with the antireserpine activity.

A synthetic analog of DOPA, 3,4-dihydroxyphenylserine, which peripherally gives rise on decarboxylation to norepinephrine, has also been tested in these brain experi-ments, but it does not seem to penetrate the blood-brain barrier in the way that DOPA can.

Some benzoquinolizines have a reserpinelike action both pharmacologically and in the storage of monoamines. These synthetic compounds have been studied with the purpose of selecting those that have desirable therapeutic features. One of them, tetrabenazine (Fig. 34-1), has a much shorter lasting action than does reserpine, but has the same type of activity qualitatively.

When reserpine is administered, it is distributed to many organs of the body without any preferential uptake by the brain. Initially, the viscera contain relatively high concentrations, but later the drug moves into adipose tissue. Careful experi-ments with radioactively labeled reserpine have demonstrated the presence of the drug in the brain during the long period over which it exerts its central depressive action. Reserpine contains two esterified groups, and both of these may be hydrolyzed

Table 34-1. Pharmacological Actions of Psychotropic Compounds[a]

Category	Typical Examples	Broad Pharmacological Effects	Broad Effects in Man	
			Clinical	Subjective
Neuroleptics, or major tranquilizers	Phenothiazines Butyrophenones Thioxanthenes Reserpine Benzoquinolizines	Decreased autonomic activity, especially of sympathetic branch; hence, antiadrenergic. Decreased motor activity. Decreased conditioned avoidance reflex. No anesthesia.	Tranquilization of overactive psychotic syndromes. Indifference to environment. In high doses, extrapyramidal symptoms produced. Calming action. Antipsychotic effect.	Drowsiness, apathy, indifference to environment, but no effect on intellect. In high doses, extrapyramidal symptoms produced.
Anxiolytic sedatives, or minor tranquilizers	Meprobamate Diazepoxides	Actions similar to those of the neuroleptics, but less potent. No anesthesia. Meprobamate in large doses causes paralysis.	Tranquilization, but without any antipsychotic effect. Sedation of anxious patients.	Muscular and mental relaxation. Less interest in environmental stimuli. Euphoria may occur.
	Barbiturates	As for other minor tranquilizers. Also, there is some blockade of parasympathetic functions. Reduced ability to undergo conditioning. In higher doses, ataxia and then anesthesia.	Tranquilization; hypnotic effect.	Sedation, sleepiness.
Antidepressants	Imipramine Chlorimipramine Desipramine Amitriptyline MAO inhibitors Pipradrol Amphetamine	Reduced autonomic activity, especially of cholinergic systems. Central sympathetic activation. Increased motor activity, with alerting behavior. Reduction of appetite and sleeping time.	Antidepressant action; stimulation; decreased fatigue. Anxiety already present may be increased. If psychotic symptoms are already present, they may be increased.	Reduced motor activity; feeling of relaxation. Increased mental energy; decreased feelings of fatigue; euphoria (but less of this with pipradrol). Greater awareness of the surroundings.

	Lithium salts		In manic-depressive psychosis there is control of the manic attack. On chronic use, there may be prophylaxis against the clinical depression.	In appropriate patients, feeling of well-being.
Psychodyslieptics, or hallucinogenic agents	Lysergide Mescaline Psilocybin Tetrahydrocannabinol	Increased motor activity. Variable autonomic effects. Catatonia.	Marked action on affect, perception, thought processes. Psychotic disturbance may be aggravated.	Autonomic phenomena (e.g., salivation, nausea), followed by psychotic reactions, with distortion of sensory perceptions, color visions, hallucinations.

aSource: based primarily on Tables 8 and 12 in [1] and Table 1 in [4].

Table 34-2. Biochemical Effects of Psychopharmacological Agents

Haloperidol
Long-lasting blockade of receptors sensitive to norepinephrine and dopamine.
Increased concentration of striatal HVA, probably by feedback-regulated overactivity of nigrostriatal fibers. May cause temporary decrease in brain catecholamines.

Chlorpromazine
Same as for haloperidol. Chlorpromazine causes an increased rate of synthesis of
dopamine in brain from tyrosine. Inhibition of some flavin (yellow) enzymes.

Reserpine
Interference with ATP- and Mg^{2+}-dependent reuptake of released monoamines across
the vesicular membrane. This results in depletion of monoamines from peripheral
sympathetic nerve endings and from central sites. Effects are long lasting; endogenous catecholamines and serotonin are released intraneuronally into the cytoplasm,
where they become subject to action of MAO. Effects of reserpinization are reversed
temporarily by administration of DOPA, but not 5-hydroxytryptophan.

Imipramine
Interference with reuptake of norepinephrine from synaptic cleft. Chlorimipramine
has a similar action on reuptake, but it is more potently expressed toward serotonin.

MAO inhibitors
Hydrazide type, tranylcypromine, and pargyline cause long-lasting, irreversible inhibition of mitochondrial MAO; monoamines are protected, and increase in concentration (more pronounced for serotonin in brain than for catecholamines); some of
conserved catecholamines may be diverted in increasing amounts into pathway of
O-methylation. Harmaline and harmine are reversible, short-acting inhibitors of MAO.
In large doses, they may deplete dopamine from central stores. Clorgyline inhibits
over a wide concentration range, yielding a double sigmoid curve.

Amphetamine
Weak inhibitor of MAO; inhibits norepinephrine reuptake in peripheral nerve (perhaps
also in CNS neurons); effective in releasing monoamines from storage vesicles into
synaptic space. Stereotypies caused by amphetamine may be mediated by striatal
dopaminergic mechanisms.

Lithium salts
Stimulate the rate of turnover of norepinephrine and serotonin in the brain; effect
is greater for serotonin than norepinephrine in chronic experiments. Inhibit the
evoked release of monoamines from brain slices. Cause increase in total body water
and extracellular water.

---

in the course of metabolism. Among the products are methyl reserpate, reserpic
acid, 3,4,5-trimethoxybenzoic acid, and other compounds.

Chlorpromazine

Pharmacologists have shown that chlorpromazine (see Fig. 34-1) acts at the postsynaptic receptor site, thereby blocking the action of transmitter substance released
into the synapse. The blockade is long lasting and not reversed by excessive amounts
of the transmitter. Chlorpromazine and haloperidol, for example, block the dopamine-sensitive receptors of the caudate nucleus. This results in increased activity
within the nigrostriatal dopaminergic system, causing an accelerated rate of turnover

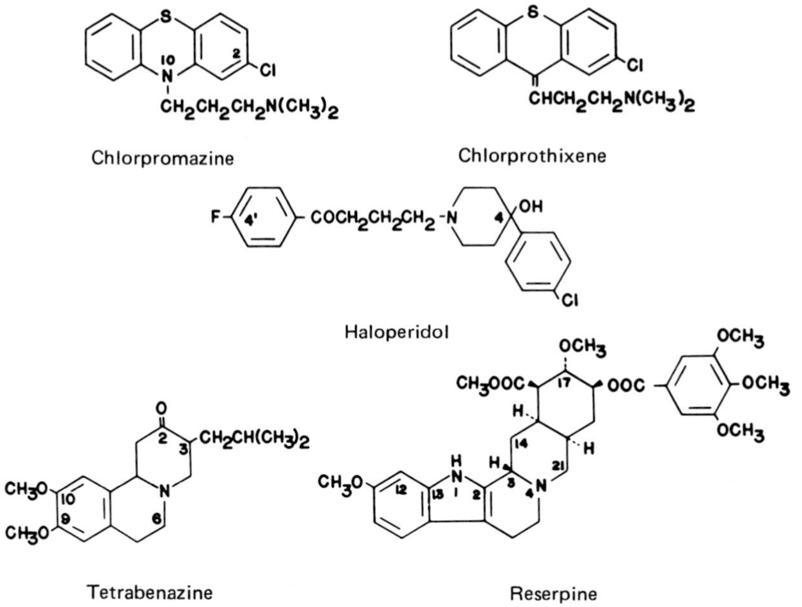

Figure 34-1
Some examples of neuroleptics.

of brain dopamine and a temporary accumulation of excessive amounts of homovanillic acid in the caudate nucleus (Chap. 32). At the same time, there is an increased output of acetylcholine from the head of the caudate nucleus (push-pull technique in the cat); the striatal cholinergic system is thought to be regulated by dopaminergic fibers. The increased impulse flow under the influence of neuroleptics is characteristic of dopamine-containing fibers and has not been detected in norepinephrine- and serotonin-containing systems of the brain. Yet such CNS fibers may also be blocked in their action, an effect that probably is the basis of the antipsychotic action of phenothiazines, butyrophenones, and thioxanthenes.

The concentration of monoamines in the brain is the steady-state result of production and utilization. The action of neuroleptics on the rates of these two processes is assessed by kinetic studies with labeled precursors. For example, if radioactive tyrosine is infused into animals, some of it reaches the brain and is synthesized into catecholamines. In animals given chlorpromazine, a substantial increase in the amount of radioactivity derived from tyrosine can be accounted for as dopamine (Table 34-3 and [8]). This is not the case with norepinephrine, the concentration of which, as measured by its radioactivity, is only slightly elevated. In this type of experiment, there is a small reduction (about 15 percent) in the concentration of endogenous tyrosine, which results in a modest increase in the specific activity of this amino acid in the brain. The apparently greater rate of utilization of tyrosine for the formation of dopamine indicates that chlorpromazine increases the activity of the neurons that synthesize dopamine, as already mentioned.

Table 34-3. Effects of Chlorpromazine and Pargyline on Rates of Formation of Catecholamines in Brain

Species[a]	Treat-ment	Compound	Precursor			Products		
			Conc. μg/g	S.A.[b] cpm/μg[c]		Dopamine cpm/g	Norepinephrine	
							μg/g	cpm/g
Rat	Control	^{14}C-tyrosine	–	6600 ± 600[d]		615 ± 30	–	183 ± 10
	CPZ	^{14}C-tyrosine	–	7500 ± 500		1262 ± 87	–	217 ± 24
Guinea pig	Control	^{14}C-tyrosine	17.7	1602		–	0.34	400
	MAOI	^{14}C-tyrosine	17.2	1600		–	0.76	180
	Control	^{3}H-DOPA	–	6253		–	0.35	876
	MAOI	^{3}H-DOPA	–	6553		–	0.64	1826

[a]Rats were injected i.p. with chlorpromazine (CPZ), 15 mg/kg, 2 hr before being killed for brain measurements. Labeled tyrosine was infused i.v. for 25 min immediately preceding death [8]. Guinea pigs were injected with labeled tyrosine or DOPA 24 or 7 hr, respectively, after pargyline (MAOI) was given, and were killed 1 hr later. The compounds were then measured in the brain stem [17].
[b]S.A. = Specific activity.
[c]Counts per minute per microgram.
[d]Mean ± S.E. Data for guinea pig are means of two or three animals.

Although little or no effect of neuroleptics on the concentration of catecholamines in the brain has been detected in acute experiments, repeated administration of chlorpromazine or haloperidol (see Fig. 34-1) decreases both striatal dopamine and cerebral norepinephrine.

Chlorpromazine, like many other tranquilizers and anesthetics, blocks excitability of the nerve membrane without affecting the resting membrane potential; at the same time, the membrane expands so that its components assume a somewhat disordered association [9]. CNS depressants, including neuroleptics, have sometimes been reported to cause an increase in the levels of ATP and phosphocreatine in the brain, but, when precautions are taken to avoid degradation of these labile esters in the course of their extraction from the brain, there proves to be no effect. Many of these drugs inhibit Na^{+}-K^{+}-ATPase but only at concentrations many times greater than needed to cause fluidization of the nerve membrane [10].

Chlorpromazine inhibits D-amino acid oxidase and some other flavoproteins, perhaps through its action as a flavin analog.

Chlorpromazine blocks secretion of certain hypothalamic neurohormones that act on the anterior pituitary gland, probably by blocking monoaminergic fibers that innervate the neuroendocrine cells. One of these hypothalamic factors causes release of growth hormone; in subjects receiving chlorpromazine, the plasma level of growth hormone declines. Another neurohormone inhibits the release of prolactin, and chlorpromazine treatment lifts this inhibition. Accordingly, neuroleptic administration results in raised concentrations of plasma prolactin (cf. Chap. 22).

The phenothiazine nucleus is a strong electron donor, and charge-transfer properties have been suggested as playing a role in the cellular actions of chlorpromazine and related compounds. Others have suggested that the molecule acts as a chelator of certain trace metals.

Chlorpromazine is readily absorbed from the gut, tissues (into which it may be injected), and blood; on the other hand, the release of the compound from most tissues is slow. Excretion by the kidney is rapid. The major portion of a single dose of chlorpromazine is excreted in the urine but some is found in the feces. With chronic administration, there seems to be some accumulation in the tissues because cessation of treatment with chlorpromazine is followed by a period of many months during which small amounts of the compound or its metabolites are found in the urine. In the brain, which is easily penetrated by chlorpromazine, the drug is found in higher concentrations in the cerebral cortex and the basal ganglia than in other parts of the CNS.

In the course of tissue metabolism, chlorpromazine undergoes a variety of reactions. These include demethylation of the amine, oxidation of the ring to yield a hydroxyl group, conjugation of this hydroxyl with glucuronic acid, and oxidation to yield chlorpromazine sulfoxide and related metabolites. After long, continued use of this drug in large doses, as in the treatment of schizophrenia, a photosensitive metabolite may accumulate in the skin; exposure to sunlight converts this substance to a purple derivative, which lends a grayish or grayish-purple cast to the complexion. Pigment may also be deposited in the cornea.

Chlorpromazine (see Fig. 34-1), a member of the thioxanthene family, is also oxidized to the sulfoxide, which represents a major metabolite of the drug. After its administration, the compound disappears from the blood with extreme rapidity.

A comprehensive collection of papers on the chemistry, metabolism, and pharmacology of phenothiazines and thioxanthenes will be found in reference [11].

ANXIOLYTICS, OR MINOR TRANQUILIZERS

Benzodiazepines

The sedative, anticonvulsant, and taming activity of such benzodiazepines as chlordiazepoxide (Fig. 34-2), along with their relaxing effect on skeletal muscle, has given these drugs a prominent role as minor tranquilizers. They seem to reduce the turnover rate of monoamines in the brain; this effect, with at least one of them, oxazepam, is more significant with respect to serotonin than to norepinephrine [12]. The metabolism of some benzodiazepines has been followed in a number of studies. Thus, a single dose of 15 mg of chlordiazepoxide, administered orally, provides a peak plasma concentration of about 1 μg per ml at 1 to 2 hr. The half-life of the radioactivity of the labeled compound in plasma is about 24 hr. Among the reactions that the molecule undergoes is a loss of the methylamino group and the formation of a lactam. The lactam ring is cleaved to yield a substituted glycine-N-oxide, which, like its precursor, is found in the urine. Half of the radioactivity of chlordiazepoxide is excreted in the urine in 3 to 3.5 days. In the case of diazepam, 10 mg taken orally provides a peak level in the blood of less than 0.1 μg per ml. Diazepam, which has 1-methyl-2-oxy substituents among others, undergoes 1-demethylation and 3-hydroxylation in

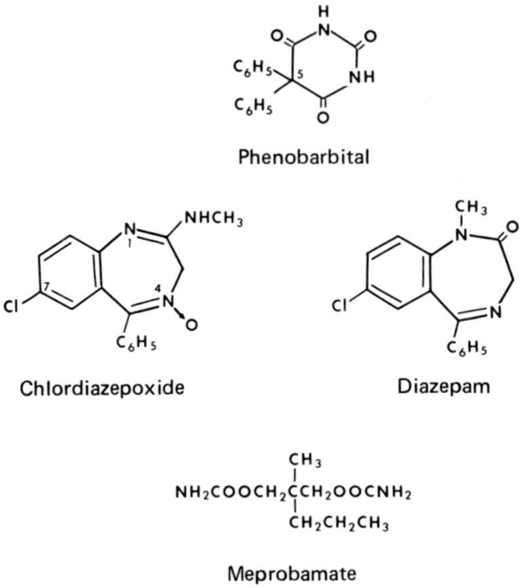

Phenobarbital

Chlordiazepoxide   Diazepam

$$NH_2COOCH_2\underset{\underset{CH_2CH_2CH_3}{|}}{\overset{\overset{CH_3}{|}}{C}}CH_2OOCNH_2$$

Meprobamate

Figure 34-2
Examples of minor tranquilizers.

metabolism. These two reactions successively result in the formation of oxazepam. The hydroxy derivative of diazepam occurs in the urine as a glucuronide.

Barbiturates
The barbiturates depress the metabolism of brain tissue in vitro to some extent; their effect is greater on electrically stimulated brain slices than on "silent" tissue. More-over, some of these compounds inhibit the respiration of brain mitochondria, whose metabolism has been stimulated by the uncoupling agent 2,4-dinitrophenol. Thio-barbiturates inhibit oxidative phosphorylation of the liver and brain mitochondria that metabolize pyruvate, and they have some action in uncoupling oxidative phos-phorylation; the oxybarbiturates affect only the former process. The site of action of the oxybarbiturates appears to be in the respiratory chain between NAD and cytochrome c; this includes a flavoprotein, which, it has been postulated, is inhibited by barbiturates (see also Chap. 31).

Barbiturates affect the neurolemma in the same manner as already described for chlorpromazine, with resulting nerve blockade. Many of them also inhibit Na^+-K^+-ATPase [9].

One of the important parameters that pharmacologists use to classify the barbitu-rates is their duration of action. This is influenced, of course, by the molecular structure. Thus, the thiobarbiturates undergo metabolism more rapidly than do the oxybarbiturates, and those with longer aliphatic side chains or an aromatic substituent are metabolized more slowly, presumably because of greater lipid solubility and longer residence in adipose tissue. These substituents are rendered more polar during metabolism through oxidations catalyzed by microsomal enzymes. In phenobarbital (Fig. 34-2), a phenolic group is inserted; in pentobarbital, the alkyl side chain of the compound undergoes hydroxylation.

The administration of phenobarbital to rats causes increased activity of the microsomal oxidizing enzymes of the liver. In addition to this induction of "drug-metabolizing enzymes," hemoprotein, which may be a component of such oxidative systems, increases.

ANTIDEPRESSANTS

Monoamine Oxidase Inhibitors

Monoamine oxidase inhibitors as a class of antidepressant drugs have had great significance for biochemical pharmacology and for the theory of mental disease (Fig. 34-3; [13, 14]). This significance remains, despite dangers inherent in their use. The first member of this class was iproniazid (N-isonicotinyl-N'-isopropylhydrazide), which was tested originally as an antitubercular drug along with isoniazid (N-isonicotinyl-hydrazide). The elation observed in patients receiving iproniazid was later put to use in the treatment of mental depression. The reversal of depression was then related to the ability of the compound to inhibit MAO powerfully and thereby to increase the concentration of monoamines in the brain. Subsequently, chemists synthesized scores of compounds on the model of iproniazid in the search for new and useful

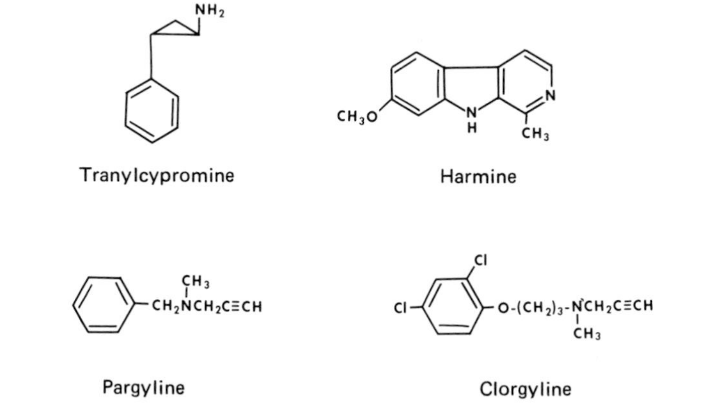

Figure 34-3
Monoamine oxidase inhibitors.

inhibitors of MAO that might also be useful therapeutically. Other structural models have yielded additional inhibitors. Representative examples are illustrated in Figure 34-3.

For many years Zeller's classification into monoamine oxidase (acting on such substrates as tyramine, isoamylamine, and epinephrine) and diamine oxidase, or histaminase, was used. As new types of amine oxidases have been discovered and as the physical and chemical properties of these enzymes have become better known, it is becoming possible to classify them not merely by the spectrum of substrates whose oxidation they catalyze, but also on the basis of their respective prosthetic groups [15, 16]. For example, classical MAO is a yellow enzyme present in the outer membrane of mitochondria. It bears flavin adenine dinucleotide (FAD), which is covalently linked through its 8-methyl group to a cysteinyl residue (in the case of the well-studied beef liver enzyme) of the peptide chain. In the plasma of pig and some other species there is a pink amine oxidase, which requires pyridoxal phosphate and cupric ions for its activity; its function is unknown. A copper-containing amine oxidase, acting upon the ϵ-amino group of lysine, has been detected in birds and other species; it is needed for the formation of desmosine in the walls of arteries.

MAO has been purified from the liver of several species and from bovine kidney and pig brain. The molecular weight of the mitochondrial enzyme in rat liver is approximately 150,000 (by ultracentrifugation measurements). This enzyme is especially sensitive to certain hydrazine derivatives, although not to hydrazine itself. It is also inhibited by quinacrine and by galactoflavin, an antimetabolite of riboflavin. Propargylamines $(RR'NCH_2CCH)$ like pargyline inhibit MAO irreversibly by forming an adduct at the N^5 position.

MAO has at least two functions. In the intestinal tract and liver, the enzyme probably serves in detoxification of certain amines formed by the gut flora. On absorption, these amines could exert pharmacological activity. In the nervous system, the enzyme serves to oxidize some of the serotonin, dopamine, and norepinephrine after their release into the synaptic space, thus terminating their action. The terminal metabolism of catecholamines may be initiated by the action of catechol O-methyltransferase (COMT), so that the substrates for MAO are 3-methoxytyramine and normetanephrine. MAO acts within the nerve fiber itself when monoamines leak from the synaptic vesicles. This occurs on a "massive" scale if reserpine is used to release the stored amines, for this takes place intraneuronally: The "free" amine within the nerve ending can then be attacked as it diffuses toward the mitochondria. MAO catalyzes the following reaction:

$$RCH_2CH_2NH_2 + O_2 + H_2O \rightarrow RCH_2CHO + H_2O_2 + NH_3$$

The aldehyde undergoes oxidation by an aldehyde dehydrogenase to form the corresponding carboxylic acid, but some reduction may also take place through the action of such enzymes as alcohol dehydrogenase (aldehyde reductase). The latter reaction is quantitatively important in the cerebral metabolism of norepinephrine, but less so for amines that lack the β-hydroxyl group.

If an animal is treated with a MAO inhibitor, the concentrations of many amines increase in the tissues, including the brain. Thus the rate of synthesis of serotonin in the brain is essentially unaffected, but, because degradation of the amine is slowed, a net increase results in the amount of the amine. This phenomenon also occurs with the catecholamines, but the increases are more modest because of two other factors: (1) Catecholamine biosynthesis is regulated by the norepinephrine feedback inhibition of tyrosine hydroxylase (Chap. 10 and [17–19]); and (2) dopamine and norepinephrine are subject to the action of COMT, and the methylated products have less biological potency than do the parent amines. These effects are illustrated by the results in Table 34-3, in which norepinephrine formation from labeled tyrosine is reduced although the tissue concentration of the amine is increased. This is not true when labeled DOPA is used.

Animals treated with a MAO inhibitor become behaviorally alert and excited. This is attributed to the increase in concentration of monoamines, but which one of them is responsible is unknown. Indeed, it is not yet known if only one specific amine mediates the response or if several act in concert. Attempts to delimit the chemistry of the response have been made by injecting precursors of specific monoamines, e.g., DOPA, tryptophan, and 5-hydroxytryptophan, but as yet there are no definitive answers. The behavior elicited by a large dose of a MAO inhibitor is not observed if the animal has been treated previously with reserpine. On the other hand, if a MAO inhibitor is administered before reserpine, the action of the latter drug is potentiated. In this case, the released amines can exert their pharmacological activity at synapses for a protracted period, unmitigated by reuptake into the vesicles of the nerve ending or catabolism by MAO.

Some investigators have cautiously tested these and other amino acids in patients on MAO-inhibitor therapy, but again the results have not been clear-cut. The profound activation of autonomic systems caused by the combination of a MAO inhibitor with a monoamine precursor has precluded simultaneous use of the two types of agent in the treatment of depression. In fact, cardiovascular and other side effects of MAO inhibitors that are used therapeutically represent autonomic reactions mediated by endogenous amines. An important precaution in the clinical use of MAO inhibitors is the avoidance of foods prepared with cheese, especially of the aged varieties, and of certain types of wine, since these contain much tyramine formed through the action of the mold or yeast on tyrosine. Ordinarily, tyramine would be metabolized to p-hydroxyphenylacetic acid before it could pass into the general circulation, but, protected by the MAO inhibitor, the tyramine reaches sympathetic postganglionic nerve endings, where it displaces stored norepinephrine stoichiometrically. The excessive release of norepinephrine to receptor sites initiates the observed, and sometimes lethal, autonomic changes.

Paradoxically, MAO inhibitors cause a lowering of blood pressure. Several mechanisms for this have been suggested, based on events that may take place centrally or peripherally. One hypothesis suggests that the inhibitor allows the accumulation of abnormal amounts of amine metabolites that are normally not detectable; an example is p-hydroxyphenylethanolamine, formed by β-oxidation of tyramine. At sympathetic nerve endings, this derivative would replace norepinephrine and then act as a substitute, or "false," transmitter, with little musculotropic action. Pargyline (see Fig. 34-3) is a MAO inhibitor that is indicated particularly in the treatment of hypertension.

Although the antidepressant action of MAO inhibitors is thought to reside in their ability to conserve monoamines and to increase the concentrations of these compounds in the brain, the drugs possess inhibitory actions on other enzymes as well. These inhibitions have not received the same thorough attention as has been given to MAO, but nevertheless are important from the toxicological point of view. For example, the metabolism of ethanol is inhibited by MAO inhibitors, so the effects of alcoholic beverages are potentiated. The MAO inhibitors act also upon the microsomal liver enzymes that catalyze the metabolic transformation, or detoxification, of other drugs. Thus additional precautionary measures must be observed when a MAO inhibitor is used therapeutically. For example, this treatment must be stopped some days before a general anesthetic is to be administered.

The long-lasting action that most MAO inhibitors exert suggests that they form a covalent link with the enzyme, as already demonstrated for pargyline. If so, the activity of the enzyme would be restored only when new enzyme has been synthesized. This has been tested experimentally in rats that have been given iproniazid to inhibit MAO and measured at suitable time intervals to determine the ability of the animals to convert exogenous dopamine to urinary dihydroxyphenylacetic acid. This ability is restored to control levels in about four days. Similar recovery times are observed for rats whose ability to catabolize pentylamine has been measured in vivo; this pathway is initiated by the action of MAO, which seems to be the reaction that limits the overall rate [16]. It is interesting that approximately the same time is needed for recovery of rats from the MAO-depressing effects of riboflavin or iron deficiency [16]. Riboflavin is needed for the synthesis of the coenzyme of MAO. Iron is required at some stage of the biosynthesis of MAO in liver (rat) and platelets (man).

There is evidence that iproniazid must undergo an oxidative reaction before it can act as an inhibitor; this may be a dealkylation, with loss of the N-isopropyl group. Tranylcypromine (see Fig. 34-3), a cyclopropylamine derivative, is thought to resemble the transient intermediate species formed by the substrate at the active site of the enzyme and, hence, to have a better fit to the enzyme.

Clorgyline (Fig. 34-3) inhibits MAO in an unusual way; the inhibition curve has two sigmoid portions at different concentration ranges. This has led to the interpretation that the enzyme consists of two major species: type A, which acts on norepinephrine and serotonin, as preferred substrates, and exhibits high sensitivity to clorgyline (as well as to harmaline and the compound known as Lilly 51641); and type B, which acts on benzylamine and phenylethylamine preferentially and has a lower sensitivity to clorgyline but is characterized by sensitivity to other propargylamine inhibitors. The question of the number of MAO species is complicated by the fact that upon electrophoresis as many as four or five bands, each reputedly with MAO activity, have been separated. Artifacts that influence the results are being detected. These include fragments of external mitochondrial membrane that remain attached to some MAO preparations; varying amounts of lipid that affect the quality of the MAO reaction and that can be removed with chaotropic agents; and the use of the nonspecific tetrazolium reduction reaction (in the presence of a monoamine) to detect "MAO activity" on electrophoretograms. Accordingly the reported multiplicity of MAO is undergoing reevaluation at the present time.

Harmine (see Fig. 34-3) and harmaline (dihydroharmine) are unique among MAO inhibitors in that they occur naturally. Their inhibitory action is short lasting and reversible. Harmine, known at one time as telepathine, is considered to have some hallucinogenic activity. The mode of action of these β-carbolines is still being explored. Along with their O-demethylated derivatives, harmol and harmalol, and certain related compounds, they antagonize peripheral actions of serotonin, although only harmine and harmaline in this group inhibit MAO. Many analogs have been synthesized in the search for useful therapeutic agents.

The metabolism of hydrazide inhibitors of MAO involves the loss of one or another of the N-substituents. For example, iproniazid yields isonicotinylhydrazide, isopropylhydrazide, isonicotinic acid, and unchanged drug in the urine. Radioactive tracer studies indicate that the isopropyl label remains in tissues considerably longer than does the isonicotinyl label. Harmaline and harmine undergo metabolism by several pathways. One of them involves O-demethylation, with subsequent glucuronidation and sulfation of this phenolic type of derivative.

Experimental and clinical uses have been found for two hydrazine derivatives that inhibit DOPA decarboxylase strongly, both in vivo and in vitro, without affecting MAO and without providing antidepressant action. They protect easily degraded substrates, such as DOPA and 5-hydroxytryptophan, from the action of decarboxylase in the peripheral tissues. Thus, when the amino acid is administered, a larger portion of it is conserved for distribution to the brain, where the amine product is needed. One of these drugs, N-seryl-N'-(2,3,4-trihydroxybenzyl)hydrazine, benserazide (Ro4-4602), inhibits cerebral decarboxylase only if it is injected in amounts much higher than those required for action on the enzyme in the liver or kidneys; hence, a differential effect can be obtained by adjustment of the dosage. The other compound is the hydrazino analog of α-methylDOPA, carbidopa; it does not pass the blood-brain barrier at all (Chap. 32).

Tricyclic Antidepressants

The tricyclic antidepressant compounds have structures based upon those of the phenothiazines (Fig. 34-4). In imipramine, the sulfur atom of promazine is replaced by an ethylene bridge, yielding a dibenzazepine or iminodibenzyl; in amitriptyline, a carbon atom is also substituted for the nitrogen in the middle ring, giving a dibenzocycloheptadiene. The use of molecular models reveals that, despite the formal resemblance of imipramine and promazine, the two benzene rings of the former compound are twisted against one another in an unsymmetrical fashion.

Some of these drugs were originally tested unsuccessfully in schizophrenic patients but ultimately were found to have important applications in the treatment of mental depression. The tricyclic antidepressants do not inhibit MAO, but their therapeutic action has been related in another way to monoamine metabolism: They appear to interfere with the process of reuptake of norepinephrine and serotonin from the synaptic cleft back into the nerve ending and its vesicles (see Chap. 10). This inhibition of the membrane "pump" provides a net increase in the concentration of free

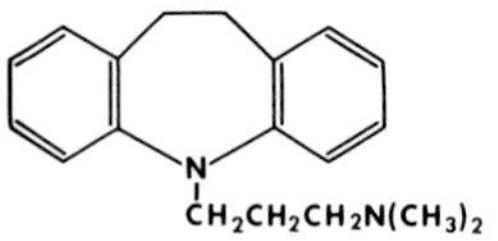

Imipramine

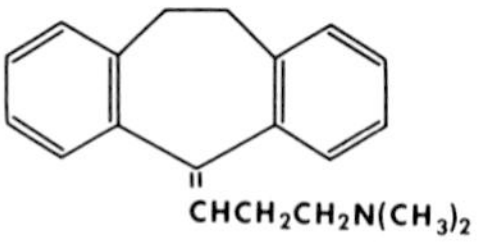

Amitriptyline

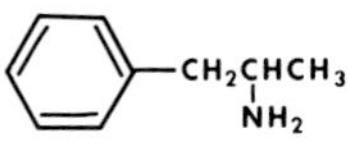

Amphetamine

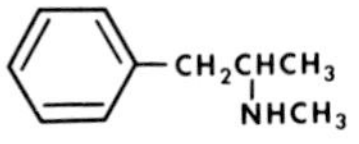

N-Methylamphetamine

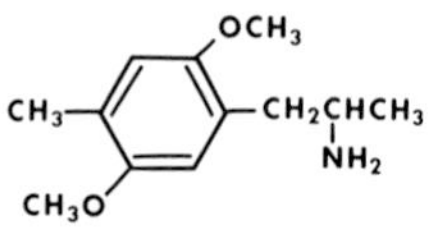

2,5-Dimethoxy-4-methylamphetamine (DOM)

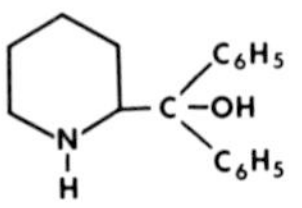

Pipradrol

Figure 34-4
Some antidepressant drugs.

amine (i.e., amine not bound in vesicles or to a macromolecule) in the synaptic cleft at any one time, with increased probabilities of stimulation of sensitive receptors on the postsynaptic neurons. There are some important differences between the tricyclic compounds in this respect; for example, imipramine, chlorimipramine, and amitriptyline strongly inhibit reuptake of serotonin, whereas desipramine, nortriptyline, and protriptyline are more effective against uptake of norepinephrine [20]. The acute administration of imipramine to experimental animals causes a decreased turnover of serotonin in neurons; this could stem from the reduced rate of reuptake of the amine and its longer residence in the synaptic space. When the drug is given to animals chronically in order to approximate the therapeutic situation, the norepinephrine content of the brain decreases and the rate of metabolism of intracisternally injected (labeled) norepinephrine increases. The reduced rate of uptake of the amine under the influence of imipramine would render it more accessible to metabolic pathways. According to several hypotheses, the inhibition of monoamine reuptake plays an important role in the antidepressant action of imipramine and its congeners.

The reuptake of dopamine is strongly affected by certain anticholinergic and antihistaminic agents, but little by the tricyclic antidepressants. It is interesting that some members of the two classes of drugs inhibiting dopamine uptake are used in the treatment of Parkinson's disease.

Imipramine is metabolized somewhat analogously to the phenothiazines: It undergoes dealkylation of the tertiary amine side chain, yielding as the first product desipramine, which also has antidepressant properties. Microsomal enzymes of the liver catalyze other oxidative reactions, as well.

Amphetamine

Amphetamine (Fig. 34-4; [21]) and some chemically related compounds have been used as therapeutic agents for many years. After the Second World War, their illicit use as psychological stimulants and, more recently, as hallucinogenic agents expanded greatly. Amphetamine abuse can lead to serious disturbances of mental behavior. The resulting "amphetamine psychosis" may be virtually indistinguishable from paranoid schizophrenia and is often diagnosed correctly only when the patient is discovered to be taking amphetamine. Some amphetamine-dependent individuals exhibit peculiar chewing movements which are undoubtedly related to the stereotypy elicited in laboratory animals by large doses of amphetamine or by apomorphine and which may be mediated by dopamine-containing fibers of the CNS.

Amphetamine was recognized early as an inhibitor of MAO. It is not especially potent in this respect in comparison with the newer synthetic derivatives discussed above, but experiments with labeled amines show that it nevertheless has some inhibitory action on MAO in vivo. It is not itself a substrate since MAO acts only on ω-amines.

The euphoriant action of amphetamines is attributable not only to its ability to inhibit MAO but also to two other actions. First, it inhibits reuptake of norepinephrine into sympathetic nerve fibers and thereby presumably causes retention of a greater supply of the amine in the synapse for stimulation of the postsynaptic fiber. Second, amphetamine can enter the nerve ending, penetrate into the amine storage vesicles, and replace the norepinephrine contained in them. This resembles the action of tyramine. The release of the amine apparently is directed outward, i.e., into the synapse. The vector of the releasing mechanism has been shown experimentally: After labeled norepinephrine is injected into an animal and time is allowed for the amine to be taken up into sympathetic nerve endings, administration of reserpine leads to a preponderance of acidic derivatives (see p. 706). On the other hand, injection of tyramine (or amphetamine) produces a preponderance of labeled amines, one fraction consisting of normetanephrine. This indicates that MAO has limited opportunity to act, but that COMT serves in the postrelease disposal of the amine; COMT occurs extraneuronally.

Despite their actions as CNS stimulants, both amphetamine and pipradrol have been used in the treatment of hyperkinetic children. Paradoxically, the drugs exert a quieting effect in this condition.

An interesting derivative of amphetamine is its p-chloro analog. This compound inhibits tryptophan hydroxylase, as does p-chlorophenylalanine. Either drug reduces the rate of formation of serotonin in the brain, with a resulting decrease in its serotonin concentration.

The excretion of amphetamine and related compounds is sharply dependent on the urinary pH. For example, a man taking about 10 mg of N-methylamphetamine (methamphetamine) orally, together with ammonium chloride, excretes 55 to 70 percent of the dose unchanged in the urine in 16 hours and an additional 6 to 7 percent as amphetamine. If he takes sodium bicarbonate, which renders the urine alkaline, only 0.6 to 2.0 percent of the total dose appears in the urine in the amine fraction because now tubular reabsorption is favored.

β-Phenylisopropylamines are metabolized along several pathways: *p*-hydroxyla-tion, with or without *O*-glucuronidation; β-hydroxylation of the side chain; *N*-dealkyla-tion if there is a substituent on the amino group; and oxidative deamination with formation of a ketone, the ketone being either reduced to the corresponding alcohol or further oxidized to an acid. Considerable species differences exist with regard to these metabolic pathways. In the dog, about one-third of a dose of *d*-amphetamine is excreted unchanged in the urine; another large portion appears as *p*-hydroxy-amphetamine, some of which is conjugated. In man and rabbit, deamination is the predominant route. Some of the phenylacetone that is formed is converted to phenyl-isopropanol; another portion is oxidized to benzoic acid, most of which appears in the urine as hippuric acid. In the rat and cat, *d*-amphetamine is hydroxylated on both the ring and the side chain, forming *p*-hydroxynorephedrine. A small amount of this metabolite is formed in man via *p*-hydroxyamphetamine.

The oxidative metabolism of amphetamine is initiated by microsomal enzymes of the liver. These increase in amount under the influence of certain drugs, so that the rate of catabolism of amphetamine is directly affected. Another pertinent type of drug interaction is simply a matter of competition between amphetamine and another drug that also is acted upon by the same hepatic enzymes [22]. The first type of interaction leads to a reduction of amphetamine content of the tissues, compared with controls dosed similarly with the compound; the second leads to increased levels.

Lithium Salts

The use of lithium salts offers considerable promise in the treatment of recurrent endogenous affective disorders. The lithium ion is therapeutically useful in the con-trol of both mania and hypomania, and it provides prophylaxis not only against the manic but also the depressive episodes of phasic illness [23].

Lithium is a monovalent cation of the alkali metal group and therefore is related to sodium and potassium; it resembles magnesium and calcium in certain aspects of its metabolism. Its determination by flame photometry in the clinical laboratory facilitates therapeutic control. By this means, dosage can be adjusted to maintain a serum level of 0.6 to 1.2 meq per liter, which can be attained with a dosage of 25 to 50 meq of lithium per day. Serum levels of 1.6 meq per liter or greater must be avoided.

Lithium salts are taken orally. There is good absorption from the gastrointestinal tract, with Li^+ passing from the blood into the tissues. Equilibration takes place quickly in liver, kidneys, and skin and less rapidly in muscle and brain. The lithium level in serum corresponds to the tissue content. The concentration of Li^+ in the cerebrospinal fluid reaches only about one-half the level in serum. Almost all the ingested Li^+ is excreted in urine.

Lithium salts appear to have some influence over the movements of norepinephrine at nerve endings and in synapses. There is evidence for a reduced availability of the amine at receptor sites as a result of Li^+ action. This may stem from increased central release of norepinephrine at a rate that outstrips biosynthesis. However, the dynamics of monoamine metabolism during lithium treatment are still being worked out, especially in chronic administration. In acute experiments, dopamine and serotonin are not affected.

Li$^+$ alters somewhat the distribution of Na$^+$, K$^+$, and water, but, with good therapeutic control of dosage, these changes are not serious. Success with lithium salts has led to research on the actions of rubidium and cesium in manic-depressive psychosis, but as yet there are no therapeutic indications for these ions.

PSYCHODYSLEPTICS, OR HALLUCINOGENIC AGENTS

The psychodysleptics represent a motley group of plant and laboratory products (Fig. 34-5). The remarkably potent *d*-LSD-25 is a synthetic derivative of a plant alkaloid, lysergic acid. The hallucinogens can be classified roughly into three major categories: (1) indole derivatives; (2) phenylalkylamine derivatives; and (3) non-nitrogenous compounds. The first two groups have lent themselves readily to biochemical theories of hallucinogenic actions based on the concept of antagonism to cerebral serotonin and catecholamines, respectively. This antagonism is thought to interfere in some way with the as-yet-unknown role of the monoamines in mental processes. Molecular orbital calculations also have been used to assess hallucinogenic potency in relation to chemical reactivity. However, our understanding of the action of these compounds at the cellular or molecular level is very limited at the present time.

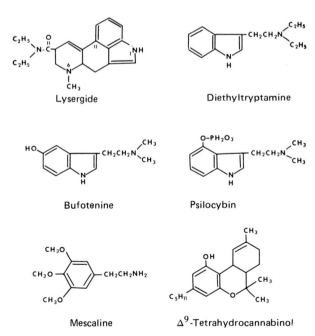

Figure 34-5
Some hallucinogens.

Indolic Hallucinogens

The simplest members of this category are the dialkyltryptamines, e.g., diethyltryptamine (see Fig. 34-5). Others are psilocybin, the O-phosphoryl ester of N-dimethyl-4-hydroxytryptamine, and d-LSD-25. Bufotenine (N-dimethyl-5-hydroxytryptamine) also is reputed to be hallucinogenic; it is found both in some plants and in the poison gland of the toad. d-LSD-25, now known as lysergide, is one of four diastereoisomers of lysergic acid diethylamide, all of which have been tested for hallucinogenic activity. Only lysergide is active to any extent, and it is exceedingly potent. While exerting its psychodysleptic action, it is estimated to be present in brain to the extent of only a few picograms per gram of tissue.

In 1953 and 1954, several investigators, including D. W. Woolley, drew attention to the possible interaction of indole alkaloids with serotonin receptors. According to this hypothesis, drugs such as lysergide would act in the brain by opposing the normal action of serotonin and thus cause a functional deficiency of serotonin at important CNS centers. This application of the antimetabolite hypothesis to psychodysleptics was soon recognized to be inadequate. For example, if lysergide is brominated to 2-bromolysergide, hallucinogenic activity is lost, but antagonism toward musculotropic actions of serotonin is retained. Thus an antiserotonin action in one context does not necessarily ensure that it will be manifested in another.

In other experiments, it has been shown that lysergide affects the metabolism of serotonin in the brain in a complex fashion, although its main known effect is to increase brain serotonin slightly and to decrease 5-hydroxyindoleacetic acid correspondingly. Studies of the distribution of serotonin in the subcellular fractions suggest that these changes probably take place within the neuron, i.e., without release of serotonin into the synaptic space, and that lysergide may cause decreased impulse activity of serotonin-containing neurons. In fact, the drug applied in very low doses iontophoretically has a direct inhibitory effect on raphe cells (rats) and specifically antagonizes the excitatory action of serotonin on certain cells of the brain stem (cats). The midline raphe nuclei are structures that contain serotonin or a serotoninlike substance, according to histofluorescence examination (see Chap. 10).

Lysergide is absorbed rapidly from the gastrointestinal tract, but there are large differences in its metabolism among species. The biological half-life of the compound in the blood of the mouse and rat is 7 minutes, but in the monkey and cat it is 1.5 to 2 hours.

Lysergide is bound to plasma proteins when transported in the blood, but it is readily released into the brain and other organs. Its highest concentration is attained in these organs soon after injection, and it declines over the next few hours. The concentration of lysergide in the brain is always less than that in the blood. Highest amounts are found in the liver.

Microsomal enzymes in the liver (but not in the brain) oxidize lysergide to its biologically inactive 2-oxy derivative. Other products, more polar than lysergide itself, are formed. The bile serves as one route of excretion of the hallucinogen, but some of it is probably then reabsorbed from the intestine.

Some investigators claim that bufotenine is a constituent of the urine of schizophrenic patients, normal subjects, or both. However, these claims have never received adequate verification.

Psilocybin, the active principle of certain mushrooms found in the southern highlands of Mexico, is unusual in being an ester of phosphoric acid. Its nonhallucinogenic hydrolytic product is psilocin, the 4-hydroxy analog of bufotenine.

An interesting and novel pathway of metabolism of some indolic compounds is by way of 6-hydroxylation. This route accounts for two-fifths of a dose of N,N'-diethyl-tryptamine but for only about one-fifth of the dose of the dimethyl and dipropyl homologs.

Mescaline

Mescaline, like lysergide, has been investigated for possible inhibitory action in many enzyme and respiratory systems in the as-yet-unsuccessful search for specific effects that might be related to its psychodysleptic action. Its structure (Fig. 34-5) suggests that it may act as an analog-inhibitor of some catecholamine system in the brain. One suggestion has been that, in schizophrenia, dopamine is doubly methylated to 3,4-dimethoxyphenylethylamine. The resemblance of this hypothetical metabolic product of a catecholamine to mescaline, which is 3,4,5-trimethoxyphenylethylamine, is clear enough. The synthetic compound has no hallucinogenic action, however, and on the few occasions when it has been identified in body fluids there has been no relation to schizophrenia.

The dose of mescaline required to elicit hallucinations is much larger than that needed for many of the other psychodysleptics. It is quite possible that some quantitatively minor metabolite of the drug is responsible for its effects.

Oral administration of labeled mescaline to humans shows that the compound is readily absorbed. It is also rapidly metabolized, for its half-life (as measured by urinary excretion of labeled products) is about six hours. Most of the tracer is excreted by the end of the second day. Among the metabolites detected are the carboxylic acid corresponding to mescaline, N-acetylmescaline, 3,4-dimethoxy-5-hydroxyphenyl-ethylamine, and the N-acetyl derivative of the last compound. Other metabolites remain to be characterized. In the rat, demethylation of the 4-methoxy group may also occur. The deamination of mescaline is sensitive to MAO inhibitors.

2,5-Dimethoxy-4-methylamphetamine (Fig. 34-4) resembles both mescaline and amphetamine in structure. It is a hallucinogen.

Tetrahydrocannabinol

As must be evident from the discussion of the nitrogen-containing psychodysleptics, research on the mechanisms of action of hallucinogens is difficult enough when chemically pure agents are available, but it is much more difficult in a galenic type of agent such as marijuana. Only recently has the chemistry of *Cannabis* been clarified, with the identification of its most important psychodysleptic constituent as $(-)$-*trans*-Δ^9-tetrahydrocannabinol (THC) [24]. It is effective in man in an oral dose of 50 to 200 μg per kg body weight and, by inhalation (smoking), in a dose of 25 to 50 μg per kg. This compound (Fig. 34-5) is now used as a standard for biological testing of other cannabinoids, i.e., C_{21} compounds characteristically found in the *Cannabis* plant and in preparations made from them, as well as related synthetic compounds and metabolites. Marijuana contains about one percent of Δ^9-THC; hashish, the resinous product of *Cannabis*, is much richer.

THC is rapidly accumulated in the liver and then metabolized to more polar products. Its half-life is about 30 minutes in humans and rats, and even less in rabbits. If taken in by smoking, it is temporarily detectable in the lungs. It is carried in the plasma in association with lipoproteins.

Hepatic microsomes act quickly on THC to convert it to the 11-hydroxy derivative; this substance is even more potent than the parent compound in regard to characteristic pharmacological actions. Further transformation leads to 8,11-dihydroxy-THC, 11-hydroxycannabinol (in which the cyclohexene ring is aromatized), and other products; side-chain oxidation also occurs. The route of excretion varies from species to species. In man, fecal excretion of radioactive Δ^9-THC exceeds urinary output after the first day, during which most of the radioactive compound eliminated is in the urine. In the rabbit, urinary excretion predominates, but in the rat the major path of excretion is by way of the bile.

BIOCHEMICAL THEORIES OF MENTAL DISORDERS

One of the aims of neurochemical research is to supply information about the substratum of mental activities. This involves integrating our knowledge of the chemical and electrical activities of the brain in a manner that will illuminate the supreme functions of that organ: thought, memory, emotional expression, and the coordination of autonomic and voluntary functions of the body. By the same token, neurochemistry is very much concerned with the question of disturbances of brain function, as in emotional disorders, various forms of mental deficiency, and the major psychoses [25–27]. One may speculate how two such disparate branches of science — the one concerned with the flow of molecules through metabolic paths, the other with the flow of ideas through neuronal circuits — may be brought together. Ultimately, one relies upon the unity of natural phenomena, including those of the brain, and in practice one adopts a pragmatic attitude toward particular problems. This pragmatism has been reflected in a large number of biochemical theories, some of them hardly more than hunches or working hypotheses, of the organic etiology of mental diseases. These theories focus on those very important, biologically active, minor constituents of the body: hormones, vitamins, and neurotransmitters. The clinical effects of the first two categories are well known and are also seen in relation to mental disorders. Replacement therapy readily overcomes psychotic manifestations in myxedema and restores the patient to his mental health. Daily provision of a few milligrams of nicotinic acid to a pellagrin clears up the dementia, along with the dermatological and gastrointestinal symptoms. Successes of these kinds in the endocrine and nutritional fields are dramatic but have not been attained in the treatment of the major psychoses.

The possibility that psychoses arise from some imbalance in neuronal pathways, i.e., an imbalance of transmitter substances, has received the greatest attention in recent years. In fact, the origins of current transmitter hypotheses of psychoses are intimately tied up with the discovery in the 1950s of the specific biochemical actions of reserpine, chlorpromazine, and iproniazid and by the elucidation of the neurochemistry of the cerebral monoamines that followed. Most recently, the working out of monoaminergic pathways that innervate the higher centers of the brain, and of others that terminate in the hypothalamic-hypophyseal systems, has provided the spur that can be expected to yield even more specific information about the roles of biogenic amines in the brain and how they are influenced by psychopharmacological agents [2, 28].

Affective Psychoses

Role of Monoamines

Various monoamine-based hypotheses have been put forward to explain mental depression. Some studies devoted to the measurement of 5HIAA and HVA in CSF have indicated that these key metabolites of serotonin and dopamine, respectively, are present in abnormally low concentrations in depressed patients, but a significant number of others report no difference at all. In comparisons of the depressive with the manic phase of manic-depressive psychosis, the concentrations of 5HIAA in lumbar CSF are significantly different; the higher mean level in the agitated patients is not caused entirely by the increased motor activity. In one study, patients receiving probenecid showed increased concentrations of 5HIAA in CSF as expected, but not as great as in normals. Thus studies of the CSF provide some support for the view that serotonin may be produced in the CNS at a subnormal rate in mental depression. Whether this is intrinsic to the depression has to be clarified. Several investigators have measured monoamines in the brains of patients who committed suicide and were therefore presumed to be depressed; generally, there is a reduced concentration of serotonin in the lower brain stem in such cases, especially in the raphe nuclei dorsalis and centralis inferior. The serotonin levels are apparently normal elsewhere in these brains [29].

Small amounts of catecholamines are excreted in the urine along with larger amounts of their metabolites, and these have been measured in depressive states. Despite the multifarious origin of the urinary compounds, it seems that the amounts of epinephrine, norepinephrine, and their metabolites increase in acute psychotic depression, as contrasted with neurotic depression. The concentrations of catecholamines have been measured in plasma, but these, even under the best of conditions, are at the very lowest level of sensitivity of the methods used, so few investigations have been conducted. There are essentially no data on the catecholamine levels in the brains of depressed patients.

According to the hypotheses that implicate a cerebral deficiency of brain biogenic amines in mental depression, there is insufficient neurohumor at certain synapses to assure normal postsynaptic activity [30–32]. One possibility whereby this might occur involves a normal rate of production or release of monoamines at crucial synapses in the brain, but an accelerated metabolism. Indeed, there is some evidence for increased levels of MAO with age, and this would correspond to the greater incidence of mental depression in older people. However, serotonin in the brain stem tends to increase with age, as does 5HIAA in the CSF, which lessens the probability that MAO is a significant factor regulating the concentration of serotonin under physiological conditions. The age factor does not apply to norepinephrine (in brain stem and hypothalamus) or dopamine (caudate nucleus), although HVA in the CSF appears to be higher in older subjects than in younger ones.

If there is central inhibition of certain neurons, with delivery of insufficient amounts of their neurohumors to the synaptic cleft, then drugs and treatments that protect the neurohumor from metabolism could be expected to have an ameliorative action in depression. This should also be true of agents that cause increased release of the neurohumor from the presynaptic fiber into the synaptic cleft. Of course, the use of MAO inhibitors in the treatment of mental depression is consistent with the first point, for these drugs cause increased amounts of monoamines to appear in the brain. In regard to release of monoamines, amphetamine, also an antidepressant drug, fits into this category. Among its several actions, it causes the release of catecholamines from the storage vesicles and it is conceivable that the increased concentration of norepinephrine in the synapse ensures increased activity of postsynaptic fibers sensitive to this monoamine. Of course, there are other possibilities. One of the most important physiological mechanisms for inactivation of the neurohumor is by the reuptake into the storage vesicle in the neuronal endings and its subsequent reutilization. Hence, drugs that prevent the reuptake of the endogenously released neurohumor would be, in effect, increasing the concentration of that neurohumor in the synapse. The tricyclic antidepressants are of this kind.

If a monoamine is deficient, then its amino acid precursor ought to reverse the symptoms of depression. With this in mind, clinical trials of several amino acids have been carried out. Levodopa, the precursor of catecholamines, has not proved useful for this purpose, nor has 5-hydroxytryptophan, the immediate precursor of serotonin. However, it has been claimed that tryptophan, with or without a MAO inhibitor, is as effective as electroshock therapy in the treatment of depression [31, 31a], so there has been some speculation as to the possible metabolites that would be responsible for this effect.

Hormonal and Metabolic Factors

The research in affective illnesses that has sought disruptions in the communications systems of the body was first directed to the endocrine systems. Changed affect resulting from endocrine dyscrasias, as in myxedema, hyperthyroidism, Addison's disease, Cushing's syndrome, and others, and restoration of mood to normal with adequate treatment have demonstrated the profound effect of hormones on the brain. This is demonstrated further by the normal swings of mood accompanying the menstrual cycle.

Many hormonal and neurohumoral factors interact with one another through the important nucleotide cyclic adenylic acid, adenosine 3',5'-monophosphate (cAMP). This compound is increased in concentration when a hormone reaches its specific target site on cell, and in some cases neuronal, membranes (see Chap. 12). For example, a dopamine-sensitive adenylate cyclase in mammalian brain increases in activity under the influence of dopamine [33, 34]. The few studies of cAMP that have thus far been initiated in psychiatry indicate that the urinary excretion of cAMP is lower in severely depressed patients than in normal persons or even moderately depressed patients [34a], and that it is higher in manic cases. The lower levels in the urine during the depressed state increase toward normal as clinical improvement occurs. It is claimed that a sharp increase in the urinary output of cAMP occurs in association with the changeover from the depressive to the manic state in bipolar patients. More information about the physiology of cAMP is needed before results of this kind can be interpreted easily.

Interruptions in the biosynthesis of the thyroid hormone result in clinical hypothyroidism, a condition that usually can be reversed by adequate treatment with a thyroid preparation. In the adult form, myxedema, many nonspecific symptoms, usually accompanied by a low BMR and elevated blood cholesterol, appear. It is thought that ischemic anoxia may ultimately be responsible for cerebral changes that result in mental abnormalities. In those myxedematous patients who develop psychotic symptoms, the personality changes sometimes are detected even before the hypothyroid state has been diagnosed. The psychosis may resemble depression, mania, schizophrenia, or paranoia. Treatment with thyroid hormone usually cures the condition rapidly, but in some cases the psychosis persists, suggesting that irreversible brain damage has occurred.

Two clinical syndromes referrable to abnormal activity of the adrenal cortex are of special interest in the present context. In Addison's disease, which represents hypofunction of the adrenal cortex, the following mental changes have been encountered: asthenia, apathy, anxiety reaction, depression, occasionally psychotic reaction. These may be reversed by adequate replacement therapy with an adrenal-cortical hormone. The second syndrome, Cushing's, represents hyperfunction of the gland and can be mimicked by chronic overdosage with cortisol or ACTH. The mental changes in this syndrome vary considerably from case to case. Depressive states are common, but mania or euphoria is seen occasionally.

Under normal conditions, stressful stimuli to the organism are funneled by way of higher nervous centers through the hypothalamus, where corticotropin releasing factor (CRF) is secreted. This is transported in the hypophyseal-portal system to the anterior pituitary, where it acts. In this way, general and environmental stimuli may be converted into an organismic endocrine response, leading to secretion of cortisol from the adrenal cortex. Much current research is being devoted to extending knowledge of the nervous pathways back from the hypothalamus to higher centers in order to delineate the "stress pathway" in anatomical terms (see Chaps. 22 and 35).

The levels of cortisol in the plasma are higher in depressed patients than in normal persons, with the greatest differences seen in the early morning. The abnormalities, which are highly specific for each individual, tend to disappear with successful treatment of the depression. The higher levels noted in plasma are apparently the result of elevated rates of production of cortisol during depressive illness. Despite these observations, the concentrations of cortisol in the lumbar CSF have not seemed to be abnormal in patients suffering from depression or mania.

In depression, there is often a higher than normal level of cortisol in the plasma, which has given rise to a special biochemical hypothesis of depression. The hypothesis states that brain serotonin is low in this mental state [32]. Tryptophan pyrrolase, a hepatic enzyme that initiates the irreversible catabolism of the essential amino acid tryptophan, is induced experimentally by administration of a glucocorticoid. As the rate of synthesis of serotonin in the brain depends upon a steady supply of trypto-phan, an accelerated rate of breakdown of that amino acid peripherally would make less available for serotonin synthesis. This hypothesis is, then, consistent with the finding of reduced levels of serotonin in some parts of the brains of depressed patients, as mentioned above. It is also consistent with the claims of successful treatment of depression by administration of tryptophan.

Voluminous literature deals with the sugar tolerance of mental patients [25]. Despite wide variability in test conditions used by investigators, there is relatively good agreement that glucose tolerance is abnormal in psychotic patients. Tolerance tends to be low in the depressive phase of manic-depressive psychosis and in endoge-nous depression. In the manic phase of manic-depressive psychosis, glucose tolerance is said to be abnormally high. In general, the curves return to normal as the patient's psychiatric condition improves. The reasons for the abnormalities that have been recorded are not known. Claims for a specific antiinsulin factor in the plasma of psychotic patients have never been confirmed. An important factor that has not been taken into consideration in the testing of glucose tolerance is the influence of diet on the concentration of many of the enzymes involved in the utilization of glu-cose. Some of these enzymes increase in amount under the influence of high-carbo-hydrate diets, so that even repetition of the test may provide sufficient stimulus to improve the tolerance. Insulin tolerance has also been measured in mental patients. Despite the availability of radioimmunoassay for this hormone in plasma, the tests have not yet been applied in psychiatry.

It has been reported that the concentration of lactic acid in the plasma becomes unusually elevated in psychotic subjects, especially after physical exercise. This abnormality may be merely a function of training, another factor that plays an important role in the regulation of carbohydrate metabolism. However, the unusual rise in lactic acid after a standard exercise test has been noted repeatedly in patients exhibiting anxiety neurosis. Conversely, the infusion of lactate into patients with anxiety neurosis regularly produces an attack of anxiety, although control infusions usually do not do so. Some normal controls also experience anxiety during such an infusion. The effects can be minimized by adding calcium ions to the infusion.

An argument, based upon theoretical grounds, is that affective psychosis involves major changes in neuronal activity with corresponding changes in cationic concentra-tions in the body. Balance studies have not been very useful, but more information has come from the use of radioactive sodium, which, on administration, equilibrates rapidly with most of the sodium in the body [31]. Thus isotopic studies are able to account for the movement of sodium from extracellular to intracellular locations, except for a portion of bone sodium that does not equilibrate within the usual time it takes for the test to be carried out. Studies with radioactive sodium, then, indicate that there is a greater retention of sodium in depressed patients; with recovery, the amount of retained sodium decreases significantly. The information available about sodium metabolism in patients suffering from mania is even more limited. There seems to be a retention of sodium in the intracellular compartments, together with that portion of the bone sodium that exchanges readily, as in depressed patients.

It has been suggested that the concentration of potassium in the tissues is some-what low in depression and increases during improvement of the mental state. If this were so, the two monovalent cations would then be inversely affected in depression, with deleterious actions on both membrane and action potentials of the neurons. The interpretation of the studies of these cations is complicated because of the obvious methodological difficulties: variations in the patient material; independent fluctuations in body water; partial binding of sodium in bone; and use of somewhat different methods from laboratory to laboratory.

The use of lithium salts in mental depression has been dealt with on pages 722 and 723.

Schizophrenia

Role of Monoamines
Direct analytical information about the brains of schizophrenic patients is even more deficient than that about affective psychoses. Again, the problem is complicated by the difficulty of deciding what one needs to measure. For many reasons, the determination of monoamines, their metabolites, and the enzymes involved in their biosynthesis and catabolism seems to be the most promising line [35]. Despite the difficulties of obtaining cadaver material and interpreting the results, there is enough encouragement from beginning studies to indicate that this is a necessary and worthwhile venture. The measurement of metabolites of dopamine, norepinephrine, and serotonin in the lumbar CSF has not yet revealed any distinctions between schizophrenic patients and healthy persons. In senile psychosis, however, the concentration of HVA is lower than in controls.

Himwich and his colleagues have claimed that the amounts of monoamines derived from tryptophan fluctuate in relation to the mental status of their schizophrenic patients. They have followed these fluctuations by measuring the urinary excretion of tryptamine, indoleacetic acid, and 5HIAA. With aggravation of psychotic symptoms, the excretion of indoles increased; with amelioration of the mental state, the excretion decreased [26]. This view stems from an earlier hypothesis that attributed to serotonin important regulatory properties in relation to psychosis; on the view that a deficiency of serotonin in the brain is associated with mental dysfunction, trials of 5-hydroxytryptophan have been undertaken. However, this precursor of serotonin does not provide a remission. Clinical experience with tryptophan, administered with a MAO inhibitor, shows that this amino acid yields metabolic products that cause the behavior of schizophrenic patients to deteriorate.

Attempts to investigate schizophrenia biochemically have often focused on the hallucinogenic drugs, which mimic, to some degree, one of the features of schizophrenia [cf. 27, 28]. An early hypothesis about this disease, or set of diseases, argued that normally part of the serum epinephrine is oxidized by ceruloplasmin to form the bicyclic compound adrenochrome and that this substance is then further metabolized to inert products. The argument went on to say that in schizophrenia there is an overproduction of adrenochrome and that it accumulates to levels that evoke hallucinations. It was proposed that this process could be prevented by diverting the methylating potential of the body away from *N*-methylation of norepinephrine into another pathway. Because nicotinic acid is normally metabolized to a methylated product, it was thought that an excess of this vitamin would serve the purpose of limiting formation of adrenochrome. Actually, large amounts of nicotinic acid do not alter the ratio of norepinephrine to epinephrine, and it is highly questionable whether adrenochrome has hallucinogenic activity [36]. Nevertheless, success has been claimed for the nicotinic acid treatment of schizophrenia. Later, this pragmatic success was explained by the assumption that certain people have unusually high requirements for this vitamin and, when the supply does not match those needs, schizophrenia results.

The reputed value of nicotinic acid in schizophrenia has become one of the prime tenets of "orthomolecular psychiatry," which has extended the hypothesis to include the argument that certain schizophrenics may have high requirements for other vitamins or trace minerals [37]; for some genetic reason, the deficiency can be made up only by supplying very large amounts of that nutrient. Carefully controlled trials with nicotinic acid and its coenzyme form, NAD, for which success in treatment has also been claimed, have failed to reveal any therapeutic benefit, and a detailed assessment of clinical results obtained with "megavitamin treatment" of schizophrenia indicates the lack of evidence for its efficacy [38].

A dopamine hypothesis in this field suggests that schizophrenic subjects are endowed with a mechanism that favors oxidation of dopamine on the ring over oxidation of the side chain. This would result in the formation of 6-hydroxydopamine, rather than norepinephrine. The 6-hydroxydopamine would then damage norepinephrine-containing neurons in those areas of the brain that are essential for normal mentation. The defect would appear as a deficiency of dopamine beta-hydroxylase in those regions and, in fact, Stein and Wise have claimed that a decrease of this kind can be detected in brain of schizophrenic patients at postmortem [39]. This claim has been contested, and the issue has not yet been resolved.

At one time, one of the primary candidates for a hypothetical methylated psychotogen was material separated partially on paper chromatograms and staining pink with one of the reagents used. A report that this substance occurred in urine of schizophrenics, but not at all or only seldom in that of normal persons, led to extensive analytical searches in urine for candidate amines and their metabolites. For a while, attention focused on DIMPEA (dimethoxyphenethylamine, dimethyldopamine); its congener mescaline has one additional methoxy group on the ring. However, DIMPEA appears to be exogenous in origin, probably coming from some plant constituent of the diet, and bears no relationship to schizophrenia. Indeed, it does not cause hallucinations. Thus, at the present time, the methylation hypothesis of schizophrenia, as applied to the catecholamine field, has generated much research without revealing a specific factor.

The discovery of small amounts of dopamine in the cerebral cortex and other parts of the brain besides the extrapyramidal system generates the possibility that there is some malfunction in this fraction of the brain dopamine. For example, there are convincing arguments that the antipsychotic action of neuroleptics is actually directed at blocking dopamine sites [40].

Many hallucinogenic substances contain the indole nucleus, derived through their biological synthesis in the plant cell from tryptophan or tryptamine. Very often the nitrogen of the side chain is alkylated, as in lysergide, psilocybin, and bufotenine. These compounds are of additional current interest because the synthetic hallucinogen *N,N*-dimethyltryptamine is said to be present in the urine and blood of schizophrenic patients. Serum and brain enzymes capable of catalyzing the *N*-methylation of tryptamine and serotonin have been described for mammalian tissues. Furthermore, clinical studies of the influence of certain amino acids in mental patients suggest that the combination of tryptophan (source of indolic compounds) and methionine (source of methyl groups) in patients previously treated with a MAO inhibitor leads to activation of the schizophrenic psychosis. It is conceivable that an indolethylamine, protected against deamination by the MAO inhibitor, is methylated to form a substance responsible for the effect. One way in which relatively large amounts of a cerebrotoxic indole might be formed is through a more effective movement of tryptophan across the blood-brain barrier, and a current hypothesis argues along these lines. It states that an alpha-2-globulin in blood plasma aids the transport of tryptophan into cells. According to the hypothesis, the globulin is present in schizophrenics in a molecular form such that the transport of tryptophan into hypothalamic neurons, among others, is increased. The result, according to this view, is that the formation of serotonin and of dimethyltryptamine is increased, the latter compound being considered the psychotogen. According to this hypothesis, the normal brain contains an enzyme that maintains the alpha-2-globulin in a less active form [40a]. There is no experimental evidence for these claims.

Hormonal and Metabolic Factors
It has already been mentioned that glucose tolerance tends to be abnormal in psychotic patients. It is low in catatonia and other forms of schizophrenia, but returns toward normal as the clinical condition improves.

Many studies have sought an abnormality in the function of the adrenal cortex in schizophrenia. Despite reports of increased adrenal activity in schizophrenic patients as compared with normals, the concentration of cortisol in the plasma of schizophrenics is within the normal range. In schizophrenics tested under a standard experimental stress, one can sometimes detect a deficiency in the adrenal-cortical response. This has been attributed either to a relative adrenal insufficiency in schizophrenics or to some defect in the connections between the higher centers and the anterior pituitary gland. However, the response level of the neuroendocrine fibers that regulate the release of the corticotropin, or their presynaptic connections, seems to be normal: Both normals and schizophrenics show the same increases in the circulating level of cortisol after they have received test doses of insulin, pyrogen, or corticotropin.

The search for hormonal deficiencies in schizophrenia has often centered on the thyroid gland, if only because of the early simplistic notion that schizophrenia is caused by defective metabolic oxidation under the control of the thyroid hormone. Overall measures of such thyroid function as plasma cholesterol, basal metabolic rate, and serum PBI concentrations are all within normal limits (when physical disease has been excluded).

About 2 to 3 percent of schizophrenic patients show the clinical picture of periodic catatonia, with its succession of phases of catatonic excitement or stupor, regularly interspersed with periods of remission, during which the patient is withdrawn and apathetic, but free from catatonia. These patients exhibit metabolic periodicity, sometimes in phase, sometimes out of phase, with the fluctuations of psychiatric state [41]. Biochemical changes occur in the metabolism of proteins, in electrolytes, and in endocrine activity. For example, in the reaction phase water and sodium chloride are retained, with a tendency to acidosis. In this phase, too, the predominant autonomic activity is adrenergic, as detected by mydriasis, salivation, pallor, and other signs, whereas in the remission phase, cholinergic activity predominates. Complete remission of periodic catatonia is brought about by the administration of thyroid hormone. However, a thyroid deficiency cannot be considered causative because the activity of this organ itself fluctuates phasically. Neuroleptics abolish the psychotic phases but do not seem to produce complete recovery. The basis for the periodicity probably resides in some higher nervous center, e.g., the hypothalamus. Recent developments in the study of hypothalamic releasing factors begin to make feasible the assessment of neuroendocrine activities in this portion of the brain.

Autoimmune Hypothesis

Heath and Krupp [42] have postulated that schizophrenia is an immunological disorder, on the basis of their detecting an IgG fraction in sera of schizophrenic patients that is said to be capable of altering brain function. This IgG antibody or antibodylike fraction, at one time called "taraxein," is said to combine with antigenic sites of neuronal nuclei in the septum. Several questions arise in connection with an autoimmune hypothesis. The first question is how a large protein molecule can penetrate to specific sites within the brain, indeed within the neurons themselves. It has been suggested that there is an altered permeability of the blood-brain barrier in schizophrenia, but experimental evidence is clearly against this. Again, taraxein has never been adequately characterized in regard to composition or even in method of preparation. In attempting to reproduce the observations with this fraction, some investigators have not been able to obtain material from schizophrenic serum that causes behavioral changes in either humans or monkeys. Others have not been able to find brain antibodies in schizophrenic patients [43].

If a circulating antibody in the plasma in schizophrenics sustains the psychosis, then blood transfusions should have an ameliorative effect. These have actually been tried, but without benefit to the patient.

ACKNOWLEDGMENT

Research in the author's laboratory in this field is supported by grants from the Medical Research Council of Canada.

REFERENCES

1. Jacobson, E. The comparative pharmacology of some psychotropic drugs. *Bull. W.H.O.* 21:411, 1959.
2. World Health Organization. *Research in Psychopharmacology: A Report of a W.H.O. Scientific Group.* W.H.O. Tech. Rep. Ser. No. 371, 1967.
3. Efron, D. (Ed.). *Psychopharmacology, a Review of Progress, 1957–1967.* Washington, D. C.: Public Health Service Publication No. 1836, 1968.
4. Von Brücke, F. T., and Hornykiewicz, O. *The Pharmacology of Psychotherapeutic Drugs.* New York: Springer-Verlag, 1969.
5. Ban, T. A. *Psychopharmacology.* Baltimore: Williams & Wilkins, 1969.
6. LeDain, G. (Chairman). *Interim Report on the Commission of Inquiry into the Non-Medical Use of Drugs.* Ottawa: Queen's Printer for Canada, 1970.
7. Usdin, E., and Efron, D. H. *Psychotropic Drugs and Related Compounds,* 2d ed. Washington, D. C.: Department of Health, Education and Welfare, Publication No. (HSM) 72-9074, 1974.
8. Nybäck, H., Sedvall, G., and Kopin, I. J. Accelerated synthesis of dopamine-C^{14} from tyrosine-C^{14} in rat brain after chlorpromazine. *Life Sci.* 6:2307, 1967.
9. Seeman, P. The membrane actions of anesthetics and tranquilizers. *Pharmacol. Rev.* 24:583, 1972.
10. Landmark, K., and Øye, I. The action of thioridazine and promazine on biological membranes: Relationship between ATPase inhibition and membrane stabilization. *Acta Pharmacol. Toxicol.* 29:1, 1971.
11. Forrest, I. S., Carr, C. J., and Usdin, E. (Eds.). *Advances in Biochemical Psychopharmacology.* Vol. 9. New York: Raven, 1974.
12. Wise, C. D., Berger, B. D., and Stein, L. Benzodiazepines: Anxiety-reducing activity by reduction of serotonin turnover in the brain. *Science* 177:180, 1972.
13. Costa, E., Gessa, G. L., and Sandler, M. Serotonin – New Vistas; Biochemistry and Behavioural and Clinical Studies. *Advances in Biochemical Psychopharmacology.* Vol. 11. New York: Raven, 1974.
14. Wolstenholme, G., and Fitzsimmons, D. W. (Eds.). *Monoamine Oxidase and Its Inhibition.* Amsterdam: Associated Scientific Publishers, in press.
15. Blaschko, H. The natural history of amine oxidases. *Rev. Physiol. Biochem. Pharmacol.* 70:83, 1974.
16. Sourkes, T. L., and Missala, K. Nutritional Requirements for Amine Metabolism *in Vivo.* In Wolstenholme, G., and Fitzsimmons, D. W. (Eds.), *Monoamine Oxidase and Its Inhibition.* Amsterdam: Associated Scientific Publishers, in press.
17. Spector, S. Regulation of Norepinephrine Synthesis. In Efron, D. (Ed.), *Psychopharmacology, A Review of Progress, 1957–1967.* Washington, D. C.: Public Health Service Publication No. 1836, 1968.
18. Cotten, M. de V. (Ed.). *Regulation of Catecholamine Metabolism in the Sympathetic Nervous System.* Baltimore: Williams & Wilkins, 1972.
19. Usdin, E., and Snyder, S. H. (Eds.). *Frontiers in Catecholamine Research.* New York: Pergamon, 1973.
20. Sulser, F., and Sanders-Bush, E. Effect of drugs on amines in the CNS. *Annu. Rev. Pharmacol.* 11:209, 1971.
21. Costa, E., and Garattini, S. (Eds.). *International Symposium on Amphetamines and Related Compounds.* New York: Raven, 1970.
22. Lal, S., Sourkes, T. L., and Missala, K. The effect of certain tranquilisers, chlorpromazine metabolites and diethyldithiocarbamate on tissue amphetamine levels in the rat. *Arch. Int. Pharmacodyn. Thér.* 207:122, 1974.

23. Schou, M. Lithium in psychiatric therapy and prophylaxis. *J. Psychiatr. Res.* 6:67, 1968.
24. Cotten, M. de V. (Ed.). Marihuana and its surrogates. *Pharmacol. Rev.* 23:259, 1971.
25. Sourkes, T. L. *Biochemistry of Mental Disease.* New York: Harper & Row, 1962.
26. Himwich, H. E. (Ed.). *Biochemistry, Schizophrenia and Affective Illnesses.* Baltimore: Williams & Wilkins, 1970.
27. Weil-Malherbe, H., and Szara, S. I. *The Biochemistry of Functional and Experimental Psychoses.* Springfield, Ill.: Thomas, 1971.
28. Snyder, S. H., Banerjee, S. P., Yamamura, H. I., and Greenberg, D. Drugs, neurotransmitters and schizophrenia. *Science* 184:1243, 1974.
29. Lloyd, K. G., Farley, I. J., Deck, J. H. N., and Hornykiewicz, O. Serotonin and 5-hydroxyindoleacetic acid in discrete areas of the brainstem of suicide victims and control patients. *Adv. Biochem. Psychopharmacol.* 11:387, 1974.
30. Schildkraut, J. J., and Kety, S. S. Biogenic amines and emotion. *Science* 156:21, 1967.
31. Coppen, A. The biochemistry of affective disorders. *Br. J. Psychiatry* 113:1237, 1967.
31a. Young, S. N., and Sourkes, T. L. Antidepressant action of tryptophan. *Lancet* 2:897, 1974.
32. Curzon, G. Tryptophan pyrrolase — a biochemical factor in depressive illness? *Br. J. Psychiatry* 115:1367, 1969.
33. Clement-Cormier, Y. C., Kebabian, J. W., Petzold, G. L., and Greengard, P. Dopamine-sensitive adenylate cyclase in mammalian brain: A possible site of action of antipsychotic drugs. *Proc. Natl. Acad. Sci. U.S.A.* 71:1113, 1974.
34. Iversen, L. L., Horn, A. S., and Miller, R. J. Actions of dopaminergic agonists on cyclic AMP production in rat brain homogenates. *Adv. Neurol.* 9:197, 1975.
*34a. Sinanan, K., Keating, A. M. B., Beckett, P. G. S., and Love, W. C. Urinary cyclic AMP in endogenous and neurotic depression. *Br. J. Psychiatry* 126:49, 1975. (A useful source of current references on urinary cyclic AMP research.)
35. Symposium on catecholamines and their enzymes in the neuropathology of schizophrenia. *J. Psychiatr. Res.* 11:1, 1974.
36. Heacock, R. A., and Powell, W. S. Adrenochrome and related compounds. *Prog. Med. Chem.* 9:275, 1973.
37. Pauling, L. Orthomolecular psychiatry. *Science* 160:265, 1968.
38. Lipton, M., Ban, T. A., Kane, F. J., Levine, J., and Mosher, L. R. *Megavitamin and Orthomolecular Therapy in Schizophrenia.* A.P.A. Task Force Report No. 7. Washington, D. C.: American Psychiatric Association, 1973.
39. Wise, C. D., Baden, M. M., and Stein, L. Post-mortem measurement of enzymes in human brain: Evidence of a central noradrenergic deficit in schizophrenia. *J. Psychiatr. Res.* 11:185, 1974.
40. Iversen, L. L. Dopamine receptors in the brain. *Science* 188:1084, 1975.
40a. Frohman, C. E., Harmison, C. R., Arthur, R. E., and Gottlieb, J. S. Conformation of a unique plasma protein in schizophrenia. *Biol. Psychiatry* 3:113, 1971.
41. Gjessing, L. R. A review of periodic catatonia. *Biol. Psychiatry* 8:23, 1974.
42. Heath, R.-G., and Krupp, I. M. Schizophrenia as a specific biological disease. *Am. J. Psychiatry* 124:1019, 1968.
43. Whittingham, S., Mackay, I. R., Jones, I. H., and Davies, B. Absence of brain antibodies in patients with schizophrenia. *Br. Med. J.* 1:347, 1968.

*Asterisk denotes key reference.

Chapter 35

Endocrine Effects on the Brain and Their Relationship to Behavior

Bruce S. McEwen

Awareness of endocrine influences on brain function is as old as the field of endocrinology itself. In 1849 Berthold described striking behavioral changes resulting from castration of roosters and the reversal of these changes after testes had been transplanted into the castrated animals [1].

Nearly 100 years later, Dr. Frank Beach published a book entitled *Hormones and Behavior* [2], which has served to instruct and motivate recent generations of investigators to explore in depth the interactions of hormones with the brain. Spectacular growth of the field of neuroendocrinology (i.e., neural control of endocrine function, described in Chap. 22 of this book) offers to the present generation of neurobiologists unparalleled opportunities to explore with great sophistication interrelated problems of the influences of neural activity on endocrine secretion and of endocrine secretion, in turn, on neural activity and behavior.

The relationships among these influences may be summarized as:

endocrine system ⇋ nervous system ⇋ behavior

This emphasizes that associations between behavioral and endocrine events are mediated via the nervous system (Chap. 22). It is the purpose of this chapter to examine chemical and molecular aspects of the influences of hormonal secretion on the nervous system and behavior. But, before doing so, it is necessary to describe briefly the behavioral events and the accompanying neural activity that can trigger hormone secretion.

BEHAVIORAL CONTROL OF HORMONE SECRETION

The brain is the producer of master hormones, the hypothalamic releasing factors, which regulate release of the anterior pituitary tropic hormones. Neurons in the hypothalamus also produce the hormones oxytocin and vasopressin, which are released by the posterior pituitary into the blood. It is therefore not surprising that certain changes in an animal's behavior are associated with the secretion of these hypothalamic releasing factors and hormones (Fig. 35-1). Consider, for example, the phenomenon of lactation, in which the suckling stimulus to the nipple triggers the release of oxytocin, which facilitates milk ejection, and of prolactin, which helps the mammary gland to replenish the supply of milk [3].

The phenomenon of stress also illustrates the behavioral control of anterior pituitary hormone secretion. Conditions associated with injury and surgical trauma, and the so-called psychic stresses of fear, novelty, and even joy, can activate the release of ACTH, which, in turn, stimulates the secretion of adrenal glucocorticoids. These behavioral stimuli are mediated by complex neural pathways and can be modified readily by learning.

737

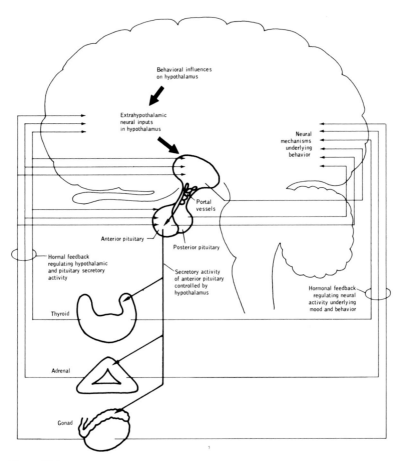

Figure 35-1
Schematic representation of possible reciprocal interactions among hypothalamic, pituitary, thyroid, adrenal, and gonadal hormones.

The secretion of the gonadotropins is also subject to behavioral modification. In the female rabbit, the act of copulation activates spinal reflex pathways that stimulate the secretion of luteinizing hormone (LH) and leads to ovulation. In the male rabbit, the act of copulation also activates the secretion of LH and increases plasma testosterone levels [4]. Social stimuli, too, modify gonadotropin secretion. In mice, olfactory cues from other females can interrupt normal estrous cycles and lead to pseudopregnancies or periods of prolonged diestrus (Lee-Boot effect), and olfactory cues from males can shorten the estrous cycle and either cause rapid attainment of estrus (Whitten effect) or terminate pregnancy in a newly impregnated mouse (Bruce effect) [3]. In male rhesus monkeys, sudden decisive defeat by other males leads to prolonged reduction in plasma testosterone levels, which can be reversed in the defeated male by the introduction of a female companion [5]. In man, the anticipation of sexual intercourse has been reported to increase beard growth, which process is under control of circulating androgen, although this finding has been disputed [6].

Several points should be stressed concerning such behavioral control by hormone secretion. First, as will become apparent in the next section of this chapter, hormones so secreted act upon target tissues, including parts of the central nervous system, to modify their function. In the brain, this feedback action includes modification of behavior and neuroendocrine function. For example, increased gonadotropin and testosterone levels in the male rabbit that result from copulation or, in the male rhesus monkey, from exposure to the female, as might be expected facilitate both sexual activity and replenishment of sperm and seminal fluid. This constitutes a form of positive feedback with respect to reproductive function.

A second aspect of behavioral control of neuroendocrine function is that the circumstances under which hypothalamic hormones are released serve to define those behavioral and physiological conditions that, it also might be expected, are subject to feedback actions of target hormones. With such gonadal steroids as testosterone, these conditions are self-evident, having to do in large part with reproductive functions. With adrenal steroids, analysis of the diverse circumstances leading to ACTH secretion is more difficult but in the end may reveal the role of adrenal steroid interactions with the brain. Each type of stress is a perturbation of the body away from a physiological norm. For example, encounter with a predator may require rapid evasive action, in which neural activity and rapidly mobilized hormones such as epinephrine play a role. Adrenal steroid secretion is slower, reaching a peak minutes after the stressful event, and as such is not expected to play a role in the immediate coping with the situation. If the evasive action is successful and the animal survives, however, it will not only have to reestablish homeostasis; presumably it will learn from its experience in order to minimize the chances of another such encounter. Adrenal steroids might be acting toward such longer adaptations. A special case of this reaction is the extinction of a conditioned avoidance response. Suppose the animal has learned to avoid a certain place where it was previously punished but then discovers that being in that place no longer results in punishment. If, for example, that place also contains a food or water supply, it is in the best interests of the animal to extinguish the avoidance response in order to take advantage of the food or water. Adrenal steroids have, in fact, been found to facilitate such extinction and thus can be said to facilitate a form of behavioral adaptation [7, 8].

We must add to this kind of analysis the fact that, in the absence of external stressors, adrenal steroids are also secreted in varying amounts according to the time of day. In nocturnally active animals such as the rat and in animals such as the human which are active during the day, the peak of this basal secretion always occurs near the end of the sleep period. Thus it is conceivable that adrenal steroid secretion may modulate behavior as a function of the time of day. Indeed it has been reported that adrenal steroids modify the detection and recognition thresholds for a variety of sensory stimuli and influence the occurrence of the so-called rapid eye movement (REM) or paradoxical phase of sleep [8].

ACTIVATIONAL AND ORGANIZATIONAL EFFECTS OF HORMONES ON THE BRAIN

Feedback effects on the brain of hormones that modify behavior and neuroendocrine function may be classified as activational or organizational. Activational effects are facilitative effects on preexisting neuroendocrine pathways or behavioral patterns that are reversible when the hormone disappears (Fig. 35-1). It is essential to consider activational effects on the behavior as facilitative or permissive because the hormones do not by themselves cause the behavior; rather, they increase its likelihood of occurrence given the proper stimuli. For example, a male rat mounting an estrogen-primed female usually will stimulate her to assume the mating (lordosis) posture, whereas a male mounting a spayed female will only infrequently elicit the lordosis response.

Organizational effects occur during a specific phase of early development and organize aspects of behavior and neuroendocrine function for a period of time beyond the presence of the hormone itself, usually for the remainder of the individual's life. Such effects will be considered below in the last sections.

Examples of Activational Effects

The specific activating or inhibiting effects of the hypothalamic, pituitary, and target gland hormones on neuroendocrine function are considered in Chapter 22. Especially noteworthy among these effects is the positive feedback action of estrogen on the brain, leading to the surge of LH that triggers ovulation. Concomitant with such hormone effects on neuroendocrine function are effects that modify appropriate behavioral responses of the animal. There are examples of behavioral effects of each class of hormone. Luteinizing hormone-releasing factor (LRF) has been observed to facilitate female mating behavior in estrogen-primed ovariectomized female rats, even in the absence of the normal LRF target gland, the pituitary [9, 10]. One suspects that other hypothalamic releasing hormones may also have behavioral effects. For example, recent reports suggest that both thyrotropin releasing factor (TRF) and melanocyte inhibiting factor (MIF) may potentiate, respectively, the noradrenergic and dopaminergic systems and exert antidepressant effects [11–13]. The posterior pituitary hormone, vasopressin, has been reported to potentiate conditioned avoidance behavior under extinction conditions in the rat, and the anterior pituitary hormone, ACTH, has been observed to have similar but less prolonged effects [14]. Melanocyte-stimulating hormone (MSH), which resembles ACTH in amino acid sequence, has been reported to have behavioral effects similar to those of ACTH [15].

The activational effects of steroid hormones have been studied the most extensively of all hormones. Estradiol facilitates mating activity in female animals of many species. In the female rate, estrogen promotes the lordosis response, increases locomotor activity, decreases food intake [16], and facilitates maternal behavior [17]. Testosterone facilitates male copulatory activity [18] and activates other behaviors, such as intermale aggression in mice [19] and territorial scent-marking activities in gerbils [20]. Progesterone appears to be important in rodents, together with estrogen, in promoting ovulation and facilitating lordosis response [21, 22]. In addition, progesterone antagonizes a variety of androgen effects [23–26] and facilitates food intake

and body weight in intact and ovariectomized adrenalectomized female rats [27].
Adrenal glucocorticoids have been reported to facilitate extinction of conditioned
avoidance behavior, modify thresholds for detection and recognition of sensory
stimuli, and influence frequency of occurrence of paradoxical sleep [8].

Biochemical Aspects of Activational Hormone Effects

Metabolism of Steroid Hormones in Brain Tissue

One important aspect of the interaction of steroid hormones with target tissues is
the metabolic transformation of the hormone to more or less active metabolites.
Such transformations appear to be of particular importance for the androgen testosterone and lead to the concept that this steroid may be a prehormone. The brain,
like the seminal vesicles, is able to convert testosterone to 5α-dihydrotestosterone
(DHT) and 3α, 5α-androstanediol (Fig. 35-2, a and b) and also, like the placenta, converts testosterone to estradiol (Fig. 35-2c). Neither conversion occurs equally in all
brain regions. Regional distribution of 5α-reductase activity toward testosterone in
rat brain is found in midbrain and brain stem; intermediate activity in hypothalamus
and thalamus; and lowest activity in cerebral cortex [28]. The pituitary has higher

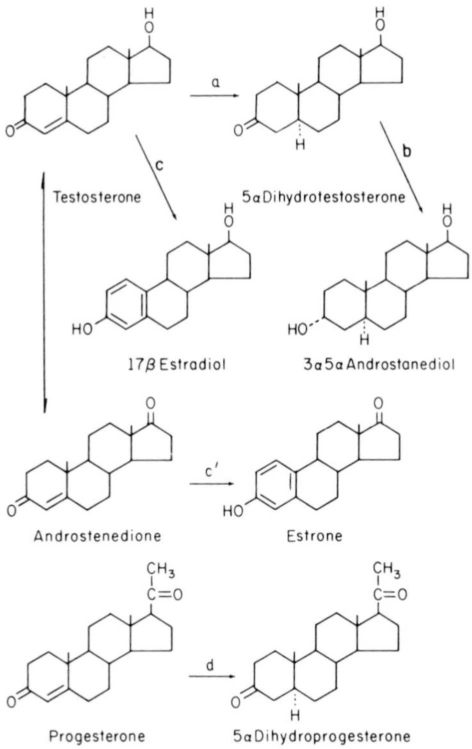

Figure 35-2
Some steroid transformations that are carried out by neural tissue.

5α-reductase activity than any region of the brain, and its activity is subject to changes as a result of gonadectomy, hormone replacement, and postnatal age. 5α-Dihydrotestosterone has been implicated in hypothalamus and pituitary as a potent regulator of gonadotropin secretion but is relatively inactive toward male rat sexual behavior [29]. Labeled metabolites with the R_f value of 5α-DHT have been detected in extracts of hypothalamic and pituitary tissue after ^{3}H-testosterone had been administered in vivo to both adult and newborn male rats. It is interesting that progesterone inhibits 5α-reductase activity toward ^{3}H-testosterone and that ^{3}H-progesterone is itself converted to ^{3}H-5α-dihydroprogesterone (Fig. 35-2d). Progesterone competition for the 5α-reductase may explain some of the antiandrogenicity of this steroid [30].

The aromatization of androgen to estrogen was first described in brain tissue and involved measuring the conversion, not of testosterone, but of androstenedione to estrone (see Fig. 35-2c'). This reaction has been described in homogenates of brain tissue from human fetuses, neonatal and adult rats, adult rabbits, and rhesus monkeys [31–33]. Aromatization is generally higher in hypothalamus and limbic structures than in cerebral cortex or pituitary gland and, from noncastrated animals, is higher in male brains than in female brains. Recent studies indicate that aromatizing activity is greater in tissue fragments than in homogenates [34] and that testosterone can be converted to estradiol (Fig. 35-2c). The physiological significance of aromatization is still unclear. Conversion of ^{3}H-androstenedione to ^{3}H-estrone has been reported to occur in isolated perfused rhesus monkey brain [31], and small amounts of ^{3}H-estradiol have been detected in neonatal rat brain after injection of ^{3}H-testosterone [35, 36]. It has been reported that estradiol will facilitate androgen-dependent behaviors in several species and also that an antiestrogenic drug, MER 25, will block androgen stimulation of female sexual behavior in the rat. Thus the suggestion has been made that some androgen effects in male and female rats (and possibly in other species as well) may involve the conversion of the androgen to estrogen. In this connection, 5α-DHT and 3α-androstanediol cannot be aromatized. This, and the fact that these steroids are much less effective in facilitating male sexual behavior in the rat [29] or in neonatally organizing the sexual differentiation of the neural control of reproductive function in the rat [37], have tended to support the aromatization hypothesis.

A number of other steroid transformations occur in brain tissue, but this metabolism does not appear to be of importance for interaction of those hormones with the putative receptor sites to be described below: Both ^{3}H-estradiol and ^{3}H-corticosterone are recovered from their binding sites in the brain tissue [38].

Cellular Mechanisms of Hormone Action

The study of hormone action has centered around the recognition of putative receptor sites for a variety of hormones and the classification of these receptor sites in terms of two fundamentally different cellular mechanisms of action, which are shown in Fig. 35-3. A number of polypeptide hormones such as glucagon, insulin, ACTH, and possibly the hypothalamic releasing factors are believed to act by way of cell-surface receptors that are associated with the enzyme adenylate cyclase. Interaction

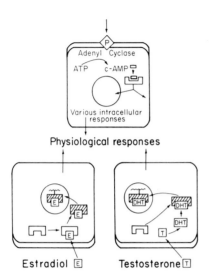

Figure 35-3
Schematic representation of interactions of hormones with target cells. *Top*: Certain protein hormones (P) and epinephrine are known to interact with cell surface receptors to stimulate adenyl cyclase. *Bottom*: Steroid hormones interact with intracellular receptors and are transferred into cell nuclear compartment.

of hormones with such receptors is believed to trigger increased formation from ATP of cyclic 3'5'-AMP, which, in turn, is an intracellular second messenger in a variety of cellular events (Fig. 35-3). This mechanism is discussed in Chapter 12. Steroid hormones, on the other hand, interact with intracellular, presumably cytoplasmic, receptor sites and are transferred to the cell nucleus where they interact with the genome to alter transcription processes (Fig. 35-3). Recent evidence suggests that thyroid hormone also may be able to act via cell nuclear receptor sites [39]. Because of the greater amount of evidence available, the remainder of this section will focus on the cellular mechanism of steroid hormone action on brain and pituitary cells.

Methods for Studying Putative Steroid Hormone Receptor Sites
Before summarizing the evidence for steroid hormone receptor sites in brain and pituitary, let us review the methods that are used to measure such sites. The availability in the early 1960s of tritium-labeled steroids of high specific activity (20 to 25 Ci per μmole at each labeled position) permitted the measurement of specific binding sites of low capacity which had previously escaped detection using ^{14}C-labeled steroids [40]. In the brain, high-resolution autoradiographic methods utilizing ^{3}H-labeled steroids have permitted mapping of steroid-hormone target cells in specific brain regions (Fig. 35-4). It should be noted that tritium permits a high degree of spatial resolution (particle range 1 to 2 μ in light-microscope autoradiography) owing to its low energy of decay [41].

Cell fractionation procedures are basic to the biochemical identification and study of steroid hormone-binding sites. Isolation of highly purified cell nuclei from small amounts of tissue from discrete brain regions is generally accomplished with the aid

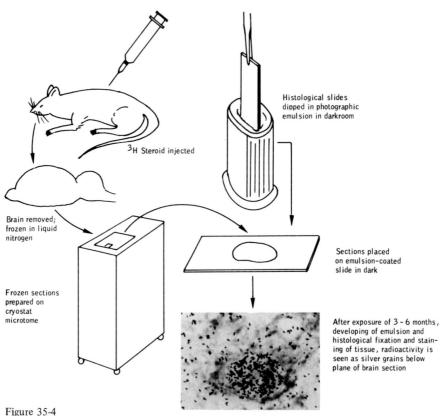

Histological slides
dipped in photographic
emulsion in darkroom

³H Steroid injected

Brain removed;
frozen in liquid
nitrogen

Sections placed
on emulsion-coated
slide in dark

Frozen sections
prepared on
cryostat
microtome

After exposure of 3 - 6 months,
developing of emulsion and
histological fixation and staining of tissue, radioactivity is
seen as silver grains below
plane of brain section

Figure 35-4
Flow diagram of a frequently used procedure for autoradiographic localization of ³H steroid
hormones in neural tissue.

of a nonionic detergent, Triton X-100, and such methods have been described in
detail elsewhere [42]. Cytosol fractions of brain tissue (prepared by centrifugation
of homogenates at 105,000 $\times$ g for 60 min) contain the soluble steroid hormone-
binding proteins, and a variety of methods intended to separate bound from unbound
steroid have been used for measuring their binding activity [40]. The most commonly
employed are gel filtration chromatography and sucrose density gradient centrifugation. Dextran-coated charcoal is frequently used because it effectively adsorbs unbound steroid and leaves intact the complexes between steroid and putative receptor.
Other methods, such as disc gel electrophoresis and precipitation of putative receptor
material by protamine sulfate, have more restricted uses.

Several general comments are in order regarding use of these methods in studies
of cytosol hormone receptors. The objective of such studies is to measure the affinity,
capacity, and specificity of the hormone-receptor interaction [40]. Measurements
of affinity and capacity are accomplished by running concentration-dependence
curves of binding and plotting the results in the form of a Scatchard plot or a reciprocal plot (Fig. 35-5). Evaluation of the x and y intercepts yields the association
constant (K_a) or its reciprocal, the dissociation constant (K_d), and the capacity of
the binding sites at saturation (B_{max}) by an infinite concentration. Problems encountered in measurements of K_d and B_{max} center around the attainment of a true equilibrium: Inasmuch as it may take hours for ligand to equilibrate with receptor, and

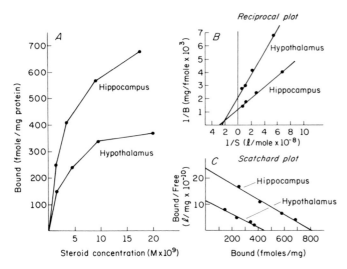

Figure 35-5
(A) Concentration-dependence curve for binding of ^{3}H-dexamethasone by cytosols from hippocampus and hypothalamus. (B) Expression of data in (A) in the form of a reciprocal plot, 1/bound vs. 1/steroid concentration. (C) Scatchard plot of same data, bound/free steroid vs. bound (fmole = 10^{-15} mole; 1 = liters).

some kinds of receptors decay measurably during this time of equilibration, there may never be attained more than a quasi equilibrium in which the K_d and B_{max} approach but never reach the true values. Specificity studies are conducted by competing for binding of ^{3}H-ligand by various unlabeled ligands. The relative ability of various unlabeled steroids to compete for binding in increasing molar excesses to ^{3}H-steroid is compared with the effectiveness of the unlabeled homologous steroid in similarly increasing excesses. A large (100 = to 1,000-fold) excess of unlabeled homologous steroid is used to estimate nonspecific binding since the techniques for separation of bound from unbound ligand measure total binding. A general criterion for nonspecific binding is that it be linear as a function of ^{3}H-ligand in the presence of a large and constant excess of unlabeled ligand, i.e., nonsaturable. Subtraction of nonspecific from total binding gives an estimate of the actual limited-capacity binding (saturable).

There are several criteria for calling a steroid hormone-binding site a putative receptor. First, it must be present in hormone-responsive tissues (or brain regions) and absent from nonresponsive ones. Second, it should bind steroids that are either active agonists or effective antagonists of the hormone effect and not bind steroids that are inactive in either sense. These two criteria will form the basis of the discussion that follows.

Properties and Topography of
Putative Steroid Receptor Sites in Brain

Let us now consider the evidence for putative steroid hormone receptor sites in brain and pituitary. In the brain and pituitary, as in other steroid hormone target tissues, there appear to be steroid hormone-binding sites for certain hormones in the cytosol and in cell nuclei. Unlike receptors in many other target tissues, however, putative brain receptors are not distributed uniformly but are concentrated in discrete brain regions.

Estradiol. (Formula in Fig. 35-2.) The first neural steroid hormone-binding sites to be recognized were those for estradiol [43]. Studies of the in vivo accumulation of ^{3}H-estradiol from the blood revealed extremely high concentrations in pituitary as well as uterus and lower but substantial uptake in the hypothalamic region of the brain. Within brain, ^{3}H-estradiol accumulation is highest in the hypothalamus and preoptic region. A substantial proportion of uptake into these brain regions and into pituitary and uterus can be blocked by concurrent administration of ^{3}H estradiol and 100- or 1,000-fold excesses of unlabeled 17β-estradiol but not by similar excesses of unlabeled testosterone or 17α-estradiol (an inactive estrogen). Such competition establishes the binding as a phenomenon of limited capacity, with specificity for active estrogens.

Autoradiography has provided more detailed information as to the distribution within the brain of binding sites for ^{3}H-estradiol [44]. Neurons, rather than glial cells, appear to contain the highest concentrations of these putative receptor sites and, in male and female rats, these neurons are concentrated within the hypophyseotropic area (medial preoptic area, anterior and medial-basal hypothalamus) and amygdala and, to a lesser extent, in the midbrain (Fig. 35-6). Not all neurons within these areas bind estradiol, but many of the cells that do bind the hormone have an intensity of labeling comparable to that found in cells of the uterus and pituitary.

The use of sedimentation rate in sucrose density gradients (approximately 8 S at low ionic strength) and specificity of binding toward active estrogens, such as 17β-estradiol and diethylstilbestrol, cell fractionation studies of the pituitary and hypothalamus, preoptic area, and amygdala demonstrated the existence of soluble (cytosol) binding sites which resemble those found in the uterus. In spite of similarities in sedimentation behavior and hormonal specificity, no one knows for sure if the estrogen-binding proteins are identical in these various estrogen target tissues. In agreement with the autoradiographic results cited above, estrogen-binding proteins are not detectable by sucrose-density gradient centrifugation in cytosols from the cerebral cortex and are demonstrable in cytosols from the pooled preoptic area, hypothalamus, amygdala, and midbrain [16]. Quantitative estimates of estradiol-binding capacity of cytosol based on Scatchard plots (see above) indicate that cerebral cortex has less than one-fourth of the capacity found in amygdala and less than one-tenth of the capacity found in hypothalamus, whereas the hypothalamus has a capacity around one-thirtieth of that found in either pituitary or uterus [43]. The dissociation constant of the binding (K_d) is between 1 and 5 $\times$ 10^{-10} M, depending on the method used to separate bound from free ^{3}H-estradiol. Based on the DNA content of these tissues, the capacity of uterine and pituitary tissues to bind ^{3}H-estradiol corresponds to 12,000 to 15,000 molecules per cell, whereas the capacity of the whole hypothalamus is around 2,000 to 3,000 molecules per cell. These estimates of binding per cell are, of course, averages which do not take into account the proportion of cells in each tissue that bind with the hormone.

The estradiol which attaches to cytosol estrogen-binding sites is transferred into the cell nuclei, and substantial amounts of the ^{3}H-estradiol in the tissue can be recovered in isolated nuclei from these target tissues [38, 43]. The relative magnitude of binding in cell nuclei closely parallels the relative magnitude of cytosol-receptor

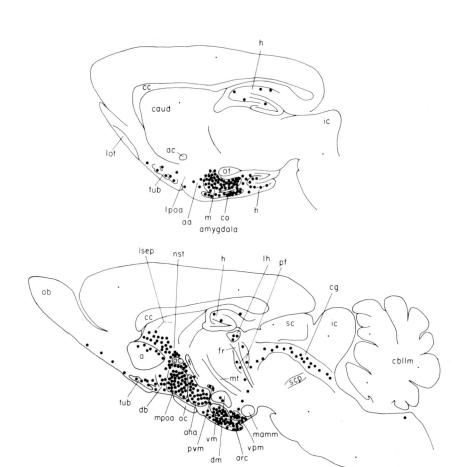

Figure 35-6
Distribution of estrogen-concentrating neurons in the brain of the female rat represented sche-matically in two sagittal sections. Most labeled neurons can be represented in a medial plane (bottom). Estradiol-concentrating neurons in the amygdala and hippocampus are represented in a more lateral plane (top). Locations of estradiol-concentrating neurons are represented by black dots (●). Figure reproduced from Pfaff and Keiner [44] by permission of the authors and The Wistar Press, Philadelphia. (Identity of the anatomical abbreviations may be found in [44].)

content in uterus, pituitary, and various brain regions. Cell nuclear-bound ^{3}H-estradiol can be extracted as a complex with protein using KCl in concentrations of 0.3 M or greater. These complexes from uterine, pituitary, and brain nuclei have lower sedimentation coefficients in sucrose-density gradient centrifugation than those of the corresponding cytosols, when the cytosols are run in low ionic-strength buffers. As reported for calf uterus [43], the 8.6 S estrogen-binding protein in cytosol is believed to be a tetrameric form of a subunit of about 61,000-Dalton molecular weight. Dissociation to a dimeric form (5.3 S) occurs at high ionic strength (0.2 M salt or greater), and further transformation to a monomer (4.5 S) is carried out by a "receptor transforming factor," believed to be an enzyme with limited proteolytic activity that is present in the cytosol. In the calf uterus, the purified cell-nuclear receptor has a sedimentation coefficient of 4.5 S and is believed to be related to the monomeric form of the cytosol receptor. Therefore, the hormone, together with the cytosol monomer, appears to be translocated into the cell nucleus and to interact with the genome.

Androgen. (Formulas in Fig. 35-2.) As noted earlier in this section, testosterone is transformed to 5α-DHT in a number of androgen target tissues, including the brain and pituitary. This transformation was first recognized in the male accessory sex glands, and subsequently receptor sites specific for 5α-DHT were recognized in these tissues [40]. In discussing putative receptor sites in the pituitary and brain, it is necessary to consider both testosterone and 5α-DHT. Measurement of tissues after either ^{3}H-testosterone or [3]H-5α-DHT has been administered to castrated male rats has shown the pituitary to have the highest concentration of radioactivity after the prostate and seminal vesicles. Hypothalamus has generally higher concentrations of radioactivity than cerebral cortex, but differences between these two structures are small, in contrast to the concentration difference between hypothalamus and cortex in ^{3}H-estradiol uptake. Owing to this low level of region-specific uptake, the detection of binding sites for androgen in rat brain has proved to be difficult. Cell fractionation of brains of castrated male rats and ring doves injected with ^{3}H-testosterone reveals low levels of cell nuclear binding in hypothalamus [38, 43]. Similar low-level cell-nuclear binding has been reported in pituitary. It is not known whether the bound radioactivity is testosterone or 5α-DHT. The most successful attempts to identify cytosol androgen-binding sites in brain have been made with [3]H-5α-DHT. An 8.6 S macromolecule, which binds [3]H-5α-DHT and not ^{3}H-testosterone or ^{3}H-progesterone, has been detected in hypothalamus cytosol by sucrose-density gradient centrifugation [43]. The binding capacity is less than one-half of that seen for cytosol binding of ^{3}H-estradiol.

Autoradiography has provided more substantial evidence for cellular localization of androgen-binding sites, although this technique does not permit identification of labeled material. ^{3}H-Testosterone uptake has been observed in 10 to 15 percent of the cells of the anterior pituitary and in selected cell groupings within the preoptic area and medial basal hypothalamus as well as in the amygdala [43]. Cellular localization of ^{3}H-testosterone has been reported in the hypothalamus, preoptic area, and midbrain of a songbird, the chaffinch *Fringilla coelebs* [43]. The midbrain localization is of particular interest because of androgen effects on bird song, motor control centers for which are located in the area of hormone uptake.

Glucocorticoids. (Formulas in Fig. 37-7.) Because of the success achieved in measuring putative estrogen receptors in the brain, attempts were made to find similar receptor molecules for adrenal steroids. Based upon the estrogen studies and on information regarding the importance of the hypothalamus in the control of ACTH secretion, it was anticipated that adrenal steroid receptors would be localized in the hypothalamic region of the brain. Such putative receptors were indeed found in hypothalamus, but the highest concentrations of binding sites for ^{3}H-corticosterone, the predominant adrenal glucocorticoid in the rat, were found in the hippocampus, amygdala, and septum [8, 38]. Cell fractionation studies revealed that this binding

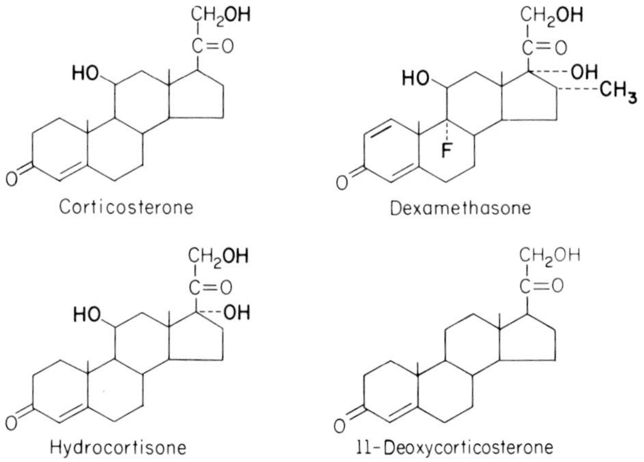

Figure 35-7
Formulas of four steroid hormones.

was due to both cytosol and cell-nuclear binding sites, in the fashion we have seen is typical for many steroid hormones and target tissues. In hippocampus, the capacity of the cytosol corticosterone-binding protein is about 850 femtomoles (fmoles per mg total cytosol protein; Fig. 35-5). A femtomole = 10^{-15}. This is large compared with the 20 fmoles per mg protein that approximates the capacity of estrogen-binding protein in the hypothalamus. Estimates of binding in cell nuclei in the hippocampus indicate that about 16,000 molecules of corticosterone are bound per cell nucleus. This is also large compared to the figure of 3,000 molecules of estradiol per cell nucleus in the hypothalamus. Differences in the binding capacity for the two steroids are undoubtedly a reflection of two factors: first, that the density of hormone-sensitive cells is different in each brain region; second, that the amount of receptor tends to reflect the amount of hormone in the blood. Physiological estrogen levels in the rat range from 3 to 30 pg per ml; corticosterone levels range from 30 to 300 ng per ml. As noted in the estrogen section, the presence together of hormone-sensitive and -insensitive cells in a given brain region makes it virtually impossible to estimate the actual hormone capacity in sensitive cells.

The properties of the soluble brain glucocorticoids that bind macromolecules toward ^{3}H-dexamethasone are illustrative of some of the characteristics of steroid hormone-binding proteins in general. These proteins have a high affinity (K_d = 3 to 4×10^{-9} M; Fig. 35-5), limited, low capacity (B_{max} = 810 fmoles per mg protein; Fig. 35-5), and a high degree of specificity for the glucocorticoids. They do, however, bind progesterone and mineralocorticoids (see below). They may be compared with transcortin, the serum protein. Unlike transcortin, the binding proteins in brain bind dexamethasone, the synthetic glucocorticoid, with high affinity. The steroid–brain receptor complex, unlike the transcortin-steroid complex, can be precipitated by the polycation, protamine sulfate. At low ionic strength, the brain receptor–steroid complex is a macromolecular aggregate of more than 600,000 molecular weight and has a low mobility in polyacrylamide gel electrophoresis; in contrast, transcortin has a molecular weight of $\approx 60,000$ and a high mobility in polyacrylamide gels, which permits it to be quantitatively separated from the binding proteins in brain.

Autoradiographic examination of the rat brain labeled with ^{3}H-corticosterone reveals that neurons in hippocampus, septum, induseum griseum, and cortico-medial amygdala are heavily labeled [8]. Most of the radioactivity appears over the region of the cell nucleus. It should be noted that there is excellent evidence that glial cells also respond to glucocorticoids and contain putative receptor sites. After adrenalectomy, the activity of the enzyme glycerolphosphatedehydrogenase (GPDH) falls in all regions of the brain and can be restored by replacement therapy with glucocorticoids [45]. In tissue culture, a rat glial-cell tumor line, RGC6, shows the glucocorticoid inducibility of GPDH and contains putative receptors for glucocorticoids, thus making it appear likely that glial cells in the nervous system are the major target for this glucocorticoid effect [46]. It is puzzling that autoradiographic studies of intact nervous tissue labeled with ^{3}H-corticosterone have failed to show specific labeling in cell nuclei associated with nonneural elements. On the other hand, glial cells are very difficult to identify in lightly stained frozen sections, and there is a substantial background of radioactive label in the neuropil surrounding the heavily labeled neurons.

Progesterone. (Formula in Fig. 35-2). As is true with testosterone, progesterone is likely to undergo one of a number of metabolic transformations in the body [43]. The conversion to 5α-dihydroprogesterone, which has been demonstrated to occur in brain (see p. 742), is one of these. This metabolite can be further reduced to 3α and 3β derivatives. 5α-Dihydroprogesterone and its 3α–OH derivative have been reported to be less active than progesterone itself in inducing lordosis behavior in estrogen-primed, spayed female rats [22]. However, 5α-DHP and not 5β-DHP appears to mimic the effect of progesterone in inhibiting ovulation under rigidly defined experimental conditions in immature rats treated with PMSG, an inducer of precocious ovulation. Another progesterone metabolite, 20α-hydroxyprogesterone, is produced by ovarian tissue and is functionally active in maintaining LH release in the rabbit after mating [43].

Studies of the fate of systemically injected ^{3}H-progesterone and ^{3}H-20α-OH-progesterone have shown that the highest uptake occurs in the midbrain region, somewhat lower uptake in hypothalamus, and still lower uptake in cerebral cortex and hippocampus of rat and guinea pig brains [43]. Midbrain and hypothalamus in guinea pig are both sites at which progesterone implants exert effects on mating that resemble physiological effects of the hormone given systemically. This pattern of uptake is, however, similar to that observed for ^{3}H-corticosterone and ^{3}H-estradiol in guinea pig brains when binding sites are saturated by either endogenous or exogenous hormones. These facts have suggested that an interaction of the steroid with lipids in the brain, and not with specific receptors, determines the pattern of uptake.

Attempts to saturate and thereby reveal limited-capacity (saturable) binding sites for ^{3}H-progesterone, using unlabeled progesterone, have proved uniformly unsuccessful and have tended to suggest that specific progesterone-binding sites either may not exist at all or be so few in number as to escape detection.

Several pieces of evidence, however, tend to support the idea of limited-capacity progesterone binding sites in brain and pituitary. First, Sar and Stumpf have reported autoradiographic localization of binding of radioactivity injected as ^{3}H-progesterone in neurons of the basomedial hypothalamus of the spayed, estrogen-primed guinea pig [43]. Unlabeled progesterone abolishes this localization, and unlabeled cortisol is without effect. According to these authors, estrogen priming is important for ^{3}H-progesterone accumulation. Similar results have been reported for cytosol binding of ^{3}H-progesterone in the pituitary and median-eminence region of spayed, estrogen-primed rats: A progesterone-saturable binding component that could not be saturated by unlabeled corticosterone has been detected [43]. Second, adrenalectomy apparently produces an increase in uptake by tissue of ^{3}H-progesterone in brain (see above). Not only does adrenalectomy remove one source of endogenous progesterone; it also unmasks glucocorticoid-binding sites in the brain to which progesterone is able to bind (see earlier discussion). One peculiarity about the interaction of progesterone with glucocorticoid receptors is that the progesterone is not extensively translocated to the cell nuclei, as was discussed in re corticosteroids. Progesterone can, in fact, block nuclear translocation of ^{3}H-corticosterone, presumably by occupying cytosol-receptor sites. Progesterone is, therefore, an antiglucocorticoid and has been shown to block a number of glucocorticoid effects on thymus, chick neural retina, and hepatoma tissue-culture cells. Thus, progesterone might exert some neural effects via an antiglucocorticoid action, although glucocorticoids are normally present in large excess over circulating progesterone and would therefore reduce the efficacy of progesterone as a competitive inhibitor.

Mineralocorticoids. (Formulas in Fig. 35-7.) Proteins capable of binding ^{3}H-deoxycorticosterone have been described in rat brain. Indeed, the brain is capable of responding to mineralocorticoids with respect to regulation of specific salt hunger [43]. A confounding factor in the study of mineralocorticoid receptors is that glucocorticoid-binding proteins in the brain have a moderate ability to bind mineralocorticoids, including aldosterone, the one that occurs naturally. It will be necessary in future work to devise means of distinguishing, both by hormonal specificity and by physical separation of binding proteins, between glucocorticoid receptors and bona fide mineralocorticoid binding sites.

Estradiol Action in the Brain and Pituitary —
Integration of Behavioral and Neuroendocrine Effects

It is evident from the previous sections that the brain and pituitary gland contain target sites for the action of steroid hormones and that these hormones activate neuroendocrine events and facilitate the occurrence of particular behaviors. Usually the neuroendocrine and behavioral effects of hormone action are coordinated with each other and together contribute to the efficient operation and survival of the individual and the species. Nowhere is this more evident than in reproduction, and this integration is particularly well illustrated by the example of reproduction in the female rat.

The estrous cycle of the female rat is usually of four days' duration (Fig. 35-8). It can be measured by means of vaginal smears, which, when examined under a light

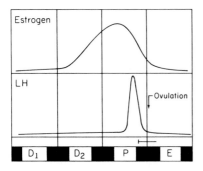

Figure 35-8
Schematic representation of plasma levels of estrogen and luteinizing hormone (LH) during the
4-day estrous cycle of the female rat. D1, D2 = diestrous days 1 and 2, respectively. P = proestrus; E = estrus. Black panel = night, white panel = day. Bar indicates onset of behavioral estrus.

microscope, reveal a characteristic pattern of cell types present in the vaginal
epithelium [16]. Indeed, the entire reproductive tract undergoes cyclic changes
under the direct control of circulating estrogen and progesterone and at about the
day of estrus becomes able to receive and maintain a fertilized egg. Plasma levels of
circulating estrogen are represented schematically in Figure 37-8. Basal levels are
below 25 picograms per ml and rise to a peak of around 40 to 80 pg per ml on the
day of proestrus [47]. Plasma levels of luteinizing hormone, the direct stimulus
for ovulation, rise to a sharp peak (the LH surge) on the afternoon of proestrus. That
the previous rise in estrogen is required for this surge can be demonstrated by administering antibodies against estrogen on diestrous day 2. In this way, the LH surge is
blocked by the effective removal of estrogen from access to target-tissue receptors [48].

   In a similar fashion, "antiestrogenic" compounds such as clomiphene or MER-25
(Formulas in Fig. 35-9) also block the LH surge when administered on diestrous day 2
[43]. These antiestrogens are known to interact with putative estrogen receptors in
the uterus, pituitary, and brain and block access of estradiol to the intracellular receptor sites. It appears, in fact, that the antiestrogens are themselves translocated into
the cell nuclei where they remain for long periods of time, unlike estradiol, which
comes out of the nucleus after a number of hours [43]. Nuclear retention of antag-
onist-receptor complexes apparently depletes cytoplasmic receptor levels and prevents
newly arriving estrogen from having an effect [49]. A *single* injection of antagonist
has been demonstrated to be as effective as estradiol itself in promoting uterine growth;
however, the antagonist-loaded tissue is rendered insensitive to further estrogen treatment [49]. The antiestrogenic action thus apparently hinges on the interference
with the "amplification" of the initial estrogenic effect that normally occurs through
further interaction of target cells with continuing supplies of estrogen. In the uterus,
this is seen as an attenuation of the growth and differentiation response to hormone.
It remains to be seen what corresponding cellular events are attenuated in brain and
pituitary by antiestrogens and what cellular events may actually be induced by antiestrogens as well as by estradiol.

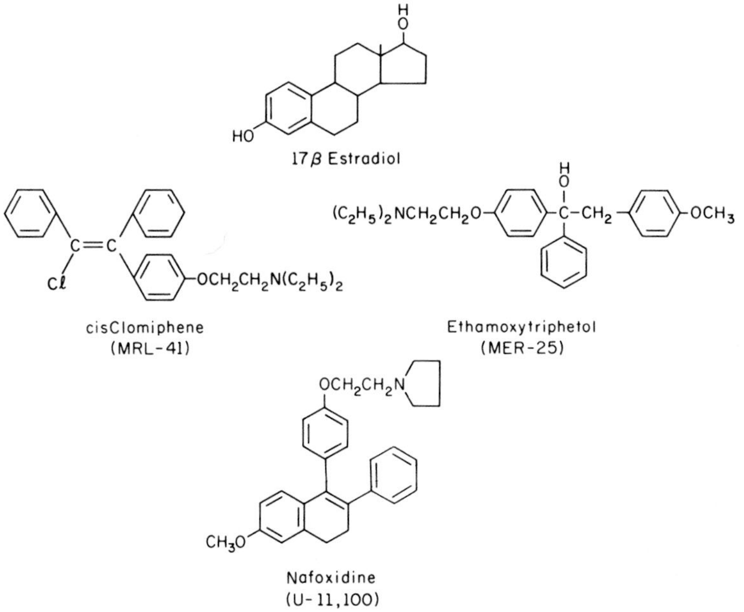

Figure 35-9
Formulas of 17β estradiol and 3 antiestrogens.

Further indication that events in the cell nucleus are required for the LH surge is provided by experiments in which actinomycin D is able to prevent the LH surge when given systemically to ovariectomized rats before the administration of a dose of estradiol, which otherwise will produce such a surge 30 hours later [43]. It is interesting to note that progesterone, another gonadal steroid, appears to be involved also in the LH surge. Actually, a substantial proportion of the progesterone in the female rat originates in the adrenal gland [50]. Thus ovariectomized rats are able to respond to estrogen alone to produce an LH surge, but ovariectomized-adrenalectomized rats require a progesterone treatment 24 hours after estrogen and 4 to 6 hours before the surge can be detected [21]. Although the latency of progesterone action is shorter than that for estradiol, it is still consistent with a mode of action in cell nuclei, but conclusive evidence on this point is lacking (see above).

The site of estrogen action on the LH surge is assumed to lie in the region of the hypothalamus and pituitary, but the exact location of these effects is unclear. As a function of time, estrogen appears first to inhibit and then to facilitate the response of the pituitary to the stimulatory effects of luteinizing hormone releasing factor (LH-RF) [43]. One of the peculiar aspects of estrogen action is that this hormone can have both negative and positive feedback effects on gonadotropin secretion. The negative effects tend to have a short latency and duration but predominate when estrogen levels are maintained at a constant level, for example, in the form of a hypothalamic implant which is left in place.

Concurrently with the neuroendocrine events described above, estrogen acts upon target sites in the hypothalamus and preoptic area to facilitate changes in the behavior of the female rat. The period of behavioral estrus is denoted by the bar in Figure 35-8, and it is clear that it is coordinated with ovulation to assure a high probability that pregnancy will result from copulation. The estrous female rat shows a number of behavioral changes: increased locomotor activity, characterized by darting motion and dragging the hind quarters so as to leave a scent; an ear quiver, which may serve as attraction (together with the very important estrous scent) for the sexual motivation of the male; and a decreased food intake, related to the fact that the female in estrus spends less time searching for food and more time searching for a mate [16]. Facilitation of lordosis, the mating posture of the female, is also apparent in estrus and is particularly striking because a female not in estrus will run away from or fight off the advances of a male. Implants of estradiol into the hypothalamus and preoptic area of the ovariectomized female rat will facilitate lordosis response, increase locomotor activity, and decrease food intake [16], suggesting that the hypophysiotropic area of the brain, which contains the highest concentration of putative estradiol receptors, is the target site for the behavioral effects. The role of these receptors is further implicated by experiments using the antiestrogens, MER-25 and cis-clomiphene (see Fig. 35-9). These agents, which interfere with estrogen binding to the receptors, block estrogen induction of behavioral receptivity and interfere with behavioral estrus when given during the normal estrous cycle [43]. Actinomycin D, an RNA-synthesis inhibitor, has been reported to block estrogen induction of lordosis. Careful implantation studies with actinomycin D have delineated the conditions for the action of this drug. It can be given up to six hours after estradiol treatment and still have an inhibitory effect over the lordosis response. When the implants are removed, a high percentage of the animals recover their responsiveness to estradiol, indicating that the drug does not simply produce a brain lesion that destroys the capacity to produce the behavior.

Both estrogen induction of the LH surge and estrogen facilitation of lordosis behavior require about 24 hours to be fully manifested. The time course for lordosis is illustrated in Figure 35-10. The ovariectomized female fails to respond to mounting attempts by the male until more than 16 hours after a single intravenous injection of the steroid. Time-course measurements after a single dose of ^{3}H-estradiol capable of promoting lordosis behavior reveal that the putative brain and pituitary receptor sites remain occupied for less than 12 hours [43]. The time difference between receptor occupation and the appearance of lordosis behavior, taken together with the actinomycin D effects noted above, suggests that intervening metabolic changes are taking place in the target cells, leading to the change in behavioral responsiveness.

A number of neurochemical markers of gonadal steroid action in brain and pituitary have been uncovered recently. For instance, in pituitary responses to estradiol, sensitivity to LH-RF is elevated and the activity of such oxidative enzymes as glucose 6-phosphate dehydrogenase is increased [43]. In female-rat hypothalamus, preoptic area, and amygdala, which contain putative estradiol receptor sites, estradiol treatment elevates activation of a number of enzymes, including choline acetyltransferase,

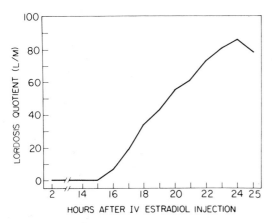

Figure 35-10
Time course of lordosis quotient (ratio of lordoses by female to number of times mounted by male) after 100 μg of estradiol was given intravenously. (Redrawn from Green, Luttge, and Whalen, *Physiol. Behav.* 5:137, 1970. Reprinted with permission of authors and Pergamon Press, copyright 1970.)

isocitrate dehydrogenase, and cysteine aminopeptidase. Estrogen treatment depresses the activity of monoamine oxidase in amygdala and hypothalamus. A number of these changes have been shown to be attenuated by the estrogen antagonist MER-25, but MER-25 by itself appears unable to produce similar changes [43]. In view of remarks above concerning the nature of the blocking effects of these antagonists, it would appear that those changes described above that like the behavioral and neuroendocrine effects of estrogen, are blocked and not mimicked by MER-25 may represent an amplification response to estrogen.

The underlying significance of alterations of activity in choline acetyltransferase and monoamine oxidase may be related to the apparent involvement of putative neurotransmitters such as norepinephrine, serotonin, dopamine, and acetylcholine in neural control of gonadotropin secretion and sexual behavior. Detailed discussion of this interesting topic is beyond the scope of this chapter, and the reader is referred to recent papers on the subject [51–54].

Organizational Effects of Hormones on Developing Brain

Gonadal Steroids and Sexual Differentiation
During the embryonic development of many vertebrates, the presumptive gonads undergo differentiation into ovaries or testes as determined by the genetic sex of the animal [55]. In lower vertebrates, especially fish, the gonad retains some of the potentiality to produce gonadal tissue of the opposite sex. As a result, sex reversals can occur under natural conditions or after hormonal administration during adult life. In mammals (and many other vertebrate classes as well), the gonadal tissue is irreversibly determined during early life, and complete sex reversals are not possible.

The secretions of the gonads determine the sex both of the reproductive tract and, apparently, of the brain as well [55–57]. In mammals and many lower vertebrate classes, the hormonally neutral sex is the female, and testicular secretions determine the differentiation of the male reproductive tract and brain from a basically feminine pattern. In other words, if the testes are removed before this differentiation occurs, the reproductive tract and brain acquire a basically feminine pattern. An important exception to this is found in many species of birds, where the hormonally neutral sex is the male, and ovarian secretion determines feminization of brain and body sex from a basically masculine pattern. We shall consider only the mammal, however, and use sexual differentiation of the rat as the principal illustration.

Shortly before the rat is born (and during fetal life in other mammals, including man), the testes secrete substantial amounts of testosterone [56]. This secretion continues for most of the first week of postnatal life (Fig. 35-11). Removal of the testes on the day of birth results in an animal that, as an adult, shows more feminine behavior after estrogen treatment than do animals castrated as adults and that will maintain a cyclic pattern of ovarian secretion when ovaries are implanted. Conversely, if testosterone is administered to female rats on the day of birth or even several days after, the results are animals that, when ovariectomized as adults and treated with gonadal steroids, show more masculine behavior from administration of androgen and less feminine behavior from estrogen than do untreated females ovariectomized as adults. In addition, these androgenized females display anovulatory sterility in the form of persistent vaginal estrus. It should be noted that the reproductive tract of the rat undergoes differentiation during the week preceding birth and is only slightly affected by postnatal manipulations of the type described above.

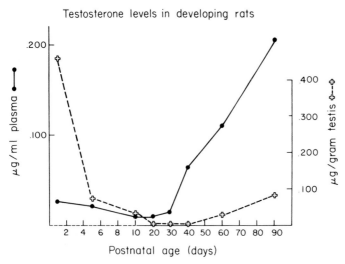

Figure 35-11
Testosterone levels in plasma (μg per ml) and testis (μg per gram) as a function of postnatal age in days. See [56].

With respect to brain differentiation, the ability to manipulate neonatally the behavioral responsiveness to gonadal hormones in the adult rat has provided endocrinologists with an excellent experimental system with which to study sexual differentiation of the brain. The two end points usually used to assess masculinization (defeminization is often a better term) are, as noted above, the presence or absence of ovulatory cycles and the frequency of occurrence of adult patterns of male or female sexual behavior under gonadal hormone treatment of gonadectomized animals. In this section, we shall consider the neurochemical aspects of sexual differentiation of the brain.

A fundamental problem has been to determine the sites of sexual differentiation within the brain. Several laboratories have shown that ovulation and female sexual behavior are suppressed most effectively by implanting pellets of testosterone in the anterior hypothalamic area and ventromedial hypothalamus; the latter site is perhaps the more effective one [58, 59]. By using implants that can be removed, one laboratory has shown that the minimal exposure time for obtaining anovulatory sterility is at least 48 hours.

Another approach in the localization of neural sites of sexual differentiation has been to measure morphological correlates of the phenomenon. A report by Raisman and Field [60] indicates that neonatal testosterone treatment, as well as the natural secretion of androgen in the neonatal male, results in a decrease in the number of presynaptic endings on dendritic spines within the preoptic area of the rat brain. The origin of these synapses is not known, but it is not the projection of the amygdala via the stria terminalis. Similar morphological differences were not found in the ventromedial hypothalamus, although it is clear from the enormous complexity of synaptic contacts with dendrites and cell bodies that other morphological differences may exist, which remain to be uncovered. Considering the heroic efforts that went into the initial work by Raisman and Field, however, it is fair to say that discovery of such correlates will not occur rapidly.

A more general conclusion is that, although the preoptic and hypothalamic areas are indeed sites of sexual differentiation, we have no basis for concluding that they are the only sites. For example, recent work has indicated possible effects of treating the neonatal cerebellum with androgen [61].

In view of the long-term organizing effects of neonatal testosterone secretion (or its administration to females) on the sexual differentiation of the brain, it is most attractive to view the primary site of action as the genome of brain cells. There is some evidence for androgen-dependent alterations in the volume of the cell nucleus and incorporation of precursors into RNA in anterior hypothalamus and amygdala during the perinatal period. The most important evidence to date for a genomic role derives, however, from studies of the long-term effects of inhibitors of protein and RNA synthesis on the defeminizing effects of testosterone on ovulatory cycles in female rats [62, 63]. Subcutaneous administration of inhibitors of RNA, protein, and DNA synthesis to 4-day-old female rats, both together with and 6 hours after testosterone propionate in varying doses, blocks or strongly attenuates the development of anovulatory sterility measured on postnatal day 80 [63]. Earlier reports indicated similar effects, although intrahypothalamic administration (in contrast to the subcutaneous route) produced variable results [62]. Thus, while the prognosis for this approach appears to be good, further work is required to delineate the intracranial sites of action of these drugs. One important technical problem is the relatively short period of inhibition produced by the drugs (usually a few hours) in relation to the longer period (at least 48 hours, as noted above) required for the defeminizing effects of testosterone. It is quite likely that a systemic dose of testosterone propionate remains in the body for at least 2 or 3 days. Because of their systemic toxicity, administration of the inhibitory drugs in larger amounts, including repeated administration, risks confounding the experiment.

The reader may have assumed up to this point that testosterone is the active hormone for producing sexual differentiation. This view has been challenged recently, however, on the basis of experiments that compared the potency of various steroids with testosterone. Surprisingly, such estrogens as estradiol and diethylstilbestrol are at least as effective as testosterone in inducing anovulatory sterility and defeminizing behavior [37]. A synthetic estrogen, 11β-methoxy-17α-ethynyl-17β-estradiol (Ru2858), is almost two orders of magnitude more effective than estradiol [64]. 5α-Dihydrotestosterone, the potent androgen, is ineffective in defeminizing behavior or inducing anovulatory sterility in the rat [65].

These findings relate to the discussion of the metabolism of testosterone in the brain and pituitary (see above): Estradiol is formed from testosterone and estrone from androstanedione in hypothalamus and limbic structures but not in pituitary; 5α-DHT is formed from testosterone in pituitary, hypothalamus, and other brain regions. These same transformations are observed to occur in newborn rats. In particular, the aromatization of testosterone to estradiol, which has been observed in vivo as well as in vitro in newborn rats [34, 35, 36], supports the concept that conversion to estrogen may be an obligatory step in sexual differentiation. It should be noted that 5α-DHT cannot be converted to estrogen. Further support for aromatization in sexual differentiation derives from experiments showing that the antiestrogen MER-25 can block testosterone-induced anovulatory sterility in female rats [65].

The search for receptors that mediate sexual differentiation must therefore consider both estrogen and testosterone. Whereas both labeled estradiol and testosterone are taken up by brain tissue in neonatal rats, these hormones are not generally observed to be concentrated in particular brain regions, as tends to be the case in adult rats (see above).

The story of estradiol-binding sites in immature brain presents a series of problems. Attempts to measure cytosol estradiol-binding sites led to the discovery of an atypical estradiol-binding protein in fetal and neonatal brains of male and female rats [65]. This protein was found equally in cerebral cortex, hypothalamus, and limbic structures. It bound estradiol but not diethylstilbestrol (which binds to the adult receptor) and disappeared from the brain by the beginning of the fourth postnatal week of life. This protein has a sedimentation coefficient of 4 S, in contrast to the 6 S to 8 S of the adult receptor. By virtue of all of these criteria and the protein's inability to bind columns of DNA cellulose, the protein was judged to be different from the adult receptor and similar to if not identical with a protein found in the blood of fetal and newborn rats. That protein is manufactured in the liver and is the same as the so-called α-fetoprotein [65]. Yet this fetoneonatal estrogen-binding protein (fEBP) remained in the tissue after the brain was perfused to remove blood contamination, and fEBP could be detected in cerebrospinal fluid [65].

The function of such a protein, found in extracellular fluid bathing the brain tissue, as well as in the blood, has been hypothesized to be protective in nature, preventing estrogen present in the maternal blood and estrogen that may be secreted in the neonatal animal from having deleterious, defeminizing effects. Testosterone does not appear to bind to this protein. Thus the hormone secreted by the testes of newborn males is free to enter the brain tissue and produce its effect. Support for the protective role of fEBP has come from studies revealing the ineffectiveness of low doses of estradiol in eliciting sexual differentiation and promoting uterine growth, compared to the potency in comparable doses of a synthetic estrogen that does not bind to the fEBP [64, 66]. Additional support comes from experiments that demonstrate that low doses of ^{3}H-estradiol do not bind in the nuclei of brain cells of female rats until after the fEBP has disappeared from the animal at the beginning of the fourth postnatal week [65].

These latter results do not by themselves indicate whether binding sites for estrogen in cell nuclei actually exist in brains of newborn rats. Further attempts are required to demonstrate such sites by overwhelming the binding capacity of the fetoneonatal protein either with large doses of labeled estradiol or with labeled diethylstilbestrol, which does not bind to this protein. Some recent autoradiographic evidence points to the existence of some estrogen-binding sites in brains of newborn rats [67]. Radioactivity injected as ^{3}H-estradiol was seen to concentrate over neurons in hypothalamus, preoptic area, and amygdala, and this labeling could be reduced by excess unlabeled estradiol and testosterone, but not by unlabeled 5α-DHT.

Biochemical experiments have supported the notion that this neuronal labeling represents, in large part, estradiol produced from testosterone [36]: Two hours after injection of testosterone-7-^{3}H into neonatal male and female rats, 50 percent of the brain-cell nuclear radioactivity was identified by crystallization as ^{3}H-estradiol. This material was concentrated $\approx$9 times per unit of protein in the cell nuclear fraction compared to the unfractionated tissue homogenate. Furthermore, the synthetic estrogen ^{3}H Ru2858, which, as noted above, does not bind to fEBP and is a highly effective agent for inducing sexual differentiation, labels cell nuclear sites 30 times more effectively than do comparable low doses of ^{3}H-estradiol [68].

Thyroid Hormone and Brain Development
The euthyroid condition during the first three postnatal weeks of life is essential for normal brain maturation. Both hypo- and hyperthyroid states, induced experimentally, lead to abnormalities in brain structure, chemistry, and behavior. Thyroid hormone deficiency from birth retards myelinization and development of the neuropil, and decreases respiratory activity of brain tissue [69]. Thyroidectomy before day 10 impairs body temperature regulation permanently [69]. Although both RNA and protein content in cerebrum and cerebellum are lower in rats made hypothyroid at birth, total DNA in each structure is normal at day 35, after cell proliferation has

ceased, suggesting that absence of thyroid hormones does not reduce total cell numbers but does reduce cell size [70]. In contrast, administration of T3 (triiodothyronine) to euthyroid rats at birth decreases the total number of cells in cerebrum and cerebellum [70].

Synapse formation in the cerebellum has been studied under both hyperthyroid and hypothyroid conditions [71]. The total number of synaptic profiles is decreased in hypothyroid rats after 10 days of age, following chemical thyroidectomy at birth. Hyperthyroidism is associated with an initial increase in synaptic density and number, followed by a reduction in number of synaptic profiles after 21 days of age [71]. Interesting parallels with synapse formation are reported for learning behavior in rats as a result of neonatal hypo- and hyperthyroidism. Learning ability is generally impaired as a result of neonatal hypothyroidism; hyperthyroid rats show an initial acceleration in learning ability but are poorer than euthyroid controls later in life [72].

Attempts to understand the mechanism of thyroid hormone action on the developing brain led one group to the observation that a mitochondrial fraction of neonatal brain (as well as adult and neonatal liver, but not adult brain) responds to thryoid hormone and increases the incorporation of labeled amino acids into protein in vitro [73]. Other evidence for adult pituitary and brain cells points to the existence of possible receptor sites in cell nuclei [39]. The existence of such sites in neonatal brain and pituitary tissue remains to be established.

ACKNOWLEDGMENTS

Research cited in this chapter from the author's laboratory was supported by USPHS Grant NS07080 and by an Institutional Grant RF70095 from the Rockefeller Foundation. The author expresses his appreciation to Ms. Winifred Berg for editorial assistance.

REFERENCES

1. Berthold, A. A. Transplantation der Hoden. *Arch. Anat. Physiol. Wiss. Med.* 16:42, 1849.
*2. Beach, F. A. *Hormones and Behavior.* New York: Hoeber, 1948.
*3. Turner, C. D., and Bagnara, J. T. *General Endocrinology.* 5th ed. Philadelphia: Saunders, 1971.
4. Saginor, M., and Horton, R. Reflex release of gonadotrophin and increased plasma testosterone concentration in male rabbits during copulation. *Endocrinology* 82:627, 1968.
5. Rose, R. M., Gordon, T. P., and Bernstein, I. S. Plasma testosterone levels in the male rhesus: Influences of sexual and social stimuli. *Science* 178:643, 1972.
6. Anon. Effects of sexual activity on beard growth in man. *Nature* 226:869, 1970. See also *Nature* 226:1277, 1970.
7. Bohus, B. Pituitary Adrenal Influences on Avoidance and Approach Behavior of the Rat. In Zimmerman, E., Gispen, W. H., Marks, B., and de Wied, D. (Eds.), *Progress in Brain Research.* Vol. 39, Drug Effects on Neuroendocrine Regulation. Amsterdam: Elsevier, 1973, p. 407.

*Asterisks denote key references.

*8. McEwen, B. S., Gerlach, J. L., and Micco, D. J., Jr. Putative Glucocorticoid Receptors in Hippocampus and Other Regions of the Rat Brain. In Isaacson, R., and Pribram, K. (Eds.), *The Hippocampus: A Comprehensive Treatise.* New York: Plenum, 1975, p. 285.

9. Moss, R. L., and McCann, S. M. Induction of mating behavior in rats by luteinizing hormone-releasing factor. *Science* 181:177, 1973.

10. Pfaff, D. W. Luteinizing hormone-releasing factor potentiates lordosis behavior in hypophysectomized ovariectomized female rats. *Science* 182:1148, 1973.

*11. Prange, A. J., Jr. *The Thyroid Axis, Drugs, and Behavior.* New York: Raven, 1974.

12. Friedman, E., Friedman, J., and Gershon, S. Dopamine synthesis: Stimulation by a hypothalamic factor. *Science* 182:831, 1973.

13. Keller, H. H., Bartholini, G., and Pletscher, A. Enhancement of cerebral noradrenaline turnover by thyrotropin-releasing hormone. *Nature* 248:528, 1974.

14. van Wimersma Greidanus, Tj. B., Bohus, B., and de Wied, D. Differential localization of the influence of lysine vasopressin and of ACTH 4-10 on avoidance behavior: A study in rats bearing lesions in the parafascicular nuclei. *Neuroendocrinology* 14:280, 1974.

15. Kastin, A. J., Dempsey, G. L., Le Blanc, B., Dyster-Aas, K., and Schally, A. V. Extinction of an appetitive operant response after administration of MSH. *Horm. Behav.* 5:135, 1974.

*16. McEwen, B. S., and Pfaff, D. W. Chemical and Physiological Approaches to Neuroendocrine Mechanisms: Attempts at Integration. In Martini, L., and Ganong, W. F. (Eds.), *Frontiers in Neuroendocrinology.* New York: Oxford University Press, 1973, p. 267.

17. Rosenblatt, J. S. Views on the Onset and Maintenance of Maternal Behavior in the Rat. In Aronson, L. R., Tobach, E., Lehrman, D. S., and Rosenblatt, J. S. (Eds.), *Development and Evolution of Behavior: Essays in Memory of T. C. Schneirla.* San Francisco: Freeman, 1970, p. 489.

18. Johnston, P., and Davidson, J. M. Intracerebral androgens and sexual behavior in the male rat. *Horm. Behav.* 3:345, 1972.

19. Owen, K., Peters, P. J., and Bronson, F. H. Effects of intracranial implants of testosterone propionate on intermale aggression in the castrated male mouse. *Horm. Behav.* 5:83, 1974.

20. Yahr, P., and Thiessen, D. D. Steroid regulation of territorial scent marking in the Mongolian gerbil (*Meriones unguicalatus*). *Horm. Behav.* 3:359, 1972.

21. Mann, D. R., and Barraclough, C. A. Role of estrogen and progesterone in facilitating LH release in 4-day cyclic rats. *Endocrinology* 93:694, 1973.

22. Whalen, R. E., and Gorzalka, B. B. The effects of progesterone and its metabolites on the induction of sexual receptivity in rats. *Horm. Behav.* 3:221, 1972.

23. Erickson, C. J., Bruder, R. H., Komisaruk, B. R., and Lehrman, D. S. Selective inhibition by progesterone of androgen-induced behavior in male ring doves (*Streptopella risoria*). *Endocrinology* 81:39, 1967.

24. Erpino, M. J. Temporary inhibition by progesterone of sexual behavior in intact male mice. *Horm. Behav.* 4:335, 1973.

25. Griffo, W., and Lee, C. T. Progesterone antagonism of androgen-dependent marking in gerbils. *Horm. Behav.* 4:351, 1973.

26. Luttge, W. G. Activation and inhibition of isolation induced inter-male fighting behavior in castrate male CD-1 mice treated with steroidal hormones. *Horm. Behav.* 3:71, 1972.

27. Ross, G. E., and Zucker, I. Progesterone and the ovarian-adrenal modulation of energy balance in rats. *Horm. Behav.* 5:43, 1974.

28. Denef, C., Magnus, C., and McEwen, B. S. Sex differences and hormonal control of testosterone metabolism in rat pituitary and brain. *J. Endocrinol.* 59:605, 1973.

29. Feder, H. H. The comparative actions of testosterone propionate and 5α androstan 17β ol 3 one propionate on the reproductive behavior, physiology, and morphology of male rats. *J. Endocrinol.* 51:241, 1971.

30. Massa, R., and Martini, L. Interference with the 5α Reductase System: A New Approach for Developing Antiandrogens. In Hubinont, P. O., Hendeles, S. M., and Preumont, P. (Eds.), *Hormones and Antagonists:* Proceedings of the 4th International Seminar on Reproductive Physiology and Sexual Endocrinology, Brussels, May 23–25, 1972. Reprinted from *Gynecol. Invest.* Vol. 2, No. 1–6, 1971/2, and Vol. 3, No. 1–4, 1972.

31. Flores, F., Naftolin, F., Ryan, K. J., and White, R. J. Estrogen formation by the isolated perfused rhesus monkey brain. *Science* 180:1074, 1973.

32. Reddy, V., Naftolin, F., and Ryan, K. J. Aromatization in the central nervous system of rabbits: Effects of castration and hormone treatment. *Endocrinology* 92:589, 1973.

33. Ryan, K. J., Naftolin, F., Reddy, V., Flores, F., and Petro, Z. Estrogen formation in the brain. *Am. J. Obstet. Gynecol.* 114:454, 1973.

34. Weisz, J., and Gibbs, C. Conversion of testosterone and androstenedione to estrogen in vitro by the brain of female rats. *Endocrinology* 94:616, 1974.

35. Weisz, J., and Gibbs, C. Metabolites of testosterone in the brain of the newborn female rat after an injection of tritiated testosterone. *Neuroendocrinology* 14:72, 1974.

36. Lieberburg, I., and McEwen, B. S. Estradiol-17β: A metabolite of testosterone recovered in cell nuclei from limbic areas of neonatal rat brains. *Brain Res.* 91:171, 1975.

*37. McEwen, B. S., Denef, C. J., Gerlach, J. L., and Plapinger, L. Chemical Studies of the Brain as a Steroid Hormone Target Tissue. In Schmitt, F. O., and Worden, F. G. (Eds.), *The Neurosciences: Third Study Program*. Boston: MIT Press, 1974, p. 599.

*38. McEwen, B. S., Zigmond, R. E., and Gerlach, J. L. Sites of Steroid Binding and Action in the Brain. In Bourne, G. H. (Ed.), *Structure and Function of Nervous Tissue.* Vol. 5. New York: Academic, 1972, p. 205.

39. Oppenheimer, J. H., Schwartz, H. L., and Surks, M. I. Tissue differences in the concentration of triiodothyronine nuclear binding sites in the rat: Liver, kidney, pituitary, heart, brain, spleen and testis. *Endocrinology* 95:897, 1974.

*40. King, R. J. B., and Mainwaring, W. I. P. *Steroid-Cell Interactions.* Baltimore: University Park Press, 1974.

*41. Feinendegen, L. E. *Tritium-Labelled Molecules in Biology and Medicine.* New York: Academic, 1967.

*42. McEwen, B. S., and Zigmond, R. E. Isolation of Brain Cell Nuclei. In Marks, N., and Rodnight, R. (Eds.), *Research Methods in Neurochemistry.* Vol. 1. New York: Plenum, 1972, p. 140.

*43. Luine, V. N., and McEwen, B. S. Steroid Hormone Receptors in Brain and Pituitary: Topography and Possible Function. In Goy, R. W., and Pfaff, D. (Eds.), *Sexual Behavior.* New York: Plenum, in press.

44. Pfaff, D. W., and Keiner, M. Atlas of estradiol-concentrating cells in the central nervous system of the female rat. *J. Comp. Neurol.* 151:121, 1973.

45. deVellis, J., and Inglish, D. Hormonal control of glycerol phosphate dehydrogenase in the rat brain. *J. Neurochem.* 15:1061, 1968.

46. deVellis, J., Inglish, D., Cole, R., and Molson, J. Effects of Hormones on the Differentiation of Cloned Lines of Neurons and Glial Cells. In Ford, D. (Ed.), *Influence of Hormones on the Nervous System,* Proc. Int. Soc. Psychoneuroendocrinology, Brooklyn, 1970. Basel: Karger, 1971, p. 25.

47. Butcher, R. L., Collins, W. E., and Fugo, N. W. Plasma concentration of LH, FSH, prolactin, progesterone, and estradiol 17β throughout the 4-day estrous cycle of the rat. *Endocrinology* 94:1704, 1974.
48. Ferin, M., Tempone, A., Zimmering, P. A., and Van de Wiele, R. I. Effect of antibodies to 17β estradiol and progesterone on the estrous cycle of the rat. *Endocrinology* 85:1070, 1969.
49. Clark, J. H., Peck, E. J., and Anderson, J. N. Oestrogen receptors and antagonism of steroid hormone action. *Nature* 251:446, 1974.
50. Fajer, A. B., Holzbauer, M., and Newport, H. M. The contribution of the adrenal gland to the total amount of progesterone produced in the female rat. *J. Physiol.* (Lond.) 214:115, 1971.
51. Gessa, G. L., and Tagliamonte, A. Role of brain monoamines in male sexual behavior. *Life Sci.* 14:425, 1974.
52. Kalra, S. P., and McCann, S. M. Effects of drugs modifying catecholamine synthesis on plasma LH and ovulation in the rat. *Neuroendocrinology* 15:79, 1974.
53. Lindström, L. H., and Meyerson, B. J. The effect of pilocarpine, oxotremorine and arecoline in combination with methyl-atropine or atropine on hormone activated oestrous behavior in ovariectomized rats. *Psychopharmacologia* 11:405, 1967.
54. Zemlan, F. P., Ward, I. L., Crowley, W. R., and Margules, D. L. Activation of lordotic responding in female rats by suppression of serotonergic activity. *Science* 179:1010, 1973.
*55. Burns, R. K. Role of Hormones in the Differentiation of Sex. In Young, W. C. (Ed.), *Sex and Internal Secretions,* 3d ed. Vol. I. Baltimore: Williams & Wilkins, 1961, p. 76.
*56. Goy, R. W. Early Hormonal Influences on the Development of Sexual and Sex-Related Behavior. In Schmitt, F. O. (Ed.), *The Neurosciences: Second Study Program.* New York: Rockefeller University Press, 1970, p. 196.
57. Jost, A. Hormonal factors in the sex differentiation of the mammalian foetus. *Philos. Trans. R. Soc. Lond.* [Biol.] 259:119, 1970.
58. Hayashi, S., and Gorski, R. A. Critical exposure time for androgenization by intracranial crystals of testosterone propionate in neonatal female rats. *Endocrinology* 94:1161, 1974.
59. Nadler, R. D. Intrahypothalamic exploration of androgen-dependent brain loci in neonatal female rats. *Trans. N.Y. Acad. Sci.* 34:572, 1972.
60. Raisman, G., and Field, P. M. Sexual dimorphism in the neuropil of the preoptic area of the rat and its dependence on neonatal androgen. *Brain Res.* 54:1, 1973.
61. Litteria, M., and Thorner, M. W. Inhibition of the incorporation of ^{3}H lysine in the Purkinje cells of the adult female rat after neonatal androgenization. *Brain Res.* 69:170, 1974.
62. Gorski, R. A., and Shryne, J. Intracerebral antibiotics and androgenization of the neonatal female rats. *Neuroendocrinology* 10:109, 1972.
63. Salaman, D. F., and Birkett, S. Androgen-induced sexual differentiation of the brain is blocked by inhibitors of DNA and RNA synthesis. *Nature* 247:109, 1974.
64. Doughty, C., Booth, J. E., McDonald, P. G., and Parrott, R. F. Effects of oestradiol 17β, oestradiol benzoate, and the synthetic oestrogen, Ru2858, on sexual differentiation in the neonatal female rat. *J. Endocrinol.* 67:419, 1975.
65. Plapinger, L., and McEwen, B. S. Gonadal Steroid-Brain Interactions in Sexual Differentiation. In Hutchison, J. (Ed.), *Biological Determinants of Sexual Behavior.* New York: Wiley, in press.

66. Raynaud, J. P. Influence of rat estradiol binding plasma protein (EBP) on uterotrophic activity. *Steroids* 21:249, 1973.

67. Sheridan, P. J., Sar, M., and Stumpf, W. E. Autoradiographic localization of ^{3}H estradiol or its metabolites in the central nervous system of the developing rat. *Endocrinology* 94:1386, 1974.

68. McEwen, B. S., Plapinger, L., Chaptal, C., Gerlach, J., and Wallach, G. Role of fetoneonatal estrogen binding proteins in the association of estrogen with neonatal brain cell nucleus receptors. *Brain Res.* 96:400, 1975.

69. Hamburgh, M. An analysis of the action of thyroid hormone on development based on in vivo and in vitro studies. *Gen. Comp. Endocrinol.* 10:198, 1968.

*70. Balasz, R. Effects of Hormones on the Biochemical Maturation of the Brain. In Ford, D. H. (Ed.), *Influence of Hormones on the Nervous System*, Proc. Int. Soc. Psychoneuroendocrinology, Brooklyn, 1970. Basel: Karger, 1971, p. 150.

71. Nicholson, J. L., and Altman, J. Synaptogenesis in the rat cerebellum: Effects of early hypo and hyperthyroidism. *Science* 176:530, 1972.

72. Eayrs, J. T. Endocrine influence on cerebral development. *Arch. Biol.* (Liege) 75:529, 1964.

73. Sokoloff, L., and Roberts, P. Biochemical Mechanisms of the Action of the Thyroid Hormones in Nervous and Other Tissues. In Ford, D. H. (Ed.), *Influence of Hormones on the Nervous System,* Proc. Int. Soc. Psychoneuroendocrinology, Brooklyn, 1970. Basel: Karger, 1971, p. 213.

Chapter 36

Learning and Memory: Approaches to Correlating Behavioral and Biochemical Events

Bernard W. Agranoff

BIOLOGICAL BASIS OF MEMORY

There is a common belief among biologists that the detailed mechanisms of behavioral plasticity constitute a major remaining frontier in our understanding of living systems. Certainly it is at this time the most obscure frontier and, accordingly, hypotheses run rampant. While a vast literature on human and animal learning and memory exists, results among reported studies are not as readily comparable as are usual biological experiments. Factors such as the species and even the strain of animal used, the details of construction of experimental apparatus, the sequence of training, and the conditions of animal housing appear to affect measured behavior drastically. Although there are many conventions for quantifying behavioral parameters, different methods may lead to different conclusions and no one method is more easily reconciled with known physiological facts than another. Nevertheless, such experiments have been directed at answering the question: What are the changes that take place in the brain during a few seconds of experience that will mediate the resulting altered response of an organism for long periods, even for a lifetime? That such changes are permanent and are resistant to subsequent electrical storms and silences of the brain point to a physicochemical alteration associated with learning, brought about by an underlying molecular process. The various investigations described in this chapter constitute attempts to establish, via chemical and behavioral means, the physiological changes that lead to the altered performance of an animal measured at various times after a training experience.

Major consideration in this chapter is given to correlation of neural metabolism with memory formation. One approach employs agents reported to inhibit or stimulate specific molecular processes. These agents are administered in association with training to test the hypothesis that one or another molecular process is critical in memory formation. Alternatively, in the absence of such agents, biochemical studies have been directed at gross or subtle changes in composition of the brain or on its selective incorporation of isotopically labeled precursors as a function of training.

CONSOLIDATION AND STAGES OF MEMORY FORMATION

Our present concept of memory formation and storage began with the observations in the last century that physical trauma to the human brain results in loss of memory [1]. Typically, a blow to the head may lead to a retrograde amnesia, i.e., loss of memory for events immediately preceding the trauma. The common interpretation of this phenomenon is that memory ordinarily forms after the actual experience, and the "fixation," or *consolidation,* of memory is blocked by the blow [2]. A similar conclusion has been drawn from animal experiments in which electroconvulsive shock (ECS) administered after training is shown to result in poor subsequent

765

performance, while the same treatment given sometime after training has no effect. In each instance, the growing insusceptibility is thought to reflect a physiological process of consolidation, whereby memory is converted from an unstable, or labile, form to a permanent form. This process has been reported by various investigators to take from seconds to as much as hours for completion.

Not all psychologists accept the consolidation concept, and alternate explanations for the developing strength of memory with passage of time following training have been offered. It is argued, for example, that the amnestic treatment, via ECS or some other means, is itself noxious. The subject learns that, when he performs correctly, he will be punished and he therefore does not exhibit the learned response after the treatment. The amnestic agent has been postulated to interfere with performance in other ways, for example by creating spurious electrical effects in the neuronal networks that mediate the new behavior. Both of these explanations argue that the new memory is not obliterated by the amnestic agent, but that some superimposed process prevents its retrieval. In the instance of a physical blow to the head, an argument against the consolidation hypothesis might state that fear or pain associated with the blow has repressed events closely associated in time with the accident. Whether interference is at the behavioral or electrical level, it can be argued that, in the absence of interference, memory of training very likely forms simultaneously and permanently with training experience. Although the issue has not been completely resolved, the consolidation hypothesis has gained considerable support from passive avoidance studies (see below), in which amnestic animals move in the direction that leads to the amnestic treatment [3]. If the treatment were aversive, the animals would not be expected to move in such a direction. Studies with antibiotic anti-metabolites also give considerable support for the consolidation hypothesis. Animals may be given such agents prior to training. Normal acquisition is seen but, upon retesting some time later, there is a marked memory deficit. If the injection were to have the properties of an aversive unconditioned stimulus, it is not likely that it would, in addition, exert its effect if given prior to the conditioned stimulus (backward conditioning). As discussed below, these agents may not obliterate memory, but instead may block the formation of a more stable form from a labile one

MEASUREMENT OF LEARNING AND MEMORY

Learning Paradigms

Perhaps the simplest kind of training we can easily measure is *habituation*: the decrease in some measured response seen with repeated stimulation [4]. The loss of a "startle" response after repeated tapping on an animal enclosure is an example of this phenomenon. Although it can be demonstrated in primitive organisms, habituation in higher animals is believed to involve neural networks and is not simply a reflection of muscle fatigue or receptor adaptation. Habituation has not commonly been investigated extensively as a model for biochemical correlates of behavior because by its nature it is not long-lasting, and one cannot exclude the possibility that it is mediated exclusively via short-term (and possibly nonchemical) memory mechanisms. Augmentation of responses after repeated stimuli, or sensitization, similarly does not have sufficiently long-lasting properties to serve as a model for studies on permanent alterations in the brain.

Most studies attempting to relate biochemical events to learning have employed conditioning paradigms. Commonly, a distinction is made between classical (type I) and instrumental (type II) conditioning. In classical conditioning, a neutral stimulus is conditioned by pairing it with an unconditioned stimulus. For example, an animal is exposed to a light flash (the conditioned stimulus, CS), followed by a mild punishing electrical shock (the unconditioned stimulus, UCS) which elicits a motor or autonomic response. After several pairings, the response, or one like it, is seen after the CS. In classical conditioning, the UCS presentation does not depend upon the animal's behavior. In instrumental conditioning, the response of the animal (such as pressing a bar) determines whether it will receive the UCS. Instead of a shock or other noxious stimulus, the UCS may be pleasant, such as food reward, for both type I and type II conditioning.

In general, use of a noxious UCS results in faster learning and perhaps more lasting memory than does positive reinforcement. The new behavior may be either active or passive. In a step-down task, a mouse learns that its natural tendency to step down from a small pedestal upon which it has been placed results in a mild electrical shock to the foot. It learns to remain on the pedestal (passive avoidance). Alternatively, an animal may be taught that shortly after a light or sound warning, it must leave the chamber in which it has been placed (e.g., by jumping over a hurdle) in order to avoid a punishing foot shock (active avoidance).

A further refinement in training techniques is *discrimination learning.* Animals learn to choose a given limb of an apparatus on the basis of position (right or left), illumination (light vs. dark), color, or geometrical pattern. Such tasks may or may not involve an avoidance component. For example, an animal may be placed in the starting box (the stem of a Y-maze) and be forced to make a choice of one limb of the maze by application of foot shock in the stem and the other arm of the maze. This is an escape discrimination task. In the avoidance variant of this kind of problem, a warning signal is given prior to shock, and only those trials in which animals avoid the shock are considered to be positive. Discrimination tasks have the advantage of tending to minimize the effects of illness of an experimental subject. If a subject is sick after a drug treatment, it is not likely that the illness will influence its decision in a discrimination task.

In a nondiscriminative avoidance task, in which the rate or speed of responding may be the index of learning used, the animal's state of health could easily affect recorded scores. Various avoidance tasks, including "step-down," "step-through," and "shuttlebox," are such timed paradigms. If the animal does not respond within a fixed period, it does not receive a positive response score for the given trial. Nevertheless, these paradigms have several advantages over discrimination tasks. A single trial score for a timed paradigm may be highly significant. For example, in a particular situation, a naive animal will step down within 5 sec of being placed in an apparatus whereas, after foot shock, it may take over 30 sec. In this way, measurement of one-trial learning is possible. A single two-choice discrimination task score, on the other hand, generally is not informative because on a random basis an animal will make the correct response 50 percent of the time, 25 percent of the time for two trials, and will achieve a criterion of three out of four in 12.5 percent of the sessions.

Methods of Quantitation

If we consider learning as a change in performance as a result of training, then memory is the demonstration of that new performance at later times. Loss of memory is variously referred to as extinction, forgetting, or amnesia, depending on the experimental history of a subject. When lost memory reappears, it is generally proposed that the subject had not lost memory but temporarily could not retrieve it.

A limitation to the study of biochemical correlates of behavior is the difficulty in quantifying the putative molecular and physiological factors that result in altered behavior. Molecular mechanisms have been inferred from bioassays in which the response of whole animals is measured. We do not, at present, have a simpler model for the measurement of long-term memory formation and must rely on total (usually overt) behavior.

In general, investigators tend to modify experimental conditions in the direction of maximizing an effect. They are confronted with several numbers after an experiment: These may include latency (the time between onset of stimulus and response), the tally of correct and incorrect responses for segments of a training session (first five trials, second five trials, and so forth), and various other criteria, which can be established before or even after an experiment (e.g., nine out of ten correct responses). From these, the investigator must measure the "amount" of memory. He might use raw scores (such as number of correct responses), percent of correct responses in trained subjects relative to untrained subjects, or a number of other scoring systems. None of these would seem, a priori, to be more valid than another, yet opposite conclusions sometimes can be drawn from a given set of data depending on which scoring method is used. Individual laboratories generally use only one particular method of calculation, so they are able to evaluate dependent variables, if only qualitatively. Often, however, results are not directly comparable to those of another laboratory because of a different scoring method. Given these problems, it is not surprising that conflicting reports occur.

Dosage and temporal gradients are often used in an attempt to demonstrate a partial impairment. In such instances, we are faced with the dilemma of whether some fraction of the specific components of a learned behavior is lost or whether there is a general impairment of all the components. For example, in experiments in which amnesia has been produced by ECS after training, the "lost" memory can be regained by means of a behavioral "reminder" [5]. If performance of the learned response involves a number of physiological units operating in "series," the impairment of any one could result in the loss of performance, and its introduction would permit performance of the learned response.

Despite such experimental difficulties, experiments designed to single out, by means of appropriate control experiments, the various components of behavior have proved useful. These inferred components do not yet correspond to anatomical or functional units of the brain. Hence reductionistic approaches, such as those out-lined below, have been used with increasing frequency by neurobiologists in an attempt to correlate behavior with physiological parameters.

EXPERIMENTAL PREPARATIONS

Whole Animals

For many years, the laboratory rat dominated physiological psychology. Its hardi-ness under adverse laboratory conditions is somewhat balanced by the disadvantage that much interesting innate behavior may have been bred out of this docile beast. Differences in behavior among strains have long been recognized. Much variation from experiment to experiment and laboratory to laboratory has recently been attributed to diurnal variation and seasonal effects on behavior. Inbred strains of rats can vary from supplier to supplier, as can the conditions of rearing. Strains of mice that are highly homogeneous genetically are available, reducing at least this one source of variability in behavior. Mice are generally less expensive to purchase and to maintain than rats, and they effect additional savings in radioisotope costs by virtue of their small size. The importance of genetic factors is underscored by experiments in which opposite drug effects on learning have been found in mice of different strains [6].

Evidence for learning and memory has been sought in a wide variety of animals (and even plants!). To say that any species is incapable of demonstrating learning or memory is probably unwise, since some new task can transform an animal from a "nonlearner" to a "learner." The following includes many of the better-known attempts at training among animal phylla.

Invertebrates

Touch and light pairing has been studied in protozoa, and well-documented examples of habituation have been reported in micrometazoa [7]. Coelenterates have an ele-mentary nervous system, and habituation has been reported, as well as learning. "Bait-shyness," the rejection of food previously coupled with an aversive stimulus, has been claimed for a wide variety of species, including the sea anemone [8]. That planarians can learn is generally accepted, but whether they can retain what they have learned for long periods is less apparent. This uncertainty precludes, at least for the present, conclusions regarding more complex experiments, such as chemical transfer of memory. An old observation of Yerkes (see [7]), who used a single annelid worm, reported its learning a right-left discrimination task. Behavioral studies of nematodes may be spurred by recent ultrastructural mapping of a relatively simple nematode nervous system [9]. A rather rapid life cycle suggests their suitability for genetic investigation of the nervous system.

Echinoderms have been relatively little studied, although learning has been reported in the starfish. Mollusks have been investigated more extensively. The octopus can be trained to respond to various visual or tactile stimuli and to retain (for many hours) what it has learned [10]. Occasionally it has been speculated that long-term memory and its possible chemical basis may have first appeared phylogenetically in the vertebrates. The octopus would appear to be an exception or, perhaps, to disprove this idea. A great deal of interest in mollusks, particularly *Aplysia,* relates to study of reduced systems, discussed below.

Studies with arthropods include reduced systems such as the cockroach ganglion, also discussed below, as well as of intact insects such as the social insects, in which learning has been examined. A change in flying patterns of the honeybee can be considered a manifestation of plasticity of the nervous system. Experimentally, social insects may be difficult to work with, particularly under laboratory conditions. Insects more suitable from this standpoint, such as *Drosophila,* may prove useful for the genetic dissection of memory [11]. Considering the size of species in the class Insecta, subjects may yet be discovered that are more ideal for a combined behavioral-genetic study than are those that are presently being used. Readers fortunate enough to have seen a flea circus can attest to the potential behavioral skills of insects.

Vertebrates

Among the classes that make up the phyllum Vertebrata, amphibians and reptiles are generally unpopular for behavioral studies because of the difficulty, up to the present time, of easily demonstrating reproducible, stable learning changes. The turtle, however, has proved to be a useful behavioral subject.

Fish have been studied extensively both from the standpoint of learning and of long-term memory formation [12]. Because they are poikilotherms, temperature can readily be introduced as an experimental variable. Regeneration in the central nervous system of cold-blooded animals has made possible studies on specification and regrowth. A gynogenetic strain, *Poecilia formosa* [13], permits experiments with genetically homogeneous subjects. Whereas most fish have highly developed visual systems (including color vision), many species, including the bullhead and salmon, also have highly developed olfactory systems, which have been the subject of behavioral investigation. Conditioning of the lateral-line system has been reported in weakly electrical fish [14].

Birds have been the subject of both behavioral and biochemical studies. The pigeon is a standard subject, especially for instrumental conditioning. The newly hatched chick is completely myelinated and capable of immediate behavioral responses [15]. The duckling has been studied extensively in imprinting, a behavioral phenomenon whereby lifelong patterns are specified during a critical post-hatching period [16].

Although psychologists have reported learning in virtually every common species of mammal known, the destructive nature of biochemical analyses has limited correlative studies to inexpensive, small animals. Thus, very little biochemical information is available from studies of such mammals as cats, dogs, and monkeys. Changes in the concentration of metabolites and hormones in the blood in primates, as well as in other species, during learning and stress have been studied extensively, but will not be discussed here [17]. (See also Chap. 35.)

Reduced Systems

If conditioning requires only that pairing of the CS and UCS eventually relates the CS to the unconditioned response, formal models may be constructed, using only a portion of the animal. The CS, UCS, and response may all be represented electrically and measured by means of electrodes in the central nervous system. It has been proposed that heterosynaptic facilitation in the *Aplysia* may serve as such a model [18]. The *Aplysia* has been studied also for a model habituation system, involving gill withdrawal after a mechanical stimulus [19]. This system is particularly attractive for study since the neuronal pathways that mediate the withdrawal response are known. In addition, this species has many large neurons which can be identified, impaled with electrodes, and injected. *Tritonia,* a related mollusk, exhibits a withdrawal response when a single known neuron is stimulated [20].

Neural systems have been elucidated in crayfish and lobster. In the latter, specific axons that mediate excitation and inhibition of muscle contraction have been identified, although behavioral modification of the limb has not yet been reported.

The cockroach has been extensively studied, using a model system described by Horridge and modified by Eisenstein and Cohen [21]. The thoracic ganglion of the headless cockroach mediates learning of a position habit. Separate electrodes are connected to a corresponding pair of legs. When the tip of a leg touches a saline bath beneath it, an electrical contact is made so that leg position (either in or out of the bath) is recorded. Under these conditions, the legs move independently. During the experiment, a mild punishing shock is administered to the experimental leg whenever it contacts the bath. During a 30-min trial, the experimental leg exhibits a developing avoidance of the bath. The paired, or "yoked," leg moves in and out of the bath, but is shocked whenever the experimental leg is extended. It receives as much shock as the positional leg, but not in relation to any position. No tendency for the paired leg to avoid the water bath is seen. The avoidance seen in the experimental leg appears, then, to be a specific position habit. Blocks of such learning by actinomycin D and by cycloheximide, inhibitors of RNA and protein synthesis, respectively, have been reported [22].

The reductionistic approach in a complex nervous system such as in that of a mammal might seem self-defeating. It appears, however, that in the highly evolved nervous system, specific structures – and, therefore, localized anatomical regions – may mediate specific components of the behavioral response. The effect of bilateral lesions of the hippocampus or of the mammillary bodies in man has been reported to lead to a specific memory deficit [23]. Such a phenomenon has not been demonstrated by lesioning in lower animals. Electrophysiological correlates of visual learning have been reported after visual and auditory units in the cat have been paired, using leg shock as the UCS [24]. Studies with brain explants and tissue culture, which tend to document the formation of new synapses, are being pursued actively at present [25]. The in vitro demonstration of formation of new synaptic connections could be a valuable step forward in understanding the rules that govern brain plasticity.

AGENTS AFFECTING LEARNING AND MEMORY

Two kinds of arguments exist that point to an organic basis for memory. One can be inferred based on the administration of disruptive agents that have putative specific chemical or physical effects. The second is based on the claim that specific alterations in brain composition or isotopic incorporation are functions of a behavioral variable. By their nature, blocking agents are unphysiological, so that they direct us only by inference to understanding brain mechanisms. They do, however, have the advantage of being usable in such complex systems as the whole animal and, hopefully, can tell us at what point of the training session, in what region of the brain, and in which macromolecules putative changes may be expected to take place.

Ablation

Perhaps the simplest investigative approach to the disruption of memory is to test for critical brain regions in behavior by examining the effect of their removal. Lashley concluded, from a lifetime study of the rat, that localization of learning or memory could not be demonstrated by means of lesions, and generally he rejected synaptic theories of memory. His and subsequent studies have shown that complex behaviors are, in general, more sensitive to destruction of tissue than are simple ones and that large lesions generally produce greater effects than do small ones, whereas the influence of brain regions is secondary. This latter claim has been used as an argument that memory is not localized − or is at best poorly localized − anatomically within the brain. Current ideas attempting to resolve Lashley's findings with the prevalent connectionist views of synaptic organization and, therefore, of memory are summarized as follows [26, 27]:

1. Lashley made inappropriate lesions across multiple functional areas, which were not well understood as differentiated structures at the time.
2. Infections may have created more brain damage than that due only to the surgical cut.
3. The mazes used involved too many sensory and motor variables. A variety of simpler tests might have shown behavior specifically correlated to lesions.
4. Lesions, particularly if given prior to training, are overcome by brain redundancy, i.e., an alternative pathway takes over.

Although the matter is not yet resolved, more sophisticated approaches do show promise of localizing behavioral mechanisms. These include the use of reversible injury via localized electrical stimulation, cooling probes, etc. Of biochemical interest is the radioautographic demonstration that intracerebral injections of antibiotics can produce reversible localized regions of inhibition of protein synthesis in brain [28].

Spreading Depression

The topical application of concentrated KCl to the cortical surface causes cessation of electrical activity at the site of application which then spreads over the hemicortex. Animals with spreading depression of one hemicortex can be trained and then will demonstrate learning at a later time when similarly treated, but not when KCl has been applied to the other hemicortex on retraining [29]. By varying the time after

learning at which KCl is applied, it can be shown that within a few minutes of training there is an apparent transfer of the learned response to the opposite hemisphere.
Inhibition of protein synthesis by KCl has been reported in rat, as well as in the
goldfish. In the goldfish, KCl is a convulsant and also produces memory loss.

Electroconvulsive Shock

ECS is the most commonly used experimental amnestic agent. Nevertheless, there
is no uniform way in which it is applied. Electrodes are in some instances attached
to ears, while in others convulsions are produced transcorneally or through the skin.
The amount of current, frequency of the pulse, and its duration are all varied, and
there is no agreement as to whether the agent itself or the resulting convulsion is
the amnestic effector. Amnesia has been reported after ECS was administered to
sedated animals, although they did not convulse visibly. Some claim that there is a
graded response to ECS, and others say that it is an "all-or-none" phenomenon.
The recent finding that ECS administered shortly *before* training can block memory
selectively without measurable effect on acquisition [30] suggests that it may act
much like the antibiotics (see below). It may be the depletion of metabolites or
some other disturbance in metabolic state that persists after behavioral recovery
that blocks memory. Use of gaseous convulsants, such as hexafluorodiethyl ether,
would seem to be more easily regulated [15]. Commonly used convulsant agents,
in addition to electroconvulsive shock and KCl, include hexafluorodiethylether,
strychnine, pentylenetetrazol, and picrotoxin. Subconvulsant doses of picrotoxin
and strychnine have been reported to enhance learning and memory [31]. When
amphetamine, a stimulant, is given together with cycloheximide, an inhibitor of
protein synthesis, the amnestic effect of cycloheximide is reportedly blocked, and
memory is preserved [32].

Genetic Manipulation

In 1940 Tryon reported [33] the results of breeding rats from a parent strain on the
basis of their response to maze training. Two inbred populations, "maze-bright" and
"maze-dull," were derived. Subsequent biochemical analyses for brain components
and enzymes have not revealed striking differences. Further, the "brightness" is task
specific. The ideal subject for the geneticist would be a bacterium that possesses a
demonstrable behavior. Unlikely as this might seem, recent strides in understanding
the mechanism of chemotaxis [34] show that bacteria detect an attractant molecule
gradient by means of temporal coding — a "minimemory." Koshland has discussed
the possible relevance of the chemotactic model to the nervous system [35].

Temperature

Ransmeier and Gerard [36] failed to demonstrate any interference with memory of
a previously learned maze by suddenly lowering body temperatures in the hamster,
although electrical activity of the brain was suppressed. It was subsequently demonstrated that hypothermia extended the period during which ECS could produce a
performance deficit. Temperature experiments are more easily and rigorously performed in poikilotherms. Cooling goldfish after they have been trained in a shock-
avoidance task prolongs the period during which memory of that task is susceptible

to ECS or to puromycin. Warming shortens the period of susceptibility [37, 38].
Cooling also slows protein synthesis, but cooling for long periods after training does
not appear to block memory [38]. Disruption of protein synthesis with blocking
agents in the face of otherwise normal metabolism may somehow block memory
formation, whereas a concerted slowing of all metabolism may simply delay the steps
relevant to memory formation.

Inhibitors of Protein Synthesis

Flexner reported in 1964 [39] that bilateral temporal injections into the brains of
etherized mice within 5 days after Y-maze training resulted in loss of memory retention. At later times, amnesia could be produced by injecting puromycin into six sites,
including ventricular and parietal sites, in addition to the bitemporal injections. It was
initially proposed that memory was localized in the temporal region and, after several
days, spread throughout the brain. There appeared to be no time limit to the disruption resulting from six injections of puromycin. Because puromycin is a selective
inhibitor of protein synthesis, the suggestion was made that puromycin blocked the
formation of protein required for memory function. On the basis of the amnesia
produced by inhibition of protein synthesis throughout the brain (six sites), it was
also proposed that protein synthesis was required for memory maintenance. Since
these initial proposals, Flexner's findings and conclusions have been extended as
follows [40]:

1. Injection of saline into the bitemporal sites after long intervals appears to
"restore" memory. This finding argues against a block of memory and for a chemical interference with retrieval of memory, relieved by saline.

2. In Flexner's experiments, a glutarimide antibiotic (acetoxycycloheximide,
AXM) did not block memory. Because AXM is an even more potent inhibitor of
protein synthesis than puromycin, it was concluded that some action of puromycin
other than the block of protein synthesis was responsible for its behavioral effect.
The product of the reaction of puromycin with the polysomal complex, peptidyl
puromycin, was proposed to be responsible for the behavioral effects of the drug.
Support for this argument came from the purported block of the puromycin effect
by AXM, which is also known to block the formation of peptidyl puromycin, in
addition to blocking protein synthesis.

3. Puromycin was demonstrated to cause swelling in brain mitochondria. This
effect was also thought to be mediated via peptidyl puromycin since protection was
observed when AXM was simultaneously injected.

4. When a protein synthesis blocker was injected bitemporally into mice one day
after training, the amnestic effects of this blocker could be negated by a number of
drugs, including reserpine, L-DOPA, imipramine, tranylcypromine, and D-amphetamine
[40]. It was proposed that peptidyl puromycin is absorbed to adrenergic sites in the
brain.

It appears that the effects of puromycin observed by Flexner are complex and
may not all be related to memory. Memory in the goldfish has been shown to be
blocked by antibiotics [12]. Fish were trained in a shuttlebox to swim over a barrier

upon a light signal in order to avoid a punishing electrical shock administered through the water. During 30 trials, avoidances rose from an average of about 20 percent (trials 1 to 10) to about 40 percent for trials 11 to 20 and about 60 percent for trials 21 to 30. When fish were given 20 trials on the first day of an experiment, returned to home storage tanks, and retrained 3 days later, they still averaged about 60 percent avoidance responses, demonstrating stable memory of what they had learned. Highly significant performance levels in comparison to those of untrained control groups are seen even 30 days after the day 1 training session. Findings with the fish are summarized as follows:

1. Puromycin (170 μg in 10 μl) can be injected easily into conscious goldfish by means of a 30-gauge needle and a Hamilton syringe. Protein synthesis is blocked by about 80 percent for nearly a day, but no gross neurological or behavioral abnormalities are seen.

2. If the drug is given immediately after training, no evidence of previous training is detected on the basis of response rates measured 3 days later, i.e., there is no memory of training. No effect resulted, however, when the same amount of puromycin was injected into fish that had been trained but which had been returned to their home tanks for 1 hr prior to injection. Such fish showed response rates comparable to those of uninjected control subjects. An important difference between these studies and those of Flexner was the evidence that fish have a relatively short consolidation gradient. The demonstration of memory fixation without the production of unconsciousness by convulsants or anesthetics lends support for a consolidation theory of memory. Fish are functioning normally during the time the drug is active, so the hypothesis that gross electrical disruption of the brain is necessary for the production of amnesia is not supported. In fact, the amnestic effects of ECS may not be related to the convulsions, but to metabolic sequelae [30], such as have been observed [41].

3. Puromycin injected before training has little effect on acquisition of learning, although memory, judged by performance on day 4, is blocked.

4. Groups of fish tested at various times after training and immediately after the injection of puromycin exhibit a slow decay of memory. Those tested immediately after training and injection of puromycin show normal response rates for trials 21 to 30. Other groups injected immediately after training and tested at various times thereafter exhibit a decrease in avoidance responses during the next 2 days. Thus, puromycin does not immediately obliterate memory, but rather initiates a process that takes several days for completion. Perhaps it weakens some physiological process, which then gives rise to defective intermediates that are subsequently destroyed.

5. AXM and cycloheximide produce effects similar to those of puromycin on memory in fish, in sharp contrast to the above findings in mice. Puromycin aminonucleoside (PAN), a derivative of puromycin that does not block protein synthesis, does not block memory in fish, nor does O-methyltyrosine, the amino acid moiety of puromycin [42]. In contrast, both puromycin and PAN potentiate seizures produced by pentylenetetrazol in the fish. The seizure-producing properties of puromycin would thus seem to be dissociated from its amnestic properties. Cycloheximides

do not block the seizure potentiation via puromycin, nor do they alone potentiate the pentylenetetrazol seizures. Peptidyl puromycin thus does not appear to mediate the convulsant activity of puromycin or of PAN.

6. If fish are allowed to remain in the training apparatus after a training session, they remain susceptible to the amnestic action of the antimetabolites for several hours. This unexpected result suggests that presence of the fish in the training environment delays onset of the consolidation process. When trained fish are subjected to the training environment without trials on day 2, some mild amnestic effects of the antibiotics are seen when they are retrained 5 days later [43]. This latter result underscores the complexity of learning experiments and their interpretation. In experiments with rats, in which ECS is used as the amnestic agent, it can be shown that if ECS is administered after an additional training trial, which is given after consolidation is thought to be complete, then it can activate the amnestic effect [2]. As discussed below, such results may well be consequences of the complex nature of physiological factors leading to performance.

Experiments by Barondes and co-workers [32] have used paradigms very similar to those of Flexner, although the results are, in general, more in accord with those conducted with fish. Experiments employing pretrial or posttrial injections indicate a brief (15 to 30 min) consolidation time. Short-term memory in the mouse lasts about 3 hr. Both puromycin and cycloheximide derivatives produce amnesia. Cycloheximides that do not block protein synthesis are not amnestic. Although differences in the temporal effects of puromycin as demonstrated by the Flexner and Barondes experiments have not been systematically explored, it should be pointed out that Barondes used an escape discrimination procedure whereas Flexner employed avoidance discrimination. A one-trial avoidance task in mice has also confirmed the amnestic action of cycloheximide [44].

The various agents have also been employed successfully in the newborn chick, using a one-trial passive avoidance task [45]. In an extension of these studies, a selective action of membrane agents, such as ouabain, on short-term memory was postulated. This may be related to a reported amnestic effect of amino acids that block glutamate-mediated spreading depression in the chick [46].

The inhibitors of protein synthesis have provided several valuable clues for increasing our understanding of memory formation. Short-term and long-term memory appear to be separable on the basis of sensitivity to these agents. A major question remains regarding the specificity of their action. Any generally toxic substance might produce a confusional state in such a way that acquisition is not affected, but permanent memory formation is blocked. At present, a clear-cut distinction between short-term and long-term memory is best supported by the antimetabolite treatment method. Interpretation of the effects of agents reported to block short-term memory must be considered as even more inferential. In man, several degenerative or traumatic conditions have been reported to produce a selective effect on formation of new permanent memories without effect on temporary ones [2, 47].

Flexner's studies of mice suggest that the effects of antimetabolites in this animal are localized in the entorhinal cortex and the hippocampus. At the molecular level, it is unclear whether protein inhibitors exert their amnestic effect via blocking the formation of some relatively prevalent brain protein at specific synaptic locations related to the new behavior, or whether, in addition, the proteins relevant to the block are chemically unique, like antibodies, for a particular neuronal pathway. Support is indirect for the existence of a great diversity of unique neuronal proteins. Specificity of regrowth of nerve bundles suggests the possibility of the presence of a high degree of chemical specificity within neurons [48]. Behaviorally, it has been claimed, but by no means accepted, that injection of specific proteins can alter behavior in recipient subjects. Further indirect evidence derives from a relatively high degree of hybridizable RNA found in brain as compared to other organs [49]. More direct proof would come from the identification of a change in brain protein patterns that is directly correlated with training.

If new protein is indeed formed at the synapse as a result of training, we should bear in mind the following:

1. Ribosomes are rare or absent at the presynaptic region. New protein in this region might come from some other source, e.g., presynaptic mitochondria, post-synaptic mitochondria, or postsynaptic ribosomes, the last being in relatively close proximity to the synapse. New presynaptic protein can migrate via axonal flow, in which case the time for the new protein to reach the synapse is a function of the length of the presynaptic axon.

2. The formation of novel protein sequences, if it occurs as a function of training, is assumed to be ultimately under the regulation of DNA. As in the selective mechanism for antibody formation, experience might evoke a response from a vast repertoire that arose via the natural history of the species.

3. The hypothetical behavior-specific proteins should be found only in small quantities as the result of any one training experience. However, the presence of gross alterations in brain RNA or protein is evidence for correlation with some alteration that is not specifically related to the given behavior. A block in protein synthesis that results in amnesia does not necessarily involve a specific protein. Relatively prevalent proteins in specific loci may mediate the fixation process. Because short-term memory is labile, a signal may be required in the brain to secrete a "fixative" molecule that permeates the entire organ but fixes only the labile connections. The memory process may require the presence of a critical protein with rapid turnover that catalyzes the formation of the putative fixative molecule. In this case, the protein itself does not contain information, and we might therefore see macroscopic changes in protein electrophoretic patterns or in radioactive labeling as a result of a training procedure. Whether or not a hypothetical critical protein is task specific, a further question arises as to whether the protein block that produces amnesia prevents formation of a new protein (as we have been considering up to this point), or whether the antibiotic inhibitor causes some rapidly turning over protein in the brain to fall below the effective level for normal memory formation. Although brain proteins in general turn over too slowly to support such a possibility, there are exceptions. For example, an isozyme of acetylcholinesterase in the rat brain appears to have a half-life of only 3 hr [50]. The concept of depletion of a critical protein

predicts that a protein block given two to three hours prior to training would produce a greater deficit of memory than would injections given immediately after training, since the critical protein would have reached an even lower level by the posttraining period. Experimental evidence does not support the idea [30, 51].

Inhibition of DNA and RNA Synthesis

Experiments in which amnesia is caused by protein-synthesis inhibitors lead to the speculation that RNA-synthesis blockers might also block memory. If enzyme induction mediates some aspect of the formation of long-term memory, a block at the transcriptional level (DNA-mediated synthesis of RNA) would be expected to produce a similar effect. If the protein blocker simply reduces the level of a rapidly turning over protein as discussed above, RNA blockers might or might not also cause reduction in a critical protein, depending on the stability of the appropriate messenger RNAs.

8-Azaguanine and actinomycin D, the DNA blockers, are more toxic than the protein inhibitors, and interpretation of experiments with them is often difficult. The use of these agents, as well as α-amanitin, camptothecin, and other more selective and less toxic agents, generally supports the proposal that RNA blockers are also amnestic [12].

Although DNA may constitute the reservoir of behavioral information, current concepts of cell regulation do not imply that it turns over. DNA metabolism is not required for the synthesis of novel proteins in enzyme induction. Neurons are non-dividing end cells, yet the brain is capable of enzyme induction. Other brain elements, notably glia, do divide. Memory theories involving glia have been proposed. Experimentally, arabinosyl cytosine, a DNA blocker, does not affect learning or memory in the goldfish [12].

Enzyme Inhibitors

Deutsch has reported that diisopropylphosphorofluoridate (DFP), a blocker of acetyl-cholinesterase, interferes with memory in rats [52]. The kinetics of recovery have led to the conclusion that this agent primarily affects retrieval. An optimal level of brain acetylcholinesterase has been proposed, which is such that excessively high or low amounts result in poor behavioral performance. This optimal level is employed to explain the paradoxical findings that DFP administration leads to reduction in performance in the rat, whereas in animals that have "forgotten" the task, DFP leads to improved performance. It is difficult to relate these experiments to known brain mechanisms since the role of acetylcholine in the central nervous system is poorly understood. Actual brain levels of acetylcholine after administration of DFP in these experiments have not yet been reported. An inhibitor of dopamine β-hydroxylase, bis (diethylthiocarbamoyl) disulfide, has similarly been reported to block memory formation in the rat, thus implicating catecholamine metabolism in memory [53].

Environmental Manipulation

In animal memory experiments, increasing attention is being given to the animal's surroundings before and after training [54]. An example of interaction between the amnestic agent and the environment comes from *reinstatement* experiments.

Reexposure to some elements of the training task can make rats susceptible to ECS amnesia at times after consolidation was believed to be complete [43]. The nature of these interactions is unknown but may be related in part to the general physiological state of the animal.

Brain Growth

The simplest model for brain plasticity might require that the brain grow as a result of experience. Brain components should then accrete, i.e., their amount should increase with age and experience. New synaptic endings after stimulation have been reported [55], and increased brain length [56] and cortical thickness [57] have been described. Starvation in adults generally causes negligible effects; the brain is spared. In the growing animal, however, this may not be true. It is clear that genetic abnormalities in man can produce defects during development that can be partially reversed by diet. Phenylalanine hydroxylase impairment (phenylketonuria) results in mental retardation in infants fed a normal diet. If a low phenylalanine diet is administered, the children develop relatively normally, and they are no longer very susceptible to brain damage by dietary phenylalanine. Thyroid deficiency during development results in defective brain growth, with reduction in dendritic spines in the cortex (see Chap. 18).

Administration of growth hormone to pregnant rats leads to progeny with permanently increased brain DNA [58]. Evidence that they are behaviorally superior is not convincing. Similarly, grafting a second brain into developing fish has been reported, although the behavioral significance is dubious [59].

Enhancing Agents

Several investigators have claimed that when extracts of nucleic acid or proteins from brains of donor animals that have been trained are injected into recipients, the recipient animals exhibit significantly improved performance that is specific for the original task [60]. The conditions of injection, the species injected, and the task are sufficiently different among laboratories to prevent generalizations from the results.

BIOCHEMICAL CORRELATES
OF EXCITATION, LEARNING, AND BEHAVIOR

The most direct confirmation of a role for macromolecules in behavior, inferred from the antimetabolite experiments, is the detection of changes in RNA or protein correlated with learning in the absence of the blocking agents. Such changes are indeed claimed, but they are not easily reconciled with the studies made with blocking agents. For example, changes in labeling in the brain that are great enough to be detected with relatively crude methods, such as trichloroacetic acid precipitation, would seem too large to account for putative behavior-specific macromolecules. Alterations within a small number of neurons have been studied using special methods, such as those developed and employed by Hydén. His laboratory has reported [61] changes in total RNA and base ratios of isolated RNA of Dieters' nucleus cells after training of a balancing task. Freehand isolation of these large neurons and of the associated satellite glial cells was followed by microanalytical procedures. Hydén has also examined electrophoretic patterns from the cytosol of cells in the rat hippocampus after training for handedness in an appetitive task [62]. Even with such relatively refined techniques, the alterations found seem rather gross compared to those anticipated from the learning of a single task.

Another technique for measuring regional alterations in the brain is radioautography. Glassman reported [63] increased ^{3}H-uridine incorporation into macromolecular material, presumably RNA, in the diencephalon of mice after a training experience. Changes in protein labeling were not reported. Altman found increased incorporation of ^{3}H-leucine into protein in motor neurons of the rat lumbar cord after simple exercise [64].

There are many reports of alterations in total brain RNA and protein labeling. Such global changes tend to support association with some nonspecific correlate of training, exercise, or stress. Visual stimulation has been reported to cause a change in aggregation of brain polysomes, but not of liver polysomes. This change is interpreted as reflecting increased neuronal activity in general. On the other hand, an increase in the incorporation of labeled uridine into mouse brain has been reported to be specific for training of a shock avoidance task [63]. Yoked control animals subjected to the same amount of light, buzzer, and shock as experimental animals, but which are not permitted to make the trained response, do not show the change in labeling. Overtrained animals, even when permitted to make the trained response, also do not show the increased labeling of RNA. It appears, then, that some changes in RNA metabolism take place with simple stimulation, but there may be, in addition, conditions under which only training stimulates labeling. A potentially useful way to distinguish general metabolic responses to stress, etc., from putative learning-specific changes is the use of a "split-brain" preparation, in which it can be demonstrated that the trained response can be evoked only via the trained eye. Unilateral changes in labeled uracil incorporation in the split-brain chick are reported, suggesting that these changes are related to the acquisition of an imprinted behavior [65].

Although it is difficult to draw any simple conclusion from these experiments, two sorts of issues can be delineated:

1. In none of the conditions in which RNA labeling is altered has a concomitant alteration in protein labeling been reported. Messenger, transfer, and ribosomal RNAs subserve protein synthesis, so a comparable, even greater, effect on protein labeling might have been anticipated.

2. In experiments in which increased labeling of RNA is observed, it is not certain that RNA synthesis is actually increased. An alternative explanation is a change in the size of the precursor pool. The injected radioisotopic precursor, usually a pyrimidine, must be phosphorylated intracellularly and converted to a triphosphate prior to incorporation into RNA. Efficacy of incorporation depends on the amount of endogenous intermediates present as well as upon the starting specific activity of the injected precursor. An increase in labeling might be due to reduction in the endogenous

unlabeled precursor pool, rather than to an increase in RNA synthesis. The relevant intracellular pools are most likely intranuclear (see Chap. 17).

CLINICAL IMPLICATIONS

Nowhere in biology is the animal model less appropriate for studying human disease than in the area of higher brain function, yet many basic principles established in studies on animal learning and memory may carry over into practical biomedical problems. For example, the effect of electroconvulsive shock therapy (ECT) in the treatment of depression has not been explained adequately. We do not know if the amnesia associated with ECT is related to the therapeutic effect. It has been shown in animal studies that ECS causes alterations in catecholamine turnover in the brain [41], and that catecholamines have been implicated in the depressive state. The re-instatement phenomenon described in this chapter has prompted the suggestion that ECT effects could be enhanced by temporal proximity of the treatment with pretreatment clinical evocation of a patient's psychological problems by the therapist [66].

Numerous drugs have been proposed for the aid of defective learning and memory, mainly for the young and for the aged. Other than agents having indirect effects, such as depression of overactivity, increasing attention span, and mild stimulant effects, no specific agent has gained acceptance as yet for the improvement of learning or memory. Drugs such as tricyanoaminopropene or magnesium pemoline, which have been reported to be effective in animals, have questionable biochemical, as well as behavioral, effects. In the aged, injection or ingestion of yeast RNA [2, 67] has been reported to improve memory, but this work has been seriously challenged. Specific amnestic defects in otherwise normal individuals have been reported as a result of tumors, surgery, or specific degenerative diseases such as Korsakoff's syndrome and Alzheimer's disease. Although the hippocampus, among other structures, appears to be involved in many of these conditions, the nature of their role, whether neural, humoral, or both, is unknown at present.

Finally, while relatively little is known about early dietary effects on brain function, animal experiments give us sufficient concern to recognize a potential problem in the malnourished infant, who may be destined for a life of restricted usefulness.

REFERENCES

1. Ribot, T. *Disorders of Memory*. London: Kegan, Paul, Trench, 1885.
*2. Agranoff, B. W., Springer, A. D., and Quarton, G. C. Biochemistry of Memory and Learning. In Vinken, P. J., and Bruyn, G. W. (Eds.), *Handbook of Clinical Neurology*. Vol. 29. Amsterdam: North-Holland, 1975.
3. Pearlman, C. A., Sharpless, S. K., and Jarvik, M. E. Retrograde amnesia produced by anesthetic and convulsant agents. *J. Comp. Physiol. Pyschol.* 54:109, 1961.
*4. Kimble, G. A. *Hilgard and Marquis' Conditioning and Learning,* 2d ed. New York: Appleton-Century-Crofts, 1961.

*Asterisks denote key references.

5. Schneider, A. M., and Sherman, W. Amnesia: A function of the temporal relation of footshock to electroconvulsive shock. *Science* 159:219, 1968.

6. Bovet, D., Bovet-Nitti, F., and Oliverio, A. Genetic aspects of learning and memory in mice. *Science* 163:139, 1969.

*7. Corning, W. C., Dyal, J. A., and Willows, A. O. D. (Eds.). *Invertebrate Learning.* New York: Plenum, 1975.

8. Garcia, J., and Hankins, W. G. The evolution of bitter and the acquisition of toxiphobia. In Denton, D., and Coghlin, J. P. (Eds.), *Olfaction and Taste V.* New York: Academic, 1975, p. 39.

9. Brenner, S. The genetics of *Caenorhabditis elegans.* *Genetics* 77:71, 1974.

10. Young, J. Z. Short and long memories in *Octopus* and the influence of the vertical lobe system. *J. Exp. Biol.* 52:385, 1970.

11. Quinn, W. G., Harris, W. A., and Benzer, S. Conditioned behavior in *Drosophila melanogaster. Proc. Natl. Acad. Sci. U.S.A.* 71:708, 1974.

*12. Agranoff, B. W. Biochemical Concomitants of the Storage of Behavioral Information. In Jaenicke, L. (Ed.), *25th Mosbacher Killoquium.* Berlin: Springer-Verlag, 1974, p. 597.

13. Agranoff, B. W., Davis, R. E., and Gossington, R. E. Esoteric fish. *Science* 171:230, 1971.

14. Bennett, M. V. L. Neural Control of Electric Organs. In Ingle, D. (Ed.), *The Central Nervous System and Fish Behavior.* Chicago: University of Chicago Press, 1968, p. 147.

15. Cherkin, A. Kinetics of memory consolidation: Role of amnesic treatment parameters. *Proc. Natl. Acad. Sci. U.S.A.* 63:1094, 1969.

16. Hess, E. H. Imprinting. *Science* 130:133, 1959.

17. Mason, J. W. Organization of psychoendocrine mechanisms. *Psychosom. Med.* 30(2):565, 1968.

*18. Kandel, E. R., and Tauc, L. Heterosynaptic facilitation in the giant cell of the abdominal ganglion of *Aplysia depilans. J. Physiol.* (Lond.) 181:28, 1965.

19. Kupferman, I., and Kandel, E. R. Neuronal controls of a behavioral response mediated by the abdominal ganglion of *Aplysia. Science* 164:847, 1969.

20. Willows, A. O. D. Behavioral Acts Elicited by Stimulation of Single Identifiable Nerve Cells. In Carlson, F. D. (Ed.), *Physiological and Biochemical Aspects of Nervous Integration.* Englewood Cliffs, N. J.: Prentice-Hall, 1968, p. 217.

21. Eisenstein, E. M., and Cohen, M. J. Learning in an isolated prothoracic insect ganglion. *Anim. Behav.* 13:163, 1965.

22. Glassman, E., Henderson, A., Cordle, M., Moon, H. M., and Wilson, J. E. Effect of cycloheximide and actinomycin D on the behavior of the headless cockroach. *Nature* 225:967, 1970.

23. Scoville, W. B., and Milner, B. Loss of recent memory after bilateral hippocampal lesions. *J. Neurol. Neurosurg. Psychiatry* 20:11, 1957. But see also Woolsey, R. M., and Nelson, J. S. Asymptomatic destruction of the fornix in man. *Arch. Neurol.* 32:566, 1975.

24. Morrell, F. Electrical Signs of Sensory Coding. In Quarton, G. C., Melnechuk, T., and Schmitt, F. O. (Eds.), *The Neurosciences: A Study Program.* New York: Rockefeller University Press, 1967, p. 452.

25. Olson, M. I., and Bunge, R. P. Anatomical observations on the specificity of synapse formation in tissue culture. *Brain Res.* 59:19, 1973.

26. Chow, K. L. Effects of Ablation. In Quarton, G. C., Melnechuk, T., and Schmitt, F. O. (Eds.), *The Neurosciences: A Study Program.* New York: Rockefeller University Press, 1967, p. 705.

27. Kandel, E. R., and Spencer, W. A. Cellular neurophysiological approaches in the study of learning. *Physiol. Rev.* 48:65, 1968.

28. Eichenbaum, H., Butter, C. M., and Agranoff, B. W. Radioautographic localization of inhibition of protein synthesis in specific regions of monkey brain. *Brain Res.* 61:438, 1973.

29. Bures, J., and Buresova, O. The use of Leaos spreading depression in the study of interhemispheric transfer of memory traces. *J. Comp. Physiol. Psychol.* 53:558, 1960.

30. Springer, A. D., Schoel, W. M., Klinger, P. D., and Agranoff, B. W. Anterograde and retrograde effects of electroconvulsive shock and of puromycin on memory formation in the goldfish. *Behav. Biol.* 13:467, 1975.

31. McGaugh, J. L. Time-dependent processes in memory storage. *Science* 153:1351, 1966.

*32. Barondes, S. H. Cerebral protein synthesis inhibitors block long-term memory. *Int. Rev. Neurobiol.* 12:177, 1970.

33. Tryon, R. C. Genetic differences in maze learning abilities in rats. *Yearbook Natl. Soc. Stud. Educ.* 39(1):111, 1940.

34. Adler, J. Decision making in bacteria. *Science* 184:1293, 1974.

35. Koshland, D. E., Jr. Sensory Response in Bacteria. In Agranoff, B. W., and Aprison, M. H. (Eds.), *Advances in Neurochemistry.* Vol. 2. New York: Plenum. In press.

36. Ransmeier, R. E., and Gerard, R. W. Effects of temperature, convulsion and metabolic factors on rodent memory and EEG. *Am. J. Physiol.* 179:663, 1954.

37. Davis, R. E., Bright, P. J., and Agranoff, B. W. Effect of ECS and puromycin on memory in fish. *J. Comp. Physiol. Psychol.* 60:162, 1965.

38. Neale, J. H., Klinger, P. D., and Agranoff, B. W. Temperature-dependent consolidation of puromycin-susceptible memory in the goldfish. *Behav. Biol.* 9:267, 1973.

39. Flexner, J. B., Flexner, L. B., and Stellar, E. Memory in mice as affected by intracerebral puromycin. *Science* 141:57, 1963.

*40. Roberts, R. B., and Flexner, L. B. The biochemical basis of long-term memory. *Q. Rev. Biophys.* 2:135, 1969. *See also* Roberts, R. B., Flexner, J. B., and Flexner, L. B. Some evidence of the involvement of adrenergic sites in the memory trace. *Proc. Natl. Acad. Sci. U.S.A.* 66:310, 1970.

41. Kety, S. S., Jovoy, F., Thierry, A.-M., Julou, L., and Glowinski, J. A sustained effect of electroconvulsive shock on the turnover of norepinephrine in the central nervous system of the rat. *Proc. Natl. Acad. Sci. U.S.A.* 58:1249, 1967.

42. Agranoff, B. W. Protein Synthesis and Memory Formation. In Lajtha, A. (Ed.), *Protein Metabolism of the Nervous System.* New York: Plenum, 1970, p. 533.

43. Davis, R. E., and Klinger, P. D. Environmental control of amnesic effects of various agents in goldfish. *Physiol. Behav.* 4:269, 1969.

44. Flood, J. F., Rosenzweig, M. R., Bennett, E. L., and Orme, A. E. Influence of duration of protein synthesis inhibition on memory. *Physiol. Behav.* 10:555, 1973.

45. Mark, R. F., and Watts, M. E. Drug inhibition of memory formation in chickens. *Proc. R. Soc. Lond.* 178:439, 1971.

46. Van Harreveld, A., and Fifkova, E. Involvement of glutamate in memory formation. *Brain Res.* 81:455, 1974.

*47. Whitty, C. W. M., and Zangwill, O. L. *Amnesia.* London: Butterworth, 1966.

48. Sperry, R. N. Chemoaffinity in the orderly growth of nerve fiber patterns and connections. *Proc. Natl. Acad. Sci. U.S.A.* 50:703, 1963.

49. Brown, I. R., and Church, R. B. RNA transcription from non-repetitive DNA in the mouse. *Biochem. Biophys. Res. Commun.* 42:850, 1971.

50. Davis, G. A., and Agranoff, B. W. Metabolic behaviour of isozymes of acetylcholinesterase. *Nature* 220:277, 1968.

51. Springer, A. D., Schoel, W. M., Klinger, P. D., and Agranoff, B. W. Anterograde and retrograde effects of electroconvulsive shock and of puromycin on memory formation in the goldfish. *Behav. Biol.* 13:467, 1975.

52. Deutsch, J. A., Hamburg, M. D., and Dahl, H. Anticholinesterase-induced amnesia and its temporal aspects. *Science* 151:221, 1966.

53. Randt, C. T., Goldstein, M., Anagnoste, B., and Quartermain, D. Norepinephrine synthesis inhibition: Effects on memory in mice. *Science* 172:498, 1971.

54. Davis, R. E., and Agranoff, B. W. Stages of memory formation in goldfish: Evidence for an environmental trigger. *Proc. Natl. Acad. Sci. U.S.A.* 55:555, 1966.

55. Schapiro, S., and Vukovich, K. R. Early experience effects upon cortical dendrites: A proposed model for development. *Science* 167:292, 1970.

56. Altman, J., Wallace, R. B., Anderson, W. J., and Das, G. D. Behaviorally induced changes in length of cerebrum in rats. *Dev. Psychobiol.* 1:112, 1968.

57. Diamond, M. C., Krech, D., and Rosenzweig, M. R. The effects of an enriched environment on the histology of the rat cerebral cortex. *J. Comp. Neurol.* 123:111, 1964.

58. Zamenhof, S., Mosley, J., and Schuller, E. Stimulation of the proliferation of cortical neurons by prenatal treatment with growth hormone. *Science* 152:1396, 1966.

59. Bresler, D. E., and Bitterman, M. E. Learning in fish with transplanted brain tissue. *Science* 163:590, 1969.

60. Quarton, G. C. The Enhancement of Learning by Drugs and the Transfer of Learning by Macromolecules. In Quarton, G. C., Melnechuk, T., and Schmitt, F. O. (Eds.), *The Neurosciences: A Study Program.* New York: Rockefeller University Press, 1967, p. 744.

61. Hydén, H., and Lange, P. W. A differentiation in RNA response in neurons early and late during learning. *Proc. Natl. Acad. Sci. U.S.A.* 53:946, 1965.

62. Hydén, H., and Lange, P. W. S-100 protein: Correlation with behavior. *Proc. Natl. Acad. Sci. U.S.A.* 67:1959, 1970.

*63. Glassman, E. The biochemistry of learning: An evaluation of the role of RNA and protein. *Annu. Rev. Biochem.* 38:605, 1969.

64. Altman, J. Differences in the utilization of tritiated leucine by single neurons in normal and exercised rats: An autoradiographic investigation with microdensity. *Nature* 199:777, 1963.

65. Horn, G., Rose, S. P. R., and Bateson, P. P. G. Monocular imprinting and regional incorporation of tritiated uracil into the brains of intact and split-brain chicks. *Brain Res.* 56:227, 1973.

66. Robbins, M. J., and Meyers, D. R. Motivational control of retrograde amnesia. *J. Exp. Psychol.* 84:220, 1970.

67. Cameron, D. E., and Solyom, L. Effect of RNA on memory. *Geriatrics* 16:74, 1961.

Glossary

ACh	acetylcholine
AChE	acetylcholinesterase
ACTH	adrenocorticotropic hormone
ADP	adenosine 5'-diphosphate
ALA	δ-aminolevulinic acid
AMP	adenosine 5'-phosphate (adenylic acid)
ATP	adenosine 5'-triphosphate
AXM	acetoxycycloheximide
BAL	British antilewisite (2,3-dimercaptopropanol)
BSA	bovine serum albumin
cAMP	see 3',5'-cyclic AMP
CBF	cerebral blood flow
CDP	cytidine 5'-diphosphate
Cer	ceramide
Cer Gal	galactocerebroside
Cer Glc	glucocerebroside
cGMP	cyclic 3',5'-guanosine monophosphate
ChAc	choline acetylase (choline acetyltransferase)
ChE	cholinesterase ("nonspecific" esterase)
CMP	cytidine 5'-phosphate (cytidylic acid)
CMR	cerebral metabolic rate
$CMRO_2$	cerebral metabolic rate for O_2
CNS	central nervous system
CoA	coenzyme A
COMT	catechol-O-methyltransferase
CPZ	chlorpromazine
CRF	corticotropin releasing factor
CS	conditioned stimulus
CSF	cerebrospinal fluid
CTP	cytidine triphosphate
3',5'-cyclic AMP	cyclic 3',5'-adenosine monophosphate

785

DDT	1,1,1-trichloro-2,2-bis (*p*-chlorophenyl)-ethane
DEAE	diethylaminoethyl
DFP	diisopropylphosphorofluoridate
DHT	dihydrotestosterone
DMPEA	dimethoxyphenylethylamine
DON	6-diazo-5-oxo-L-norleucine
DOPA	3,4-dihydroxyphenylalanine
DOPAC	3,4-dihydroxyphenylacetic acid
dopamine	3,4-dihydroxyphenylethylamine
EAE	experimental allergic encephalomyelitis
EAN	experimental allergic neuritis
ECS	electroconvulsive shock
ECT	electroconvulsive shock therapy
ECW	extracellular water
EDTA	ethylenediaminetetraacetate
EGTA	ethyleneglycol-bis-(β-aminoethyl ether)-N,N-tetraacetic acid
EIM	excitability inducing material
ER	endoplasmic reticulum
E_R	resting potential
FAD	flavin-adenine dinucleotide
fEBP	fetoneonatal estrogen-binding protein
fmole	femtomole (= 10^{-15} mole)
FSH	follicle-stimulating hormone
FSH-RF	follicle-stimulating hormone releasing factor
GABA	γ-aminobutyric acid
GABA-T	γ-aminobutyric acid transaminase
GAD	glutamic acid decarboxylase
GFA	glial fibrillary acidic protein
GH	growth hormone (somatotropin)
GH-RF	growth hormone releasing factor
GMP	guanosine 5′-phosphate (guanylic acid)
GPDH	glycerolphosphate dehydrogenase
GSH	glutathione (reduced form)
GTP	guanosine 5′-triphosphate
GTT	glucose tolerance test

HC	hemicholinium
HHH	hypothalamic hypophysiotropic hormone
5-HIAA	5-hydroxyindoleacetic acid
HIOMT	hydroxyindole-O-methyltransferase
HnRNA	heterogenous RNA
HPr	histidine-containing protein
5-HT	5-hydroxytryptamine (serotonin)
5-HTP	5-hydroxytryptophan
HVA	homovanillic acid (4-hydroxy-3-methoxyphenylacetic acid)
IMP	inosine 5'-phosphate
ITT	insulin tolerance test
17-KGS	17-ketogenic steroid
17-KS	17-ketosteroid
LD	lethal dose
LH	luteinizing hormone
LH-RF	luteinizing hormone releasing factor
LMM	light meromyosin
LSD	d-lysergic acid diethylamide
MAO	monoamine oxidase
MeCh	acetyl-β-methylcholine (mecholyl)
MET	metanephrine
MHPG	3-methoxy-4-hydroxyphenylglycol
MIF	migration inhibition factor
MLD	metachromatic leukodystrophy
MS	multiple sclerosis
MSH	melanocyte-stimulating hormone
MSH-IF	melanocyte-stimulating hormone inhibitory factor
NAD^+	nicotinamide-adenine dinucleotide
NADH	reduced form of NAD^+
$NADP^+$	nicotinamide-adenine dinucleotide phosphate
NADPH	reduced form of $NADP^+$
Na_E	exchangeable sodium
NANA	*see* Neu-NAc

NE	norepinephrine
NEM	N-ethylmaleimide
Neu-NAc	N-acetylneuraminic acid (NANA)
NGF	nerve growth factor
NMET	normetanephrine
NMR	nuclear magnetic resonance
NVP	naphthyl vinyl pyridyl
11-OHCS	11-hydroxycorticosteroid
17-OHCS	17-hydroxycorticosteroid
OMP	orotidine 5'-phosphate
PAM	pyridine-2-aldoxime methiodide
PAN	puromycin aminonucleoside
PAPS	3'-phosphoadenosine 5'-phosphosulfate
PEP	phosphoenolpyruvate
PG	prostaglandin
P_i	inorganic phosphate
PIF	prolactin inhibitory factor
PNMT	phenylethanolamine-N-methyltransferase
PNS	peripheral nervous system
PPD	N,N'-dimethyl-p-phenylenediamine
PSD	postsynaptic density
RER	rough endoplasmic reticulum
RF	releasing factor
RQ	respiratory quotient
SER	smooth endoplasmic reticulum
SME	stalk-median eminence tissue
SSA-D	succinic semialdehyde dehydrogenase
SSPE	subacute sclerosing panencephalitis
TBW	total body water
TCA	tricarboxylic acid cycle; trichloracetic acid
TEC	triethylcholine
THC	Δ^1-tetrahydrocannabinol
TMB-4	N,N-trimethylene-bis-(pyridinium-4-aldoxime) dibromide

TnT, TnC, TnI	protein components of troponin
TRF	thyrotropin releasing factor
Tris	tris-(hydroxymethyl)-aminomethane
TSH	thyroid-stimulating hormone (thyrotropin)
TSP	total soluble protein
UCS	unconditioned stimulus
UDP	uridine $5'$-diphosphate
UMP	uridine $5'$-phosphate (uridylic acid)
UTP	uridine $5'$-triphosphate
VMA	vanillylmandelic acid

Index

Index